Reservoir Quality of Clastic and Carbonate Rocks: Analysis, Modelling and Prediction

Geological Society books refereeing procedures

The Society makes every effort to ensure that the scientific and production quality of its books matches that of its journals. Since 1997, all book proposals have been refereed by specialist reviewers as well as by the Society's Books Editorial Committee. If the referees identify weaknesses in the proposal, these must be addressed before the proposal is accepted.

Once the book is accepted, the Society Book Editors ensure that the volume editors follow strict guidelines on refereeing and quality control. We insist that individual papers can only be accepted after satisfactory review by two independent referees. The questions on the review forms are similar to those for *Journal of the Geological Society*. The referees' forms and comments must be available to the Society's Book Editors on request.

Although many of the books result from meetings, the editors are expected to commission papers that were not presented at the meeting to ensure that the book provides a balanced coverage of the subject. Being accepted for presentation at the meeting does not guarantee inclusion in the book.

More information about submitting a proposal and producing a book for the Society can be found on its website: www.geolsoc.org.uk.

It is recommended that reference to all or part of this book should be made in one of the following ways:

ARMITAGE, P.J., BUTCHER, A.R., CHURCHILL, J.M., CSOMA, A.E., HOLLIS, C., LANDER, R.H., OMMA, J.E. & WORDEN, R.H. (eds) 2018. *Reservoir Quality of Clastic and Carbonate Rocks: Analysis, Modelling and Prediction*. Geological Society, London, Special Publications, **435**.

GRIFFITHS, J., FAULKNER, D.R., EDWARDS, A.P. & WORDEN, R.H. 2018. Deformation band development as a function of intrinsic host-rock properties in Triassic Sherwood Sandstone. *In*: ARMITAGE, P.J., BUTCHER, A.R., CHURCHILL, J.M., CSOMA, A.E., HOLLIS, C., LANDER, R.H., OMMA, J.E. & WORDEN, R.H. (eds) *Reservoir Quality of Clastic and Carbonate Rocks: Analysis, Modelling and Prediction*. Geological Society, London, Special Publications, **435**, 161–176. First published online January 19, 2016, https://doi.org/10.1144/SP435.11

GEOLOGICAL SOCIETY SPECIAL PUBLICATION NO. 435

Reservoir Quality of Clastic and Carbonate Rocks: Analysis, Modelling and Prediction

EDITED BY

P. J. ARMITAGE
BP Upstream Technology, UK

A. R. BUTCHER
Geological Survey of Finland, Finland

J. M. CHURCHILL
Shell UK Ltd, UK

A. E. CSOMA
MOL Group Exploration, Hungary

C. HOLLIS
University of Manchester, UK

R. H. LANDER
Geocosm, USA

J. E. OMMA
Rocktype, UK

and

R. H. WORDEN
University of Liverpool, UK

2018
Published by
The Geological Society
London

THE GEOLOGICAL SOCIETY

The Geological Society of London (GSL) was founded in 1807. It is the oldest national geological society in the world and the largest in Europe. It was incorporated under Royal Charter in 1825 and is Registered Charity 210161.

The Society is the UK national learned and professional society for geology with a worldwide Fellowship (FGS) of over 10 000. The Society has the power to confer Chartered status on suitably qualified Fellows, and about 2000 of the Fellowship carry the title (CGeol). Chartered Geologists may also obtain the equivalent European title, European Geologist (EurGeol). One fifth of the Society's fellowship resides outside the UK. To find out more about the Society, log on to www.geolsoc.org.uk.

The Geological Society Publishing House (Bath, UK) produces the Society's international journals and books, and acts as European distributor for selected publications of the American Association of Petroleum Geologists (AAPG), the Indonesian Petroleum Association (IPA), the Geological Society of America (GSA), the Society for Sedimentary Geology (SEPM) and the Geologists' Association (GA). Joint marketing agreements ensure that GSL Fellows may purchase these societies' publications at a discount. The Society's online bookshop (accessible from www.geolsoc.org.uk) offers secure book purchasing with your credit or debit card.

To find out about joining the Society and benefiting from substantial discounts on publications of GSL and other societies worldwide, consult www.geolsoc.org.uk, or contact the Fellowship Department at: The Geological Society, Burlington House, Piccadilly, London W1J 0BG: Tel. +44 (0)20 7434 9944; Fax +44 (0)20 7439 8975; E-mail: enquiries@geolsoc.org.uk.

For information about the Society's meetings, consult *Events* on www.geolsoc.org.uk. To find out more about the Society's Corporate Affiliates Scheme, write to enquiries@geolsoc.org.uk.

Published by The Geological Society from:
The Geological Society Publishing House, Unit 7, Brassmill Enterprise Centre, Brassmill Lane, Bath BA1 3JN, UK

The Lyell Collection: www.lyellcollection.org
Online bookshop: www.geolsoc.org.uk/bookshop
Orders: Tel. +44 (0)1225 445046, Fax +44 (0)1225 442836

British Library Cataloguing in Publication Data

A catalogue record for this book is available from the British Library.
ISBN 978-1-78620-139-3
ISSN 0305-8719

Distributors
For details of international agents and distributors see:
www.geolsoc.org.uk/agentsdistributors

Typeset by Nova Techset Private Limited, Bengaluru & Chennai, India
Printed and bound by CPI Group (UK) Ltd, Croydon CR0 4YY

Contents

Acknowledgements

The Geological Society thanks the following companies for their generous sponsorship of the Petroleum Group and the conference that led to this volume. The Petroleum Group is thanked for its contribution towards the colour printing in this volume.

Petroleum Group corporate sponsors at the time of the conference:

Sponsors of Reservoir Quality of Clastic and Carbonate Rocks: Analysis, Modelling and Prediction conference:

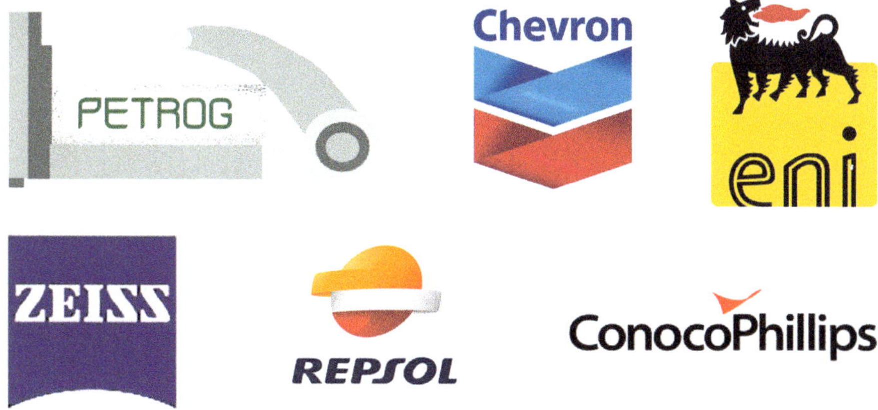

Petroleum reservoir quality prediction: overview and contrasting approaches from sandstone and carbonate communities

R. H. WORDEN[1]*, P. J. ARMITAGE[2], A. R. BUTCHER[3], J. M. CHURCHILL[4], A. E. CSOMA[5], C. HOLLIS[6], R. H. LANDER[7] & J. E. OMMA[8]

[1]*Department of Earth, Ocean and Ecological Sciences, University of Liverpool, 4 Brownlow Street, Liverpool L18 1LX, UK*

[2]*BP Upstream Technology, Chertsey Road, Sunbury, Middlesex TW16 7LN, UK*

[3]*Geological Survey of Finland, Espoo, 02151 Finland*

[4]*Shell UK Limited, 1 Altens Farm Road, Aberdeen AB12 3FY, UK*

[5]*MOL Group Exploration, Október Huszonharmadika u. 18, Budapest, H-1117, Hungary*

[6]*University of Manchester, Manchester, M13 9PL, UK*

[7]*Geocosm, Town Plaza 233, Durango, CO 81301, USA*

[8]*Rocktype, Oxford OX4 1LN, UK*

**Correspondence: r.worden@liv.ac.uk*

Abstract: The porosity and permeability of sandstone and carbonate reservoirs (known as reservoir quality) are essential inputs for successful oil and gas resource exploration and exploitation. This chapter introduces basic concepts, analytical and modelling techniques and some of the key controversies to be discussed in 20 research papers that were initially presented at a Geological Society conference in 2014 titled 'Reservoir Quality of Clastic and Carbonate Rocks: Analysis, Modelling and Prediction'. Reservoir quality in both sandstones and carbonates is studied using a wide range of techniques: log analysis and petrophysical core analysis, core description, routine petrographic tools and, ideally, less routine techniques such as stable isotope analysis, fluid inclusion analysis and other geochemical approaches. Sandstone and carbonate reservoirs both benefit from the study of modern analogues to constrain the primary character of sediment before they become a hydrocarbon reservoir. Prediction of sandstone and carbonate reservoir properties also benefits from running constrained experiments to simulate diagenetic processes during burial, compaction and heating. There are many common controls on sandstone and carbonate reservoir quality, including environment of deposition, rate of deposition and rate and magnitude of sea-level change, and many eogenetic processes. Compactional and mesogenetic processes tend to affect sandstone and carbonate somewhat differently but are both influenced by rate of burial, and the thermal and pressure history of a basin. Key differences in sandstone and carbonate reservoir quality include the specific influence of stratigraphic age on seawater composition (calcite v. aragonite oceans), the greater role of compaction in sandstones and the greater reactivity and geochemical openness of carbonate systems. Some of the key controversies in sandstone and carbonate reservoir quality focus on the role of petroleum emplacement on diagenesis and porosity loss, the role of effective stress in chemical compaction (pressure solution) and the degree of geochemical openness of reservoirs during diagenesis and cementation. This collection of papers contains case study-based examples of sandstone and carbonate reservoir quality prediction as well as modern analogue, outcrop analogue, modelling and advanced analytical approaches.

Porosity and permeability (reservoir quality) exert fundamental controls on the economic viability of a petroleum accumulation (Blackbourn 2012). They need to be quantified from basin access and exploration, via appraisal and field development through secondary and tertiary recovery in order to

From: ARMITAGE, P. J., BUTCHER, A. R., CHURCHILL, J. M., CSOMA, A. E., HOLLIS, C., LANDER, R. H., OMMA, J. E. & WORDEN, R. H. (eds) 2018. *Reservoir Quality of Clastic and Carbonate Rocks: Analysis, Modelling and Prediction.* Geological Society, London, Special Publications, **435**, 1–31.
First published online May 01, 2018, https://doi.org/10.1144/SP435.21

minimize cost and maximize return on investment. Consequently, the question of reservoir quality prediction is becoming ever more critical with exploration and production of petroleum in increasingly challenging conditions and from less conventional reservoirs. The porosity of a reservoir affects the volume of oil or gas in place and permeability affects the rate at which oil and gas flow from the reservoir to the wellbore (Gluyas & Swarbrick 2004). Measuring porosity and permeability using core analysis techniques (Emery & Robinson 1993), or using wireline logs via a combination of density, sonic, neutron and NMR logs (Rider & Kennedy 2011), is essential for reservoir characterization but these approaches do not directly reveal the controls on reservoir quality and thus do not facilitate prediction away from the wellbore. Defining grain types and their mineralogy, grain–grain contacts, intergranular volume (a proxy for the extent of compaction), matrix type and amount, cement mineralogy and morphology, and pore types, their genesis and proportions is vital in order to establish the dominant controls on reservoir quality (Fig. 1; Primmer *et al.* 1997). Quantifying the conditions of effective stress, temperature and thermal history as well as the composition of fluids under which a sedimentary rock evolved, from deposition through burial and uplift, is also essential to achieve a holistic understanding and make credible forward predictions of reservoir quality. Process-oriented diagenetic models that consider these depositional and burial history characteristics may provide deeper insights into the controls on reservoir quality while also having the potential to serve as powerful tools for prediction, particularly when integrated with other disciplines such as sedimentology, structural geology, geomodelling and reservoir engineering.

Despite the importance of understanding the controls on porosity and permeability of reservoirs, numerous fundamental issues lack consensus. Reservoir quality is controlled by interdependent sedimentary and diagenetic factors, including the origin of the sediment (provenance), depositional environment, weathering and climatic conditions in the hinterland and at the site of deposition, compaction, recrystallization and dissolution, authigenic mineral growth, petroleum charge, depth of burial, extent and rate of heating, fluid pressure and effective stress and structural deformation (Morad *et al.* 2012). Some of these factors are interlinked but they are typically considered separately, thus hampering reservoir quality prediction. Yet further issues arise owing to some fundamental unresolved controversies that thwart the development of universally accepted models and approaches for reservoir quality prediction. These include the effect of petroleum on mineral growth, the question of whether effective stress or temperature controls pressure solution (chemical compaction) and the question of open v. closed systems during the diagenesis of reservoirs. These controversies will be addressed at the end of this paper.

Traditionally, carbonate and clastic reservoir quality prediction has been undertaken by what are effectively separate communities, although there are many apparent overlaps (Figs 1–3). There are also many assumptions adopted by one community for which contradictory assumptions have been made by the other community. Bringing these scientific communities together provides an opportunity to take a leap into bigger advancements in reservoir

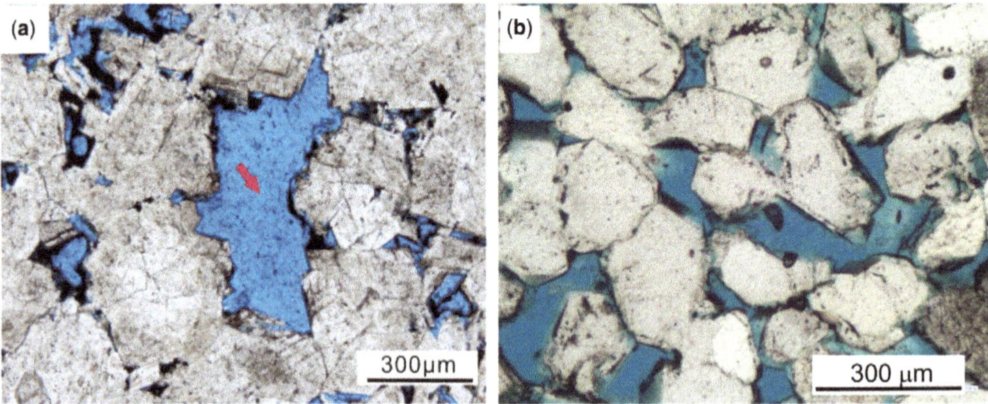

Fig. 1. Images of high-porosity carbonate and sandstone reservoir samples. (**a**) Fabric-obliterative oolitic dolomite, 3200 m Fiexianguan Fm, Sichuan Basin, China. (**b**) Quartz cemented sandstone, 3400 m, Ula Formation, North Sea Basin. (Image (a) courtesy of Lei Jiang and image (b) courtesy of Mohammed Bukar.)

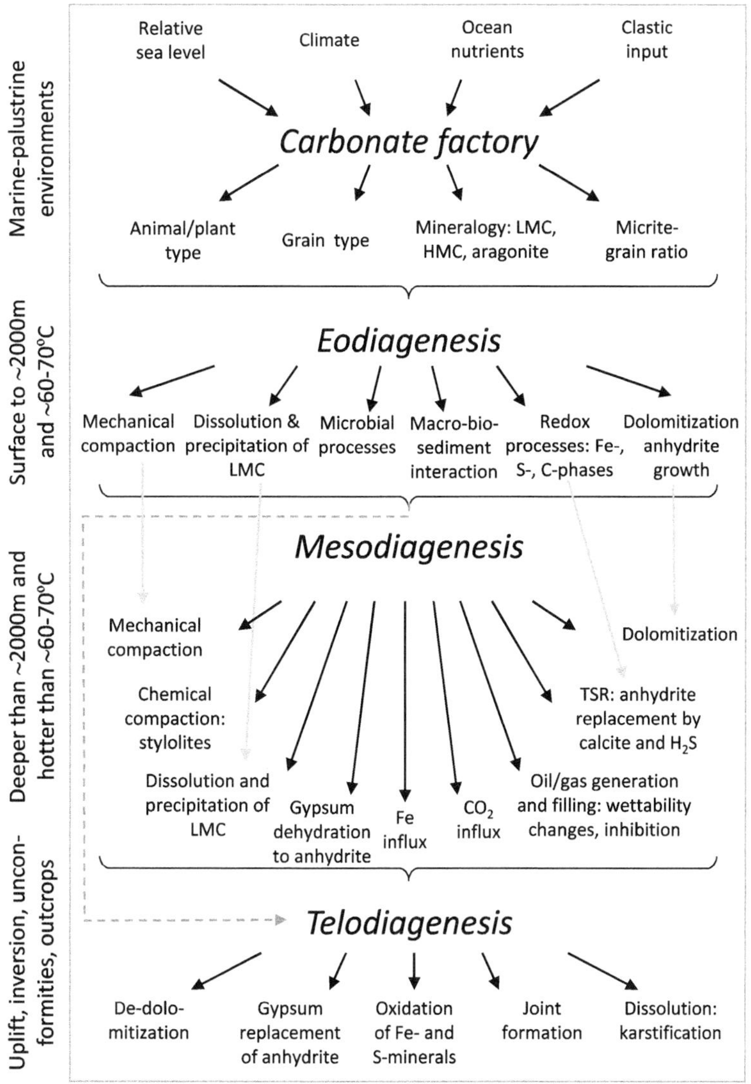

Fig. 2. Controls on carbonate reservoir quality split between the major controls of the origin of the carbonate grains: the 'carbonate factory', eogenetic (early/shallow burial diagenetic) controls, mesogenetic (burial diagenetic) controls and even telogenetic (uplift-related) influences. The carbonate factory is controlled by sea-level variations, climate, ocean currents and supply of nutrients and the amount of clastic material fed into the systems. Together these control the types and abundances of different life forms, grain types and mineralogy and the amount of matrix-micrite. If the split between eodiagenetic and mesogenetic is accepted to be about 2000 m and 60–70°C then mechanical compaction and the replacement of unstable mineral grains and matrix start in the eogenetic realm and continue into the earlier stages of the mesogenetic realm. Similarly, dolomitization and redox processes, such as sulphate reduction, occur in both the eogenetic and mesogenetic realms although under quite different conditions of temperature, stress and fluid composition. LMC, low magnesium calcite; HMC, high magnesium calcite.

quality prediction for clastic, carbonate and unconventional reservoirs. The aim of this paper, and the accompanying volume, is to achieve some level of consensus as to our true current understanding of the controls on reservoir quality across both communities.

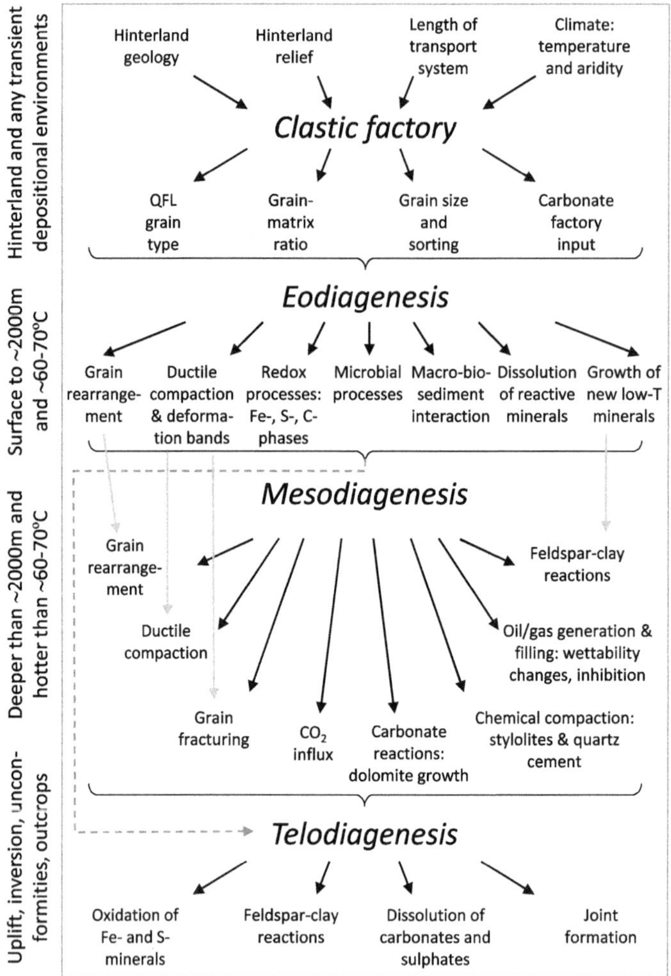

Fig. 3. Controls on sandstone reservoir quality split between the major controls on clastic grains: the 'clastic factory', eogenetic (early/shallow burial diagenetic) controls, mesogenetic (burial diagenetic) controls, and even telogenetic (uplift-related) influences. The clastic factory is controlled by geology in the source sediment area, climate, relief and the length of the sediment transport systems. Together these control the composition of the sand (QFL), the amount of matrix, sediment texture and the extent of *in-situ* (autochthonous) carbonate generation. If the split between eodiagenetic and mesogenetic is accepted to be about 2000 m and 60–70°C then the mechanical compaction processes of grain rearrangement and ductile grain compaction start in the eodiagenetic realm and continue into the earlier stages of the mesogenetic realm. Grain fracturing can start in deformation bands during shallow burial (eodiagenesis) and continue during burial into the mesogenetic realm. Similarly reactive grain dissolution and/or replacement by minerals stable at low temperature occurs in both the eodiagenetic and mesogenetic realms although under quite different temperature, stress and fluid composition conditions.

Overview of reservoir quality controls in carbonates and sandstones

Depositional characteristics of primary sediments

Even before diagenesis starts, the primary make-up of the sediment has a major role in influencing the history and reservoir quality evolution of the subsequent rock (**Nguyen *et al.* 2016**; Figs 2 & 3). In carbonates, the mineralogy of primary $CaCO_3$ is of fundamental importance to the subsequent reactivity of the sediment. The composition of seawater and global sea-level have fluctuated through time, leading to preferential precipitation of aragonite and/or high-Mg calcite at specific times (Sandberg

1983). Aragonite and high-Mg calcite are metastable and therefore the resultant carbonate is more reactive than limestone that is deposited during periods when low-Mg calcite is preferentially precipitated (Fig. 2). In sandstones the supply of sediment involves a wide variety of depositional minerals (quartz, feldspars and the spectrum of clay, carbonate and ferromagnesian minerals; Fig. 3). A new perspective and improved database of mineral distribution in modern clastic sedimentary environments may serve to help understand where minerals accumulate in specific environments and so help to develop appropriate analogues to be used to understand and predict mineral distribution in ancient and deeply buried rocks deposited in an equivalent depositional environment (Wooldridge *et al.* 2017*a*, *b*). Furthermore, some new carbonate plays involve Mg-clay minerals and their partial dissolution, raising important questions about the origin and diagenetic importance of siliceous minerals in carbonate systems (**Tosca & Wright 2015**).

Eodiagenesis (early/shallow burial diagenesis)

Diagenesis starts as soon as sediment has been deposited and the eventual reservoir quality is at least partly a function of the conditioning that the sediment experienced in the near-surface realm (Morad *et al.* 2000; Worden & Burley 2003; Figs 2–4). Early/shallow burial diagenesis, also known as eodiagenesis (Choquette & Pray 1970; Morad *et al.* 2000; Worden & Burley 2003), is the result of a combination of (1) changing conditions (e.g. pH, redox state and CO_2-saturation) compared with those in which sediment was initially deposited or formed; and (2) continued progress towards thermodynamic equilibrium. Eogenetic processes include meteoric infiltration, groundwater flow, bioturbation, microbial activity and seawater reflux leading to dolomite and anhydrite precipitation (Figs 2 & 3). The pivotal role of eodiagenesis is widely acknowledged in the carbonate community (Tucker & Wright 1990; Moore & Wade 2013; Tosca & Wright 2015) although the sandstone community seems to have taken longer to acknowledge the universal importance of the processes that condition a sediment before it undergoes mesodiagenesis. One of the most important eogenetic reactions in carbonates is the dissolution, or replacement by low-Mg calcite, of aragonite and high-Mg calcite. This can be facilitated by the flux of seawater within carbonate sediments above and below the sediment–water interface, and meteoric water during platform emergence. All of these processes tend to happen relatively soon after deposition and can result in widespread cementation and lithification of the primary sediment as well as a drastic redistribution of porosity prior to burial.

The problem of defining eodiagenesis. The definition of eodiagenesis and mesodiagenesis (i.e. burial diagenesis) is somewhat academic but it raises questions about the primary controls on diagenesis: depth, temperature, effective stress, fluid composition and mineral composition. Depth of burial is typically treated as a master control although rate of burial, depth-related heating and depth-related effective stress are the extrinsic variables that primarily need to be taken into account (Figs 2 & 3). Fluid composition is somewhat more important for carbonates than sandstones simply because carbonate minerals tend to be more soluble and have faster dissolution and precipitation rates than silicate minerals (see later section on 'Open v. closed diagenetic systems and secondary porosity'). Eodiagenesis in sandstones has been defined in terms of sediment that is (1) within the realm of subaerially influenced water (oxidized and with CO_2 influenced by surface biological activity), (2) buried to less than about 2000 m or (3) buried to depths where the sediment is at less than about 60 or 70°C (Figs 2 & 3; Choquette & Pray 1970; Morad *et al.* 2000; Worden & Morad 2003). In sandstones, temperature is probably the most important delimiter between eodiagenesis and mesodiagenesis (Fig. 3) since few mineral reactions happen in sandstones during burial between exiting the realm where microbial processes are dominant and moving to depths/temperatures where a semi-predictable suite of reactions (e.g. illitization of smectite, feldspar dissolution, quartz cement, dolomite growth) starts to occur. Elapsed time also plays a role, given that most diagenetic reactions are kinetically controlled (Ehrenberg *et al.* 2009). Thus rocks exposed to lower temperatures for longer periods may have comparable extents of diagenetic alteration to younger, hotter counterparts.

Why eodiagenesis matters to deeply buried sandstone and carbonate reservoirs. Eogenetic processes fundamentally condition the sediment prior to its burial and heating. Eodiagenesis can lead to total or partial occlusion of pores owing to mass cementation (e.g. by calcite) and it can lead to alteration (replacement or dissolution) of unstable detrital components (aragonite, feldspars, opaline silica, etc.) owing to interaction with groundwater. Eodiagenesis can lead to accumulation of normally minor components such as Fe-minerals during times of reduced sedimentation. The role of eodiagenesis in sandstone reservoir quality is becoming increasingly acknowledged, for example in terms of the possible creation of coats on detrital grains that

Fig. 4. Effects of eodiagenesis: nodules, nodular beds and concretions. (**a**) Upper Jurassic Portland Limestone with chert beds, Wessex Basin, UK (Maliva *et al.* 1999). (**b**) Lower Jurassic Bridport Sandstone, Wessex Basin, UK, with bioclast-derived calcite cemented nodules and nodular layers, Wessex Basin, UK (Bryant *et al.* 1988). (Images (a) and (b) courtesy of Richard Worden.)

then may inhibit later quartz cement (Dowey *et al.* 2017; Wooldridge *et al.* 2017*a*, *b*; **Daneshvar & Worden 2017**; Fig. 5). Eodiagenesis can lead to the early growth of carbonate and sulphate cements that either may inhibit compaction or be widely disseminated and so greatly damage porosity. Subsequent mesogenetic processes are contingent upon both the primary sediment character and the combination of eogenetic processes that were imposed on the primary sediment.

Dominant eogenetic minerals in sandstones and carbonates. The main pore-filling and replacing cementing minerals that develop during eodiagenesis of sandstones and carbonates are calcite, dolomite, siderite, pyrite, gypsum, kaolinite, smectite, berthierine (Fe-rich 7 Å clay), glauconite and amorphous or cryptocrystalline silica (Figs 2 & 3; Morad *et al.* 2012). These minerals can be found as eogenetic phases in both carbonates and sandstones. Eodiagenesis leads to locally pore-filling cements (commonly carbonate, less commonly sulphate) as nodules or concretions (Fig. 4); grain coats (Fig. 5), replacement of unstable detrital or diagenetic components such as aragonite or high-Mg calcite clasts, feldspar grains or lithic grains; and a suite of (typically biologically mediated) redox processes involving carbon, iron and sulphur compounds and minerals. Eodiagenesis of carbonates is dominated by the growth of calcite and dolomite (Tucker &

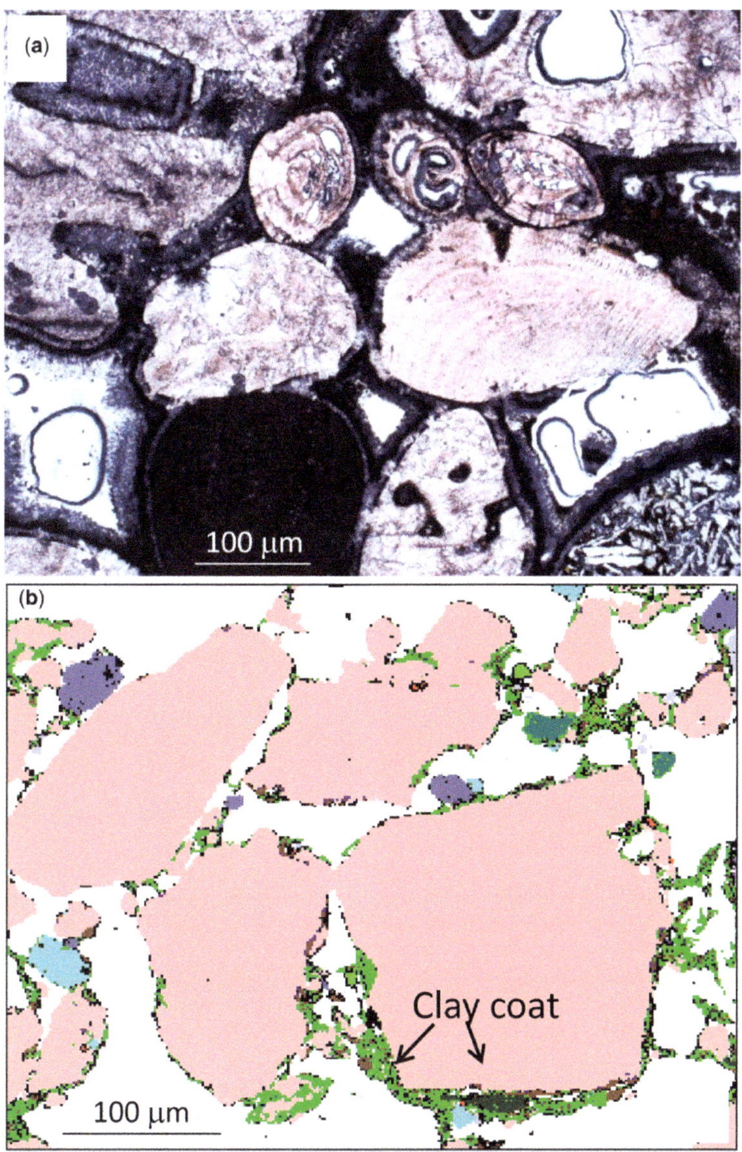

Fig. 5. Grain coatings on carbonates and sandstones formed during eodiagenesis: (**a**) light optical image of a modern carbonate sand, Bahamas, lime mud coating isopachous fringing aragonite cement. (**b**) QEMSCAN image of a modern sand, Ravenglass Estuary, NW England, UK, with clay minerals (green) coating quartz (pink), plagioclase (turquoise), K-feldspar (dark green) and dolomite (purple) grains. (Image (a) courtesy of Cathy Hollis, image (b) courtesy of Richard Worden, James Utley and Luke Wooldridge.)

Wright 1990; Moore & Wade 2013). Eodiagenesis of sandstones is dominated by the alteration of feldspars and other labile grain types to kaolinite and smectite and growth of Fe-minerals such as pyrite, siderite or Fe-clay depending on water composition (Worden & Burley 2003). Eogenetic growth of siderite, berthierine and glauconite can occur in carbonates and sandstones but only when there is an abundant supply of iron. Eogenetic silica cement as chert, grain-coating amorphous silica or microquartz grain coats (French *et al.* 2012; Worden *et al.* 2012; French & Worden 2013) can grow in carbonates and sandstones, typically requiring a biogenic source of amorphous silica such as diatoms, radiolaria or sponge spicules (Vagle *et al.* 1994; Haddad *et al.* 2006). Mg-clay growth and dissolution during eodiagenesis is now recognized as an important component in the reservoir quality evolution of some lacustrine carbonate reservoirs (Tosca & Wright 2015).

Meteoric and surface water influences during eodiagenesis. Meteoric water has extremely low salinity (e.g. 10–20 mg l^{-1} in temperate coastal regions), is saturated with atmospheric oxygen and is mildly acidic (pH 5.6) owing to dissolved atmospheric CO_2. As a consequence, the flow of meteoric water through recently deposited sediment will tend to cause a predictable series of diagenetic processes including: alteration of feldspars to kaolinite (or even gibbsite if there is sufficient meteoric water throughput), dissolution of relatively soluble minerals such as calcite or gypsum and oxidation of any reduced Fe-mineral (ferrous minerals such as pyrite, biotite or chlorite) (Figs 2 & 3). Meteoric water that percolates through soils and peats may develop high bicarbonate concentrations as it dissolves soil gas-derived CO_2 owing to respiration, microbial decay or fermentation. Such waters effectively become weak solutions of carbonic acid that are capable of promoting dissolution, forming leached sandstone horizons known as ganisters in coaly, deltaic sequences (Percival 1983, 1992) and karstification in carbonates (Dewever *et al.* 2010). Microbiological processes may convert sulphide minerals into dissolved sulphate (via bacterial sulphate reduction). These waters effectively contain sulphuric acid, which is capable of exacerbating dissolution processes, especially of carbonate minerals, in both limestones and sandstones.

Meteoric water will continue to dissolve carbonate minerals until the water is saturated with carbonate; if flow is accompanied by evaporation of the meteoric water, then the water can become supersaturated with calcite and lead to calcite growth within the sediment. In both clastic and carbonate sediments, this results in calcrete growth first, followed by dolocrete and then ultimately gypcrete growth (Schmid *et al.* 2003). The ability of flowing meteoric water to alter feldspars to clay minerals will continue until the meteoric-derived groundwater develops high metal concentrations, for example during evaporation (Jacobson *et al.* 1988). Under humid conditions, vast cave systems can develop in carbonate rocks with dissolution taking place from percolating groundwater in the vadose zone, and also at the top of the water table (phreatic caves). Karstic terrain will not always develop during emergence of a carbonate platform, however; there has to be sufficient time and water of the correct composition for dissolution to take place and sediment permeability has to be high enough to facilitate fluid flux.

If surface water holds fine suspended sediment and percolates through a porous, unsaturated zone, e.g. a vadose zone of an unconsolidated, typically low salinity aquifer or into intertidal granular sediments at low tide in estuaries or deltas where the water typically approaches seawater salinity (Barrie *et al.* 2015), then the fine suspended sediment may be retained within the granular matrix. The matrix acts as a filter and traps the fine suspended sediment; this is known as mechanical infiltration (Matlack *et al.* 1989; Moraes & De Ros 1990; Herringshaw & McIlroy 2013). Infiltration is one way to add fine sediment to previously clean, clay-free sediment. Infiltration only happens when turbid water soaks into unsaturated permeable (typically granular) sediment and is thus of importance to granular sediment with a water table (i.e. deltaic, fluvial or intertidal clastic sediment). Infiltration is widely cited as a control on the presence of clay minerals in granular sediments. Although there have been few studies of modern sedimentary environments where this has been proved, there is a burgeoning soil-science literature where infiltration is accepted as a major control on redistribution of fine materials in a granular matrix (Dowey *et al.* 2017).

Animal–sediment interaction and eodiagenesis. Bioturbation occurs when organisms live within sediment and disturb primary sediment structures; this occurs in both carbonate and clastic sediment (Taylor & Goldring 1993). Organisms live in recently deposited sediment to find shelter from predators, a home and/or nutrition within the sediment itself (Bromley 1996). Bioturbation physically moves sediment around and tends to mix primary sedimentary layers together. This mixing can lead to significant redistribution of porosity, which can result in the degradation of reservoir quality (Needham *et al.* 2005), enhanced vertical permeability and differential diagenesis, for example, preferential cementation of burrows (Walter & Burton 1990). Some animals can select the minerals they ingest; for example *Macaronichnus* preferentially ingests quartz, leading to quartz-rich back-filled burrows and the potential for locally enhanced permeability paths (Gingras *et al.* 2002). Microfauna, such as diatoms, as well as burrowing macrofauna, live in sediment (Vos *et al.* 1988), creating biofilms that result in a bulk sediment that is more coherent than expected (Schindler *et al.* 2015). Biofilms lead to the creation of thin coats of fine-grained minerals (i.e. clay and silt) on detrital grains (Wooldridge *et al.* 2017*a*, *b*) and the organically mediated precipitation of minerals, such as dolomite (Gingras *et al.* 2004).

Microbial influences on eodiagenesis. Microbial activity occurs in all near-surface sediments and needs to be accounted for in developing a holistic understanding of eodiagenesis and the early stages of reservoir quality evolution of sedimentary rocks (McMahon *et al.* 1992). Assuming there is sufficient biomass, different populations of bacteria utilize different sources of oxygen to survive in the sediment (Irwin *et al.* 1977). These sources include dissolved oxygen in water, oxygen derived from the reduction of nitrate to N_2, oxygen derived from the reduction of ferric to ferrous iron and oxygen derived from the reduction of sulphate to sulphide. The metabolic activities of bacteria typically convert organic forms of carbon into CO_2; this CO_2 is the ultimate fate of the oxygen that has been removed from solution as well as from nitrate, ferric iron, oxidized manganese and sulphate. Bacterial processes thus lead to a series of chemically reduced species of iron, sulphur, manganese and nitrogen and an abundance of CO_2 (McMahon *et al.* 1992). Framboidal pyrite, common in many types of sediment including sandstones and carbonates, is the result of eogenetic microbial activity. In non-marine settings, or where marine aqueous sulphate has been exhausted and where ferric iron is abundant, the reduced form of iron is free to make eogenetic minerals other than pyrite, such as siderite or Fe-clay minerals, such as berthierine. Even in fully marine sediment, where detrital ferric iron minerals are abundant, there may not be enough sulphate (and thus sulphide) available to convert all the newly reduced iron into pyrite, thus allowing for the growth of other Fe minerals such as siderite or Fe-clay minerals including berthierine or glauconite (Worden & Burley 2003). Microbially created carbonate minerals may also result from the metabolic production of CO_2; these are best identified by a combination of textural, compositional and stable isotope studies. It is now recognized that some carbonate reservoirs only exist owing to microbial activity (Lipinski *et al.* 2013; Bahniuk *et al.* 2015).

Dolomitization and eodiagenesis. On a global basis, many carbonate reservoirs comprise a proportion of dolostone. Although many conceptual models of dolomitization have been presented, it is now widely concluded that platform-scale dolomitization is most likely to occur from the reflux of evaporated seawater, since this is the most volumetrically abundant source of fluids with a high-Mg/Ca ratio (Adams & Rhodes 1960; Jiang *et al.* 2016). Reflux is interpreted to occur owing to the formation of brine pools on the platform top, particularly within arid environments. Evaporation leads to precipitation of gypsum, thereby decreasing the Ca/Mg of the fluid, and increases brine density. The brine sinks and refluxes seaward through the sediment pile, with dolomitization being facilitated by the slight increase in temperature and occurring preferentially in sediments with a high reactive surface area and/or a 'seed' (nucleation point) for dolomitization, such as high-Mg calcite (Jones & Xiao 2005; Jiang *et al.* 2014). It is not always necessary for brines to have reached gypsum saturation for reflux dolomitization to occur since there simply needs to be repeated flux of seawater (Garcia-Fresca *et al.* 2012; Newport *et al.* 2017). Typical features of platform-scale dolomitization by reflux include stratabound, fabric-selective and sometimes fabric retentive dolomitization, all at temperatures of less than *c.* 60°C. Basin- and field-scale analysis of carbonate reservoirs shows that dolomitization may improve permeability through a redistribution of porosity, creating a more effective pore network, although over-dolomitization (i.e. replacement and cementation) close to the source of the dolomitizing fluids may degrade reservoir quality (Lucia & Major 1994; Saller & Henderson 1998). Although dolomitization has been commonly reported in carbonate reservoirs, it has also been reported to occur in sandstones (Morad *et al.* 1992, 2012), especially in marginal marine depositional environments.

Mesodiagenesis (burial diagenesis)

The problem of defining mesodiagenesis. Burial diagenesis, sometimes known as mesodiagenesis (Choquette & Pray 1970; Morad *et al.* 2000; Worden & Burley 2003), has been defined in terms of sediment that (1) was physically separated from atmospherically influenced water, (2) was buried to greater than about 2000 m or (3) was buried to depths where the sediment is greater than *c.* 60 or 70°C. For sandstones, temperature is probably the most important factor differentiating between eodiagenesis and mesodiagenesis since few mineral reactions happen in sandstones during burial between relatively shallow depths where microbial processes are dominant and moving to depths/temperatures where a range of chemical reactions (e.g. illitization of smectite, feldspar dissolution, quartz cement, dolomite growth) start to occur. Elapsed time also plays a role, given that most diagenetic reactions are kinetically controlled. Thus rocks exposed to lower temperatures for longer periods may have comparable extents of diagenetic alteration with younger, hotter counterparts.

Typical mesogenetic processes in sandstones and carbonates. Typical, but not universal, mesogenetic processes, as sandstones are buried from 60 to >150°C, are (in approximate order) development of K-feldspar overgrowths, chloritization of berthierine and chlorite growth, smectite illitization, K-feldspar and plagioclase albitization and dissolution and replacement by illite, quartz cementation, K-feldspar reaction with kaolinite to create illite,

Fe-dolomite growth, stylolite formation, rare fracturing owing to shear, minor barite growth, minor pyrite growth and other minor or localized processes such as anhydrite growth. Most of these processes lead to a profound decrease in porosity although it is noteworthy that the creation of grain-coating chlorite or microquartz can impede quartz cementation and thus lead to better than expected reservoir quality (Ehrenberg 1993; Aase *et al.* 1996). In very deeply buried sandstones, the dominant remaining porosity type tends to be secondary pores (Fig. 6) within dissolved feldspar grains (Dutton & Loucks 2010).

Typical burial processes in carbonates include complete dissolution or replacement of aragonite and high-Mg calcite, often leading to the local creation of secondary porosity (Fig. 6), gypsum dehydration to anhydrite, pressure dissolution (chemical compaction), dolomitization, growth of calcite and/or dolomite spar that may become more Fe-rich with time, fracturing, secondary porosity generation and, locally, anhydrite reaction

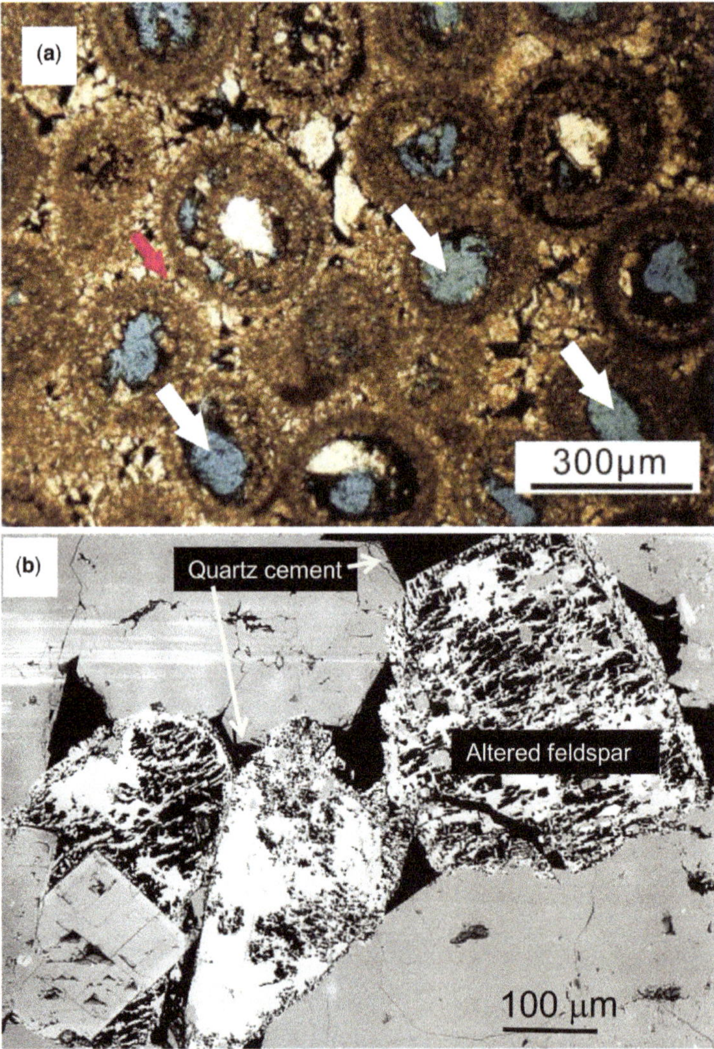

Fig. 6. Secondary porosity in sandstones and carbonates. (**a**) Lower Triassic Fiexianguan Formation, Sichuan Basin, China (Jiang *et al.* 2014). Secondary pores within ooids comprise a substantial fraction of the effective porosity. (**b**) Upper Triassic Chaunoy Formation, Paris Basin, France (Worden *et al.* 1999). Secondary porosity after feldspar dissolution led to ineffective porosity and clay mineral growth. (Image (a) courtesy of Lei Jiang and image (b) courtesy of Richard Worden.)

with petroleum via thermochemical sulphate reduction.

Mesogenetic carbonate and sulphate cements in sandstones and carbonates. Calcite and dolomite are, naturally enough, the dominant cements in carbonate reservoirs (**Barnett *et al.* 2015**; **John 2015**; Tosca & Wright 2015), but they are also common in sandstone reservoirs. Ferroan calcite and dolomite grow in carbonates during mesodiagenesis in rocks that contain, or have an influx of, iron. Since sandstones typically have a more complex detrital mineralogy than carbonates, they have a greater likelihood of containing abundant iron. Consequently, ferroan dolomite is a common mesogenetic cement in sandstones. If there is a very large amount of iron, exceeding the available amount of calcium and magnesium, then siderite can develop.

Anhydrite replaces eogenetic gypsum during burial and so is found in marginal marine carbonates deposited under relatively hot and arid conditions (Warren 1999). Anhydrite is an important cement in carbonate reservoirs since it reacts with petroleum fluids at elevated temperature producing toxic, corrosive and environmentally damaging H_2S via thermochemical sulphate reduction (Worden *et al.* 1995, 2000*b*; Worden & Smalley 1996). Anhydrite is also a relatively common, but minor, mesogenetic cement in sandstones, especially those that are interbedded with evaporite lithologies (Sullivan *et al.* 1994).

Quartz cementation during mesodiagenesis of reservoirs. Quartz is the single most important pore-occluding cement in deeply buried sandstones (McBride 1989; Bjørlykke & Egeberg 1993; Worden & Morad 2000; **Chudi *et al.* 2016**; **King & Goldstein 2016**; Nguyen *et al.* 2016; **Stricker & Jones 2016**; **Wells *et al.* 2015**). Quartz cement grows as a function of time, temperature and grain size (see later section on 'The use of forward diagenetic models to study reservoir quality'; Walderhaug 1994*a*, *b*, 1996). Effective stress has been proposed to influence quartz cementation (and see later in the 'Controversies' section on the controls on pressure solution; Bjørkum 1996; Sheldon *et al.* 2003). Fluid composition has also been proposed to influence quartz cementation (see later section on 'The effect of petroleum emplacement on sandstone and carbonate diagenesis'; Worden *et al.* 1998; Taylor *et al.* 2010; Worden *et al.* 2018). Quartz cement in sandstones can be inhibited by the presence of grain-coating materials such as microcrystalline quartz and chlorite (Ehrenberg 1993; Aase *et al.* 1996). Quartz can also be present in small amounts in carbonates, especially those associated with evaporitic lithologies.

Clay cementation during mesodiagenesis of reservoirs. Detrital clay minerals in primary sand deposits tend to be a complex mixture of illite, chlorite, kaolinite and smectite with less common accumulations of gibbsite, goethite, berthieriene, glauconite and palygorskite (Houseknecht & Pittman 1992; Worden & Morad 2003). In sandstones, mesodiagenesis lead to the conversion of smectite into illite and, to a lesser extent, into chlorite. During eo- and meso-diagenesis, gibbsite, goethite, berthieriene and palygorskite tend to react with other (detrital and eogenetic) minerals and aqueous species creating illite and chlorite. Kaolinite can be generated during mesogenetic feldspar alteration (Ehrenberg & Jakobsen 2001), but kaolinite tends to react with detrital K-feldspar to produce illite at elevated temperatures (Bjørkum *et al.* 1993). In the absence of K-feldspar, kaolinite transforms into dickite (a polymorph of kaolinite) at elevated temperature (Beaufort *et al.* 1998). Mesodiagenesis therefore leads to sandstones developing a simplified clay mineral assemblage that is typically dominated by illite and chlorite.

Illite is a very important pore-occluding mineral in clay-bearing sandstones (Nguyen *et al.* 2016), but can also be found in small quantities in carbonates. Where present, illite typically has a huge effect on reservoir properties such as permeability, water saturation (S_w), resistivity and wettability (Worden & Morad 2003; Worthington 2003).

Chlorite represents a complex group of Fe–Mg–Al-rich clay minerals that are seldom of much importance to carbonates but can be of paramount importance to sandstones since the Fe-rich variety of chlorite can coat detrital grains and inhibit quartz cementation (Dowey *et al.* 2012; Stricker & Jones 2016). Since Fe-rich chlorite limits the growth of quartz cement and so preserves porosity in deeply buried sandstones, it can be an important mineral in Fe-rich sandstones. Fe-chlorite can develop where there was insufficient supply of dissolved sulphate during eodiagenesis (to create sulphide and thus lock the iron up as pyrite) and insufficient supply of dissolved, typically biogenic, carbonate (to lock the iron up by creating siderite). Mg-rich chlorite can also be present in sandstones, typically from arid and semi-arid environments, but typically occurs as a pore-filling cement rather than a grain-coat phase (Worden & Morad 2003; Morad *et al.* 2010).

Other mesogenetic cements in sandstone and carbonate reservoirs. Cements of lesser importance to sandstone reservoir quality include pyrite, barite, albite, K-feldspar, halite and zeolites. Pyrite is a ubiquitous trace diagenetic mineral in carbonates and sandstones. Barite is a relatively common, but volumetrically minor, late diagenetic mineral in deeply buried sandstones (Khalifa & Morad 2012); barite is typically interpreted to be derived diagenetically altered detrital K-feldspar. Barite is relatively rare in carbonates, except those that have experienced hydrothermal mineralization (Neilson & Oxtoby 2008). Sandstones rich in detrital feldspar are routinely found to contain a few per cent feldspar cement, including albite and K-feldspar (Schmid *et al.* 2004). Feldspar cements are also found, albeit rarely, in carbonates (Spotl *et al.* 1996). Halite cement is common in sandstones and carbonates that are stratigraphically or structurally adjacent to evaporite beds or salt diapirs. Rarely, halite can fill pores and destroy reservoir quality when present in large volumes (Schroder *et al.* 2005). Zeolite cements are common in volcaniclastic sediments and can fill a significant proportion of the porosity. However, volcaniclastic sandstones tend to have low permeability owing to their large quantities of clay minerals and ductile grains (Zhu *et al.* 2012) and therefore are of relatively little economic significance in most basins.

Trace mesogenetic cements in sandstones and carbonates include base metal sulphides (galena, sphalerite), Fe-oxides, ilmenite, rutile and apatite. These minerals seldom influence reservoir quality in sandstone or carbonate hydrocarbon reservoirs.

Compaction and structural diagenesis

Compaction includes the family of processes that lead to porosity loss via grain reorganization (repacking), ductile grain deformation, brittle grain fracture and pressure solution (see later section on 'Pressure solution or chemical compaction'). Structural diagenesis broadly refers to the interaction of deformation and geochemical processes in rocks (Laubach *et al.* 2010). This term is generally applied where diagenetic reactions take place in association with deformation structures such as fractures, faults, and deformation bands. Structural controls on reservoir quality have received much interest from both sandstone and carbonate communities in recent years. Carbonates tend to become relatively hard and brittle soon after deposition, owing to the precipitation of eogenetic framework-strengthening cements (Fig. 2), and they subsequently tend to undergo fracturing in response to changes in stress. Sandstones can be relatively weak and capable of substantial mechanical compaction (by grain reorientation, grain re-packing, and ductile deformation) and large-scale (bed-scale) ductile deformation owing to changes in stress until mesodiagenesis commences. Early cementation of sandstones, e.g. by carbonate cement, naturally strengthens the rock and inhibits mechanical compaction (Fig. 3).

Normal compactional processes occur in response to the progressive increase in vertical effective stress associated with increasing overburden thickness during normal sedimentation and are important since they lead to porosity loss. Loose granular materials respond to increased effective stress by realigning grains to adopt a close-packed (lower porosity) structure, weak-brittle grains undergo fracturing and disaggregation while weak-ductile grains undergo ductile deformation (Worden *et al.* 2000*a*). The magnitude of compactional porosity loss in the absence of pore-filling materials is a function of the mechanical properties, size distributions and abundances of the constituent clastic particles together with the magnitude and orientation of the 3D stress distribution (Pittman & Larese 1991; Chuhan *et al.* 2003; Karner *et al.* 2005). Stabilization of sediment by cement precipitation can inhibit compaction by creating a strong framework (Pittman & Larese 1991). Carbonates and sandstones typically have quite different compactional responses to increasing effective stress since carbonates tend to lithify (cement) and strengthen soon after deposition.

Under certain circumstances, stylolites (chemical-dissolution/pressure solution seams; see later) may develop. In carbonates, this can happen at depths greater than *c.* 300–600 m (30–40°C; Ebner *et al.* 2009) whereas in sandstones this only happens in rocks at greater than *c.* 2500 m (>90°C; Bjørkum 1996). This difference in depth and temperature probably reflects the much lower solubility and dissolution rates of silicate minerals than carbonate minerals. In both carbonates and sandstones, stylolites may nucleate on clay, or organic-rich, laminae or at grain contacts with enhanced solubility (e.g. quartz-muscovite grain contacts).

Fracturing and deformation bands in sandstone reservoirs. Fracture-related porosity is viewed as paramount and beneficial to reservoir quality in many carbonate reservoirs. Fractures can form in response to burial (vertical loading), in the vicinity of salt diapirs and displacive faults or in response to folding. Fractures have also been recognized as critical for production in many tight gas sandstone reservoirs. In contrast, shear stresses in porous sandstones commonly result in fine-grained gouge in the axis of a fault. Away from the axis of a fault, cataclastic deformation bands (**Busch *et al.* 2015**; **Griffiths *et al.* 2016**) are common in poorly lithified sandstones (i.e. not deep or hot enough to have experienced mesogenetic cementation) with deformation bands being the result of localized shear, sand grain disaggregation and locally enhanced rates of quartz cementation. Deformation bands may also form from vertical loading that is not related to tectonic stress (Fossen 2010). Deformation bands undergo localized quartz cementation far earlier (at lower temperatures) than their undeformed host owing to locally increased surface area and generation of highly reactive quartz fracture surfaces (Lander *et al.* 2008*a*, *b*; Griffiths *et al.* 2016).

Reservoir quality can be affected by deformation bands owing to the locally reduced porosity and permeability. Deformation bands are less commonly reported from carbonates, probably because the host sediment is stronger and not capable of the creation of shear bands. It remains to be seen whether deformation band fracturing mechanisms are truly dependent on host-reservoir lithology.

Fracture porosity and carbonate reservoirs. The form and distribution of fractures in carbonates has been widely studied over the last 10 years. In particular, the mechanical stratigraphic layering of carbonate platforms has been well described; dissipation of stress varies according to bed thickness, such that more numerous (i.e. more narrowly spaced), shorter fractures tend to occur in thinner beds while thicker beds have more widely spaced, longer fractures (Nelson 2001). Some fractures cut across bed boundaries, while others terminate or slip at bed contacts, particularly where there is a contrast in competency between rigid carbonate beds and ductile, clay-rich beds. Fracture densities also vary between lithologies (e.g. limestone and dolomite), between facies with different petrophysical properties and with distance from faults (Korneva *et al.* 2017 – online). Nearly all carbonate reservoirs are fractured, but the importance of those fractures to reservoir performance depends on matrix porosity, fracture pore volumes and connectivity of the fracture network.

Telodiagenesis (uplift-related diagenesis)

The diagenesis of sedimentary successions that have been exhumed during uplift or basin inversion is referred to as uplift-related diagenesis, or telodiagenesis (Choquette & Pray 1970; Abdel-Wahab & Turner 1991; Morad *et al.* 2000; Worden & Burley 2003). The most extreme case of telodiagenesis is when the overburden to a given suite of sediment rock is totally eroded so that these rocks now sit at the Earth's surface. Under these circumstances, telodiagenesis is effectively the same as weathering and includes the processes of (1) oxidation of reduced phases such as pyrite, Fe-clay minerals and Fe-carbonates, and (2) dissolution or alteration of minerals by low pH groundwater including alteration of feldspars to clay minerals and dissolution of carbonate and sulphate minerals. Note that weathering reactions of sedimentary rocks tend to reverse what happened during the previous burial diagenesis event. There are also physical consequences of uplift and removal of overburden, including the development of decompaction joints (Neuzil 2003) that serve to allow the fluids that cause weathering to penetrate metres or even 10's of metres below the surface.

During the earlier stages of uplift, before sedimentary rocks re-emerge at the Earth's surface, it is possible that other processes may start to alter the mineralogy and fabric of a sandstone or a carbonate. Dedolomitization may occur in carbonates (Fig. 2); this is a process whereby dolomite is replaced by calcite during uplift, especially in aquifers (Jacobson *et al.* 2010). Calcite in carbonates and sandstones can be dissolved leaving enhanced porosity (**Mahdi & Aqrawi 2017**). Anhydrite can be replaced by gypsum if there is sufficient water available (Fig. 2). Silicate reactions, typically orders of magnitude slower than carbonate and sulphate reactions, are less likely to occur until the sediment is very close to the Earth's surface. Note that shales do not typically decompact during uplift (until the weathering zone is reached) and that shale porosity values can be used to estimate the degree of uplift (Law 1998): mesogenetic compaction processes are considered to be largely irreversible although decompaction joints are a common attribute of brittle rocks at outcrop (Figs 2 & 3). Finally, bacterial sulphate reduction can occur, leading to dissolution of sulphate and its local replacement by calcite, and biodegradation of hydrocarbons, significantly reducing oil quality (Saller *et al.* 2014).

When sedimentary rocks have been exhumed but have not been exposed at the Earth's surface, then the weathering reactions of oxidation, feldspar alteration to clay and carbonate and sulphate mineral dissolution can still occur if the overlying rocks allow the ingress of surface-related fluids either through a permeable matrix or via fractures and joints. The telodiagenesis realm in low-permeability non-brittle rocks may extend to no more than a few metres below the Earth's surface (Armitage *et al.* 2016). The telodiagenesis realm in sedimentary rocks with considerable matrix permeability (which to hydrogeologists are known as aquifers) can extend for hundreds of metres or more below the Earth's surface.

It is worth reflecting on the fact that every unconformity surface, in all sedimentary successions, has probably experienced telodiagenesis. It is also worth adding a note of caution that for all analogue studies of sedimentary rocks at outcrop, the mineralogy and reservoir quality of the outcrop are unlikely to represent what would be found in equivalent rocks that have not undergone uplift.

Methodologies applied to reservoir quality prediction for sandstone and carbonates

Analytical techniques

There is almost total overlap in the range of analytical tools and techniques that have been used for reservoir quality prediction in sandstones and carbonates (Fig. 7). Reservoir quality is often first considered using routine techniques such as seismic interpretation, wireline logging and core analysis, irrespective of host reservoir type. Sedimentological core description, to reveal rock type, lithofacies, sedimentary and tectonic structures and macroporosity, is also fundamental to both sandstones and carbonates. There is also increasing emphasis on the application of trace fossil types and abundance

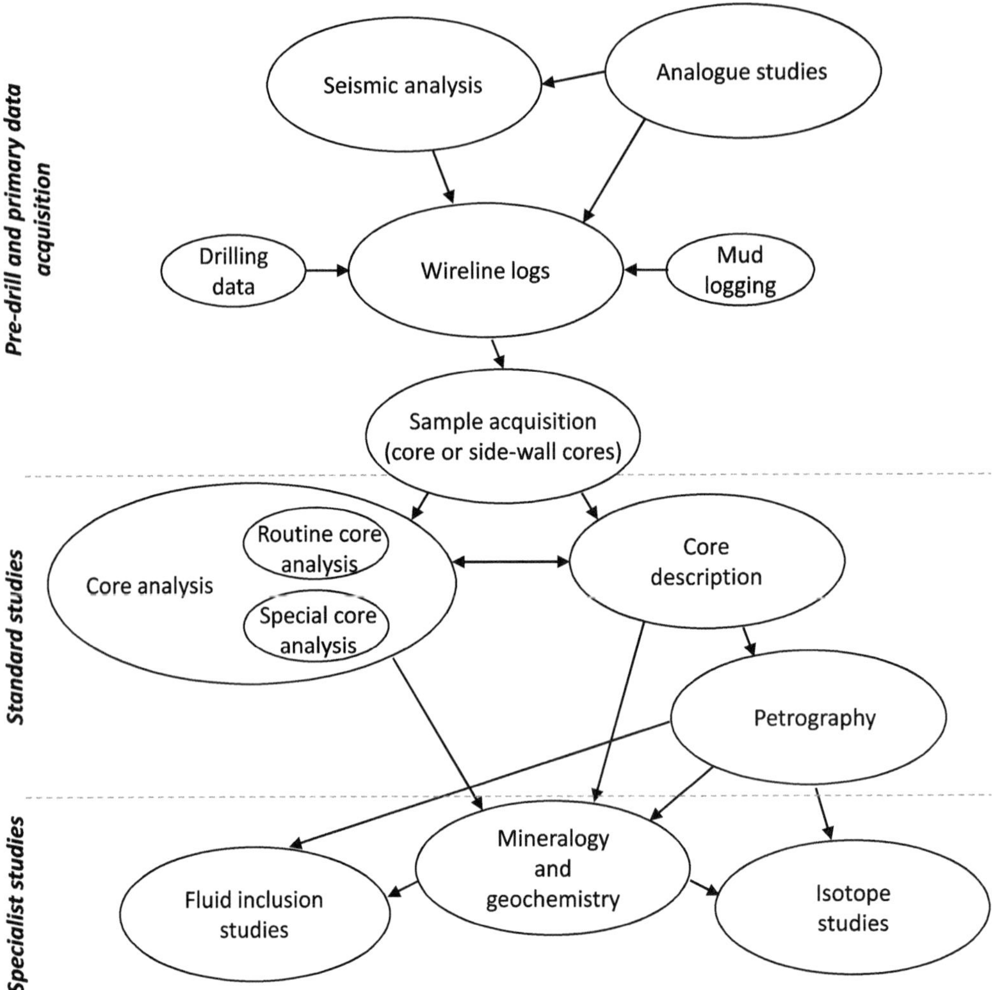

Fig. 7. Carbonate and clastic reservoir quality communities use many of the same techniques and data types and could learn much from each other. The first available data are seismic (e.g. attribute) data although analogue studies also help inform decisions. Drilling wells provides data via mud-logging (cuttings analysis) and drilling conditions (e.g. rate of penetration data). Wireline and subsurface logging provide high-quality indirect reservoir-quality data. If core is collected then core analysis (routine and special) follows. Routine core analysis includes porosity, permeability, grain density and fluid saturation measurements. Special core analysis includes wettability, relative permeability and pore size distribution measurements. Detailed core description should come next with the production of a detailed sedimentary log. In order to understand the link between core description and core analysis and log data, petrographic techniques need to be applied, preferably to core plugs. After petrographic analysis, there is a range of specialist techniques (e.g. mineralogy analysis using SEM-EDS, fluid inclusions studies and radiogenic and stable isotope techniques) that can be applied to answer questions about the timing of cement growth, the timing of petroleum filling, the effect that petroleum had on reservoir quality, the origin of the materials that caused cementation, the type of water that facilitated diagenesis and reservoir quality evolution.

(ichnology) to help reveal subtleties of the environment of deposition as well as the influence of bioturbation on reservoir quality.

A wide variety of petrological techniques are routinely used to describe sandstones and carbonates to help make sense of core analysis data in light of rock fabric and diagenetic history. Standard petrological techniques include transmitted and reflected light optics, secondary electron imaging of broken rock chips, back-scattered electron microscopy (BSEM) of polished sections, cathodoluminescence techniques integrated with light optics or BSEM (SEM-CL) images (Griffiths *et al.* 2016; Nguyen *et al.* 2016; Wells *et al.* 2015), and point chemical analysis of secondary X-rays in the SEM using energy dispersive spectral (EDS) detectors. Optical and compositional analysis of detrital heavy minerals has been used as a provenance tool while trace element analysis of diagenetic cements has been used to determine the sources of cement and interpret formation water origin and migration (**Götte 2016**; Kraishan *et al.* 2000).

Mineralogical analyses (to give proportions of component minerals) of cuttings and core samples is applied to both sandstone and carbonate successions using X-ray diffraction (**Gier *et al.* 2015**) and infrared techniques, supplementing point counting with a petrographic microscope. Chemical analysis of cuttings and core samples is less frequently applied to sandstone and carbonate successions but can be achieved using benchtop X-ray fluorescence (Gier *et al.* 2015), portable X-ray fluorescence and inductively coupled plasma-based tools used to solve questions of lithology determination, sediment provenance and chemical stratigraphy specifically to help build reservoir models through improved correlation.

A new generation of integrated SEM-EDS tools, such as QEMSCAN (Armitage *et al.* 2010, 2013*b*; Daneshvar & Worden 2017), now permits automated mineral identification and mapping of areas from standard polished sections (6 cm^2), or larger if desired (Fig. 5). QEMSCAN analysis primarily provides detailed, repeatable, quantitative mineralogy based on chemical analysis of micron-sized spots collected at pre-defined points across a polished section. If the electron beam spacing is small enough (i.e. approaching 1 μm between the analytical points) then a high-resolution mineral map of the thin section is produced. The image of the rock reveals the distribution of mineral and porosity across the selected area; this provides a valuable stepping stone between thin section description and core description as well as providing information about porosity and its distribution. QEMSCAN analysis cannot routinely differentiate different forms of the same mineral (e.g. calcite cement from calcite grains, quartz cement from quartz grains), but future integration of CL detectors may lead to increased functionality of these automated approaches.

A suite of geochemical techniques are also routinely applied to both sandstones and carbonates. These include a range of fluid inclusion techniques (Wells *et al.* 2015) such as homogenization thermometry to reveal trapping temperature, freezing point depression to reveal aqueous salinity and fluid inclusion compositional analysis using thermal phase changes as well as microbeam techniques. Conventional light element stable isotope analyses (H, C, O, S) have also been applied to sandstones and carbonates for many years with C and O stable isotopes proving especially useful for constraining the origin, timing and conditions of carbonate cement growth (King & Goldstein 2016). More recently clumped C–O isotopes have proved invaluable to determination of the temperature of carbonate mineral growth in both sandstones and carbonates (John 2015). Radiogenic isotope analysis (e.g. $^{87}Sr/^{86}Sr$) is also used, more in research than routine cases, to constrain the timing and origin of carbonate cements (King & Goldstein 2016). Radiometric analysis for dating (e.g. using K–Ar techniques) has been applied to sedimentary and diagenetic problems with interesting and thought-provoking results, but this is not routine in reservoir quality studies. The analysis of less conventional stable isotopes (Mg, Fe, etc.; Geske *et al.* 2015) is now possible although the science of metal isotopes is in its relative infancy and has not yet found routine application for reservoir quality analysis.

The use of modern analogues to study reservoir quality

Carbonate sedimentology and reservoir quality prediction has developed in major ways based upon the study of modern (Holocene) sedimentary analogues (Dravis 1979; McClain *et al.* 1992; Whittle *et al.* 1993; Macintyre *et al.* 1996; Harris 2010). The sandstone community has more recently started using modern analogues to help develop predictive models of reservoir quality with some focus on understanding carbonate cement distribution, especially in aquifers, marginal marine settings and arid intracontinental soils (Arakel 1986; McBride *et al.* 1995; McBride & Parea 2001). New focus on the use of clastic depositional analogues to help with reservoir quality prediction has arisen with an imperative to understand the origin and distribution of grain-coating clay minerals that serve to inhibit quartz cement and thus preserve porosity, in deeply buried sandstone reservoirs (Dowey *et al.* 2017; Wooldridge *et al.* 2017*a*, *b*; Daneshvar & Worden 2017), but with much work yet to be done to extend the approach to carbonate cemented marine sandstones and their relationship to modern clastic sedimentary systems. There is much still to learn about reservoir quality prediction from the study of modern clastic environments.

Experimental simulation of diagenesis

A useful way to develop new understanding about sandstone and carbonate reservoir quality and diagenesis is to perform experiments in laboratories. No experiment can fully reproduce natural conditions, not least since most geological processes take much longer than the time available to researchers. There are a variety of ways to speed up processes, including using higher temperatures than found in diagenetic systems; an example of this are routine rock-eval pyrolysis measurements of source rock maturity and richness (Hunt 1995), simulation of dolomitization (Kaczmarek & Sibley 2011) and measurements of the rate of redox reactions during thermochemical sulphate reduction (Cross *et al.* 2004). Reaction rates can also be increased by using analogue materials that are more soluble than silicates or carbonates (Sathar *et al.* 2012) and fluid compositions that are exceptionally far from equilibrium with the host rock (Heald & Renton 1966; Chermak & Rimstidt 1990; Lander *et al.* 2008*a*).

Experiments on compaction and mineral diagenesis have, thus far, more commonly been applied to sandstones than carbonates, this possibly owing to (1) the greater importance of mechanical compaction for sandstones than carbonates and (2) the more complex mineralogy and geochemistry of sandstones than carbonates being more amenable to experimental simulation. However, chemo-mechanical compaction of fine-grained carbonates has been successfully reproduced during experiments under low temperature conditions, proving the importance of water composition (**Wang *et al.* 2016**), as well as the effect of the presence of oil v. water on compaction, pressure solution and porosity loss (Risnes *et al.* 2003). Much work has recently been undertaken on the experimental simulation of reactions between CO_2 and host rocks to determine the effects of carbon capture and storage on reservoir and caprock quality, utilizing realistic low temperatures and relatively long-duration experiments (Armitage *et al.* 2013*a*).

Experiments on the eogenetic evolution of clastic sedimentary systems have revealed interesting roles for animal–sediment interactions (McIlroy *et al.* 2003; Needham *et al.* 2005, 2006; Worden *et al.* 2006), as well as more conventional physical processes such as infiltration (Matlack *et al.* 1989). As the recognition of the importance of the biological origin of grain-coating materials develops (Wooldridge *et al.* 2017*a*, *b*), then new questions emerge about the durability of these grain coats that demand to be tested under experimental, as well as empirical (i.e. from modern analogues), conditions. The role of biological grain coats on the mechanical properties of sediment also needs to be determined to assess with matrix stabilization resulting from sediment–biological interactions.

The use of forward diagenetic models to study reservoir quality

Over the past 20 years, reservoir quality assessment has expanded from a focus on characterization to encompass process-oriented models. These models have been designed to predict rock properties in undrilled locations and reconstruct properties through geological time. Augmenting the discipline with models has reinforced, and not detracted from, the importance of rock characterization given that such data are essential for constraining parameters and testing performance. In addition to their obvious utility in diagenetic research and hydrocarbon exploration and production, diagenetic models have important applications in geothermal energy production and waste disposal and are vital components for constraining rock properties through time in fields as diverse as palaeohydrology, structural deformation and seismology. Notable model types include reactive transport models, coupled effect-oriented compaction and cementation models, and grain/pore scale geometrical models. Diagenetic models of all types benefit from integration with depositional models that constrain the composition and texture of sediments and from 3D basin models that constrain thermal, stress and fluid flux histories.

Reactive transport models (RTMs) consider a comprehensive set of aqueous geochemical reactions while accounting for mass transport by fluid advection and diffusion (Bethke 1996, 2008). These models account for the thermodynamic driving forces for reactions but also consider reaction kinetics. The grid cells in such models generally have dimensions on the orders of metres, with time scales that may range from hundreds to millions of years. RTMs have been widely applied to carbonate systems, especially in the area of dolomitization (Jones *et al.* 2002, 2003, 2004; Xiao & Jones 2006; Whitaker & Xiao 2010; **Consonni *et al.* 2016**). They also have been used to simulate silicate reactions associated with sandstone and mudstone diagenesis (Czernichowski-Lauriol *et al.* 1996; Thyne *et al.* 2001; Brosse *et al.* 2003; Sanjuan *et al.* 2003; Xiao & Jones 2006; Yuan *et al.* 2017; **Geloni *et al.* 2015**). Geloni *et al.* (2015) incorporated compaction into their RTM simulations, thereby addressing an important factor that has been missing from most previous RTM studies. Although RTMs have been widely applied to waste disposal and CO_2 injection problems (Pollyea & Fairley 2012; Steefel *et al.* 2015), to date we are unaware of published studies where results from RTM simulations served as a basis for making pre-drill porosity and permeability predictions that were subsequently evaluated by post-drill sample characterization. Challenges associated with using RTMs for reservoir quality prediction studies include providing appropriate boundary conditions for fluid flow through geological time, defining grain-scale reactive surface areas and their evolution with diagenetic alteration, applying the appropriate reaction kinetics (particularly for silicate reactions), incorporating compaction and linking diagenetic alteration to permeability and other rock properties (Bartels *et al.* 2005; Xie *et al.* 2015).

Coupled effect-oriented compaction and cementation models, unlike RTMs, do not have a thermodynamic foundation, although they do incorporate theoretically robust algorithms (Ajdukiewicz & Lander 2010). To date the focus of such models has been on sandstones, where compaction is an important mechanism of porosity loss and where geochemical reactions mainly involve relatively insoluble aluminosilicates. The model scale of reference is comparable with a core plug and simulations are designed to be directly comparable with petrographic and core plug data (Lander & Walderhaug 1999). Simulation timescales typically span millions of years. Compaction simulations generally consider mean grain size and sorting, variations in the mechanical properties of framework grains and pore-filling materials, as well as effective stress. Additionally, coupled effect-oriented compaction and cementation models incorporate kinetic models for quartz overgrowth cements (Walderhaug 1994*b*, 1996), plagioclase albitization (Perez & Boles 2005), the reaction of kaolinite and K-feldspar to form fibrous illite (Franks & Zwingmann 2010; Lander & Bonnell 2010), and the reaction of volcanic rock fragments and other labile mineral grains to form grain-coating chlorite (Bonnell *et al.* 2014). Other diagenetic processes such as carbonate cementation and replacement may be considered in a simulation but currently are not predicted *a priori* except where they occur as a by-product of albitization and volcanic rock fragment alteration. A prerequisite to applying models of this type involves an optimization procedure where model parameter values are obtained that provide an optimal match with samples that have associated petrographic, core analysis and burial history data. Pre-drill reservoir quality predictions may be made by accounting for the compositions, textures and burial histories of the prediction locations. This modelling approach reproduces wide ranges in observed extents of diagenetic alteration in sandstones (Walderhaug *et al.* 2000; Taylor *et al.* 2004; Perez & Boles 2005; Lander *et al.* 2008*a*; Tobin *et al.* 2010; English *et al.* 2017) and has proven to be an accurate basis for pre-drill prediction of porosity, permeability, seismic velocities and other rock properties (Taylor *et al.* 2010, 2015; Wojcik *et al.* 2016; Chudi *et al.* 2016). Current limitations of the approach include its applicability only to sandstones and siltstones, consideration of a restricted set of geochemical reactions and the lack of mass transfer simulation. Challenges in applying the approach for reservoir quality prediction include obtaining adequate analogue datasets for parameter optimization and assessing when diagenetic processes not considered by the modelling approach (e.g. carbonate cementation) may materially impact reservoir quality.

A final class of diagenetic models simulates processes at the grain and pore scale. Such models are enticing because they have the potential to predict textures and morphologies that may be compared directly with data from natural samples and laboratory experiments and because they serve as a more rigorous basis for predicting a wide range of rock properties compared with the model types discussed above. To date, most grain/pore scale simulation studies either have involved imposing a predefined diagenetic state as a means to rigorously simulate the impact of diagenetic alteration on bulk rock properties (Bakke & Øren 1997; Øren & Bakke 2002, 2003; Øren *et al.* 2007; Jin *et al.* 2012; Mousavi & Bryant 2013; Prodanovic *et al.* 2013; van der Land *et al.* 2013; Hosa & Wood 2017) or consider a subset of the diagenetic processes that affect reservoir quality (Abe & Mair 2005; Lander *et al.* 2008a; Marketos & Bolton 2009; Gale *et al.* 2010; Cheung *et al.* 2013; Ankit *et al.* 2015; Lander & Laubach 2015; Wendler *et al.* 2016). Although the promise of models of this class is enormous, considerable work is needed to develop comprehensive modelling approaches that are capable of making accurate pre-drill predictions.

Improvements in our ability to accurately predict reservoir quality in both clastics and carbonates probably will arise not only from new developments within each of the model types discussed above but also by linking these types into integrated modelling systems that take advantage of the strengths that each has to offer. For instance, the reaction and mass transport capabilities of an RTM could be enhanced by coupling with a grain/pore scale model that rigorously simulates compaction and dissolution/precipitation in 3D and that predicts reactive surface areas in addition to fluid transport and mechanical properties.

Controversies

Pressure solution or chemical compaction?

Pressure solution, now routinely called chemical compaction by some sandstone diagenesis practitioners, is controversial because it is unclear whether it is actually driven by pressure or simply by elevated solubility owing to increased temperature or even by specific mineral–chemical conditions such as at mica–quartz interfaces (Oelkers *et al.* 1996; Sheldon *et al.* 2003; Walderhaug *et al.* 2006; Kristiansen *et al.* 2011). In fact, it is not pressure that is the main variable; rather it is effective (or contact) stress that needs to be accounted for where effective stress is equal to the lithostatic pressure minus the fluid pressure. Contact stresses show major variations along a given grain–grain contact with the stress gradient, causing a chemical potential gradient and thus leading to diffusion out of the contact zone (Sheldon *et al.* 2003). Pressure solution happens more readily (shallower) in carbonates than in sandstones but there seems to be little debate about the role of effective stress v. temperature in carbonates. Some relatively shallow-buried (<600 m) carbonate successions contain horizontal stylolites owing to normal burial (Ebner *et al.* 2009) that are cut by vertical stylolites owing to thrust tectonics (Starmer 1995); this suggests that the direction of principal effective stress (pressure) is the master control, at least in carbonate successions (Fig. 8). However, note that the role of thermally mediated processes involved in chemical compaction, so widely acknowledged in the sandstone world, may have lessons for carbonate reservoir quality prediction. If pressure solution is indeed driven by increasing effective stress then high degrees of (early) overpressure could lead to anomalously elevated reservoir quality owing to restricted amounts of cement (Stricker & Jones 2016). In contrast, if pressure solution is not driven by increasing effective stress then early overpressure cannot be the cause of reduced amounts of cement and enhanced reservoir quality.

Open v. closed systems and secondary porosity

Some older models of sandstone diagenesis invoked large-scale water flow systems as the main controls on reservoir quality (Giles *et al.* 2000); these tended to assume relatively open geochemical systems. Some invoked a vital role for organic acids derived from source rocks (e.g. enhancing the concentration

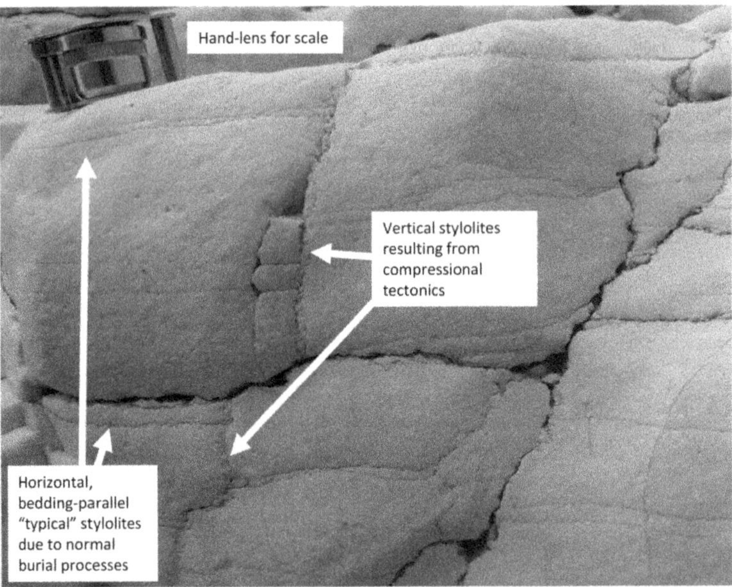

Fig. 8. Upper Cretaceous Chalk outcrop from Selwicks Bay, Flamborough Head, UK (Starmer 1995). The Chalk developed bedding-parallel stylolites during normal burial (to *c.* 1000 m). The region underwent compressional tectonics with the main Flamborough Head Fault Zone, about 50 m from this site; compression resulted in a transient change in direction of the principal stress to horizontal and led to the development of vertical stylolites. The structures in this rock suggest that pressure (or more correctly, effective stress) plays a major part in the chemical compaction of rocks, thus apparently contradicting Bjørkum (1996) in which it was speculated that pressure plays a negligible role in chemical compaction. (Image courtesy of Richard Worden.)

of scarcely soluble Al and Si; Barth & Riis 1992). Alternative models have recognized the limited complexing ability of organic acids and their scarcity in formation waters and now tend to assume a relatively closed system for sandstone diagenesis (Lander & Walderhaug 1999). In fact the degree of 'openness' is likely to be partly a function of the aqueous solubility of a given species in a sandstone with alumina, silica and TiO_2 being effectively immobile (and thus part of a closed system) while CO_2 is demonstrably part of an open diagenetic system in sandstones (**Király *et al.* 2016**). Observations from cemented veins routinely show limited penetration of the cementing fluid from the vein into the matrix, suggesting that mass-import of material cannot be a process that affects the bulk properties of a porous rock unit (reservoir; Fig. 9). Nonetheless, there are detailed, large-scale studies of diagenetic systems that seem to require loss and gain of species such as calcium and magnesium (Fig. 9; Consonni *et al.* 2016; **Poteet *et al.* 2016**), potassium (Gier *et al.* 2015) and even aluminium to explain the distribution of minerals in reservoirs (Nguyen *et al.* 2016).

Fundamental proof of the openness of CO_2 in diagenetic systems comes from (1) pyrolysis experiments that show that CO_2 is generated from source rocks in large volumes (Andresen *et al.* 1994; Hunt 1995) and (2) the $\delta^{13}C$ values of carbonate cements reflecting external input in many systems (Macaulay *et al.* 2000). Carbonate rocks probably involve open system diagenesis during dolomitization (Fig. 9; Whitaker & Xiao 2010; Jiang *et al.* 2014; Consonni *et al.* 2016) and are also probably susceptible to influxes of CO_2 during kerogen maturation (Heasley *et al.* 2000).

Secondary porosity in carbonates, previously accepted as a routine cause of enhanced reservoir quality, has recently become controversial with issues surrounding whether it happens, what it looks like and whether it actually enhances overall reservoir quality (Ehrenberg *et al.* 2013). Despite published scepticism, mesogenetic porosity has proved to be essential to the effectiveness of reservoirs in some carbonate systems (Barnett *et al.* 2015; Poteet *et al.* 2016). Interestingly, the perception of the importance of secondary porosity in sandstones has risen and fallen but now seems to be recognized as crucial in some tight gas sandstones and other deeply buried sandstones where secondary porosity (after feldspar dissolution) is just about the only remaining porosity (Dutton *et al.* 2012; Nguyen *et al.* 2016).

The question of open v. closed systems is probably too simplistic for both sandstones and carbonates. Most cases probably fall somewhere between being at least partly a function of the species under consideration, the scale of observation, and palaeohydrology of the system. Better application of geochemical considerations, more routine use of high-precision quantitative mineral analyses and high-density sampling, and more rigorous palaeohydrological studies need to be made by the reservoir quality community to better assess the role of material flux for reservoir quality analysis of carbonates and sandstones.

The effect of petroleum emplacement on sandstone and carbonate diagenesis

The question of the role of petroleum emplacement on sandstone diagenesis (and the possible preservation of porosity) has fired up much excitement with numerous papers seeking to address the problem from a combination of theoretical, oil-field data and experimental approaches (Marchand *et al.* 2001; Bloch *et al.* 2002; Taylor *et al.* 2010; Sathar *et al.* 2012; Wells *et al.* 2015). Some have concluded that petroleum emplacement has no influence on sandstone diagenesis (Taylor *et al.* 2010). Some have concluded that petroleum emplacement stops sandstone diagenesis (Gluyas *et al.* 1993), while others have concluded that sandstone diagenesis can be inhibited (slowed down) by petroleum emplacement (Worden *et al.* 1998; Marchand *et al.* 2001; Worden *et al.* 2018).

The equivalent debate has been somewhat more muted in the world of carbonate diagenesis (Heasley *et al.* 2000; Kolchugin *et al.* 2016; **Morad *et al.* 2016**) with many practitioners simply assuming that addition of petroleum will, in one way or another, seriously impede the diagenetic processes that influence reservoir quality. There is an overlap of interests in the case of carbonate cemented sandstone reservoirs which have, in some cases, been shown to display differences in the amount of cement above and below oil–water contacts (De Souza & Silva 1998).

Progress can be made by careful examination of high-quality petrographic data, in conjunction with detailed analysis of fluid saturation, specific to each core plug, derived from petrophysical wireline logs and an analysis of the timing of petroleum filling. It is also important to take account of primary compositional and textural variation between individual samples ensuring that assessment of the effect of varying water saturation is made between samples that were originally similar (Worden *et al.* 2018).

A special case of the effect of petroleum emplacement on diagenesis is thermochemical sulphate reduction (TSR), during which oxidized sulphate (typically anhydrite) reacts with reduced carbon in petroleum fluids producing calcite, H_2S and several other phases such as water, elemental

Fig. 9. Role of open systems in reservoir diagenesis. (**a**) Termination of fault/fracture controlled dolomite body, Lower Carboniferous Limestone, North Wales Platform. Dolomitized beds are located beneath thin, low permeability beds and terminate sharply against fractures. (**b**) Fracture-controlled calcite cementation in Upper Devonian Balnagown Group (aeolian facies of the Gaza Formation) at Tarbert Ness in NE Scotland. The fracture has led to localized calcite cementation of the sandstone owing to an influx of cementing species but the effect is limited to the near-fracture (proximal) region since the new cement closed pores and limited advective influx into the more distal parts of the sandstone. (Image (a) courtesy of Cathy Hollis, image (b) courtesy of Richard Worden.)

sulphur, CO_2 and mineral sulphides. TSR is common in deeply buried, anhydrite-bearing carbonates (Worden & Smalley 1996; Cai *et al.* 2003), where it leads to toxic and corrosive sour (H_2S-rich) gas but may also lead to small increases in reservoir porosity (Jiang *et al.* 2018). TSR is more rarely considered or reported in sandstones (Worden & Smalley 2001; Worden *et al.* 2003), although it may be a common but minor process in which the sour gas is ameliorated by reaction with Fe-minerals, resulting in commonly reported late-stage pyrite cement.

Burial dolomitization

Although most platform-scale dolomitization takes place in the shallow subsurface (see section on 'Dolomitization and eodiagenesis'), significant volumes of dolomite can also form during mesodiagenesis, especially in the vicinity of faults. There has been much focus on the genesis and importance of so-called hydrothermal dolomite bodies as reservoir rocks in recent years (Davies & Smith 2006). Typical features include non-stratabound geometry, with bodies forming irregular and discontinuous 'halos' around faults, usually in the hanging wall of normal faults or within the uppermost array of faults (Reidal shears) within strike-slip faults. Dolomite is usually non-fabric selective and fabric destructive, forming complex characteristic textures (e.g. zebra dolomite) with large vuggy pores connected by fractures (Davies & Smith 2006). Fluid inclusion data and associated mineralization (e.g. quartz, fluorite, barite and galena) point to elevated temperatures of dolomitization in many cases, explaining why this type of dolomite has been described as hydrothermal in origin. Although these bodies have been well described from outcrop, their importance as reservoir targets has not been proven. Furthermore, there is no satisfactory model for their formation; whereas some bodies appear to have formed from compactional dewatering of adjacent clastic basins (Frazer *et al.* 2014), in many cases identification of a satisfactory source of brine and Mg is lacking. It has been shown, primarily using clumped isotopes, that some reportedly fault-related, 'hydrothermal' dolomite bodies formed initially at lower temperature by normal marine reflux processes and were then altered (recrystallized), after deformation, by high-temperature fluids (Swart *et al.* 2016). Certainly, the study of these bodies continues to provide important information on fluid flux and reaction in the burial realm and is leading to a new generation of conceptual models, including fault-controlled convection of seawater, sometimes via basal sandstone aquifers (Hollis *et al.* 2017).

Conclusions

(1) Sandstones and carbonates are typically studied using the same suite of analytical tools and techniques starting with log analysis and petrophysical core analysis, core description, then into routine petrographic tools and even into the less routine realms of stable isotopes, fluid inclusions and other geochemical approaches.

(2) Reservoir quality in sandstones and carbonates is influenced by many of the same factors including: depositional environment, relative sea-level change, grain types, grain size and sorting, matrix proportion, biogenic input (e.g. from a nearby shelf), rate of deposition and isolation from surface conditions, palaeoclimate, soil formation (only in exposed sediment), near-surface water composition, eogenetic water flux (type, amount), rate of burial and heating (T_{max}), timing of oil filling, timing and magnitude of fluid overpressure development, fracturing and the importance of redox conditions for element minerals that contain Fe and S (typically Fe-silicates, pyrite, anhydrite), mass transfer associated with diffusion and fluid advection which impacts carbonate reactions in both carbonate and sandstone reservoirs owing to the high solubility of the associated minerals.

(3) Sandstones and carbonates have some significant differences in the controls on their reservoir quality including: carbonates are influenced by the depositional age of sediment (on the type of $CaCO_3$: aragonite v. calcite seas); carbonates tend to experience limited porosity loss from compaction owing to the strengthening effect of cements that form at low effective stresses; most sandstones undergo considerable porosity loss owing to compaction; silicate minerals are much less soluble than carbonate minerals, leading to a greater tendency for local conservation of mass for diagenetic reactions in deeply buried sandstones than carbonates; and sour gas (H_2S) and CO_2 may substantially modify reservoir quality in carbonates but are less important in sandstones.

(4) Reactive transfer models provide important insights into understanding the spatial variations in diagenetic reactions and have great promise for predicting a broad range in diagenetic reactions involving mass transfer in both carbonates and sandstones. To date the primary focus of these models is on process understanding; they have seen limited use for pre-drill reservoir quality prediction. On the other hand, process-oriented diagenesis models have had impressive success in accurately predicting reservoir quality of sandstones in a variety of basins prior to drilling where compaction, quartz cementation and reactions involving certain framework grains and clay minerals are the main diagenetic processes. Current models of this type, however, do not consider a number of processes that may be important in some reservoirs such as carbonate cementation and zeolite

formation processes that RTMs could be used to address.

(5) Some of the key controversies in sandstone and carbonate reservoir quality focus on the role of petroleum emplacement on diagenesis and porosity loss, the role of effective stress in chemical compaction (pressure solution), the degree of geochemical openness of reservoirs during diagenesis and cementation and the origin of hydrothermal dolomite.

(6) Improved dialogue between sandstone and carbonate reservoir quality practitioners and researchers may help to resolve some of the research questions and problems. There are interesting and exciting possibilities for overlap between these two previously relatively isolated communities.

We would like to credit reviewers Ian Billing and Sadoon Morad for their insightful and useful comments. The final version of Figure 7 is largely thanks to the Ian Billing's suggested improvements. Series editor Dave Hodgson is also thanks for his helpful and positive comments. Finally we would like to thank all the hard work by the many reviewers of the 20 papers in this special publication.

References

Aase, N.E., Bjørkum, P.A. & Nadeau, P.H. 1996. The effect of grain-coating microquartz on preservation of reservoir porosity. *American Association of Petroleum Geologists* Bulletin, **80**, 1654–1673.

Abdel-Wahab, A.A. & Turner, P. 1991. Diagenesis of the Nubia Formation, Central Eartern Desert, Egypt. *Journal of African Earth Sciences*, **13**, 343–358.

Abe, S. & Mair, K. 2005. Grain fracture in 3D numerical simulations of granular shear. *Geophysical Research Letters*, **32**, https://doi.org/10.1029/2004gl022123

Adams, J.E. & Rhodes, M.L. 1960. Dolomitization by seepage refluxion. *American Association of Petroleum Geologists Bulletin*, **44**, 1912–1921.

Ajdukiewicz, J.M. & Lander, R.H. 2010. Sandstone reservoir quality prediction: the state of the art. *American Association of Petroleum Geologists Bulletin*, **94**, 1083–1091, https://doi.org/10.1306/Intro060110

Andresen, B., Throndsen, T., Barth, T. & Bolstad, J. 1994. Thermal generation of carbon dioxide and organic acids from different source rocks. *Organic Geochemistry*, **21**, 1229–1242.

Ankit, K., Urai, J.L. & Nestler, B. 2015. Microstructural evolution in bitaxial crack-seal veins: a phase-field study. *Journal of Geophysical Research – Solid Earth*, **120**, 3096–3118, https://doi.org/10.1002/2015jb011934

Arakel, A.V. 1986. Evolution of calcrete in paleodrainages of the Lake Napperby area, central Australia. *Palaeogeography Palaeoclimatology Palaeoecology*, **54**, 283–303, https://doi.org/10.1016/0031-0182(86)90129-x

Armitage, P.J., Worden, R.H., Faulkner, D.R., Aplin, A.C., Butcher, A.R. & Iliffe, J. 2010. Diagenetic and sedimentary controls on porosity in Lower Carboniferous fine-grained lithologies, Krechba field, Algeria: a petrological study of a caprock to a carbon capture site. *Marine and Petroleum Geology*, **27**, 1395–1410, https://doi.org/10.1016/j.marpetgeo.2010.03.018

Armitage, P.J., Faulkner, D.R. & Worden, R.H. 2013*a*. Caprock corrosion. *Nature Geoscience*, **6**, 79–80, http://www.nature.com/ngeo/journal/v6/n2/abs/ngeo1716.html#supplementary-information

Armitage, P.J., Worden, R.H., Faulkner, D.R., Aplin, A.C., Butcher, A.R. & Espie, A.A. 2013*b*. Mercia Mudstone Formation caprock to carbon capture and storage sites: petrology and petrophysical characteristics. *Journal of the Geological Society, London*, **170**, 119–132, https://doi.org/10.1144/jgs2012-049

Armitage, P.J., Worden, R.H., Faulkner, D.R., Butcher, A.R. & Espie, A.A. 2016. Permeability of the Mercia Mudstone: suitability as caprock to carbon capture and storage sites. *Geofluids*, **16**, 26–42, https://doi.org/10.1111/gfl.12134

Bahniuk, A.M., Anjos, S., Franca, A.B., Matsuda, N., Eiler, J., McKenzie, J.A. & Vasconcelos, C. 2015. Development of microbial carbonates in the Lower Cretaceous Codo Formation (north-east Brazil): implications for interpretation of microbialite facies associations and palaeoenvironmental conditions. *Sedimentology*, **62**, 155–181, https://doi.org/10.1111/sed.12144

Bakke, S. & Øren, P.-E. 1997. 3-D pore-scale modelling of sandstones and flow simulations in the pore networks. *SPE Journal*, **2**, 136–149.

Barnett, A.J., Wright, V.P., Chandra, V.S. & Jain, V. 2015. Distinguishing between eogenetic, unconformity-related and mesogenetic dissolution: a case study from the Panna and Mukta fields, offshore Mumbai, India. *In*: Armitage, P.J., Butcher, A. *et al.* (eds) *Reservoir Quality of Clastic and Carbonate Rocks: Analysis, Modelling and Prediction*. Geological Society, London, Special Publications, **435**. First published online December 18, 2015, https://doi.org/10.1144/SP435.12

Barrie, G.M., Worden, R.H., Barrie, C.D. & Boyce, A.J. 2015. Extensive evaporation in a modern temperate estuary: stable isotopic and compositional evidence. *Limnology and Oceanography*, **60**, 1241–1250, https://doi.org/10.1002/lno.10091

Bartels, J., Clauser, C., Kuhn, M., Pape, H. & Schneider, W. 2005. Reactive flow and permeability prediction – numerical simulation of complex hydrogeothermal problems. *In*: Harvey, P.K., Brewer, T.S., Pezard, P.A. & Petrov, V.A. (eds) *Petrophysical Properties of Crystalline Rocks*. Geological Society, London, Special Publications, **240**, 133–151.

Barth, T. & Riis, M. 1992. Interactions between organic-acid anions in formation waters and reservoir mineral phases. *Organic Geochemistry*, **19**, 455–482.

Beaufort, D., Cassagnabere, A., Petit, S., Lanson, B., Berger, G., Lacharpagne, J.C. & Johansen, H. 1998. Kaolinite-to-dickite reaction in sandstone reservoirs. *Clay Minerals*, **33**, 297–316.

Bethke, C.M. 1996. *Geochemical Reaction Modeling, Concepts and Applications*. Oxford University Press, New York.

Bethke, C.M. 2008. *Geochemical and biogeochemical reaction modelling*, 2nd edn. Cambridge University Press, Cambridge.

Bjørkum, P.A. 1996. How important is pressure in causing dissolution of quartz in sandstones? *Journal of Sedimentary Research*, **66**, 147–154.

Bjørkum, P.A., Walderhaug, O. & Aase, N.E. 1993. A model for the effect of illitization on porosity and quartz cementation of sandstones. *Journal of Sedimentary Petrology*, **63**, 1089–1091.

Bjørlykke, K. & Egeberg, P.K. 1993. Quartz cementation in sedimentary basins. *American Association of Petroleum Geologists Bulletin*, **77**, 1538–1548.

Blackbourn, G.A. 2012. *Cores and Core Logging for Geoscientists*. Whittles, Dunbeath.

Bloch, S., Lander, R.H. & Bonnell, L. 2002. Anomalously high porosity and permeability in deeply buried sandstone reservoirs: origin and predictability. *American Association of Petroleum Geologists Bulletin*, **86**, 301–328.

Bonnell, L.M., Lander, R.H. & Larese, R.E. 2014. Prediction of the formation of grain coating chlorite by the in situ alteration of volcanic rock fragments. (Abstract.) *Reservoir Quality of Clastic and Carbonate Rocks: Analysis, Modelling and Prediction*. Geological Society, London, 112–113.

Bromley, R.G. 1996. *Trace Fossils: Biology, Taphonomy and Applications*. Chapman & Hall, London.

Brosse, E., Margueron, T. et al. 2003. The formation and stability of kaolinite in Brent sandstone reservoirs: a modelling approach. *In*: Worden, R.H. & Morad, S. (eds) *Clay Mineral Cements in Sandstones. Special Publication of the International Association of Sedimentologists*. Blackwell Science Ltd, Oxford, 383–410.

Bryant, I.D., Kantorowicz, J.D. & Love, C.F. 1988. The origin and recognition of laterally continuous carbonate cemented horizons in the Upper Lias sands of Southern England. *Marine and Petroleum Geology*, **5**, 108–133.

Busch, B., Winkler, R., Osivandi, K., Nover, G., Amann-Hildenbrand, A. & Hilgers, C. 2015. Evolution of small-scale flow barriers in German Rotliegend siliciclastics. *In*: Armitage, P.J., Butcher, A. et al. (eds) *Reservoir Quality of Clastic and Carbonate Rocks: Analysis, Modelling and Prediction*. Geological Society, London, Special Publications, **435**. First published online November 18, 2015, https://doi.org/10.1144/SP435.3

Cai, C.F., Worden, R.H., Bottrell, S.H., Wang, L.S. & Yang, C.C. 2003. Thermochemical sulphate reduction and the generation of hydrogen sulphide and thiols (mercaptans) in Triassic carbonate reservoirs from the Sichuan Basin, China. *Chemical Geology*, **202**, 39–57, https://doi.org/10.1016/s0009-2541(03)00209-2

Chermak, J.A. & Rimstidt, J.D. 1990. The hydrothermal transformation rate of kaolinite to muscovite/illite. *Geochimica et Cosmochimica Acta*, **54**, 2979–2990, https://doi.org/10.1016/0016-7037(90)90115-2

Cheung, L.Y.G., O'Sullivan, C. & Coop, M.R. 2013. Discrete element method simulations of analogue reservoir sandstones. *International Journal of Rock Mechanics and Mining Sciences*, **63**, 93–103, https://doi.org/10.1016/j.ijrmms.2013.07.002

Choquette, P.W. & Pray, L. 1970. Geologic nomenclature and classification of porosity in sedimentary carbonates. *American Association of Petroleum Geologists Bulletin*, **54**, 207–250.

Chudi, O.K., Lewis, H., Stow, D.A.V. & Buckman, J.O. 2016. Reservoir quality prediction of deep-water Oligocene sandstones from the west Niger Delta by integrated petrological, petrophyscial and basin modelling. *In*: Armitage, P.J., Butcher, A. et al. (eds) *Reservoir Quality of Clastic and Carbonate Rocks: Analysis, Modelling and Prediction*. Geological Society, London, Special Publications, **435**. First published online December 14, 2016, https://doi.org/10.1144/SP435.8

Chuhan, F.A., Kjeldstad, A., Bjørlykke, K. & Hoeg, K. 2003. Experimental compression of loose sands: relevance to porosity reduction during burial in sedimentary basins. *Canadian Geotechnical Journal*, **40**, 995–1011.

Consonni, A., Frixa, A. & Maragliulo, C. 2016. Hydrothermal dolomitization: simulation by reaction transport modelling. *In*: Armitage, P.J., Butcher, A. et al. (eds) *Reservoir Quality of Clastic and Carbonate Rocks: Analysis, Modelling and Prediction*. Geological Society, London, Special Publications **435**. First published online June 14, 2016, https://doi.org/10.1144/SP435.13

Cross, M.M., Manning, D.A.C., Bottrell, S.H. & Worden, R.H. 2004. Thermochemical sulphate reduction (TSR): experimental determination of reaction kinetics and implications of the observed reaction rates for petroleum reservoirs. *Organic Geochemistry*, **35**, 393–404, https://doi.org/10.1016/j.orggeochem.2004.01.005

Czernichowski-Lauriol, I., Sanjuan, B., Rochelle, C., Bateman, K., Pearce, J. & Blackwell, P. 1996. *Inorganic Geochemistry*. Final Report of JOULE II Project Number ct 92-0031.

Daneshvar, E. & Worden, R.H. 2017. Feldspar alteration and Fe-minerals: origin, distribution and implications for sandstone reservoir quality in estuarine sediments. *In*: Armitage, P.J., Butcher, A. et al. (eds) *Reservoir Quality of Clastic and Carbonate Rocks: Analysis, Modelling and Prediction*. Geological Society, London, Special Publications, **435**. First published online April 13, 2017, https://doi.org/10.1144/SP435.17

Davies, G.R. & Smith, L.B. 2006. Structurally controlled hydrothermal dolomite reservoir facies: an overview. *American Association of Petroleum Geologists Bulletin*, **90**, 1641–1690.

De Souza, R.S. & Silva, D.A. 1998. Origin and timing of carbonate cementation of the Namorado Sandstones (Cretaceous), Albacora Field Brazil: implication for oil recovery. *In*: Morad, S. (ed.) *Carbonate Cementation in Sandstones*. International Association of Sedimentologists, Special Publications, **26**, 309–325.

Dewever, B., Berwouts, I., Swennen, R., Breesch, L. & Ellam, R.M. 2010. Fluid flow reconstruction in karstified Panormide platform limestones (north-central Sicily): implications for hydrocarbon prospectivity in the Sicilian fold and thrust belt. *Marine and Petroleum Geology*, **27**, 939–958, https://doi.org/10.1016/j.marpetgeo.2009.10.018

Dowey, P.J., Hodgson, D.M. & Worden, R.H. 2012. Prerequisites, processes, and prediction of chlorite grain coatings in petroleum reservoirs: a review of subsurface examples. *Marine and Petroleum Geology*, **32**, 63–75, https://doi.org/10.1016/j.marpetgeo.2011.11.007

Dowey, P.J., Worden, R.H., Utley, J. & Hodgson, D.M. 2017. Sedimentary controls on modern sand grain coat formation. *Sedimentary Geology*, **353**, 46–63, https://doi.org/10.1016/j.sedgeo.2017.03.001

Dravis, J. 1979. Rapid and widespread generation of recent oolitic hardgrounds on a high-energy Bahamian platform, Eleuthera Bank, Bahamas. *Journal of Sedimentary Petrology*, **49**, 195–207.

Dutton, S.P. & Loucks, R.G. 2010. Diagenetic controls on evolution of porosity and permeability in lower Tertiary Wilcox sandstones from shallow to ultradeep (200–6700 m) burial, Gulf of Mexico Basin, USA. *Marine and Petroleum Geology*, **27**, 69–81, https://doi.org/10.1016/j.marpetgeo.2009.08.008

Dutton, S.P., Loucks, R.G. & Day-Stirrat, R.J. 2012. Impact of regional variation in detrital mineral composition on reservoir quality in deep to ultradeep lower Miocene sandstones, western Gulf of Mexico. *Marine and Petroleum Geology*, **35**, 139–153, https://doi.org/10.1016/j.marpetgeo.2012.01.006

Ebner, M., Koehn, D., Toussaint, R., Renard, F. & Schmittbuhl, J. 2009. Stress sensitivity of stylolite morphology. *Earth and Planetary Science Letters*, **277**, 394–398, https://doi.org/10.1016/j.epsl.2008.11.001

Ehrenberg, S.N. 1993. Preservation of anomalously high-porosity in deeply buried sandstones by grain coating chlorite – examples from the Norwegian continental shelf. *American Association of Petroleum Geologists Bulletin*, **77**, 1260–1286.

Ehrenberg, S.N. & Jakobsen, K.G. 2001. Plagioclase dissolution related to biodegradation of oil in Brent Group sandstones (Middle Jurassic) of Gullfaks Field, northern North Sea. *Sedimentology*, **48**, 703–721.

Ehrenberg, S.N., Nadeau, P.H. & Steen, O. 2009. Petroleum reservoir porosity v. depth: influence of geological age. *American Association of Petroleum Geologists Bulletin*, **93**, 1281–1296, https://doi.org/10.1306/06120908163

Ehrenberg, S.N., Walderhaug, O. & Bjørlykke, K. 2013. Carbonate porosity creation by mesogenetic dissolution: reality or illusion?: reply. *AAPG Bulletin*, **97**, 347–349, https://doi.org/10.1306/07111212089

Emery, D. & Robinson, A.G. 1993. *Inorganic Geochemistry: Application to Petroleum Geology*. Blackwell, Oxford.

English, K.L., English, J.M. *et al.* 2017. Controls on reservoir quality in exhumed basins an example from the Ordovician sandstone, Illizi Basin, Algeria. *Marine and Petroleum Geology*, **80**, 203–227, https://doi.org/10.1016/j.marpetgeo.2016.11.011

Fossen, H. 2010. Deformation bands formed during soft-sediment deformation: observations from SE Utah. *Marine and Petroleum Geology*, **27**, 215–222, https://doi.org/10.1016/j.marpetgeo.2009.06.005

Franks, S.G. & Zwingmann, H. 2010. Origin and timing of late diagenetic illite in the Permian-Carboniferous Unayzah sandstone reservoirs of Saudi Arabia. *American Association of Petroleum Geologists Bulletin*, **94**, 1133–1159, https://doi.org/10.1306/04211 009142

Frazer, M., Whitaker, F. & Hollis, C. 2014. Fluid expulsion from overpressured basins: implications for Pb–Zn mineralisation and dolomitisation of the East Midlands platform, northern England. *Marine and Petroleum Geology*, **55**, 68–86, https://doi.org/10.1016/j.marpetgeo.2014.01.004

French, M.W. & Worden, R.H. 2013. Orientation of microcrystalline quartz in the Fontainebleau Formation, Paris Basin and why it preserves porosity. *Sedimentary Geology*, **284**, 149–158, https://doi.org/10.1016/j.sedgeo.2012.12.004

French, M.W., Worden, R.H., Mariani, E., Larese, R.E., Mueller, R.R. & Kliewer, C.E. 2012. Microcrystalline quartz generation and the preservation of porosity in sandstones: evidence from the Upper Cretaceous of the Sub-Hercynian Basin, Germany. *Journal of Sedimentary Research*, **82**, 422–434, https://doi.org/10.2110/jsr.2012.39

Gale, J.F.W., Lander, R.H., Reed, R.M. & Laubach, S.E. 2010. Modeling fracture porosity evolution in dolostone. *Journal of Structural Geology*, **32**, 1201–1211, https://doi.org/10.1016/j.jsg.2009.04.018

Garcia-Fresca, B., Lucia, F.J., Sharp, J.M., Jr, & Kerans, C. 2012. Outcrop-constrained hydrogeological simulations of brine reflux and early dolomitization of the Permian San Andres Formation. *American Association of Petroleum Geologists Bulletin*, **96**, 1757–1781, https://doi.org/10.1306/02071210123

Geloni, C., Ortenzi, A. & Consonni, A. 2015. Reactive transport modelling of compacting siliciclastic sediment diagenesis. *In*: Armitage, P.J., Butcher, A. *et al.* (eds) *Reservoir Quality of Clastic and Carbonate Rocks: Analysis, Modelling and Prediction*. Geological Society, London, Special Publications **435**. First published online December 10, 2015, https://doi.org/10.1144/SP435.7

Geske, A., Goldstein, R.H. *et al.* 2015. The magnesium isotope (delta Mg-26) signature of dolomites. *Geochimica Et Cosmochimica Acta*, **149**, 131–151, https://doi.org/10.1016/j.gca.2014.11.003

Gier, S., Worden, R.H. & Krois, P. 2015. Comparing clay mineral diagenesis in interbedded sandstones and mudstones, Vienna Basin, Austria. *In*: Armitage, P.J., Butcher, A. *et al.* (eds) *Reservoir Quality of Clastic and Carbonate Rocks: Analysis, Modelling and Prediction*. Geological Society, London, Special Publications **435**. First published online November 20, 2015, https://doi.org/10.1144/SP435.9

Giles, M.R., Indrelid, S.L., Beynon, G.V. & Amthor, J. 2000. The origin of large-scale quartz cementation: evidence from large data sets and coupled heat-fluid mass transport modelling. *In*: Worden, R.H. & Morad, S. (eds) *Quartz Cement in Sandstones*. Blackwells, Oxford, 21–38.

Gingras, M.K., MacMillan, B., Balcom, B.J., Saunders, T. & Pemberton, S.G. 2002. Using magnetic resonance imaging and petrographic techniques to understand the textural attributes and porosity distribution in Macaronichnus-burrowed sandstone. *Journal of Sedimentary Research*, **72**, 552–558, https://doi.org/10.1306/122901720552

GINGRAS, M.K., PEMBERTON, S.G., MUELENBACHS, K. & MACHEL, H.G. 2004. Conceptual models for burrow-related, selective dolomitization with textural and isotopic evidence from the Tyndall Stone, Canada. *Geobiology*, **2**, 21–30, https://doi.org/10.1111/j.1472-4677.2004.00022.x

GLUYAS, J. & SWARBRICK, R. 2004. *Petroleum Geoscience*. Blackwell, Oxford.

GLUYAS, J.G., ROBINSON, A.G., EMERY, D., GRANT, S.M. & OXTOBY, N.H. 1993. The link between petroleum emplacement and sandstone cementation. *In*: PARKER, J.R. (ed.) *Petroleum Geology of Northwest Europe: Proceedings of the 4th Conference*. The Geological Society, London, 1395–1402.

GÖTTE, T. 2016. Trace element composition of authigenic quartz in sandstones and its correlation with fluid–rock interaction during diagenesis. *In*: ARMITAGE, P.J., BUTCHER, A. *ET AL*. (eds) *Reservoir Quality of Clastic and Carbonate Rocks: Analysis, Modelling and Prediction*. Geological Society, London, Special Publications, **435**. First published online January 19, 2016, https://doi.org/10.1144/SP435.2

GRIFFITHS, J., FAULKNER, D.R., EDWARDS, A.P. & WORDEN, R.H. 2016. Deformation band development as a function of intrinsic host-rock properties in Triassic Sherwood Sandstone. *In*: ARMITAGE, P.J., BUTCHER, A. *ET AL*. (eds) *Reservoir Quality of Clastic and Carbonate Rocks: Analysis, Modelling and Prediction*. Geological Society, London, Special Publications, **435**. First published online January 19, 2016, https://doi.org/10.1144/SP435.11

HADDAD, S.C., WORDEN, R.H., PRIOR, D.J. & SMALLEY, P.C. 2006. Quartz cement in the Fontainebleau sandstone, Paris Basin, France: crystallography and implications for mechanisms of cement growth. *Journal of Sedimentary Research*, **76**, 244–256, https://doi.org/10.2110/jsr.2006.024

HARRIS, P.M. 2010. Delineating and quantifying depositional facies patterns in carbonate reservoirs: insight from modern analogs. *American Association of Petroleum Geologists Bulletin*, **94**, 61–86, https://doi.org/10.1306/07060909014

HEALD, M.T. & RENTON, J.J. 1966. Experimental study of sandstone cementation. *Journal of Sedimentary Petrology*, **36**, 977–991.

HEASLEY, E.C., WORDEN, R.H. & HENDRY, J.P. 2000. Cement distribution in a carbonate reservoir: recognition of a palaeo oil–water contact and its relationship to reservoir quality in the Humbly Grove field, onshore, UK. *Marine and Petroleum Geology*, **17**, 639–654, https://doi.org/10.1016/s0264-8172(99) 00057-4

HERRINGSHAW, L.G. & MCILROY, D. 2013. Bioinfiltration irrigation-driven transport of clay particles through bioturbated sediments. *Journal of Sedimentary Research*, **83**, 443–450, https://doi.org/10.2110/jsr.2013.40

HOLLIS, C., BASTESEN, E. *ET AL*. 2017. Fault controlled dolomitization in a rift basin. *Geology*, **45**, 219–222, https://doi.org/10.1130/G38s394.1

HOSA, A. & WOOD, R. 2017. Quantifying the impact of early calcite cementation on the reservoir quality of carbonate rocks: a 3D process-based model. *Advances in Water Resources*, **104**, 89–104, https://doi.org/10.1016/j.advwatres.2017.02.019

HOUSEKNECHT, D.W. & PITTMAN, E.D. 1992. *Origin, Diagenesis and Petrophysics of Clay Minerals in Sandstones*. SEPM Special Publications, **47**, Society of Economic Paleontologists and Mineralogists, Tulsa, OK.

HUNT, J.M. 1995. *Petroleum Geochemistry and Geology*. W. H. Freeman and Company, New York.

IRWIN, H., CURTIS, C. & COLEMAN, M.L. 1977. Isotopic evidence for source of diagenetic carbonates formed during burial of organic rich sediments. *Nature*, **269**, 209–213.

JACOBSON, A.D., ZHANG, Z.F., LUNDSTROM, C. & HUANG, F. 2010. Behavior of Mg isotopes during dedolomitization in the Madison Aquifer, South Dakota. *Earth and Planetary Science Letters*, **297**, 446–452, https://doi.org/10.1016/j.epsl.2010.06.038

JACOBSON, G., ARAKEL, A.V. & CHEN, Y.J. 1988. The central Australian groundwater discharge zone – evolution of associated calcrete and gypcrete deposits. *Australian Journal of Earth Sciences*, **35**, 549–565, https://doi.org/10.1080/08120098808729469

JIANG, L., WORDEN, R.H., CAI, C.F., LI, K.K., XIANG, L., CAI, L.L. & HE, X.Y. 2014. Dolomitization of gas reservoirs: the Upper Permian Changxing and Lower Triassic Feixianguan Formations, northeast Sichuan Basin, China. *Journal of Sedimentary Research*, **84**, 792–815, https://doi.org/10.2110/jsr.2014.65

JIANG, L., CAI, C.F., WORDEN, R.H., CROWLEY, S.F., JIA, L.Q., ZHANG, K. & DUNCAN, I.J. 2016. Multiphase dolomitization of deeply buried Cambrian petroleum reservoirs, Tarim Basin, north-west China. *Sedimentology*, **63**, 2130–2157, https://doi.org/10.1111/sed.12300

JIANG, L., WORDEN, R.H. & YANG, C. 2018. Thermochemical sulphate reduction can improve carbonate petroleum reservoir quality. *Geochemica et Cosmochimica Acta*, **223**, 127–140.

JIN, G.D., PATZEK, T.W. & SILIN, D.B. 2012. Modeling the impact of rock formation history on the evolution of absolute permeability. *Journal of Petroleum Science and Engineering*, **100**, 153–161, https://doi.org/10.1016/j.petrol.2012.03.005

JOHN, C.M. 2015. Burial estimates constrained by clumped isotope thermometry: example of the Lower Cretaceous Qishn Formation (Haushi-Huqf High, Oman). *In*: ARMITAGE, P.J., BUTCHER, A. *ET AL*. (eds) *Reservoir Quality of Clastic and Carbonate Rocks: Analysis, Modelling and Prediction*. Geological Society, London, Special Publications **435**. First published online November 18, 2015, https://doi.org/10.1144/SP435.5

JONES, G.D. & XIAO, Y. 2005. Dolomitization, anhydrite cementation, and porosity evolution in a reflux system: insights from reactive transport models. *American Association of Petroleum Geologists, Bulletin*, **89**, 577.

JONES, G.D., WHITAKER, F.F., SMART, P.L. & SANFORD, W.E. 2002. Fate of reflux brines in carbonate platforms. *Geology*, **30**, 371–374, https://doi.org/10.1130/0091-7613(2002)030<0371:forbic>2.0.co;2

JONES, G.D., SMART, P.L., WHITAKER, F.F., ROSTRON, B.J. & MACHEL, H.G. 2003. Numerical modeling of reflux dolomitization in the Grosmont platform complex (Upper Devonian), Western Canada sedimentary basin. *American Association of Petroleum Geologists Bulletin*, **87**, 1273–1298, https://doi.org/10.1306/03260302007

Jones, G.D., Whitaker, F.F., Smart, P.L. & Sanford, W.E. 2004. Numerical analysis of seawater circulation in carbonate platforms: II. The dynamic interaction between geothermal and brine reflux circulation. *American Journal of Science*, **304**, 250–284, https://doi.org/10.2475/ajs.304.3.250

Kaczmarek, S.E. & Sibley, D.F. 2011. On the evolution of dolomite stoichiometry and cation order during high-temperature synthesis experiments: an alternative model for the geochemical evolution of natural dolomites. *Sedimentary Geology*, **240**, 30–40, https://doi.org/10.1016/j.sedgeo.2011.07.003

Karner, S.L., Chester, J.S., Chester, F.M., Kronenberg, A.K. & Hajash, A. 2005. Laboratory deformation of granular quartz sand: implications for the burial of clastic rocks. *American Association of Petroleum Geologists Bulletin*, **89**, 603–625, https://doi.org/10.1306/12200404010

Khalifa, M. & Morad, S. 2012. Impact of structural setting on diagenesis of fluvial and tidal sandstones: the Bahi Formation, Upper Cretaceous, NW Sirt Basin, North Central Libya. *Marine and Petroleum Geology*, **38**, 211–231, https://doi.org/10.1016/j.marpetgeo.2011.05.006

King, B. & Goldstein, R.H. 2016. History of hydrothermal fluid flow in the midcontinent, USA: the relationship between inverted thermal structure, unconformities and porosity distribution. *In*: Armitage, P.J., Butcher, A. *et al.* (eds) *Reservoir Quality of Clastic and Carbonate Rocks: Analysis, Modelling and Prediction.* Geological Society, London, Special Publications **435**. First published online August 17, 2016, https://doi.org/10.1144/SP435.16

Király, C., Sendula, E. *et al.* 2016. The relevance of dawsonite precipitation in CO_2 sequestration in the Mihályi-Répcelak area, NW Hungary. *In*: Armitage, P.J., Butcher, A. *et al.* (eds) *Reservoir Quality of Clastic and Carbonate Rocks: Analysis, Modelling and Prediction.* Geological Society, London, Special Publications **435**. First published online June 20, 2016, https://doi.org/10.1144/SP435.15

Kolchugin, A.N., Immenhauser, A., Walter, B.F. & Morozov, V.P. 2016. Diagenesis of the palaeo-oil–water transition zone in a Lower Pennsylvanian carbonate reservoir: constraints from cathodoluminescence microscopy, microthermometry, and isotope geochemistry. *Marine and Petroleum Geology*, **72**, 45–61, https://doi.org/10.1016/j.marpetgeo.2016.01.014

Korneva, I., Bastesen, E. *et al.* 2017. The effects of dolomitization on petrophysical properties and fracture distribution within rift-related carbonates (Hammam Faraun Fault Block, Suez Rift, Egypt). *Journal of Structural Geology*, https://doi.org/10.1016/j.jsg.2017.06.005

Kraishan, G.M., Rezaee, M.R. & Worden, R.H. 2000. Significance of trace element composition of quartz cement as a key to reveal the origin of silica in sandstones: an example from the Cretaceous of the Barrow sub-basin, Western Australia. *In*: Worden, R.H. & Morad, S. (eds) *Quartz Cementation in Sandstones.* International Association of Sedimentologists, Special Publications, **29**, 317–332.

Kristiansen, K., Valtiner, M., Greene, G.W., Boles, J.R. & Israelachvili, J.N. 2011. Pressure solution – the importance of the electrochemical surface potentials. *Geochimica et Cosmochimica Acta*, **75**, 6882– 6892, https://doi.org/10.1016/j.gca.2011.09.019

Lander, R.H. & Bonnell, L.M. 2010. A model for fibrous illite nucleation and growth in sandstones. *American Association of Petroleum Geologists Bulletin*, **94**, 1161–1187, https://doi.org/10.1306/04211009121

Lander, R.H. & Laubach, S.E. 2015. Insights into rates of fracture growth and sealing from a model for quartz cementation in fractured sandstones. *Geological Society of America Bulletin*, **127**, 516–538, https://doi.org/10.1130/b31092.1

Lander, R.H. & Walderhaug, O. 1999. Predicting porosity through simulating sandstone compaction and quartz cementation. *American Association of Petroleum Geologists Bulletin*, **83**, 433–449.

Lander, R.H., Larese, R.E. & Bonnell, L.M. 2008*a*. Toward more accurate quartz cement models: the importance of euhedral v. noneuhedral growth rates. *American Association of Petroleum Geologists Bulletin*, **92**, 1537–1563, https://doi.org/10.1306/07160808037

Lander, R.H., Solano-Acosta, W., Thomas, A.R., Reed, R.M., Kacewicz, M., Bonnell, L.M. & Hooker, J.N. 2008*b*. Simulation of fault sealing from quartz cementation within cataclastic deformation zones. *AAPG Hedberg Conference Basin and Petroleum Systems Modeling: New Horizons in Research and Applications*, Napa Valley, CA.

Laubach, S.E., Eichhubl, P., Hilgers, C. & Lander, R.H. 2010. Structural diagenesis. *Journal of Structural Geology*, **32**, 1866–1872, https://doi.org/10.1016/j.jsg.2010.10.001

Law, A. 1998. Regional uplift in the English Channel: quantification using sonic velocity. *In*: Underhill, J.R. (ed.) *Development, Evolution and Petroleum Geology of the Wessex Basin.* The Geological Society, London, 187–197.

Lipinski, C.J., Franseen, E.K. & Goldstein, R.H. 2013. Reservoir analog model for oolite-microbialite sequences, Miocene terminal carbonate complex, Spain. *American Association of Petroleum Geologists Bulletin*, **97**, 2035–2057, https://doi.org/10.1306/06261312182

Lucia, F.J. & Major, R.P. 1994. Porosity evolution through hypersaline reflux dolomitization. *In*: Purser, B.H., Tucker, M.E. & Zenger, D.H. (eds) *Dolomites: A Volume in Honour of Dolomieu.* International Association of Sedimentologists, Special Publications. Blackwells, Oxford, 325–341.

Macaulay, C.I., Fallick, A.E., Haszeldine, R.S. & McAulay, G.E. 2000. Oil migration makes the difference: regional distribution of carbonate cement delta C-13 in northern North Sea Tertiary sandstones. *Clay Minerals*, **35**, 69–76.

Macintyre, I.G., Reid, R.P. & Steneck, R.S. 1996. Growth history of stromatolites in a Holocene fringing reef, Stocking Island, Bahamas. *Journal of Sedimentary Research*, **66**, 231–242.

MAHDI, T.A. & AQRAWI, A.A.M. 2017. Role of facies diversity and cyclicity on the reservoir quality of the mid-Cretaceous Mishrif Formation in Southern Mesopotamian Basin, Iraq. *In*: ARMITAGE, P.J., BUTCHER, A. ET AL. (eds) *Reservoir Quality of Clastic and Carbonate Rocks: Analysis, Modelling and Prediction*. Geological Society, London, Special Publications, **435**. First published online February 22, 2017, https://doi.org/10.1144/SP435.19

MALIVA, R.G., DICKSON, J.A.D., SCHIAVON, N. & FALLICK, A.E. 1999. Self-organization origin of wood-grained chert, Portland Limestone Formation (Upper Jurassic), southern England. *Geological Magazine*, **136**, 413–421, https://doi.org/10.1017/s0016756899002629

MARCHAND, A.M.E., HASZELDINE, R.S., SMALLEY, P.C., MACAULAY, C.I. & FALLICK, A.E. 2001. Evidence for reduced quartz-cementation rates in oil-filled sandstones. *Geology*, **29**, 915–918.

MARKETOS, G. & BOLTON, M.D. 2009. Compaction bands simulated in Discrete Element Models. *Journal of Structural Geology*, **31**, 479–490, https://doi.org/10.1016/j.jsg.2009.03.002

MATLACK, K.S., HOUSEKNECHT, D.W. & APPLIN, K.R. 1989. Emplacement of clay into sand by infiltration. *Journal of Sedimentary Petrology*, **59**, 77–87.

MCBRIDE, E.F. 1989. Quartz cement in sandstones: a review. *Earth Science Reviews*, **26**, 69–112.

MCBRIDE, E.F. & PAREA, G.C. 2001. Origin of highly elongate, calcite-cemented concretions in some Italian coastal beach and dune sands. *Journal of Sedimentary Research*, **71**, 82–87.

MCBRIDE, E.F., MILLIKEN, K.L., CAVAZZA, W., CIBIN, U., FONTANA, D., PICARD, M.D. & ZUFFA, G.G. 1995. Heterogeneous distribution of calcite cement at the outcrop scale in Tertiary sandstones, Northern Apennines, Italy. *American Association of Petroleum Geologists Bulletin*, **79**, 1044–1063.

MCCLAIN, M.E., SWART, P.K. & VACHER, H.L. 1992. The hydrogeochemistry of early meteoric diagenesis in a Holocene deposit of biogenic carbonates. *Journal of Sedimentary Petrology*, **62**, 1008–1022.

MCILROY, D., WORDEN, R.H. & NEEDHAM, S.J. 2003. Faeces, clay minerals and reservoir potential. *Journal of the Geological Society*, **160**, 489–493.

MCMAHON, P.B., CHAPELLE, F.H., FALLS, W.F. & BRADLEY, P.M. 1992. Role of microbial processes in linking sandstone diagenesis with organic rich clays. *Journal of Sedimentary Petrology*, **62**, 1–10.

MOORE, C.H. & WADE, W.J. 2013. *Carbonate Reservoirs: Porosity, Evolution and Diagenesis in a Sequence Stratigraphic Framework*. Elsevier, Amsterdam.

MORAD, D., PAGANONI, M. ET AL. 2016. Origin and evolution of microporosity in packstones and grainstones in a Lower Cretaceous carbonate reservoir, United Arab Emirates. *In*: ARMITAGE, P.J., BUTCHER, A. ET AL. (eds) *Reservoir Quality of Clastic and Carbonate Rocks: Analysis, Modelling and Prediction*. Geological Society, London, Special Publications **435**. First published online December 21, 2016, https://doi.org/10.1144/SP435.20

MORAD, S., MARFIL, R., ALAASM, I.S. & GOMEZGRAS, D. 1992. The role of mixing zone dolomitization in sandstone cementation – evidence from the Triassic Buntsandstein, the Iberian Tange, Spain. *Sedimentary Geology*, **80**, 53–65.

MORAD, S., KETZER, J.M. & DE ROS, L.F. 2000. Spatial and temporal distribution of diagenetic alterations in siliciclastic rocks: implications for mass transfer in sedimentary basins. *Sedimentology*, **47**, 95–120.

MORAD, S., AL-RAMADAN, K., KETZER, J.M. & DE ROS, L.F. 2010. The impact of diagenesis on the heterogeneity of sandstone reservoirs: a review of the role of depositional facies and sequence stratigraphy. *American Association of Petroleum Geologists Bulletin*, **94**, 1267–1309, https://doi.org/10.1306/04211009178

MORAD, S., KETZER, J.M. & DE ROS, L.F. 2012. Linking diagenesis to sequence stratigraphy: an integrated tool for understanding and predicting reservoir quality distribution. *In*: MORAD, S., KETZER, J.M. & DEROS, L.F. (eds) *Linking Diagenesis to Sequence Stratigraphy*. Special Publication of the International Association of Sedimentologists, **45**. Chichester, John Wiley & Sons, 1–36.

MORAES, M.A.S. & DE ROS, L.F. 1990. Infiltrated clays in fluvial Jurassic sandstones of Recôncavo Basin, northeastern Brazil. *Journal of Sedimentary Petrology*, **60**, 809–819.

MOUSAVI, M.A. & BRYANT, S.L. 2013. Geometric models of porosity reduction by ductile grain compaction and cementation. *American Association of Petroleum Geologists Bulletin*, **97**, 2129–2148, https://doi.org/10.1306/05171311165

NEEDHAM, S.J., WORDEN, R.H. & MCILROY, D. 2005. Experimental production of clay rims by macrobiotic sediment ingestion and excretion processes. *Journal of Sedimentary Research*, **75**, 1028–1037, https://doi.org/10.2110/jsr.2005.078

NEEDHAM, S.J., WORDEN, R.H. & CUADROS, J. 2006. Sediment ingestion by worms and the production of bio-clays: a study of macrobiologically enhanced weathering and early diagenetic processes. *Sedimentology*, **53**, 567–579, https://doi.org/10.1111/j.1365-3091.2006.00781.x

NEILSON, J.E. & OXTOBY, N.H. 2008. The relationship between petroleum, exotic cements and reservoir quality in carbonates – a review. *Marine and Petroleum Geology*, **25**, 778–790, https://doi.org/10.1016/j.marpetgeo.2008.02.004

NELSON, R.A. 2001. *Geologic Analysis of Naturally Fractured Reservoirs*. 2nd edn. Gulf, Houston, TX.

NEUZIL, C.E. 2003. Hydromechanical coupling in geologic processes. *Hydrogeology Journal*, **11**, 41–83, https://doi.org/10.1007/s10040-002-0230-8

NEWPORT, R., HOLLIS, C., BODIN, S. & REDFERN, J. 2017. Examining the interplay of climate and low amplitude sea-level change on the distribution and volume of massive dolomitisation. *Depositional Record*, **3**, 38–59, https://doi.org/10.1002/dep2.25

NGUYEN, D.T., HORTON, R.A. & KAESS, A.B. 2016. Diagenesis, plagioclase dissolution and preservation of porosity in Eocene and Oligocene sandstones at the Greeley oil field, southern San Joaquin basin, California, USA. *In*: ARMITAGE, P.J., BUTCHER, A. ET AL.

(eds) *Reservoir Quality of Clastic and Carbonate Rocks: Analysis, Modelling and Prediction*. Geological Society, London, Special Publications **435**. First published online June 28, 2016, https://doi.org/10.1144/SP435.14

OELKERS, E.H., BJØRKUM, P.A. & MURPHY, W.M. 1996. A petrographic and computational investigation of quartz cementation and porosity reduction in North Sea sandstones. *American Journal of Science*, **296**, 420–452.

ØREN, P.E. & BAKKE, S. 2002. Process based reconstruction of sandstones and prediction of transport properties. *Transport in Porous Media*, **46**, 311–343.

ØREN, P.E. & BAKKE, S. 2003. Reconstruction of Berea sandstone and pore-scale modelling of wettability effects. *Journal of Petroleum Science and Engineering*, **39**, 177–199, https://doi.org/10.1016/s0920-4105(03)00062-7

ØREN, P.E., BAKKE, S. & HELD, R. 2007. Direct pore-scale computation of material and transport properties for North Sea reservoir rocks. *Water Resources Research*, **43**, https://doi.org/10.1029/2006wr005754

PERCIVAL, C.J. 1983. The Firestone Sill Ganister, Namurian, Northern England – the A2 horizon of a Podzol or podzolic palaeosol. *Sedimentary Geology*, **36**, 41–49, https://doi.org/10.1016/0037-0738(83)90020-9

PERCIVAL, C.J. 1992. The Harthope ganister – a transgressive barrier-island to a shallow-marine sand-ridge from the Namurian of Northern England. *Journal of Sedimentary Petrology*, **62**, 442–454.

PEREZ, R. & BOLES, A.R. 2005. An empirically derived kinetic model for albitization of detrital plagioclase. *American Journal of Science*, **305**, 312–343, https://doi.org/10.2475/ajs.305.4.312

PITTMAN, E.D. & LARESE, R.E. 1991. Compaction of lithic sands – experimental results and application. *American Association of Petroleum Geologists Bulletin*, **75**, 1279–1299.

POLLYEA, R.M. & FAIRLEY, J.P. 2012. Implications of spatial reservoir uncertainty for CO2 sequestration in the east Snake River Plain, Idaho (USA). *Hydrogeology Journal*, **20**, 689–699, https://doi.org/10.1007/s10040-012-0847-1

POTEET, J.E., GOLDSTEIN, R.H. & FRANSEEN, E.K. 2016. Diagenetic controls on location of reservoir spots reltive to paleotopgraphic and structural highs. *In*: ARMITAGE, P.J., BUTCHER, A. *ET AL.* (eds) *Reservoir Quality of Clastic and Carbonate Rocks: Analysis, Modelling and Prediction*. Geological Society, London, Special Publications, **435**. First published online December 21, 2016, https://doi.org/10.1144/SP435.18

PRIMMER, T.J., CADE, C.A. *ET AL.* 1997. Global patterns in sandstone diagenesis: their application to reservoir quality prediction for petroleum exploration. *In*: KUPECZ, J.A., GLUYAS, J. & BLOCH, S. (eds) *Reservoir quality prediction in sandstones and carbonates*. AAPG, Memoirs, Tulsa, OK, **69**, 61–78.

PRODANOVIC, M., BRYANT, S.L. & DAVIS, J.S. 2013. Numerical simulation of diagenetic alteration and its effect on residual gas in tight gas sandstones. *Transport in Porous Media*, **96**, 39–62, https://doi.org/10.1007/s11242-012-0072-3

RIDER, M. & KENNEDY, M.J. 2011. *The Geological Interpretation of Well Logs*. Rider-French Consulting, Cambridge.

RISNES, R., HAGHIGHI, H., KORSNES, R.I. & NATVIK, O. 2003. Chalk–fluid interactions with glycol and brines. *Tectonophysics*, **370**, 213–226, https://doi.org/10.1016/s0040-1951(03)00187-2

SALLER, A.H. & HENDERSON, N. 1998. Distribution of porosity and permeability in platform dolomites: insight from the Permian of west Texas. *American Association of Petroleum Geologists Bulletin*, **82**, 1528–1550.

SALLER, A.H., POLLITT, D. & DICKSON, J.A.D. 2014. Diagenesis and porosity development in the First Eocene reservoir at the giant Wafra Field, Partitioned Zone, Saudi Arabia and Kuwait. *American Association of Petroleum Geologists Bulletin*, **98**, 1185–1212.

SANDBERG, P.A. 1983. An oscillating trend in Phanerozoic non-skeletal carbonate mineralogy. *Nature*, **305**, 19–22, https://doi.org/10.1038/305019a0

SANJUAN, B., GIRARD, J.P., LANINI, S., BOURGUINON, A. & BROSSE, E. 2003. Geochemical modelling of diagenetic illite and quartz cement formation in Brent sandstone reservoirs: example of the Hild Field, Norwegian North Sea. *In*: WORDEN, R.H. & MORAD, S. (eds) *Clay Mineral Cements in Sandstones*. International Association of Sedimentologists, Special Publications. Blackwell Science, Oxford, 425–452.

SATHAR, S., WORDEN, R.H., FAULKNER, D.R. & SMALLEY, P.C. 2012. The effect of effect of oil saturation on the mechanism of compaction mechanism of compaction in granular materials: higher oil saturations lead to more more grain fracturing and less pressure solution. *Journal of Sedimentary Research*, **82**, 571–584, https://doi.org/10.2110/jsr.2012.44

SCHINDLER, R.J., PARSONS, D.R. *ET AL.* 2015. Sticky stuff: redefining bedform prediction in modern and ancient environments. *Geology*, **43**, 399–402.

SCHMID, S., WORDEN, R.H. & FISHER, Q.J. 2003. The origin and regional distribution of dolomite cement in sandstones from a Triassic dry river system, Corrib Field, offshore west of Ireland. *Journal of Geochemical Exploration*, **78-9**, 475–479, https://doi.org/10.1016/s0375-6742(03)00118-3

SCHMID, S., WORDEN, R.H. & FISHER, Q.J. 2004. Diagenesis and reservoir quality of the Sherwood Sandstone (Triassic), Corrib Field, Slyne Basin, west of Ireland. *Marine and Petroleum Geology*, **21**, 299–315, https://doi.org/10.1016/j.marpetgeo.2003.11.015

SCHRODER, S., GROTZINGER, J.P., AMTHOR, J.E. & MATTER, A. 2005. Carbonate deposition and hydrocarbon reservoir development at the Precambrian–Cambrian boundary: the Ara Group in South Oman. *Sedimentary Geology*, **180**, 1–28, https://doi.org/10.1016/j.sedgeo.2005.07.002

SHELDON, H.A., WHEELER, J., WORDEN, R.H. & CHEADLE, M.J. 2003. An analysis of the roles of stress, temperature, and pH in chemical compaction of sandstones. *Journal of Sedimentary Research*, **73**, 64–71.

SPOTL, C., KRALIK, M. & KUNK, M.J. 1996. Authigenic feldspar as an indicator of paleo-rock water interactions in Permian carbonates of the northern Calcareous Alps, Austria. *Journal of Sedimentary Research*, **66**, 139–146.

STARMER, I.C. 1995. Deformation of the Upper Cretaceous Chalk at Selwicks Bay, Flamborough Head, Yorkshire: its significance in the structural evolution of the northeast England and North Sea Basin. *Proceedings of the Yorkshire Geological Society*, **50**, 213–228, first published online June 1, 1995, https://doi.org/10.1144/pygs.50.3.213

Steefel, C.I., Yabusaki, S.B. & Mayer, K.U. 2015. Reactive transport benchmarks for subsurface environmental simulation. *Computational Geosciences*, **19**, 439–443, https://doi.org/10.1007/s10596-015-9499-2

Stricker, S. & Jones, S.J. 2016. Enhanced porosity preservation by pore fluid overpressure and chlorite grain coatings in the Triassic Skagerrak, Central Graben, North Sea, UK. *In*: Armitage, P.J., Butcher, A. et al. (eds) *Reservoir Quality of Clastic and Carbonate Rocks: Analysis, Modelling and Prediction*. Geological Society, London, Special Publications **435**. First published online January 5, 2016, https://doi.org/10.1144/SP435.4

Sullivan, M.D., Haszeldine, R.S., Boyce, A.J., Rogers, G. & Fallick, A.E. 1994. Late anhydrite cements mark basin inversion – isotopic and formation water evidence, Rotliegend Sandstone, North Sea. *Marine and Petroleum Geology*, **11**, 46–54, https://doi.org/10.1016/0264-8172(94)90008-6

Swart, P.K., Cantrell, D.L., Arienzo, M.M. & Murray, S.T. 2016. Evidence for high temperature and O-18-enriched fluids in the Arab-D of the Ghawar Field, Saudi Arabia. *Sedimentology*, **63**, 1739–1752, https://doi.org/10.1111/sed.12286

Taylor, A.M. & Goldring, R. 1993. Description and analysis of bioturbation and ichnofabrics. *Journal of the Geological Society*, **150**, 141–148, https://doi.org/10.1144/gsjgs.150.1.0141

Taylor, T.R., Stancliffe,, Macaulay, C. & Hathon, L.A. 2004. High temperature quartz cementation and the timing of hydrocarbon accumulation in the Jurassic Norphlet Sandstone, offshore Gulf of Mexico, USA. *In*: Cubbit, J.M., England, W.A. & Larter, S.R. (eds) *Understanding Petroleum Reservoirs; Toward an Integrated Reservoir Engineering and Geochemical Approach*. Geological Society, London, Special Publications, **237**, 257–278. First published January 1, 2004, https://doi.org/10.1144/GSL.SP.2004.237.01.15

Taylor, T.R., Giles, M.R. et al. 2010. Sandstone diagenesis and reservoir quality prediction: models, myths, and reality. *American Association of Petroleum Geologists Bulletin*, **94**, 1093–1132, https://doi.org/10.1306/04211009123

Taylor, T.R., Kittridge, M.G., Winefield, P., Bryndzia, L.T. & Bonnell, L.M. 2015. Reservoir quality and rock properties modeling – Triassic and Jurassic sandstones, greater Shearwater area, UK Central North Sea. *Marine and Petroleum Geology*, **65**, 1–21, https://doi.org/10.1016/j.marpetgeo.2015.03.020

Thyne, G., Boudreau, B.P., Ramm, M. & Midtbo, R.E. 2001. Simulation of potassium feldspar dissolution and illitization in the Statfjord Formation, North Sea. *American Association of Petroleum Geologists Bulletin*, **85**, 621–635.

Tobin, R.C., McClain, T., Lieber, R.B., Ozkan, A., Banfield, L.A., Marchand, A.M.E. & McRae, L.E. 2010. Reservoir quality modeling of tight-gas sands in Wamsutter field: integration of diagenesis, petroleum systems, and production data. *American Association of Petroleum Geologists Bulletin*, **94**, 1229–1266, https://doi.org/10.1306/04211009140

Tosca, N.J. & Wright, V.P. 2015. Diagenetic pathways linked to labile Mg-clays in lacustrine carbonate reservoirs: a model for the origin of secondary porosity in the Cretaceous pre-salt Barra Velha Formation, offshore Brazil. *In*: Armitage, P.J., Butcher, A. et al. (eds) *Reservoir Quality of Clastic and Carbonate Rocks: Analysis, Modelling and Prediction*. Geological Society, London, Special Publications **435**. First published online November 20, 2015, https://doi.org/10.1144/SP435.1.

Tucker, M.E. & Wright, V.P. 1990. Chapter 7: Diagenetic processes, products and environments. *In*: Tucker, M.E. & Wright, V.P. (eds) *Carbonate Sedimentology*. Blackwell Science, Oxford, 314–364.

Vagle, G.B., Hurst, A. & Dypvik, H. 1994. Origin of quartz cement in some sandstone from the Jurassic of the Inner Moray Firth (UK). *Sedimentology*, **41**, 363–377.

van der Land, C., Wood, R., Wu, K., van Dijke, M.I.J., Jiang, Z., Corbett, P.W.M. & Couples, G. 2013. Modelling the permeability evolution of carbonate rocks. *Marine and Petroleum Geology*, **48**, 1–7, https://doi.org/10.1016/j.marpetgeo.2013.07.006

Vos, P.C., De Boer, P.L. & Misdorp, R. 1988. Sediment stabilization by benthic diatoms in intertidal sandy shoals; qualitative and quantitative observations. *In*: de Boer, P.L., van Gelder, A. & Nio, S.D. (eds) *Tide-influenced Sedimentary Environments and Facies*. Reidel Publishing, Dordrecht, 511–526.

Walderhaug, O. 1994*a*. Precipitation rates for quartz cement in sandstones determined by fluid inclusion microthermometry and temperature history modelling. *Journal of Sedimentary Research Section a-Sedimentary Petrology and Processes*, **64**, 324–333.

Walderhaug, O. 1994*b*. Temperatures of quartz cementation in Jurassic sandstones from the Norwegian continental shelf – evidence from fluid inclusions. *Journal of Sedimentary Research Section a-Sedimentary Petrology and Processes*, **64**, 311–323.

Walderhaug, O. 1996. Kinetic modeling of quartz cementation and porosity loss in deeply buried sandstone reservoirs. *American Association of Petroleum Geologists Bulletin*, **80**, 731–745.

Walderhaug, O., Lander, R.H., Bjørkum, P.A., Oelkers, E.H., Bjørlykke, K. & Nadeau, P.H. 2000. Modelling quartz cementation and porosity in reservoir sandstones: examples from the Norwegain continental shelf. *In*: Worden, R.H. & Morad, S. (eds) *Quartz Cementation in Sandstones*. International Association of Sedimentologists, Special Publications, Oxford, **29**, 39–50.

Walderhaug, O., Bjørkum, P.A. & Aase, N.E. 2006. Kaolin-coating of stylolites, effect on quartz cementation and general implications for dissolution at mineral interfaces. *Journal of Sedimentary Research*, **76**, 234– 243, https://doi.org/10.2110/jsr.2006.015

Walter, L.M. & Burton, E.A. 1990. Dissolution of Recent platform carbonate sediments in marine pore fluids. *American Journal of Science*, **290**, 601–643.

WANG, W.Y., MADLAND, M.V., ZIMMERMAN, U., NERMOEN, A., KORSNES, R.I., BERTOLINI, S.R.A. & HILDEBRAND-HABEL, T. 2016. Evaluation of porosity change during chemo-mechanical compaction in flooding experiments on Liège outcrop chalk. *In*: ARMITAGE, P.J., BUTCHER, A. ET AL. (eds) *Reservoir Quality of Clastic and Carbonate Rocks: Analysis, Modelling and Prediction*. Geological Society, London, Special Publications, **435**. First published online October 26, 2016, https://doi.org/10.1144/SP435.10

WARREN, J.K. 1999. *Evaporites: Their Evolution and Economics*. Blackwell Science, Oxford.

WELLS, M., HIRST, P., BOUCH, J., WHEAR, E. & CLARK, N. 2015. Deciphering multiple controls on reservoir quality and inhibition of quartz cement in a complex reservoir: Ordovician glacial sandstones, Illizi Basin, Algeria. *In*: ARMITAGE, P.J., BUTCHER, A. ET AL. (eds) *Reservoir Quality of Clastic and Carbonate Rocks: Analysis, Modelling and Prediction*. Geological Society, London, Special Publications **435**. First published online December 11, 2015, https://doi.org/10.1144/SP435.6

WENDLER, F., OKAMOTO, A. & BLUM, P. 2016. Phase-field modeling of epitaxial growth of polycrystalline quartz veins in hydrothermal experiments. *Geofluids*, **16**, 211–230, https://doi.org/10.1111/gfl.12144

WHITAKER, F.F. & XIAO, Y.T. 2010. Reactive transport modeling of early burial dolomitization of carbonate platforms by geothermal convection. *American Association of Petroleum Geologists Bulletin*, **94**, 889–917, https://doi.org/10.1306/12090909075

WHITTLE, G.L., KENDALL, C.G.S., DILL, R.F. & ROUCH, L. 1993. Carbonate cement fabrics displayed – a traverse across the margin of the Bahamas Platform near Lee Stocking Island in the Exuma Cays. *Marine Geology*, **110**, 213–243, https://doi.org/10.1016/0025-3227(93)90086-b

WOJCIK, K.M., ESPEJO, I.S., KALEJAIYE, A.M. & UMAHI, O.K. 2016. Bright spots, dim spots: geologic controls of direct hydrocarbon indicator type, magnitude, and detectability, Niger Delta Basin. *Interpretation: a Journal of Subsurface Characterization*, **4**, SN45–SN69, https://doi.org/10.1190/int-2016-0062.1

WOOLDRIDGE, L.J., WORDEN, R.H., GRIFFITHS, J. & UTLEY, J.E.P. 2017*a*. Clay-coated sand grains in petroleum reservoirs: understanding their distribution via a modern analogue. *Journal of Sedimentary Research*, **87**, 338–352, https://doi.org/10.2110/jsr.2017.20

WOOLDRIDGE, L.J., WORDEN, R.H., THOMPSON, A., GRIFFITHS, J. & CHUNG, P. 2017*b*. Biofilm origin of clay-coated sand grains. *Geology*, **45**, 875–878, https://doi.org/10.1130/G39161.1

WORDEN, R.H. & BURLEY, S.D. 2003. Sandstone diagenesis: the evolution from sand to stone. *In*: BURLEY, S.D. & WORDEN, R.H. (eds) *Sandstone Diagenesis, Recent and Ancient*. International Association of Sedimentologists Reprint Series, Oxford, 3–44.

WORDEN, R.H. & MORAD, S. 2000. Quartz cementation in sandstones: a review of the key controversies. *In*: WORDEN, R.H. & MORAD, S. (eds) *Quartz Cementation in Sandstones*. International Association of Sedimentologists, Special Publications, oxford, 1–20.

WORDEN, R.H. & MORAD, S. 2003. Clay minerals in sandstones: controls on formation, distribution and evolution. *In*: WORDEN, R.H. & MORAD, S. (eds) *Clay Mineral Cements in Sandstones*. International Association of Sedimentologists, Special Publications, **34**, Blackwells, Oxford, 3–41.

WORDEN, R.H. & SMALLEY, P.C. 1996. H_2S-producing reactions in deep carbonate gas reservoirs: Khuff Formation, Abu Dhabi. *Chemical Geology*, **133**, 157–171, https://doi.org/10.1016/s0009-2541(96) 00074-5

WORDEN, R.H. & SMALLEY, P.C. 2001. H_2S in North Sea oil fields: importance of thermochemical sulphate reduction in clastic reservoirs. *In*: CIDU, R. (ed.) *Water-Rock Interaction 2001*. Balkema, Lisse, 659–662.

WORDEN, R.H., SMALLEY, P.C. & OXTOBY, N.H. 1995. Gas souring by thermochemical sulfate reduction at 140 degrees C. *American Association of Petroleum Geologists Bulletin*, **79**, 854–863.

WORDEN, R.H., OXTOBY, N.H. & SMALLEY, P.C. 1998. Can oil emplacement prevent quartz cementation in sandstones? *Petroleum Geoscience*, **4**, 129–137, https://doi.org/10.1144/petgeo.4.2.129

WORDEN, R.H., COLEMAN, M.L. & MATRAY, J.M. 1999. Basin scale evolution of formation waters: a diagenetic and formation water study of the Triassic Chaunoy Formation, Paris Basin. *Geochimica Et Cosmochimica Acta*, **63**, 2513–2528.

WORDEN, R.H., MAYALL, M. & EVANS, I.J. 2000*a*. The effect of ductile–lithic sand grains and quartz cement on porosity and permeability in Oligocene and lower Miocene clastics, South China Sea: prediction of reservoir quality. *American Association of Petroleum Geologists Bulletin*, **84**, 345–359.

WORDEN, R.H., SMALLEY, P.C. & CROSS, M.M. 2000*b*. The influence of rock fabric and mineralogy on thermochemical sulfate reduction: Khuff Formation, Abu Dhabi. *Journal of Sedimentary Research*, **70**, 1210–1221.

WORDEN, R.H., SMALLEY, P.C. & BARCLAY, S.A. 2003. H_2S and diagenetic pyrite in North Sea sandstones: due to TSR or organic sulphur compound cracking? *Journal of Geochemical Exploration*, **78–79**, 487–491, https://doi.org/10.1016/s0375-6742(03)00072-4

WORDEN, R.H., NEEDHAM, S.J. & CUADROS, J. 2006. The worm gut; a natural clay mineral factory and a possible cause of diagenetic grain coats in sandstones. *Journal of Geochemical Exploration*, **89**, 428–431, https://doi.org/10.1016/j.gexplo.2005.12.011

WORDEN, R.H., FRENCH, M.W. & MARIANI, E. 2012. Amorphous silica nanofilms result in growth of misoriented microcrystalline quartz cement maintaining porosity in deeply buried sandstones. *Geology*, **40**, 179–182, https://doi.org/10.1130/g32661.1

WORDEN, R.H., BUKAR, M. & SHELL, P. 2018. Does oil emplacement inhibit quartz cementation and preserve reservoir quality? *American Association of Petroleum Geologists Bulletin*, **102**, 49–75.

WORTHINGTON, P.F. 2003. Effect of clay content upon some physical properties of sandstone reservoirs. *In*: WORDEN, R.H. & MORAD, S. (eds) *Clay Mineral Cements in Sandstones*. International Association of Sedimentologists, Special Publications, **34**, Blackwell Science, Oxford, 191–211.

Xiao, Y. & Jones, G.D. 2006. Reactive transport modeling of carbonate and siliciclastic diagenesis and reservoir quality prediction. *SPE Journal*, Paper 101669, 1–10.

Xie, M.L., Mayer, K.U. *et al.* 2015. Implementation and evaluation of permeability-porosity and tortuosity–porosity relationships linked to mineral dissolution-precipitation. *Computational Geosciences*, **19**, 655–671, https://doi.org/10.1007/s10596-014-9458-3

Yuan, G.H., Cao, Y.C., Gluyas, J. & Jia, Z.Z. 2017. Reactive transport modeling of coupled feldspar dissolution and secondary mineral precipitation and its implication for diagenetic interaction in sandstones. *Geochimica et Cosmochimica Acta*, **207**, 232–255, https://doi.org/10.1016/j.gca.2017.03.022

Zhu, S.F., Zhu, X.M., Wang, X.L. & Liu, Z.Y. 2012. Zeolite diagenesis and its control on petroleum reservoir quality of Permian in northwestern margin of Junggar Basin, China. *Science China – Earth Sciences*, **55**, 386–396, https://doi.org/10.1007/s11430-011-4314-y

Diagenetic pathways linked to labile Mg-clays in lacustrine carbonate reservoirs: a model for the origin of secondary porosity in the Cretaceous pre-salt Barra Velha Formation, offshore Brazil

NICHOLAS J. TOSCA[1,2]* & V. PAUL WRIGHT[3]

[1]*Department of Earth & Environmental Sciences, University of St Andrews, St Andrews KY16 9AL, UK*

[2]*Present address: Department of Earth Sciences, University of Oxford, Oxford OX1 3AN, UK*

[3]*PW Carbonate Geoscience Ltd and Natural Sciences (Geology), National Museum of Wales, Cathays Park, Cardiff CF10 3NP, UK*

**Correspondence: nickt@earth.ox.ac.uk*

Abstract: The lacustrine carbonate reservoirs of the South Atlantic host significant accumulations of chemically reactive and Al-free Mg-silicate minerals (e.g. stevensite, kerolite and talc). Petrographic data from units such as the Cretaceous Barra Velha Formation in the Santos Basin suggest that Mg-silicate minerals strongly influenced, and perhaps created, much of the observed secondary porosity. The diagenetic interactions between reactive Mg-silicate minerals and carbonate sediments are, however, poorly known. Here we develop a conceptual model for the origin of secondary porosity in the Barra Velha Formation guided by considerations of the chemistry that triggers Mg-silicate crystallization, as well as the geochemical and mineralogical factors that act as prerequisites for rapid Mg-silicate dissolution during early and late diagenesis. We conclude that sub-littoral zones of volcanically influenced rift lakes would have acted as the locus for widespread Mg-silicate accumulation and preservation. Organic-rich profundal sediments, however, would be especially prone to Mg-silicate dissolution and secondary porosity development. Here, organic matter diagenesis (especially methanogenesis) plays a major role in modifying the dissolved inorganic carbon budget and the pH of sediment porewaters, which preferentially destabilizes and then dissolves Mg-silicates. Together, the sedimentological, stratigraphic and geochemical predictions of the model explain many enigmatic features of the Barra Velha Formation, providing a novel framework for understanding how Mg-silicate–carbonate interactions might generate secondary porosity more broadly in other lacustrine carbonate reservoirs across the South Atlantic.

The proposal that carbonate-hosted porosity in the large oil fields in the Barra Velha Formation of the Santos Basin, offshore Brazil, is a result of the dissolution of Mg-silicates, such as stevensite (Wright & Barnett 2015), raises the possibility that an unstudied diagenetic system strongly influences the properties of lacustrine carbonates. To address this issue, this paper reviews the distribution and diagenetic behaviour of Mg-silicates, especially those associated with rift basins such as those currently the focus of intense exploration in the South Atlantic. Based on these considerations, we develop a model for understanding how Mg-silicate–carbonate interactions lead to the generation of secondary porosity in lacustrine reservoirs such as the Barra Velha Formation.

Lake systems developed on rifted margins, specifically those that are volcanically influenced, produce a wide variety of carbonate sediments (Renaut *et al.* 2002; Gierlowski-Kordesch 2010; Wright 2012). In these settings, the lithology of the surrounding catchment influences lake-water chemistry, which, in turn, controls the style of carbonate production (Jones & Bowser 1978; Wright 2012). Lakes developed on volcanic catchments tend to feature waters rich in Ca, Mg, $SiO_{2(aq)}$ and HCO_3^-, commonly reaching alkaline pH values (Yuretich & Cerling 1983; Cerling 1994). This chemistry commonly triggers the formation of Mg-rich clay minerals (here collectively referred to as Mg-silicates) that are abundant products of alkaline lake systems. Nevertheless, despite their frequent occurrences in these systems, Mg-silicates have received little attention from the standpoint of carbonate reservoir properties.

Mg-silicates are the most geochemically reactive of all of the clay mineral groups; they form very early during peri-marine and lacustrine sediment

From: Armitage, P. J., Butcher, A. R., Churchill, J. M., Csoma, A. E., Hollis, C., Lander, R. H., Omma, J. E. & Worden, R. H. (eds) 2018. *Reservoir Quality of Clastic and Carbonate Rocks: Analysis, Modelling and Prediction.* Geological Society, London, Special Publications, **435**, 33–46.
First published online November 20, 2015, https://doi.org/10.1144/SP435.1

diagenesis, and are known to precipitate directly (and very rapidly) from the water column, setting them apart from other clay minerals (Millot 1970; Jones 1986). Because they exhibit marked variation in response to changes in pH, cation chemistry, salinity and the presence/absence of Al-bearing detrital material, Mg-rich clays have long been recognized as valuable tools in developing facies models for lacustrine environments. Most notably, Millot (1970) developed a facies model for the distribution of Mg-rich clays in closed-basin settings that has largely withstood the test of time. The model places the most Al- and Fe-rich clay minerals at the basin margins, as aluminosilicate detritus becomes more important, with clay mineral chemistry progressively becoming more Mg- and Si-rich (and Al-poor) towards the basin centre. Eventually, clays formed in the centre of the basin are dominated by Mg-rich smectite (i.e. stevensite), kerolite and sepiolite. This model, and the decades of subsequent work refining it (see more expanded discussions in Jones 1986; Calvo *et al.* 1999; Deocampo *et al.* 2009; Tosca 2015), places lacustrine clay mineral formation and Mg-silicates within an easily understood depositional framework. Nevertheless, predicting the distribution of Mg-rich clays on a thermodynamic basis has been hindered by gaps in the understanding of the Mg-silicate crystallization process (Jones 1986; Galán & Pozo 2011; Tosca 2015).

Although the formation of lacustrine Mg-silicates has received considerable attention, the diagenesis of Mg-silicate clay minerals and the processes that drive dissolution and generate porosity in lacustrine carbonates have received very little. In the following sections, we explore this process from a number of perspectives. We start by introducing the major structural characteristics of the most common lacustrine Mg-rich clay minerals and examine the processes that lead to their formation. This background makes it easier to understand why Mg-silicates are so susceptible to destruction during early and later diagenesis of lacustrine carbonates, as we discuss below.

Structural factors influencing Mg-silicate formation and diagenesis

Although there are a number of Mg-rich silicate phases that are known to occur in lacustrine carbonates (Pozo & Casas 1999; Galán & Pozo 2011; Bristow *et al.* 2012), they all share many structural characteristics and have nearly identical chemical compositions. The most common Mg-silicates deposited with lacustrine carbonates are composed almost exclusively of Mg and Si, with only stevensite containing additional cations, most commonly Na. As a consequence, the crystal structures of these phases are closely related but differ significantly in detail (Brindley *et al.* 1977; Bailey 1980; Guggenheim & Eggleton 1987; Guggenheim & Krekeler 2011). This overall structural similarity offers a possible explanation for the common co-occurrence and broadly similar geochemical affinity of Mg-silicates for saline alkaline environments. However, the subtle structural details that distinguish one phase from another mean that each Mg-silicate has specific genetic significance that lends detailed insight into the sedimentary and geochemical environment of deposition.

The lacustrine Mg-silicates are all built loosely on permutations of the talc structure. The crystal structure of talc is composed of a 'TOT' (also referred to as a 2:1 type) structure, whereby layers of Mg-octahedra ('O' layers) are effectively sandwiched by two layers of Si-tetrahedra ('T' layers), attached at the top and bottom of the octahedral layer (Fig. 1). Because talc contains an octahedral sheet composed only of Mg with no substitution in the tetrahedral or octahedral layer, the layers have no permanent charge and therefore no expandability or interlayer cations can be accommodated (Bailey 1980; Evans & Guggenheim 1988; Drits *et al.* 2012). Kerolite, a common lacustrine Mg-silicate, is most intuitively viewed as a hydrated variety of the talc structure. Here, interlayer and surface adsorbed H_2O lead to a slightly larger interlayer repeat distance (Fig. 1), although the details of how this is accommodated are not well known (Brindley *et al.* 1977; Miller *et al.* 1991). Because of these close similarities, kerolite in practice is commonly mistaken for talc (and vice versa). However, in most sedimentary environments that give rise to Mg-silicates, kerolite precipitates initially and talc (*sensu stricto*) formation usually results from prolonged dehydration: for example, upon burial diagenesis (Mitsuda & Taguchi 1977; Noack *et al.* 1989; Tosca *et al.* 2011). Stevensite, yet another common lacustrine Mg-silicate, is broadly similar to talc and kerolite, except that the Mg-octahedral layer contains a number of cation vacancies that impart a low but permanent negative charge to the structure: this must be accommodated by positively charged interlayer cations (Fig. 1) (Faust & Murata 1953; Faust *et al.* 1959; Wilson 2013). This classifies stevensite as an expandable clay and member of the smectite group, but overall its structure still shares many structural similarities with talc and kerolite. Finally, sepiolite is composed of 'TOT ribbons' of Mg-octahedra and Si-tetrahedra (Guggenheim & Krekeler 2011). These ribbons run parallel with the crystallographic *x*-axis and, when linked together at their corners, form voids or channels that accommodate molecular H_2O (Fig. 1).

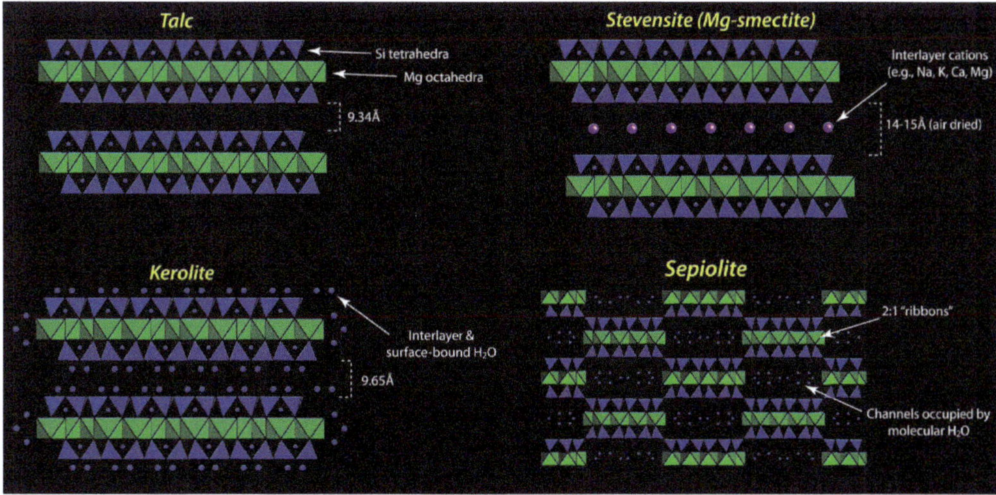

Fig. 1. A comparison of major structural features of common lacustrine Mg-silicate minerals, showing that all of the minerals under consideration are built from permutations of a 2:1 arrangement of Si tetrahedra and Mg octahedra, and are nominally free of Al.

From the standpoint of reservoir quality, the lacustrine Mg-silicate minerals discussed above share one critical feature that sets them apart from the rest of the clay mineral family: they do not contain appreciable levels of Al^{3+}. The lack of Al^{3+} allows these phases to be precipitated with relative ease directly from most natural waters (because Al^{3+} is normally insoluble under the chemical conditions of most surface waters). Also, the absence of Al^{3+} has direct consequences for Mg-silicate dissolution rates and mechanisms, which in turn places strict limits on the range of secondary products that might be formed upon their destruction.

Without Al^{3+}, the structures of the lacustrine Mg-silicates are composed entirely of Mg–O and Si–O bonds, and this strongly influences their dissolution kinetics compared to Al-bearing counterparts (Oelkers 2001*a*, *b*; Saldi *et al.* 2007). The dissolution of silicate minerals typically proceeds through the breaking of metal–oxygen bonds, which can occur at different rates depending on the bond types present (Oelkers 2001*a*, *b*). Thus, as a silicate mineral dissolves, metals are extracted from the structure in the relative order of their bond disruption rates. Of course, to maintain charge balance within the structure, proton (H^+) exchange necessarily accompanies metal release, causing H–O bonds to be formed to replace broken metal–O bonds (Oelkers 2001*a*, *b*).

One consequence of the structures discussed above is that Mg–O bonds are destroyed faster than both Al–O and Fe(III)–O bonds, both of which tend to be common in other aluminosilicates such as illite or montmorillonite (Casey & Westrich 1992; Oelkers 2001*a*, *b*). Mg–O bonds are also destroyed faster than Si–O bonds. There is abundant evidence to support these trends, including the more rapid release of Mg from serpentine, forsterite and enstatite (Casey & Ludwig 1995; Oelkers 2001*a*, *b*; Oelkers & Schott 2001), as well as antigorite, talc and phlogopite dissolution (Saldi *et al.* 2007; Brantley 2008). In drawing analogies with silicate mineral dissolution and ligand exchange reactions that occur around a particular cation, Casey & Ludwig (1995) noted that bond destruction rates strongly correlate with H_2O exchange rates around a given cation (which also correlate with bond strength). Indeed, H_2O exchange rates for Mg^{2+} are orders of magnitude faster than for Al^{3+} and Fe^{3+}, providing a mechanistic basis for observed differences in dissolution rates among silicates.

Given the variation in metal–O bond disruption rates, it follows that the slowest metal–oxygen bond to break that holds the mineral structure together therefore becomes rate-limiting (Oelkers 2001*a*, *b*; Saldi *et al.* 2007). The overall mineral dissolution rate is then a function of whether this bond disruption rate is accelerated or decelerated under specific conditions. Importantly, however, not all bonds need to be broken to completely destroy a mineral structure. For example, in the Mg-olivine mineral forsterite, Mg–O bonds are the rate-limiting step simply because their destruction liberates silica tetrahedra without needing to break Si–O bonds (because the SiO_4 tetrahedra are isolated from one another: Oelkers 1999, 2001*a*, *b*).

Mg-rich clay minerals appear to dissolve in a similar fashion, with a mechanism dictated by the specifics of the crystal structures discussed above. For example, under acidic conditions, the rates of proton exchange for Mg^{2+} in the octahedral sheet are so rapid that as this proceeds inwards from the edges, partially detached silica species are liberated in the process, in turn leading to increased surface area and further accelerated rates of dissolution (Saldi *et al.* 2007).

Together, the distinctly Al-free composition and similar structures of the Mg-rich silicates leads to dissolution rates that are much more rapid than the other aluminosilicate minerals. Most importantly, Al-free compositions lead to congruent dissolution insofar as all mineral components are liberated into solution without the formation of secondary aluminosilicates. This makes specific predictions for the range of possible products produced once Mg-silicates dissolve. However, predicting the occurrence of secondary porosity resulting from Mg-rich clay dissolution logically involves understanding the processes responsible for their original distribution within lacustrine facies.

Chemical controls on Mg-silicate formation in lacustrine sediments

A brief inspection of the chemical formulas of the Al-free Mg-silicates indicates that their stabilities will be affected by dissolved Mg^{2+}, $SiO_{2(aq)}$, pH and salinity. Indeed, natural occurrences of Al-free Mg-silicates in saline alkaline lake settings, combined with laboratory synthesis studies, indicate that these variables are responsible for the formation of one particular phase over another in most environments (Jones 1986; Jones & Galan 1988; Galán & Pozo 2011; Jones & Conko 2011). However, although the most important chemical variables in driving Mg-rich clay formation have long been recognized, predicting the occurrence of Mg-silicates from natural waters has historically been difficult and difficulties are often attributed sluggish precipitation kinetics.

The chemical and physical processes that lead to Mg-rich clay formation are easier to understand if placed within the broader context of mineral nucleation and crystal growth kinetics. Like most other minerals, clay precipitation first involves nucleation (the *de novo* generation of crystal nuclei) and then crystal growth. Mg-silicates may nucleate directly from water (termed homogeneous nucleation) or on the surface of a pre-existing solid (termed heterogeneous nucleation: Stumm 1992; Tosca 2015). Classical nucleation theory states that the homogeneous nucleation of Mg-silicates from water necessarily requires a critical level of supersaturation to be overcome (Stumm 1992). This critical supersaturation boundary predicts that spontaneous nucleation will not occur unless a given water is many times more supersaturated than equilibrium solubility predicts.

Until recently, the critical supersaturation boundary that triggers Mg-silicate nucleation from water was not well known, but a recent review of experimental synthesis data (Tosca 2015) delineated a clear boundary in more quantitative terms. This boundary can be expressed in terms of a 'Mg-silicate solubility diagram' that captures the three major chemical controls on Mg-silicate formation (i.e. $a_{SiO_2(aq)}$, $a_{Mg^{2+}}$ and pH, or a_{H^+}) (Jones 1986; Jones & Galan 1988) (Fig. 2). Figure 2 shows that this nucleation boundary lies well above the equilibrium solubilities for Mg-silicate minerals, implying that significant supersaturation is required to initiate nucleation. Figure 2 also shows that a number of different combinations of pH, Mg^{2+} and $SiO_{2(aq)}$ may result in the formation of Mg-silicates directly from water, but the individual phase that grows from such nuclei (e.g. stevensite, kerolite or sepiolite) is dependent on the kinetics of subsequent crystal growth (Jones 1986; Tosca & Masterson 2014; Tosca 2015).

Most importantly, the nucleation boundary in Figure 2 indicates why the evaporative concentration of some lake waters leads to the formation of Mg-silicates from solution and some do not. Unless this boundary is crossed, Mg-silicates will not precipitate directly from water, regardless of the phase. As Figure 2 shows, at higher $SiO_{2(aq)}$ concentrations (or activities), lower Mg and/or pH is required to overcome the nucleation boundary. This boundary also provides further support for why the most common formation pathway for stevensite is now widely recognized as authigenic precipitation from lake water (Wilson 2013). Critically, any process that results in a rapid or sudden supersaturation above this boundary will result in Mg-silicate precipitation directly from water. Such a process could include porewater modification by biological activity, mixing of water sources (e.g. hot spring input) or, much more commonly, evaporative concentration. The type of Mg-silicate formed is sensitive to the pH, Mg/Si ratio, salinity, $SiO_{2(aq)}$ input and temperature under which crystal growth occurs.

The nucleation boundary discussed above also agrees with geochemical and mineralogical analyses of evaporating saline lakes. Combined mineralogical and lake-water analyses reported in Gac *et al.* (1977) and Darragi & Tardy (1987) offers a clear example. These authors showed that in Lake Yoa of Northern Chad, authigenic stevensite precipitates with aragonite. Although aragonite is commonly formed from the Lake Yoa waters, the authigenic carbonate phase associated with

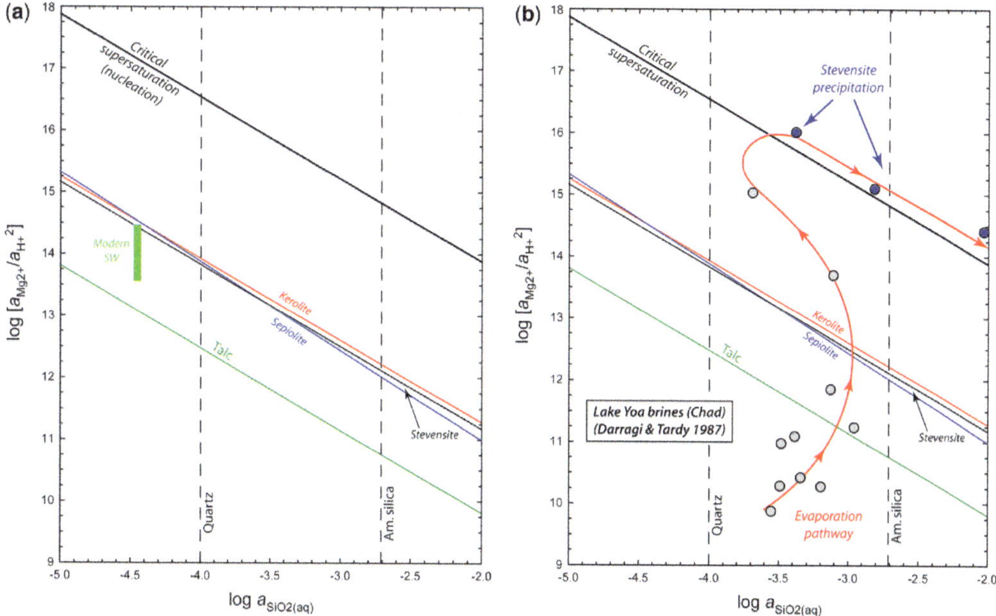

Fig. 2. **(a)** Solubility diagram for the Mg–SiO_2–H_2O system at 25°C. The diagram illustrates the solubility of Mg-silicates for water compositions that have increasing Mg^{2+} concentrations and/or pH (larger *y*-axis values) or increasing $SiO_{2(aq)}$ concentrations (larger *x*-values). Water compositions plotting on or above a given boundary indicate supersaturation with respect to the Mg-silicate phase of interest. Note that the 'critical supersaturation' boundary delineated by experimental studies (Tosca 2015) corresponds to the point at which nucleation of Mg-silicates will occur directly from water with no available substrate. Modern seawater (SW) is plotted for reference, and shows why nucleation of Mg-silicates does not occur. Vertical dashed lines indicate solubilities for quartz and amorphous silica. **(b)** Chemical evolution of waters from Lake Yoa indicating the effects of evaporative concentration (data from Darragi & Tardy 1987). Waters from which stevensite has precipitated are plotted as blue circles. Sources of thermodynamic data are detailed in Tosca (2015).

Mg-silicate mineralization may vary even within the Lake Chad system, illustrating the varying impact of Mg-silicates on the resulting Ca/Mg ratio of solution (e.g. Gac *et al.* 1977; Eugster & Jones 1979; Darragi & Tardy 1987). For example, authigenic low-Mg calcite has been observed to crystallize with Mg-silicates in the Pantanal wetlands of Brazil (Barbiéro *et al.* 2002), and, indeed, Mg-silicates and low-Mg calcitic spherules both crystallized early in the history of the Barra Velha precursor sediment (Wright & Barnett 2015). Regardless of the identify of the main carbonate phase, however, the chemical evolution of the evaporating water in Lake Yoa follows a pathway illustrated in Figure 2. This pathway agrees with the supersaturation levels predicted from synthesis studies, and represents an important step in understanding when and where Mg-silicates are likely to form. This also provides some of the most convincing evidence that the main genetic pathway for widespread stevensite formation is through evaporative concentration.

The relationships discussed above can be used to make predictions regarding the distribution of Mg-silicates in lacustrine systems. Clearly, homogeneous nucleation of Mg-silicates directly from water is one of the only mechanisms that will result in the most laterally extensive accumulations of Mg-silicates. This is simply because climatic controls (e.g. evaporative hydrological regimes) act over regionally extensive areas. In contrast, the nucleation of Mg-silicates on pre-existing surfaces (e.g. heterogeneous nucleation, which requires lower levels of supersaturation) is critically dependent on the nature and distribution of the substrate itself (Tosca 2015). A number of substrates may act as favourable surfaces that cause Mg-silicate nucleation (e.g. amorphous silica, detrital certain clay minerals and biological materials, such as extracellular polymeric substances (EPS)), but this process is arguably most important when favourable chemistry and favourable surface properties (e.g. binding properties that decrease Mg-silicate–solution interfacial energy: Tosca 2015) coincide in more locally confined environments, such as microbial mats (Leveille *et al.* 2002; Souza-Egipsy *et al.* 2005; Bontognali *et al.* 2010). For example,

Arp *et al.* (2003) noted some biofilm laminae in subfossil microstromatolites from a marine-influenced crater lake in Indonesia had undergone replacement by Mg–Si. Bontognali *et al.* (2010) identified the potential for Mg and silica to be concentrated in EPS in microbial mats in recent microbial mats from Abu Dhabi. Burne *et al.* (2014) recorded Mg-silicates (stevensite) in modern thrombolites from Lake Clifton in Western Australia, which provides the structural rigidity for thrombolite growth, produced by biofilm mineralization, with aragonite growing in the stevensite matrix. In a more ancient example, Tosca *et al.* (2011) have explained the association of microbialites and early talc in Neoproterozoic carbonates as potentially reflecting microbial activity in influencing porewater chemistry. Wright & Barnett (2015) have identified multi-metre-scale cyclothems in the Barra Velha Formation of the Santos Basin, offshore Brazil, in which stevensite was a common matrix, but direct evidence for a microbial influence is certainly absent from the associated carbonates: these authors invoked evaporation as the likely trigger for large-scale stevensite precipitation.

Sedimentological implications of Mg-silicate nucleation and gel formation

Once the nucleation boundary is crossed, nucleation of Mg-silicates from water does not immediately produce a crystalline clay mineral but, instead, produces a Mg-silicate 'gel'. This precipitation pathway undoubtedly influences early and later diagenetic reactions and sedimentary textures in many non-marine carbonates.

There is abundant experimental and field-based evidence to support this pathway. For example, in their experimental study of Mg-silicate (including stevensite) precipitation from water at 25°C, Tosca & Masterson (2014) showed that the initial stages of stevensite precipitation occur rapidly (e.g. minutes to hours) upon crossing the critical supersaturation, producing a 'gel-phase' characterized by stevensite crystallites that exhibit limited stacking order and a high degree of hydration (e.g. between 20 and 30 wt% H_2O). The 'gel' phase, then, is simply defined as a hydrated aggregate of Mg-silicate nanoparticles, and is evidently persistent through the early diagenetic stages of lacustrine and peri-marine sediments. For example, a number of authors have reported the observation of Mg-silicate 'gels', 'poorly developed clays' or 'poorly crystalline silicates' (Millot 1970; Jeans 1971; Brindley *et al.* 1977; Gac *et al.* 1977; Calvo *et al.* 1999; Pozo & Casas 1999; Leveille *et al.* 2000, 2002; Polyak & Guven 2000; De Santiago Buey *et al.* 2000; Miller & James 2012). The environments in which these gel phases are most often found commonly involve fluids at high solute concentration (i.e. strongly supersaturated with respect to Mg-silicates), high salinity and pH, and most often include speleothems, evaporitic lacustrine environments and saline porewater. The results reported in these studies are in general agreement with those of Tosca & Masterson (2014) in terms of structural aspects detectable from X-ray diffraction (XRD), Fourier transform infrared spectroscopy (FT-IR) and thermal analysis. Indeed, the precipitation of a hydrated 'poorly crystalline' precursor appears to be a common pathway involved in the *de novo* formation of Mg-silicates from supersaturated solutions, thereby providing a link between experimental results and numerous occurrences of Mg-silicate 'gel'. In fact, because of the viscous nature of the gel and because of its role in removing Mg^{2+} from water (especially if the Mg/Si ratio in the fluid is less than that of stevensite), Wright (2012) suggested that stevensite gel might act as an ideal substrate for spherulitic calcite precipitation, which would explain some of the unusual textures observed in Santos Basin carbonates. Because the gel phase has been reported in lacustrine and peri-marine subsurface sediments in addition to those exposed to overlying bottom water (e.g. Pozo & Casas 1999; De Santiago Buey *et al.* 2000; Leveille *et al.* 2000, 2002; Polyak & Guven 2000; Miller & James 2012), the gel phase evidently persists in the subsurface, with the kinetics of dehydration dependent on the factors discussed above. Thus, Mg-silicate gels provide an important source of late diagenetic water to buried sediment, as well as a unique chemical and physical substrate to mediate further diagenetic reactions upon burial.

This Mg-silicate precipitation pathway predicts that a number of diagenetic reactions will take place between the initially precipitated gel and the surrounding sediments. For example, dehydration will ultimately take place upon the deposition of Mg-silicate gel, and this may occur in response to wetting–drying cycles, prolonged exposure to high-salinity solutions (i.e. low a_{H_2O}) or upon even shallow burial and mild heating (as most H_2O is released at very low temperatures during thermal analysis: see Tosca & Masterson 2014). Indeed, 'gel-dehydration' has been invoked by a number of authors as an important driver to explain the development of Mg-silicate minerals in evaporative environments and during sediment diagenesis (Gac *et al.* 1977; Stoessell & Hay 1978; Darragi & Tardy 1987; Pozo & Casas 1999). Gel dehydration was invoked as a main driver for the formation of kerolite associated with microbial mats in speleothems in basaltic caves found in Hawaii (Leveille *et al.* 2000, 2002). A similar dehydration pathway was suggested to explain stevensite associated with evaporitic deposits from saline lakes (Gac *et al.* 1977; Stoessell & Hay 1978; Darragi & Tardy 1987; Pozo & Casas 1999; De Santiago Buey *et al.* 2000). Indeed, gel dehydration has also been implicated in explaining sedimentary textures associated with palygorskite- and sepiolite-bearing ooids formed in palustrine settings influenced by wetting–drying cycles (Miller & James 2012). The common feature of all of these environments involves initially high supersaturation states with respect to Mg-silicate minerals and environments driving the prolonged and, perhaps, cyclic dehydration of initially formed products. Importantly, because the initial gel phase may incorporate between 20 and 30 wt% H_2O even after drying at room temperature (Tosca & Masterson 2014), this may represent an important basinal fluid source to drive the destabilization of other minerals upon burial.

The sedimentological aspects of gel formation require additional consideration. Such gels are likely to accumulate in specific low-energy settings and where they will not be exposed to desiccation. They are unlikely to accumulate around lake margins where exposure and wave reworking will occur, so should be more characteristic of deeper sublittoral areas, although, as discussed below, not necessarily in organic-rich profundal settings. In small lakes or ponds with limited fetch, Mg-silicate gels have the potential to accumulate in relatively shallow settings, although the gels would then be prone to exposure.

Early diagenesis, Mg-silicate dissolution and porosity formation

Once Mg-silicates are deposited, there are a number of factors that can lead to their destabilization and subsequent dissolution during diagenesis. However, evidence of Mg-silicate (e.g. stevensite) dissolution, as documented from field settings, and the underlying reasons for this rather unusual phenomenon have not received much attention in the literature. From first principles, a simple shift in the saturation state of a given porewater with respect to Mg-silicates will establish a thermodynamic drive towards dissolution. Once this drive is established, dissolution and removal will take place at rapidly, controlled in part by the structure and bonding characteristics discussed above, and also by the extremely high specific surface area inherent to Mg-silicates. Under what conditions would such a drive for dissolution be established? Once this happens, how long would Mg-silicates persist in lacustrine sediments, and would this process affect surrounding carbonate sediments?

Lake Turkana, a brackish $Na–HCO_3$ lake of the East African rift system, offers important insight into how a thermodynamic drive for Mg-silicate dissolution could be established and also how labile Mg-silicate phases can be under certain chemical conditions. Stevensite is an important early precipitate from the lake water at Turkana, as demonstrated by cation mass balance and mineralogical analyses (Cerling 1979, 1996; Yuretich & Cerling 1983). Stevensite forms directly from evaporating lake water, which in turn exerts a strong control on Mg^{2+} depletion of the overall basin (Yuretich & Cerling 1983; Cerling 1996). In a detailed study of porewater chemistry at Lake Turkana, Cerling (1996) found that a number of sediment cores showed chemical variations with depth, indicating a diagenetic process not previously identified in lacustrine sediments. Specifically, Cerling (1996) observed increases in excess alkalinity with depth (mostly dominated by carbonate alkalinity), and further showed that these alkalinity increases were accompanied by releases in both Mg^{2+} and $SiO_{2(aq)}$ corresponding to a Mg/Si ratio stoichiometrically equivalent to stevensite. Because dissolved SO_4 is depleted in the Turkana porewaters and the $\delta^{13}C$ composition of the dissolved inorganic carbon (DIC) from these waters progressively increases with depth, Cerling (1996) argued that methanogenesis (only efficient in the near-absence of dissolved sulphate: Hoehler *et al.* 2010) triggered these changes. Methanogenesis, or the microbial generation of methane, typically occurs via one of two possible pathways (Canfield *et al.* 2005): (1) the reduction of CO_2 by H_2; or (2) the fermentation of organic matter, specifically acetate or other organic substrates, to produce methane and dissolved inorganic carbon. The second pathway is typically more important in non-marine sediments (Canfield *et al.* 2005) and results in CH_4 that is highly enriched in ^{12}C, and DIC that is highly enriched in ^{13}C. In addition to down-core increases in carbonate alkalinity (from organic matter fermentation), Mg^{2+} and $SiO_{2(aq)}$, Cerling (1996) also observed a significant increase in the $\delta^{13}C$ of DIC, supporting the idea that methanogenesis drove stevensite dissolution in lake Turkana sediments with depth.

The latter observation suggests that the carbon isotope composition of lacustrine carbonates may, in fact, be a useful guide in identifying processes that may have influenced the alkalinity budget of sediments during early diagenesis, perhaps initiating the corrosion of Mg-silicates (e.g. methanogenesis). However, the $\delta^{13}C$ composition of lacustrine carbonates may be influenced by a number of factors, including the composition of input waters, atmospheric CO_2 exchange and the mixing of gas sources produced by microbial respiration (e.g. Gierlowski-Kordesch 2010). Particularly relevant to Mg-silicate-rich systems, however, are changes in the relative amounts of authigenic dolomite and calcite, which may be driven by changes in Ca and Mg availability (among many other factors). This may in turn affect the bulk $\delta^{13}C$ composition because the mineral-specific fractionation factors for dolomite and calcite are offset from one another (e.g. Bristow *et al.* 2012). These caveats indicate that an integrated mineralogical, geochemical and isotopic approach is needed to establish reliable fingerprints of early diagenetic Mg-silicate corrosion. Nevertheless, the basic observation from Lake Turkana that a silicate mineral such as stevensite shows rapid dissolution behaviour in preference to calcium carbonate sediments is, at first glance, a rather unusual one: it raises the broader question of why such a phenomenon might occur in lacustrine sediments and whether it might be a more general process to which Mg-silicates are particularly susceptible.

To illustrate why stevensite dissolution should occur in the Lake Turkana sediments, we have constructed a simple thermodynamic model exploring how porewater chemistry is modified by methanogenesis. Here, we use a thermodynamic model detailed in Tosca & Masterson (2014) that uses the Pitzer ion interaction model for calculating ion activity coefficients and, therefore, mineral saturation states in high ionic strength waters.

As a starting point, our calculations begin with the initial chemical composition of Lake Turkana water, as reported in Cerling (1996). Here the water has a pH of approximately 9.2 and is initially supersaturated with respect to stevensite, despite being somewhat depleted in Mg^{2+} (from widespread stevensite formation, as discussed above). The most important chemical change occurring in response to methanogenesis is the addition of excess DIC from the fermentation of organic substrates. These changes can be very simply modelled by adding dissolved inorganic carbon to the system as HCO_3^-. In the model, this is also equivalent to adding CO_2 in dissolved form (Fig. 3a) because, regardless of the form of DIC added to the water (e.g. $CO_{2(aq)}$, HCO_3^- or CO_3^{2-}), the distribution of carbonate species is dictated by both the equilibrium thermodynamics and the prevailing pH of the lake Turkana system.

As HCO_3^- is added to the Lake Turkana water, the increase in carbonate alkalinity rapidly shifts the pH from 9.2 to approximately 8.5 as about 60 mmol/kg HCO_3^- is added. At this endpoint in our calculations, the pH and alkalinity values of the simulated porewater are nearly identical to those compositions reported in Cerling (1996). This simple change has marked effects on the saturation state of Mg-silicate minerals. Because Mg-silicate phases display a thermodynamic and kinetic sensitivity to pH, the decrease from pH 9.2 to 8.5 causes stevensite to shift from supersaturated (in the initial Lake Turkana water) to undersaturated (in the porewater once methanogenesis is established) (Fig. 3b). At the same time, because porewater chemistry is effectively buffered by the carbonate system, the saturation states with respect to carbonate minerals actually show slight increases in saturation state and, therefore, remain unsusceptible to dissolution. This provides convincing evidence that small changes in carbonate alkalinity, driven, for example, by the microbial diagenesis

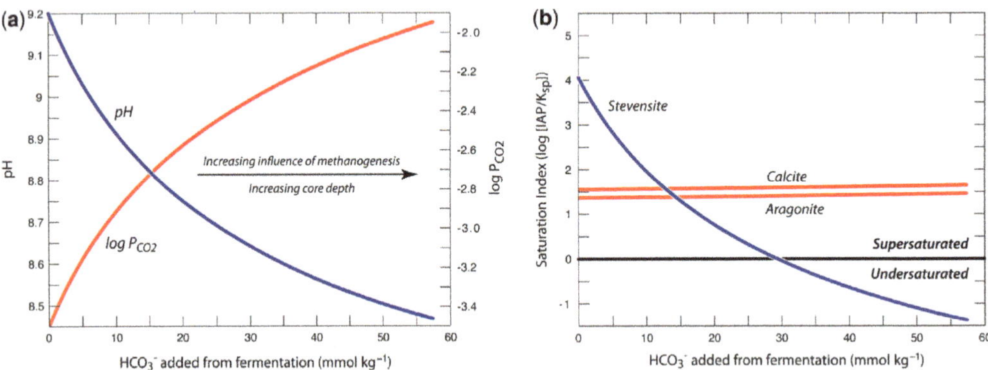

Fig. 3. Thermodynamic model calculations illustrating the effect of methanogenesis on Lake Turkana porewater chemistry. **(a)** Evolution of porewater pH (left axis) and dissolved CO_2 (right axis) during methanogenesis. As methanogenesis produces dissolved inorganic carbon (DIC), the pH of the original water (corresponding to Lake Turkana water reported in Cerling 1996) decreases from 9.2 to less than 8.5, closely corresponding to the porewater chemistry reported by Cerling (1996). **(b)** Calculated saturation states of stevensite, calcite and aragonite during methanogenesis in Lake Turkana porewater. The resulting pH decreases and DIC increases result in negligible effects on carbonate mineral stability, but result in porewaters that are undersaturated with respect to stevensite, enabling dissolution to proceed. IAP, ion activity product; K_{sp}, solubility product.

of organic matter, are effective in pushing Mg-silicates outside of their thermodynamic stability windows, establishing a thermodynamic drive for dissolution.

At first glance, it is somewhat unexpected that the pH values associated with stevensite dissolution in the Lake Turkana cores reach a minimum of around 8.5, yet are able to dissolve stevensite. This indicates that acidic conditions are not necessarily required to dissolve stevensite, so long as thermodynamic undersaturation is established. A related question, however, is why stevensite appears to dissolve so rapidly in the Lake Turkana cores once this undersaturation is established. We can understand this better by returning to the effects of the stevensite crystal structure and particle size on dissolution rate and the lifetime of stevensite in undersaturated sediment porewater.

Aside from the bonding types and crystal structure of stevensite as outlined above, one additional factor causing rapid dissolution lies in its high specific surface area. Like most clay minerals that form under Earth surface conditions, stevensite occurs in a variety of morphological crystal forms that are, at a maximum, of the order of hundreds of nanometres in size (Wilson 2013). In fact, neoformed stevensite has been shown to occur as distinctive spheres ranging from 50 to 1000 Å in diameter (or 100 nm maximum: De Santiago Buey *et al.* 2000; Benhammou *et al.* 2009). As a consequence, the specific surface area of stevensite has been reported to lie between 150 and 400 $m^2\ g^{-1}$, roughly equivalent to the area of a tennis court contained in 1 g. Freshly precipitated stevensite, expected upon initial precipitation from evaporating lake water, undoubtedly lies at the upper end of this range and, perhaps, beyond. Regardless of exact values, the very small size of discrete stevensite particles means that in the presence of a thermodynamic drive for dissolution, the lifetime of these particles within a sediment might be expected to be far lower than many other minerals. This can be quantitatively illustrated if we assume a dissolution rate for stevensite using laboratory dissolution rates measured for crystalline talc as a structural analogue (Saldi *et al.* 2007). If we assume far-from-equilibrium dissolution rates for talc at pH 8.5 (from Saldi *et al.* 2007) and a discrete spherical particle size of 100 nm (e.g. De Santiago Buey *et al.* 2000), then by using the molar volume of stevensite and a simple kinetic relationship between dissolution rate and mineral lifetime (Lasaga 1998), we estimate that, under the Lake Turkana porewater conditions, individual stevensite particles would last no longer than about 70 years. Although there are large uncertainties inherent in this calculation, this is broadly consistent with the indications of rapid stevensite dissolution from the Lake Turkana cores (Cerling 1996), and also with reports of near-instantaneous removal of Mg-silicate scale at circum-neutral pH (e.g. Gunnlaugsson & Einarsson 1989). Indeed, these calculations show that, once a thermodynamic undersaturation for stevensite exists, complete dissolution occurs in a geological instant.

Mg-silicates may also be prone to destabilization during later diagenetic reactions initiated upon burial. For example, the thermal energy supplied upon burial often leads to the migration of cations from interlayer (exchangeable) positions to vacant octahedral sites (Jaynes & Bigham 1987; Jaynes *et al.* 1992; Komadel *et al.* 2005; Petit *et al.* 2008). This process, known as the Hofmann–Klemen effect, occurs mainly in montmorillonite-group clay minerals, to which stevensite belongs. Stevensite itself would be particularly vulnerable to this process because its structure demands that at least 10% of octahedral sites are vacant (Faust & Murata 1953). Univalent cations (e.g. Na^+ and Li^+) are most susceptible to migration in the stevensite structure, and divalent cations may also migrate (Jaynes & Bigham 1987; Jaynes *et al.* 1992; Komadel *et al.* 2005; Petit *et al.* 2008). Importantly, the migration of cations within the structure produces a loss of layer charge and expandability, and the generation of 'talc-like' species. Along with these changes, the loss of a proton from structural hydroxyl (-OH) groups occurs (Jaynes *et al.* 1992; Komadel *et al.* 2005; Petit *et al.* 2008), which would produce acidic porewater. In this way, stevensite may initiate its own self-destruction and dissolution resulting from cation migration driven by shallow burial.

Discussion and summary

Based on the considerations above, Mg-silicates formed in lacustrine settings should display a series of what might be termed 'behaviours' that can help in understanding and predicting their distribution and subsequent diagenesis. In terms of distribution, they not only require highly alkaline conditions, commonly met by the carbonate alkalinity provided in rifts with high volcanic activity, but also a specific chemistry. As discussed above, this chemistry may be generated through a number of means, but the most rapid and laterally extensive mechanism is through evaporative concentration. In the case of the Barra Velha Formation of Brazil, the abundance of calcite and Mg-silicates and the absence of the more typical evaporite minerals such as chlorides and sulphates, supports the view that the catchment areas were draining basic volcanic terrains. Wright (2012) also highlighted these controls, following the proposal from Cerling (1994) that lakes where volcanic terrains predominate tend to produce calcite,

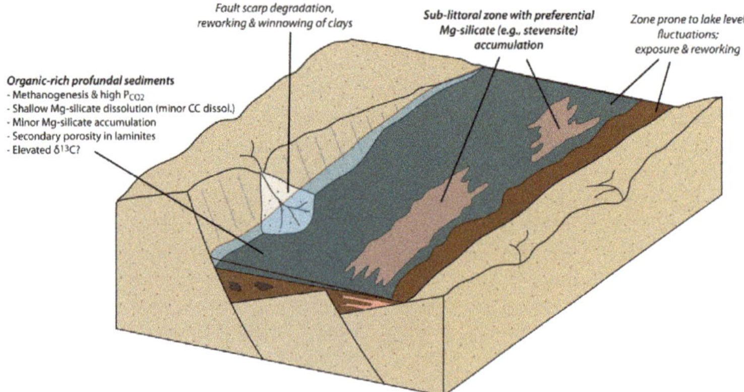

Fig. 4. Modelled distribution of Mg-silicates in an idealized rift-basin setting. Shallower but sub-littoral waters prone to evaporative concentration commonly cross the supersaturation boundary required for nucleation directly from water, and are less prone to the reworking and exposure that results in the largest and most widespread accumulations. Sediments deposited in deeper profundal settings are strongly influenced by organic matter diagenesis. Here methanogenesis, for example, would result in higher DIC (and dissolved CO_2 partial pressure (P_{CO_2})) conditions, where Mg-silicates would commonly destabilize and dissolve, imparting heavy $\delta^{13}C$ compositions to DIC that may be transferred to the carbonate sediments. CC, calcite.

tri-octahedral smectite, analcime and various bicarbonate–carbonate minerals. What remains unclear is whether a sufficient source of both Mg and $SiO_{2(aq)}$ was available to fuel the extensive precipitation of Mg-silicates in these Cretaceous lakes. An unusual and alternative source of solutes might be from the serpentinization of the mantle. Zalan *et al.* (2011) have shown that the Santos, Campos and Espírito Santo basins are all expressions of a magma-poor passive margin, and argue for the presence of a continuous belt of exhumed mantle at the transition between continental and oceanic crusts that flanks all three basins. They infer that serpentinization of exhumed mantle, deduced from gravimetric modelling, occurred down to depths of 6–8 km. This suggestion notwithstanding, direct evidence of serpentinization and/or the presence of exhumed mantle in the offshore Brazilian basins has remained elusive.

In summary, given suitable geochemical conditions, Mg-silicate gels are predicted to accumulate in lower-energy settings below wave base, and not prone to exposure, desiccation and reworking (Fig. 4). However, the effects of methanogenesis in profundal areas where organic matter could be concentrated implies that Mg-silicates would be more prone to destabilization and dissolution, so reducing the likelihood of accumulations in such settings. Although methanogenesis is demonstrably effective in tipping the porewater alkalinity balance in favour of Mg-silicate dissolution and carbonate precipitation, any process that disrupts this balance (i.e. decreases pH and Mg-silicate saturation, but increases carbonate alkalinity) will produce the same effect. Thus, from a facies model standpoint, we are left with a 'Goldilocks' situation where Mg-silicates would not accumulate in sediments deposited in waters that were too shallow or potentially too deep. Shallow closed-system lakes are, in reality, characteristically dynamic and so the actual distribution of Mg-silicates is expected to exhibit more complexity, with controls on chemistry and also with accommodation space playing a critical role. In rift settings, the updip regions of

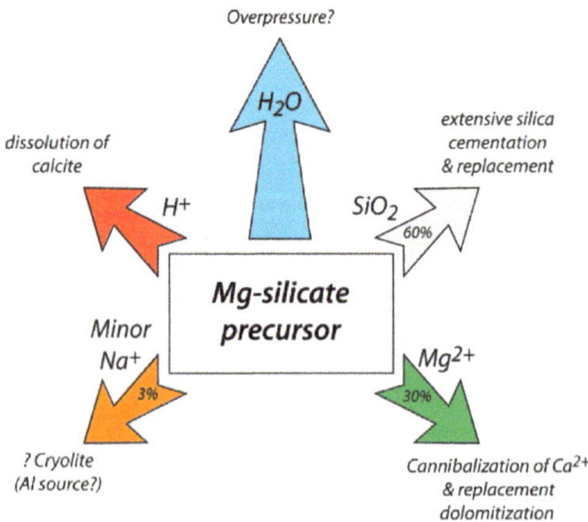

Fig. 5. Schematic showing the relative proportions of chemical components released from stevensite dissolution, and their primary diagenetic products.

half-graben are prone both to differential tilting and to footwall uplift, and so these might be seen as poor sites for Mg-silicate gel accumulation. Deeper settings (e.g. in lows adjacent to main faults), however, are expected to result in the highest levels of organic matter accumulation in profundal zones, promoting, for example, methanogenesis (Fig. 4) (Bosence 1998).

The congruent dissolution of Mg-silicates, leaving no direct *in situ* product, is likely to produce porosity. Wright & Barnett (2015) have identified petrographical criteria for identifying porosity formed after the dissolution of stevensite in the Barra Velha Formation of the Santos Basin. One of the most distinctive manifestations of this effect is seen in the widespread pseudo-fenestral porosity seen in the spherulitic facies. The resulting pore systems are illustrated by Terra *et al.* (2010) (their figs 20 & 21, especially fig. 20b).

Complete decay of the Mg-silicates would result in the release of Mg (30%), silica (60%) and minor Na (3%), in addition to bound water (Fig. 5). Thus, expected diagenetic products would include replacement dolomites (as the Mg^{2+} cannibalized any existing calcite), extensive silica cementation and replacement, and Na-silicates. The release of bound water might contribute to local overpressure, which would limit physical and chemical compaction. These products all fit with anecdotal evidence that the Barra Velha shows replacement dolomitization, extensive silicification, the presence of cryolite (although the source of Al is unresolved), and the survival of open-packed pore systems and the rarity of stylolites.

Jones & Xiao (2013) have proposed that geothermal convection linked to localized conduits might also be a critical factor in enhancing porosity and permeability in lacustrine carbonates sealed by salt, as in the specific case of the Barra Velha Formation in the Santos Basin. The study implies an awareness that late-stage corrosion has affected these carbonates (as is also our experience), but an alternative process that could cause later-stage dissolution could also reflect an inherent property of the decay on Mg-silicates, specifically the production of protons by the Hofmann–Klemen effect, discussed earlier. If this operated in the Barra Velha Formation, it would produce a more widespread dissolution effect and not one linked to major fluid conduits such as faults – a testable alternative model if and when more data become available from the region.

In summary, although both the occurrence and diagenetic consequences of Mg-silicates in carbonate reservoirs are somewhat unfamiliar from the standpoint of petroleum geology, many enigmatic features can be understood from first-principle considerations rooted in geochemistry, mineralogy and sedimentology. The model presented here provides a novel framework for understanding and predicting Mg-silicate–carbonate interactions and their geological consequences, and should provide a fertile starting point to develop and test new hypotheses for porosity evolution as data continue to emerge.

The authors thank Andrew Barnett, Frances Abbotts, and Rachel Wood for helpful discussion, and Jeff Lukasik and an anonymous reviewer for thoughtful reviews.

References

Arp, G., Reimer, A. & Reitner, J. 2003. Microbialite formation in seawater of increased alkalinity, Satonda Crater Lake, Indonesia. *Journal of Sedimentary Research*, **73**, 105–127.

Bailey, S. W. 1980. Structures of layer silicates. *In*: Brindley, G. W. & Brown, G. (eds) *Crystal Structures of Clay Minerals and their X-ray Diffraction*. Mineralogical Society, London, 1–124.

Barbiéro, L., de Queiroz Neto, J. P., Ciornei, G., Sakamoto, A. Y., Capellari, B., Fernandes, E. & Valles, V. 2002. Geochemistry of water and ground water in the Nhecolândia, Pantanal of Mato Grosso, Brazil: variability and associated processes. *Wetlands*, **22**, 528–540.

Benhammou, A., Tanouti, B., Nibou, L., Yaacoubi, A. & Bonnet, J.-P. 2009. Mineralogical and physicochemical investigation of Mg-smectite from Jbel Ghassoul, Morocco. *Clays and Clay Minerals*, **57**, 264–270.

Bontognali, T. R. R., Vasconcelos, C., Warthmann, R. J., Bernasconi, S. M., Dupraz, C., Strohmenger, C. J. & McKenzie, J. A. 2010. Dolomite formation within microbial mats in the coastal sabkha of Abu Dhabi (United Arab Emirates). *Sedimentology*, **57**, 824–844.

Bosence, D. W. J. 1998. Stratigraphic and sedimentological models of rift basins. *In*: Purser, B. H. & Bosence, D. W. J. (eds) *Sedimentation and Tectonics of Rift Basins: Red Sea–Gulf of Aden*. Chapman & Hall, London, 9–25.

Brantley, S. L. 2008. Kinetics of mineral dissolution. *In*: Brantley, S. L., Kubicki, J. & White, A. F. (eds) *Kinetics of Water–Rock Interaction*. Springer, Berlin, 151–210.

Brindley, G. W., Bish, D. L. & Wan, H. M. 1977. Nature of kerolite and its relation to talc and stevensite. *Mineralogical Magazine*, **41**, 443–452.

Bristow, T. F., Kennedy, M. J., Morrison, K. D. & Mrofka, D. D. 2012. The influence of authigenic clay formation on the mineralogy and stable isotopic record of lacustrine carbonates. *Geochimica et Cosmochimica Acta*, **90**, 64–82.

Burne, R. V., Moore, L. S., Christy, A. G., Troitzsch, U., King, P. L., Carnerup, A. M. & Hamilton, P. J. 2014. Stevensite in the modern thrombolites of Lake Clifton, Western Australia: a missing link in microbialite mineralization? *Geology*, **42**, 575–578, https://doi.org/10.1130/G35484.1

CALVO, J. P., BLANC-VALLERON, M. M., RODRIGUEZ-ARANDIA, J. P., ROUCHY, J. M. & SANZ, M. E. 1999. Authigenic clay minerals in continental evaporitic environments. *In*: THIRY, M. & SIMON-COINÇON, R. (eds) *Palaeoweathering, Palaeosurfaces and Related Continental Deposits*. International Association of Sedimentologists, Special Publications, **27**, 129–151.

CANFIELD, D. E., KRISTENSEN, E. & THAMDRUP, B. 2005. *Aquatic Geomicrobiology*. Elsevier, Amsterdam.

CASEY, W. H. & LUDWIG, C. 1995. Silicate mineral dissolution as a ligand exchange reaction. *In*: WHITE, A. F., BRANTLEY, S. L. & RIBBE, P. H. (eds) *Chemical Weathering Rates of Silicate Minerals*. Reviews in Mineralogy and Geochemistry, **31**. Mineralogical Society of America, Washington, DC, 87–118.

CASEY, W. H. & WESTRICH, H. R. 1992. Control of dissolution rates of orthosilicate minerals by divalent metal-oxygen bonds. *Nature*, **355**, 157–159.

CERLING, T. E. 1979. Paleochemistry of plio-pleistocene lake Turkana, Kenya. *Palaeogeography, Palaeoclimatology, Palaeoecology*, **27**, 247–285.

CERLING, T. E. 1994. Chemistry of closed basin lake waters: a comparison between African Rift Valley and some central North American rivers and lakes. *In*: GIERLOWSKI-KORDESCH, E. H. & KELTS, K. (eds) *The Global Geological Record of Lake Basins*, **1**. Cambridge University Press, Cambridge, 29–30.

CERLING, T. E. 1996. Pore water chemistry of an alkaline lake: Lake Turkana, Kenya. *In*: JOHNSON, T. C. & ODADA, E. O. (eds) *Limnology, Climatology and Paleoclimatology of the East African Lakes*. Gordon and Breach, Amsterdam, 225–240.

DARRAGI, F. & TARDY, Y. 1987. Authigenic trioctahedral smectites controlling pH, alkalinity, silica and magnesium concentrations in alkaline lakes. *Chemical Geology*, **63**, 59–72.

DEOCAMPO, D. M., CUADROS, J., WING-DUDEK, T., OLIVES, J. & AMOURIC, M. 2009. Saline lake diagenesis as revealed by coupled mineralogy and geochemistry of multiple ultrafine clay phases: Pliocene Olduvai Gorge, Tanzania. *American Journal of Science*, **309**, 834–858.

DE SANTIAGO BUEY, C., BARRIOS, M. S., ROMERO, E. G. & MONTOYA, M. D. 2000. tMg-rich smectite 'precursor' tphase in the Tagus basin, Spain. *Clays and Clay Minerals*, **48**, 366.

DRITS, V. A., GUGGENHEIM, S., ZVIAGINA, B. B. & KOGURE, T. 2012. Structures of the 2:1 layers of pyrophyllite and talc. *Clays and Clay Minerals*, **60**, 574–587.

EUGSTER, H. P. & JONES, B. F. 1979. Behavior of major solutes during closed-basin brine evolution. *American Journal of Science*, **279**, 609–631.

EVANS, B. W. & GUGGENHEIM, S. 1988. Talc, pyrophyllite, and related minerals. *In*: BAILEY, S. W. (ed.) *Hydrous Phyllosilicates*. Reviews in Mineralogy, **19**. Mineralogical Society of America, Washington, DC, 225–294.

FAUST, G. T. & MURATA, K. J. 1953. Stevensite, redefined as a member of the montmorillonite group. *American Mineralogist*, **38**, 973–987.

FAUST, G. T., HATHAWAY, J. C. & MILLOT, G. 1959. A restudy of stevensite and allied minerals. *American Mineralogist*, **44**, 342–370.

GAC, J. Y., DROUBI, A., FRITZ, B., & TARDY, Y. 1977. Geochemical behaviour of silica and magnesium during the evaporation of waters in Chad. *Chemical geology*, **19**, 215–228.

GALÁN, E. & POZO, M. 2011. Palygorskite and sepiolite deposits in continental environments. Description, genetic patterns and sedimentary settings. *In*: GALAN, E. & SINGER, A. (eds) *Developments in Palygorskite–Sepiolite Research*. Developments in Clay Science, **3**. Elsevier, Amsterdam, 125–173.

GIERLOWSKI-KORDESCH, E. H. 2010. Lacustrine carbonates. *In*: ALONSO-ZARZA, A. M. & TANNER, E. H. (eds) *Carbonates in Continental Settings: Facies, Environments and Processes*. Developments in Sedimentology, **61**. Elsevier, Amsterdam, 1–101.

GUGGENHEIM, S. & EGGLETON, R. A. 1987. Modulated 2:1 layer silicates; review, systematics, and predictions. *American Mineralogist*, **72**, 724–738.

GUGGENHEIM, S. & KREKELER, M. P. 2011. The structures and microtextures of the palygorskite–sepiolite group minerals. *In*: GALAN, E. & SINGER, A. (eds) *Developments in Palygorskite–Sepiolite Research. A New Outlook on these Nanomaterials*. Developments in Clay Science, **3**. Elsevier, Amsterdam, 1–32.

GUNNLAUGSSON, E. & EINARSSON, A. R. 1989. Magnesium-silicate scaling in mixture of geothermal water and deaerated fresh water in a district heating system. *Geothermics*, **18**, 113–120.

HOEHLER, T., GUNSALUS, R. P. & MCINERNEY, M. J. 2010. Environmental constraints that limit methanogenesis. *In*: TIMMIS, K. N. (ed.) *Handbook of Hydrocarbon and Lipid Microbiology*. Springer, Berlin, 636–654.

JAYNES, W. F. & BIGHAM, J. M. 1987. Charge reduction, octahedral charge, and lithium retention in heated, Li-saturated smectites. *Clays and Clay Minerals*, **35**, 440–448.

JAYNES, W. F., TRAINA, S. J., BIGHAM, J. M. & JOHNSTON, C. T. 1992. Preparation and characterization of reduced-charge hectorites. *Clays and Clay Minerals*, **40**, 397–404.

JEANS, C. V. 1971. The neoformation of clay minerals in brackish and marine environments. *Clays and Clay Minerals*, **9**, 209–217.

JONES, B. F. 1986. Clay mineral diagenesis in lacustrine sediments. *In*: MUMPTON, F. A. (ed.) *Studies in Diagenesis*. United States Geological Survey, Bulletin, **1578**, 291–300.

JONES, B. F. & BOWSER, C. J. 1978. The mineralogy and related chemistry of lake sediments. *In*: LERMAN, A. (ed.) *Lakes: Chemistry, Geology and Physics*. Springer, Berlin, 179–227.

JONES, B. F. & CONKO, K. M. 2011. Environmental influences on the occurrences of sepiolite and palygorskite: a brief review. *In*: GALAN, E. & SINGER, A. (eds) *Developments in Palygorskite–Sepiolite Research*. Developments in Clay Science, **3**. Elsevier, Amsterdam, 69–83.

JONES, B. F. & GALAN, E. 1988. Sepiolite and Palygorskite. *In*: BAILEY, S. W. (ed.) *Hydrous Phyllosilicates*, Reviews in Mineralogy, **19**. Mineralogical Society of America, Washington, DC, 631–674.

Jones, G. D. & Xiao, Y. 2013. Geothermal convection in South Atlantic subsalt lacustrine carbonates: developing diagenesis and reservoir quality predictive concepts with reactive transport models. *Bulletin American Association of Petroleum Geologists*, **97**, 1249–1271.

Komadel, P., Madejova, J. & Bujdak, J. 2005. Preparation and properties of reduced-charge smectites – a review. *Clays and Clay Minerals*, **53**, 313–334.

Lasaga, A. C. 1998. *Kinetic Theory in the Earth Sciences*. Princeton University Press, Princeton, NJ.

Leveille, R. J., Fyfe, W. S. & Longstaffe, F. J. 2000. Unusual secondary Ca-Mg-carbonate-kerolite deposits in basaltic caves, Kauai, Hawaii. *Journal of Geology*, **108**, 613–621.

Leveille, R. J., Longstaffe, F. J. & Fyfe, W. S. 2002. Kerolite in carbonate-rich speleothems and microbial deposits from basaltic caves, Kauai, Hawaii. *Clays and Clay Minerals*, **50**, 514–524.

Miller, A. K., Guggenheim, S. & Koster van Groos, A. F. 1991. Bond energy of adsorbed and interlayer water: kerolite dehydration at elevated pressures. *Clays and Clay Minerals*, **39**, 127–130.

Miller, C. R. & James, N. P. 2012. Autogenic microbial genesis of middle miocene palustrine ooids; Nullarbor Plain, Australia. *Journal of Sedimentary Research*, **82**, 633–647.

Millot, G. 1970. *Geology of Clays*. Springer, London.

Mitsuda, T. & Taguchi, H. 1977. Formation of magnesium-silicate hydrate and its crystallization to talc. *Cement and Concrete Research*, **7**, 223–230.

Noack, Y., Decarreau, A., Boudzoumou, F. & Trompette, R. 1989. Low-temperature oolitic talc in upper Proterozoic rocks, Congo. *Journal of Sedimentary Research*, **59**, 717.

Oelkers, E. H. 1999. A comparison of forsterite and enstatite dissolution rates and mechanisms. *In*: Jamtveit, B. & Meakin, P. (eds) *Growth, Dissolution and Pattern Formation in Geosystems*. Kluwer, Dordrecht, 253–268.

Oelkers, E. H. 2001*a*. An experimental study of forsterite dissolution rates as a function of temperature and aqueous Mg and Si concentrations. *Chemical Geology*, **175**, 485–494.

Oelkers, E. H. 2001*b*. General kinetic description of multioxide silicate mineral and glass dissolution. *Geochimica et Cosmochimica Acta*, **65**, 3703– 3719.

Oelkers, E. H. & Schott, J. 2001. An experimental study of enstatite dissolution rates as a function of pH, temperature, and aqueous Mg and Si concentration, and the mechanism of pyroxene/pyroxenoid dissolution. *Geochimica et Cosmochimica Acta*, **65**, 1219–1231.

Petit, S., Righi, D. & Decarreau, A. 2008. Transformation of synthetic Zn-stevensite to Zn-talc induced by the Hofmann-Klemen effect. *Clays and Clay Minerals*, **56**, 645–654.

Polyak, V. J. & Guven, N. 2000. Authigenesis of trioctahedral smectite in magnesium-rich carbonate speleothems in Carlsbad Cavern and other caves of the Guadalupe Mountains, New Mexico. *Clays and Clay Minerals*, **48**, 317.

Pozo, M. & Casas, J. 1999. Origin of kerolite and associated Mg clays in palustrine-lacustrine environments; the Esquivias Deposit (Neogene Madrid Basin, Spain). *Clay Minerals*, **34**, 395.

Renaut, R. W., Morley, C. K. & Jones, B. 2002. Fossil hot-spring travertine in the Turkana basin, northern Kenya: structure, facies, and genesis. *In*: Renaut, R. W. & Ashley, G. M. (eds) *Sedimentation in Continental Rifts*. Society for Sedimentary Geology (SEPM), Special Publications, **73**, 123–141.

Saldi, G. D., Köhler, S. J., Marty, N. & Oelkers, E. H. 2007. Dissolution rates of talc as a function of solution composition, pH and temperature. *Geochimica et Cosmochimica Acta*, **71**, 3446–3457.

Souza-Egipsy, V., Wierzchos, J., Ascaso, C. & Nealson, K. H. 2005. Mg-silica precipitation in fossilization mechanisms of sand tufa endolithic microbial community, Mono Lake (California). *Chemical Geology*, **217**, 77–87.

Stoessell, R. K. & Hay, R. L. 1978. Geochemical origin of sepiolite and kerolite at Amboseli, Kenya. *Contributions to Mineralogy and Petrology*, **65**, 255–267.

Stumm, W. 1992. *Chemistry of the Solid–Water Interface: Processes at the Mineral–Water and Particle–Water Interface in Natural Systems*. John Wiley, New York.

Terra, G. J. S., Spadini, A. R. *et al.* 2010. Classificação de rochas carbonáticas aplicável às bacias sedimentares [Carbonate rock classifcation applied to Brazilian sedimentary basins]. *Boletin Geociencias Petrobras*, **18**, 9–29.

Tosca, N. J. 2015. Geochemical pathways to Mg-clay formation. *In*: Pozo, M. & Galán, E. (eds) *Magnesian Clays: Characterization, Origins and Applications*. AIPEA (Association Internationale pour l'Etude des Argiles), Special Publications, **2**, 283–329.

Tosca, N. J. & Masterson, A. L. 2014. Chemical controls on incipient Mg-silicate crystallization at 25°C: implications for early and late diagenesis. *Clay Minerals*, **49**, 165–194.

Tosca, N. J., Macdonald, F. A., Strauss, J. V., Johnston, D. T. & Knoll, A. H. 2011. Sedimentary talc in Neoproterozoic carbonate successions. *Earth and Planetary Science Letters*, **306**, 11–22.

Wilson, M. J. 2013. *Rock-Forming Minerals, Volume 3C, Sheet Silicates: Clay Minerals*. Geological Society, London.

Wright, V. P. 2012. Lacustrine carbonates in rift settings: the interaction of volcanic and microbial processes on carbonate deposition. *In*: Garland, J., Neilson, J. E., Laubach, S. E. & Whidden, K. J. (eds) *Advances in Carbonate Exploration and Reservoir Analysis*. Geological Society, London, Special Publications, **370**, 39–47, https://doi.org/10.1144/SP370.2

Wright, V. P. & Barnett, A. 2015. An abiotic model for the development of textures in some South Atlantic early Cretaceous lacustrine carbonates. *In*: Bosence, D. W. J., Gibbons, K. A., Le Heron, D. P., Morgan, W. A., Pritchard, T. & Vining, B. A. (eds) *Microbial Carbonates in Space and Time: Implications for Global Exploration and Production*. Geological Society, London, Special Publications, **418**, 209–219, https://doi.org/10.1144/SP418.3

Yuretich, R. F. & Cerling, T. E. 1983. Hydrogeochemistry of Lake Turkana, Kenya: mass balance and mineral reactions in an alkaline lake. *Geochimica et Cosmochimica Acta*, **47**, 1099–1109.

ZALAN, P. V., DO CARMO, M., SEVERINO, G., RIGOTI, C. A., MAGNAVITA, L. P., BACH DE OLIVEIRA, J. A. & VIANNA, A. R. 2011 An entirely new 3D-view of the crustal and mantle structure of a South Atlantic passive margin – Santos, Campos and Espírito Santo basins, Brazil. *American Association of Petroleum Geologists Annual Convention and Exhibition*, Houston, TX, USA, 10–13 April 2011, Search and Discovery Article #30177, http://www.searchanddiscovery.com/documents/2011/30177zalan/ndx_zalan.pdf

Origin and evolution of microporosity in packstones and grainstones in a Lower Cretaceous carbonate reservoir, United Arab Emirates

DANIEL MORAD[1], MATTEO PAGANONI[2*], AMENA AL HARTHI[3], SADOON MORAD[3], ANDREA CERIANI[3], HOWRI MANSURBEG[4], AISHA AL SUWAIDI[3], IHSAN S. AL-AASM[5] & STEPHEN N. EHRENBERG[3]

[1]*Department of Earth Sciences, University of Oslo, N-0316 Oslo, Norway*

[2]*Department of Earth Sciences, University of Oxford, South Parks Road, Oxford OX1 3AN, UK*

[3]*School of Geosciences, Grant Institute, University of Edinburgh, Edinburgh EH9 3FE, UK*

[4]*Department of Petroleum Geosciences, Soran University, Soran, Kurdistan Region, Iraq*

[5]*Department of Earth and Environmental Sciences, University of Windsor, Windsor, Ontario, Canada*

**Correspondence: matteo.paganoni@spc.ox.ac.uk*

Abstract: Microporosity in carbonate reservoirs is generated by the complex interplay between depositional and diagenetic processes. This petrographical, SEM, fluid-inclusion and isotopic study of a Lower Cretaceous carbonate reservoir, Abu Dhabi, UAE, revealed that: (1) micritization of ooids and skeletal fragments, which resulted in spheroidal (rounded) micrite, accounts for most microporosity in peloidal packstones and grainstones; and (2) transformation of spheroidal micrite into subhedral/euhedral micrite and microspar, known as aggrading neomorphism, could happen via precipitation of syntaxial calcite overgrowths around micrite (micro-overgrowths) and not only, as suggested previously in the literature, by recrystallization involving the dissolution (of micrite) and reprecipitation (of microspar). Precipitation of calcite cement around micrite (i.e. destruction of microporosity) is more extensive in the water zone than in the oil zone, which is possibly contributing to the lower porosity and permeability of the carbonate reservoir in the water zone. Similarity in bulk oxygen isotopic values of micritized packstones and grainstones in the water and oil zones (average $\delta^{18}O_{V\text{-}PDB} = -7.2‰$ and $-7.8‰$, respectively) is attributed to: (1) a small difference in temperatures between the crest (oil zone) and the flanks (water zone); and (2) calcite precipitation around micrite occurred prior and subsequent to oil emplacement. Bulk carbon and strontium isotopic compositions of micritized packstones and grainstones in the water and oil zones (average $\delta^{13}C_{V\text{-}PDB} = +3.7‰$ and average $^{87}Sr/^{86}Sr$ ratios = 0.707469) indicate that calcite cement was derived from marine porewaters and/or dissolution of the host limestones. The minimum formation temperatures of bulk micrite/microspar, which are inferred based on paragenetic relationships, fluid-inclusion microthermometry and oxygen isotope data, are around 58–78°C.

The importance of microporosity in reservoir properties and heterogeneity distribution has received increasing regional and global attention during the past four decades (e.g. Wilson 1975; Budd 1989; Lambert *et al.* 2006; Volery *et al.* 2010; Deville de Periere *et al.* 2011). Microporous reservoirs are characterized by high capillary forces, which increase the water saturation and thicken the transition zone, resulting in significant volumes of oil in place. Furthermore, an increase in the amounts of micrite and related microporosity induces changes to the acoustic and transport properties of peloidal packstones and grainstones (Vanorio & Mavko 2011). The origin and diagenetic evolution of microporosity in limestones is a subject of long-standing debate and controversy (Ahr 1989; Budd 1989; Moshier 1989*a*; Saller & Moore 1989; Al-Aasm & Azmy 1996; Richard *et al.* 2007; Volery *et al.* 2010; Deville de Periere *et al.* 2011). Several origins of microporosity have been suggested, including:

- partial dissolution of aragonite without significant subsequent precipitation of low-Mg calcite (LMC) cement in a mixing zone early diagenetic environment (Saller & Moore 1989);

From: Armitage, P. J., Butcher, A. R., Churchill, J. M., Csoma, A. E., Hollis, C., Lander, R. H., Omma, J. E. & Worden, R. H. (eds) 2018. *Reservoir Quality of Clastic and Carbonate Rocks: Analysis, Modelling and Prediction.* Geological Society, London, Special Publications, **435**, 47–66.
First published online December 21, 2016, https://doi.org/10.1144/SP435.20

- mineralogical stabilization of the precursor material and subsequent recrystallization of coarser micrite/microspar crystals in an early diagenetic meteoric or fluctuating environment (Budd 1989; Kaldi 1989; Richard *et al.* 2007; Volery *et al.* 2010; Léonide *et al.* 2014) – the recrystallization process is thought to happen through a hybrid Ostwald ripening (Léonide *et al.* 2014);
- mineralogical stabilization in a marine environment within a relatively closed diagenetic system and subsequent limited matrix cementation (Moshier 1989*b*);
- burial diagenetic precipitation of microcrystalline calcite driven by regional fluid flow (Vahrenkamp *et al.* 2014);
- burial diagenetic dissolution driven by aggressive fluids of either allochems (Dravis 1989) or euhedral micrite/microspar, resulting, in the latter case, in the formation of rounded micrite (referred to as spheroidal herein: Lambert *et al.* 2006).

Micritization and conversion of micrite into microspar, which is referred to in the literature as recrystallization or aggrading neomorphism, are ubiquitous diagenetic processes in carbonate sediments (Friedman 1964; Bathurst 1972; Brand & Veizer 1980, 1981; Tucker 2001; Madden & Wilson 2012) and their role in reservoir quality evolution is debated in the literature (Deville de Periere *et al.* 2011). Micritization, which involves the transformation of allochems into peloids composed of microcrystalline micrite without internal structure, has been suggested to occur on the seafloor in warm quiet-water environments (Bathurst 1966, 1975; Burgess 1979; Perry 1999; Waite *et al.* 2008). Micritization is thought to be accomplished by endolithic algae, bacteria and fungi through microboring (Bathurst 1966, 1975; Kobluck & Risk 1977; Morita 1980; Novitski 1981; Gunther 1990; Chafetz & Buczynski 1992; Monty 1995; Perry 1999; Sherman *et al.* 1999; Reid & Macintyre 2000; Berkyova & Munnecke 2010). These authors suggested that repetition of these processes resulted in complete obliteration of the original grain texture and preservation of the grain margins. Additionally, Reid & Macintyre (2000) proposed that micritization might take place through recrystallization (aggrading neomorphism) (*sensu* Folk 1965; Purdy 1968) on the seafloor. Other researchers (e.g. Tucker & Marshall 2004; Spadafora *et al.* 2010) suggested that calcite precipitation during micritization occurs through microbially mediated nanoglobule formation and subsequent merging into larger micrite crystals. Further investigation of this process is thus required to better understand the characteristics of carbonate crystals and mechanisms of crystal growth within peloids (Reid & Macintyre 1998).

Recrystallization or aggrading neomorphism is suggested by Folk (1965) to indicate a change in size and/or shape of a carbonate grain/crystal (e.g. micrite into microspar) with little or no change in chemical composition or mineralogy. Neomorphism also includes the conversion/stabilization of aragonite and high-Mg calcite (HMC) into LMC (Folk 1965). These neomorphic processes have been proposed to occur in both meteoric and burial-diagenetic environments (Tucker & Wright 2009). The conversion of micrite into microspar is suggested to follow the Ostwald ripening phenomenon (Volery *et al.* 2010; Deville de Periere *et al.* 2011). According to this phenomenon, small anhedral micrite crystals, which have a large surface area (i.e. kinetically less stable), can dissolve and precipitate as coarser, euhedral crystals with a smaller surface area (i.e. kinetically more stable) (Al-Aasm & Veizer 1986; Moshier 1989*a*; Al-Aasm & Azmy 1996; Munnecke *et al.* 1997).

Here we present an integrated petrographical, geochemical and fluid-inclusion study of peloidal packstones and grainstones reservoir rocks of the Lower Cretaceous Upper Kharaib Formation, in a giant oil field located onshore Abu Dhabi, in the United Arab Emirates (UAE) (Fig. 1). This study aims to constrain the origin and diagenetic evolution of microporosity in packstones and grainstones of the reservoir in the crest (i.e. the oil zone) v. the flanks (i.e. the water zone) of the anticline that forms the trap for hydrocarbon accumulation. Microporosity in mudstones and wackestones was not included in this study. The lower portion (about 18 m) of the studied reservoir (i.e. the Upper Kharaib Formation) in the crest lies within the transition zone.

Geological setting of the Kharaib Formation

The Thamama Group, which comprises thick Lower Cretaceous, shallow-marine carbonates, includes the Shuaiba, Kharaib, Lekhwair and Habshan formations (Alsharhan 1990) (Fig. 2). Oil production in the studied field is mainly from the Upper Lekhwair to the Lower Shuaiba Formation (Dewever *et al.* 2011). The structure of the oil field is a faulted and fractured anticline with a length of approximately 25 km and a width of 7.5 km, and was the result of compression induced by convergent tectonics at the Arabian Plate boundary in Oman (92 Ma: Sharland *et al.* 2001). This orogenesis also triggered the formation of number of other anticlines in the UAE, resulting in the development of a foreland basin and initiation of hydrocarbon maturation and

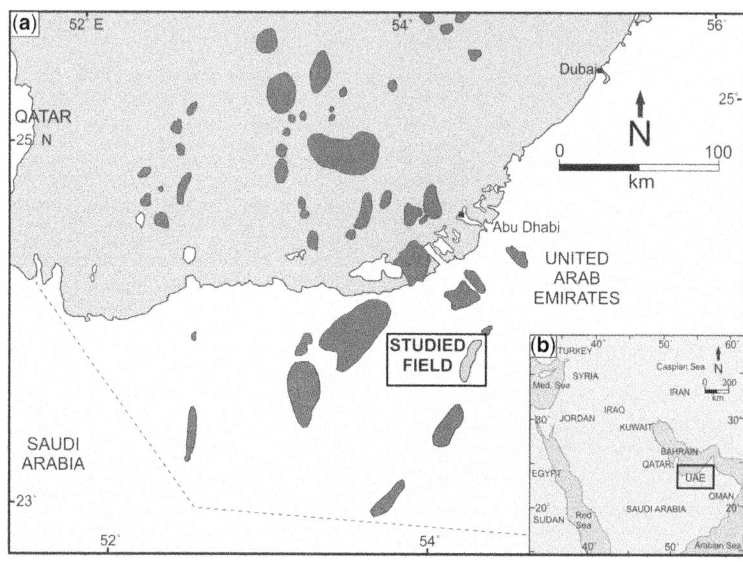

Fig. 1. (**a**) Location map of the study area. The analysed field is located onshore Abu Dhabi and is shown in light grey (black rectangle). The other major hydrocarbon fields are in dark grey. (**b**) Small-scale map of the Arabian Peninsula.

migration. Oil charging into the Thamama Group reservoirs occurred dominantly during the late Cretaceous–Eocene (Taher 1997; Vahrenkamp *et al.* 2014). The oil source rocks have been suggested to be either the Late Jurassic Diyab Formation or the Shuaiba-equivalent distal facies (Bab Formation), which were presumably matured during the late Cretaceous (Murris 1980; Milner 1998; Al-Suwaidi *et al.* 2000).

Thickness variation of the Cretaceous succession between the crest and flanks is attributed by Oswald *et al.* (1995) to differences in the extent of chemical compaction (i.e. stylolitization) above and below the oil–water contact (OWC). Deformation in the Oman Mountains caused southwards tilting of the field, which was accompanied by faulting in the Campanian. The faults in the field are orientated perpendicular to the southwards tilting. The Zagros

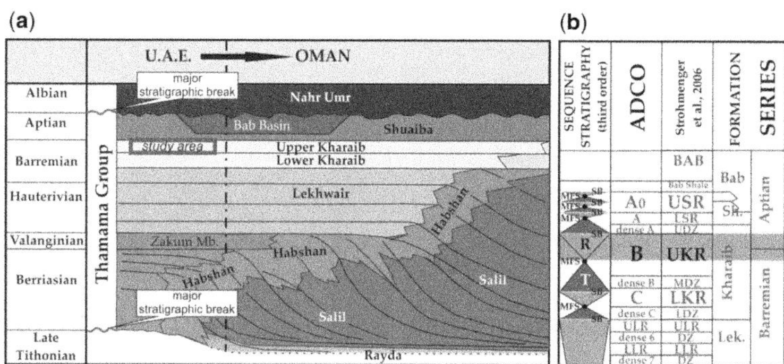

Fig. 2. (**a**) Regional sequence stratigraphy diagram of the Lower Cretaceous showing that the Upper Thamama Group in the UAE has been consistently located far from the palaeo-shelf-break. (**b**) Sequence stratigraphy model for the Upper Thamama Group (Barremian and Aptian) showing third-order transgression periods (T) and regression periods (R), as well as sequence boundaries (SB) and maximum flooding surfaces (MFS). The nomenclatures used by Strohmenger *et al.* (2006) and ADCO are shown (Lek., Lekhwair Formation; Sh., Shuaiba Formation; DZ, Dense Zone; LLR, Lower Lekhwair Reservoir; ULR, Upper Lekhwair Reservoir; LDZ, Lower Dense Zone; LKR, Lower Kharaib Reservoir; MDZ, Middle Dense Zone; UKR, Upper Kharaib Reservoir; UDZ, Upper Dense Zone; LSR, Lower Shuaiba Reservoir; USR, Upper Shuaiba Reservoir). Modified after Vahrenkamp *et al.* (2014).

orogenesis resulted in a later northwards tilting of the structure in Oligo-Miocene times, establishing a new OWC (Oswald *et al.* 1995).

The Thamama Group comprises two second-order supersequences: the older Berriasian–early Valanginian Habshan Formation; and the younger Valanginian–Aptian, comprising the Lekhwair, Kharaib (Barremian and early Aptian) and Shuaiba formations (Strohmenger *et al.* 2006). The Kharaib Formation is interpreted as a second-order late transgressive sequence, which comprises two third-order composite sequences. The lower third-order composite sequence represents the interval from a sequence boundary (SB) at the base of a lower dense zone (i.e. having very low porosity and permeability) to a SB about 6 m below a middle dense zone. The upper third-order composite sequence, which includes the reservoir studied here, directly overlies the previous sequence and is capped by sequence boundary located below an upper dense zone, in the lowermost part of the Shuaiba Formation. The two third-order composite sequences consist of 14 fourth-order parasequence sets exhibiting mainly aggradation (mudstones/wackestones to grainstones) and progradation patterns, which are attributed to climatic cycles (Strohmenger *et al.* 2006).

In the studied field, the Kharaib Formation consists of cyclical shallow-water platform carbonates, which were deposited on a broad distal carbonate ramp under the influence of sea level fluctuations. The water depth at which the carbonates were deposited varied between a few metres and a few tens of metres. The formation consists of lime mudstones, wackestones, packstones, grainstones and rudstones variably stacked as fourth- and fifth-order aggradation parasequences and parasequence sets. The base of the formation is dominated by *Orbitolina* mudstones, which pass upwards into oncoidal and algal packstones/grainstones. The rudist-rich packstones/grainstones and miliolid grainstones are abundant in the upper part of the formation (Neilson *et al.* 1998). Rudist-rich, high-energy facies are not dominant, probably as the location of the field is far from the palaeo-shelf-break (Van Buchem *et al.* 2002). Bed-parallel stylolites, formed by chemical compaction at significant burial depth, are common in the formation, particularly on the flanks of the studied anticline (Paganoni *et al.* 2016).

Samples and methods

Four hundred samples were collected from four wells, two of which are located in the oil zone in the crest, with the other two in the water zone on the flanks of the reservoir. Polished thin sections were prepared for all samples, subsequent to impregnation with blue epoxy, and were subjected to petrographical examination. Modal analyses of the allochems, cement types, dolomite and macroporosity types (intergranular, intragranular and moldic) were performed on 124 samples from the crest and 68 samples from the flanks by counting 300 points in each thin section. Forty-seven thin sections were used to conduct backscattered electron (BSE) imaging and electron microprobe analyses (EMP). Small sample chips obtained from 24 samples in the crest wells and 27 samples in the flank wells were gold coated and examined using a scanning electron microscope (SEM) equipped with a BSE detector. A field emission electron probe microanalyser (FE-EPMA), JXA-8530F JEOL SUPERPROBE, equipped with four crystal spectrometers (WDS), secondary (SE) and backscattered electron detectors (BSE), at the Department of Earth Sciences, Uppsala University, Sweden was used to analyse the carbonate cements. The operation conditions during the analyses were an accelerating voltage of 20 kV, a measured beam current of 8 nA and a beam diameter of 1–5 μm. The standards and count times used were wollastonite (Ca, 10 s), (Mg, 10 s), (Mn, 20 s), hematite (Fe, 20 s) and strontianite (Sr, 20 s). Detections limits were 48 ppm Mg, 90 ppm Fe, 135 Mn and 259 ppm Sr.

Small sample chips were obtained from dominantly peloidal packstones and grainstones for carbon, oxygen and strontium isotope analyses, subsequent to petrographical examination. The samples come from both the crest (i.e. the oil zone) and the flanks (i.e. the water zone) of the studied oil field. Stable oxygen and carbon isotope ratios were measured for 396 samples, and strontium isotopes were analysed for 88 samples. The samples for stable O and C isotopes were reacted with pure H_3PO_4 *in vacuo* at 25°C for 4 h (Al-Aasm *et al.* 1990). The isotope analyses were performed on a Thermo Finnigan Delta Plus mass spectrometer at the Department of Earth and Environmental Sciences, University of Windsor, Canada. Analytical precision was ±0.05‰. Results are reported in ‰ as $\delta^{18}O_{V\text{-}PDB}$ and $\delta^{13}C_{V\text{-}PDB}$. The $^{87}Sr/^{86}Sr$ isotopic ratios were measured at Institut für Geologie, Mineralogie und Geophysik, Bochum, Germany, using an automated Finnigan 261 mass spectrometer equipped with nine Faraday collectors. Correction for isotopic fractionation during the analyses was made by normalization to $^{86}Sr/^{88}Sr = 0.1194$. The mean standard error of mass spectrometer performance was ±0.00003 for standard NBS-987.

Fluid-inclusion microthermometry was performed on representative samples of coarse crystalline calcite filling moldic pores in 23 samples from both the oil and water zones. Both homogenization (T_h) and final ice melting temperatures ($T_{m_{ice}}$) were

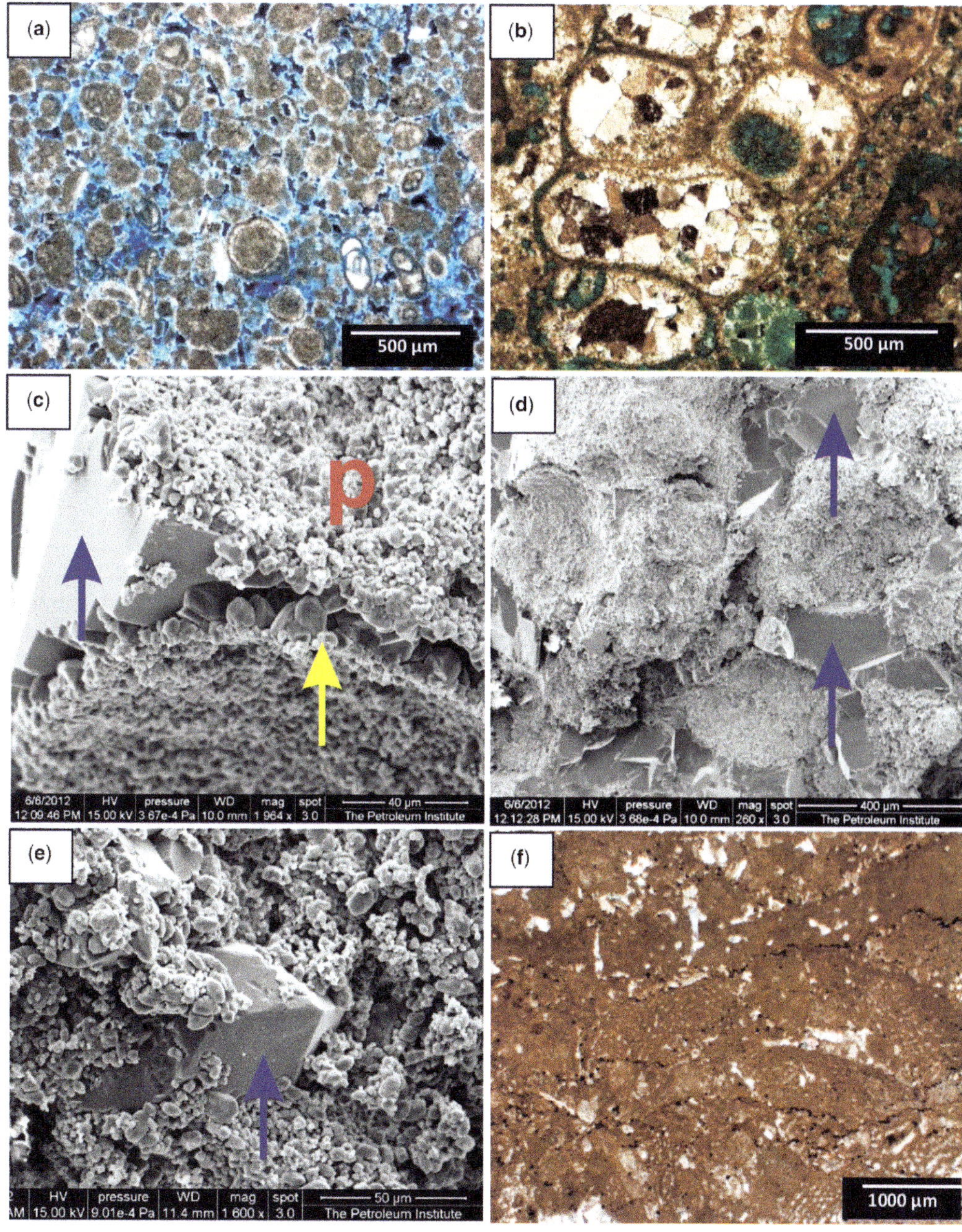

Fig. 3. Optical photomicrographs (XPL: a, b & f) and SEM images (c–e). (**a**) Peloidal grainstone with abundant and well-connected intergranular pores. Oil zone 2756.0 m (9042.0 ft). (**b**) Micritized bioclasts in which the chambers (i.e. primary intragranular pores) are filled with equant drusy calcite cement. Water zone, depth 2917.5 m (9572.0 ft). (**c**) Peloids (p) coated with circumgranular calcite (yellow arrow), which is engulfed by intergranular equant calcite cement (blue arrow). Oil zone, 2756.3 m (9094.3 ft). (**d**) Peloidal grainstones cemented by intergranular blocky calcite cement (blue arrows). Oil zone, 2771.9 m (9094.3 ft). (**e**) Blocky calcite crystal (blue arrow) engulfing a micrite/microspar crystals. Water zone, depth 2923.1 m (9590.2 ft). (**f**) Mechanically compacted peloidal packstone containing small amounts of circumgranular calcite cement. Mechanical compaction was then followed by chemical compaction (i.e. stylolitization). Oil zone, depth 2770.0 m (9088.0 ft).

Fig. 4. (**a**) Optical photomicrograph (XPL) of peloidal packstone showing dolomite (white crystals in the upper right-hand side of the figure), which has replaced the mud matrix and peloids. Oil zone, depth 2769.1 m (9085.0 ft). (**b**) BSE image showing a rhombic dolomite crystal, which has replaced and engulfed micrite. The bright patch is pyrite engulfed by the dolomite crystal. Water zone, depth 2936.3 m (9633.6 ft). (**c**) SEM image of rhombic dolomite, filling an intragranular pore in a peloid and engulfing micrite. Oil zone (transition zone), depth 2794.9 m (9169.6 ft).

measured. The former is the inclusion's minimum entrapment temperature (i.e. the temperature of mineral precipitation) (Goldstein & Reynolds 1994). The latter is the temperature at which the last ice melts in an aqueous fluid inclusion and reflects the trapped diagenetic fluid's salinity (Goldstein & Reynolds 1994). Measured $T_{m_{ice}}$ can be linked with wt% NaCl eq. using the relationships of Hall *et al.* (1988) and Bodnar (1993). The microthermometric measurements were performed at the Department of Earth and Environmental Sciences, University of Pavia, Italy, using a Linkam THMS600G heating–freezing stage, coupled with an Olympus BX60. The thermocouple was calibrated using synthetic pure water and CO_2 inclusions. In order to avoid overheating, low heating rates were used for measuring homogenization temperatures (T_h), measured in order of increasing T_h.

Porosity and permeability data were obtained for 290 core plugs (diameter of 3.8 cm) from three wells (two in the crest and one in the flank). Prior to measurements, the core plugs were examined for microfractures, cleaned by oil extractor and dried in vacuum oven at 60°C for 24 h. A helium porosimeter was used for porosity measurements and the permeability was measured by helium permeameter by applying (6.89×10^5–2.76×10^6 Pa). These measurements were performed by PanTerra Geoconsultants for the Abu Dhabi Company for Onshore Oil Operations (ADCO).

Results

Petrography

Limestones of the Kharaib Formation comprise mainly peloidal grainstones, packstones, mudstones and wackestones. This study is focused on comparison of packstones and grainstones in the oil and water zone samples. In addition to micritization and conversion of micrite into micrite/microspar crystals, diagenetic alterations of the studied limestones include dissolution of allochems, cementation by primarily circumgranular and equant calcite, dolomitization, mechanical compaction, and chemical compaction. The latter event resulted in the widespread development of stylolites and dissolution seams in the succession. Stylolitization is generally more extensive in the water than in the oil zone (Oswald *et al.* 1995; Paganoni *et al.* 2016).

Micritized bioclasts and, less commonly, ooids occur in all the studied limestones in both the oil and water zones. The pervasive to complete conversion of allochems into micritized grains (i.e. the formation of peloids), resulted in the complete destruction of internal fabric (Fig. 3a). However, in many cases, the presence of vague concentric laminae and of chambers within peloids helps in

Fig. 5. SEM images showing peloids with various micrite shapes. (**a**) Peloid with dominantly spheroidal micrite having smooth surfaces. Oil zone, depth 2760.0 m (9055.2 ft). (**b**) Peloid with mixed spheroidal and subhedral micrite. Water zone, depth 2923.1 m (9590.2 ft). (**c**) Higher magnification image showing sub-spheroidal micrite crystals. The crystals surfaces are smooth, corroborating our idea of calcite micro-overgrowths around micrite. Oil zone, depth 2759.2 m (9052.5 ft). (**d**) Subhedral micrite, typical of the water zone (arrow). Water zone, depth 2923.1 m (9590.2 ft). (**e**) Peloid composed of mainly euhedral micrite. Water zone, depth 2928.5 m (9608.0 ft). (**f**) A peloid composed of micro-rhombic, subhedral and spheroidal micrite (continuous arrow). The micro-rhombic micrite varies in size and is overall tightly interlocked with each other. Water zone, depth 2928.5 m (9608.0 ft). Note the presence of rounded 'holes' (dotted arrows) within the micro-rhombic micrite possibly created due to the detachment of subhedral/spheroidal micrite during sample preparation.

the identification of ooids and bioclasts (Fig. 3b), respectively. The chambers are occluded, partly to completely, by equant and blocky calcite cement or, less commonly, by micrite mud which shows less microporosity than the peloids. The peloid limestones are cemented by circumgranular rims of tiny (3–10 μm; average 5 μm across) scalenohedral and/or equant calcite crystals (Fig. 3c). The moldic, intergranular and intragranular pores in peloidal grainstones are filled with coarse (50–400 μm across) equant calcite crystals, which engulf the circumgranular calcite cement rim (Fig. 3c, d) and micrite/microspar crystals (Fig. 3e). Coarse intergranular crystals occur as syntaxial calcite overgrowths around echinoderm fragments and engulf microporosity-rich peloids. Where circumgranular rims are less developed and the limestones lack scattered equant calcite cement, both peloids and bioclasts have suffered severe mechanical compaction (Fig. 3f).

Locally, in the peloidal packstones and grainstones, the peloids are slightly dolomitized (dolomite crystals are 10–50 μm across) (Fig. 4a). BSE examination shows that microporosity between the micrite particles is preserved within the replacive rhombic dolomite crystals (Fig. 4b). SEM examination reveals that these dolomite crystals engulf, as well as fill, space between the micrite particles (Fig. 4c). SEM examination of peloidal packstones and grainstones indicates that the micrite has spheroidal (rounded) and anhedral forms intermixed with variable amounts and sizes of subhedral to euhedral/subhedral micrite/microspar (Fig. 5a–f). Spheroidal micrite displays smooth surfaces (Fig. 5a, c), even at magnifications of up to ×50 000. These ≤2 μm micrite spheroids are commonly merged/aligned and, at times, engulfed by subhedral/euhedral micrite/microspar (Fig. 5b). Visual estimations indicate that the subhedral to euhedral micrite/microspar is relatively more abundant and more interlocked in the water than oil zone (Fig. 5d, e). Micro-rhombic micrite is relatively rare in the studied limestones, and varies widely in size and degree of interlocking crystals even within the same peloid. This micrite is closely associated with, and engulfs, spheroidal and subhedral micrite. Under the SEM, rounded 'holes' within the micro-rhombic micrite have been observed, which are presumably due to the detachment of rounded (spheroidal) and subhedral micrite during sample preparation (Fig. 5f).

BSE examinations revealed that rarely the rounded micrite displays thin rims, which are brighter than the cores, suggesting the presence of differences in chemical composition (Fig. 6). However, the small size of such rims precludes performing quantitative EMP analyses because the smallest spot size by such analyses is 5 μm. Nevertheless,

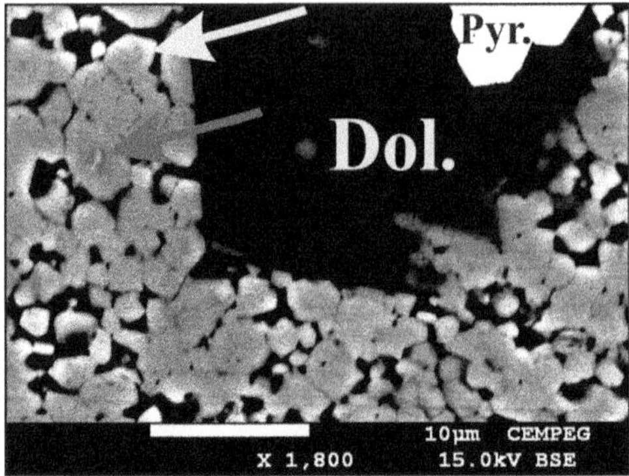

Fig. 6. BSE image with enhanced contrast showing the presence of thin calcite growth around micrite (upper arrow), and that microspar has darker core and brighter (richer in Mn) rim (lower arrow). Dol., dolomite; Pyr., pyrite. Water zone, depth 2936.3 m (9633.6 ft).

EMP analysis revealed consistently higher Mn concentrations in the rims compared to the inner micrite particle (around 80–100 and 0–55 ppm, respectively). However, such Mn concentrations are within the analytical errors and/or detection limits of the microprobe. In many cases, it is difficult to distinguish between micrite/microspar formed by calcite cement around spheroidal micrite and discrete, pore-filling equant calcite cement. EMP analysis revealed the consistent presence of Mg (744–2787 ppm; average 1792 ppm), with concentrations of Sr, Fe and Mn varying from below detection limit to 596 ppm (average 245 ppm), 428 ppm (average 106 ppm) and 137 ppm (average 84 ppm), respectively.

The peloids display variable amounts of microporosity (Fig. 7a) and are, in some cases, pervasively to completely dissolved, which has thus resulted in the formation of macro-intragranular and moldic pores in grainstones (Fig. 7b). These peloidal grainstones are extensively cemented by intergranular, circumgranular and/or equant calcite. Few of the moldic and large intragranular pores within peloids are partly or completely filled by equant calcite occluding space between and engulfing subhedral and euhedral micrite crystals (Fig. 7a). Most commonly, the peloids in both the oil and water zones lack macro-intragranular porosity (Fig. 7c, d). Pore throats in the micrite/microspar within the peloids are less than 2.5 μm. Many of the peloids contain scattered round- to oval-shaped micropores (4–10 μm across), which are lined by microcrystalline scalenohedral and equant calcite. Some authors (e.g. Moshier 1989*a*) have called these microvugs.

Modal analyses revealed that macroporosity in packstones and grainstones is most abundant in the

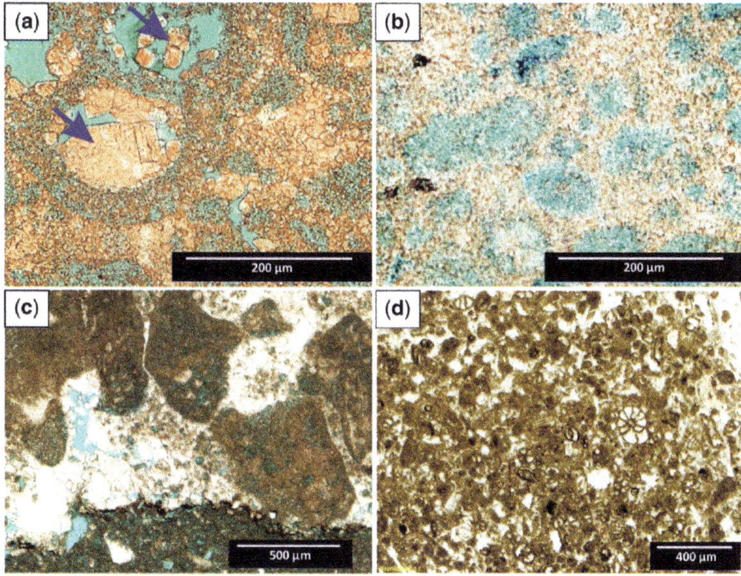

Fig. 7. Optical photomicrographs (PPL). (**a**) Abundant microporosity in micritized allochems. Note the presence of equant calcite cement filling intragranular pores (blue arrows). Oil zone, depth 2771.3 m (9092.3 ft). (**b**) Peloidal grainstone heavily cemented by microspar. Note that the peloids have been subjected to extensive dissolution, provoking a porosity inversion. Oil zone, depth 2749.6 m (9021 ft). (**c**) Peloidal packstone cemented locally by equant calcite. Oil zone, depth 2767.9 m (9081.0 ft) (**d**) Peloidal packstone that lacks macroporosity. Water zone, depth 2931.0 m (9616.0 ft).

crest dominated by intragranular pores (average 4%), whereas moldic and intergranular pores are least abundant (average 1.2%). Total macroporosity in the flanks is 0–4%. Core-plug porosity and permeability measurements in the packstones and grainstones overall have higher values in the oil zone than in the water zone. Porosity and permeability are markedly higher in the crest (oil zone: porosity 22–23% and permeability 1.1×10^{-10} cm^2 or 11.2 mD) than in flanks (water zone: porosity 12–13% and permeability 5.0×10^{-12} cm^2 or 0.5 mD). The dense zones in the crest and flanks have very low porosity (typically <1%) and permeability (around $1–10 \times 10^{-13}$ cm^2 or 0.1–0.01 mD) (Paganoni *et al.* 2016). Porosity is positively correlated with permeability in both the oil and water zones. A cross-plot of porosity v. permeability displays positive correlation for grainstone- and packstone-dominated samples from both the oil and water zones (Fig. 8).

Stable oxygen, carbon and strontium isotopes

Bulk isotopes values for packstones and grainstones from the oil and water zones show partial overlap in the $\delta^{13}C$ (+2.7 to +4.3‰, average +3.8‰ and +2.9 to +4.7‰, average +3.7‰, respectively) and $\delta^{18}O$ (−8.6 to −6.7‰, average −7.8‰ and −8.8 to −5.8‰, average −7.2‰, respectively). The cross-plot of carbon v. oxygen isotopes, in the oil and water zones, reveals a crudely negative correlation (Fig. 9). The $\delta^{13}C$ values are enriched with depth, in both the oil and water zones (Fig. 10).

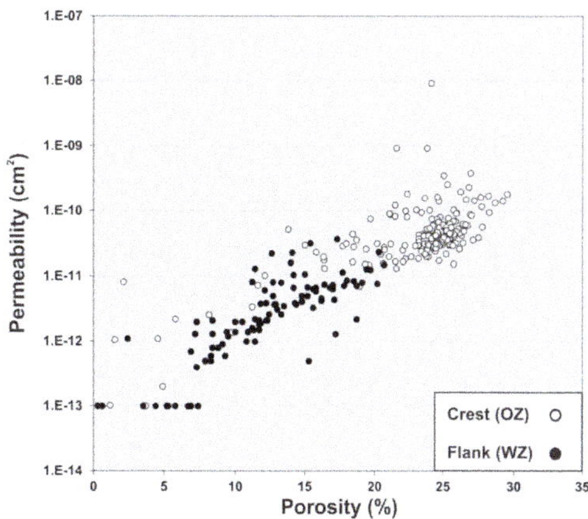

Fig. 8. Cross-plot of core-plug porosity v. permeability for dominantly peloidal grainstones and packstones in the oil and water zones, showing a positive correlation. Note that both porosity and permeability are higher in the oil than in the water zone.

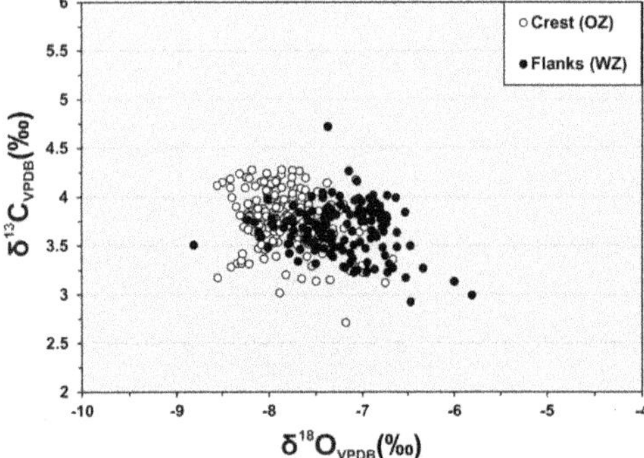

Fig. 9. Cross-plot of oxygen v. carbon isotopes of bulk samples (grainstones and packstones) in the Kharaib Formation for the water and oil zone, showing a crudely negative correlation trend. Note that oxygen values are lower than expected for calcite precipitated in equilibrium with Lower Cretaceous seawater. The Lower Cretaceous isotopic values for the best preserved signature of calcite in equilibrium with seawater are inferred to be between about −4.5 and −1.5‰ for $\delta^{18}O_{VPDB}$, and between about +1.0 and +4.0‰ for $\delta^{13}C_{VPDB}$ for carbon.

Instead, the $\delta^{18}O$ values show a rather non-systematic fluctuation, with the higher values being generally encountered within the fine-grained intervals (Fig. 11). The bulk $^{87}Sr/^{86}Sr$ ratios for packstones and grainstones in the oil zone are similar to those in the water zone (average 0.707469) (Fig. 12).

Microthermometry

The T_h and fluid salinity of the equant coarse-crystalline calcite, which fills moldic/intragranular pores, were obtained using fluid-inclusion microthermometry (Fig. 13a). The T_h of this calcite in the oil zone is 75–130°C (average 95°C: Fig. 13b) and salinity is 3.0–7.5 wt% eq. NaCl (average 5.2%). In the water zone, T_h is 85–135°C (average 110°C: Fig. 13c) and salinity is 4.5–7.9 wt% eq. NaCl (average 6.2%).

Discussion

Constraints on micritization

The complete micritization of ooids and skeletal fragments in the Lower Cretaceous limestone

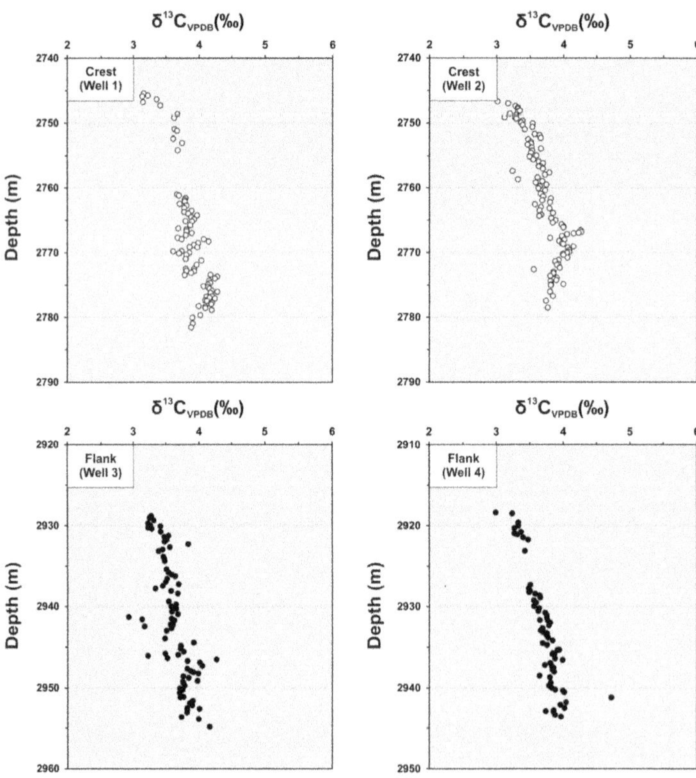

Fig. 10. Carbon isotopic composition of limestones *vs* depth from crest (i.e. the oil zone; wells 1 and 2) and flanks wells (i.e. the water zone; wells 3 and 4), showing a general trend of $\delta^{13}C_{VPDB}$ increase with depth.

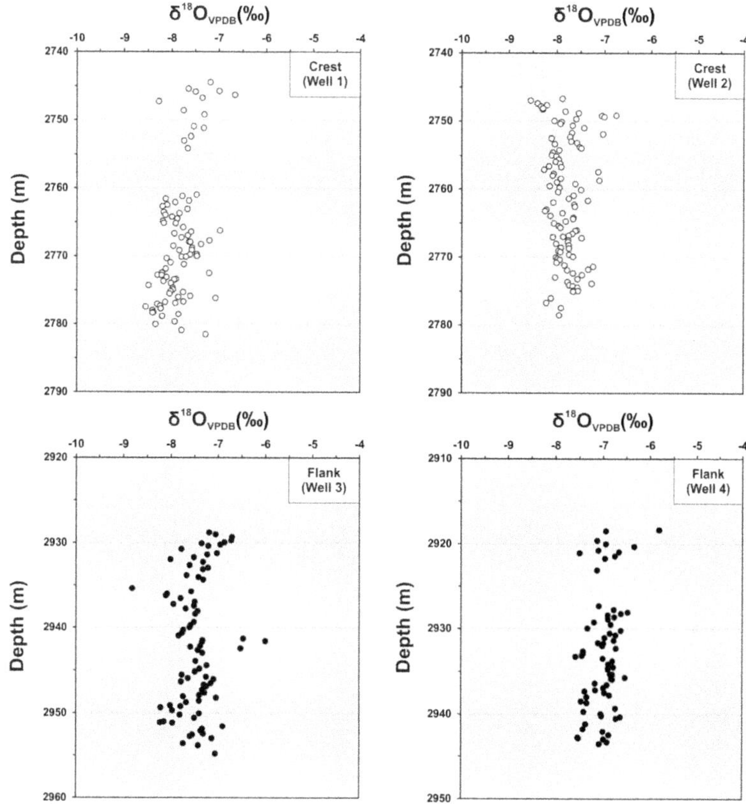

Fig. 11. Oxygen isotopic composition of limestones v. depth from the crest (i.e. the oil zone: wells 1 and 2) and flanks wells (i.e. the water zone: wells 3 and 4), showing a non-systematic distribution of $\delta^{18}O_{VPDB}$ values with depth.

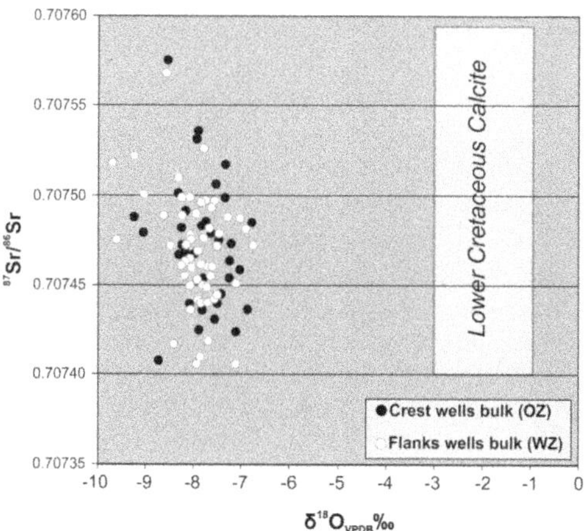

Fig. 12. Cross-plot showing that there is no correlation between strontium ($^{87}Sr/^{86}Sr$) v. oxygen isotopes for bulk samples (dominated by peloidal grainstones and packstones) in the Kharaib Formation, for oil and water zones.

reservoir could lead to the formation of microporous peloids, a process that was introduced by Bathurst (1966), but the diagenetic environment of this process is still a controversial issue (Budd 1989; Moshier 1989*b*; Reid & Macintyre 2000; Léonide *et al.* 2014). Micritization is likely to have occurred in marine water (Bathurst 1987) as we have no convincing evidence of meteoric water diagenesis, such as meniscus and gravity cements. Extensive micritization of the allochems suggests slow sedimentation rates: that is, extended exposure of these allochems to microbial micritization (e.g. Husinec & Read 2006). The spheroidal shape of the micrite composing the peloids (Fig. 5a, c) is considered here to represent evidence of a microbial precipitation of $CaCO_3$. This interpretation contradicts the suggestion of Lambert *et al.* (2006), who interpreted the spheroidal/rounded micrite as a product of mesogenetic dissolution due to fluxes of CO_2 and/or organic acids (derived from thermal maturation of organic matter) into limestones of the oil zone. Lines of

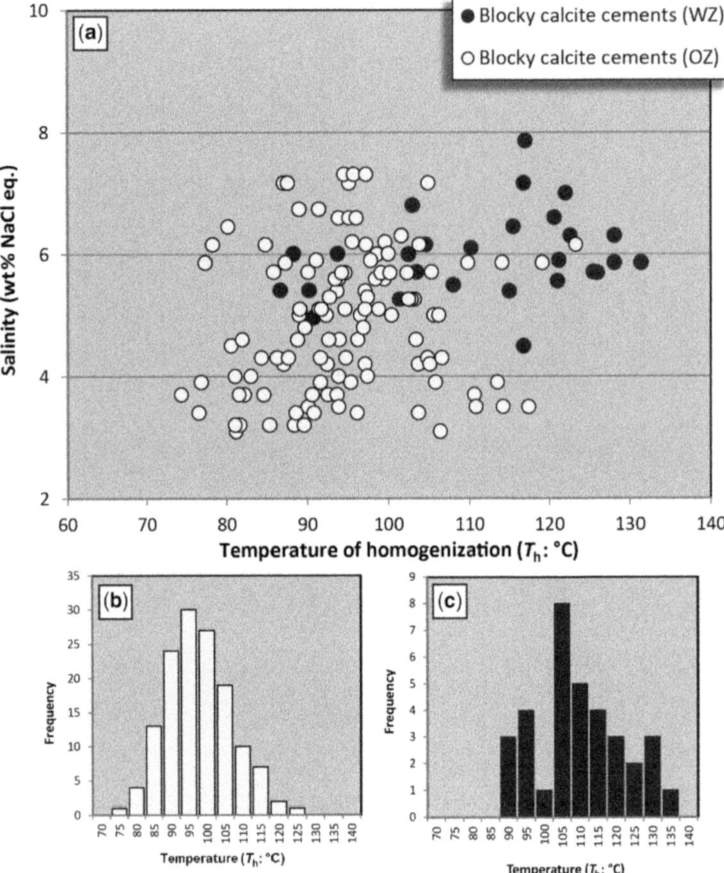

Fig. 13. Fluid-inclusion microthermometry results. (**a**) Cross-plot of the homogenization temperature (T_h, °C) v. salinity (wt% NaCl eq.) for blocky calcite cements in the oil and water zones. (**b**) & (**c**) T_h frequency histograms for blocky calcite cement fluid inclusions in the oil (white) and water (black) zones.

evidence supporting our rejection of this interpretation include:

- Several studies have concluded that microbial carbonates typically have spheroidal shapes with smooth surfaces (Folk 1993; Perri & Tucker 2007; Maruthamuthu *et al.* 2010; Spadafora *et al.* 2010), although experimental work has revealed that spheroidal calcium carbonate can be precipitated inorganically (Cheng *et al.* 2014).
- There is no reason that a flux of organic acids would be restricted to the oil zone, where most of the rounded (spheroidal) micrite has been observed, because such acids are expected to migrate through the flanks too.
- Dissolution of micrite would result in etched and pitted surfaces and irregular grains shapes (Davis *et al.* 2007) rather than spheroids with smooth surfaces. Nevertheless, we neither exclude mesogenetic dissolution nor the flux of aggressive waters into the limestones, but we only presume that these processes have probably not created rounded (spheroidal) micrite with smooth surfaces.

The conversion of allochems into peloids (Fig. 3a) is likely to be associated with the formation of micropores between the micrite particles (Moshier 1989*b*; Fischer *et al.* 2014). Moreover, formation of microporosity by microbially driven dissolution of the allochems cannot be excluded (Ehrlich 1998; Davis *et al.* 2007; Jacobson & Wu 2009). Severe compaction of peloids in packstones and grainstones that lack cement (Fig. 3f) suggests that micritization is early diagenetic in origin (Geslin 1994). The presence of variable amounts of microvugs within the peloids (Fig. 7) is attributed to the presence of empty or partly cemented primary intragranular pores (i.e. chambers) in the bioclasts (e.g. *Orbitolina*). The variation in the amounts of micro- and macro-intragranular pores within peloids in the studied packstones and grainstones depends on the degree of cementation of these pores. We found no convincing evidence for the formation of microporosity by partial dissolution of aragonitic bioclasts during early diagenesis as was suggested for other limestone successions (Volery *et al.* 2009; Deville de Periere *et al.* 2011).

It is uncertain whether the spheroidal micrite, which is suggested here to be formed during microbial micritization of the allochems, was initially calcite (HMC or LMC), vaterite and/or aragonite. The original mineralogy would have important implications for the diagenetic evolution of the peloids (Saller & Moore 1989). For example, microcrystalline aragonite is strongly soluble in porewaters, and may explain the presence of dissolved peloids embedded in extensively calcite-cemented grainstones. The presence of moldic and large intragranular pores embedded in calcite-cemented grainstones (Fig. 7b) is a feature that is commonly reported in the literature and attributed to concomitant complete dissolution of aragonite allochems and cementation by sparry calcite (i.e. porosity inversion) (James & Choquette 1984; Swirydczuk 1988; Melim *et al.* 2002; Tucker & Wright 2009). It is unlikely that the metastable aragonite bioclasts were dissolved and reprecipitated as micrite, as was suggested for mudstones by Munnecke & Samtleben (1996).

Constraints on the formation of microspar cement (micro-overgrowth)

There are several lines of evidence suggesting that the formation of subhedral to euhedral crystals of micrite and microspar occurs by the precipitation of calcite around the micrite. In other words, the micrite acted as nuclei around which the growth of an extremely thin layer of syntaxial calcite overgrowths occurred. Owing to the large number of micrite nuclei, the precipitation of calcite micro-overgrowths would result in a mosaic of equant, micrite-sized crystals (Moshier 1989*a*). Micro-rhombic micrite is rarely observed in the studied limestones, while it was reported to be widely occurring in other microporous limestones, including the Cretaceous of the UAE (Budd 1989; Moshier 1989*b*). A possible explanation is that our investigation is mainly related to peloids, whereas the other studies focused on micrite mud.

The cementation mechanism is similar to the well-known phenomenon of syntaxial growth of calcite around echinoderm fragments. Evidence supporting this interpretation includes:

- the intimate association of spheroidal micrite with subhedral and euhedral micrite/microspar;
- the relatively greater amounts of euhedral and subhedral micrite in the water zone than in the oil zone, which suggests that calcite precipitation around micrite has continued in the water zone whereas it was halted in the oil zone;
- the presence of petrographical (BSE and SEM) evidence for the engulfment of micrite by a 1–2 μm-thick layer of euhedral calcite cement (e.g. Fig. 6);
- the observation of rounded 'holes' within the micro-rhombic micrite, possibly derived from detachment of subhedral/spheroidal micrite (Fig. 5f);
- the consistently higher Mn content in the thin cement layer compared to the micrite core. This observation, although within the analytical microprobe errors, suggests that precipitation occurred from porewaters with a different chemistry compared to the formation of micrite nucleus. The Mn increase in carbonate cement is typically attributed to the establishment of reducing conditions during burial below the seafloor (Grover & Read 1983; Emery 1987; Barnaby & Rimstidt 1989).

The more common presence of subhedral and euhedral micrite/microspar than spheroidal micrite in the water zone compared to the oil zone, a phenomenon observe by other authors (Lambert *et al.* 2006; Deville de Periere *et al.* 2011), has been shown to result in lower $\delta^{18}O$ values in the water zone (Oswald *et al.* 1995; Cox *et al.* 2010). The lower oxygen isotopic values were attributed to an increase in precipitation temperature in the water zone, assuming no changes occurred to the oxygen isotopic composition of the porewaters (Dickson & Coleman 1980). Unlike these previous studies (Oswald *et al.* 1995; Cox *et al.* 2010), we found that the $\delta^{18}O$ values for micrite/microspar in the water zone are relatively similar to those in the oil zone (Figs 9 & 11). Rather, isotopic data indicated that $\delta^{18}O$ values are slightly lower in the oil than in the water zone, which can be attributed to the formation of early diagenetic spheroidal micrite in both zones and the initiation of calcite precipitation around spheroidal micrite throughout the formation prior to oil emplacement in micropores of both the oil and water zones. Micro-cement formation in the oil zone could have been promoted by the presence of irreducible water in the micropores (e.g. Cox *et al.* 2010; Deville de Periere *et al.* 2011). Continuation of diagenesis in the micropores in the oil zone is expected because oil is preferentially emplaced in macropores (cf. Cox *et al.* 2010).

The precipitation of calcite in the crest by diffusion in irreducible porewaters from sites of high Ca^{2+} and $CO_3{}^{2-}$ concentration, such as along stylolites, into the micropores is expected to be retarded due to oil emplacement (cf. Heasley *et al.* 2000; Cox *et al.* 2010). Simultaneously, higher rates of cementation of micro- and macropores are expected to occur in the water zone owing to more efficient mass flux by diffusion and advection, leading to a more rapid loss of porosity and permeability in the water zone than in the oil zone. This scenario would imply that cementation continued further, but at slow rates, in the oil zone, resulting in the addition of a late generation of calcite cement with low $\delta^{18}O$ values (Cox *et al.* 2010).

Based on the results presented and discussed above, we conclude that recrystallization/aggrading neomorphism (Richard *et al.* 2007; Volery *et al.* 2010; Deville de Periere *et al.* 2011; Léonide *et al.* 2014) is not the only possible mechanism for the conversion of micrite to euhedral micrite/microspar. Here, recrystallization/aggrading neomorphism is considered less likely because: (1) where dissolution is part of the conversion process, the micrite particles would display irregular shapes and etched/pitted surfaces (Davis *et al.* 2007), and no such surfaces have been observed in the studied rocks; instead, the micrite particles have spheroidal shapes with smooth surfaces; (2) it is difficult to envisage why widespread calcite dissolution would be occurring in calcite-saturated, low-temperature, diagenetic environments. Calcite dissolution and reprecipitation is possible with a HMC or aragonite precursor, it appears to be unlikely for a LMC micrite. In summary, although we cannot exclude that micrite can grow larger by dissolution and reprecipitation as microspar (i.e. becomes more stable following Ostwald ripening), bigger crystals can also grow by precipitation of calcite micro-overgrowths around micrite. The precipitation of a mineral around a similar mineral is thermodynamically the easiest step in crystallization (Hillner *et al.* 1992; Nesse 2000; Teng *et al.* 2000).

Various processes can source calcium-carbonate cement during diagenesis, including: (1) dissolution of aragonitic and HMC allochems in marine porewaters (Saller & Moore 1989); (2) meteoric dissolution of HMC and aragonite particles (Budd 1989); (3) an increase in alkalinity and/or lower P_{CO_2} in marine porewaters with enhanced calcium carbonate saturation during burial diagenesis (Moshier 1989*a*, *b*); and/or (4) stylolitization, which occurs during burial diagenesis (Moshier 1989*b*; Alsharhan & Sadd 2000; Paganoni *et al.* 2016). Sources (1)–(3) are typical for near-surface to shallow burial diagenesis (eogenetic regime of Choquette & Pray 1970), whereas source (4) applies for burial diagenesis (mesogenetic regime of Choquette & Pray 1970). Additionally, Vahrenkamp *et al.* (2014) suggested that cementation in the Upper Kharaib Formation occurred during burial diagenesis and was driven by an east to west fluid-flow system, providing an externally derived source of calcium carbonate. Furthermore, the tectonic history of the area suggests a change in the OWC through time. This change has displaced oil where there was water, and vice versa, complicating the rock–fluid interactions and, possibly, homogenizing the isotopic signal.

The carbon and strontium isotopic compositions (Figs 9, 10 and 12) are similar to the ranges inferred for ambient seawater (Scholle & Arthur 1980; Veizer *et al.* 1999). These isotopic compositions suggest that calcite cementation of the studied limestones occurred under a relatively closed diagenetic system, with dissolved carbon and strontium being derived from connate marine porewaters and dissolution of the host limestones (Budd 1989; Moshier 1989*a*). The data presented here show no evidence of a flux of fluids charged with dissolved ^{12}C-rich carbon, which is typically derived from the oxidation of organic matter and hydrocarbons (Morad 1998). Alternatively, an eventual flux of such fluids, which could be meteoric or basinal in origin, was buffered in terms of carbon and strontium isotopes by interaction with the host limestones under a relatively low water/rock ratio (Wigley *et al.* 1978; Budd 1989).

Formation temperature of calcite micro-overgrowths

The lower than expected $\delta^{18}O$ values for calcite precipitated from Lower Cretaceous seawater in the studied bulk micrite/microspar (Fig. 9) is attributed to the precipitation of cement around the micrite during progressive burial (i.e. increase in temperature) under conditions of a high water/rock ratio (Moshier 1989*a*; Al-Aasm 2000). Micrite formed on the seafloor is expected to have $\delta^{18}O$ values similar to those of calcite precipitated from ambient seawater (e.g. Maliva *et al.* 2009). The precise burial depths and temperatures at which the precipitation of calcite around micrite occurred are difficult to constrain because this cement is too small to contain measurable fluid inclusions for microthermometry or to obtain precise stable isotope data. Based on the discussion and interpretations above, the stable isotope data probably represent mixtures of micrite formed by micritization of allochems on the seafloor and their thin layer of calcite overgrowths (higher temperature), thus only minimum temperatures of micrite/microspar crystals can be obtained using bulk $\delta^{18}O$ values.

Another difficulty associated with inferring the cement temperatures is the unknown oxygen isotopic composition of porewaters from which precipitation occurred. Using the oxygen isotopic compositions of micrite/microspar in peloidal packstones and grainstones ($\delta^{18}O = -5.8$ to -8.8‰), the oxygen isotope fractionation equation of Friedman & O'Neil (1977), and assuming that precipitation occurred from Lower Cretaceous seawater ($\delta^{18}O_{V\text{-}SMOW} = 0$‰: Veizer *et al.* 1999), then temperatures of about 45–65°C are obtained. Higher temperatures (58–78°C) are obtained if precipitation is assumed to have occurred from evolved marine porewaters ($\delta^{18}O_{V\text{-}SMOW} = +2$‰: Egeberg & Aagaard 1989). These latter temperatures are closer to those obtained for micrite/microspar in the studied Cretaceous reservoir by Vahrenkamp *et al.* (2014) using clumped oxygen isotope analyses techniques. However, these authors concluded that precipitation of micrite/microspar was from a geochemically more evolved ($\delta^{18}O_{V\text{-}SMOW} = +3.3$ to $+4.9$‰) fluid-flow system, which would result here in higher temperatures (68–108°C) than those obtained using geochemically evolved marine porewaters with $\delta^{18}O_{V\text{-}SMOW} = +2$‰.

Accepting that calcite has precipitated around spheroidal micrite that was formed at around 30°C, the inferred 58–78°C for bulk micrite/microspar would then be minimum precipitation temperatures of the calcite micro-overgrowths. Thus, isotopic and fluid-inclusion analyses suggest that the calcite macro-cements occurred partly simultaneously to and partly after the precipitation of calcite micro-overgrowths. Furthermore, the inferred minimum temperature range (58–78°C) suggests that calcite micro-overgrowths around micrite could be partly derived from initial phases of stylolitization of the host limestone (Oswald *et al.* 1995; Ehrenberg 2006; Paganoni *et al.* 2016).

Consequently, using these inferred temperatures and the average homogenization temperatures recorded in primary fluid inclusions entrapped in blocky calcite cements in the oil and water zones (95 and 110°C, respectively: Fig. 13), which are close to the maximum burial temperatures reached by the reservoir (Oswald *et al.* 1995; Vahrenkamp *et al.* 2014), we can roughly estimate the volumetric proportion of micrite and their calcite micro-overgrowth. This estimation can be achieved with: $x + y = 1$, which implies that $xT_m + yT_o = T_b$, where x is micrite proportion; T_m is seafloor micrite precipitation temperature, assumed to be 30°C; y is proportion of calcite micro-overgrowths around micrite; T_o is temperature of precipitation of calcite micro-overgrowths; and T_b is bulk temperature of micrite plus calcite micro-overgrowths. Thus, for an inferred bulk temperature of 58°C, the proportions of micrite:calcite micro-overgrowths are 57%:43% in the oil zone and 70%:35% in the water zone. For an inferred bulk temperature of 78°C, the proportions micrite:calcite micro-overgrowths are 26%:74% in the oil zone and 40%:60% in the water zone.

Possible variations in the proportion of micrite core and micro-overgrowths can partly explain the non-systematic variation in the $\delta^{18}O$ values of peloidal packstones and grainstones. Alternatively, the scattered $\delta^{18}O$ values v. depth (Fig. 11), with higher values in the limestone intervals having a higher micrite matrix (i.e. having lower permeability), can also be attributed to diagenesis under a variable water:rock ratio (Al-Aasm 2000; Morad *et al.* 2012). Low permeability implies limited fluid flow and, hence, limited temperature oxygen isotope fractionation between calcite and porewaters. The reason behind the crude negative correlation between $\delta^{18}O$ and $\delta^{13}C$ values (Fig. 9) is not immediately clear, but could indicate a higher input of ^{12}C derived from the oxidation of more abundant organic matter in the muddy than in the grainy intervals. This interpretation may explain the crude trend of the progressive increase in $\delta^{13}C$ with depth and towards the transition zone (Fig. 10). Cementation by macro-calcite continued mainly in the water zones owing to greater diffusive and advective mass transfer than in the oil zone (Cox *et al.* 2010). The crude trend of the increase in fluid salinity suggests precipitation of late diagenetic coarse calcite cements at higher temperatures and from more evolved formation waters (Fig. 13a).

Constrains on the formation of macropores

The presence of partly dissolved peloids and moldic pores (Fig. 7a, b) suggests dissolution of unstable grains during early diagenesis and, possibly, later the flux of aggressive fluids into the limestones. In addition to a possible role of organic acids and CO_2-charged mesogenetic fluids, carbonate dissolution can occur by the flux of meteoric waters, which commonly promotes the formation of secondary porosity during near-surface to shallow-burial diagenesis (Choquette & Pray 1970; James & Choquette 1984; Harris *et al.* 1985). Granier (2000) argued that, like the other formations of the Thamama Group, the Kharaib Formation is bounded by unconformities. However, strong evidence of subaerial exposure has not been recognized during this and other studies of the Upper Kharaib Formation in this oil field (Dewever *et al.* 2011). Specifically, Dewever *et al.* (2011) only found limited exposure evidence in the uppermost part of the reservoir (i.e. red-coloured and reworked horizons), which was related to the limited flux of meteoric waters. Conversely, other authors have observed evidence of more intense exposure in adjacent areas (Alsharhan 1990; Van Buchem *et al.* 2002, 2010; Hillgärtner *et al.* 2003; Strohmenger *et al.* 2006). Thus, the studied succession may have passed through a meteoric water lens, but only for a limited period of time, allowing the dissolution of unstable allochems and concomitant calcite cementation. Furthermore, Volery *et al.* (2009) concluded that low-intensity and transient exposure events during a higher-order transgressive systems tract (TST)/highstand systems tract (HST) enhance the formation and redistribution of microporosity. Conversely, long-term exposure events related to a major fall in the relative sea level can result in extensive meteoric water diagenesis, including the destruction of microporosity, pervasive cementation and development of karstic reservoirs (Volery *et al.* 2009). The absence of unequivocal subaerial exposure isotope signatures is probably due to limited/temporary meteoric water incursion into the Upper Kharaib sediments in dry climate conditions and/or overprint by burial-diagenetic rock–water interaction.

Impact of calcite micro-overgrowths on reservoir quality

Carbonate reservoirs are inherently heterogeneous owing to strong spatial and temporal variations in depositional facies (Esrafili-Dizaji & Rahimpour-Bonab 2009; Harris 2010). Diagenesis can accentuate the spatial and temporal variations in the reservoir porosity and permeability (Budd 2001). The origin and diagenetic evolution of microporosity in depositional mudstones and wackestones differ from those in the packstones and grainstones. The lime-mud-dominated facies can undergo pervasive to complete destruction of microporosity by rearrangement of the micrite particles by mechanical compaction. Nevertheless, porosity can be preserved in these facies due to limited compaction (Maliva *et al.* 2009), re-developed through secondary porosity formation (Bruna *et al.* 2013) or simply enhanced/redistributed by dissolution–recrystallization processes (Budd 1989). Conversely, microporosity in the grainy limestones, which is the focus of this study, is mainly initially formed by the micritization of allochems. The degree of microporosity retention in the peloids is influenced by the physical and geochemical conditions prevailing during burial diagenesis, including the rock:water ratio, the extent of stylolitization and the flux of basinal fluids. Micritization is expected to alter the physical and chemical properties of the resulting peloidal packstones and grainstones. The micropores within the peloids increase the water saturation and the capillary pressure.

The formation of calcite micro-overgrowths around micrite within peloids impacts the reservoir quality of the packstones and grainstones (Budd 1989; Moshier 1989*a*, *b*). The higher porosity and permeability of grainstones and packstones in the oil compared to the water zone is attributed mainly to differences in chemical compaction and subsequent macro-cement precipitation (e.g. Oswald *et al.* 1995; Neilson *et al.* 1998; Paganoni *et al.* 2016). Nevertheless, the presence of larger amounts of loosely packed spheroidal micrite without or with incomplete nanometre- to micrometre-thick calcite overgrowths in the oil than in the water zone might partly contribute to the observed porosity and permeability differences (Lambert *et al.* 2006; Deville de Periere *et al.* 2011). The generally substantially greater core-plug porosity than thin-section macroporosity suggests that microporosity is dominant in the studied reservoir, and that the differential development of calcite micro-overgrowth between oil and water zones may impact reservoir quality. Moreover, Ehrenberg & Walderhaug (2015) have suggested that macropores are more readily cemented than micropores. A possible source for the calcite micro-overgrowths could derive from the initial phases of stylolitization of the limestones.

The precipitation of calcite micro-overgrowths around micrite makes the peloids more rigid and may explain the subsequent preservation of microporosity during burial diagenesis (cf. Moshier 1989*b*; Volery *et al.* 2010). Peloids in which the micrite lacks calcite micro-overgrowths are ductile and, hence, were severely compacted. Deformation of the ductile peloids was most severe where there is no circumgranular and intergranular equant calcite cement supporting the framework of the host limestone (cf. Moshier 1989*a*; Munnecke *et al.* 1997; Bruna *et al.* 2013). These cements, together with the partial cementation of peloids by micro-overgrowths, helped in the preservation of microporosity in the oil zone (cf. Moshier 1989*b*; Morad *et al.* 2012). Conversely, massive precipitation of micro-overgrowths around the micrite would ultimately result in substantial destruction of microporosity within the peloids, which is the case in many samples from the water zone. The low rock:water ratio in the water zone would favour progressive cementation and destruction of reservoir quality.

The higher core-plug porosity and permeability in the oil zone compared to the water zone (Fig. 8) is thus attributed to the precipitation of larger amounts of cements in the micro- and macropores in packstones and grainstones of the water zone. The higher porosity in peloidal packstones and grainstones from the oil zone compared to the water zone (Fig. 8) may indicate that such porosity is preserved rather than created, because oil emplacement is expected to retard or even stop cementation (Oswald *et al.* 1995; Neilson *et al.* 1998; Heasley *et al.* 2000). Conversely, micro- and macropores in the water zone were subjected to more extensive cementation.

Conclusions

Based on this integrated petrographical, geochemical and fluid-inclusion study, it is concluded that the conversion of allochems in packstones and grainstones of the studied Lower Cretaceous Kharaib limestone reservoir into peloids has resulted in the formation of substantial amounts of microporosity. Micritization has also resulted in the formation of spheroidal (rounded) micrite, which acted as nuclei for the growth of thin (<1 μm) syntaxial calcite overgrowths and resulting in the formation of subhedral to euhedral micrite and microspar. Such cementation of micropores in peloidal packstones and grainstones is more extensive in the water zone than in the oil zone. Therefore, microporosity is better preserved in the oil zone. The paragenetic sequence, fluid-inclusion microthermometry of calcite macro-cement, which engulfs micrite/microspar, and the lack of evidence of meteoric water influence suggest that the precipitation of micrite/microspar has occurred prior to, during and probably even after oil emplacement. This conclusion is supported by the fairly similar oxygen isotopic compositions of micrite/microspar in the oil and water zones. Using these oxygen isotopic values and assuming that the precipitation of micrite/microspar occurred from geochemically evolved marine porewaters ($\delta^{18}O_{SMOW} = +2‰$), we infer minimum temperatures of around 58–78°C for calcite micro-overgrowth formation. This temperature range suggests that calcite cement could be partly derived from the initial phases of stylolitization of the host limestone.

We are greatly indebted for the extensive and constructive comments and suggestions made by David Budd, three anonymous referees and the volume editor Anita Csoma, which helped us improve the manuscript. We are grateful to the Oil Subcommittee for funding this project and to ADCO for giving us access to samples and necessary data, as well as for permission to publish the paper.

References

AHR, W.M. 1989. Early diagenetic microporosity in the Cotton Valley Limestone of east Texas. *Sedimentary Geology*, **63**, 275–292.

AL-AASM, I.S. 2000. Chemical and isotopic constraints for recrystallization of sedimentary dolomites from the Western Canada Sedimentary Basin. *Aquatic Geochemistry*, **6**, 229–250.

AL-AASM, I.S. & AZMY, K. 1996. Diagenesis and evolution of microporosity of Middle–Upper Devonian Kee Scarp Reefs, Norman Wells, Northwest Territories, Canada: petrographic and geochemical aspects. *American Association of Petroleum Geologists Bulletin*, **80**, 82–100.

AL-AASM, I.S. & VEIZER, J. 1986. Diagenetic stabilization of aragonite and low-Mg calcite, II. Stable isotopes in rudists. *Journal of Sedimentary Petrology*, **56**, 763–770.

AL-AASM, I.S., TAYLOR, B.E. & SOUTH, B. 1990. Stable isotope analysis of multiple carbonate samples using selective acid extraction. *Chemical Geology*, **80**, 119–125.

ALSHARHAN, A.S. 1990. Geology and reservoir characteristics of Lower Cretaceous Kharaib Formation in Zakum Field, Abu Dhabi, United Arab Emirates. *In*: BROOKS, J. (ed.) *Classic Petroleum Provinces*. Geological Society, London, Special Publications, **50**, 299–316, https://doi.org/10.1144/GSL.SP.1990.050.01.16

ALSHARHAN, A.S. & SADD, J.L. 2000. Stylolites in Lower Cretaceous carbonate reservoirs, U.A.E. *In*: ALSHARHAN, A.S. & SCOTT, R.W. (eds) *Middle East Models of Jurassic/Cretaceous Carbonate System*. Society of Economic Paleontologists and Mineralogists (SEPM), Special Publications, **69**, 185–207.

AL-SUWAIDI, A.S., TAHER, A.K., ALSHARHAN, A.S. & SALAH, M.G. 2000. Stratigraphy and geochemistry of Upper Jurassic Diyab Formation, Abu Dhabi, U.A.E. Middle East models of Jurassic/Cretaceous carbonate systems. *In*: ALSHARHAN, A.S. & SCOTT, R.W. (eds) *Middle East Models of Jurassic/Cretaceous Carbonate System*. Society of Economic Paleontologists and Mineralogists (SEPM), Special Publications, **69**, 249–271.

BARNABY, R.J. & RIMSTIDT, J.D. 1989. Redox conditions of calcite cementation interpreted from Mn and Fe contents of authigenic calcites. *Geological Society of America Bulletin*, **101**, 795–804.

BATHURST, R.G.C. 1966. Boring algae, micrite envelopes and lithification of molluscan biosparites. *Geological Journal*, **5**, 15–32.

BATHURST, R.G.C. 1972. *Carbonate Sediments and Their Diagenesis*. Developments in Sedimentology, **12**. Elsevier, Amsterdam, 549–593.

BATHURST, R.G.C. 1975. *Carbonate Sediments and Their Diagenesis*. 2nd edn. Elsevier, Amsterdam.

BATHURST, R.G.C. 1987. Diagenetically enhanced bedding in argillaceous platform limestones: stratified cementation and selective compaction. *Sedimentology*, **34**, 749–778.

BERKYOVA, S. & MUNNECKE, A. 2010. 'Calcispheres' as a source of lime mud and peloids – evidence from the early Middle Devonian of the Prague Basin, the Czech Republic. *Bulletin of Geosciences*, **85**, 585–602.

BODNAR, R.J. 1993. Revised equation and table for determining the freezing point depression of H_2O–NaCl solutions. *Geochimica et Cosmochimica Acta*, **57**, 683–684.

BRAND, U. & VEIZER, J. 1980. Chemical diagenesis of a multicomponent carbonate system – 1: trace elements. *Journal of Sedimentary Petrology*, **50**, 1219–1236.

BRAND, U. & VEIZER, J. 1981. Chemical diagenesis of a multicomponent carbonate system – 2: stable isotopes. *Journal of Sedimentary Petrology*, **51**, 987–997.

BRUNA, P.O., GUGLIELMI, Y. ET AL. 2013. Porosity gain and loss in unconventional reservoirs: example of rock typing in Lower Cretaceous hemipelagic limestones, SE France (Provence). *Marine and Petroleum Geology*, **48**, 186–205.

BUDD, D.A. 1989. Micro-rhombic calcite and microporosity in limestones: a geochemical study of the Lower

Cretaceous Thamama Group, UAE. *Sedimentary Geology*, **63**, 293–311.

Budd, D.A. 2001. Permeability loss with depth in the Cenozoic carbonate platform of west-central Florida. *American Association of Petroleum Geologists Bulletin*, **85**, 1253–1272.

Burgess, C.J. 1979. The development of a Lower Jurassic carbonate tidal flat, Central High Atlas, Morocco. 2: diagenetic history. *Journal of Sedimentary Petrology*, **49**, 413–427.

Chafetz, H.S. & Buczynski, C. 1992. Bacterially induced lithification of microbial mats. *Palaeos*, **7**, 277–293.

Cheng, H., Zhang, X. & Song, H. 2014. Morphological investigation of calcium carbonate during ammonification–carbonization process of low concentration calcium solution. *Journal of Nanomaterial*, **2014**, Article 503696.

Choquette, P.W. & Pray, L.C. 1970. Geologic nomenclature and classification of porosity in sedimentary carbonates. *American Association of Petroleum Geologists Bulletin*, **54**, 207–250.

Cox, P.A., Wood, R.A., Dickson, J.A.D., Al Rougha, H.B., Shebl, H. & Corbett, P.W.M. 2010. Dynamics of cementation in response to oil charge: evidence from a cretaceous carbonate field, U.A.E. *Sedimentary Geology*, **228**, 246–254.

Davis, K.J., Nealson, K.H. & Luttge, A. 2007. Calcite and dolomite dissolution rates in the context of microbe-mineral surface interactions. *Geobiology*, **5**, 191–205.

Deville de Periere, M., Durlet, C., Vennin, E., Lambert, L., Bourillot, R., Caline, B. & Poli, E. 2011. Morphometry of micrite particles in Cretaceous microporous limestones of the Middle East: influence on reservoir properties. *Marine and Petroleum Geology*, **28**, 1727–1750.

Dewever, B., Fugel, F., Van Steenwinkel, M. & Ehrenberg, S.N. 2011. *Sedimentological, Diagenetic and Reservoir Rock Typing Study of Upper Thamama Reservoirs A, B, C and Unit 6 & 7 in the Sahil Field, U.A.E.* Panterra Geoconsultants Report prepared for the Abu Dhabi Company for Onshore Oil Operations (ADCO).

Dickson, J.A.D. & Coleman, M.L. 1980. Changes in carbon and oxygen isotope composition during limestone diagenesis. *Sedimentology*, **27**, 107–118.

Dravis, J.J. 1989. Deep-burial microporosity in Upper Jurassic Haynesville oolitic grainstones, East Texas. *Sedimentary Geology*, **63**, 325–341.

Egeberg, P.K. & Aagaard, P. 1989. Origin and evolution of formation waters from oil fields on the Norwegian shelf. *Applied Geochemistry*, **4**, 131–142.

Ehrenberg, S.N. 2006. Porosity destruction in carbonate platforms. *Journal of Petroleum Geology*, **29**, 41–52.

Ehrenberg, S.N. & Walderhaug, O. 2015. Preferential calcite cementation of macropores in microporous limestones. *Journal of Sedimentary Research*, **85**, 780–793.

Ehrlich, H.L. 1998. Geomicrobiology: its significance for geology. *Earth Science-Reviews*, **45**, 45–60.

Emery, D. 1987. Trace-element source and mobility during limestone burial diagenesis – an example from the Middle Jurassic of eastern England. *In*: Marshall, J.D. (ed.) *Diagenesis and Sedimentary Sequences*. Geological Society, London, Special Publications, **36**, 201–217, https://doi.org/10.1144/GSL.SP.1987.036.01.16

Esrafili-Dizaji, B. & Rahimpour-Bonab, H. 2009. Effects of depositional and diagenetic characteristics on carbonate reservoir quality: a case study from the South Pars gas field in the Persian Gulf. *Petroleum Geoscience*, **15**, 325–344, https://doi.org/10.1144/1354-079309-817

Fischer, C., Kurganskaya, I., Schafer, T. & Luttge, A. 2014. Variability of crystal surface reactivity: what do we know? *Applied Geochemistry*, **43**, 132–157.

Folk, R.L. 1965. Some aspects of recrystallization in ancient limestones. *In*: Pray, L.C. & Murray, R.C. (eds) *Dolomitization and Limestones Diagenesis*. Society of Economic Paleontologists and Mineralogists (SEPM), Special Publications, **13**, 14–48.

Folk, R.L. 1993. SEM imaging of bacteria in carbonate sediments and rocks. *Journal of Sedimentary Petrology*, **63**, 990–999.

Friedman, G.M. 1964. Early diagenesis and lithification in carbonate sediments. *Journal of Sedimentary Petrology*, **34**, 777–813.

Friedman, I. & O'Neil, J.R. 1977. Compilation of stable isotope fractionational factors of geochemical interest. *In*: Fleischer, M. (ed.) *Data of Geochemistry*. 6th edn. United States Geological Survey, Professional Paper, **440**.

Geslin, J.K. 1994. Carbonate pseudomatrix in siliciclastic-carbonate turbidites from the Oquirrh-Wood River Basin, Southern Idaho. *Journal of Sedimentary Research*, **64**, 55–58.

Goldstein, R.H. & Reynolds, T.J. 1994. *Systematics of Fluid Inclusions in Diagenetic Minerals*. Society of Economic Paleontologists and Mineralogists (SEPM), Short Course Notes, **31**.

Granier, B.R. 2000. Lower Cretaceous stratigraphy of Abu Dhabi and the United Arab Emirates – A reappraisal. *Paper presented at the Abu Dhabi International Petroleum Exhibition and Conference*, 13–15 October 2000, Abu Dhabi, United Arab Emirates.

Grover, G., Jr & Read, J.F. 1983. Paleoaquifer and deep-burial related cements defined by regional cathodoluminescent patterns, Middle Ordovician carbonates, Virginia. *American Association of Petroleum Geologists Bulletin*, **67**, 1275–1303.

Gunther, A. 1990. Distribution and bathymetric zonation of shell boring endoliths in recent reef and shelf environments: Cozumel, Yucatan (Mexico). *Facies*, **22**, 233–262.

Hall, D.L., Sterner, S.M. & Bodnar, R.J. 1988. Freezing point depression of $NaCl-KCl-H_2O$ solutions. *Economic Geology*, **83**, 197–202.

Harris, P.M.M. 2010. Delineating and quantifying depositional facies patterns in carbonate reservoirs: insight from modern analogs. *American Association of Petroleum Geologist Bulletin*, **94**, 61–86.

Harris, P.M.M., Kendall, C.G.S.C. & Lerche, I. 1985. Carbonate cementation – a brief review. *In*: Schneidermann, N. & Harris, P.M. (eds) *Carbonate Cements*. Society of Economic Paleontologists and Mineralogists, Special Publications, **36**, 79–95.

HEASLEY, E.C., WORDEN, R.H. & HENDRY, J.P. 2000. Cement distribution in a carbonate reservoir: recognition of a palaeo oil–water contact and its relationship to reservoir quality in the Humbley Grove field, Onshore, UK. *Marine and Petroleum Geology*, **17**, 639-654.

HILLGÄRTNER, H., VAN BUCHEM, F.S.P., GAUMENT, F., RAZIN, P., PITTET, B., GRÖTSCH, J. & DROSTE, H. 2003. The Barremian–Aptian evolution of the eastern Arabian carbonate platform margin (northern Oman). *Journal of Sedimentary Research*, **73**, 756–773.

HILLNER, P.E., MANNE, S., GRATZ, A.J. & HANSMA, P.K. 1992. AFM images of dissolution and growth on a calcite crystal. *Ultramicroscopy*, **42–44**, 1387–1393.

HUSINEC, A. & READ, J.F. 2006. Transgressive oversized radial ooid facies in the Late Jurassic Adriatic Platform interior: low-energy precipitates from highly supersaturated hypersaline waters. *Geological Society of America Bulletin*, **118**, 550–556.

JACOBSON, A.J. & WU, L. 2009. Microbial dissolution of calcite at $T = 28°C$ and ambient pCO_2. *Geochmica et Cosmochimica Acta*, **73**, 2314–2331.

JAMES, N.P. & CHOQUETTE, P.W. 1984. Diagenesis 9. Limestones – the meteoric diagenetic environment. *Geoscience Canada*, **11**, 161–194.

KALDI, J. 1989. Diagenetic microporosity (chalky porosity), middle Devonian Kee Scarp reef complex, Norman wells, northwest territories, Canada. *Sedimentary Geology*, **63**, 241–252.

KOBLUCK, D.R. & RISK, M.J. 1977. Calcification of exposed filaments of endolithic algae, micrite envelope formation and sediment production. *Journal of Sedimentary Petrology*, **47**, 517–528.

LAMBERT, L., CHRISTOPHE, D., JEAN-PAUL, L. & GERARD, M. 2006. Burial dissolution of micrite in Middle East carbonate reservoirs (Jurassic–Cretaceous): keys for recognition and timing. *Marine and Petroleum Geology*, **23**, 79–92.

LÉONIDE, P., FOURNIER, F. ET AL. 2014. Diagenetic patterns and pore space distribution along a platform to outer-shelf transect (Urgonian limestone, Barremian–Aptian, SE France). *Sedimentary Geology*, **306**, 1–23.

MADDEN, R.H.C. & WILSON, M.E.J. 2012. Diagenesis of Neogene delta-front patch reefs: alteration of coastal, siliciclastic-influenced carbonates from humid equatorial regions. *Journal of Sedimentary Research*, **82**, 871–888.

MALIVA, R.G., MISSIMER, T.M., CLAYTON, E.A. & DICKSON, J.A.D. 2009. Diagenesis and porosity preservation in Eocene microporous limestones, South Florida, USA. *Sedimentary Geology*, **217**, 85–94.

MARUTHAMUTHU, S., DHANDAPANI, P., PONMARIAPPAN, S., SATHIYANARAYANAN, S., MUTHUKRISHNAN, S. & PALANISWAMY, N. 2010. Scale formation by calcium-precipitating bacteria in cooling water system. *Journal of Failure Analysis and Prevention*, **10**, 416–426.

MELIM, L.A., WESTPHAL, H., SWART, P.K., EBERLI, G.P. & MUNNECKE, A. 2002. Questioning carbonate diagenetic paradigms: evidence from the Neogene of the Bahamas. *Marine Geology*, **185**, 27–53.

MILNER, P.A. 1998. Source rock distribution and thermal maturity in the Southern Arabian Peninsula. *GeoArabia*, **3**, 339–356.

MONTY, C.L.V. 1995. The rise and nature of carbonate mud-mounds: an introductory actualistic approach. *In*: MONTY, C.L.V., BOSENCE, D.W.J., BRIDGES, P.H. & PRATT, B.R. (eds) *Carbonate Mud-mounds, Their Origin and Evolution*. International Association of Sedimentologists, Special Publications, **23**, 11–48.

MORAD, S. 1998. Carbonate cementation in sandstones: distribution patterns and geochemical evolution. In MORAD, S. (ed.) Carbonate Cementation in Sandstones. International Association of Sedimentologists, Special Publications, **26**, 1–26.

MORAD, S., AL-AASM, I.S., NADER, F.H., CERIANI, A. & GASPARRINI, M. 2012. Impact of diagenesis on the spatial and temporal distribution of reservoir quality in the Jurassic Arab C and D members, offshore Abu Dhabi oilfield, United Arab Emirates. *GeoArabia*, **17**, 17–54.

MORITA, R.Y. 1980. Calcite precipitation by marine bacteria. *Geomicrobiology Journal*, **2**, 63–82.

MOSHIER, S.O. 1989*a*. Microporosity in micritic limestones: a review. *Sedimentary Geology*, **63**, 191–213.

MOSHIER, S.O. 1989*b*. Development of microporosity in a micritic limestone reservoir, Lower Cretaceous, Middle East. *Sedimentary Geology*, **63**, 217–240.

MUNNECKE, A. & SAMTLEBEN, C. 1996. The formation of micritic limestones and the development of limestone-marl alternations in the Silurian of Gotland, Sweden. *Facies*, **34**, 159–176.

MUNNECKE, A., WESTPHAL, H., REIJMER, J.J.G. & SAMTLEBEN, C. 1997. Microspar development during early marine burial diagenesis: a comparison of Pliocene carbonates from the Bahamas with Silurian limestones from Gotland (Sweden). *Sedimentology*, **44**, 977–990.

MURRIS, R.J. 1980. Hydrocarbon habitat of the Middle East. *In*: MIALL, A.D. (ed.) *Facts and Principles of World Petroleum Occurrence*. Canadian Society of Petroleum Geologists, Memoirs, **6**, 765–800.

NEILSON, J.E., OXTOBY, N.H., SIMMONS, M.D., SIMPSON, I.R. & FORTUNATOVA, N.K. 1998. The relationship between petroleum emplacement and carbonate reservoir quality: examples from Abu Dhabi and the Amu Darya Basin. *Marine and Petroleum Geology*, **15**, 57–72.

NESSE, W.D. 2000. *Introduction to Mineralogy*. Oxford University Press, New York.

NOVITSKI, J.A. 1981. Calcium carbonate precipitation by marine bacteria. *Geomicrobiology Journal*, **2**, 375–388.

OSWALD, E.J., MUELLER, H.W., III, GOFF, D.F., AL-HABSHI, H. & AL-MATROUSHI, S. 1995. Controls on porosity evolution in Thamama Group carbonate reservoirs in Abu Dhabi, UAE. *In*: *Middle East Oil Show*, 11–14 March 1995, Bahrain, Conference Proceedings. Society of Petroleum Engineers, Richardson, TX, 251–265.

PAGANONI, M., AL HARTHI, A. ET AL. 2016. Impact of stylolitization on diagenesis of a lower cretaceous carbonate reservoir from a giant oilfield, Abu Dhabi, United Arab Emirates. *Sedimentary Geology*, **335**, 70–92.

PERRI, E. & TUCKER, M.E. 2007. Bacterial fossils and microbial dolomite in Triassic stromatolites. *Geology*, **35**, 207–210.

PERRY, C.T. 1999. Biofilm-related calcification, sediment trapping and constructive micrite envelopes: a criterion for the recognition of ancient grass-bed environments? *Sedimentology*, **46**, 33–45.

Purdy, E.G. 1968. *Carbonate Diagenesis: An Environmental Survey*. Istituto di geologia e paleontologia dell'Università di Roma, Rome.

Reid, R.P. & Macintyre, I.G. 1998. Carbonate recrystallization in shallow marine environments: a widespread diagenetic process forming micritized grains. *Journal of Sedimentary Research*, **68**, 928–946.

Reid, R.P. & Macintyre, I.G. 2000. Microboring v. recrystallization: further Insight into the micritization process. *Journal of Sedimentary Research*, **1**, 24–28.

Richard, J., Sizun, J.P. & Machhour, L. 2007. Development and compartmentalization of chalky carbonate reservoirs: the Urgonian Jura-Bas Dauphiné platform model (Génissiat, southeastern France). *Sedimentary Geology*, **198**, 195–207.

Saller, A.H. & Moore, C.H. 1989. Meteoric diagenesis, marine diagenesis, and microporosity in Pleistocene and Oligocene limestones, Enewetak Atoll, Marshall Islands. *Sedimentary Geology*, **63**, 253–272.

Scholle, P.A. & Arthur, M.A. 1980. Carbon isotope fluctuations in Cretaceous pelagic limestones: potential stratigraphic and petroleum exploration tool. *American Association of Petroleum Geologists Bulletin*, **64**, 67–87.

Sharland, P.R., Archer, R. *et al.* 2001. *Arabian Plate Sequence Stratigraphy*. GeoArabia, Special Publications, **2**.

Sherman, C.E., Fletcher, C.H. & Rubin, K.H. 1999. Marine and meteoric diagenesis of Pleistocene carbonates from a nearshore submarine terrace, Oahu, Hawaii. *Journal of Sedimentary Research*, **69**, 1083–1097.

Spadafora, A., Perri, E., Mckenzie, J.A. & Vasconcelos, C. 2010. Microbial biomineralization processes forming modern Ca:Mg carbonate stromatolites. *Sedimentology*, **57**, 27–40.

Strohmenger, C.J., Weber, L.J. *et al.* 2006. High-resolution sequence stratigraphy and reservoir characterization of Upper Thamama (Lower Cretaceous) Reservoirs of a Giant Abu Dhabi Oil Field, United Arab Emirates. *In*: Harris, P.M. & Weber, L.J. (eds) *Giant Hydrocarbon Reservoirs of the World: From Rocks to Reservoir Characterization and Modeling*. American Association of Petroleum Geologists, Memoirs, **88**, 139–171.

Swirydczuk, K. 1988. Mineralogical control on porosity type in Upper Jurassic Smackover ooid grainstones, southern Arkansas and northern Louisiana. *Journal of Sedimentary Research*, **58**, 339–347.

Taher, A.A. 1997. Delineation of organic richness and thermal history of the Lower Cretaceous Thamama Group, East Abu Dhabi: a modeling approach for oil exploration. *GeoArabia*, **2**, 65–88.

Teng, H.H., Dove, P.M. & De Yoreo, J.J. 2000. Kinetics of calcite growth: surface processes and relationships to macroscopic rate laws. *Geochimica et Cosmochimica Acta*, **64**, 2255–2266.

Tucker, M.E. 2001. *Sedimentary Petrology: An Introduction to the Origin of Sedimentary Rocks*. Wiley-Blackwell, Hoboken, NJ.

Tucker, M.E. & Marshall, J. 2004. Diagenesis and Geochemistry of Upper Muschelkalk (Triassic) Buildups and Associated Facies in Catalonia (NE Spain): a paper dedicated to Francesc Calvet. *Geologica Acta*, **2**, 257–269.

Tucker, M.E. & Wright, V.P. 2009. *Carbonate Sedimentology*. John Wiley, Chichester, UK.

Vahrenkamp, V., Barata, J., Van Laer, P.J., Swart, P. & Murray, S. 2014. Micro rhombic calcite of a giant Barremian (Thamama B) reservoir onshore Abu Dhabi – clumped isotope analyses fix temperature, water composition and timing of burial diagenesis. *Paper presented at the Abu Dhabi International Petroleum Exhibition and Conference*, 10–13 November 2014, Abu Dhabi, UAE.

Van Buchem, F.S.P., Pittet, B. *et al.* 2002. High resolution sequence stratigraphic architecture of Barremian/Aptian carbonate systems in northern Oman and the United Arab Emirates (Kharaib and Shu'aiba formations). *GeoArabia*, **7**, 461–500.

Van Buchem, F.S.P., Al-Husseini, M.I., Maurer, F., Droste, H.J. & Yose, L.A. 2010. Sequence-stratigraphic synthesis of the Barremian–Aptian of the eastern Arabian Plate and implications for the petroleum habitat. *In*: Van Buchem, F.S.P., Al-Husseini, M.I., Maurer, F. & Droste, H. (eds) *Barremian–Aptian Stratigraphy and Hydrocarbon Habitat of the Eastern Arabian Plate*. GeoArabia, Special Publications, **4**, 9–48.

Vanorio, T. & Mavko, G. 2011. Laboratory measurements of the acoustic and transport properties of carbonate rocks and their link with the amount of microcrystalline matrix. *Geophysics*, **76**, E105–E115.

Veizer, J., Ala, D. *et al.* 1999. $^{87}Sr/^{86}Sr$, $\delta^{13}C$ and $\delta^{18}O$ evolution of Phanerozoic seawater. *Chemical Geology*, **161**, 50–88.

Volery, C., Davaud, E., Foubert, A. & Caline, B. 2009. Shallow-marine microporous carbonate reservoir rocks in the Middle East: relationship with seawater Mg/Ca ratio and eustatic sea level. *Journal of Petroleum Geology*, **32**, 313–326.

Volery, C., Davaud, E., Foubert, A. & Caline, B. 2010. Lacustrine microporous micrites of the Madrid Basin (Late Miocene, Spain) as analogues for shallow-marine carbonates of the Mishrif reservoir Formation (Cenomanian to Early Turonian, Middle East). *Facies*, **56**, 385–397.

Waite, R., Wtzel, A., Meyer, A.A. & Strasser, A. 2008. The paleoecological significance of nerineoid mass accumulations from the Kimmeridgian of the Swiss Jura Mountains. *Palaios*, **23**, 548–558.

Wigley, T.M., Plummer, L.N. & Pearson, F.J. 1978. Mass transfer and carbon isotope evolution in natural water systems. *Geochimica et Cosmochimica Acta*, **42**, 1117–1139.

Wilson, J.L. 1975. *Carbonate Facies in Geologic History*. Springer, New York.

Distinguishing between eogenetic, unconformity-related and mesogenetic dissolution: a case study from the Panna and Mukta fields, offshore Mumbai, India

A.J. BARNETT[1]*, V.P. WRIGHT[2], V.S. CHANDRA[3] & V. JAIN[4]

[1]*BG Group, 100 Thames Valley Park, Reading, RG6 1PT, UK*

[2]*PW Carbonate Geoscience, 18 Llandennis Avenue, Cardiff CF23 6JG, UK*

[3]*Institute of Petroleum Engineering, Heriot-Watt University, Edinburgh, EH14 4AS, UK*

[4]*BG India, BG House, Lake Boulevard, Hiranandani Business Park, Mumbai, Maharashtra 400076, India*

**Correspondence: andrew.barnett@bg-group.com*

Abstract: The Panna–Mukta fields host hydrocarbons in the Bassein Formation Eocene–Oligocene ramp limestones. The pore system is almost wholly secondary, comprising microporosity, mouldic porosity, vugs, solution-enlarged stylolites and fractures. Although petrographical evidence points to dissolution after extensive late cementation, the presence of a high permeability layer close to a palaeokarstic surface at the Eocene–Oligocene boundary has raised the possibility that this secondary porosity could be related to subaerial exposure. However, the Panna–Mukta reservoirs show a strong correlation between secondary matrix porosity and stylolite density measured from cores. Stylolites only developed in 'clean' limestones lacking argillaceous material, whereas more argillaceous limestones in the succession are characterized by dissolution seams and have poor reservoir quality. These cleaner limestones occur preferentially below the Eocene–Oligocene boundary, representing an upwards-shallowing sequence, whereas the argillaceous limestones occur further below the Eocene–Oligocene boundary in the lower part of the same shallowing-upwards sequence and in the transgressive limestones at the base of the Bassein A. This secondary porosity distribution suggests movement of corrosive fluids along pre-existing stylolites. Despite an apparent link between porosity distribution and an unconformity, secondary porosity development was mesogenetic and related to the distribution of facies that favoured stylolites that acted as conduits for the flow of corrosive fluids. The Bassein Formation reservoirs show unequivocal evidence of significant porosity development by mesogenetic dissolution but the exact process or processes by which such porosity creation occurs requires further work.

Secondary porosity is of considerable importance to the prospectivity of carbonates and often determines the presence or absence of reservoir-grade porosity. Most limestones, by virtue of having been deposited in shallow water, are prone to secondary porosity development as a consequence of relative sea-level falls and the effects of acidic meteoric fluids producing early (eogenetic) unconformity-related dissolution (Tucker & Wright 1990; Moore 2001; Ahr 2008). However, secondary porosity formation due to dissolution at depth related to burial fluids is increasingly being invoked, and Wright & Harris (2013) have attempted to define such dissolution (also referred to as burial corrosion) as:

> porosity formation caused by fluids unrelated to recharge from the overlying land surface, or adjacent water bodies, but related typically to a confined aquifer, where the source of the fluid is ultimately from below the formation (hypogene).

Recognizing the relative importance of eogenetic v. mesogenetic dissolution in generating the final preserved pore system is commonly difficult. The exact mechanisms by which burial dissolution proceeds are often difficult to establish and remain controversial. This debate has been reinvigorated by some workers (e.g. Ehrenberg *et al.* 2012) who not only question the veracity of the commonly cited mechanisms of burial dissolution, but also suggest that many of the petrographical characteristics used to identify burial dissolution can be explained without invoking mesogenetic processes. Thus, the aim of this paper is to assess the relative roles of eogenetic v. mesogenetic processes in creating secondary porosity in the Bassein Limestone of the

From: Armitage, P. J., Butcher, A. R., Churchill, J. M., Csoma, A. E., Hollis, C., Lander, R. H., Omma, J. E. & Worden, R. H. (eds) 2018. *Reservoir Quality of Clastic and Carbonate Rocks: Analysis, Modelling and Prediction.* Geological Society, London, Special Publications, **435**, 67–84.
First published online December 18, 2015, https://doi.org/10.1144/SP435.12

Panna–Mukta fields, offshore India. In so doing, several key questions are addressed: Can significant amounts (>10%) of secondary porosity be produced by mesogenetic processes, as defined above, or is such porosity development restricted to eogenetic or telogenetic diagenesis? If significant secondary porosity development can occur during mesogenesis, what are the exact mechanisms by which this occurs? These questions are not just of academic significance. If the possibility of large-scale porosity creation by mesogenesis can be ruled out, then exploration for carbonate plays where prospectivity is probably dependent on secondary porosity development should be limited to areas where eogenetic or telogenetic porosity enhancement is likely to have occurred.

Location and geological setting

The Panna and Mukta fields are located in the Bombay Basin in central west offshore India (Fig. 1). The Panna Field is located *c.* 50 km east of the giant Bombay High Field and 95 km west of Bombay. It lies immediately north of the giant Bassein gas field, separated by a shallow syncline. Present-day seawater depth ranges from 45 to 55 m. The Panna Field is a broad, low-relief anticlinal trap (Fig. 1). The oil column is *c.* 20 m thick and is largely in transition to water. It is sandwiched between a gas cap and a 40–60 m-thick water leg. The gas column is typically 50 m thick. The Mukta Field is located 15 km west of the Panna Field. Present seawater depth is *c.* 65 m. Production is from three stacked oil pools within several anticlinal closures.

The principal hydrocarbon-bearing formations in both the Panna and Mukta fields are the Eocene Bassein B and Early Oligocene Bassein A limestones. However, porosity development is less continuous in the Mukta area. The Bassein reservoirs are sourced from the Paleocene–Early Eocene deltaic clastics of the Panna Formation. The seal comprises the Late Oligocene Alibag Shale (Fig. 2a).

Sedimentation in the Bombay Basin occurred in a passive margin setting of Late Cretaceous–recent age. The Tapti–Surat Depression (NE) and the adjacent Mahim Graben (east) comprise the Late Cenozoic clastic depocentre, as well as the probable source areas for hydrocarbons. The Mahim Graben is separated from the Panna Field by the Panna East Fault Zone (Fig. 1). This is the major structural feature in the Panna–Mukta blocks. It was an extensional fault during the Late Eocene and has undergone an episode of transpression during the post-mid Miocene (Basu *et al.* 1982). Despite this phase of inversion, the Panna–Mukta area has been in a continuously marine setting from the Early Oligocene to the onset of glacio-eustasy in the mid-Miocene (i.e it has not been subject to an influx of meteoric water during that period).

The deposition of the Eocene–Early Oligocene limestones of the Panna–Mukta area took place in an unusual platform configuration on what appears to be an isolated (detached) southerly dipping ramp. This interpretation is supported by the absence of a shelf margin or resedimentation as sediment gravity flows. However, shoreline facies appear to be absent. Approximately 95% of the component limestones are dominated by foraminifera (Fig. 3;

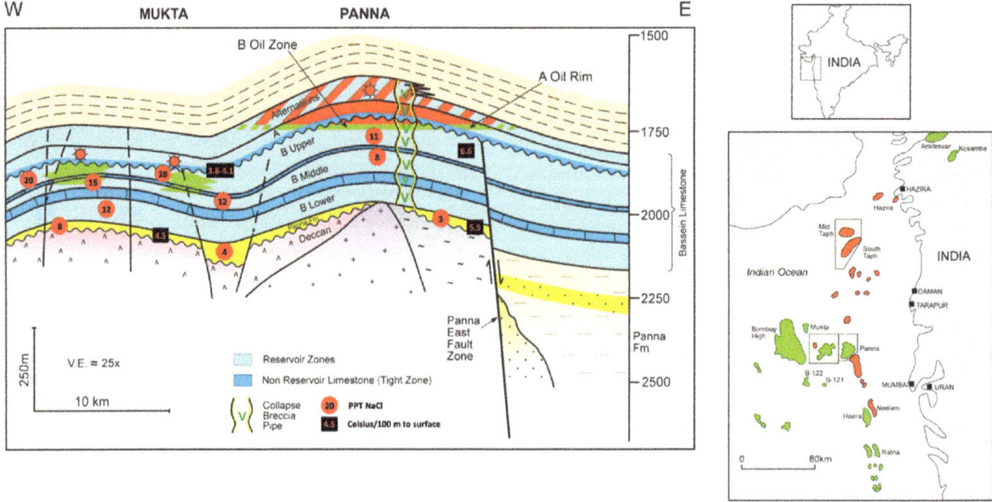

Fig. 1. Location map and generalized structural cross-section across the Panna and Mukta fields. B, Bassein.

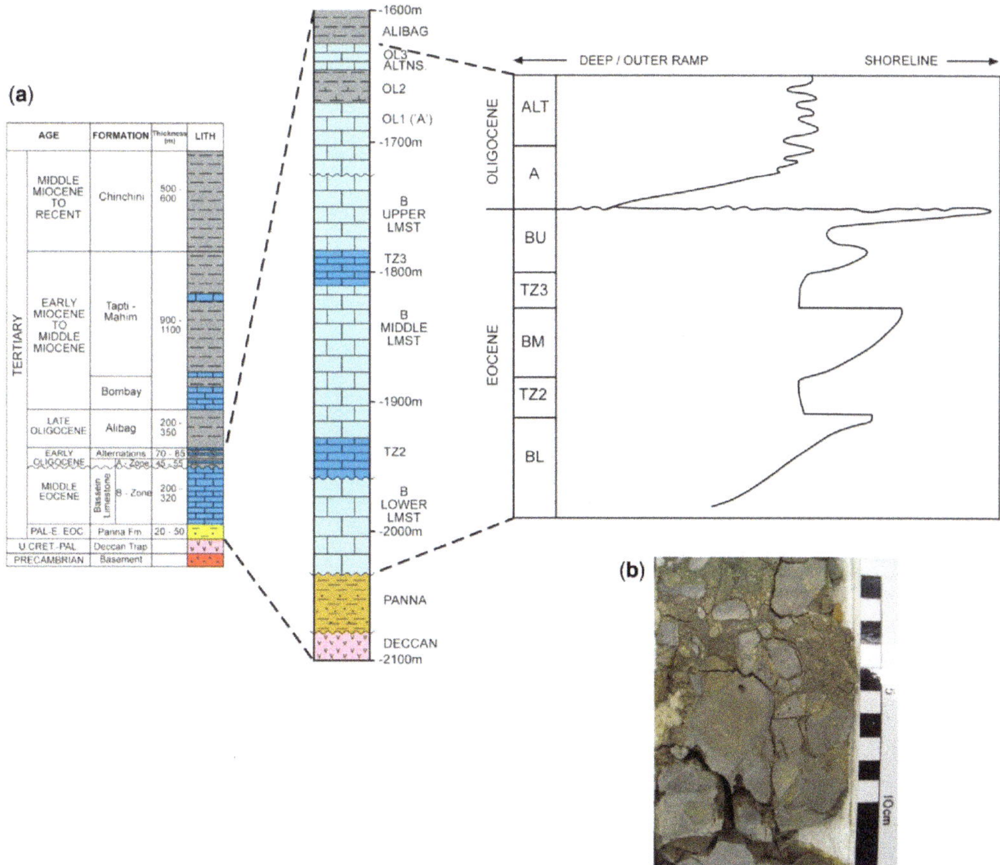

Fig. 2. (**a**) Stratigraphic column and interpreted sea-level history for the Panna and Mukta fields. ALT, Alternations; A, Bassein A; BU, Bassein B Upper; TZ3, Tight Zone 3; BM, Bassein B Middle; TZ2, Tight Zone 2; BL, Bassein B Lower. The Tight Zones correspond to deeper-water facies. (**b**) Rubbly palaeokarstic horizon at the top of the Bassein B. Smaller scale divisions are 1 cm.

Table 1). Sedimentary structures are rarely observed. Consequently, the facies scheme presented below is largely based on the presence, abundance and preservation state of foraminifera (Wright & Barnett 2011). Comparison can be made with well-established facies models for Eocene, foram-dominated successions from outcrop studies that integrate extant forms and associated biota such as calcareous algae (Buxton & Pedley 1989; Racey 1994; Beavington-Penney & Racey 2004). Although using such studies allows broad relationships between depositional depth and facies to be established and relative sea-level changes to be identified (Figs 2a & 3), the bulk of the limestones in the Bassein B are packstones dominated by a *Coskinolina* assemblage (Sedimentary Facies C: Table 1; Figs 2 & 3) and identifying clear depth trends is difficult. The component subfacies cover a wide range of depths from above fair-weather wave base in inner ramp settings to probable mid-ramp depths. The complex interfingering of the different *Coskinolina* subfacies suggests that the facies pattern was more a mosaic than one with shore-parallel facies belts. The shallowest parts of the ramp are represented by miliolid packstones and grainstones (Sedimentary Facies M: Table 1; Figs 2 & 3), but these are extremely rare.

Separating the Bassein B units are the 'Tight Zones', which are composed of more argillaceous, facies consisting of rotalid (Sedimentary Facies R) and bioclastic packstone and wackestones (Sedimentary Facies H: Table 1; Figs 2 & 3). These are characterized by extensive clay seams.

The Bassein A limestones are more mud-rich (wackestone–packstone textures predominate) and heterogeneous than the Bassein B, and are dominated by Sedimentary Facies G and N (Table 1; Figs 2 & 3). Nummulitids are common as is

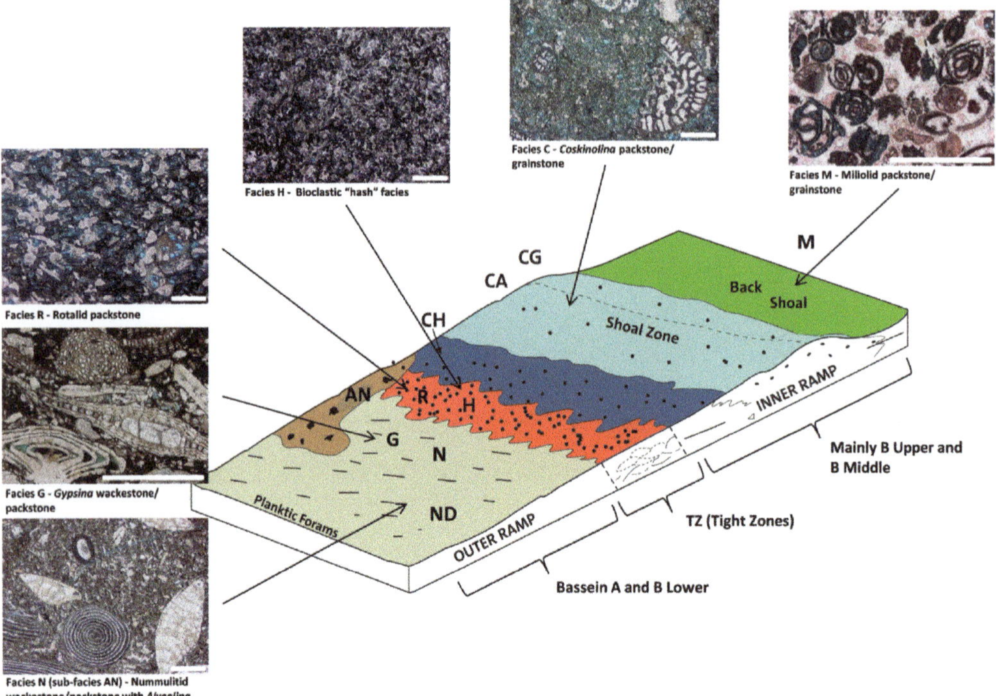

Fig. 3. Depositional model for the Bassein A and Bassein B limestones. For facies abbreviations, see Table 1. All scale bars are 1 mm.

Gypsina. Rotalids, miliolids and diverse small benthic foraminifera occur along with rare *Orbitolites* and *Peneropolis*. Planktonic foraminifera also occur rarely, as do minor corals. The overall depositional setting is likely to represent mid-ramp–outer-ramp depths, probably below storm-weather wave base but within the photic zone (Wright & Barnett 2011).

The Bassein A and B are separated by a palaeokarstic surface at the Eocene–Oligocene boundary and, based on biostratigraphic data, we estimate that this represents a hiatus of about 3.5 myr. The palaeokarst profile is thin, comprising <0.3 m of limestone clasts in a matrix of green clay (Fig. 2b). The subsequent marine transgression represents a major flooding event and a backstepping of facies belts (Fig. 2a). Consequently, the facies of the Bassein A interval are noticeably deeper water in character than those of the Bassein B (Fig. 3; Table 1).

The current reservoir temperature in the B Upper is 109–117°C at 1715 m TVDSS (total vertical depth subsea). The formation waters of the Bassein Limestone have low salinity. Immediately west of the Panna East Fault (Fig. 1), a salinity of approximately 11 000 ppm is observed in the B Upper, decreasing to around 8000 ppm in the B Middle and about 3000 ppm in the Panna Formation. The salinity of formation water in Mukta Field is up to 8000 ppm in the Panna Formation and 20 000 ppm in the B Upper. The most likely source of this fresh- to brackish water is the Deccan Formation, which forms a highly fractured aquifer onshore, 60 km to the east of Panna Field. The produced gas comprises 7.3% CO_2 and 500–600 ppm H_2S.

Tracer test

On 7 March 2012, tritiated water (^{3}H-labelled water) was injected into the upper part of the Bassein B at well PB-7HZ. Monitoring of the surrounding wells shows that tritiated water bypassed some of the closer wells, breaking through earlier at other, more distant wells (Fig. 4). This suggests that a high-permeability conduit (e.g. palaeokarst-related cavernous porosity) is preferentially connecting these wells. More surprising, however, was the actual breakthrough time, which was less than 1 day in some cases. The authors do not know of any examples of similarly rapid breakthrough. Breakthrough of this rapidity is unequivocal evidence of the presence of a high-permeability conduit, or conduits, in

Table 1. *Summary of sedimentary facies in the Bassein Limestone of the Panna and Mukta fields*

Sedimentary facies	Subfacies	Component grains	Palaeoenvironmental interpretation	Dominant interval of occurrence
M	M – Miliolid facies	Mainly small miliolids	Restricted inner ramp (back shoal)	Bassein B
C	CC – *Coskinolina*-dominated facies	Major: large specimens of *Coskinolina* and large rotalids Minor: small miliolids, rotalids, textularids and alveolinids	Inner ramp (shoal)	Bassein B
	CG – *Coskinolina* grainstone facies	Major: commonly fragmented *Coskinolina,* small and large rotalids, miliolids and textularids Minor: alveolinids, rare *Orbitolites* and *Peneropolis*	Inner ramp (shoal)	
	CA – *Coskinolina and Aiveolina* facies	Major: *Coskinolina* and *Alveolina* Minor: *Orbitolites,* rare undifferentiated soritids and green algae	Inner ramp (shoal)	
	CP– *Coskinolina* packstone facies	Major: *Coskinolina*	Inner ramp (shoal)	
	CH – *Coskinolina* 'hash' packstone to wackestone facies	Major: coskinolinid debris	Proximal mid-ramp	
H	H – 'Hash' facies with fine skeletal debris	Major: finely comminuted bioclasts, rotalid and echinoderm debris	Mid-ramp	Tight zones in Bassein B
R	R – Rotalid packstone facies	Major: rotalids (e.g. *Lockhartia*), echinoderm debris	Mid-ramp	Bassein A and Tight Zones in Bassein B
	RH – Rotalid 'hash' packstone facies	Major: rotalid fragments, finely comminuted undifferentiated bioclasts, echinoderm debris	Mid-ramp	
G	G – *Gypsina* wackestone to packstone facies	Major: *Gypsina* Minor: nummulitids, nummulithoclastic debris, *Lepidocyclina, Operculina*	Outer ramp	Bassein A and B Lower
N	N – Nummulitid grainstone to wackestone facies	Major: Nummulitids	Outer ramp	Bassein A and B Lower
	AN – Nummulitid wackestone to packstone facies with alveolinids	Major: nummulitids, alveolinids Minor: *Operculina*	Outer ramp	
	ND – Nummulitid facies with discocyclinids	Major: nummulitids, discocyclinids	Outer ramp	

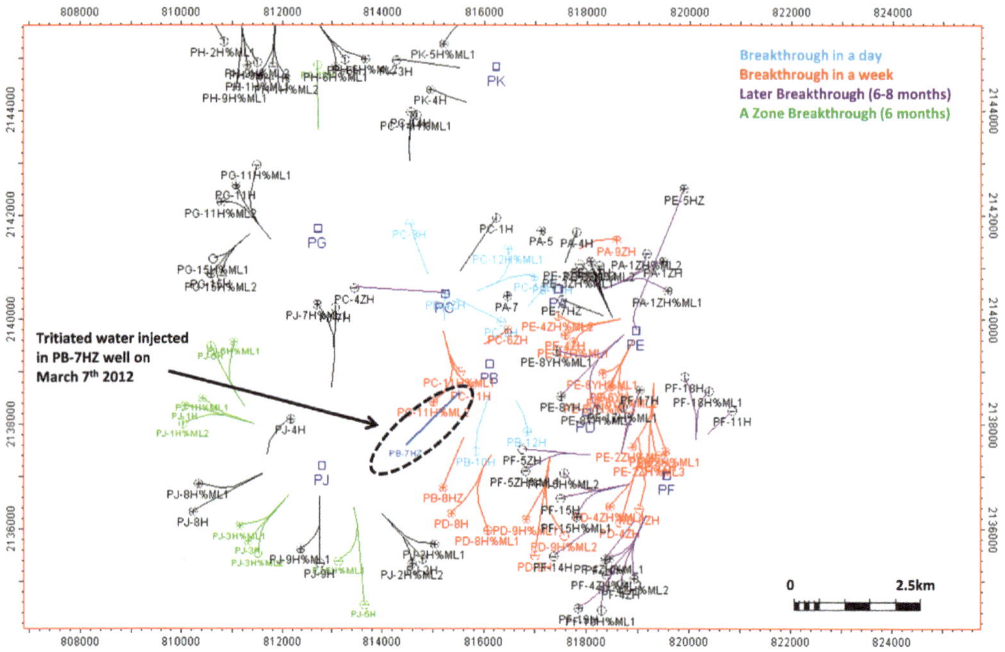

Fig. 4. Map of the Panna Field showing the breakthrough times for tritiated water injected into the PB-7HZ well.

Table 2. *Dual-range Fourier transform infrared (FT-IR) spectroscopy for 25 samples from the Bassein Formation*

Sample No.	Calcite (%)	Dolomite (%)	Pyrite (%)	Dickite (%)	Other (%)
1	95.7	0.8	0.5	1.8	1.2
2	96.2	0.0	0.5	1.0	2.3
3	95.1	0.9	0.6	1.3	2.1
4	94.4	1.3	0.1	1.5	2.8
5	99.0	0.0	0.3	0.0	0.7
6	99.2	0.0	0.4	0.0	0.4
7	98.6	0.0	0.7	0.5	0.3
8	99.2	0.0	0.4	0.0	0.4
9	99.3	0.0	0.4	0.0	0.3
10	99.4	0.0	0.4	0.0	0.1
11	99.3	0.0	0.4	0.0	0.3
12	99.0	0.0	0.4	0.0	0.6
13	99.4	0.0	0.4	0.0	0.2
14	99.5	0.0	0.3	0.0	0.2
15	99.4	0.0	0.3	0.0	0.3
16	99.3	0.0	0.4	0.0	0.3
17	99.3	0.0	0.3	0.0	0.4
18	99.3	0.0	0.4	0.0	0.2
19	99.6	0.0	0.4	0.0	0.0
20	99.2	0.0	0.5	0.0	0.3
21	99.4	0.0	0.4	0.0	0.3
22	96.8	0.0	1.0	1.2	0.9
23	92.2	0.0	3.0	3.9	0.9
24	97.5	0.0	0.5	0.9	1.2
25	75.4	0.0	23.7	0.0	0.9

the upper Bassein B close to the Eocene–Oligocene unconformity described above.

Methods and materials

The observations described in this paper are based on the description of >1200 m of slabbed core and the semi-quantitative examination of *c.* 500 thin sections. Fluid inclusion ($n = 3$) analyses from non-ferroan calcite cements have been performed (Table 2), as have Fourier transform infrared (FT-IR) spectroscopy analyses ($n = 25$: Table 3). Mercury injection capillary pressure (MICP) analyses are also available ($n = 22$).

Table 3. *Two-phase fluid inclusion homogenization temperatures and measured* $\delta^{18}O$‰ *(PDB) in calcite micro-drilled samples*

Sample	Cement type	Mean homogenization temperature (°C)	$\delta^{18}O$‰ (PDB) in calcite
A	Calcite	70	−5.8
B	Calcite	115	−6.9
C	Calcite	130	−8.1

Samples A–C are from successively older cement generations. Data from Esteban & Taberner (2000).

Core samples were described for sedimentary facies and pore types (including a visual estimate of porosity), including fractures, stylolites and solution seams. A similar approach was applied to describe petrographical samples, but particular attention was given to the identification and abundance of foraminifera because of their importance in palaeoenvironmental interpretation. Because the initial core description suggested that there was an association between zones of higher stylolite density and secondary matrix macroporosity, a small subset (40 m) of core material immediately below the Eocene–Oligocene boundary was examined in greater detail by measuring the percentage proportion of rock with visible secondary matrix porosity and the number of stylolites per metre. Following Wanless (1979), the term 'stylolite' is herein used to refer to sutured-seam solution, whereas 'solution seam' is used to denote non-sutured-seam solution (synonymous with microstylolite, microstylolite swarms, clay seams).

Fluid-inclusion microthermometry was performed on successive generations of pre-dissolution, pore-filling, non-ferroan calcite cement from a single sample. The results are shown in Table 2.

Dual-range FT-IR spectroscopy was performed as detailed in Herron *et al.* (1997). Samples were prepared by cutting a disk from one end of each plug. The portion for mineralogy was ground to less than 2.5 μm, then mixed with KBr and pressed into a pellet. Transmittance FTIR spectra were collected with a Perkin Elmer System 2000 FT-IR spectrometer. Both far-IR and mid-IR spectra were collected. Quantitative mineralogy is achieved because minerals have characteristic properties that can be effectively used in quantitative analysis. Among these properties are chemical bond vibrational energies that lead to characteristic infrared (IR) absorbance spectra. Mineralogy can be quantitatively solved from the sample spectrum and the spectra of the mineral constituents. Of course, many variations exist that can affect the mineral properties, including their IR spectra. To account for natural variability in minerals, the FT-IR spectroscopy procedure employed herein uses 50 different mineral standards representing 28 minerals. We include multiple standards for quartz, calcite, dolomite, kaolinite, illite, smectite and chlorite. The use of multiple mineral standards is an important factor in attaining such a high degree of accuracy. In the case of feldspars, we have solved for the Na-, K- and Ca-feldspar end-member spectra, and solved for the fractions of the end members in the samples. The use of Fourier Transform produces smooth spectra with reduced noise content that are suitable for full spectrum analysis. The procedure used also combines the mid-IR spectrum with the far-IR spectrum prior to data processing. The resulting dual-range FT-IR mineralogy is more accurate than possible from either mid-IR or far-IR alone. The mineralogy detection limits are generally approximately 1–2 wt%. The results are shown in Table 3.

Mercury porosimetry measurements were carried out on a Micrometritic AutoPore II 9220 instrument. Plugs, 0.5 inch in diameter and approximately 0.7 inch long, were cut from one of the end pieces. These plugs were then dried in a vacuum oven at 105°C for 12 h. The plugs were placed in individual penetrometers. The penetrometers were loaded into four low-pressure ports. During the low-pressure run, the penetrometers (plus sample) were evacuated then filled with mercury, the pressure was then increased to 14 psia. After the low-pressure run, the penetrometers were transferred to two high-pressure ports. During the high-pressure intrusion (drainage) run, the pressure was incrementally increased from 14 to 60 000 psia, at each increment the pressure was allowed to equilibrate prior to collecting a measurement.

Results

Pore system

The porosity in the Bassein A and B is almost wholly secondary (the only exception being intraparticle porosity within the foraminifera tests), and dissolution has created a range of pore types at various scales. The finest scale of porosity development is matrix and intraparticle microporosity. These are the volumetrically dominant pore types in the Bassein reservoirs, making up, on average, nearly 70% of the total pore system (41.3–94.4%; mean = 69.9%; $n = 22$) based on petrography and MICP analyses (Figs 5a, b & 6). The microporosity grades into mouldic pores and submillimetre-scale vugs, some of which have subsequently been filled by dickite (Fig. 5a, d, e). The majority of the moulds are the sites of former calcitic foraminifera but dissolution was highly selective, and miliolid and textularid forams were preferentially dissolved compared to rotalids. This must reflect the susceptibility of the fine crystal sizes of the miliolids and textularids and their perforate test walls. At a slightly larger scale, millimetre-scale vugs occur and attain sizes of up to about 8 mm. Some of these cross-cut pre-existing carbonate cements, including saddle dolomite and demonstrably late, fracture-filling calcite (see the section 'Paragenesis and the timing of secondary porosity formation': Fig. 5c). Larger, centimetre-scale vugs (up to *c.* 5 cm) are observed in core, but only occur locally and are extremely rare. Vugs may be completely occluded by dickite (Fig. 5c) or be partially filled by non-ferroan calcite and/or saddle dolomite or remain open (Fig. 5f). Note that these pore types

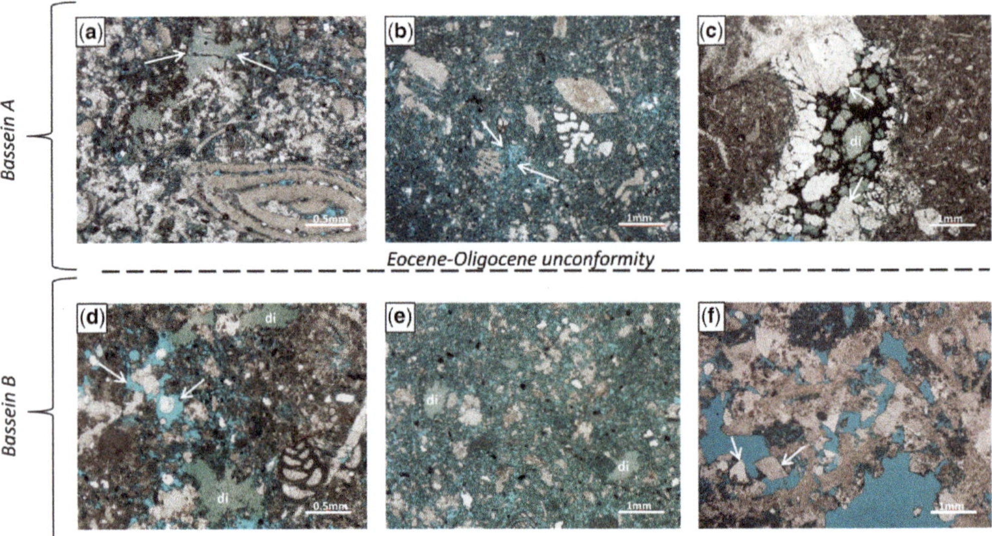

Fig. 5. Thin-section photomicrographs showing the commonly occurring pore types in the Bassein A and B intervals. Note that the porosity is almost wholly secondary in nature and that similar pore types occur both above and below the Eocene–Oligocene unconformity. (**a**) Nummulitic packstone with well-developed matrix microporosity (note the greenish coloration of the matrix) and a dickite-filled vug (arrowed). Minor primary porosity occurs preserved within nummulite tests. (**b**) Nummulitic wackestone with well-developed matrix microporosity. Leached micrite matrix locally grades into incipient vugs highlighted by the brighter blue colour (arrowed). (**c**) Nummulitic wackestone with coral fragments showing a vug filled by calcite, dickite (di) and bitumen. Note the irregular, corroded margins of the calcite crystals (arrowed), indicating that a phase of dissolution occurred prior to the precipitation of dickite. (**d**) *Coskinolina* packstone with matrix and intraparticle microporosity and vugs. Vugs are locally open (arrowed), but elsewhere are occluded by dickite (di). (**e**) *Coskinolina* packstone with well-developed matrix microporosity locally grading into small vugs, some of which are filled by dickite (di) while others remain open. (**f**) Vugs cross-cutting a coral fragment. The vugs are partially lined by saddle dolomite (arrowed).

are characteristic of both the Bassein A and Bassein B (i.e. the same pore types occur above and below the palaeokarst at the Eocene–Oligocene boundary).

At a larger scale, stylolites and associated fractures or tensions gashes occur, both of which show evidence of having been solution-enlarged (Fig. 7). The stylolite-associated fractures are generally normal or at a high angle to the stylolites, and are <10 cm in length. They are commonly heavily corroded and locally lined with non-ferroan calcite or saddle dolomite cements, which also show evidence of corrosion. Dickite-filled fractures are also common. En echelon fractures are rare, as are fractures that are not associated with stylolites. Although not observed in core owing to poor recovery, borehole image logs show the very local presence of thin (<1 m), heavily leached rubbly zones that presumably represent a more well-developed end member of the millimetre- to centimetre-scale vugs seen in thin section and core (cf. Chandra *et al.* 2015*a*, *b*).

At the coarsest scale, the stratigraphy of the entire field, fault-related collapse breccia pipes occur, which extend from the top to the base of the reservoir section – a vertical distance of *c.* 350 m. The collapse breccia pipes only seem to occur in those parts of the field where the Panna Formation pinches out against the Deccan Formation (Fig. 8). It should be noted that, similar to the finer-scale porosity, the larger pore types occur either side of the Eocene–Oligocene unconformity, suggesting that there is no genetic link between the unconformity and the majority of porosity generation. Furthermore, the preferential occurrence of the collapse breccia pipes above the pinch out of the Panna Formation suggests that their development was controlled by the underlying structure and stratigraphic architecture, and not by an overlying karstic surface.

Visible secondary matrix porosity and stylolite density

A qualitative examination of >1200 m of core suggested that there is an association between zones of higher stylolite density and secondary matrix

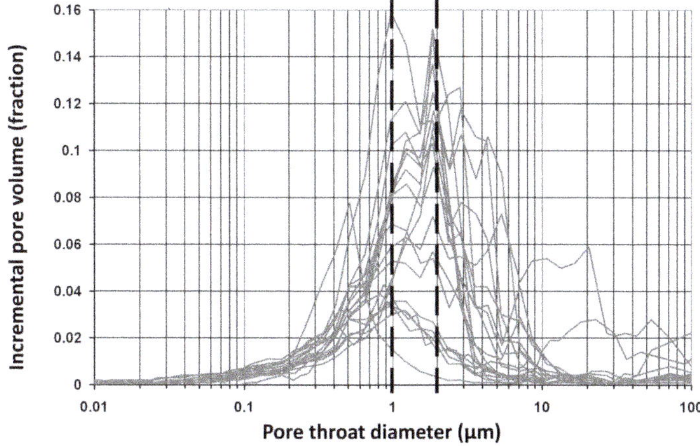

Reservoir	*N*	*Min Microporosity (%)*	*Max Microporosity (%)*	*Mean Microporosity (%)*
Bassein B *(1 µm cut-off)*	22	26.0	90.6	47.1
Bassein B *(2 µm cut-off)*	22	41.3	94.4	69.9

Fig. 6. Mercury porosimetry plot showing the abundance of microporosity. The table gives minimum, maximum and mean microporosity values using both a 1 µm (after Cantrell & Hagerty 1999) and 2 µm (after Moshier 1989) pore-throat diameter cut-off to define microporosity.

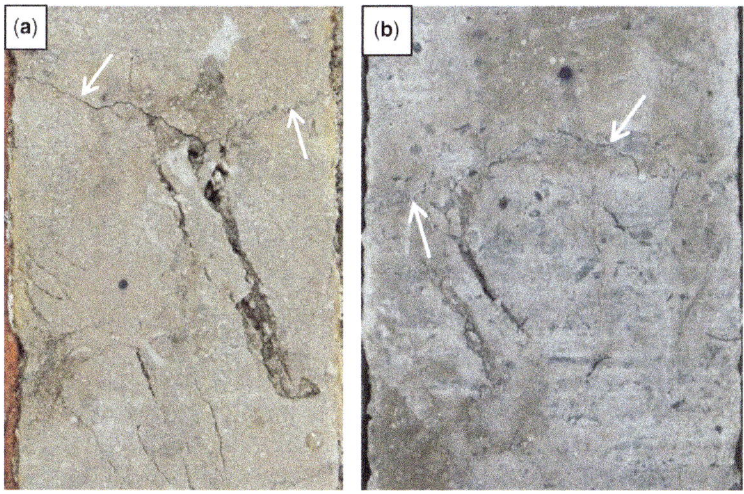

Fig. 7. Core photographs showing stylolites (arrowed) and associated solution-enlarged tension gashes and fractures. (**a**) An example from the Bassein A. (**b**) An example from the Bassein B. As in Figure 5, note that a similar pattern of porosity development occurs both above and below the Eocene–Oligocene unconformity. The core pieces in each photograph are 10 cm wide.

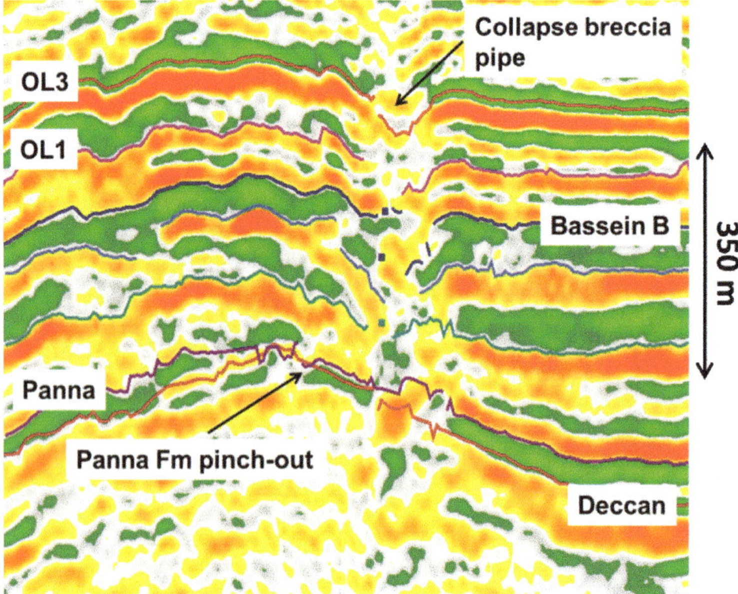

Fig. 8. Seismic section showing a collapse breccia pipe. A number of these features occur on the Panna Field and their occurrence is apparently limited to those areas where the Panna Formation pinches out onto Deccan Traps. Note that the collapse breccia pipes cross-cut the Eocene–Oligocene unconformity. OL1, top Bassein A; OL3, top Alternations/base Alibag Shale.

macroporosity (e.g. mouldic pores and vugs). This association is especially clear in the Mukta Field and in the oil rim of the Panna Field, where clusters of stylolites are associated with zones of heavier oil staining. Note that this pattern is quite distinct from the association of stylolites and tensions gashes/fractures shown in Figures 6 and 8, and normally occurs as centimetre- to decimetre-thick horizons of mouldic pores and small vugs immediately adjacent to swarms of stylolites. The apparent positive correlation between stylolite density and secondary matrix macroporosity has been further examined by measuring the proportion of rock with visible secondary matrix porosity and the number of stylolites per metre from two cored Panna wells. Both cores come from the upper part of the Bassein B immediately below the Eocene–Oligocene unconformity. The resulting cross-plot shows a positive correlation between zones of increased stylolite density and zones of increased secondary matrix macroporosity development (cf. Chandra *et al.* 2015*a*, *b*). To test the robustness of this correlation, a Spearman's rank correlation coefficient test was performed. For both wells, calculated r' values exceed critical values and the null hypothesis is rejected (i.e. there is a positive correlation between secondary matrix macroporosity and stylolite density). For the red well, the calculated r' value is 0.57, whereas the critical value (r' 0.05, 15) is 0.44. For the green well, the calculated r' value is 0.84, while the critical value (r' 0.05, 11) is 0.53.

Discussion

Paragenesis and the timing of secondary porosity formation

The principal diagenetic phases are shown in Figure 9 and illustrated in Figure 10. The presence of burrow-related, nodular fabrics (Ricken 1986) suggests that extensive early cementation took place during shallow-marine burial. Interparticle and intraparticle porosity was extensively occluded by non-ferroan calcite spar during this phase of diagenesis. These calcite cements clearly predate any major corrosion and record progressively higher temperature fluids, with progressive burial from 70°C to >130°C based on the study of fluid inclusions from successive cement generations (samples A–C in Table 3). Ongoing and subsequent burial led to extensive compaction, pressure solution and porosity loss. Stylolites and solution seams developed, enhancing earlier formed nodular fabrics. Many stylolites developed fractures and tension gashes. This was followed by a major phase of dissolution.

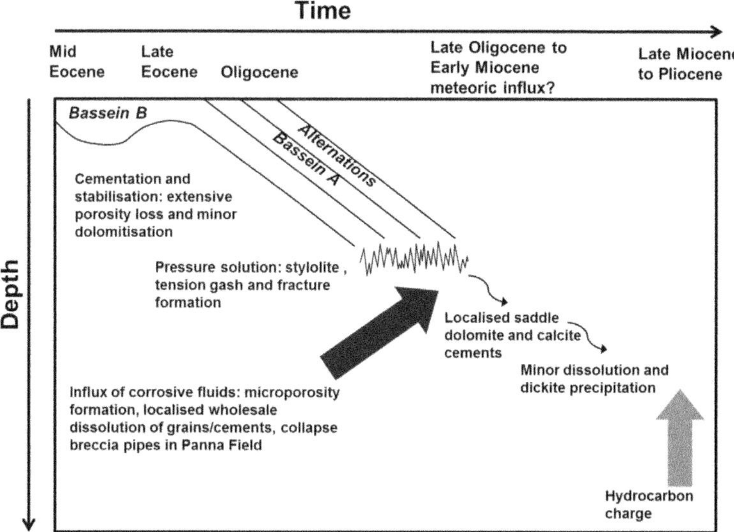

Fig. 9. Paragenetic sequence for the Bassein Formation in the Panna and Mukta fields.

The general criteria for recognizing mesogenetic dissolution have been summarized by Mazullo & Harris (1992) and include the following:

- dissolution of saddle dolomite and other late cements (especially in fractures);
- dissolution of cements with hydrocarbon inclusions;
- dissolution along stylolites and solution-enlargement of stylolite-related fractures;
- dissolution of compacted grains;
- an association with dickite cements – dickite is a high-temperature (>90°C) polymorph of kaolinite, and is regarded as evidence that acidic, organic-rich fluids have affected the host rock, leaching Al and SiO_2 (Maliva *et al.* 1999);
- an association with metal sulphides.

Many of the above features are readily identified in the Bassein Limestone. Dissolution along stylolites and the solution-enlargement of associated fractures is observed both in thin section and in core (Figs 7 & 10). Solution-enlarged stylolites and associated fractures cannot, therefore, be attributed to the plucking of rock material during the thin-section preparation process (cf. Ehrenberg *et al.* 2012). Dissolution of vug- and fracture-filling saddle dolomite and late non-ferroan calcite occurs. As already noted, pre-corrosion calcite cements have homogenization temperatures of 70–130°C (Table 3; Fig. 10c). Saddle dolomite and late non-ferroan calcite are commonly associated with dickite, which represents up to 3.9% of some core plug samples based on FT-IR spectroscopy (Table 2; Fig. 10f). Dickite commonly 'sits' on etched dolomite and calcite crystal surfaces, indicating that it both post-dates those phases and their dissolution. Metal sulphides occur in the form of pyrite, which is commonly observed to partially replace saddle dolomite (Table 2; Fig. 10e). There is some ambiguity over the relative timing of some fractures and saddle dolomite precipitation. In some fractures, with apparently uncorroded margins, there is a distinct generation of bladed, uncorroded non-ferroan calcite, commonly encased in saddle dolomite that was later etched, locally partly covered with pyrite, followed by dickite precipitation. It is possible that these fractures were not affected by the main phase of corrosion but were subject to a later phase of corrosion prior to dickite precipitation. Alternatively, they could represent a second phase of fracturing after the main corrosion event. Saddle dolomite also occurs in fractures that were already clearly corroded before dolomite precipitation, and the dolomite was itself clearly affected by corrosion followed by dickite precipitation.

The role of top Eocene palaeokarst v. mesogenesis in porosity generation

The Eocene–Oligocene boundary in the Panna and Mukta fields is marked by a clear palaeokarstic surface, a biostratigraphically resolvable hiatus (*c.* 3.5 myr) and high-permeability conduits proven by production data. These observations could be interpreted to indicate a genetic link between all of the secondary porosity beneath this unconformity and dissolution related to subaerial exposure. Furthermore, the volumetrically dominant pore

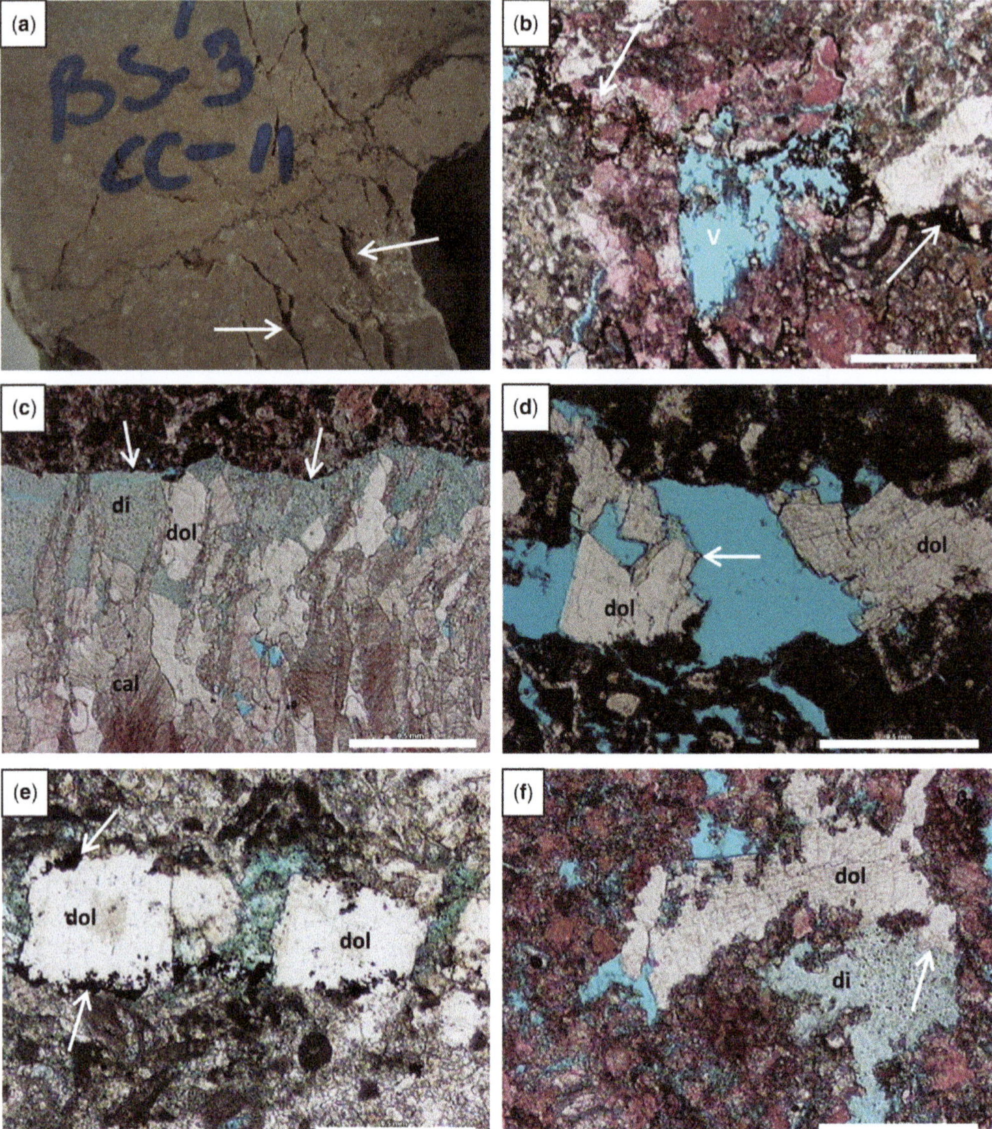

Fig. 10. Core photographs and thin-section photomicrographs showing evidence of late dissolution or burial corrosion. The scale bar for all of the photomicrographs is 1 mm. (**a**) Core photograph showing stylolites and associated solution-enlarged tension gashes (arrowed). Note the mouldic and vuggy porosity either side of the stylolite (top right-hand side). (**b**) Solution-enlarged stylolite (arrowed) and associated vug (v). (**c**) Fracture-filling bladed, non-ferroan calcite (cal: pink coloration) and saddle dolomite (dol). The margins of both the calcite and dolomite crystals are corroded, and the remaining pore space is occluded by dickite (di). Note the irregular, solution-enlarged fracture margin (arrowed). (**d**) Vug partially filled by saddle dolomite (dol). Note the etched margins of the dolomite crystals (arrowed). (**e**) Fracture-filling saddle dolomite (dol) partially replaced by pyrite (arrowed). (**f**) Vug-filling saddle dolomite (dol) and dickite (di). Note the irregular etched contact between the two phases.

types (matrix and intraparticle microporosity, mouldic macropores and vugs) are not specifically diagnostic of either eogenetic or mesogenetic dissolution. The evidence for some late dissolution is unequivocal (e.g. dissolution of saddle dolomite and other high-temperature carbonate cements, solution-enlargement of stylolites). However, this simply means that some, possibly very limited, late dissolution took place and does not exclude the possibility that the vast majority of the pore system was created in a near-surface environment.

Ehrenberg *et al.* (2012), in their critique of the concept of mesogenetic dissolution, quote Bob Loucks who, in a 2003 presentation, stated that 'what is in a hole might not have anything to do with the origin of the hole'. That criticism can be fairly applied to several of the criteria for recognizing mesogenetic dissolution cited by Mazullo & Harris (1992). If a crystal of saddle dolomite, for example, occurs in a vug and the saddle dolomite itself is corroded, it does not necessarily follow that the host vug was produced by the same phase of dissolution that etched the saddle dolomite. The vug may owe its origin to eogenetic or mesogenetic dissolution or to a combination of the two.

Based on several lines of evidence, we conclude that it is improbable that meteoric diagenesis at the Eocene–Oligocene unconformity is the dominant control on porosity development in the Bassein Limestone. First, very similar pore types at a variety of scales occur both above and below this unconformity (Figs 5, 7 & 8). It could be argued that collapse breccia pipes are simply the products of upwards propagation of collapse related to the Eocene–Oligocene karstic event. Upwards propagation from underlying karst systems of up to 760 m has been documented by Loucks (1999) and McDonnell *et al.* (2007). However, this would not explain why the collapse breccia pipes only occur where the Panna Formation pinches out against the Deccan Formation, suggesting that the fluids responsible for their formation were derived from the basin and not from the near-surface environment. Furthermore, it is noticeable that the fill of the pipes is chaotic (i.e. reflectors cannot easily be traced across the pipe) in the reservoir section (up to OL3 on Fig. 8), but take on more of a sag appearance with traceable reflectors in the Alibag Formation. This sag-like morphology is similar to the upwards propagation of underlying palaeokarst described by Loucks (1999) and McDonnell *et al.* (2007). Secondly, contrary to what we might normally expect, zones of increased stylolite density correlate with zones of increased secondary macroporosity (Fig. 11). It is difficult to explain this pattern of porosity development without invoking the movement of corrosive fluids along reopened stylolites and their associated fracture systems. It follows that the associated matrix porosity is, at least, post-stylolitic in age (i.e. porosity was generated after the cessation of active pressure dissolution). The only other way that this relationship could develop is if the Bassein Limestone was buried sufficiently to develop both the stylolites and the late cements described above, and then uplifted sufficiently to be leached by surface-related meteoric fluids. A phase of late Miocene inversion is known from the Bombay Basin. However, a telogenetic origin for the secondary porosity remains unlikely because, regardless of the amount of erosion caused by this erosion event, the Bassein Limestone is still overlain by 1–1.5 km of late Oligocene–mid-Miocene deposits, mainly marine mudrocks.

As in other limestone successions, stylolites (sutured-seam solution *sensu* Wanless 1979) in the Bassein Formation only develop in relatively 'clean' limestones lacking argillaceous material. This is the case with the Bassein B *Coskinolina*-dominated limestones and also with the deeper-water nummulitic facies in the Bassein A, which, although having been probably deposited in mid- to outer-ramp settings, are in fact relatively clean in lacking an argillaceous matrix. The more argillaceous limestones, such as those in the Tight Zones in the Bassein B, are characterized by solution seams or 'horsetails' (non-sutured-seam solution *sensu* Wanless 1979) rather than stylolites. For the Bassein B, the cleaner facies occur preferentially in the upper part of upwards-shallowing sequences: the more argillaceous facies in the lower part of the sequence (Fig. 12). Because stylolite density is correlated with secondary porosity development (Fig. 11), there is a facies control on reservoir quality, with the solution-seam-dominated intervals forming the lower-quality rock or even the Tight Zones within the Bassein B. The stylolite-dominated intervals are characterized by better secondary porosity development and, therefore, better reservoir quality as virtually all of the porosity is secondary. As a result of this vertical organization of facies, porosity distribution in the Bassein B mimics that expected where porosity development is eogenetic, unconformity-related and occurring in association with a shallowing-upwards unit, when, in fact, a type of mechanical stratigraphy that governs stylolite v. solution-seam development controls porosity distribution.

Mechanism of dissolution

Based on the above, a viable mechanism of dissolution for the creation of porosity in the Panna and Mukta fields needs to explain:

- The reopening (or, at least, the transmissibility) of stylolites to allow the influx of corrosive fluids.
- The origin and characteristics of the fluid(s) that could generate a widespread mean porosity of approximately 16% in the Bassein B Upper and *c.* 10% in the Bassein A.

Following deposition, the Bassein Limestone underwent extensive porosity loss by calcite

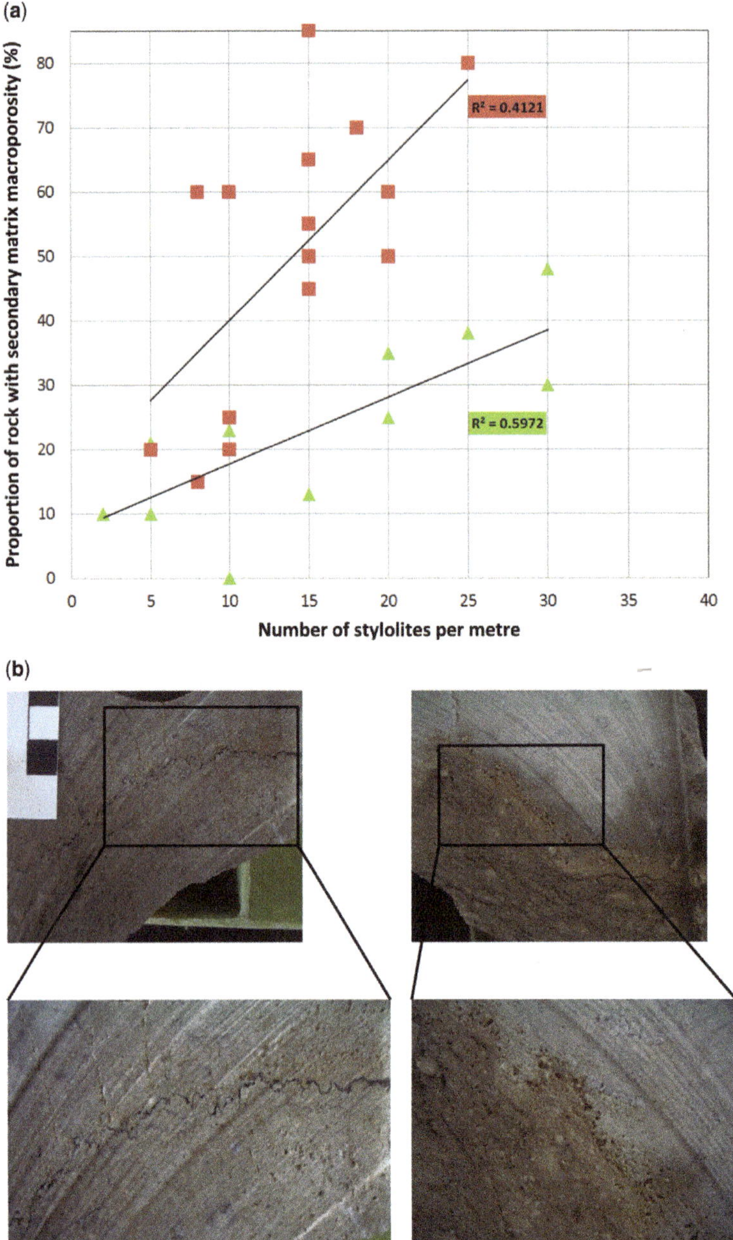

Fig. 11. (**a**) Cross-plot showing the relationship between visible secondary porosity and stylolite density from two Panna wells. Red, well A; green, well B. (**b**) Core photograph showing the association of stylolites with secondary macroporosity. Note the preferential development of visible mouldic and vuggy macroporosity in the immediate vicinity of the stylolite planes.

cementation and pressure dissolution, including the development of stylolites and solution seams. Pre-corrosion calcite cements record progressively higher temperature fluids from 70°C to >130°C based on the study of fluid inclusions (Table 3). This cementation and compaction probably would have reduced depositional porosities of about 40% (Moore 2001) to as little as 5–10%. It is likely

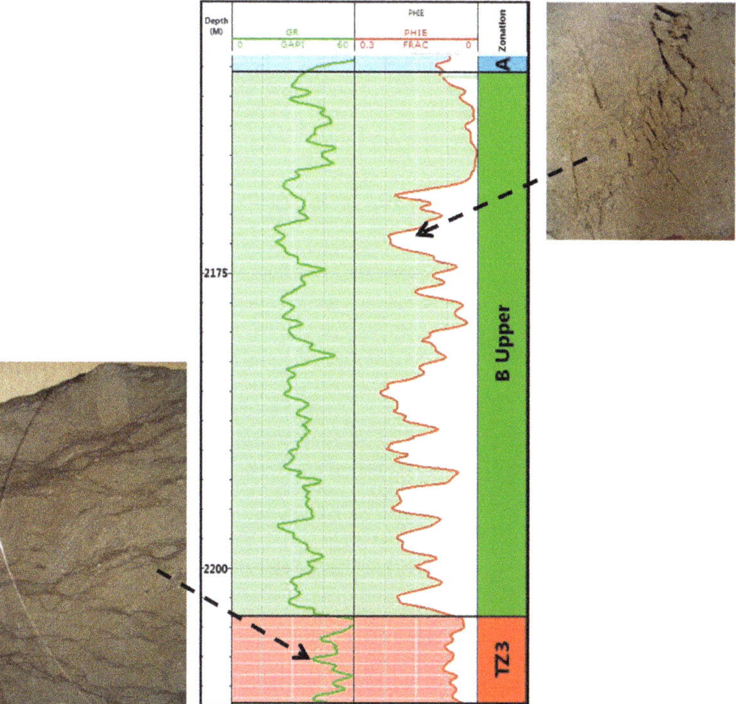

Fig. 12. The control of sedimentary facies on stylolite development and the generation of secondary porosity. From left to right, the log tracks are measured depth in metres, gamma ray (GR), effective porosity (PHIE) and reservoir zone. Porosity is significantly higher in the 'clean', clay-free limestones compared to argillaceous limestones such as those in the Tight Zones and the lowermost part of the Bassein A. This appears to be due to the fact that clay-free limestones develop stylolites that are associated with secondary porosity development. In contrast, argillaceous limestones develop dissolution seams that are not associated with secondary porosity enhancement (see the text for more details).

that the more argillaceous limestones of the Tight Zones experienced even more severe porosity destruction, such that they were not enhanced by later dissolution (i.e. they were too tight to allow the ingress of corrosive fluids). Furthermore, they lack stylolites, the apparently preferred pathway for fluid movement. In order to reopen stylolites and other conduits for fluid flow, this phase of progressive porosity loss must have been followed by a phase of uplift and unloading. Unloading of stylolites and associated fractures and tension gashes has been documented by Nelson (1981) from the Tain Creek Limestone of the western Wyoming thrust belt. In this example, all tension gashes and unloading fractures are now filled with late-stage calcite cement testifying to the fact that they were temporarily open to fluid flow. An episode of post-mid Miocene transpression has been documented in the Bombay Basin (Basu *et al.* 1982) and the uplift associated with this event could have caused unloading. However, it should also be noted that recent laboratory studies (Heap *et al.* 2014) demonstrate that Jurassic limestone samples with stylolites orientated parallel to the direction of fluid flow exhibit permeabilities that are an order of magnitude higher than the porosity–permeability relationships in similar samples lacking stylolites (cf. fig. 2 of Heap *et al.* 2014). While the findings of Heap *et al.* (2014) are contrary to the widely held view that stylolites act as permeability barriers, the result of these authors may explain the positive correlation between stylolite density and secondary matrix porosity observed in the Bassein Limestone (Fig. 11).

In terms of the character of the fluids responsible, the following processes are commonly invoked to explain mesogenetic dissolution (Giles & Marshall 1986; Giles & de Boer 1989; Esteban & Taberner 2003; Ehrenberg *et al.* 2012):

- CO_2 and organic acids produced during thermal maturation of organic matter;
- H_2S created by the maturation of sulphur-rich source rocks;

- mixing corrosion – a process involving the mixing of two solutions that are both in equilibrium with a given mineral but are of different compositions: the resulting solution can be either under- or oversaturated with respect to the mineral in question and, consequently, this process could result in dissolution or cementation;
- H_2S created by thermochemical sulphate reduction (TSR) in evaporite-bearing carbonate successions;
- thermal convection – owing to the retrograde solubility of carbonate minerals, ascending warm water will be undersaturated when it enters a cooler reservoir and could cause dissolution.

Several of these mechanisms can probably be immediately eliminated based on simple geological considerations. Large-scale carbonate dissolution can occur by the oxidation of H_2S, as has occurred at Carlsbad Caverns (Hill 1995). However, as Ehrenberg *et al.* (2012) pointed out, this particular mechanism requires an abundance of oxygen, a condition that is not generally satisfied in the deeper subsurface. TSR is probably only capable of producing extremely localized dissolution (Machel 2001) and could not explain the widespread secondary porosity generation that we observe in the Bassein A and B, although some authors do attribute the enlargement of previously formed secondary porosity to this process (cf. Ma *et al.* 2008). In any case, there is a complete lack of evaporite sulphates in the studied succession and, hence, none of the normal indicators of this process (e.g. calcitization of anhydrite) are seen in core or thin section.

Dissolution by acidic fluids generated from CO_2 during thermal maturation of organic matter has the potential to produce some secondary porosity. The Bassein reservoirs are sourced from a Type III source rock that could, theoretically, generate a certain amount of CO_2 to produce corrosive fluids (Giles & De Boer 1989). The mass balance calculations of Giles & De Boer (1989) demonstrate that Type III source rocks would require a minimum organic carbon content of 5% to generate sufficient CO_2 to produce dissolution exterior to the source-rock interval, assuming that the inorganic composition of the source rock had only 30% of the reactive minerals of the typical shale. The mean total organic content (TOC) of the organic-rich shales that source the Bassein Limestone is 11.5% ($n = 7$; mean = 11.5; 0.3–60.1). Furthermore, coal horizons also occur locally in the Panna Formation ($n = 4$; mean = 31.9; 16.1–45.9) and they would be able to expel virtually all of the CO_2 they generate. Nevertheless, the majority of workers conclude that the amount of CO_2 generated by this process cannot account for the large volumes of secondary porosity (Giles & Marshall 1986; Taylor *et al.* 2010; Ehrenberg *et al.* 2012), such as is present in the Bassein Limestone, owing to the neutralization of the generated acids within the source rock (many source rocks contain considerably more carbonate than the 'typical' shale cited above: cf. table 2 in Giles & Marshall 1986) or along migration pathways. Indeed, isotopic evidence summarized by Giles & Marshall (1986) suggests that much of the CO_2 generated from organic matter is ultimately incorporated into carbonate concretions within the source rock itself. Furthermore, the produced gas from the Panna–Mukta fields is relatively low (<8%).

Carbonate minerals show retrograde solubility (i.e. the solubility of calcite falls with increasing temperature at a constant partial pressure of CO_2) and, as a result, warmer water moving upwards will be undersaturated when it enters a cooler reservoir and could, therefore, cause dissolution (Giles & de Boer 1989). As already noted, the observation that the collapse breccia pipes form above the onlap point and pinch out of the Panna Formation suggests that the dissolution that created these features at least was produced by ascending basinal fluids.

Mixing corrosion has been proposed as a mechanism for burial corrosion (Esteban & Taberner 2003), whereby two fluids, potentially saturated with respect to the calcite mix and because of the non-linear solubility curve for calcite, produce a fluid that can be undersaturated with respect to calcite. In terms of mixing corrosion, the low salinities of the formation water of the Bassein Limestone raises the likelihood of some later ingress of non-connate fluids, the most likely source being the onshore Deccan Formation, 60 km to the east of the Panna Field. It is also notable that the formation water of the Bassein Limestone has very low salinity. Immediately west of the Panna East Fault, a salinity of *c.*11 000 ppm is observed in the B Upper, decreasing to *c.* 8000 ppm in the B Middle and to *c.* 3000 ppm in the Panna Formation (Fig. 1). The most likely source of this fresh- to brackish water is meteoric water that originally percolated through the Deccan Formation which forms a highly fractured aquifer onshore, 60 km to the east of the Panna Field. This conclusion is supported not only by an upsection increase in salinity from the Panna Formation to the top of the Bassein B but also by a westwards (i.e. offshore) increase in salinity within both formations. The salinity of the formation water in the Mukta Field is up to 8000 ppm in the Panna Formation and 20 000 ppm in the B Upper (Fig. 1). Based on this pattern, it seems reasonable to conclude that the current formation water of the Bassein Limestone was sourced from the Deccan Formation via the Panna

Formation. This fluid pathway would allow ample potential for the ingress of warmer, ascending fluids. The current reservoir temperature in the B Upper is 109–117°C at 1715 m TVDSS. The geothermal gradient across the Panna structure is 60°C km^{-1} and, as a result, water entering the Panna Formation and the Panna East Fault Zone on the downthrown side of this fault zone at depths of >2500 m TVDSS would be expected to have temperatures in excess of 160°C. In terms of the cooling formation water model of Giles & de Boer (1989), the decrease in temperature associated with the ascension of these waters could have considerable potential for dissolution. However, for this mechanism to be effective, the initial pH of the cooling fluid must be <6.9 (Giles & de Boer 1989). The pH of the water in the present-day Deccan aquifer is 7.0–7.5 (Lunkad & Raymahashay 1978) and there is no reasonable basis to assume that it was significantly lower in the late Miocene. It has already been noted that CO_2 and carboxylic acids generated during the maturation of organic matter are unlikely to generate significant volumes of secondary porosity. However, this mechanism does provide a source of weak acids that could shift the pH of the formation water towards an acidic value and into the range in which cooling would result in significant undersaturation (Giles & de Boer 1989). Furthermore, the common occurrence of dickite in the Bassein Limestone is clear evidence of the past passage of acidic, organic-rich fluids (Maliva *et al.* 1999).

Conclusions

Despite the ongoing controversy as to the exact mechanism by which burial dissolution proceeds, the present study strongly supports the possibility of major (>10%) secondary porosity generation by mesogenetic processes. The occurrence of similar secondary pores, including the collapse breccia pipes, both above and below the Eocene–Oligocene unconformity and the positive correlation between secondary matrix porosity and stylolite density is impossible to explain by an eogenetic mechanism. A telogenetic mechanism (i.e. stylolitization and burial cementation followed by sufficient uplift for leaching by surface-related meteoric fluids) is equally unlikely because, although a phase of late Miocene inversion is known to have occurred in the Bombay Basin, the Bassein Limestone is overlain by 1–1.5 km of late Oligocene–mid-Miocene strata.

Nevertheless, it is clear, as highlighted by Ehrenberg *et al.* (2012), that identifying a viable mechanism by which large-scale mesogenetic dissolution can occur remains problematic, and a number of key questions remain unresolved: Where, for example, does all the resulting dissolved calcite go in the subsurface?

The case study presented herein facilitates the recognition of several criteria for discriminating between secondary porosity generation by eogenetic and telogenetic processes v. mesogenetic processes:

- Secondary porosity distribution cross-cuts unconformities and/or biostratigraphically resolvable hiatuses, and is more readily related to structural features.
- A positive correlation exists between secondary porosity distribution and stylolites or other demonstrably late discontinuities.
- Subsidence history precludes the possibility of telogenetic porosity enhancement.

The authors would like to thank reviewers Benjamin Brigaud and Yasin Makhloufi, and editors Peter Armitage and Richard Worden, for their comments, which greatly improved the final manuscript. BG India is thanked for logistical support and for granting permission to publish.

References

Ahr, W.M. 2008. *Geology of Carbonate Reservoirs: The Identification, Description and Characterization of Hydrocarbon Reservoirs in Carbonate Rocks.* John Wiley, Chichester.

Basu, N.D., Banerjee, A. & Tamhane, D.M. 1982. Facies distribution and petroleum geology of the Bombay offshore basin, India. *Journal of Petroleum Geology*, **5**, 51–75.

Beavington-Penney, S.J. & Racey, A. 2004. Ecology of extant nummulitids and other larger benthic foraminifera: applications in palaeoenvironmental analysis. *Earth-Science Reviews*, **67**, 219–265.

Buxton, M.W.N. & Pedley, H.M. 1989. A standardised model for Tethyan Tertiary carbonate ramps. *Journal of the Geological Society, London*, **146**, 746–748, https://doi.org/10.1144/gsjgs.146.5.0746

Cantrell, D.L. & Hagerty, R.M. 1999. Microporosity in Arab Formation carbonates, Saudi Arabia. *GeoArabia*, **2**, 129–154.

Chandra, V., Barnett, A.J., Corbett, P., Geiger, S., Wright, V.P., Steele, R. & Milroy, P.G. 2015*a*. Effective integration of reservoir rock-typing and simulation using near-wellbore upscaling. *Marine and Petroleum Geology*, **67**, 307–326.

Chandra, V., Wright, V.P. *et al.* 2015*b*. Evaluating the impact of a late-burial corrosion model on reservoir permeability and performance in a mature carbonate field using near-wellbore upscaling. *In*: Agar, S.M. & Geiger, S. (eds) *Fundamental Controls on Fluid Flow in Carbonates*. Geological Society, London, Special Publications, **406**, 427–445, https://doi.org/10.1144/SP406.11

Ehrenberg, S.N., Walderhaug, O. & Bjørlykke, K. 2012. Carbonate porosity creation by mesogenetic dissolution: reality or illusion? *American Association of Petroleum Geologists Bulletin*, **96**, 217–233.

Esteban, M. & Taberner, C. 2000. *The Origin of Porosity in the Panna-Mukta Reservoirs, Bombay Offshore,*

India. Phase 2 Confidential Report for Enron Oil & Gas India.

Esteban, M. & Taberner, C. 2003. Secondary porosity development during late burial in carbonate reservoirs as a result of mixing and/or cooling of brines. *Journal of Geochemical Exploration*, **78–79**, 355–359.

Giles, M.R. & de Boer, R.B. 1989. Secondary porosity: creation of enhanced porosities in the subsurface from the dissolution of carbonate cements as a result of cooling formation waters. *Marine and Petroleum Geology*, **6**, 261–269.

Giles, M.R. & Marshall, J.D. 1986. Constraints on the development of secondary porosity in the subsurface: re-evaluation of processes. *Marine and Petroleum Geology*, **3**, 243–255.

Heap, M.J., Baud, P., Reuschlé, T. & Meredith, P.G. 2014. Stylolites in limestones: barriers to fluid flow. *Geology*, **42**, 51–54.

Herron, M.M., Matteson, A. & Gustavson, G. 1997. Dual-range FT-IR mineralogy and the analysis of sedimentary formations. *SCA paper*, **9729**.

Hill, C.A. 1995. H_2S-related porosity and sulfuric acid oilfield karst. *In*: Budd, D.A., Saller, A.H. & Harris, P.M. (eds) *Unconformities in Carbonate Strata-Their Recognition and the Significance of Associated Porosity*. American Association of Petroleum Geologists, Memoirs, **61**, 301–306.

Loucks, R.G. 1999. Paleocave carbonate reservoirs: origins, burial-depth Modifications, spatial complexity and reservoir implications. *American Association of Petroleum Geologists Bulletin*, **83**, 1795–1834.

Lunkad, S.K. & Raymahashay, B.C. 1978. Ground water quality in weathered Deccan Basalt of Malwa Plateau, India. *Quarterly Journal of Engineering Geology and Hydrogeology*, **11**, 273–277, https://doi.org/10.1144/GSL.QJEG.1978.011.04.01

Ma, Y., Zhang, S., Guo, T., Zhu, G., Cai, X. & Li, M. 2008. Petroleum Geology of the Puguang sour gas field in the Sichuan Basin, SW China. *Marine and Petroleum Geology*, **25**, 357–370.

Machel, H.G. 2001. Bacterial and thermochemical sulphate reduction in diagenetic settings – old and new insights. *Sedimentary Geology*, **140**, 143–175.

Maliva, R.G., Dickson, J.A.D. & Fallick, A.E. 1999. Kaolin cements in limestones: potential indicators of organic-rich pore waters during diagenesis. *Journal of Sedimentary Research*, **69**, 158–163.

Mazullo, S.J. & Harris, P.M. 1992. Mesogenetic dissolution: its role in porosity development in carbonate reservoirs. *American Association of Petroleum Geologists Bulletin*, **76**, 607–620.

McDonnell, A., Loucks, R.G. & Dooley, T. 2007. Quantifying the origin and geometry of circular sag structures in northern Fort Worth Basin, Texas: paleocave collapse, pull-apart fault systems, or hydrothermal alteration? *American Association of Petroleum Geologists Bulletin*, **91**, 1295–1318.

Moore, C.H. 2001. *Carbonate Reservoirs: Porosity Evolution and Diagenesis in a Sequence Stratigraphic Framework*. Developments in Sedimentology, **55**. Elsevier, Amsterdam.

Moshier, S.O. 1989. Microporosity in micritic limestones: a review. *Sedimentary Geology*, **63**, 191–213.

Nelson, R.A. 1981. Significance of fracture sets associated with stylolite zones. *American Association of Petroleum Geologists Bulletin*, **65**, 2417–2425.

Racey, A. 1994. Biostratigraphy and palaeobiogeographic significance of Tertiary nummulitids (foraminifera) from northern Oman. *In*: Simmons, M.D. (ed.) *Micropalaeontology and Hydrocarbon Exploration in the Middle East*. Chapman & Hall, London, 343–370.

Ricken, W. 1986. *Diagenetic Bedding*. Lecture Notes in Earth Sciences, **6**. Springer, Berlin.

Taylor, T.R., Giles, M.R. *et al.* 2010. Sandstone diagenesis and reservoir quality prediction: models, myths and reality. *American Association of Petroleum Geologists Bulletin*, **94**, 1093–1132.

Tucker, M.E. & Wright, V.P. 1990. *Carbonate Sedimentology*. Blackwell Science, Oxford.

Wanless, H.R. 1979. Limestone response to stress: pressure solution and dolomitisation. *Journal of Sedimentary Petrology*, **49**, 437–462.

Wright, V.P. & Barnett, A.J. 2011. Burial corrosion and porosity formation: from the seismic to micropore scale, but what processes do we blame? *In*: Engel, A.S., Engel, S.A., Moore, P.J. & DuChene, H. (eds) *Carbonate Geochemistry: Reactions and Processes in Aquifers and Reservoirs*. Karst Waters Institute, Special Publications, **16**, 81–83.

Wright, V.P. & Harris, P.M. 2013. Carbonate dissolution and porosity development in the burial (mesogenetic) environment. *Search and Discovery Article 50860. AAPG Annual Convention and Exhibition*, 19–22 May 2013, Pittsburgh, PA, http://www.searchanddiscovery.com/documents/2013/50860wright/ndx_wright.pdf

Role of facies diversity and cyclicity on the reservoir quality of the mid-Cretaceous Mishrif Formation in the southern Mesopotamian Basin, Iraq

THAMER A. MAHDI[1]* & ADNAN A. M. AQRAWI[2]

[1]*Department of Earth Science, College of Science, University of Baghdad, Al-Jadria, Baghdad, Iraq*

[2]*Statoil ASA, Forusbeen 50, Stavanger N-4035, Norway*

**Correspondence: tamgeology@gmail.com*

Abstract: An integrated sedimentological and petrophysical approach was implemented to define the role of facies diversity and cyclicity on the reservoir quality of the Mishrif Formation in several oil fields in southern Iraq. The reservoir quality in most regressive cycles was enhanced upwards from deep-marine facies towards the shallower shelf-margin facies. The change in reservoir quality could be detected in the facies stacked systematically within the regressive cycles, which was also easily recognized using the porosity logs. The impact of early diagenetic overprints was quite obvious in developing both reservoir and non-reservoir rock types within the Mishrif Formation in the study area. A simple rock-typing nomenclature was proposed based on the available data in order to classify the existing reservoir (R) and non-reservoir (S) rock types.

The best-recognized reservoir rock types were rudistid microfacies with grain-dominated fabrics (including both grainstone (R1) and grain-dominated packstone (R2)), which were subjected to an early diagenetic dissolution process, usually located beneath discontinuity surfaces. Such reservoir units or rock types have a regional extent within the southern Mesopotamian Basin, as they have often developed during the Mishrif shelf-margin progradation. In addition, the other important reservoir rock type was a microbialite (i.e. peloidal mud-dominated packstone (R3)), which was additionally characterized by micropores within the mud-dominated portion of the facies. However, owing to the variable intensity of the diagenetic effects and differences in the depositional texture components, the reservoir quality in this rock type could vary regionally.

The regional distribution of the rudistid grainstone and grain-dominated packstone reservoir rock types (R1 and R2) was mostly related to the palaeogeographical highs that existed during deposition. However, such reservoir rock types could pinch out within the depositional sequences, showing their potential to become stratigraphic traps outside the structural crest of the field. The delineation of the reservoir rock types within a sequence-stratigraphic framework can be quite beneficial for reservoir prediction and exploration within and outside of the field.

Introduction and general geology

The Cretaceous geological history of the eastern passive margin of the Arabian Plate was dominated by shallow-water carbonate sedimentation that included the Mishrif Formation carbonates deposited in the NE (Sharland *et al.* 2001; Ziegler 2001).

The oil fields of southern Iraq (i.e. the study area: Fig. 1) host almost two-thirds of the proven oil reserves of Iraq (Aqrawi *et al.* 2010*a*). The oil was usually entrapped in giant structures. For example, the Rumaila (north and south), together with West Qurna anticline, form a super-giant anticlinorium that hosts around 110 billion barrels of oil in place (Daly 2010). These reserves are mostly in the Cretaceous reservoirs, and particularly within the Zubair and Mishrif formations (Aqrawi *et al.* 2010*a*).

The carbonate reservoirs of the Mishrif Formation (Middle Cenomanian–Early Turonian) are part of the Albian–Turonian prolific petroleum system of the Arabian Plate (e.g. van Buchem *et al.* 2011). The carbonate reservoirs of this petroleum system were subjected to a spectrum of post-depositional diagenetic processes that increased their reservoir heterogeneity (Hollis 2011). Some of these diagenetic processes were broadly recognized in the Mishrif Formation facies, such as dissolution, cementation and compaction (Aqrawi *et al.* 1998; Lambert *et al.* 2006; Volery *et al.* 2009; Periere *et al.* 2011). In addition, it was also observed and reported previously by Aqrawi *et al.* (1998) that most of the porous and permeable rudistid facies are restricted to the shallowing-upwards cycles. However, such facies were not integrated with

From: Armitage, P. J., Butcher, A. R., Churchill, J. M., Csoma, A. E., Hollis, C., Lander, R. H., Omma, J. E. & Worden, R. H. (eds) 2018. *Reservoir Quality of Clastic and Carbonate Rocks: Analysis, Modelling and Prediction.* Geological Society, London, Special Publications, **435**, 85–105.
First published online February 22, 2017, https://doi.org/10.1144/SP435.19

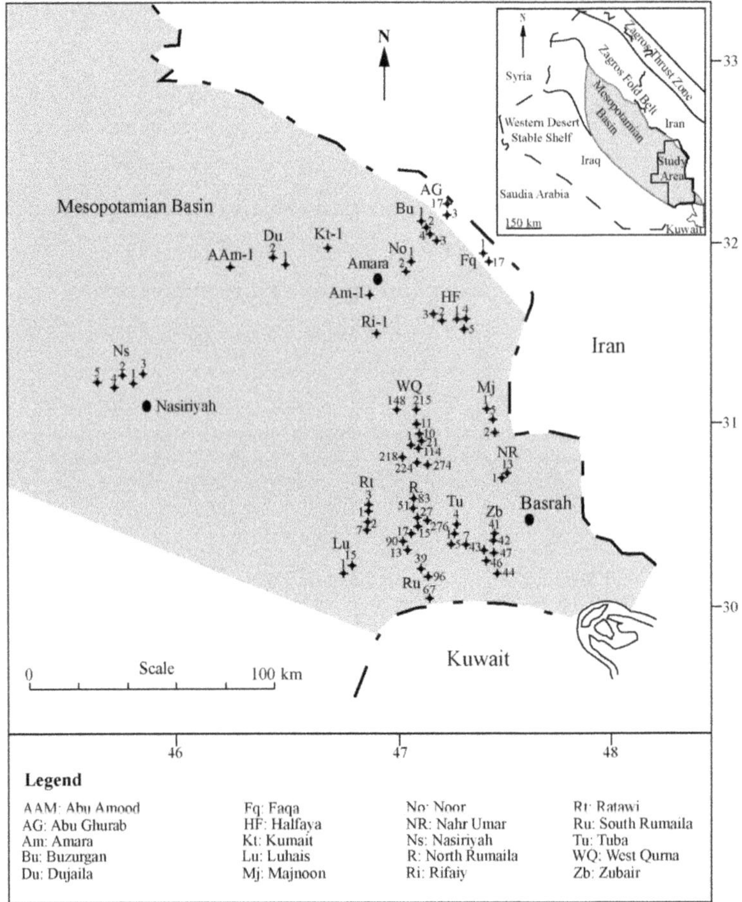

Fig. 1. Schematic maps of the southern Mesopotamian Basin illustrating the location of the major structural elements and the locations of the 36 studied wells (well and oil field names are cited in the legend) (modified from Mahdi & Aqrawi 2014).

diagenesis and petrophysical properties within the framework of each depositional cycle. Such data integration is considered as a prerequisite for the carbonate reservoir modelling and performance prediction (e.g. Kerans & Tinker 1997).

The Iraq South Oil Company (SOC) had subdivided the Mishrif Formation into three major units for the purpose of reservoir engineering and production activities: namely, A, BI and BII. These Mishrif units also exhibit distinct log patterns, and can be further subdivided into petrophysical units (Fig. 2). The latter units were used by Iraqi oil companies to correlate and map the continuity of the reservoir units across several oil fields by ignoring the vertical and lateral genetic facies changes that existed across unit boundaries. The petrophysical heterogeneity of these reservoir units is due to variations in porosity, water saturation and fluid volume (Fig. 2). Microfacies and petrographical analysis, combined with core porosity and permeability data, of units A, BI, and BII show that variations in reservoir quality also occur within a single unit, and are related to the heterogeneity of depositional facies and the effects of diagenetic overprints. An example of this relationship can be observed in unit A. The middle and lower parts of this unit are characterized by good reservoir quality, as shown by a high porosity (15–34%), and a large volume of both moveable and residual hydrocarbons (Fig. 2). This is due to the occurrence of microporous, vuggy wackestones and packstones of back-shoal and lagoonal environments. In contrast, the upper part has a poorer reservoir quality owing to the effect of the compaction and stylolitization of the tight tidal flat mudstones and wackestones (Fig. 2). Therefore, these facies display lower porosity values and have fewer

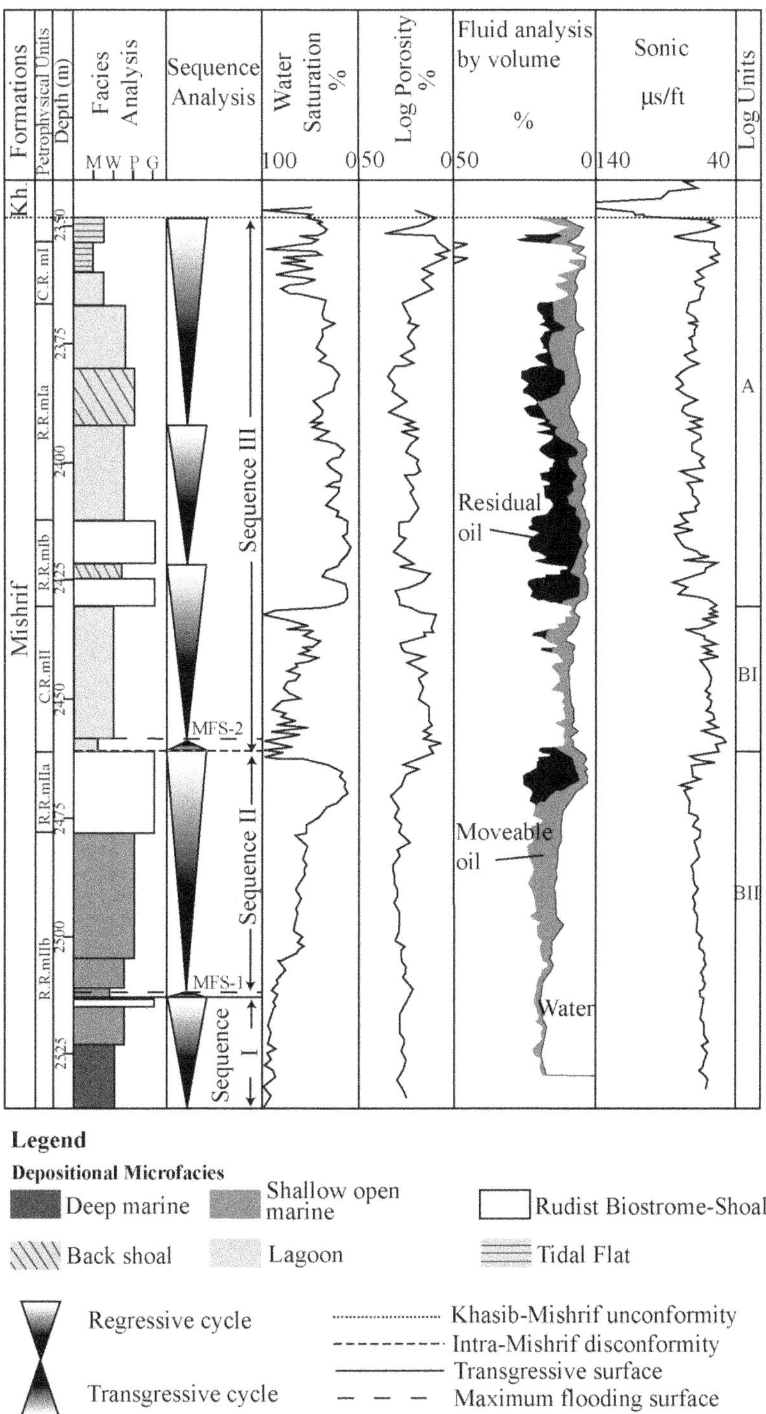

Fig. 2. Sedimentological composite log of the Mishrif Formation at well WQ-224, showing the recognized microfacies within each transgressive–regressive sequence. MFS, maximum flooding surface.

hydrocarbons. It is clear that a more robust rock-typing classification that integrates diagenesis, facies and petrophysical properties in a sequence-stratigraphic framework is needed to fully describe vertical and lateral heterogeneities. Therefore, the main purposes of this study are to:

- Characterize the reservoir units or main rock types of the Mishrif Formation in the southern Mesopotamian Basin, taking into account their depositional facies and the impact of diagenetic overprints within a sequence-stratigraphic framework. It has been reported previously by Aqrawi *et al.* (2010*b*) that a rock type is the final fabric of the depositional facies after having been subjected to diagenetic overprints.
- Show the vertical and lateral distribution of these reservoir units throughout the study area in order to reduce the risk of their prediction during future exploration activities.

Referring to the regional studies of Sharland *et al.* (2001) and Ziegler (2001), the study area within the Mesopotamian Basin is located in the NE of the Arabian Plate (Fig. 1). The southern Mesopotamian Basin is also a part of a foreland basin. The latter basin passes eastwards into the Zagros Fold and Thrust Belt, and also gives way westwards to the stable shelf of the western desert of Iraq (Fig. 1). The field structures in the study area are completely buried with no or limited surface expression on LandSat images. In addition, the seismic data show that these structures are gently dipping anticlines, with high angles to the Zagros Fold Belt (Yokoi & Sato 2004; Daly 2010). Regional studies of the southern Mesopotamian Basin indicated that some field structures in the Basra region appear to follow a north–south trend generated by Cambrian salt and basement complex activities of the Arabian Plate (e.g. Jassim & Goff 2006; Aqrawi *et al.* 2010*a*). Further regional tectonic activities within the study area included the late Cretaceous Alpine Orogeny, which was associated with the closure of the Neo-Tethys and thrust emplacement of the Hawasina and Semail nappes in the Oman Mountains (Sharland *et al.* 2001). This last event had some influence during the deposition of the Mishrif Formation and its equivalents along the eastern Arabian Plate (e.g. van Buchem *et al.* 2002; Al-Zaabi *et al.* 2010).

Data and methodology

Sedimentological and petrophysical data from 36 wells in southern Iraq were analysed and integrated (Fig. 1). The sedimentological data included logged and sampled cores at a 1–2 m scale. Available cores were originally slabbed in order to study and identify the large-scale features, such as macropores and rudists. More than 3000 standard thin sections (30 μm thick) were prepared from the collected samples of both cores and cuttings. The samples were selected whenever a change in porosity occurred along the well successions.

The petrographical analyses revealed some key information about the texture, fabric, pore types and diagenetic overprints in each facies. The depositional textures were classified according to the classification of Dunham (1962), and microfacies analysis and interpretation were based on the classification and concepts given in Flügel (1982), which were published recently by Mahdi *et al.* (2013). The thin sections were also impregnated with blue-stained epoxy in order to identify pore types visually under polarizing microscope.

The sequence-stratigraphic principles and scheme of the Mishrif Formation by Mahdi *et al.* (2013) and Mahdi & Aqrawi (2014) were also followed in this study. As a result, both microfacies and wireline log data were employed to delineate the transgressive–regressive cycles and their boundaries.

Helium porosity and air permeability measurements were made on 570 core plugs (of 1.5 × 2.5 inches) using a Core Lab Helium Porosimeter and Core Lab Micropermeameter. Mercury injection capillary pressure (MICP) curves were also obtained for 11 core samples in order to support the reservoir characterization results. The capillary pressure was measured directly from core plugs, which were mostly sampled every 1–0.3 m, and adequately represented distinct depositional textures and diagenetic fabrics. Different degrees of cementation and compaction are taken into consideration during sampling. The core plugs were selected horizontally parallel to the bedding planes.

Computer-processed interpretation (CPI) logs provided valuable data about the effective porosity (PHIE), water saturation, bulk volume oil (BVO) and bulk volume water (BVW) of the recognized reservoir units, and further assisted in characterizing them as reservoir and non-reservoir rock types.

The impact of diagenetic processes on the depositional facies textures has resulted in the delineation of various reservoir and non-reservoir (i.e. seal) rock types. A simple rock-typing nomenclature was proposed based on the available sedimentological and petrophysical data in order to classify the existing reservoir (R) and non-reservoir (S) rock types.

Depositional facies characterization within a sequence-stratigraphic framework

Facies associations of the Mishrif Formation consist of various types of grains, which are dominated by bioclasts, foraminifera, algae and rudist fragments (e.g. Mahdi *et al.* 2013). The detailed microfacies analysis revealed that such facies associations were deposited within a range of depositional settings from deep-marine to tidal-flat environments (e.g. Aqrawi *et al.* 1998; Mahdi *et al.* 2013) (Fig. 2). Deposition of the Mishrif Formation took place initially along a ramp that prograded into a low-angle carbonate shelf through time (Aqrawi *et al.* 1998; Sherwani 1998). In addition, some patchy rudist biostromes nucleated on the crestal parts of some giant fields, such as West Qurna, Rumaila and Zubair (Sadooni & Aqrawi 2000; Sadooni 2005).

Vertical stacking of the Mishrif Formation facies exhibits deposition in cyclic patterns reflecting periodic oscillations in sea level, which were due to transgression and regression phases (Figs 2 & 3). However, such phases were also affected by tectonic events within the study area (Mahdi *et al.* 2013). Such tectonic and sea-level change interactions were recorded in the form of regional discontinuity and maximum flooding surfaces recognized along the studied successions of the Mishrif Formation (Figs 2 & 3). Within the transgressive successions, the identified maximum flooding surfaces and transgressive surfaces can be correlated with the global Middle Cenomanian and Late Cenomanian–Early Turonian eustatic sea-level rises (e.g. Aqrawi *et al.* 2010*a*; Mahdi *et al.* 2013; Mahdi & Aqrawi 2014). However, exposure surfaces occur at the top of regressive cycles (Figs 2 & 3). Exposure surfaces resulted from the combination of both a Middle Cenomanian eustatic sea-level fall and a Middle Turonian inversion and uplift event (Sharland *et al.* 2001). In addition, a Late Cenomanian exposure surface was observed in some oil fields, such as Zubair and Rumaila (Fig. 3). More details about the sequence-stratigraphic framework of the Mishrif Formation in the southern Mesopotamian Basin can be found in a recent publication by Mahdi & Aqrawi (2014).

The rudistid facies

Rudists have a particular importance in hydrocarbon exploration in the Middle East, because large volumes of hydrocarbons have been produced from the rudist-bearing facies (e.g. Videtich *et al.* 1988; Al-Sharhan & Nairn 1997; Aqrawi *et al.* 2010*a*).

In the southern Mesopotamian Basin, data for characterizing the rudistid reservoir facies were mainly derived from core samples. An examination of the Mishrif rudists from these samples showed two distinct fabrics preserved: rudist shells embedded in a micritic matrix (Fig. 4a); and fragmented rudist shells and bioclasts with vuggy pores (Fig. 4b).

Rudists and their derived bioclasts form different depositional facies and textures based on their location across the shelf margin of the Mishrif carbonate platform (Mahdi *et al.* 2013), such as rudstone, rudistid-floatstone, packstone and grainstone (Fig. 5). The gradual transition of rudistid facies from mud-dominated (e.g. floatstone) to grain-dominated (e.g. grainstone) fabric might also reflect a gradual increase in water energy from the deeper slope towards the shallower shelf margin (Fig. 5). Under the effects of wave action, a large amount of rudist bioclasts were produced (Scott 1988), and their size gradually increased upwards within the identified regressive cycles of the Mishrif Formation (Mahdi *et al.* 2013). Such trends were also important when delineating the reservoir zones within such cycles, because these zones are topped by rudistid units (Mahdi & Aqrawi 2014). The reservoir units were probably subjected to strong vadose and meteoric dissolution, particularly along the upper cycle succession, where some karstified surfaces were recognized. The main karstified surfaces in the Mishrif Formation include the intra-Mishrif disconformity, the Late Cenomanian–Early Turonian boundary and the upper unconformable Mishrif contact (Fig. 2). Most porosity of the rudistid facies resulted from the dissolution of the aragonitic components of rudist shells. Owing to the dissolution of the rudist shells, the typical resultant pore type is moldic, which reflects some of the morphological features of rudist shells, such as accessory cavities, canals and cellular-prismatic microstructures (Fig. 5).

The lateral variation in rudistid facies deposited across the shallow carbonate platform towards the shelf margin displays a change from rudistid floatstone to grainstone (Fig. 5). This lateral facies change was also associated with the later dissolution that enhanced the vuggy porosity of these reservoir units (Fig. 5). In most studied successions, the same facies change also displays a general increase in the core porosity and permeability values upwards towards the rudistid packstone–grainstone facies within each regressive cycle (Figs 2 & 3). This trend might also support the positive relationship between enhanced reservoir properties and rudistid facies. Having defined the vertical distribution of rudistid facies of the Mishrif Formation, it was possible to predict their lateral distribution based on Walther's law (Middleton 1973), as shown in Figure 5. Such a vertical variation in reservoir properties implies the same lateral variation, and, thus, palaeogeographical effect on reservoir quality distribution from shelf margin to shallow open marine. The lateral changes in rudistid facies shown in Figure 5 are the result of a relative sea-level fall (Mahdi & Aqrawi 2014), which resulted in an increase in the wave action that consequently destroyed rudist

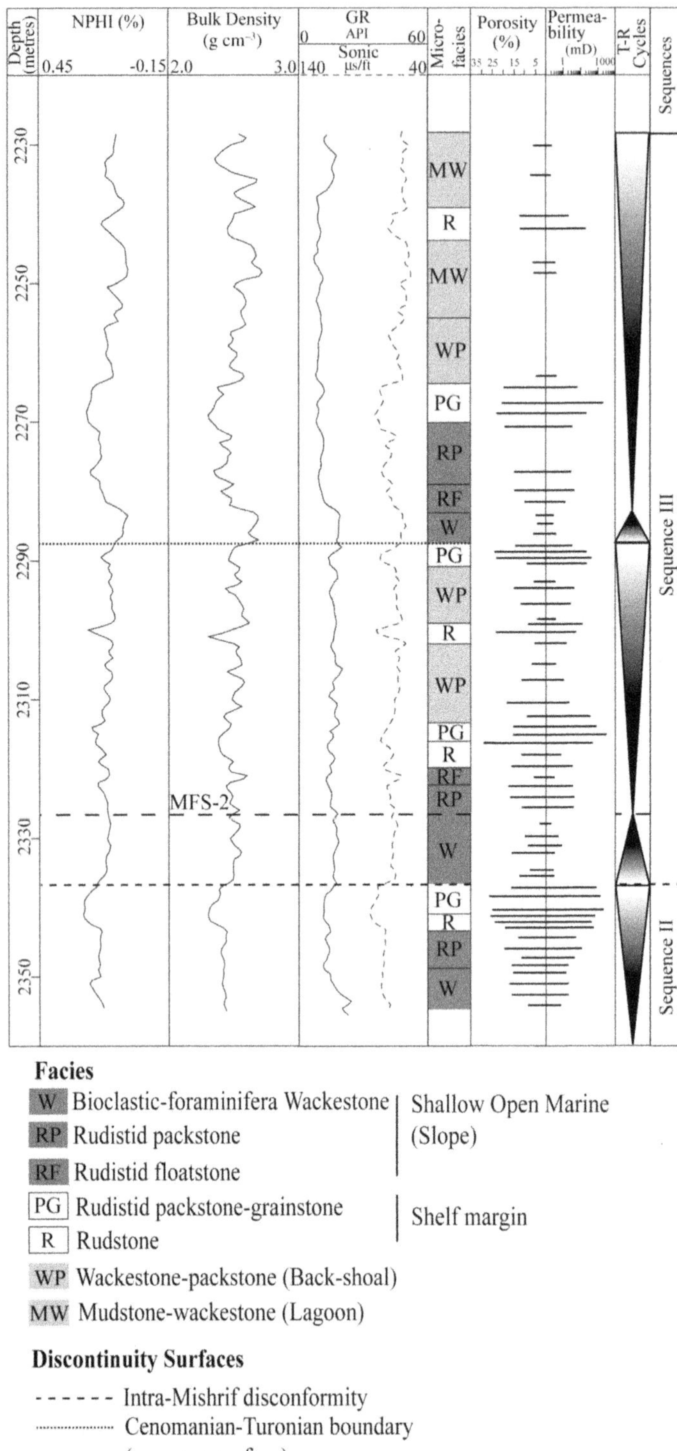

Fig. 3. Stratigraphic succession of the Mishrif Formation at well R-15. Measured core-plug porosity and permeability are also displayed for some facies. T–R, transgressive–regressive.

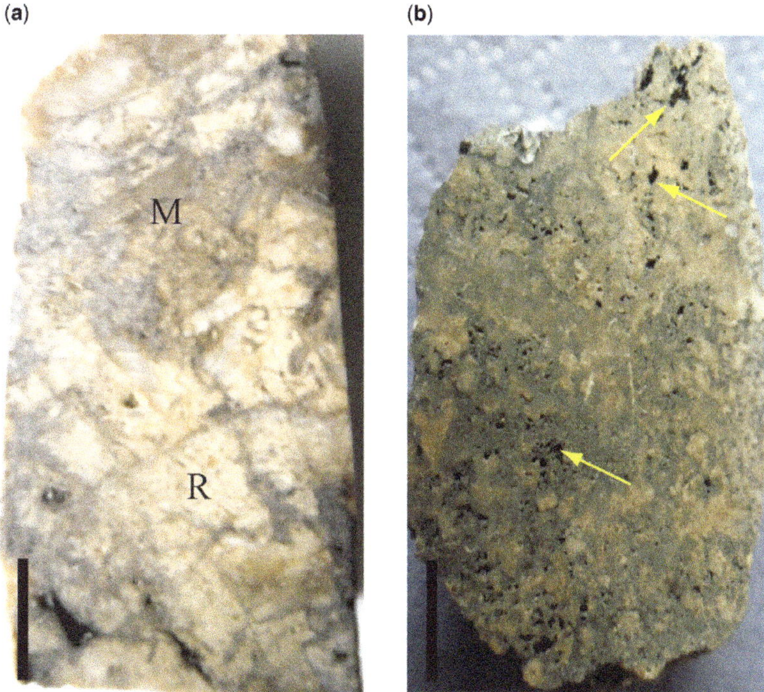

Fig. 4. Core sample photographs of the Rudistid facies. (**a**) Rudstone with preserved radiolitid rudist (R) and micritic matrix (M) (well Ns-5 at 2011 m depth). (**b**) Oil-stained rudistid limestone consisting of rudist bioclasts with vuggy pores indicated by arrows (well Fq-17 at 4061 m depth). Scale bar is 3 cm.

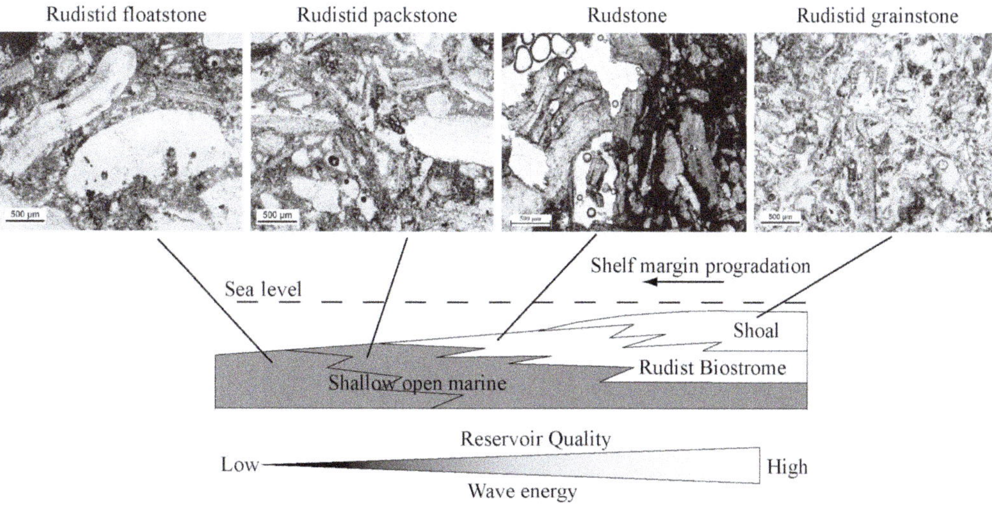

Fig. 5. A schematic conceptual model showing the lateral distribution of rudistid facies deposited across the shelf margin of the Mishrif carbonate platform and their reservoir quality. (Note the relative increase in dissolution of the rudist bioclasts from rudistid floatstone to rudistid grainstone. Also, some remnants of the rudist microstructure are quite clear within the rudstone microfacies.)

congregations at the shelf margin and then transported them basinwards as dispersed debris in the muddy deposits. This process also caused a basinwards progradation of rudistid facies due to relative sea-level changes.

As a result, the continuity of rudistid reservoir units of the Mishrif Formation was mostly controlled by the geometry of rudist biostromes because the rudists of the Mishrif Formation can be either autoparabiostrome or parabiostrome, meaning that they lack visible topographical relief and also framework (e.g. Kershaw 1994). The geometry of a rudist biostrome can be either tabular or lenticular (Gilli *et al.* 1995). Through the correlation of rudist biostrome and shoal facies that form major reservoir units, a high lateral continuity and apparently 'layer-cake'-like geometry for these units are recorded (Fig. 6). More specifically, the reservoir units of the Mishrif Formation have a tabular shape, and are laterally persistent at the field scale. Such a high degree of continuity was also supported by the consistency of porosity log signatures between the correlated well logs (Fig. 6). However, lateral variations in the core porosity and permeability values of the same rudist body were common, because any biostrome body can contain different abundances of shell fragments and matrix (Kershaw 1994), and this will consequently cause variations in both pore types and porosity values (Fig. 6).

The microbial facies

Microbial facies have not been recognized previously in the Mishrif Formation as most of the reservoir studies were focused on the rudistid facies (Mahdi & Aqrawi 2014). In addition, oil-field exploration and development activities concentrate on mapping structural highs that are usually dominated by rudistid facies, with the microbalite facies absent. Mahdi & Aqrawi (2014) identified microbial facies in the Mishrif Formation dominated by peloidal packstones that displayed very good reservoir potential. These facies were interpreted to have been deposited in shallow-water areas with low wave energy. Most of the porous and permeable microbial reservoir units were found within the third-order Sequence II of Mahdi *et al.* (2013). Some of these units were overlain by the intra-Mishrif disconformity surface. Such facies were developed at different depositional settings within the Mishrif depositional basin. Also, it has been well documented that microbial facies may represent an important contributor in carbonate platforms

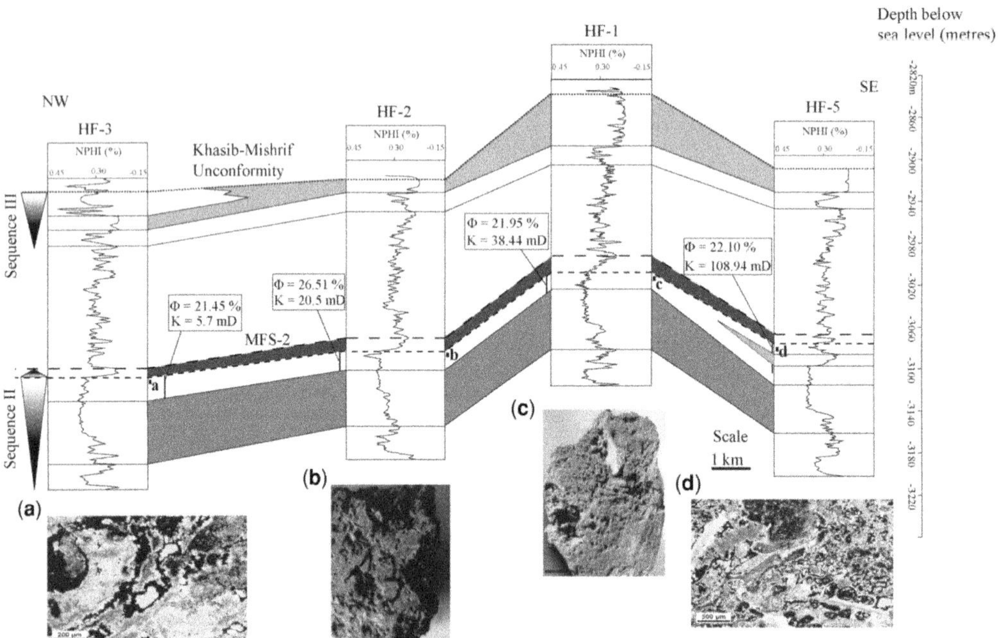

Fig. 6. Sequence-stratigraphic cross-section across the Halfaya Field structure correlating the rudistid reservoir units within sequences II (SII) and III (SIII). (Note that the average core porosity and permeability values are also shown for these reservoir units in Sequence II, and the continuity of rudistid facies within the regressive cycles of both sequences.)

throughout the geological record, including the Cretaceous (e.g. Riding & Awramik 2000).

The microbial facies of the Mishrif Formation are characterized by packstones dominated by peloids (Fig. 7). Other components include skeletal grains, such as echinoderm, benthonic foraminifera and rudist bioclasts (Fig. 7d). Intraclasts are also common as non-skeletal grains in such facies (Fig. 7d). The peloidal texture of the microbial facies shows various degrees of heterogeneity in the size of peloids (Fig. 7). The peloidal fabric is one of the building block arrays that determine the pore characteristics in the microbial facies (Ahr *et al.* 2011). The peloidal microfabric displayed interparticle pores (Fig. 7a) and solution-enhanced fenestral-like vugs (Fig. 7b–d). The interparticle pores are primarily formed during deposition, but they are rarely preserved without some effect from dissolution, which formed the fenestral-like vugs (Fig. 7b). This hybrid pore type could be considered as a distinctive character of the peloidal-dominated microbial facies of the Mishrif Formation (Mahdi & Aqrawi 2014).

The occurrence of fenestral-like vugs in the microbial facies below the intra-Mishrif disconformity might also indicate a dissolution process that took place during the sea-level fall. Another possible mechanism for creating such pore types in a similar facies could be related to the degradation of organic matter that resulted in the shrinkage and creation of fenestral vugs between micritic clots and peloids (e.g. Riding & Tomás 2006). The association of both primary interparticle and secondary vuggy pores increased the reservoir quality of these microbial facies.

In the regressive cycles of the Mishrif Formation, microbial facies might form stratigraphic traps that grade laterally and interfinger with tight shallow, open-marine wackestones and mud-dominated packstones. Potential examples of such a stratigraphic trap may be present in the area between the Kumait and Amara oil fields, where the porous microbial facies pinched out in an up-dip direction towards the shallow open-marine facies around the Amara oil field (Fig. 8). In addition, the porosity log motifs of these facies show high porosity, which was then abruptly sealed by less porous transgressive deep-marine facies of Sequence III (Fig. 8). These stratigraphic traps could be expected along the Mishrif shelf slope

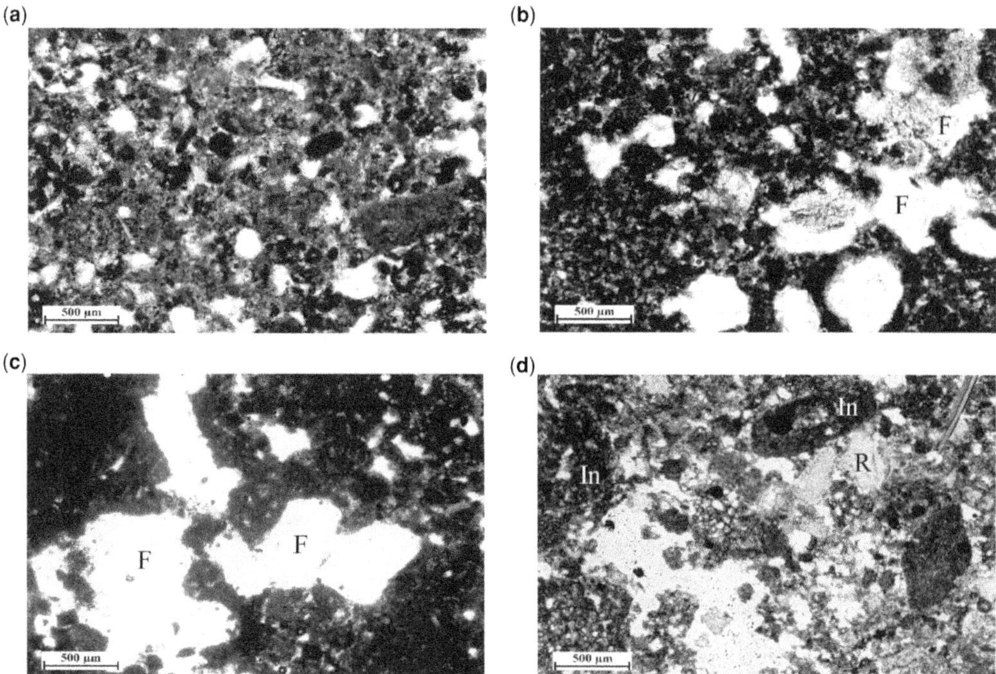

Fig. 7. Thin-section photomicrographs of microbial facies that show good reservoir potential. (**a**) Peloidal packstone–grainstone with interparticle pores (well Kt-1 at 3200 m depth). (**b**) Same as the previous facies with fenestral-like vugs (F) (well Kt-1 at 3199 m depth). (**c**) Fenestral-like vugs (F) in peloidal wakestone–packstone microfacies (well WQ-215 at 2666.5 m depth). (**d**) Peloidal packstone–grainstone with intraclasts (In) and rudist bioclasts (R). This microfacies shows interparticle pores and extensive dissolution vugs (well WQ-21at 2591.5 m depth).

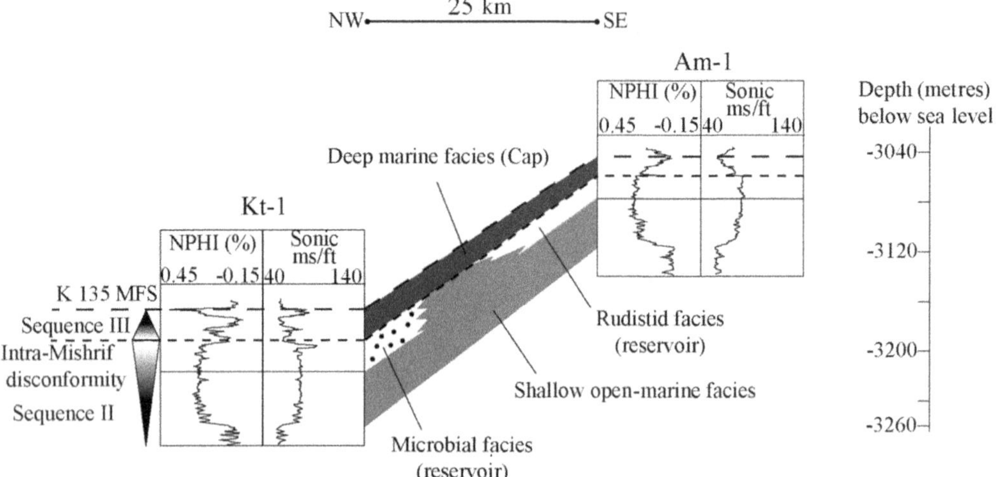

Fig. 8. A stratigraphic cross-section between the Kumait (well Kt-1) and Amara (well Am-1) fields showing the pinch out of microbial facies present in well Kt-1, which are not present in well Am-1, and which have been replaced by rudistid facies. This may indicate the deposition of the microbial facies in deeper open-marine areas of the Mishrif platform during the regressive depositional period of Sequence II. The overlying transgressive deep-marine facies of Sequence III may act as a local seal unit for the microbial reservoir facies. The key for the symbols is in the legend in Figure 3.

successions, where some facies of Sequence II were deposited.

Diagenetic framework

The diagenesis of the Mishrif Formation in the study area has been described in detail by Aqrawi *et al.* (1998). Results of that paper are summarized here, since diagenesis forms one of the building blocks of the rock-typing methodology presented in this paper (e.g. Aqrawi *et al.* 2010*b*).

Based on petrographical characteristics, three diagenetic stages are recognized. The early stage belongs to the marine and meteoric phreatic environments. Diagenetic processes that operated in the marine phreatic environment include the formation of micrite envelopes rimming the skeletal grains, and the partial dissolution of skeletal grains that are composed of metastable minerals (e.g. aragonite). The effect of dissolution was more aggresive in the meteoric phreatic environment, especially in rudistid facies, and was responsible for creating secondary porosity. However, meteoric isopachous and syntaxial rim cements partially or completely occluded interparticle and secondary pores, particularly in cemented grainstone–packstone facies that turned them into a non-reservoir rock type (i.e. S3).

The second diagenetic stage represents a burial environment, which affected the Mishrif Formation at deeper intervals under higher temperatures relative to the previous environments. The output of this stage was mainly cementation and compaction. Precipitation of granular, drusy and blocky cements occurred after a period of dissolution that affected rudists and other skeletal grains. Stylolites and increasing grain compaction reduced the porosity in lime mud-dominated and grain-dominated facies, respectively. Fracturing occurred at an earlier stage, before the deeper part of the burial. As a result, the fractures were either cemented by blocky calcite or remained partially open.

The late diagenetic stage is related to the periodic exposure of the Mishrif's carbonate platform as a result of sea-level emergence. Dissolution is the major observed process that took place when the platform was flushed by meteoric waters along discontinuity surfaces. This process formed a complex pore system of interconnected vugs, which were particularly found below the Mishrif–Khasib unconformity and the intra-Mishrif disconformity surface.

Capillary pressure curves and the impact of diagenesis on the depositional facies

The capillary pressure curves show a clear increase in entry pressure from grain-dominated to mud-dominated microfacies. In the same trend, there is also a consistent decrease in permeability and an increase in microporosity. In addition, the existence of small separate vugs in the mud-dominated fabric has caused an increase in the entry pressure (e.g. Lucia 1995, 2007).

Also, different capillary pressure curves are observed for the same facies due to diagenetic overprints. The main diagenetic processes observed include cementation in reservoir facies, particularly grainstone and grain-dominated packstone facies (Fig. 9a, b), and compaction and neomorphism in non-reservoir facies, such as wackestone and mud-dominated packstone (Fig. 9c, d). The overall result of these processes was an overall decrease in permeability and pore-throats sizes that, in turn, increased entry pressures.

In most analysed samples, the capillary pressure curves exhibited a large transition zone with no clear plateau, which points to a wide range of pore-throat size distributions (Fig. 9). However, only the grainstone facies displays a clear plateau, with the lowest observed entry pressures (Fig. 9a) indicating a rather uniform large pore-throat size distribution with higher permeability.

The combined results of both capillary pressure analysis and porosity and permeability trends show that grainstone has the highest reservoir quality and represents the best flow units with the highest production capability. Reservoir quality decreases in the mud-dominated facies (i.e. mudstone, wackestone and mud-dominated packstone). Nevertheless, there was an absence of capillary pressure results and porosity and permeability data for the microporous facies. As a result, the current evaluation cannot be generalized over the entire Mishrif Formation reservoirs unless more samples are analysed in future evaluations.

Mishrif rock types

Based on the Mishrif facies heterogeneity, diagenetic overprint, and the available porosity and permeability data, six rock types were identified that have reservoir and non-reservoir characteristics, using the rock type definition of Aqrawi *et al.* (2010*b*). All of these rock types may occur in every single Mishrif unit (i.e. A, BI and BII)

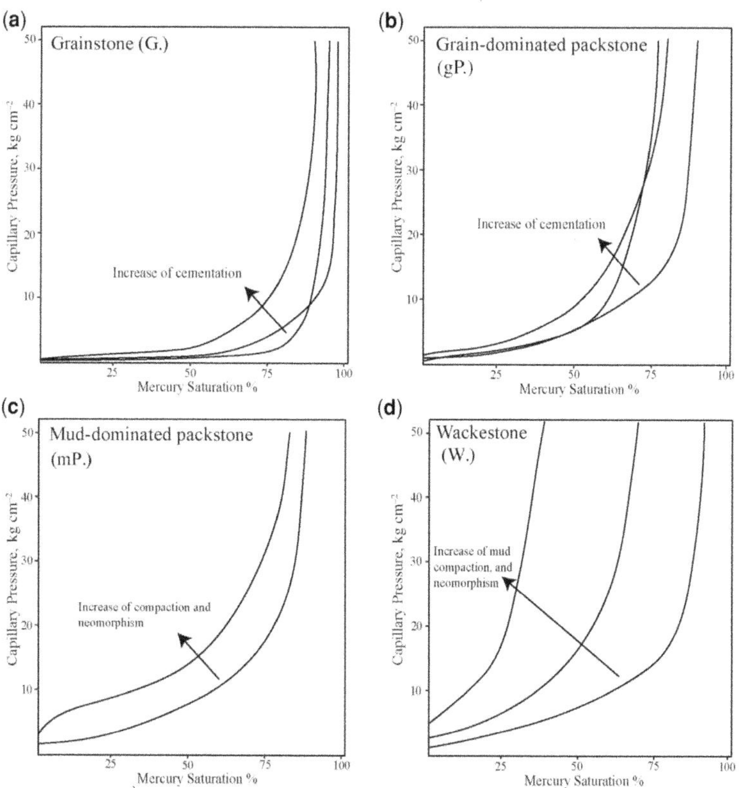

Fig. 9. MICP curves for four selected microfacies of the Mishrif Formation. The effect of diagenesis can be observed through the variation in pressure for each rock fabric.

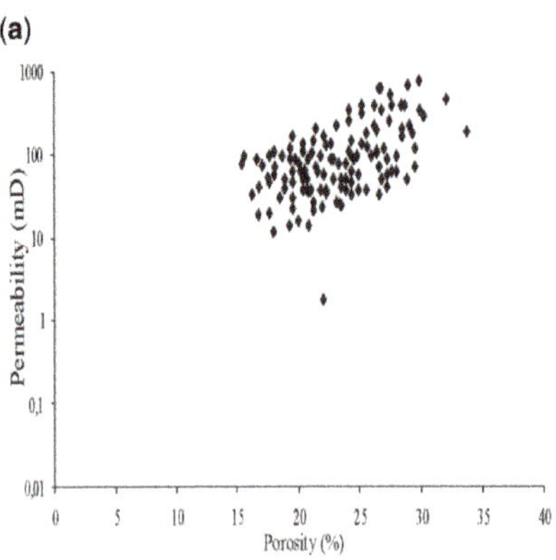

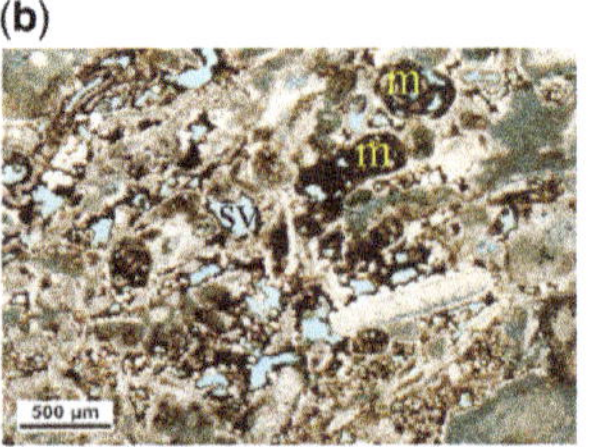

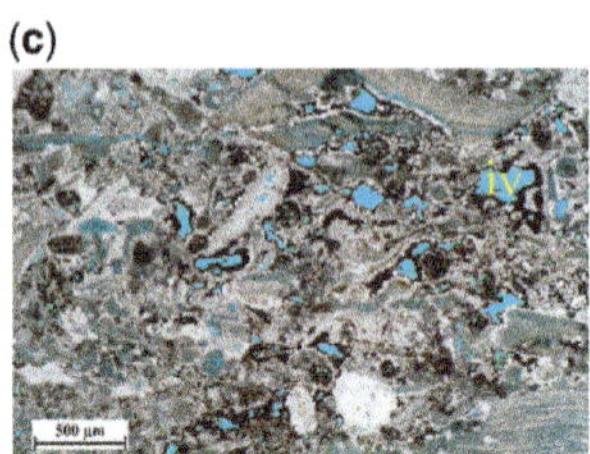

2- R2. Grain-dominated Packstone (gP.)

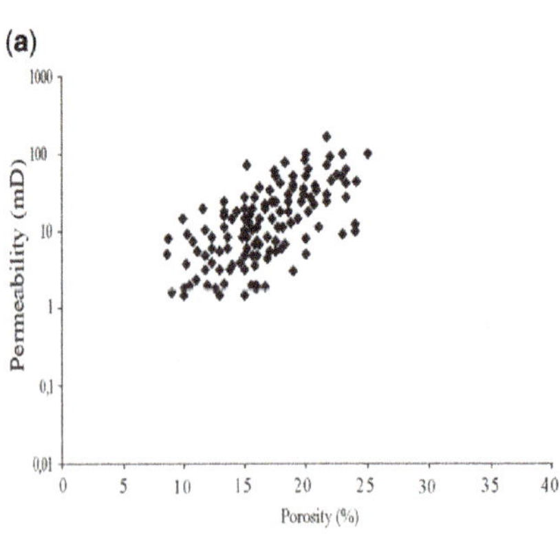

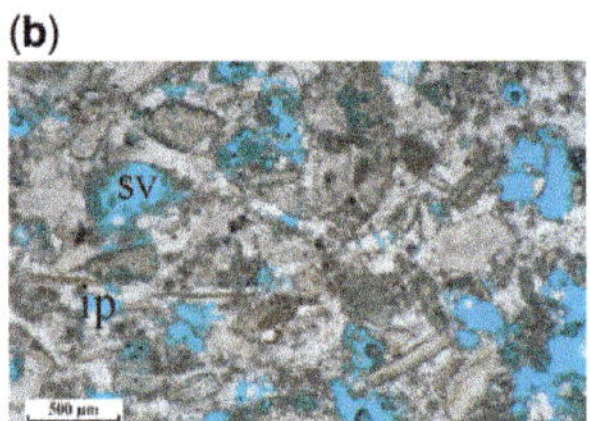

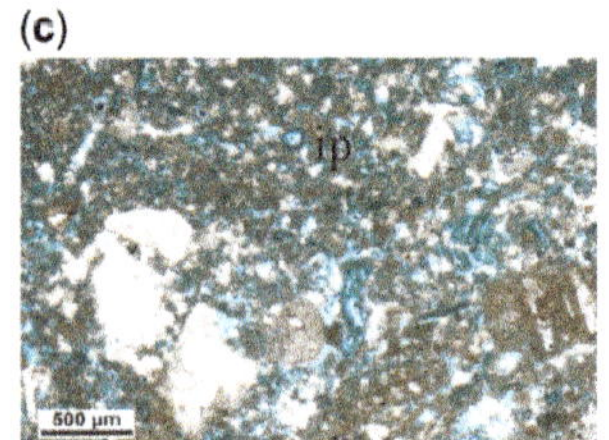

3- R3. Mud-dominated Packstone (mP.)

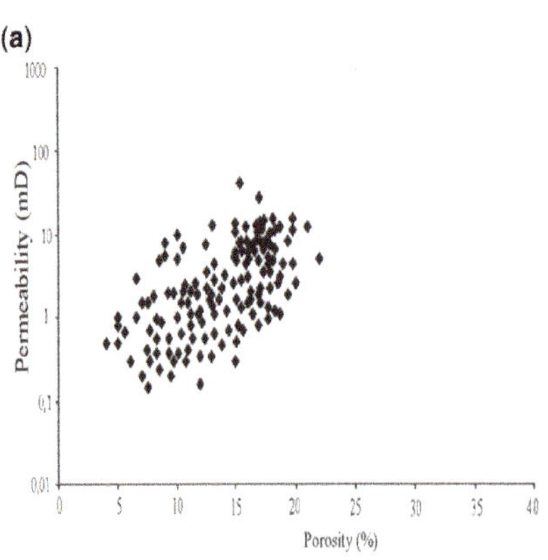

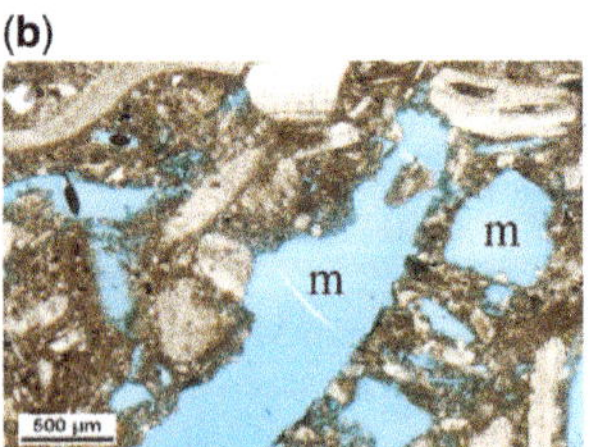

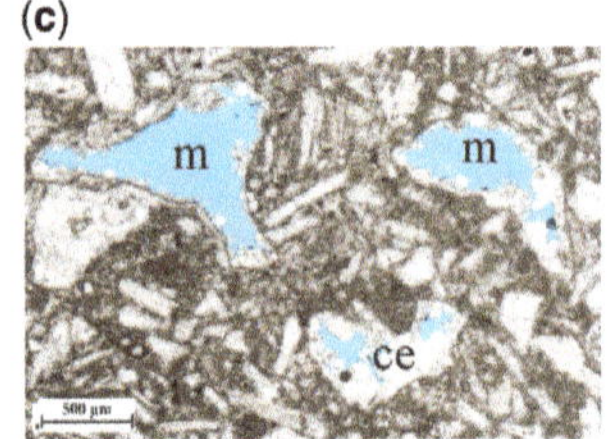

identified previously by SOC (Fig. 2). The reservoir and non-reservoir rock-type classification presented in this study does not involve pore-throat size distribution or dynamic data. However, once these data are available, this classification can be modified and further detailed, including more rock types or petrophysical classes *sensu* Lucia (1995).

Based on rock fabric, the non-reservoir and reservoir rock types include mud- and grain-dominated depositional microfacies, respectively, after having been subjected to diagenesis (Aqrawi *et al.* 2010*b*). The mud-dominated microfacies are mostly deposited in low-energy environments. The size of the micrite particles, which are commonly recrystallized to microrhombic crystals, and the amount of interparticle porosity in these mud-dominated facies control their pore-size distribution (Lucia 2007). Depending on their diagenetic history, the micrite-rich facies of the Mishrif Formation rarely form potential reservoirs unless they have been fractured and/or dolomitized. They usually form dense layers characterized by very low porosity and permeability values, and are typically not fractured (e.g. Lambert *et al.* 2006; Periere *et al.* 2011).

The grain-dominated microfacies and their associated pore types result from high-energy depositional conditions with variable diagenetic overprints. The skeletal and non-skeletal grains are major components that form porous and permeable reservoir units, particularly when they are dissolved. However, they can also be heavily cemented, forming tight non-reservoir units, if the geochemical solutions were supersaturated with $CaCO_3$.

Reservoir rock types

R1. Grainstones (G.). This reservoir rock type is classified as grainstones that are deposited in high-energy, depositional environments, such as shoal and rudist biostromes. They include rudistid grainstone and peloidal or bioclastic grainstone, in addition to the occasional rudstone (Fig. 10.1b, c). The absence of lime mud is the main feature in this rock type. Therefore, grain size controls pore size (Lucia 2007). This relationship is hard to define in the Mishrif Formation due to the alteration of interparticle pores by early diagenetic dissolution. This produces a complex pore system of interconnected and separate vugs, as dissolution also enlarges the depositional interparticle pores (Fig. 10.1b, c).

Based on the observed pore system of the grainstone facies, and their high porosity and permeability trend (16–34%, 2–900 mD: Fig. 10.1a), this rock type forms the best reservoir units of the Mishrif Formation. The reservoir potential of this rock type might be slightly decreased by the occurrence of meteoric isopachous, mosaic and syntaxial rim cement (Fig. 10.1b, c).

R2. Grain-dominated packstones (gP.). These include both depositional rudistid packstones and peloidal or bioclastic-rich packstones (Fig. 10.2b, c), which are deposited in moderate- to high-energy environments near rudist biostromes or shoals. In this grain-dominated packstone texture, the intergranular spaces are occupied by interparticle pores (Lucia 2007). Lime mud is a very minor component of this rock type, which may reflect the overall energy of the system in which the rock type was deposited. In the Mishrif Formation, the original interparticle pores of grain-dominated packstones are largely altered by early diagenetic dissolution, which enlarges the pore spaces and increases their interconnectivity (Fig. 10.2a). Therefore, grain-dominated packstones form the second-best reservoir rock type, with abundant interconnected vugs and interparticle pores, in addition to the moldic pores (Fig. 10.2b, c).

The core porosity–permeability measurements for this rock type are also high (Fig. 10.2a), with the lower values attributed to the occurrence of different cement types, which can completely or partially fill some pores (Fig. 10.2b, c).

R3. Mud-dominated packstones (mP.). The mud-dominated packstones were originally deposited in shallow, open-marine and back-shoal environments. Low-energy depositional conditions led to an abundance of lime mud in this microfacies,

Fig. 10. Porosity v. permeability cross-plots and thin-section photomicrographs of reservoir rock types (R1, R2 and R3). (**1a**) Porosity v. permeability cross-plot of grainstone rock type (G). Oil-filled interconnected (iv), separate vugs (sv) and moldic pores (m) in rudistid grainstone facies (G). The facies might have been extensively affected by dissolution, as indicated by the presence of meteoric isopachous cement in (**1b**) and (**1c**) (1b is from well Tu-4 at 2344 m depth and 1c is from well Tu-5 at 2402 m depth). (**2a**) Porosity v. permeability cross-plot of the grain-dominated packstone rock type (gP). The rudistid packstone is characterized by separate vugs (sv) and interparticle pores (ip) (**2b**), which are partially enlarged by later dissolution. Also, the grains are partially micritized (2b). The peloidal grain-dominated packstone is characterized by micritized skeletal grains (**2c**) and also consists of interparticle and isolated vuggy pores due to the dissolution and cementation. (well Zb-43 at 2403 m depth in 2b and well R-27 at 2234 m depth in 2c). (**3a**) Porosity v. permeability cross-plot of the mud-dominated packstone rock type (mP). Large moldic pores (m) result from the dissolution of rudist fragments in the rudistid packstone facies in (**3b**). The moldic pores are of various sizes mostly due to partial filling by meteoric cement (ce) in (**3c**) (well Fq-17 at 4076 m depth in 3b and well R-51 at 2276.5 m depth in 3c).

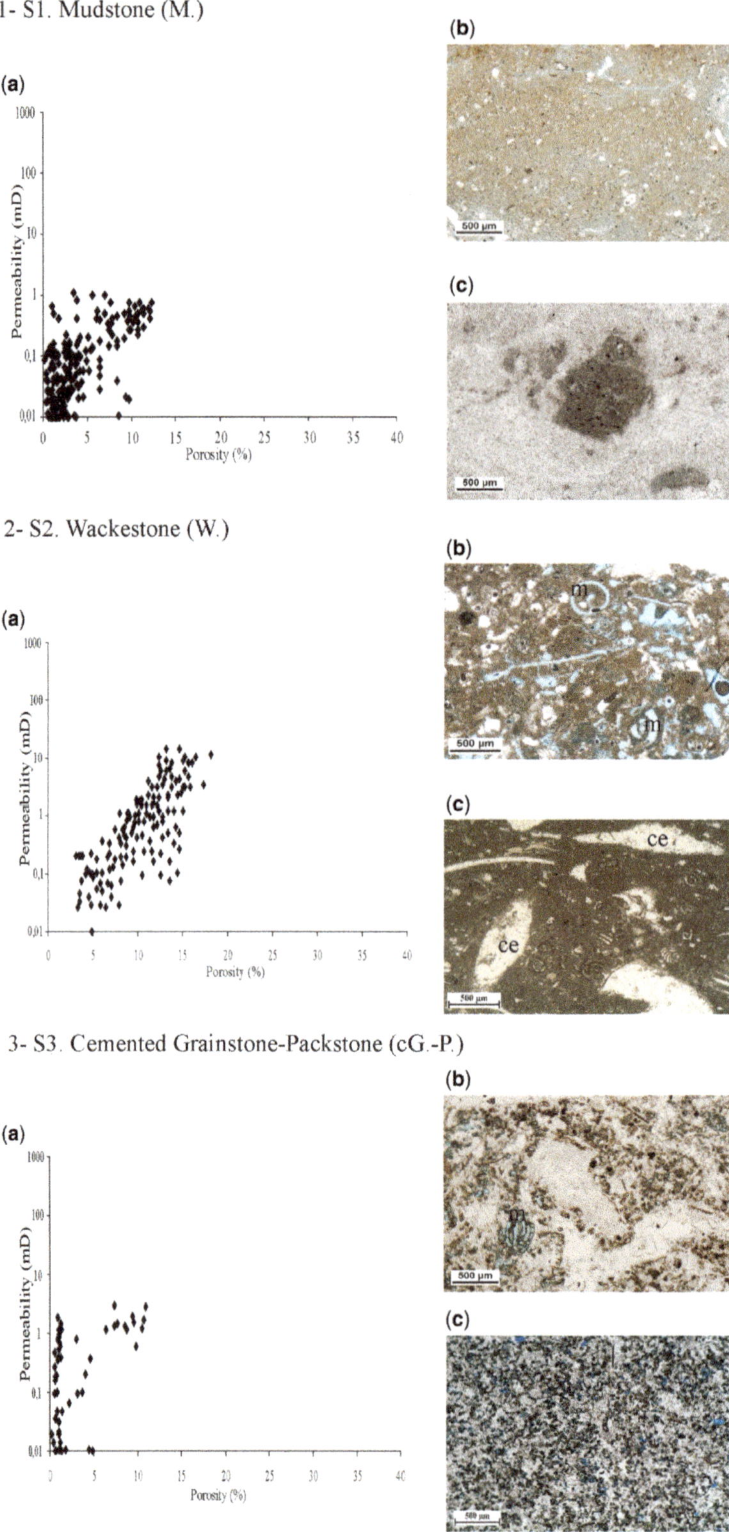
1- S1. Mudstone (M.)
(a)
(b)
(c)
Permeability (mD)
Porosity (%)
500 µm
2- S2. Wackestone (W.)
(a)
(b)
(c)
m
ce
Permeability (mD)
Porosity (%)
500 µm
3- S3. Cemented Grainstone-Packstone (cG.-P.)
(a)
(b)
(c)
m
Permeability (mD)
Porosity (%)
500 µm

with micritization of skeletal grains also common (Fig. 10.3b, c).

The porosity of this microfacies can be related to both the intergrain matrix micropores and the isolated vugs (Fig. 10.3b, c). The higher porosity and permeability of this rock type was due to early diagenetic dissolution of skeletal grains, which increased through the vug interconnectivity (Fig. 10.3b). Therefore, this microfacies is a reservoir rock type that is characterized by higher porosity and permeability compared to other mud-dominated facies, such as wackestone and mudstone fabrics (Fig. 10.3a). It is important to mention that the microbial peloid-rich packstone facies are also included within this rock type class, as they are characterized by a high reservoir quality.

Non-reservoir rock types

S1. Mudstones (M.). Some depositional mudstones recognized within the Mishrif Formation were slightly argillaceous, suggesting deposition in a deep-marine environment. Other mudstone microfacies were deposited within lagoonal or tidal-flat environments (Fig. 11.1b, c). The matrix micropores were the dominant pore type observed in this rock type. They were considered as interparticle or intercrystalline pores (Lucia 2007). Examination of thin sections has shown a high porosity in the mudstone rock fabric due to matrix micropores (Fig. 11.1b) that can be patchily distributed. Also, the same samples had a high core porosity (Fig. 11.1a). However, it was not possible to determine the origin of these intercrystalline micropores without utilizing a scanning electron microscope, which was outside the scope of this study. The preservation of micropores and, hence, their resistance to compaction might have resulted from early cementation of low-Mg calcite mud before deeper burial, which was reported as a common mechanism in the Cretaceous microporous carbonate facies (e.g. Volery *et al.* 2009). The effect of compaction and neomorphism on this microfacies is evident in the loss of microporosity, as indicated by the lower values of core porosity (Fig. 11.1a, b). As a result, most mudstones formed non-reservoir rock types, and may form sealing units overlying the grainstone and grain-dominated packstone reservoirs.

S2. Wackestones (W.). The wackestone microfacies were deposited in different depositional settings and below fair-weather wave base. The high porosity of some of wackestone rock fabrics can be attributed to the occurrence of both separate vugs and micropores (Fig. 11.2b). Such vugs can be moldic, intrafossil or shelter, and they were mostly isolated in a tight matrix. The vug size differs according to the dissolved grain types, which include rudists, foraminifera and algae (Fig. 11.2b). However, these vugs can either be not connected or completely occluded by cement, which decreases both porosity and permeability (Fig. 11.2c). The potential reservoirs of wackestone rock types were characterized by bimodal porosity with the occurrence of meso- and micropores. The average trend of porosity and permeability values in wackestones is lower than the mud-dominated packstones due to the higher amount of dissolved skeletal components in the packstones (Figs 10.3a & 11.2a).

S3. Cemented grainstone–packstone (cG.-P.). The large amount of cementation in these grainstones and packstones decreases their reservoir quality (Fig. 11.3a), and alters them into tight, non-porous and impermeable units. The original components of these facies are similar to reservoir grainstone and grain-dominated packstone rock types (i.e. R1 and R2). The origin of the cemented fabric is related to the early meteoric diagenesis and submarine cementation of the skeletal grains and pores shortly after deposition, which led to early lithification of the sediments. In addition, the residual porosity can be occluded further during late burial cementation and compaction (Fig. 11.3b, c).

Role of cyclicity in the quality of the reservoir facies

The sequence-stratigraphic framework has a strong influence on variations in reservoir architecture, petrophysical properties and producibility (Ruppel

Fig. 11. Porosity v. permeability cross-plots and thin-section photomicrographs of non-reservoir rock types (S1, S2 and S3). (**1a**) Porosity v. permeability cross-plot of the mudstone rock type (M). Thin-section photomicrograph of deep-marine mudstone with micropores (**1b**); and the same mudstone fabric can be highly neomorphosed, leaving relicts of the original texture observed in (**1c**) (well WQ-11 at 2425 m depth in 1b and well Ri-1 at 2992 m depth in 1c). (**2a**) Porosity v. permeability cross-plot of the wackestone rock type (W). Separate moldic vugs (m) with micropores in the foraminiferal–bioclastic wackestone microfacies (**2b**). Large moldic pores of dissolved shell fragments, which are completely occluded by cement (ce), in the lagoonal wackestone (**2c**) (well R-83 at 2344 m depth in 2b and well Ru-96 at 2305 m depth in 2c). (**3a**) Porosity v. permeability cross-plot of the cemented grainstone–packstone rock type (cG-P). The low reservoir quality of this grainstone fabric is mostly related to the extensive cementation (**3b**) and (**3c**). However, remnants of moldic pores (m) (3b), as well as interparticle pores (3c), can be observed (well WQ-215 at 2623 m depth in 3b, and well WQ-21 at 2490 m depth in 3c).

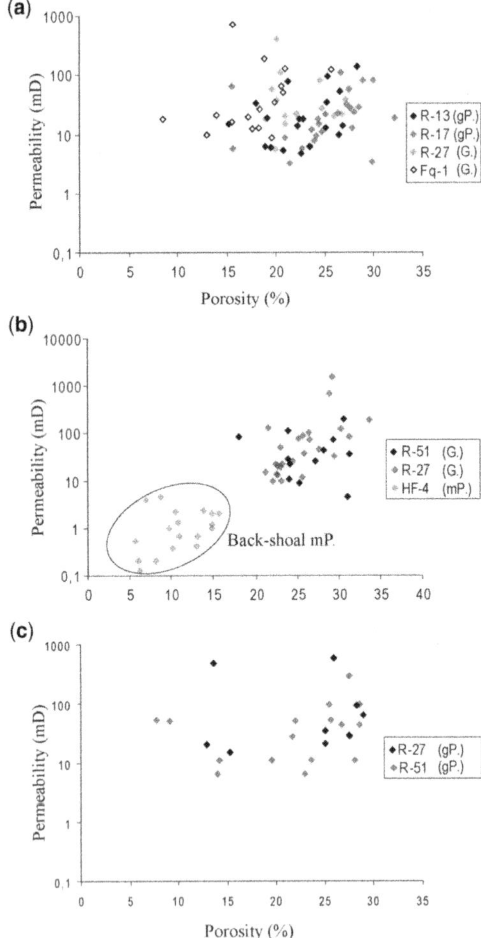

Fig. 12. Porosity v. permeability cross-plots of two selected microfacies of the Mishrif Formation located a few centimetres below the boundary of a regressive cycle. The analysed samples consist of rudistid grainstones and grain-dominated packstones that are directly below the intra-Mishrif disconformity (**a**), the Cenomanian–Turonian boundary (exposure surface) (**b**) and the Khasib-Mishrif unconformity (**c**). Note the lower porosity and permeability values of the back-shoal mud-dominated packstones that topped the Cenomanian–Turonian boundary (b).

2004; Lucia 2007). This relationship could be deduced for the Mishrif Formation in the study area by examining the vertical stacking of reservoir and non-reservoir facies (or rock types) within the transgressive–regressive cycles. The sequence-stratigraphic framework established by Mahdi *et al.* (2013) and Mahdi & Aqrawi (2014) provided regionally recognized sequence boundaries and other key surfaces. The sedimentological, and porosity and permeability, data show that reservoir facies are mostly contained within the regressive cycles (Figs 3 & 6). The majority of the porous and permeable units, which are the recognized reservoir rock types (R1, R2 and R3), occur within the rudistid facies that are concentrated at the top of regressive cycles (Figs 3 & 6). They are typically developed below exposure surfaces that represent the upper boundaries of the regressive cycles. These surfaces were distinguished as top contacts or sequence boundaries, such as Sequence II (i.e. intra-Mishrif disconformity) (Fig. 2) and Sequence III (i.e. Cenomanian–Turonian boundary) (Fig. 3). The rudistid facies below these surfaces are characterized by the highest porosity and permeability values of all the rock types largely due to diagenetic dissolution (Fig. 12). However, poor-quality or non-reservoir facies (or rock types S1 and S2) occurred in the muddy facies at the basal parts of the same cycles. The inner shelf mudstones and wackestones that occupy the uppermost part of the same regressive cycles appear to act as non-reservoir units, which are non-reservoir rock types S1 and S2, owing to their very low porosity and permeability values (Fig. 12b). The lowest porosity and permeability values occurred in the mud-dominated facies of transgressive cycles (Fig. 13), turning them into local seals. However, the microbial mud-dominated packstones, which were subjected to early diagenetic dissolution, can also make good reservoir units with a reasonable porosity and permeability, which is why they are included in reservoir rock type R3.

The rudistid facies are hosted in the transgressive–regressive sequences that were defined by Mahdi *et al.* (2013). However, their thickness,

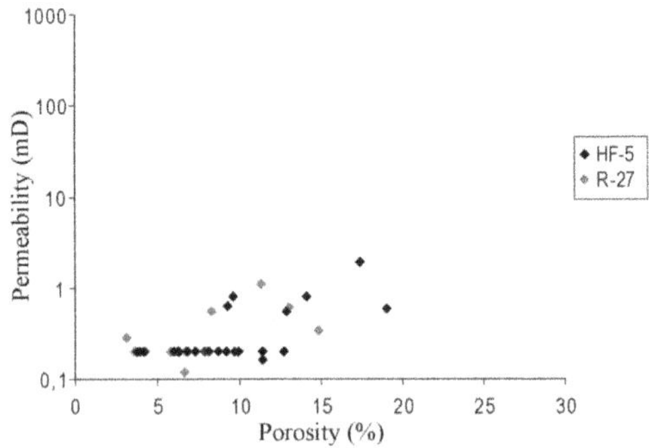

Fig. 13. Porosity v. permeability cross-plot of the wackestone microfacies showing poor reservoir quality. The samples are selected from the transgressive succession of Sequence III in well HF-5 of the Halfaya Field and well R-27 of the North Rumaila Field.

lateral continuity and relationship with other facies differ in these sequences (Mahdi & Aqrawi 2014). In each defined third-order sequence (Mahdi & Aqrawi 2014), the best reservoir units were found in reservoir rock types: R1; grainstones (G.) and R2; and grain-dominated packstones (gP.), which were bounded upwards by discontinuity surfaces. Most hydrocarbon production comes from these two reservoir rock types of sequences II and III, where thick units of coarse rudists were leached during relative sea-level falls due to early diagenetic dissolution. During the deposition of Sequence II, high rates of shelf-margin progradation were recorded in the southern Mesopotamian Basin (Mahdi *et al.* 2013; Mahdi & Aqrawi 2014): this may explain the widespread distribution and occurrence of rudistid reservoirs in many oil fields (Figs 3 & 6). The Middle Cenomanian sea-level fall led to the enhancement of the reservoir characteristics of rudistid facies through the dissolution of a large amount of skeletal components during the exposure of the Mishrif carbonate platform. Therefore, the regressive cycle of Sequence II represents an important exploration target that can be easily determined through mapping the intra-Mishrif disconformity surface. Below this sequence boundary, thick units of reservoir rock types – coarse-grained and leached grainstones (R1) and grain-dominated packstones (R2) characterized by high porosity, permeability and hydrocarbon saturation – are usually present (Figs 3 & 6). In addition, the potential reservoir units below the same boundary usually occur on the crest and flanks of anticlinal structures (Fig. 6), which form most structural traps in the study area.

Sequence III exhibits more variations in the distribution of reservoir and non-reservoir rock types. The reservoir rock types were found in several stratigraphic units, which are separated by shallow, open-marine wackestones (S2) or lagoonal-tidal flat mudstones (S1) that act as seals (Fig. 2). In contrast to Sequence II, the tops of Sequence III regressive cycles are not usually dominated by rudistid

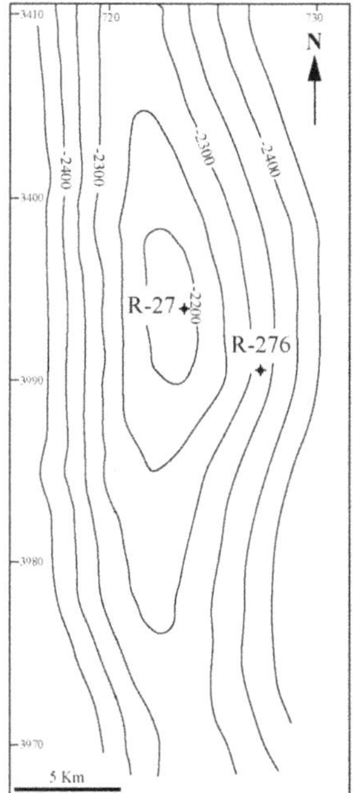

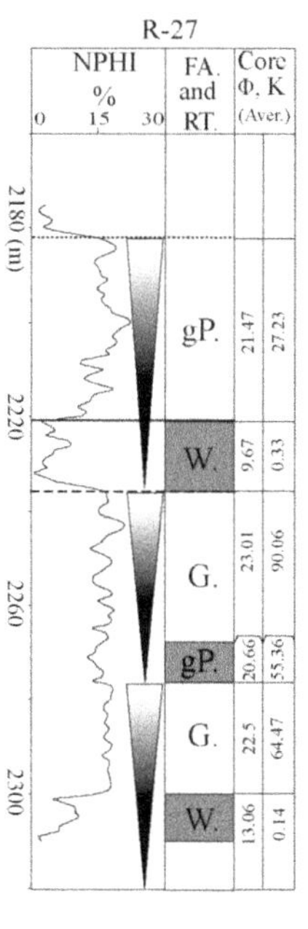

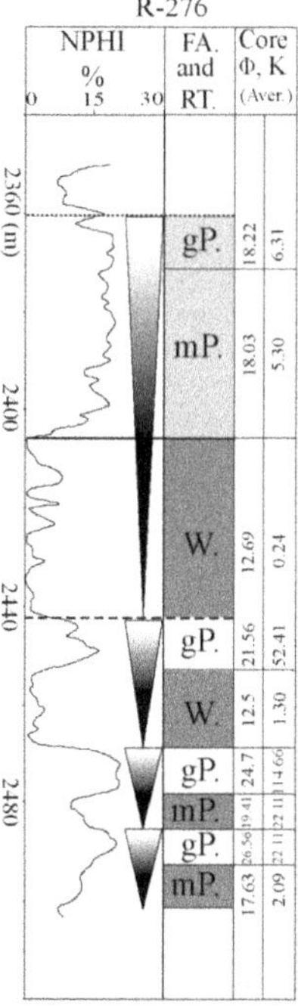

Forced regressive surface
FA. : Facies Associations
RT. : Rock Types

Fig. 14. Vertical distribution of the reservoir and non-reservoir rock types of the Mishrif Formation within two wells of the North Rumaila oil field. The different locations of the wells, R-27 and R-276, exhibit two different vertical and lateral distributions of facies along with associated changes in porosity and permeability values. In well R-27, which penetrates the crest of the Rumaila anticlinal structure, the regressive cycles are capped by grainstone reservoir facies, which have higher average porosity and permeability values. Towards the flank well R-276, these facies are grade into grain-dominated packstone reservoir facies and thicker wackstones of lower reservoir quality. The key for the symbols is in the legend in Figure 2.

reservoir facies. They are often replaced by reservoir rock type R3 (i.e. mud-dominated packstones) (Figs 3 & 6).

During the Late Cenomanian–Middle Turonian, differential uplift and subsidence took place in the Mesopotamian Basin (Sherwani 1998), which formed some important palaeo-structures, such as the Rumaila and Tuba oil fields. Rudists, their derived grainstones and grain-dominated packstones (i.e. reservoir rock types R1 and R2) could reach up to 40 m in thickness, as seen in well R-27 (of the North Rumaila Field) (Fig. 14). The maximum thickness of these facies with the highest porosity and permeability values (R1) occupy the crestal part of the North Rumaila anticline (Fig. 14). Towards the flanks of the same structure, these facies grade laterally into grain-dominated packstones microfacies (R2), as seen in well R-276. Such a lateral facies shift is associated with a corresponding decrease in the porosity and permeability values (Fig. 14). Another distinct change occurs in the mud-dominated microfacies (non-reservoir rock types: S1 and S2) that are thicker in well R-276, and also have lower porosity and permeability values (Fig. 14). The capping units above the Cenomanian–Turonian exposure surface consist of shallow, open-marine wackestones (S2) that were deposited during the Early Turonian flooding. They represent sealing facies that thin towards well R-27 (Fig. 14). In both wells, the rudistid reservoir facies form the tops of the regressive cycles, with a remarkable change in thickness that was controlled by palaeobathemetry and the available accommodation space. The thick reservoir units in well R-27 were deposited when accommodation space was available over the shallowest portions of the Rumaila structure. The preferred nucleation of rudists was on the crestal part of this structure,

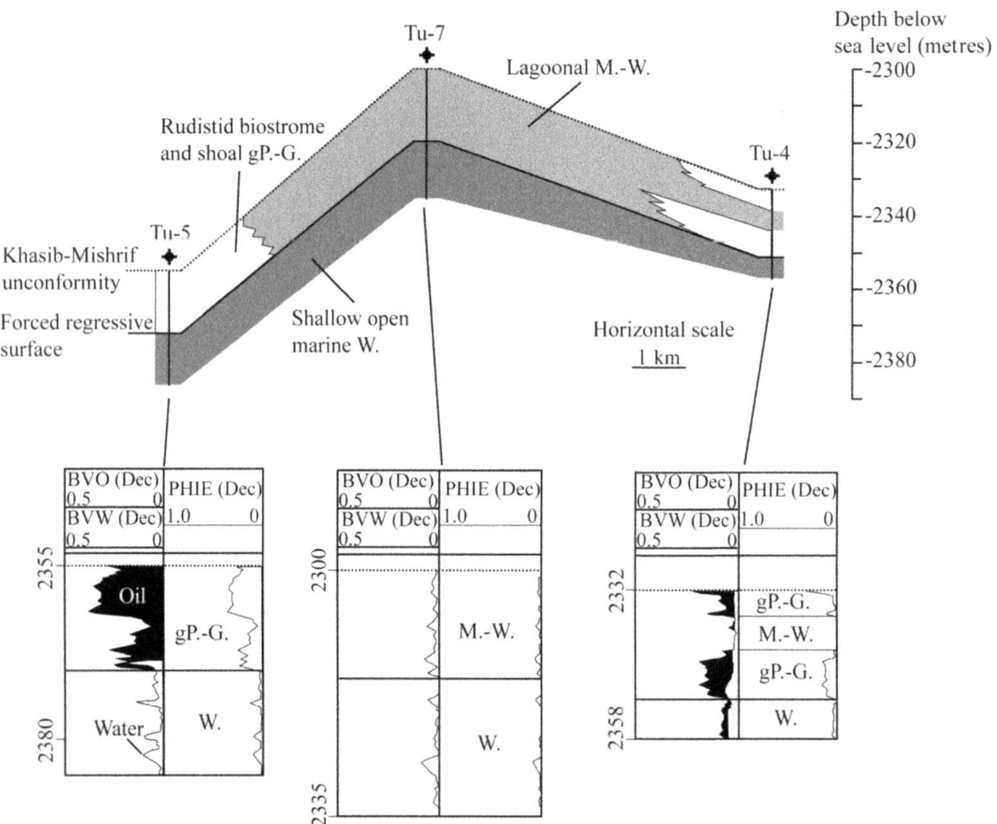

Fig. 15. Lateral facies changes from reservoir to non-reservoir rock types, forming stratigraphic traps along the flanks of the Tuba Field anticline in southern Iraq. Reservoir quality is higher in these facies at the flanks compared to the other facies along the crest of the structure. This is shown clearly in the well log succession of flank wells Tu- 4 and Tu-5 compared to the crestal well Tu-7. The displayed succession represents the uppermost part of Sequence III of the Mishrif Formation in the Tuba Field.

which continued with the development of non-reservoir lagoonal facies towards well R-276. This sharp lateral facies shift is caused by tectonic movements that took place towards the end of the deposition of the Mishrif Formation in the Early Turonian. Evidence of this assumption is indicated by the forced regressive and the upper unconformable surface contact of the Mishrif Formation that bounds both rudistid and lagoonal facies successions in the two wells (Fig. 14).

Remarkable lateral facies changes occur between the upper unconformable Mishrif contact and the forced regressive surfaces, which can form potential stratigraphic traps in some oil fields. The best example of such traps occurs in the Tuba oil field, where local rudist biostrome and shoal reservoir facies were found in wells Tu-4 and Tu-5. Both wells are located on the flanks of the Tuba anticline (Fig. 15). The reservoir facies in these two wells consist of grainstones and grain-dominated packstones (i.e. reservoir rock types R1 and R2, respectively), which pinch out updip towards well Tu-7 that is located at the crest of the Tuba anticline (Fig. 15). In well Tu-7, non-reservoir facies consist of lagoonal mudstones and wackestones (i.e. non-reservoir rock types S1 and S2) that display very low porosity and lack hydrocarbons, which may act as local seals. In contrast, high porosity and large hydrocarbon volumes are present in both the Tu-4 and Tu-5 wells. This type of stratigraphic trap is expected in the last regressive cycle of Sequence III. Similar lateral facies changes with potential stratigraphic traps may also occur in the Halfaya oil field between wells HF-2 and HF-3 (Fig. 6).

Conclusions

The reservoir quality of the Mishrif Formation is largely related to the existence of the rudistid facies, in which the reservoir properties are further enhanced by dissolution. As in other Albian–Turonian rudistid formations of the Arabian Plate region, these facies are characterized by high porosity and permeability. Their characteristics display a significant upwards improvement in reservoir quality within the regressive cycles, which are usually capped by discontinuity surfaces, such as the intra-Mishrif disconformity and the Khasib–Mishrif unconformity. Along such major sequence boundaries, meteoric fluids and karstification enhanced the reservoir properties of rudistid facies by the dissolution process, which produced an interconnected vuggy-pore system. In the same regressive cycles, pore types of the Mishrif Formation show a systematic development from interparticle, isolated pores to interconnected vuggy pores of grain-dominated facies. This implies reservoir partition into mud- and grain-dominated facies with a different quality that depends on the effect of later diagenetic overprints and the original rock fabric. Therefore, this study suggests a new classification of reservoir and non-reservoir facies, in which their vertical and lateral distributions differ in the sequence-stratigraphic framework of the Mishrif Formation.

Based on rock-type identification and correlation in selected oil fields, a tabular-like geometry of reservoirs is predicted. These reservoirs are hosted within long-extended prograding rudistid lithosomes, particularly in Sequence II (i.e. the Lower Mishrif unit), which exhibit a pinch-out trend downdip of some anticline structures. Therefore, most of the traps in this sequence are combined stratigraphic-structural type similar to the trap discovered recently in the lower Mishrif unit in the Gharaf Field (Embong *et al.* 2012). Such large-distance progradation can be observed in Oman in the Cenomanian Natih E intraplatform carbonates that spread laterally more quickly due to the limited accommodation space to build upwards (Droste & van Steenwinkel 2004). In the Mishrif Formation, pinch-out traps and clinoform geobodies can be associated with prograding rudistid facies (e.g. Farzadi 2006; Avu *et al.* 2011; Mahdi & Aqrawi 2014). Owing to facies partitioning in the prograding clinoforms, reservoir properties in such geobodies may be changed in the dip direction and along strike. Therefore, they can have a big influence on the reservoir behaviour (Adams *et al.* 2011). Another type of stratigraphic traps occurs in Sequence III, where the reservoir rock types pinch out updip of the anticlinal structures (e.g. the Tuba Field). Hence, their discovery is not related to the occurrence of structural closure. The reservoir facies in these traps have less lateral continuity, and they are capped by thick inner-shelf, non-reservoir rock types that were deposited during the availability of much accommodation space.

The first author would like to acknowledge Statoil ASA for full sponsorship of his PhD study at the University of Bergen, Norway. Thanks are due to Jeroen Kenter (Statoil) and Professor William Helland-Hansen (University of Bergen) for their useful comments after reviewing the first draft of this paper. Also, the authors acknowledge Anita Csoma (editor) and two anonymous reviewers for providing lots of insights that improved the original manuscript.

References

Adams, E.W., Grélaud, C., Pal, M., Csoma, A.E., Omar, S., Al-Ja'aidi, O.S. & Al-Hinai, R. 2011. Improving reservoir models of Cretaceous carbonates with digital outcrop modelling (Jabal Madmar, Oman): static modelling and simulating clinoforms. *Petroleum*

Geoscience, **17**, 309–332, https://doi.org/10.1144/1354-079310-031
AHR, W.M., MANCINI, E.A. & PARCELL, W.C. 2011. Pore characteristics in microbial carbonate reservoirs. AAPG Search & Discovery Article 30167, *AAPG Annual Convention and Exhibition*, April 10–13, 2011, Houston, Texas, USA.
AL-SHARHAN, A.S. & NAIRN, A.E.M. 1997. *Sedimentary Basins and Petroleum Geology of the Middle East*. Elsevier, Amsterdam.
AL-ZAABI, M., TAHER, A., AZZAM, I. & WITTE, J. 2010. Geological overview of the Middle Cretaceous Mishrif Formation in Abu Dhabi. *Paper presented at the Abu Dhabi International Petroleum Exhibition and Conference*, 1–4 November 2010, Abu Dhabi, UAE, https://doi.org/10.2118/137894-MS
AQRAWI, A., THEHNI, G., SHERWANI, G. & KAREEM, B. 1998. Mid-Cretaceous rudist-bearing carbonates of the Mishrif Formation: an important reservoir sequence in the Mesopotamian Basin. *Journal of Petroleum Geology*, **21**, 57–82.
AQRAWI, A.A.M., GOFF, J.C., HORBURY, A. & SADOONI, F.N. 2010*a*. *The Petroleum Geology of Iraq*. Scientific Press, Beaconsfield, UK.
AQRAWI, A.A.M., MAHDI, T.A., SHERWANI, G.H. & HORBURY, A.D. 2010*b*. Characterization of the mid-Cretaceous Mishrif reservoir of the southern Mesopotamian Basin, Iraq. AAPG Search & Discovery Article 50264, *AAPG GEO 2010 Middle East Geoscience Conference & Exhibition*, March 7–10, 2010, Manama, Bahrain.
AVU, A., PATON, G. & VAN KLEEF, F. 2011. Stratigraphic analysis in carbonate zones: an investigation using 3D seismic analysis techniques on an offshore UAE field. *In*: *31st Annual GCSSEPM Foundation Bob F Perkins Research Conference 2011: New Views on Seismic Imaging – Their Use in Exploration and Production*. Gulf Coast Section SEPM (GCSSEPM), Houston, TX, USA, 828–847.
DALY, M.C. 2010. *BP in Rumaila*. BP, London, http://www.bp.com/en/global/corporate/press/speeches/bp-in-rumaila.html
DROSTE, H. & VAN STEENWINKEL, M. 2004. Stratal geometries and patterns of platform carbonates: the Cretaceous of Oman. *In*: EBERLI, G., MASSEFERRO, J.-L. & SARG, J.F.R. (eds) *Seismic Imaging of Carbonate Reservoirs and Systems*. American Association of Petroleum Geologists, Memoirs, **81**, 185–206.
DUNHAM, R.J. 1962. Classification of carbonate rocks according to depositional texture. *In*: HAM, W.E. (ed.) *Classification of Carbonate Rocks. A Symposium*. American Association of Petroleum Geologists, Memoirs, **1**, 108–121.
EMBONG, M., HIGHASHI, M. ET AL. 2012. Reservoir charactersation of Mishrif Formation of Garraf field, Iraq, using 3D seismic and AI inversion. *Paper presented at the Petroleum Geoscience Conference and Exhibition (PGCE) 2012*, 23–24 April 2012, Kuala Lumpur, Malaysia.
FARZADI, P. 2006. The development of Middle Cretaceous carbonate platforms, Persian Gulf, Iran: constraints from seismic stratigraphy, well and biostratigraphy. *Petroleum Geoscience*, **12**, 59–68, https://doi.org/10.1144/1354-079305-654
FLÜGEL, E. 1982. *Microfacies Analysis of Limestones*. Springer, Berlin.
GILLI, E., MASSE, J.P. & SKELETON, P.W. 1995. Rudist as gregarious sediment dwellers, not reef builders, on Cretaceous carbonate platforms. *Palaeogeography, Palaeoclimatology, Palaeoecology*, **118**, 245–267.
HOLLIS, C. 2011. Diagenetic controls on reservoir properties of carbonate successions within the Albian–Turonian of the Arabian Plate. *Petroleum Geoscience*, **17**, 223–241, https://doi.org/10.1144/1354-079310-032
JASSIM, S.Z. & GOFF, J.C. (eds). 2006. *Geology of Iraq*. Dolin, Prague and Moravian Museum, Brno, Czech Republic.
KERANS, C. & TINKER, C. 1997. *Sequence Stratigraphy and Characterization of Carbonate Reservoirs*. Society of Economic Paleontologists and Mineralogists, Short Course Notes, **40**.
KERSHAW, S. 1994. Classification and geological significance of biostromes. *Facies*, **31**, 81–92.
LAMBERT, L., DURLET, C., LOREAU, J.P. & MARNIER, G. 2006. Burial dissolution of micrite in Middle East carbonate reservoirs (Jurassic–Cretaceous): keys for recognition and timing. *Marine and Petroleum Geology*, **23**, 79–92.
LUCIA, F.J. 1995. Rock fabric/petrophysical classification of carbonate pore space for reservoir characterization. *American Association of Petroleum Geologists Bulletin*, **79**, 1275–1300.
LUCIA, F.J. 2007. *Carbonate Reservoir Characterization*. 2nd edn. Springer, New York.
MAHDI, T.A. & AQRAWI, A.A.M. 2014. Sequence stratigraphic Analysis of the mid-Cretaceous Mishrif Formation in southern Mesopotamian Basin, Iraq. *Journal of Petroleum Geology*, **37**, 287–312.
MAHDI, T.A., AQRAWI, A.A.M., HORBURY, A. & SHERWANI, G.H. 2013. Sedimentological characterization of the mid-Cretaceous Mishrif reservoir in southern Mesopotamian Basin, Iraq. *GeoArabia*, **18**, 139–174.
MIDDLETON, G.V. 1973. Johannes Walther's Law of the Correlation of Facies. *Geological Society of America Bulletin*, **84**, 979–988.
PERIERE, M.D., DURLET, C., VENNIN, E., LAMBERT, L., BOURILLOT, R., CALINE, B. & POLI, E. 2011. Morphometry of micrite particles in cretaceous microporous limestones of the Middle East: influence on reservoir properties. *Marine and Petroleum Geology*, **28**, 1727–1750.
RIDING, R. & TOMÁS, S. 2006. Stromatolite reef crusts, Early Cretaceous, Spain: bacterial origin of in situ precipitated peloid microspar? *Sedimentology*, **53**, 23–34.
RIDING, R. & AWRAMIK, S.M. (eds). 2000. *Microbial Sediments*. Springer, Berlin.
RUPPEL, S.C. (ed.). 2004. *Multidisciplinary Imaging of Rock Properties in Carbonate Reservoirs for Flow-Unit Targeting*. The University of Texas at Austin, Bureau of Economic Geology, Final Technical Report.
SADOONI, F.N. 2005. The nature and origin of Upper Cretaceous basin margin rudist buildups of the Mesopotamian Basin, southern Iraq, with consideration of possible hydrocarbon stratigraphic entrapment. *Cretaceous Research*, **26**, 213–224.
SADOONI, F.N. & AQRAWI, A.A.M. 2000. Cretaceous sequence stratigraphy and petroleum potential of the

Mesopotamian Basin, Iraq. *In*: SCOTT, B. & ALSHARHAN, A.S. (eds) *Middle East Models of Jurassic/Cretaceous Carbonate Systems*. Society of Economic Paleontologists and Mineralogists, Special Publications, **69**, 315–334.

SCOTT, R.W. 1988. Evolution of the late Jurassic and early Cretaceous reefs biotos. *Palaios*, **3**, 184–193.

SHARLAND, P.R., ARCHER, R. ET AL. 2001. *Arabian Plate Sequence Stratigraphy*. GeoArabia, Special Publications, **2**.

SHERWANI, G.H. 1998. *Sequence stratigraphy and depositional systems of Cenomanian–Early Turonian formations in southern Iraq*. PhD thesis, University of Baghdad.

VAN BUCHEM, F., RAZIN, P., HOMEWOOD, P., OTTERBOOM, H. & PHILIP, J. 2002. Stratigraphic organization of carbonate ramps and organic intrashelf basins: Natih Formation (middle Cretaceous) of northern Oman. *American Association of Petroleum Geologists Bulletin*, **86**, 21–54.

VAN BUCHEM, F.S.P., SIMMONS, M.D., DROSTE, H.J. & DAVIES, R.B. 2011. Late Aptian to Turonian stratigraphy of the eastern Arabian Plate – depositional sequences and lithostratigraphic nomenclature. *Petroleum Geoscience*, **17**, 211–222, https://doi.org/10.1144/1354-079310-061

VIDETICH, P.E., MCLIMANS, P.K., WATSON, H.K.S. & NAGY, R.M. 1988. Depositional, diagenetic, thermal, and maturation histories of Cretaceous Mishrif Formation, Fateh Field, Dubai. *American Association of Petroleum Geologists Bulletin*, **7**, 1143–1159.

VOLERY, C., DAVAUD, E., FOUBERT, A. & CALINE, B. 2009. Shallow-marine microporous carbonate reservoir rocks in the Middle East: relationship with seawater Mg/Ca ration and eustatic sea level. *Journal of Petroleum Geology*, **32**, 313–326.

YOKOI, K. & SATO, R. 2004. Petroleum potential of Iraq – Outline of petroleum geology and undeveloped giant oil fields in southern Iraq. *Journal of the Japanese Association for Petroleum Technology*, **69**, 12–22.

ZIEGLER, M.A. 2001. Late Permian to Holocene paleofacies evolution of the Arabian Plate and its hydrocarbon occurrences. *GeoArabia*, **6**, 445–504.

Burial estimates constrained by clumped isotope thermometry: example of the Lower Cretaceous Qishn Formation (Haushi-Huqf High, Oman)

CÉDRIC M. JOHN

Department of Earth Science and Engineering and Qatar Carbonate and Carbon Storage Research Centre, Imperial College London, London SW7 2AZ, UK
Cedric.john@imperial.ac.uk

Abstract: Accurate determination of the thermal history of sedimentary basins is critical to constrain the timing of diagenetic processes. Here, clumped isotope palaeothermometry is used to estimate minimum burial depth of the Lower Cretaceous Qishn Formation in east central Oman. Fossil oysters were collected in a soft argillaceous unit, thin-sectioned, and studied under petrographical and cathodoluminescence microscopy, revealing a variable state of shell preservations with early silica replacement of carbonate and late-stage calcite cementation. Clumped isotopes values varied from 0.602 to 0.666‰, with the best-preserved oyster shell yielding a temperature of 37 ± 4°C and a calculated oxygen isotope ratio in seawater ($\delta^{18}O_{seawater}$) compatible with Cretaceous seawater having experienced moderate evaporation (1.0–1.5‰ VSMOW). The new minimum estimates for the burial depth of the Qishn Formation is 1.0–1.2 km, based on the temperature difference between the well-preserved oyster and the partially neomorphosed oyster recording the highest burial temperature (63 ± 4°C). This is in excess of what was predicted by previous studies (<400 m), but compatible with conodonts alteration index temperatures (<80°C). This study highlights the potential of clumped isotopes as a quantitative tool to estimate temperature and burial depth in fine-grained carbonate succession where fluid inclusions are absent, and offers a new tool to constrain the thermal histories of sedimentary basins.

Two essential parameters in reconstructing the pathway from sediment to rock are the depth to which the sediment was buried, and the maximal burial temperature attained. These parameters are crucial because they influence organic and inorganic diagenesis, thus impacting porosity and permeability in the reservoir, organic matter maturation, and the history of hydrocarbon migration. Hence, being able to better constrain the thermal history of sedimentary basins is of great interest not only to understand fundamental Earth science problems, but also to improve Play Fairway Analysis and the relative timing of porosity evolution and hydrocarbon charging.

Traditional palaeothermometers have been used for decades, but every tool suffers from intrinsic limitations. Fluid-inclusion palaeothermometry provides a mean of reconstructing the maximum burial temperature and, potentially, the temperature of initial entrapment of the inclusion, but inclusions are prone to resetting (Goldstein & Reynolds 1994). In addition, generating the data can be very time intensive, and the method only works when crystals large enough to contain visible fluid inclusions are present in the sample. The alteration colour of conodonts is another classical approach to assessing maximum burial depth (Epstein *et al.* 1976), but the method only gives a broad quantitative range for temperatures and is only applicable in Cambrian–early Triassic sediments. Oxygen isotope ratios in carbonates ($\delta^{18}O_{carbonate}$) are sensitive to the temperature of the fluid at time of mineralization, and can thus be used as an additional palaeotemperature proxy (Kim & O'Neil 1997). However, $\delta^{18}O_{carbonate}$ also depends on the isotopic composition of the fluid from which the carbonate precipitates ($\delta^{18}O_{water}$) and thus it is very difficult to disentangle the compositional information from the temperature component of $\delta^{18}O_{carbonates}$ (Kim & O'Neil 1997).

This paper focuses on using the carbonate 'clumped isotope' palaeothermometer (Δ_{47}), a relatively new stable isotope technique that yields the temperature of precipitation (and subsequent neomorphism and/or recrystallization) of carbonates (Eiler 2007), which can be a very useful addition to the current toolkit to reconstruct the thermal history of basins. The Δ_{47} parameter is mathematically defined as:

$$\Delta_{47} = \left[\frac{R^{47}}{2R^{13}R^{18} + 2R^{17}R^{18} + R^{13}(R^{17})^2} - \frac{R^{46}}{2R^{18} + 2R^{13}R^{17} + (R^{17})^2} - \frac{R^{45}}{R^{13} + 2R^{17}} + 1\right]1000$$

From: Armitage, P. J., Butcher, A. R., Churchill, J. M., Csoma, A. E., Hollis, C., Lander, R. H., Omma, J. E. & Worden, R. H. (eds) 2018. *Reservoir Quality of Clastic and Carbonate Rocks: Analysis, Modelling and Prediction.* Geological Society, London, Special Publications, **435**, 107–121.
First published online November 18, 2015, https://doi.org/10.1144/SP435.5

where R^i is the measured ratios of CO_2 of masses 45, 46 and 47 to mass 44, and R^{13} is the ratio of $^{13}C/^{12}C$ and R^{18} is the ratio of $^{18}O/^{16}O$ calculated from R^{45} and R^{46} assuming random distribution. R^{17} is calculated from R^{18} assuming a mass-dependent relationship between ^{18}O and ^{17}O.

At low temperatures, 'clumping' of heavy isotopes ($^{13}C-^{18}O$) is favoured because the vibrational energy in a heavy–heavy bond is less than half that of the corresponding light–light bond ($^{12}C-^{16}O$). The heavier molecule, consequently, is more stable and its abundance deviates from a stochastic distribution following a predictable law that is a function of temperature (Schauble *et al.* 2006; Eiler 2007). At high temperatures, the effects of entropy mask the effects of clumping, giving rise to a stochastic distribution of isotopologues. Therefore, clumped isotopes offer a single-phase palaeothermometer applicable to all carbonates and to carbon-fluoro apatite (Schauble *et al.* 2006; Eagle *et al.* 2010; Eiler 2011; Huntington *et al.* 2011). The relationship between Δ_{47} in carbonate and temperature was established for Earth surface temperatures by earlier work (Ghosh *et al.* 2006; Dennis & Schrag 2010), and has recently been calibrated in the 25–250°C range (Kluge *et al.* 2015). The high-temperature calibration significantly reduces uncertainties in applications of the proxy to subsurface environments where temperatures are often in excess of 70°C.

The goal of this study is to demonstrate how clumped isotopes can be used to gain insight into the thermal history of a sedimentary basin. The case study selected is the Lower Cretaceous Qishn Formation, a series of shallow-water tidal carbonates that crop out in the region of the Haushi-Huqf tectonic high (see Ries & Shackleton 1990) in east central Oman (Fig. 1). The Qishn Formation is an ideal candidate for applying this technique for several reasons. First, it is a surface analogue to the oil-bearing Kahraib and Shuaiba formations in Oman and the Middle East (Immenhauser *et al.* 2004). Second, the tectonic history of the Haushi-Huqf High is relatively sparsely constrained (Ries & Shackleton 1990): estimates of maximum burial depth of the Cretaceous outcrops are of 400 m based on post-Cretaceous sediment coverage (Immenhauser *et al.* 2004; Sattler *et al.* 2005), a plausible but loosely constrained estimate that seems at odds with evidence of pressure solution at the outcrop (Sena & John 2013; Sena *et al.* 2014). Hence, obtaining better constraints in this region through the application of novel methods is desirable. Finally, because of the low burial and temperature regime of the Qishn Formation, and the fine grained-nature of the sediment deposited, no fluid-inclusion data can be extracted. Thermal constraints do exist in the Permian Gharif Formation that sits unconformably immediately below the Lower Cretaceous series, and which contains conodonts with a colour alteration index (CAI) of 1.0–1.5 that indicates very little post-burial heating (less than 50–80°C: Angiolini *et al.* 2003). However, conodonts only provide a maximum temperature limit (<80°C), and cannot provide a good minimum estimate to help bracket the maximum burial depth more accurately (Epstein *et al.* 1976).

The approach taken in this study is to measure the clumped isotope signature of Cretaceous oysters collected in the Qishn Formation, and to derive temperatures for palaeo-environments and post-deposition diagenetic transformation of the shells. The latter also yields quantitative minimum estimates of burial depth for the Haushi-Huqf structural high.

Material and methods

Background on the Qishn Formation and samples

A recent detailed facies description and palaeoenvironmental interpretation of the Qishn Formation can be found in Sena & John (2013), and the broad depositional framework of the formation is described earlier works by Immenhauser *et al.* (2004) and Sattler *et al.* (2005). The Qishn Formation (Beydoun & Bamahmoud 1993) is composed of carbonate sediments deposited during the Barremian–Aptian stages, and is equivalent to the Upper Kharaib and Lower Shu'aiba formations of northern Oman (Immenhauser *et al.* 2004), also known as the Kahmah Group in Oman and the Thamama Group in the United Arab Emirates (UAE). The Khamah Group is interpreted as one large second-order sea-level cycle, beginning with the aggrading–retrograding pelagic Rayda marls and the prograding Habshan–Lekhwair platform carbonates, and terminating with the aggradation of shallow-marine platform carbonates of the Kharaib and Shu'aiba formations (see Grelaud *et al.* 2006 and references therein). Some of the best examples of Cretaceous tidal flat deposits are found in the Qishn Formation of Oman (Sattler *et al.* 2005; Sena & John 2013), which contain laminated sediment with rhyzocretions, tidal channels or scours, desiccation cracks, and generally fine-grained facies rich in Baccinella algae, as well as packstone–grainstone facies rich in rudist fragments, Orbitolina and corals. The Cretaceous oysters investigated here were sampled at Wadi Jarrah (Fig. 1) within an argillaceous wackestone–packstone unit interpreted as a deep subtidal, high-diversity fauna assemblage and located immediately above omission surface 'OS1' (Sattler *et al.* 2005; Sena &

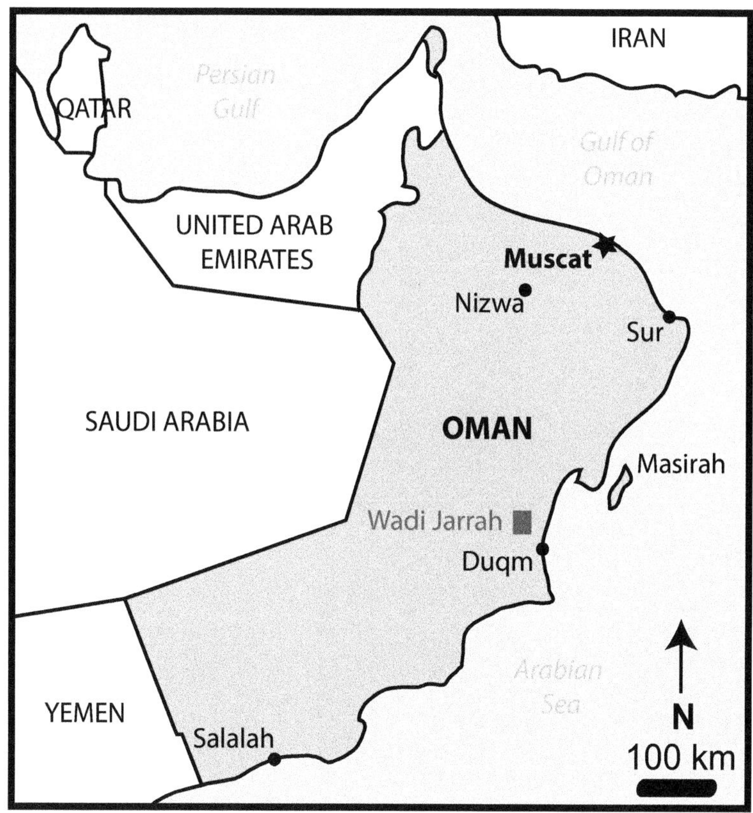

Fig. 1. Location of the Haushi-Huqf High and the Wadi Jarrah outcrops where oysters were extracted (square). Modified from Sena & John (2013).

John 2013) (Fig. 2). The wackestone–packstone unit is bioturbated, and contains (alongside oysters) peloids, echoinoids, some porifers, benthic foraminifers and various other skeletal debris (Sena & John 2013). A total of five oysters were sampled by hand and measured in this study, and are named samples 4-1, 4-2, 5-1, 7-2 and 7-4 (Fig. 3).

Analytical procedures

Each oyster shell was rinsed under reverse-osmosis water and sonicated to remove clays, and then cleaned with 10% hydrogen peroxide to remove any remnant of organic material. The shells were sawed in two along the plane of symmetry of each valve, and doubly polished thin sections were made that comprised the entire cross-section of the shell. Thin section were first examined under a Zeiss Axioskop 40 polarization microscope and then under a CITL Cathodoluminescence Mk5-2 stage (operating conditions of about 280 μA and 15 kV) mounted on a Nikon Eclipse 50i microscope. Carbonate powers were extracted from the centre of the cross-sections close to the shell hinges using a dental drill equipped with a tungsten-carbide bit, using medium speed. To determine the mineralogy of the shell, an aliquot of the carbonate powder was analysed by attenuated total reflectance Fourier transform infrared (ATR-FTIR). Spectra were recorded using a Thermo Electron Nicolet 5700 spectrometer equipped with a deuterated triglycine sulphate (DTGS) detector and a Smart Orbit diamond ATR. Spectra was measured at a resolution of 4 cm^{-1} in the range 4000–400 cm^{-1} (32 scans) and were background subtracted.

Clumped isotope analyses were performed in the Qatar Stable Isotope Laboratory at Imperial College following laboratory procedures previously described in Dale *et al.* (2014). About 5 mg of carbonate powder were reacted on a manual vacuum line in individual reaction vessels with approximately 2 ml of 105% orthophosphoric acid at 90°C. The reactant CO_2 was cleaned using a procedure initially described by Dennis & Schrag (2010), where the gas is passively passed through a hydrocarbon trap densely packed with Porapak Q (filled length

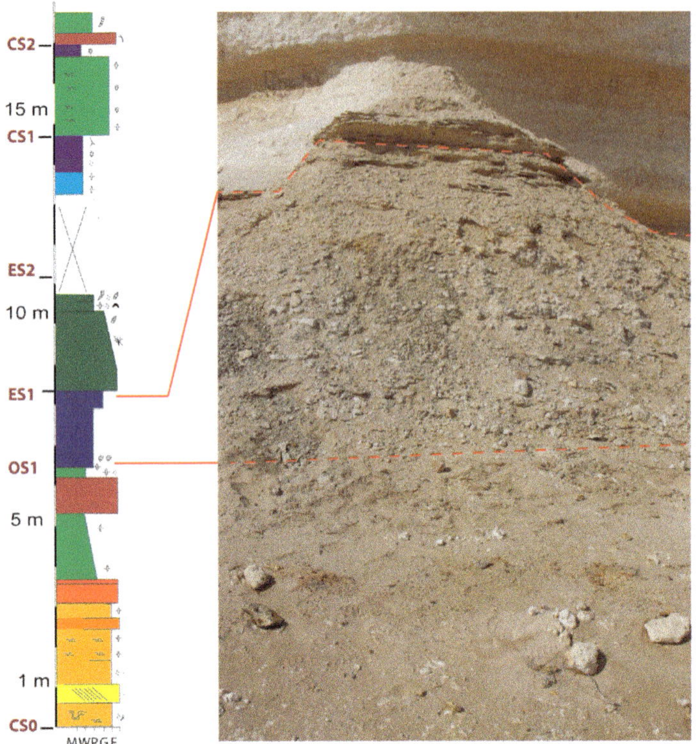

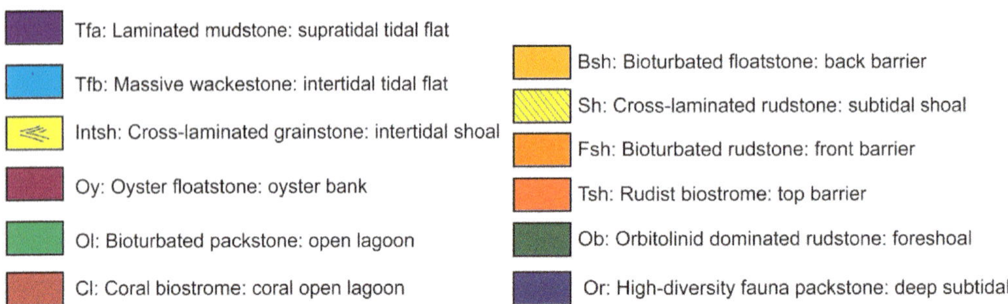

Fig. 2. Simplified stratigraphic log of the Jurf and Qishn formations (modified from Sena & John 2013), and photograph of the packstone bed from which oysters were extracted, immediately on top of flooding surface OS1 (Sattler *et al.* 2005).

13 cm, inner diameter *c.* 8 mm) held at −35°C. Two mass spectrometers ('Pinta' and 'Nina') with distinct reference frames (Table 1) were used for the analysis of the gas, following procedures described by Huntington *et al.* (2009) and Dennis *et al.* (2011). Each measurement consists of eight cycles with seven sequences per cycle, and include a peak centre, background measurements and an automatic bellows pressure adjustment aimed at a 15 V signal at mass 44. The sample gas is measured against an Oztech reference gas standard ($\delta^{13}C = -3.63‰$ VPDB, $\delta^{18}O = -15.79‰$ VPDB for Niña; $\delta^{13}C = -3.62‰$ VPDB, $\delta^{18}O = -15.73‰$ VPDB for Pinta). Heated gases (1000°C) were used to correct for non-linearity in the mass spectrometer following Huntington *et al.* (2009). Heated gas was combined with an in-house Carrara Marble (ICM) and an interlaboratory carbonate standard ('ETH3')

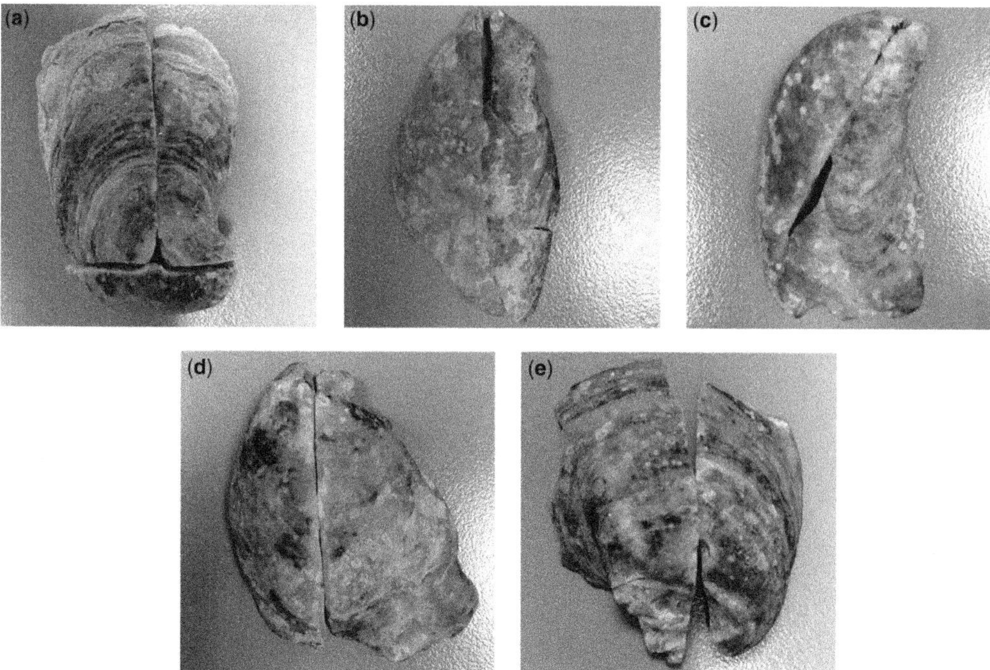

Fig. 3. Pictures of the five oyster shells selected for this study: (**a**) sample 4-1; (**b**) sample 4-2; (**c**) sample 5-1; (**d**) sample 7-3; and (**e**) sample 7-4.

to build a day-to-day secondary transfer function alongside a series of CO_2 gases equilibrated at various temperatures ('EQ') (Table 1) to transfer the measured values into the 'Carbon dioxide equilibrated scale' reference frame following Dennis *et al.* (2011). The Δ_{47} values for ICM (0.312‰) and ETH3 (0.634‰) reacted at 90°C were previously determined in-house using a primary reference frame consisting of equilibrated CO_2 at 1000, 80, 50 and 25°C following Dennis *et al.* (2011). Sample contamination was monitored using the mass 48 and mass 49 signals, the Δ_{47} values are linearity corrected using heated gas data (Huntington *et al.* 2009) and corrected for isotope fractionation during phosphoric acid digestion using a value of 0.069 (Guo *et al.* 2009; Wacker *et al.* 2013). Carbonate $\delta^{18}O$ values are calculated using the acid fractionation factors of Kim & O'Neil (1997). The Δ_{47} values were converted to temperature using the recent calibration developed at Imperial College (Kluge *et al.* 2015), and the $\delta^{18}O_{water}$ was calculated using the $\delta^{18}O_{calcite}$, clumped isotope temperatures and the equation reported in Friedman & O'Neil (1977). The standard error for Δ_{47} values of standard was ±0.002‰, and the standard deviation for $\delta^{18}O$ and $\delta^{13}C$ were 0.2 and 0.1‰, respectively.

Results

Mineralogy and petrography

FTIR results indicate that all samples are composed of calcite and silica, and that aragonite is absent (Fig. 4). Petrographical observations confirm the presence of silica replacing carbonates and forming cauliflower structure in all samples (Fig. 5e, f). The silicification is fabric selective, with silica replacement mimicking the structure of the carbonate shell (Fig. 5e, f). The abundance of silica in the shell is, however, not proportional to the magnitude of the FTIR absorbance band in each sample: sample 4-1, for instance, has a large number of silica replacement structures but a low absorbance for silica; whilst, conversely, sample 7-4 has a strong absorbance for silica, but relatively few silica replacement structures. Under cathodoluminescence (CL) microscopy, the silica structure in samples 4-1, 4-2, 5-1 and 7-1 display a dull orange luminescence. By contrast, silica replacement structures in sample 7-4 are non-luminescent. The un-silicified shell material is diversely and variously preserved. Sample 7-4 shows well-preserved detailed structure of the shell with very fine laminations (Fig. 5a) and a complete lack of luminescence under CL (Fig. 6c1,

Table 1. *Summary of all standards used to correct the data and project it in the absolute reference frame of Dennis* et al. *(2011)*

Reference frame for mass spectrometer 'Pinta'					Reference frame for mass spectrometer 'Nina'					
Date	Δ_{47} heated (raw)	Δ_{47} heated (raw)	Date	Δ_{47} linearity corrected	Date	Δ_{47} heated (raw)	Δ_{47} heated (raw)	Date	Δ_{47} linearity corrected	
Heated gas			**Carrara Marble**		**Heated gas**			**Carrara Marble**		
30 May 2014	−0.85	16.49	1 June 2014	−0.582	20 February 2013	−0.621	12.802	12 April 2013	0.188	
30 May 2014	−1.02	−48.95	3 June 2014	−0.557	21 February 2013	−0.997	−13.566	16 April 2013	−0.564	
2 June 2014	−0.87	16.58	7 June 2014	−0.516	22 February 2013	−1.016	−12.074	17 April 2013	−0.529	
6 June 2014	−0.88	16.54	10 June 2014	−0.542	25 February 2013	−1.243	−27.428	20 April 2013	−0.562	
9 June 2014	−0.74	16.59	16 June 2014	−0.546	26 February 2013	−1.604	−48.978	22 April 2013	−0.517	
12 June 2014	−0.99	−48.98	19 June 2014	−0.549	28 February 2013	−1.544	−48.583	14 May 2013	−0.587	
18 June 2014	−0.83	17.53	27 June 2014	−0.597	11 April 2013	−0.475	−47.680	26 May 2013	−0.566	
27 June 2014	−0.88	17.42	1 July 2014	−0.552	12 April 2013	0.187	6.791	27 May 2013	−0.556	
27 June 2014	−1	−49.08	6 July 2014	−0.587	15 April 2013	−0.031	−0.623	4 June 2013	−0.453	
30 June 2014	−0.87	17.44	9 July 2014	−0.570	16 April 2013	−0.438	−15.752	7 June 2013	−0.554	
4 July 2014	−0.86	17.05	14 July 2014	−0.577	16 April 2013	−1.371	−34.672	8 June 2013	−0.592	
8 July 2014	−0.88	16.96	24 July 2014	−0.559	17 April 2013	−1.391	−34.675	15 June 2013	−0.535	
11 July 2014	−1	−48.78	29 July 2014	−0.578	17 April 2013	−0.498	21.778	17 June 2013	−0.546	
17 July 2014	−0.85	16.99	3 August 2014	−0.573	18 April 2013	−0.936	−10.017	18 June 2013	−0.566	
21 July 2014	−1.01	−49.01	2 September 2014	−0.524	18 April 2013	−0.942	−10.016	21 June 2013	−0.486	
28 July 2014	−0.83	17.02	Change of reference frame		19 April 2013	−0.714	6.318	22 June 2013	−0.543	
30 July 2014	−0.98	−48.88	17 May 2013	−0.622	25 April 2013	−0.937	−10.754	24 June 2013	−0.549	
1 August 2014	−0.85	14.46	26 May 2013	−0.633	26 April 2013	−0.704	7.874	29 June 2013	−0.539	
30 August 2014	−0.83	14.11	26 May 2013	−0.597	29 April 2013	−0.677	8.692	6 July 2013	−0.562	
30 August 2014	−0.97	−48.58	27 May 2013	−0.622	6 May 2013	−0.908	−7.030	9 July 2013	−0.437	
30 August 2014	−0.83	15.24	27 May 2013	−0.622	8 May 2013	−1.137	−18.975			
30 August 2014	−0.86	13.9	30 May 2013	−0.579	9 May 2013	−0.714	2.958			
Change of reference frame			29 June 2013	−0.574	16 May 2013	−0.711	5.576			
9 May 2013	−0.78	−1.27	6 July 2013	−0.609	18 May 2013	−0.779	1.817			
10 May 2013	−0.83	6.33	17 July 2013	−0.587	19 May 2013	−1.641	−49.251	Date	Δ_{47} heated (raw)	Δ_{47} heated (raw)
22 May 2013	−0.93	−48.91	19 July 2013	−0.616	20 May 2013	−1.610	−48.990			
23 May 2013	−0.9	−47.95	23 July 2013	−0.600	22 May 2013	−0.673	9.507	**Equilibrated gas 25°C**		
23 May 2013	−0.81	5.77	4 August 2013	−0.582	24 May 2013	−0.681	5.924	22 April 2013	−0.14	−8.478

29 May 2013	−0.927	−48.807
25 June 2013	−0.930	−48.653
25 June 2013	−0.803	7.890
26 June 2013	−0.930	−48.653
3 July 2013	−0.826	7.730
8 July 2013	−0.786	6.666
12 July 2013	−0.986	−48.701
16 July 2013	−0.816	−4.486
18 July 2013	−0.788	9.784
24 July 2013	−0.895	2.631
25 July 2013	−0.730	6.787
26 July 2013	−0.812	−32.078
28 July 2013	−0.731	1.139
29 July 2013	−0.811	−48.902
30 July 2013	−0.754	−27.247
31 July 2013	−0.958	−48.811
31 July 2013	−0.845	10.918
1 August 2013	−0.942	−49.132
1 August 2013	−0.885	8.699
1 August 2013	−0.842	9.662
2 August 2013	−0.811	9.152
3 August 2013	−0.824	11.166
7 August 2013	−0.866	7.279
12 August 2013	−0.830	11.910
14 August 2013	−0.959	−48.789
19 August 2013	−0.861	8.328
21 August 2013	−0.858	7.961
28 August 2013	−0.855	7.222
8 September 2013	−0.826	12.997
9 September 2013	−0.815	15.202

6 August 2013	−0.594
10 August 2013	−0.605
13 August 2013	−0.604
20 August 2013	−0.624
28 August 2013	−0.529
3 September 2013	−0.569
7 September 2013	−0.605
10 September 2013	−0.555
10 September 2013	−0.545
ETH3	
1 June 2014	−0.278
5 June 2014	−0.240
8 June 2014	−0.265
11 June 2014	−0.321
17 June 2014	−0.242
27 June 2014	−0.316
2 July 2014	−0.269
7 July 2014	−0.286
10 July 2014	−0.282
16 July 2014	−0.314
19 July 2014	−0.289
25 July 2014	−0.245
31 July 2014	−0.289
3 September 2014	−0.261
Change of reference frame	
30 June 2013	−0.273
19 July 2013	−0.325
19 July 2013	−0.313
19 July 2013	−0.314
20 July 2013	−0.327
26 July 2013	−0.238
26 July 2013	−0.183
26 July 2013	−0.145

29 May 2013	−1.656	−49.265
3 June 2013	−0.764	3.964
5 June 2013	−0.627	11.644
6 June 2013	−0.644	8.853
12 June 2013	−0.619	
20 June 2013	−0.618	11.665
25 June 2013	−1.651	−49.092
28 June 2013	−0.648	10.992
1 July 2013	−0.643	11.083
2 July 2013	−0.686	7.631
5 July 2013	−0.622	12.577
8 July 2013	−0.681	6.874
10 July 2013	−0.696	7.745
12 July 2013	−1.747	−49.283

Equilibrated gas 50°C		
7 May 2013	−0.287	−10.952
25 May 2013	0.462	33.77
Equilibrated gas 80°C		
15 April 2013	−0.215	−37.294
3 May 2013	−0.464	−15.698
13 May 2013	−0.485	−15.129
31 May 2013	−0.418	−15.922
10 June 2013	0.202	25.814

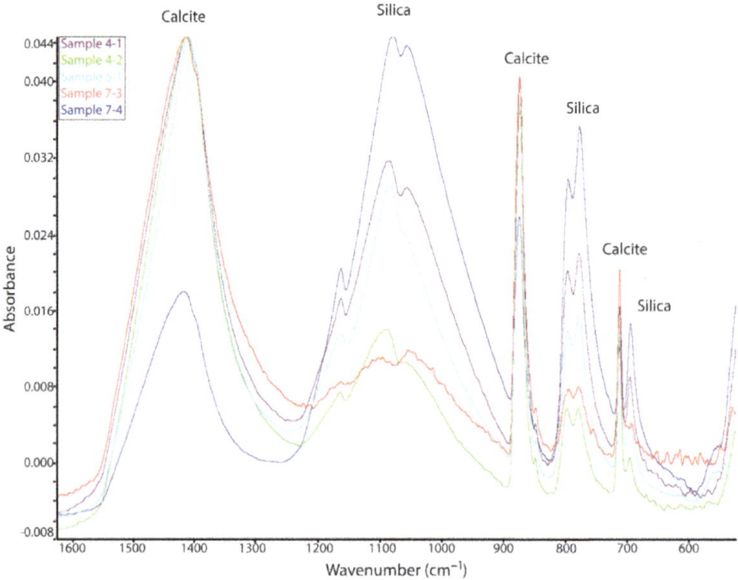

Fig. 4. FTIR results for all samples. The phases identified are marked on the figure, and are limited to silica and calcite.

c2). The other shell samples (4-1, 4-2, 5-1 and 7-3) show evidence of calcite neomorphose, resulting in a loss of the original shell texture (Fig. 5b–d). In addition, these samples also yield a bright luminescent cement phase under CL, either as replacement of original shell material or within microfractures (Fig. 6a1, a2). Late-stage calcite is limited to one microfracture in sample 7-4, and this was not sampled during drilling.

Stable isotope results

The Δ_{47} values of oyster shells range from 0.666 $\pm$ 0.008‰ (sample 7-4) to 0.602 $\pm$ 0.009‰ (sample 4-2: Table 2). These Δ_{47} values correspond to temperatures of 37$\pm$3 and 63 $\pm$ 4°C using the calibration of Kluge *et al.* (2015) (Fig. 7). The $\delta^{18}O_{calcite}$ of all samples is relatively similar, with an average of 2.5 $\pm$ 0.16‰ VPDB (1 SD). The calculated $\delta^{18}O_{water}$ based on Friedman & O'Neil (1977) varies from +1.7‰ VSMOW for sample 7-4 to 5.9‰ VSMOW for sample 4-2. All other samples have values intermediate between these two, and fall on a line that follows the 2.5‰ calcite line in a clumped isotope temperature v. water isotope composition space (Fig. 7). The carbon isotope data clusters into two groups: the first is around 0.15‰ and the second is around 1.4‰. One sample, 7-4, has a large standard deviation in $\delta^{13}C$, representing different parts of the shell being sampled and each falling into a different cluster for $\delta^{13}C$.

Discussion

Diagenetic history of the oyster shells

Sample 7-4 has the best-preserved shell, with the fine lamellar original structure of the shell still largely intact at the location were powders were extracted (Fig. 5a). A phase of silica replacement during early burial is visible within this shell, as well as in other shells, and it is apparent that this replacement was mimetic as details of the shell growth bands can still be seen in the silica material (Fig. 5e, f). It would appear that other than this early phase of silicification, sample 7-4 was not heavily altered during burial. For the other samples, evidence from petrography points to a more complex diagenetic history following silicification. The detailed petrographical structure of the shell in samples 4-1, 4-2, 5-1 and 7-3 is variable, with zones of fair preservation (Fig. 5c, d) ranging to zones that are clearly heavily recrystallized (Fig. 5b), and, thus, these samples are referred below as 'altered shell samples'. The altered shell samples show a greater area of cauliflower 'silicified' structures in the shell (see Fig. 6b1), but less overall silica content in FTIR than sample 7-4 (Fig. 4). However, the silica structures within samples such as 4-1 show a dull orange luminescence under CL, whereas opal should not luminesce under the cold CL used in this study and should show a blue luminescence under hot CL (Boggs & Krinsley 2006). The dull orange luminescence of silicified

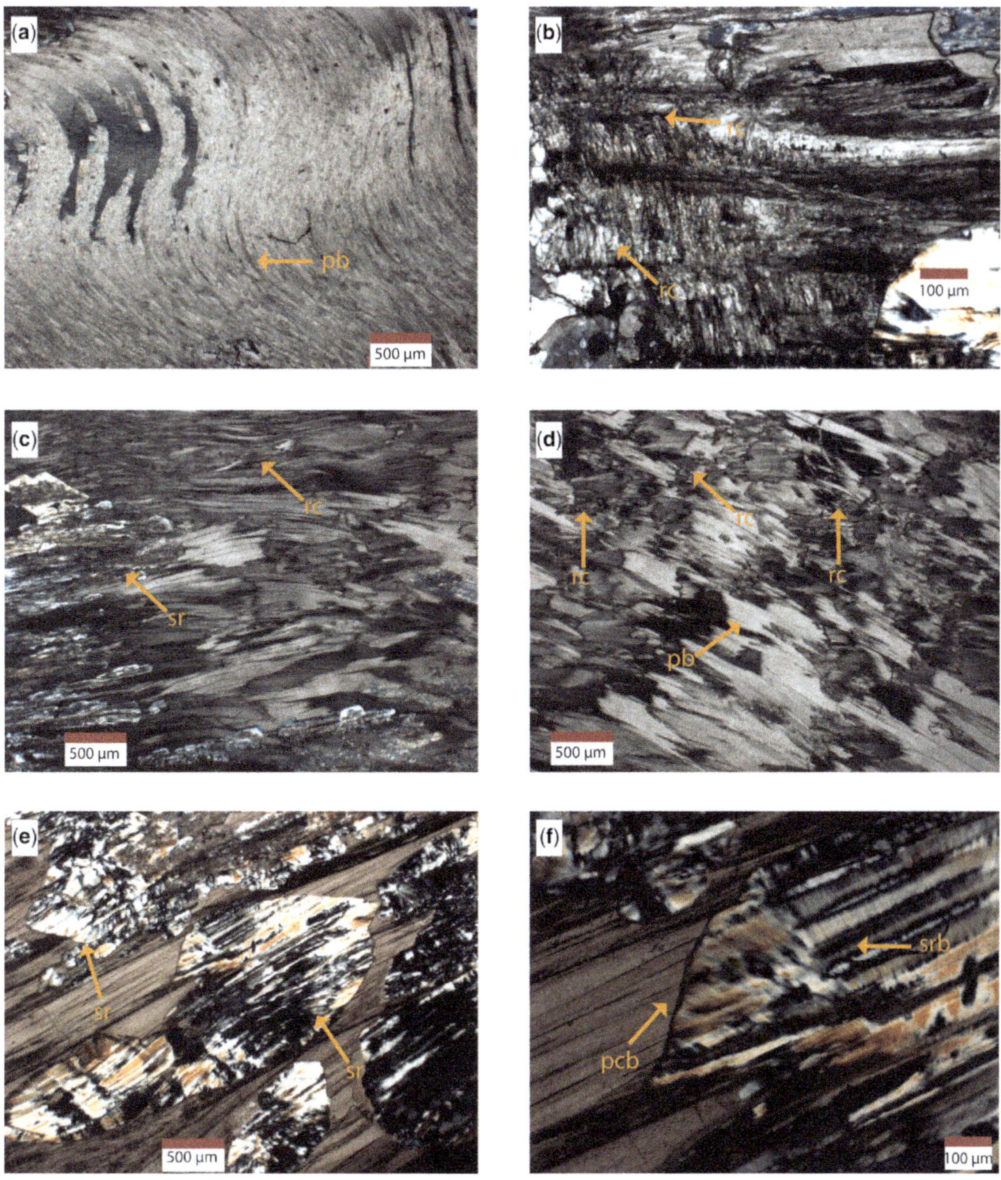

Fig. 5. Petrographical microscope images. (**a**) Sample 7-4, note the well-preserved growth bands of the oyster (pb). (**b**) Sample 4-1, zone of largely neomorphosed calcite within the shell, with recrystallized calcite (rc). (**c**) Sample 4-2, the shell structure is preserved, but some areas are replaced by diagenetic silica (sr), whereas other areas are composed of recrystallized calcite (rc). (**d**) Sample 4-2, same as (c), this area of the shell shows some preserved original banding (pb), but detailed analysis reveals that the structure of the shell has been altered and zones of recrystallized calcite (rc) exist. (**e**) Silica replacement structures within sample 4-1. (**f**) Same as (e), but at $\times 100$ total magnification. Note how the mimetic silica replacement (srb) preserves the banding of the calcitic shell (pcb).

zones in the altered shell samples can be interpreted as subsequent replacement of diagenetic silica by burial calcite, a luminescent phase (Fig. 6b2). This luminescence of silicified structure is absent or much less pronounced in sample 7-4 (Fig. 6c), indicating minimum late-stage calcification. Early silica replacement of calcite followed by (re)calcification of silica later in the diagenetic history is common: for instance, as reported for the Lower Cretaceous calcareous turbidites of the Eastern

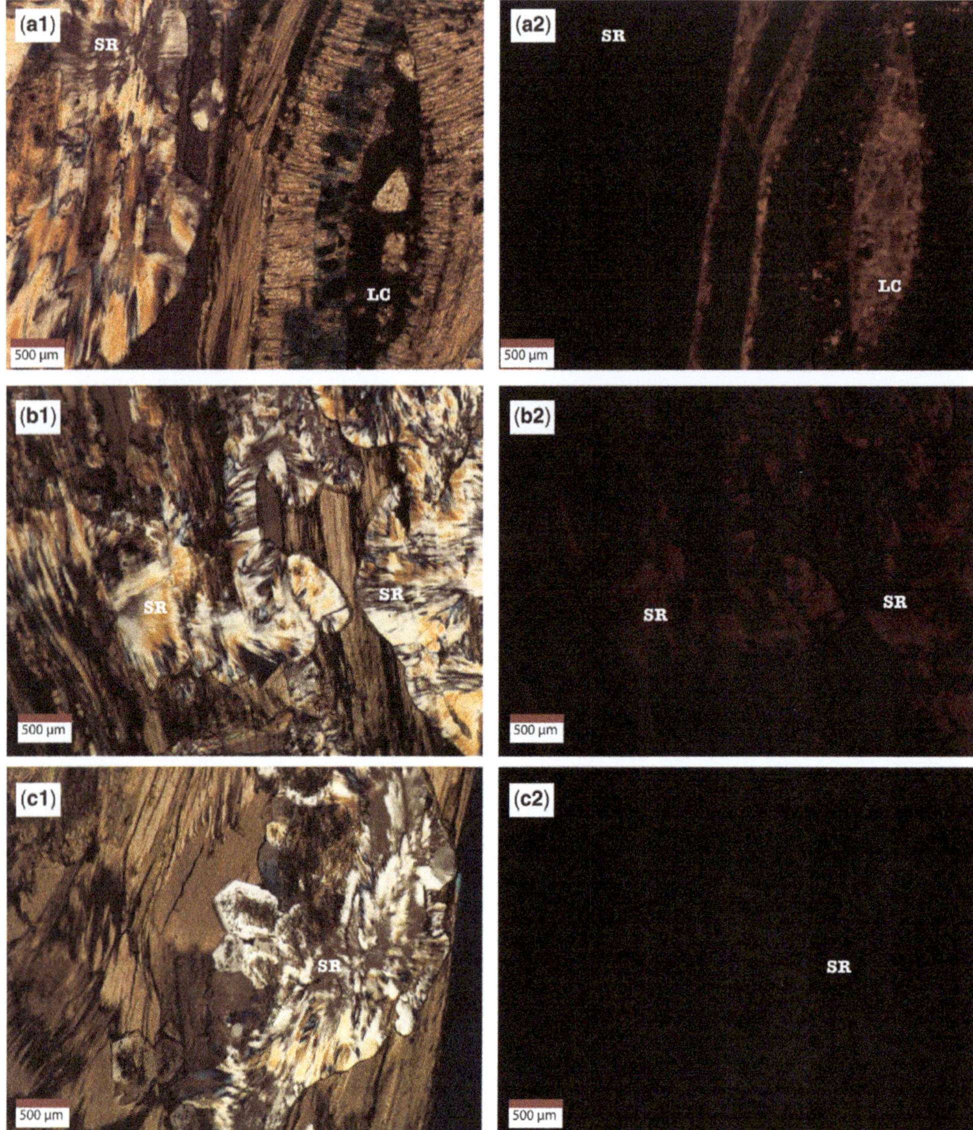

Fig. 6. Cathodoluminescence microscope images. (**a**) Sample 4-1, showing silica replacement (sr) and a late-stage calcite (lc) that luminesce under CL (**a1**). (**b**) Sample 4-1, showing silica replacement (sr) that has a dull luminesce under CL (**b1**), interpreted as late-stage calcification of the early diagenetic silica. (**c**) Sample 7-4, showing silica replacement (sr) and well-preserved shell structure, and a non-luminescence under CL (**c1**).

Alps (Hesse 1987). We propose that the same mechanism applied to our sample material, at least for the altered shell samples. Irrespective of the luminescence of silica structure, a bright-luminescent calcitic cement, interpreted as a burial Mn^{2+}-rich cement, is visible in all of the altered shell samples (Fig. 6a1, a2). The combination of the presence of late-stage cements, potential calcification of initially silicified structures and calcite shell neomorphose explain why sample 7-4 is the only one to preserve a low-temperature, near-surface signal.

Palaeo-environmental implications of sample 7-4

Sample 7-4 yields temperatures that are compatible with a Lower Cretaceous shallow, tropical sea (sea-surface temperature as low as 34°C), although

Table 2. *Clumped isotope results for the oyster shells measured in this study*

Sample	Mass spectrometer	Run date	49 parameter	Δ_{48} offset	$\delta^{18}O$ PDB (corrected)	$\delta^{18}C$ (PDB)	Δ_{47} corrected	±1 SE	Temperature (°C)	Error	$\delta^{18}O$ fluid (‰ VSMOW)
4-1	Nina	14 August 2013	0.07	−0.04	−2.54	0.19	0.623				
	Pinta	14 August 2013	0.05	0.52	−2.73	0.19	0.615				
	Pinta	19 July 2014	−0.10	0.04	−2.65	0.06	0.599				
				Averages:	**−2.64**	**0.15**	**0.612**	**0.007**	**59**	**3**	**5.5**
				SD:	**0.10**	**0.08**					
4-2	Nina	21 August 2013	0.09	0.49	−2.84	1.33	0.607				
	Pinta	21 August 2013	0.07	2.70	−2.92	1.48	0.585				
	Pinta	21 August 2013	0.15	0.87	−2.90	1.43	0.615				
				Averages:	**−2.88**	**1.41**	**0.602**	**0.009**	**63**	**4**	**5.9**
				SD:	**0.04**	**0.08**					
5-1	Nina	14 August 2013	0.05	−0.29	−2.62	1.19	0.610				
	Pinta	14 August 2013	0.07	0.44	−2.56	1.16	0.581				
	Nina	21 August 2013	0.06	−0.03	−2.50	1.14	0.636				
				Averages:	**−2.56**	**1.16**	**0.609**	**0.016**	**60**	**7**	**5.8**
				SD:	**0.06**	**0.03**					
7-3	Nina	19 July 2014	−0.04	1.48	−2.39	1.30	0.628				
	Pinta	20 July 2013	0.01	−0.60	−2.55	1.80	0.594				
	Pinta	19 July 2014	−0.05	−0.45	−2.73	1.10	0.623				
				Averages:	**−2.56**	**1.40**	**0.615**	**0.011**	**58**	**5**	**5.6**
				SD:	**0.17**	**0.36**					
7-4	Nina	4 September 2013	0.28	0.34	−2.15	1.03	0.674				
	Pinta	3 August 2014	0.07	0.49	−2.80	0.13	0.659				
				Averages:	**−2.47**	**0.58**	**0.666**	**0.008**	**37**	**3**	**1.7**
				SD:	**0.46**	**0.64**					

Sample name, mass spectrometer and dates of each runs are indicated, as well as the two parameters used to monitor contaminations (49 parameter and Δ_{48} offset), $\delta^{18}O$ and $\delta^{13}C$ of each shell, and the Δ_{47} value of the shell. Values of each acquisition are averaged per sample, the temperature calculated using Kluge *et al.* (2015), and fluid $\delta^{18}O_{SMOW}$ calculated using the average $\delta^{18}O_{calcite}$ value, clumped temperature and the equation of Friedman & O'Neil (1977).

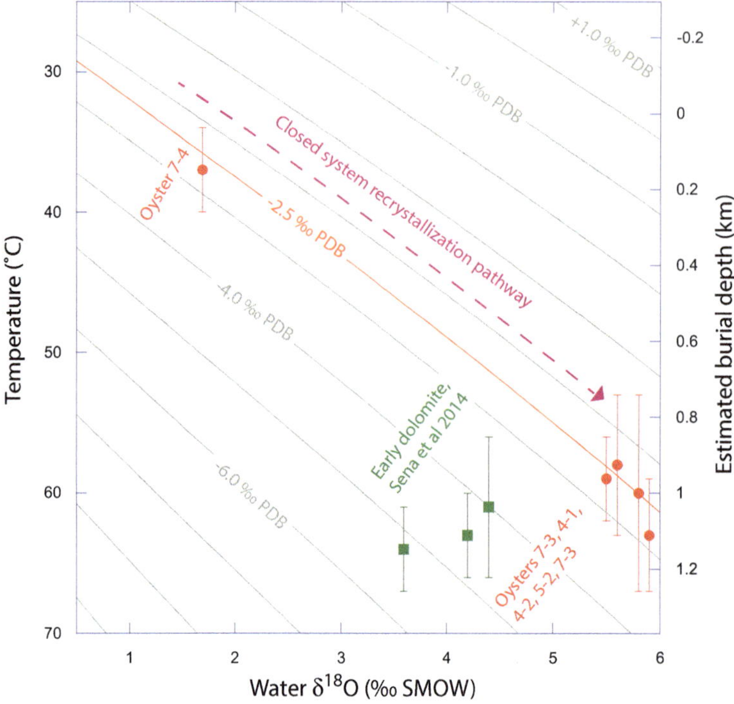

Fig. 7. Clumped isotope results for the Cretaceous oysters analysed in this study (red data points) and from legacy data on dolomite (Sena *et al.* 2014) recalibrated using the temperature calibration of Kluge *et al.* (2015). The vertical axis on the left is clumped isotope temperatures in °C; the horizontal axis is calculated fluid $\delta^{18}O$ in ‰ SMOW. The vertical axis on the right corresponds to burial depths calculated using a geothermal gradient of 27°C km^{-1}, and are thus minimum estimates (see the text for details). The grey lines in the plot represent lines of equal $\delta^{18}O$ composition for the calcite, in ‰ VPDB. The orange line represents the average $\delta^{18}O$ composition of all oysters in ‰ VPDB. As can be seen on the figure, the oyster values follow a closed-system recrystallization trend, with the heavier calculated fluid $\delta^{18}O$ being an artefact of higher temperatures with a virtually unchanged calcite $\delta^{18}O$. Temperatures obtained from dolomite are compatible with the maximum temperatures obtained from oysters.

partial neomorphose of the shell cannot be completely ruled out. Assuming that sample 7-4 records a temperature very close to pristine depositional conditions, surface water $\delta^{18}O$ in the Haushi-Huqf region during the Cretaceous can be calculated at around +1.1–1.7‰ VSMOW (Fig. 4). Given a $\delta^{18}O_{seawater}$ value estimated as 0.0–0.2‰ for the Cretaceous (Zhou *et al.* 2008), this implies some degree of isolation and evaporation on the Arabian Platform surrounding the Haushi-Huqf High because $\delta^{18}O_{seawater}$ values have been shifted by at least 1–1.7‰ relative to open-marine Cretaceous seawater. By comparison, modern sea-surface waters in the Mediterranean have values about 0.8–1.5‰ heavier than SMOW (Gat *et al.* 2003), and the Dead Sea has surface $\delta^{18}O_{seawater}$ values between 3.5 and 4.5‰ heavier than SMOW (Gat 1984). So the Haushi-Huqf $\delta^{18}O_{seawater}$ values point, by analogy, to some degree of evaporation and slightly elevated salinity close to the modern Mediterranean Sea, but much less than in regions experiencing high degrees of evaporation, such as the Dead Sea. The interpretation of isolation and high salinity is compatible with evidence from the biogenic assemble in the Qishn and equivalent Shuaiba formations, with abundant Baccinella mounds and Orbitolina beds indicative of slightly elevated salinity conditions around the relatively isolated Haushi-Huqf High (Immenhauser *et al.* 2004; Sena & John 2013). In addition, the carbon isotope values of the shell are compatible with deposition within marine conditions. The fact that $\delta^{13}C$ varies between shells, but also between different parts of a same shell (sample 7-4), points to heterogeneities in the carbon isotopes. Since these heterogeneities do not affect clumped isotope values, it is likely that they represent an intra-shell vital effect for carbon isotopes as opposed to burial diagenesis. Interestingly, this disequilibrium effect for carbon isotope has not impacted clumped isotope, something that has been observed before (see, for example, Henkes *et al.* 2013).

Assessing the burial history of the Haushi-Huqf High

The shell samples recording a higher temperature offer a regional window into the burial history of the Lower Cretaceous Qishn. The data points for all shells follow a typical closed-system recrystallization trend in a temperature–$\delta^{18}O_{water}$ space: that is, the $\delta^{18}O_{carbonate}$ values for all of the shells remains constant (at *c.* 2.5‰ VPDB) even if the temperature and back-calculated $\delta^{18}O_{water}$ values vary (Fig. 7). This behaviour is best understood if one considers that the stratigraphic unit in which the oysters are contained is a low-permeability shale, and thus recrystallization will occur with minimum fluid exchange with the surrounding rock. Conversely, the calculated $\delta^{18}O_{water}$ (Fig. 7) will appear very large, but this is an artefact of a constant $\delta^{18}O_{carbonate}$ but more elevated temperature used to back-calculate fluid composition. The temperature range recorded in oysters (34–67°C, accounting for measurement error) is compatible with the data from conodonts from the Gahrif Formation that indicates low burial temperatures of below 50–80°C (Angiolini *et al.* 2003). The combination of clumped isotope temperatures and CAI temperatures offers a chance to extract quantitative burial estimates by providing both a minimum and maximum estimate of burial depth. The difference between the maximum temperature recorded in oysters and the Earth surface temperature (34°C, sample 7-4) is a function of: (a) the geothermal gradient in Oman; and (b) the depth at which the sample was buried when it recrystallized. Sample 4-2 records the highest temperature and thus yields a minimum burial estimate for the Qishn Formation, because the temperature recorded by this oyster is intermediate between low depositional temperature of the well-preserved area of the shell and the high-temperature late-stage cements present in the shell. Recent work on heat flux in the Arabian Shield suggest that the geothermal gradient in Oman is best approximated by a range between 21 and 27°C km^{-1} (Rolandone *et al.* 2013). The temperature difference between samples 4-2 and 7-4 is 33°C, which, depending on assumptions related to the geothermal gradient, translates into 1.0–1.2 km of minimum burial (Fig. 4, right vertical axis where the maximum geothermal gradient is used for depth estimates). Combined with the maximum temperature provided by the CAI (80°C), this brackets nicely the range of maximum burial of the Qish Formation in this region to between 1.0 and 1.7 km. This result is larger by at least 800 m compared to previous sediment cover estimates of <400 m burial (Immenhauser *et al.* 2004). This explains why well-developed horizontal stylolites are present at outcrop (Sena & John 2013), as these were developed under a more significant overburden stress. The immediate implication is that the total post-Cretaceous sediment cover of the Haushi-Huqf High was previously underestimated, and thus that denudation rates must also have been higher than previously estimated during uplift.

Mention should be made that previous work has demonstrated that the clumped isotope value of calcites exposed to temperatures in excess of 120°C would start to slowly equilibrate to burial temperatures (Dennis & Schrag 2010; Passey & Henkes 2012; Henkes *et al.* 2014). Hence, an alternative interpretation of the data presented here could be that the formation was exposed to much greater burial depth, but without visible recrystallization/neomorphism. This 'solid state' reordering of clumped isotopes is, however, a very slow process, and it would only be initiated in the case of the Hausi-Huqf High if the sediments were buried for 60 myr or more to a depth of >4 km (120°C) based on reported geothermal gradients in Oman. This contradicts temperature data from conodonts in the Gharif Formation (Angiolini *et al.* 2003), and would be much more difficult to explain given known constraints on post-Cretaceous sediment cover. A solid-diffusion interpretation would also ignore petrographical evidence that shows less-well-preserved shell structure in samples recording higher temperatures (Figs 5 & 6). Furthermore, previous clumped isotope data on early dolomite of the underlying Jurf Formation (Sena *et al.* 2014), reinterpreted using the calibration of Kluge *et al.* (2015), yields temperatures of 56–67°C (Fig. 7, green data points), very similar to the temperatures obtained from the altered oyster shells (Fig. 7, red data points). The similar temperatures between calcite and dolomite in the Qishn and Jurf formations would be harder to explain with solid-state reordering because different mineralogies (calcite and dolomite) should have very different Arhenius parameters and activation energies for solid-state reordering, and thus different final temperatures of equilibration (Passey & Henkes 2012). Thus, a simpler scenario with shallow burial and diagenetic transformation of the oysters best fits the data.

Conclusions and perspectives

Clumped isotope data proves very useful in reconstructing and quantifying the burial depth of the Lower Cretaceous Qishn Formation in the Haushi-Huqf High region. It yields temperatures of neomorphic transformation in fine-grained biogenic carbonates where fluid-inclusion data are not available, and at the low range of temperatures where typically fluid inclusions are difficult to interpret. The temperatures obtained from clumped isotopes when combined with previous conodonts estimates bracket the maximum temperature reached by the sediment to in-between 67°C (from clumped isotope palaeothermometry) and 80°C (conodonts alteration index).

This approach offers many advantages in carbonate rocks and clastic rocks containing carbonate cements. From a practical point of view, the method can be applied to multiple components from wells (oysters, micrite, cements and other skeletal components) to constrain the burial history of a reservoir unit. The temperatures of some components will result from mineralogical transformation and diagenesis, as is the case with the oysters of the Qishn Formation, but other components will display a more complex solid-state re-ordering of clumped isotopes. By exploiting the richness of this thermal record, one can piece together the temperature–depth history of a given basin. One clear direction of research for the method will be to overcome the sample size limitation (currently 24 mg) in order to address more focused questions: for instance, using new analytical protocols (Hu *et al.* 2014) or improved preparation lines and analysers. Ultimately, the goal will be to pair temperature measurements from clumped isotopes with radiometric dating of carbonates in order to unlock the burial histories of sedimentary basins in areas of interest.

I gratefully acknowledge funding from the Qatar Carbonates and Carbon Storage Research Centre (QCCSRC), provided jointly by Qatar Petroleum, Shell, and Qatar Science and Technology Park. This paper was presented as part of an invited presentation at the Geological Society of London 'Reservoir Quality of Clastic and Carbonate Rocks: Analysis, Modelling and Prediction', and the conveners are thanked for giving me the opportunity to present this work. Finally, I thank Shuram Oil and Gas for great logistic help during our Oman fieldwork, Leticia Bombieri for sample preparation, and the members of my research group for fruitful discussion and for helping in making our clumped isotope laboratory a success. Peter Armitage (editor), Cathy Holis (associate editor) and two anonymous reviewers are thanked for their comments that helped to improve the manuscript.

References

Angiolini, L., Balini, M., Garzanti, E., Nicora, A., Tintori, A., Crasquin, S. & Muttoni, G. 2003. Permian climatic and paleogeographic changes in Northern Gondwana: the Khuff Formation of Interior Oman. *Palaeogeography, Palaeoclimatology, Palaeoecology*, **191**, 269–300, https://doi.org/10.1016/S0031-0182(02)00668-5

Beydoun, Z. R. & Bamahmoud, M. O. 1993. The Qishn Formation, Yemen: lithofacies and hydrocarbon habitat. *Marine and Petroleum Geology*, **10**, 364–372.

Boggs, S. & Krinsley, D. 2006. *Application of Cathodoluminescence Imaging to the Study of Sedimentary Rocks*. Cambridge University Press, Cambridge

Dale, A., John, C. M., Mozley, P. S., Smalley, P. C. & Muggeridge, A. H. 2014. Time-capsule concretions: unlocking burial diagenetic processes in the Mancos Shale using carbonate clumped isotopes. *Earth and Planetary Science Letters*, **394**, 30–37, https://doi.org/10.1016/j.epsl.2014.03.004

Dennis, K. J. & Schrag, D. P. 2010. Clumped isotope thermometry of carbonatites as an indicator of diagenetic alteration. *Geochimica et Cosmochimica Acta*, **74**, 4110–4122.

Dennis, K. J., Affek, H. P., Passey, B. H., Schrag, D. P. & Eiler, J. M. 2011. Defining an absolute reference frame for 'clumped' isotope studies of CO_2. *Geochimica et Cosmochimica Acta*, **75**, 7117–7131, https://doi.org/10.1016/j.gca.2011.09.025

Eagle, R. A., Schauble, E. A., Tripati, A. K., Tütken, T., Hulbert, R. C. & Eiler, J. M. 2010. Body temperatures of modern and extinct vertebrates from ^{13}C-^{18}O bond abundances in bioapatite. *Proceedings of the National Academy of Sciences of the United States of America*, **107**, 10 377–10 382, https://doi.org/10.1073/pnas.0911115107

Eiler, J. M. 2007. 'Clumped-isotope' geochemistry – The study of naturally-occurring, multiply-substituted isotopologues. *Earth and Planetary Science Letters*, **262**, 309–327, https://doi.org/10.1016/j.epsl.2007.08.020

Eiler, J. M. 2011. Paleoclimate reconstruction using carbonate clumped isotope thermometry. *Quaternary Science Reviews*, **30**, 3575–3588, https://doi.org/10.1016/j.quascirev.2011.09.001

Epstein, A. G., Epstein, J. B. & Harris, L. D. 1976. *Conodont Color Alteration – An Index to Organic Metamorphism*. United States Geological Survey, Professional Papers, **995**, http://pubs.usgs.gov/pp/0995/report.pdf

Friedman, I. & O'Neil, J. R. 1977. Chapter KK. Compilation of stable isotope fractionation factors of geochemical interest. *In*: Fleisher, M. (ed.) *Data of Geochemistry*. 6th edn. United States Geological Survey, Professional Papers, **440-KK**

Gat, J. R. 1984. The stable isotope composition of dead-sea waters. *Earth and Planetary Science Letters*, **71**, 361–376, https://doi.org/10.1016/0012-821X(84)90103-1

Gat, J. R., Klein, B., Kushnir, Y., Roether, W., Wernli, H., Yam, R. & Shemesh, A. 2003. Isotope composition of air moisture over the Mediterranean Sea: an index of the air-sea interaction pattern. *Tellus*, **55**, 953–965, https://doi.org/10.1034/j.1600-0889.2003.00081.x

Ghosh, P., Adkins, J. et al. 2006. ^{13}C-^{18}O bonds in carbonate minerals: a new kind of paleothermometer. *Geochimica et Cosmochimica Acta*, **70**, 1439–1456.

Goldstein, R. H. & Reynolds, T. J. 1994. *Systematics of Fluid Inclusions in Diagenetic Minerals*. SEPM Short Course, **31**. Society for Sedimentary Geology (SEPM), Tulsa, OK.

GRELAUD, C., RAZIN, P., HOMEWOOD, P. W. & SCHWAB, A. M. 2006. Development of incisions on a periodically emergent carbonate platform (Natih Formation, Late Cretaceous, Oman). *Journal of Sedimentary Research*, **76**, 647–669, https://doi.org/10.2110/jsr.2006.058

GUO, W., MOSENFELDER, J. L., GODDARD, W. A. & EILER, J. M. 2009. Isotopic fractionations associated with phosphoric acid digestion of carbonate minerals: insights from first-principles theoretical modeling and clumped isotope measurements. *Geochimica et Cosmochimica Acta*, **73**, 7203–7225, https://doi.org/10.1016/j.gca.2009.05.071

HENKES, G. A., PASSEY, B. H., WANAMAKER, A. D., JR., GROSSMAN, E. L., AMBROSE, W. G., JR. & CARROLL, M. L. 2013. Carbonate clumped isotope compositions of modern marine mollusk and brachiopod shells. *Geochimica et Cosmochimica Acta*, **106**, 307–325, https://doi.org/10.1016/j.gca.2012.12.020

HENKES, G. A., PASSEY, B. H., GROSSMAN, E. L., BROCK, W. S., PEREZ-HUERTA, A. & YANCEY, T. E. 2014. Temperature limits for preservation of primary calcite clumped isotope paleotemperatures. *Geochimica et Cosmochimica Acta*, **139**, 362–382, https://doi.org/10.1016/j.gca.2014.04.040

HESSE, R. 1987. Selective and reversible carbonate – silica replacements in Lower Cretaceous carbonate-bearing turbidites of the Eastern Alps. *Sedimentology*, **34**, 1055–1077, https://doi.org/10.1111/j.1365-3091.1987.tb00592.x

HU, B., RADKE, J., SCHLÜTER, H.-J., HEINE, F. T., ZHOU, L. & BERNASCONI, S. M. 2014. A modified procedure for gas-source isotope ratio mass spectrometry: the long-integration dual-inlet (LIDI) methodology and implications for clumped isotope measurements. *Rapid Communications in Mass Spectrometry*, **28**, 1413–1425, https://doi.org/10.1002/rcm.6909

HUNTINGTON, K. W., EILER, J. M. ET AL. 2009. Methods and limitations of 'clumped' CO_2 isotope (Δ 47) analysis by gas-source isotope ratio mass spectrometry. *Journal of Mass Spectrometry*, **44**, 1318–1329, https://doi.org/10.1002/jms.1614

HUNTINGTON, K. W., BUDD, D. A., WERNICKE, B. P. & EILER, J. M. 2011. Use of clumped-isotope thermometry to constrain the crystallization temperature of diagenetic calcite. *Journal of Sedimentary Research*, **81**, 656–669, https://doi.org/10.2110/jsr.2011.51

IMMENHAUSER, A., HILLGARTNER, H. ET AL. 2004. The Barremian–lower Aptian Qishn Formation (Huqf Area, Oman): a new outcrop analogue for Kharaib/Shuaiba Reservoirs. *GeoArabia*, **9**, 153–194.

KIM, S.-T. & O'NEIL, J. R. 1997. Equilibrium and nonequilibrium oxygen isotope effects in synthetic carbonates. *Geochimica et Cosmochimica Acta*, **61**, 3461–3475, https://doi.org/10.1016/S0016-7037(97)00169-5

KLUGE, T., JOHN, C. M., JOURDAN, A. L., DAVIS, S. & CRAWSHAW, J. 2015. Laboratory calibration of the calcium carbonate clumped isotope thermometer in the 25–250°C temperature range. *Geochimica et Cosmochimica Acta*, **157**, 213–227, https://doi.org/10.1016/j.gca.2015.02.028

PASSEY, B. H. & HENKES, G. A. 2012. Carbonate clumped isotope bond reordering and geospeedometry. *Earth and Planetary Science Letters*, **351–352**, 223–236, https://doi.org/10.1016/j.epsl.2012.07.021

RIES, A. C. & SHACKLETON, R. M. 1990. Structures in the Huqf-Haushi Uplift, east Central Oman. *In*: ROBERTSON, A. H. F., SEARLE, M. P. & RIES, A. C. (eds) *The Geology and Tectonics of the Oman Region*. Geological Society, London, Special Publications, **49**, 653– 663, https://doi.org/10.1144/GSL.SP.1992.049.01.39

ROLANDONE, F., LUCAZEAU, F., LEROY, S., MARESCHAL, J.-C., JORAND, R., GOUTORBE, B. & BOUQUEREL, H. 2013. New heat flow measurements in Oman and the thermal state of the Arabian Shield and Platform. *Tectonophysics*, **589**, 77–89, https://doi.org/10.1016/j.tecto.2012.12.034

SATTLER, U., IMMENHAUSER, A., HILLGÄRTNER, H. & ESTEBAN, M. 2005. Characterization, lateral variability and lateral extent of discontinuity surfaces on a Carbonate Platform (Barremian to Lower Aptian, Oman). *Sedimentology*, **52**, 339–361, https://doi.org/10.1111/j.1365-3091.2005.00701.x

SCHAUBLE, E., GHOSH, P. & EILER, J. 2006. Preferential formation of $^{13}C-^{18}O$ bonds in carbonate minerals, estimated using first-principles lattice dynamics. *Geochimica et Cosmochimica Acta*, **70**, 2510–2529, https://doi.org/10.1016/j.gca.2006.02.011

SENA, C. M. & JOHN, C. M. 2013. Impact of dynamic sedimentation on facies heterogeneities in Lower Cretaceous peritidal deposits of central east Oman. *Sedimentology*, **60**, 1156–1183, https://doi.org/10.1111/sed.12026

SENA, C. M., JOHN, C. M., JOURDAN, A. L., VANDEGINSTE, V. & MANNING, C. 2014. Dolomitization of lower cretaceous peritidal carbonates by modified seawater: constraints from clumped isotopic paleothermometry, elemental chemistry, and strontium isotopes. *Journal of Sedimentary Research*, **84**, 552–566, https://doi.org/10.2110/jsr.2014.45

WACKER, U., FIEBIG, J. & SCHOENE, B. R. 2013. Clumped isotope analysis of carbonates: comparison of two different acid digestion techniques. *Rapid Communications in Mass Spectrometry*, **27**, 1631–1642, https://doi.org/10.1002/rcm.6609

ZHOU, J., POULSEN, C. J., POLLARD, D. & WHITE, T. S. 2008. Simulation of modern and middle Cretaceous marine $\delta^{18}O$ with an ocean–atmosphere general circulation model. *Paleoceanography*, **23**, PA3223, https://doi.org/10.1029/2008PA001596

Feldspar alteration and Fe minerals: origin, distribution and implications for sandstone reservoir quality in estuarine sediments

EHSAN DANESHVAR & RICHARD H. WORDEN*

Department of Earth, Ocean and Ecological Sciences, University of Liverpool, 4 Brownlow Street, L69 3GP, UK

**Correspondence: r.worden@liv.ac.uk*

Abstract: The occurrence and distribution of minerals in modern sedimentary systems hold many clues to help unravel the origin and distribution of reservoir quality-controlling minerals in ancient and deeply buried sandstones, but few quantitative studies have been undertaken. Here we have used a range of techniques including X-ray diffraction, scanning electron microscopy and fully automated mineralogical QEMSCAN analysis to provide a comprehensive understanding of mineral composition and distribution within the post-glacial, clastic sediments of the Ravenglass Estuary, NW England. The Ravenglass Estuary is fed by two main rivers: one drains a granite-dominated hinterland, the other drains a hinterland that contains andesite and Triassic red bed sandstones. The granite-supplied arm has slightly more quartz-rich and Fe mineral-poor sediment than the andesite- and red bed-supplied sediment. The provenance signals are muted for feldspar and mica minerals heavy-mineral garnet populations seem to be sensitive to provenance. Detrital K-feldspar grains are preferentially associated with illite-dominated clay mineral coats, whereas all plagioclase mineral grains are preferentially associated with kaolinite-dominated clay mineral coats. This can be explained by rapid early diagenesis in the sediment with K-feldspar grain surfaces replaced by illite and plagioclase grain surfaces replaced by kaolinite. The andesite- and red bed-supplied sediment contains twice the amount of Fe minerals, which are dominated by chlorite, than the granite-supplied sediment. Chlorite rarely is associated with grain coatings on feldspar grains, possibly because it is predominantly a detrital mineral. Detrital Fe minerals seem to be locally replaced by pyrite due to bacterial sulphate reduction, suggesting that some early diagenetic processes may serve to lock away iron and prevent it from creating Fe-rich clay minerals.

Sedimentary provenance studies are generally based on geological parameters such as mineralogy, geochemistry and lithology and are designed to determine the source of the sand grains. Diagenetic processes such as cementation, compaction and dissolution during burial influence reservoir quality by changing the primary mineralogy and fabric (Ramm 2000; Worden & Burley 2003). Sediment provenance impacts reservoir quality as it controls the primary sand grain type and strongly influences the types of detrital clay mineral (Morton 1987; Haughton *et al.* 1991). For example, the alteration of feldspars is common during the burial diagenesis of sandstones but different types of feldspar, from different provenance areas, are found under different diagenetic evolution paths for a given water composition of the formation (Boles 1982; Morad 1988; Aagaard *et al.* 1990; Morad *et al.* 1990; Parsons *et al.* 2005; Tyrrell *et al.* 2011).

This research was designed to study detrital feldspars and their alteration to clay minerals in a modern, marginal marine (estuarine) environment, and to investigate the relationship between feldspar alteration patterns and provenance. The estuary selected was the Ravenglass Estuary in Cumbria (UK) (Fig. 1). This westward-opening estuary has two main arms that join near the sea with each arm having hinterlands of rather different geologies. The northern arm drains Triassic sandstones and Palaeozoic andesite, the southern arm drains Palaeozoic granite (Bousher 1999; Daneshvar 2012).

The main origins of clay minerals in estuarine environments have been documented (Bokuniewicz 1995; Aagaard *et al.* 2000; Tucker 2001; Burley & Worden 2003; Worden & Morad 2003; Sionneau *et al.* 2008) and can be summarized as: (1) clay minerals can be directly deposited after originating in weathering profiles or from soil and drift deposits in a fluvial hinterland (called inherited clay minerals), (2) clay minerals can be neo-formed from other aluminosilicate minerals such as feldspars, and (3) clay minerals can be transformed from precursor clay minerals.

Estuaries are highly efficient sediment traps for inherited clay minerals, and during marine transgression estuarine deposits have high preservation potentials (Boyle *et al.* 1974; Boyle *et al.* 1977; Dalrymple *et al.* 1992). In estuarine systems,

From: ARMITAGE, P. J., BUTCHER, A. R., CHURCHILL, J. M., CSOMA, A. E., HOLLIS, C., LANDER, R. H., OMMA, J. E. & WORDEN, R. H. (eds) 2018. *Reservoir Quality of Clastic and Carbonate Rocks: Analysis, Modelling and Prediction.* Geological Society, London, Special Publications, **435**, 123–139.
First published online April 13, 2017, https://doi.org/10.1144/SP435.17

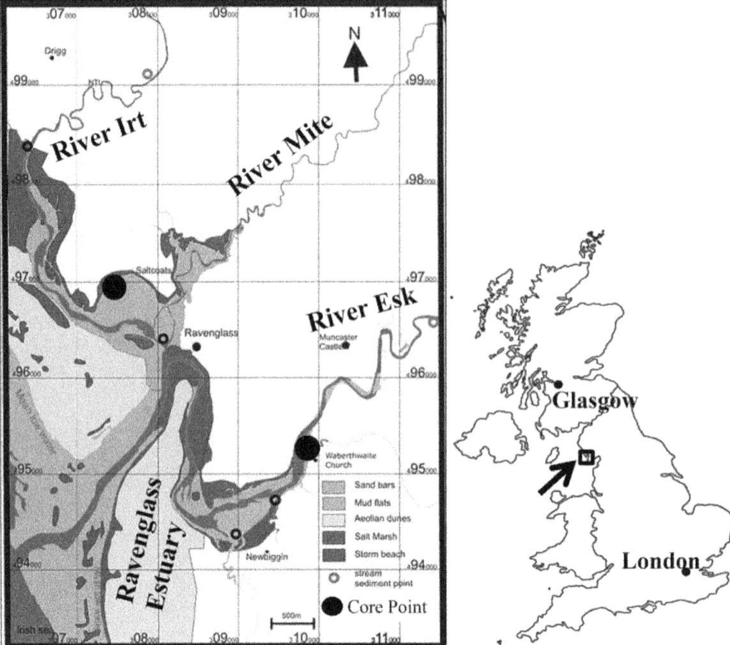

Fig. 1. Map of the study area, showing the location of the Esk arm of the Ravenglass Estuary and coring point. Sketch map showing lithological composition in Ravenglass catchment area.

clay minerals can also form *in situ* either by direct growth in pore waters or from the transformation of detrital clay minerals. In terms of transformation, inherited clay minerals tend to be modified by ion exchange or cation rearrangement (Berner 1980; Aller & Aller 1998; Deer *et al.* 2013). Clay mineral neo-formation and transformation can take place soon after deposition directly within the depositional environment (Berner 1980; Baker *et al.* 2000; Burley & Worden 2003; Worden & Morad 2003; McIlroy *et al.* 2003). For example, processes that have been previously proposed to occur in estuarine environments include the formation of kaolinite and/or illite from weathered feldspars, the transformation of mixed-layer chlorite to Fe-rich chlorite and the precipitation of gibbsite after kaolinite dissolution and alteration (Velde 1985; Drever & Hurcomb 1986; Drever & Zobrist 1992; Huang 1993; Velde & Church 1999; Worden & Morad 2003; Daneshvar 2012).

Inherited clay minerals, which formed in the hinterland, can be transported as suspended load through the rivers to the estuary before settling out (Zhang *et al.* 1990; Fan *et al.* 2008). The distribution of clay minerals in modern sediments is typically considered to reflect the weathering pattern (Thiry 2000; Tucker 2001) whereas neoformed clay minerals are the products of alteration during diagenesis (Bjølykke 1998; Worden & Morad 2003). The distribution of clay minerals and sedimentary depositional environments in the Ravenglass Estuary were previously investigated (Daneshvar 2012) in order to identify the sources of clay minerals in the estuary, and to study alteration and early diagenetic processes. In this study, the main research questions are:

(1) What are the dominant mineralogical assemblages, including clays and detrital minerals, in this modern sedimentary environment?
(2) Are there any spatial and stratigraphic variations in mineral composition, especially in Fe minerals and various clay minerals, in the estuary that reflect variations in provenance?
(3) What are the lessons for the spatial distribution of reservoir quality in deeply buried, ancient estuarine sandstones given any spatial and stratigraphic variations in feldspar provenance in the studied modern estuary?

Background geology

The northern part of the UK (including Cumbria) is presently undergoing limited isostatic recovery following the last glacial maximum (Bousher 1999). The west Cumbrian region has been affected by glaciation during the Quaternary on at least three occasions. The last glaciation occurred in the Late

Devensian, at about 28–13 ka (Moseley 1978; McDougall 2001). The Ravenglass area has benefitted from substantial geological and geomorphological research due to the location of the nearby nuclear reprocessing plant at Sellafield and the low-level waste repository at Drigg. Glacial deposits have been largely removed from the land surface following the last glaciation (Merritt & Auton 2000). Merritt & Auton (2000) reported that the upper group of the Quaternary sediments within the area around Sellafield and Ravenglass includes estuarine, alluvial, organic and aeolian sequences sitting on top of the final glacial deposits. The estuarine sediments have a maximum thickness of about 10–15 m in this area (Bousher 1999).

The Ravenglass Estuary in NW England (Fig. 1) lies near the small town of Ravenglass, located on the west coast of Cumbria. The Ravenglass Estuary, which encompasses the tidal reaches of the River Esk, River Irt and River Mite, occupies an area of 5.6 km^2 of which 86% is intertidal (Bousher 1999). The Ravenglass Estuary is fed by two main arms; the River Esk drains the Palaeozoic Eskdale granite in the south; the River Irt drains the Triassic Sherwood Sandstone Group and the Ordovician Borrowdale Volcanic Group in the north (Fig. 1).

The Triassic Sherwood Sandstone is locally known as the St Bees Sandstone (SBSF) (Colter & Ebbern 1978; Barnes *et al.* 1994; Strong *et al.* 1994) and has extensive coastal outcrop in the cliffs at the St. Bees Head, 32 km north of Ravenglass village. The St Bees Sandstone is a feldspathic sandstone dominated by major quantities (>10%) of detrital quartz, K-feldspar and albite and minor mica (between 1% and 10%, mainly muscovite and less abundant biotite) and trace (<1%) hematite and ilmenite with major amounts of disseminated carbonate cement (Barnes *et al.* 1994). The carbonate cement is composed of calcite and dolomite with a minor amount of ferroan dolomite (Mildowski *et al.* 1998).

The Eskdale Granite, the largest exposed intrusion in Cumbria, was emplaced during the late Ordovician during the final stages of the closure of the Iapetus Ocean and the Caledonian Orogeny (Soper 1987). There are two main types of granite; an older biotite-granodiorite in the south and a younger pink muscovite-granite in the north (Rundle 1979). The River Esk mostly drains the pink granite (Simpson 1934), locally called Cumberland Pink Granite, which is composed of major quartz, feldspar (K-feldspar and plagioclase) and muscovite; and minor biotite and hematite with trace quantities of apatite (Simpson 1934; Strong *et al.* 1994). Tourmaline occurs as a joint-coating minerals and as a replacement for feldspar. Biotite, where present, is typically chloritized or replaced by hematite. Overall, the Eskdale Granite shows intense chloritization (Brown *et al.* 1964).

The Borrowdale Volcanic Group (BVG) is a group of late Ordovician igneous rocks that consists of two main exposed outcrops in the region (the Lake District) including northern and southern outcrops. The older NW outcrops are drained through the River Irt into the Ravenglass Estuary and are mainly comprised of andesite along with less abundant tuffs and agglomerates. In contrast to the NW outcrops, the younger SE part of the BVG lithologically comprises basaltic compositions. The upper boundary of the BVG in the SE is an unconformity with the overlying Windermere Supergroup, while the lower boundary in the NW is an unconformity with the underlying Skiddaw Group (Fitton 1972). Apart from the primary andesitic and basaltic mineral composition, secondary minerals such as hematite and chlorite, the result of weathering of the mother rock, are also reported across the outcrops (Entwisle *et al.* 2005).

The estuary contains several depositional environments including tidal channels (with mixed wave and river influences), tidal bars, tidal flats (sandy and muddy), aeolian dunes and storm deposits. Subaqueous gravel dune bedforms, made up of shell-rich fine to medium gravels, are found at bends in the intertidal part of the rivers. The major part of the estuary is intertidal and of this just under half is fine-grained sediment (silts and sandy silts) with the rest comprising sand-dominated sediments with some surfaces that are armoured with gravel.

Methodology

Coring and sampling

Several short cores were taken across the estuary, providing up to 1 m of undisturbed modern sediments. Two cores are dealt with in detail here, one from the northern arm of the estuary (from the River Irt) and one from the southern arm of the estuary (from the River Esk). A Van Walt window sampler was used to acquire the cores. This is a pneumatic drill-driven corer that recovers *c.* 1 m of largely undisturbed sediment in a thin film of plastic sheath. The window sampler was driven into the substrate with an Atlas Copco Cobra TT percussion hammer. Once the polythene sheath around the core was sliced open, samples were taken every 5–10 cm, preserved in sealed plastic jars and then stored in a refrigerator before preparation for mineral and petrological analysis. Surface samples were also collected from around the estuary; all surface samples were preserved in sealed plastic bags and stored at 4°C, pending mineralogical analysis.

Analytical approaches

Mineral analysis was performed using X-ray diffraction (XRD) and automated mineralogy for a fully quantitative mineral assemblage analysis (QEMSCAN). XRD was used to examine the samples in bulk, as well as the clay fractions. Bulk fractions were crushed using a micromill for 10 minutes. Randomly oriented samples were analysed by XRD using a PANalytical X'Pert Pro MPD X-ray diffractometer. Clay fractions from portions of the cored sediment sample were suspended by ultrasonic disaggregation in distilled water. Suspended materials were then separated in three steps (grain size separation) using a centrifuge at 300, 500 and 3500 rpm for 3, 5 and 30 minutes respectively. Finally, solids were dried in an oven at 55°C overnight. Clay fractions of sediment samples were prepared for XRD analysis in four ways: (1) air dried, (2) Mg-saturated and then glycolated, (3) heated for one hour at 400°C, and (4) heated for one hour at 550°C. Scans covered the 2θ (two times the angle between the sample and the incident X-ray beam) range of 3.66–70.00° over a scan time of 13 min 21 s, with 0.04 rad Soller slits in both the incident and diffracted beam paths. HighScore Plus software operated the X-ray diffractometer; the same software was used to interpret the XRD traces with reference patterns taken from the International Centre for Diffraction Data, Powder Diffraction File 2 Release 2008. Semi quantification of the clay minerals was based on the intensity of the XRD peaks at *c.* 14 Å (*c.* 6.2°), 9.9 Å (8.9°), *c.* 7.15 Å (*c.* 12.3°) and 7.07 Å (12.5°).

Samples were prepared as grain mounts by dispersing the freeze-dried original sediment in thermally stable resin. The resin blocks were then sliced and polished in preparation for electron optical examination. Backscattered electron (BSE) and scanning electron microscopy (SEM) imaging along with energy-dispersive X-ray spectroscopy (EDS) analyses were carried out on a Philips XL 30 fitted with an Oxford Instruments secondary X-ray detector. BSE images were collected at 25 kV.

QEMSCAN analysis was used to define the mineralogy and texture in fully quantitative terms (Pirrie *et al.* 2003). The system comprises a scanning electron microscope combined with four energy dispersive spectrometers, a microanalyser and an electronic processing unit that integrates the scanned data using a powerful software suite (iDiscover) to provide information about the chemical and mineral compositions of the samples. Polished thin sections were carbon-coated before being transferred to QEMSCAN for analysis. Each sample was divided by the machine into a series of 2 mm × 2 mm fields that are analysed individually. An electron beam of 15 kV was directed at the sample. Initially, primary and secondary backscatter electrons were measured, with the brightness indicating the sample density and the BSE surface signal equated to atomic weight. A backscatter cut-off was used to differentiate sediment from the mounting medium (resin) and the mounted sediment was further analysed with X-rays collected from each point, at a selected step size, by the four EDS detectors. The step size selected in this case was 10 μm to ensure that the fine fraction in the sediment was analysed as well as sand grains. The EDS spectra from each point analysed were compared to a library of spectra using a Species Identification Protocol (SIP) and each point was assigned to a specific mineral. The output is a mineral map of each sample with the resolution limited by the step size selected at the time of data collection. The data were subsequently manipulated to remove contamination and a summary of the fully quantitative mineralogical content of the rock was produced, in addition to textural information.

Results

Core description

Core 1 was obtained on a sand bar near the channel bank in the Esk estuary (Fig. 2a). The core has been divided into six units based on grain size, colour, degree of bioturbation and bioclast population. Unit A (90–60 cm) comprises relatively coarse-grained sediments at the base and fines upwards over 40 cm. Unit A is a light-coloured, moderately to well-sorted, medium-grained sand containing shell fragments. In unit B, the sediment coarsens upwards over the 25 cm thickness (60–35 cm). This unit contains dark grey silt and clay-dominated sediments with organic material and roots. Unit C (35–25 cm) is dark, moderately-sorted, medium- to fine-grained sand that contains plant material and rootlets. Unit D (25–20 cm) is lighter than C, and is a poorly sorted, medium-grained sand with a low degree of bioturbation. Unit E (20–5 cm) is a dark, bioturbated, poorly sorted, medium-grained sand with shell fragments and some plant material. Unit F (0–5 cm) is a poorly sorted and bioturbated medium- to fine-grained sand with clay and silt.

The basal, well-sorted bioclastic sand of Unit A in the Esk core suggests a strong marine influence at the environment of deposition, possibly representing an offshore deposit or a storm-beach deposit. The overlying silt- and clay-dominated units with plant material suggest a low-energy environment, either a muddy tidal flat and/or an overbank setting. The upper bioturbated sand-prone units are interpreted as being part of an in-channel tidal bar deposit, as could be expected from the present environment of deposition.

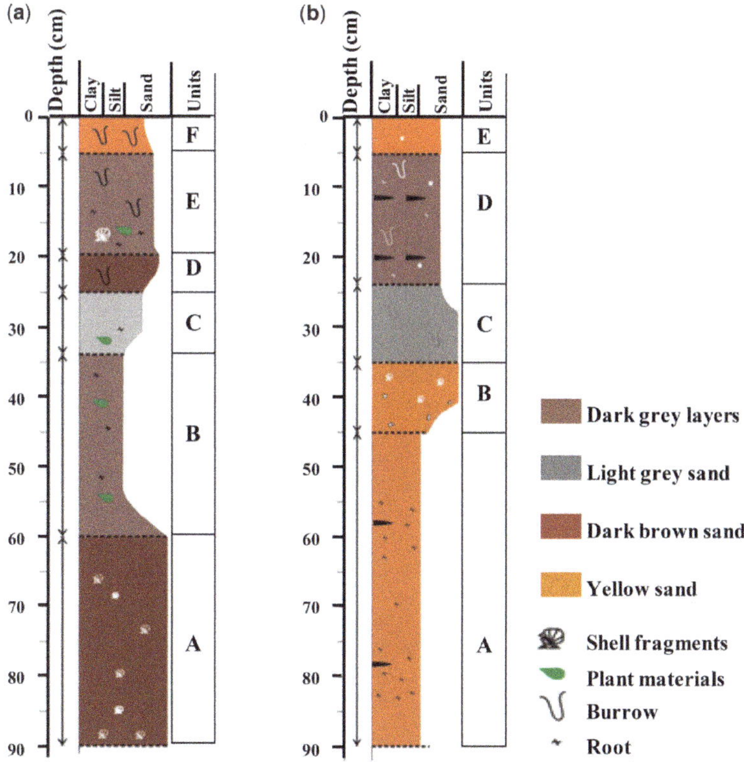

Fig. 2. Detailed stratigraphy of the core from the Ravenglass Estuary. (**a**) Core no. 1 in sand bar close to the channel on the Esk estuary (**b**) Core no.2 at the immediate marine environment close to the Ravenglass estuary mouth on Irt estuary.

Core 2 was collected near the Saltcoats area of the Irt estuary in the intertidal zone (Fig. 2b). This core has been divided into five units (A to E). Unit A (95–45 cm) is a light-coloured silt- and clay-dominated unit. Unit B (45–35 cm) is light-coloured, moderately sorted coarse sand with shell fragments. A thin gravel layer marks an abrupt contact with the silty clay of Unit A. Unit C (35–25 cm) contains poorly sorted, medium to coarse sand, devoid of mud laminae and shell fragments, which is lighter in colour than Unit B. Unit D (25–5 cm) is a poorly sorted, medium-grained sand with silt and clay. This unit is dark in colour and includes several millimetre-thick, dark mud laminae, sub-vertical burrow-fills and thin shell fragments. Unit D fines upwards over 20 cm and is capped by a sharply defined dark layer. Unit E (5–0 cm) is a clean, light-coloured, moderately to poorly sorted medium sand. At depths of 65–55 cm and 90–75 cm multiple millimetre-thick dark mud laminae that contain organic material are present.

The mud laminae in the lower part of core 2 indicate low-energy conditions. The overall coarsening upwards in this core could be interpreted as progressive shallowing of the environment of deposition (regressive). However, the intra-estuary position of the core suggests that the lower, fine-grained deposits may be muddy tidal flat deposits. The shell fragments in the upper core indicate higher-energy environments. The coarser sediment is not well sorted, as might be expected in a shoreface setting, and is likely to mark the interplay of river-derived and marine-derived deposits as befits the present position within the estuary.

SEM analysis

Clay minerals that coat sand grains are too fine-grained to produce high-quality, single-mineral EDS spectra in SEM analysis. EDS spectra represent mixtures of minerals and must be interpreted with care. Kaolinite has been identified in places (Fig. 3), with approximately equal Al and Si peaks and no K or any other cations. Illite has also been identified with abundant Si, somewhat less Al and a sizeable K peak (Fig. 3b). Some EDS spectra of coating minerals show high Si, somewhat lower

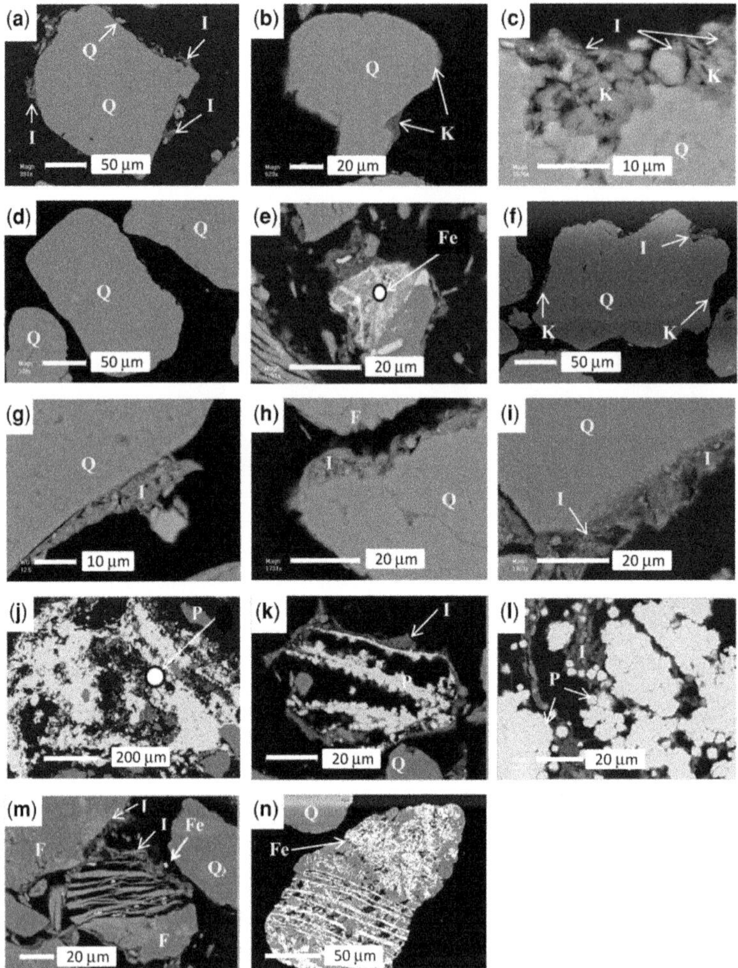

Fig. 3. BSE images of polished section grain mounts. (**a**) Sample at a depth of 30–40 cm (Esk estuary), the quartz sand grain in the middle is coated with clay-grade material and flake-shaped illite is noted as the grain coating. (**b**) Sample at 61–70 cm depth, the quartz grain in the middle is surrounded by very fine grained kaolinite. (**c**) Sample at 50–58 cm depth showing sand grains with two different age coatings of different compositions. Older booklet-shape pore-filling kaolinite is seen, along with coating flake-shaped illite. (**d**) Image of a grain at 70–80 cm depth that shows no visible coating around the sand grain. (**e**) BSE image of a coated sand grain with iron oxide at a depth of 50–58 cm. (**f**) BSE image of a quartz grain in the middle with a fine-grained kaolinite coating. (**g**) The quartz sand grain in the middle is coated with clay-grade material, mainly flakes of illite, at a depth of 10–20 cm (Irt estuary). (**h**) Quartz grain with a thick coating of illite–chlorite at a depth of 30–40 cm. (**i**) BSE image of a sand grain at a depth of 40–50 cm, which EDS analyses show to be illite–chlorite. (**j**) BSE image of a sand grain at a depth of 70–80 cm that is dominated by pyrite. (**k**) Residual grain coated with iron oxides. (**l**) Pyrite crystals around the grains, which show that sulphate was supplied and bacterially reduced to sulphide that then reacted with the available Fe to create pyrite in the Irt estuary sediments. (**m**) A K-feldspar grain torn to strings to form illite and Fe oxide particles between the strings. (**n**) A quartz grain covered with a secondary deposition of Fe oxide in the Esk estuary sediments. Q, quartz; K, kaolinite; I, illite; Fe, Fe oxides; P, pyrite.

Al, a large Fe-peak and rather less Mg (Fig. 3); such spectra are indicative of chlorite. However, most spectra seem to contain some K and Fe suggesting either the presence of intergrown illite and a chlorite-like material or the occurrence of some sort of K-bearing smectite phase.

BSE microscope images of Esk estuary samples from core 1 reveal that the sediment contains a

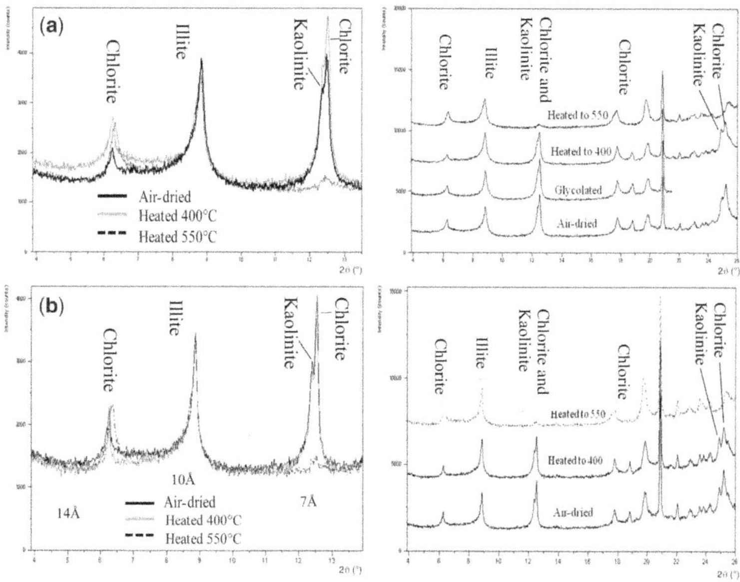

Fig. 4. (**a**) Representative XRD patterns of the <2 μm fraction of the cored sediment samples in the Esk estuary taken at a depth of 5–10 cm. The patterns for each sample are stacked (right). Overlapping the XRD data (left) from the low-angle end of the pattern illustrates the drop in intensity of the 7.15 and 7 Å peaks during heating to 400°C and their near total collapse at 550°C. Glycolation seems to have little effect on the sample. Note also the broadening of the chlorite (001) peak. The sediment seems to be composed of illite, chlorite and kaolinite with negligible smectite present. (**b**) Representative XRD patterns of the <2 μm fraction of the cored sediment sample at Irt estuary, taken at a depth of 10–20 cm. The patterns for each sample are stacked (right). Overlapping the XRD data (left) from the low-angle end of the pattern illustrates the drop in intensity of the 14, 7.15 and 7 Å peaks during heating to 400°C and peaks at 7.15 and 7 Å show near total collapse at 550°C and a significant shift from 14 Å to 12 Å while 10 Å peak is unchanged at 550°C. Glycolation seems to have little effect on the sample. There was a slight collapse of the 14 Å trace on heating to 550°C with a broad hump developing at about 12 Å. This is characteristic of vermiculite, suggesting that the 14 Å peak represents a combination of chlorite and vermiculite. The sediments thus seem to be composed of illite, chlorite, dioctahedral chlorite and vermiculite and kaolinite with negligible smectite present.

mixture of clastic grains and marine (mostly siliceous) bioclasts such as diatoms in association with some iron oxide minerals. Core 1 sediments are moderately sorted with sand grains that are variably coated with clay minerals. Grains in the lower parts of the core have a notably low degree of clay coating (Fig. 3). The BSE and EDS investigations revealed that the grain-coating clay minerals include kaolinite, illite, chlorite and combination of chlorite and illite. The clay minerals present are very fine-grained, tending to be less than a few micrometres in size. Fe oxides are present in core samples. EDS spectra reveal that most grains at the base of the core are predominantly coated with the chlorite with minor illite in comparison to grains from the shallower section, which are predominantly coated with illite with minor chlorite.

BSE images of the Irt estuary samples from core 2 reveal poorly sorted sediments with sand grains that are variably coated with clay minerals. The sediments contain a combination of clastic grains and marine microfauna, such as diatoms. The clay coat coverage generally decreases up-section. The BSE and EDS investigations revealed that the clay minerals present are kaolinite, illite, chlorite and combinations of illite and chlorite. Pyrite is an important authigenic mineral that decreases in abundance upwards in the core. The clay minerals present are fine-grained and in some cases occur as single-phase accumulations. EDS spectra reveal that most grains at the base of the core are coated with illite with minor amounts of mixed illite and chlorite, in contrast to grains in the shallow part, which are coated more with mixed illite and chlorite.

XRD analysis

X-ray diffraction confirmed the unsurprising dominance of quartz with somewhat smaller quantities of feldspar. Clay minerals are present in relatively small quantities in core 1 from the Esk sediments (Fig. 4). The relative heights of the chlorite (001) and (002) peaks (ratio of 1:4) suggest that the chlorite in the sediment core is relatively Fe-rich (Hillier & Velde 1992; Hillier 2003). Scans of the glycolated sample showed no discernable changes in peaks intensity suggesting that smectite is negligible in these sediments (Moore & Reynolds 1997). A scan of the sample heated up to 400°C revealed a distinct decrease in intensity for the peaks at 12.3° and 12.5° (equivalent to 7.15 Å and 7.07 Å, respectively). Chlorite (002) peaks are reportedly unaffected when heated to 400°C (Starkey *et al.* 1984). In contrast, kaolinite (001) peaks typically show a decrease in intensity on heating to 400°C. This confirms the presence of kaolinite. For the 400°C scan, the intensities of the peaks at 6.2° and 8.9° remain unchanged, further confirming that illite and chlorite are present. A scan of a sample heated up to 550°C shows a significant collapse for both the 12.3° and 12.5° peaks, while the intensity of the peak at 6.2° increased and slightly sharpened, behaviour that is typical of chlorite. Figure 3 shows that the peaks for the chlorite (002) at 12.5° and chlorite (004) at 25.3° decrease in intensity up-section. This suggests that the overall quantity of Fe-rich clay minerals decreases up-section.

QEMSCAN mineralogy and texture

The SEM and EDS analysis revealed a range of clay minerals including illite, chlorite and kaolinite. Some EDS spectra displayed mixtures of these minerals that proved hard to unravel. QEMSCAN, with its automated and repeatable approach to mineral identification (Grauch *et al.* 2008; Haberlah *et al.* 2011; Allen *et al.* 2012), resolved this issue and allowed detailed and robust quantification of the amounts of the various clay minerals. The quantitative data from QEMSCAN report all values as mass per cent. Where abundances are greater than 1%, QEMSCAN results are quoted to one decimal place. Where abundances are less than 1%, QEMSCAN results are quoted to two decimal places.

Esk estuary samples (core 1). QEMSCAN imaging shows that sediments from the Esk estuary core are composed of the subangular to very angular fine-grained, quartz-dominated sands (Fig. 5) suggesting that these are mineralogically fairly mature. QEMSCAN confirms that quartz is the most abundant mineral in the samples, ranging from 68.4 to 84.8% of the bulk mineralogy (Figs 5 & 6a; Table 1). Mica, including muscovite and biotite, makes up to 2.8% of the bulk mineralogy (Fig. 6a). K-feldspar is the most type of common feldspar across all samples ranging from 5.0 to 7.3%. Albite ($Na_{1.0-0.9}$ $Ca_{0.0-0.1}$) is the second most abundant feldspar (2.6–4.7%), with andesine ($Na_{0.5-0.7}Ca_{0.5-0.3}$), labradorite ($Na_{0.3-0.5}Ca_{0.7-0.5}$) and anorthite ($Na_{0.0-0.1}Ca_{1.0-0.9}$) all in trace

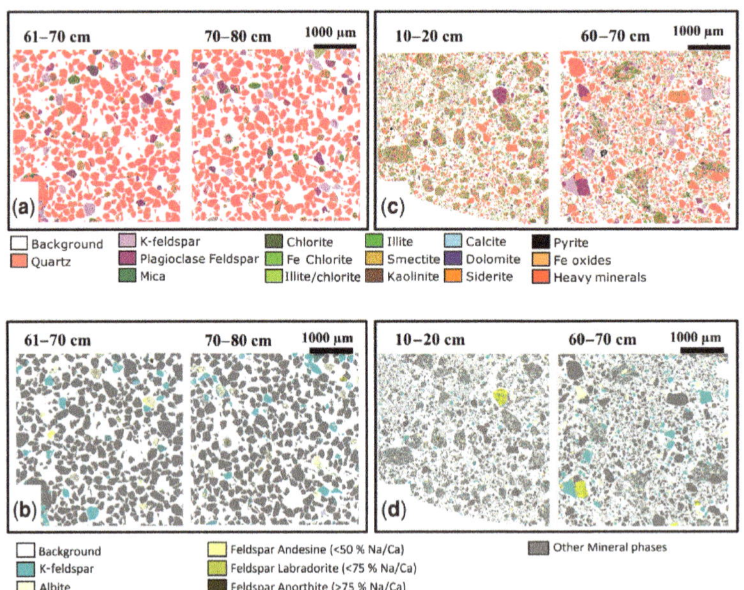

Fig. 5. Representative automated mineral maps for sediment samples from (**a**, **c**) the Esk and (**b**, **d**) Irt estuaries. The false-coloured images (on the right) show feldspar differentiated distribution.

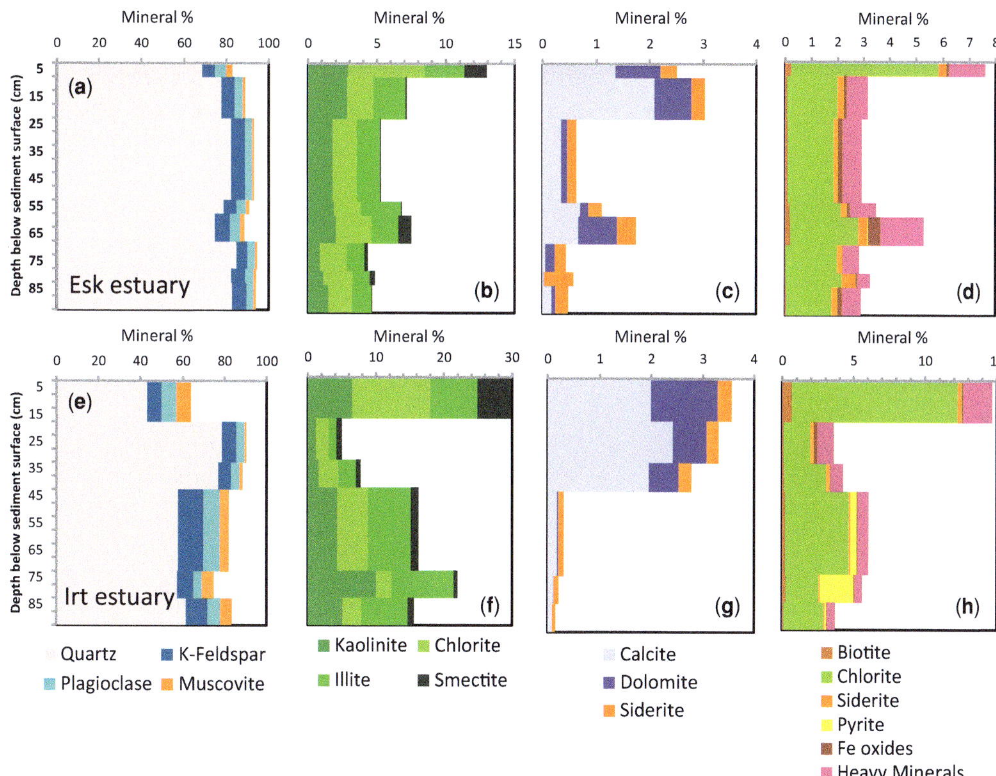

Fig. 6. QEMSCAN-derived mineral variations with depth for the Esk (top) and Irt (bottom) sediment cores. (**a**, **e**) The main detrital minerals. (**b**, **f**) Clay minerals. (**c**, **g**) Carbonate minerals. (**d**, **h**) Fe-bearing minerals. All data are reported as mass per cent.

quantities, making up 0.38, 0.45 and 0.62% respectively (Table 1).

The overall abundance of clay minerals varies from 5 to 13% with an average of about 6% (Fig. 6b). Overall clay minerals are dominated by chlorite (average 2.4%), kaolinite (average 1.8%) and illite (average 1.9%) while QEMSCAN-defined smectite is the least abundant clay (average 0.43%).

Carbonate minerals vary from <1 to 3% (Fig. 6c). They are dominated by calcite with less abundant dolomite and siderite. The minor amounts of calcite make up between 0.02 and 2.1% while dolomite is a trace phase and comprises up to 0.84%. Siderite, unlike other carbonates, is consistently distributed throughout the core. Notably, calcite and dolomite are more prevalent within the top part of core 1 (up to 20 cm depth).

The pyrite content of the Esk core is very low throughout (<0.04%) and the typically Fe-bearing heavy minerals are sporadically distributed, comprising between 0.5 and 1.6% (Fig. 6d). The garnet types correspond to dominant almandine–spessartine (Fe and Mn rich garnet) (Fig. 7a).

Irt estuary samples (core 2). QEMSCAN shows that Irt estuarine sediments in the core are composed of subangular to angular, very fine to fine sand grains that are categorized as arkosic to subarkosic (Fig. 6e). Lithic fragments are common throughout Irt estuary sediments.

Although the quartz content is lower in Irt than Esk sediments, it is the most common mineral throughout the Irt, ranging from 43.3 to 78.7% of the bulk mineralogy. There are smaller quantities of K-feldspar (the most abundant feldspar; from 6.2 to 12.0%) and albite ($Na_{1.0-0.9}Ca_{0.0-0.1}$) (from 2.7% to 6.0%) (Table 1). Andesine ($Na_{0.5-0.7}Ca_{0.5-0.3}$), labradorite ($Na_{0.3-0.5}Ca_{0.7-0.5}$) and anorthite ($Na_{0.0-0.1}$ $Ca_{1.0-0.9}$) all are trace phases making up to 0.66, 0.75 and 0.83%, respectively (Table 1). Flakes of mica, including muscovite and biotite, are present throughout (4.3% total mica) (Fig. 6e).

QEMSCAN shows that the clay mineral content is higher in the Irt core than in the Esk core and is generally dominated by illite (average 5.5%), kaolinite (average 4.8%) and chlorite (average 4.3%) with smaller amounts of smectite (average 1.5%)

Table 1. *Bulk mineralogy analysis using QEMSCAN*

Depth (cm)	Quartz	K-feldspar	Albite	Andesine	Labradorite	Anorthite	Muscovite	Bictite	Kaolinite	Chlorite	Illite	Smectite	Calcite	Dolomite	Siderite	Pyrite	Fe oxides	Heavy minerals
Esk estuary mineralogy																		
0–5	68.45	6.13	4.7	0.17	0.15	0.43	2.79	0.24	2.9	5.58	2.88	1.59	1.36	0.84	0.3	0	0.09	1.4
10–20	77.83	6.19	3.22	0.31	0.21	0.06	1	0.08	2.84	1.92	2.34	0.06	2.09	0.69	0.24	0.01	0.1	0.81
20–30	82.33	6.51	3.01	0.18	0.24	0.07	0.74	0.1	1.79	1.76	1.69	0.05	0.35	0.11	0.16	0	0.17	0.74
45–50	78.83	6.02	3.77	0.28	0.32	0.16	1.43	0.18	1.89	1.94	2.92	0.06	0.7	0.14	0.25	0	0.1	1.01
50–58	81.87	5.02	4.02	0.2	0.45	0.58	1.08	0.06	1.3	2.34	1.22	0.48	0.23	0.29	0.17	0	0.01	0.68
58–61	74.55	7.25	3.76	0.24	0.32	0.41	1.91	0.2	2.02	2.61	1.93	0.95	0.66	0.73	0.35	0.04	0.45	1.62
61–70	84.84	5.24	2.71	0.26	0.04	0.53	0.93	0.03	0.87	1.95	1.27	0.25	0.05	0.18	0.2	0	0.01	0.64
70–80	82.47	6.47	3.09	0.09	0.1	0.62	1.14	0.04	1.12	2.13	1.24	0.37	0.02	0.01	0.54	0	0.05	0.5
80–90	82.74	6.91	2.55	0.38	0.15	0.11	1.09	0.06	1.48	1.72	1.41	0.04	0.17	0.07	0.23	0	0.16	0.73
Irt estuary Mineralogy																		
10–20	43.31	6.67	5.01	0.6	0.79	0.51	6.84	0.58	6.52	11.54	6.83	4.97	2	1.28	0.27	0.04	0.09	2.05
20–30	78.66	6.95	3.03	0.18	0.24	0.35	0.84	0.12	1.23	1.86	1.12	0.74	2.42	0.65	0.23	0	0.24	1.14
30–40	76.93	6.21	2.74	0.37	0.36	0.48	1.38	0.15	1.6	2.92	2.53	0.63	1.95	0.58	0.24	0.01	0.03	0.89
60–70	57.79	12.04	5.96	0.66	0.75	0.49	4.23	0.22	4.35	4.47	6.36	1.11	0.18	0.02	0.1	0.46	0.06	0.75
70–80	57.32	7.69	3.77	0.16	0.07	0.02	5.55	0.27	10.05	2.28	9.17	0.49	0.1	0.01	0.09	2.39	0.06	0.51
80–90	61.53	10.19	5.14	0.15	0.08	0.83	5.5	0.12	5.12	2.78	6.82	0.87	0.08	0	0.06	0.13	0.02	0.58

All data are reported in mass per cent.

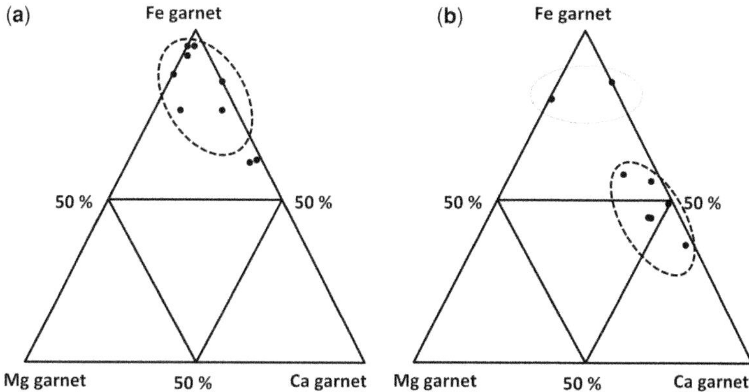

Fig. 7. Ternary diagrams showing the garnet speciation for samples from the (**a**) Esk estuary and (**b**) Irt estuary. Note the difference in the proportions of Ca and Fe garnet between the two locations. Dashed lines represent definite groups and faint solid line represents tentative group.

(Fig. 6f). Notably, the top of core 2 is made up of 30.1% of clay minerals that are dominated by chlorite.

In the Irt core, pyrite and carbonates are more abundant than in the Esk core, averaging 0.51 and 1.7% respectively (Fig. 6g, h). Calcite is the main carbonate mineral (up to 2.4%) and is more prevalent within the top part of the core (between 10 and 40 cm) (Fig. 6g). Other carbonates including dolomite and siderite are less abundant making up to 1.3 and 0.27%, respectively.

Heavy minerals are more prevalent in the upper samples and represent up to 2.1% of the sediment (Fig. 6h) with garnet speciation analysis showing the presence of grossular (Ca-rich) along with almandine and spessartine (Fe- and Mn-rich varieties) (Fig. 7b).

Discussion

The modern Ravenglass Estuary sediments studied here have been bioturbated and have sand-filled burrows that are visible in the cores. At the two core sites the lack of gravel suggests that the evolving depositional environments have not been fully fluvial. The reported presence of diatoms implies that there has been a persistent marine influence during the deposition of the 5–10 m of estuarine sediment (Bousher 1999) since the end of the most recent glaciation at approximately 17 ka. If we make a simple assumption that the rate of sedimentation has been fairly consistent then the 1 m cores represent between 1.7 and 3.4 kyr of deposition. Overall, the sedimentary characteristics of the metre-long cores in the Esk and the Irt suggest that these sites have been estuarine for at least 1.7–3.4 kyr as the cores do not contain evidence of fully marine or fully fluvial conditions.

Mineralogy of the sediment in the two arms of the estuary

The mineralogical and chemical composition of clastic sediments is the product of various factors including the hinterland geology, climate and length of the fluvial transport system. Hinterland lithology has a primary control on sediment supplied. The intensity of physical and chemical weathering is governed by humidity, temperature and temperature ranges. Although the rate of sediment supply is controlled by the overall slope of the basin and climate, the rate of supply, in turn, regulates sorting during transportation and deposition (Roddaz *et al.* 2006; Bhuiyan *et al.* 2011; Caracciolo *et al.* 2012). The mineralogical composition of clastic sediments can also be influenced by post-depositional *in-situ* weathering and mineral–biological interaction such as bioturbation, microbial activity and various soil-forming processes.

SEM petrology, QEMSCAN and XRD analyses identified a range of minerals that is present in the estuarine sediment in both arms of the estuary (Figs 3, 4 & 5). The quantitative data from QEMSCAN and XRD analyses, where all values are reported as mass per cent, revealed that the modern sediment in the southern arm of the estuary (Esk) is dominated by quartz (79.3%) with subordinate K-feldspar (6.2%), plagioclase (4.2%), chlorite 2.4%), illite (1.9%), muscovite (1.3%) and kaolinite (1.8%) alongside trace (<1%) quantities of carbonates, biotite, Fe oxides and pyrite (Fig. 6, Table 1). QEMSCAN analysis suggested that there was 0.43% smectite; XRD analysis suggested only trace swelling characteristics for the clay, thus supporting the quantitative QEMSCAN assessment.

The granite-sourced sediments of the southern arm of the estuary, drained by the River Esk, are thus quartz-dominated with an average quartz/(quartz + total feldspar) (Q/QF) index of 88 ± 2%. Treating the lithic population (L) as being equal to the sum of muscovite, biotite, chlorite and heavy minerals, then the southern arm of the estuary has a Q/(QFL) index of 84 ± 4%.

The quantitative data from QEMSCAN and XRD analyses revealed that the modern sediment in the northern arm of the estuary (Irt) is dominated by quartz (62.6%) with subordinate K-feldspar (8.3%), plagioclase (5.5%), chlorite 4.3%), illite (5.5%), muscovite (4.1%), kaolinite (4.8%) and calcite (1.1%) with trace (<1%) quantities of other carbonates, biotite, Fe oxides and pyrite (Fig. 6, Table 1). QEMSCAN analysis suggested that there was 1.5% smectite. XRD analysis (Fig. 3) also suggested minor swelling characteristics for the clay, thus supporting this quantitative QEMSCAN assessment.

The arkosic sandstone- and andesite-sourced sediments of the northern arm of the estuary, drained by the River Irt, are quartz rich with an average Q/QF index of 81 ± 6% and a Q/(QFL) index of 72 ± 11%.

Overall, therefore, the granite-provenanced sediment is somewhat cleaner and more quartz-rich than the sediment of andesite- and arkose-sandstone provenance.

Feldspar populations in the estuary

Plagioclase and K-feldspar are found in sediment in both arms of the estuary (Table 1). SEM and QEMSCAN analyses revealed that the feldspars occur as detrital grains (Figs 4 & 5).

Feldspar in the southern (Esk) arm of the estuary is dominated by K-feldspar with a K-feldspar/(K-feldspar + plagioclase) index of 55 ± 5% (Table 1). Feldspar in the northern (Irt) arm of the estuary is also dominated by K-feldspar with a K-feldspar/(K-feldspar + plagioclase) index of 57 ± 5%. The feldspar populations are thus effectively the same in both arms of the estuary, irrespective of provenance. The Triassic sandstone in the hinterland is dominated by K-feldspar with the Borrowdale Volcanic Group andesite dominated by plagioclase. In contrast the granite contains both K-feldspar and plagioclase. The similarity in the overall feldspar population appears to be a coincidence.

A range of types of plagioclase (between the sodic and calcic end-members) are found in sediment in both arms of the estuary (Table 1). The southern arm of the estuary contains a range of types of plagioclase, with sodium-dominated albite the most abundant type. Detrital andesine, labradorite and anorthite are found in measurable quantities as well as albite. Albite represents 81 ± 4% of the plagioclase population.

The northern arm of the estuary contains different types of plagioclase, including andesine, labradorite and anorthite, and is also dominated by albite, which represents about 79 ± 9%. Thus the specific plagioclase populations are similar in both the Esk and Irt arms of the estuary.

Heavy mineral indications of provenance differences in the estuary

QEMSCAN data revealed that the heavy minerals in the southern Esk arm of the estuary include hornblende, garnet, apatite, rutile, zircon, ilmenite and epidote. Rutile and hornblende are the dominant heavy minerals. The heavy-mineral garnet population in the southern arm is dominated by Fe–Al almandine (Fig. 7).

Heavy minerals in the northern Irt arm of the estuary also include hornblende, garnet, apatite, rutile, zircon, ilmenite and epidote. Rutile and hornblende are the dominant heavy minerals. The Irt sediments have a similar heavy mineral population to the Esk sediments in the southern arm. However, the garnet population of the northern arm is dominated by Ca-rich grossular (Fig. 7). Heavy mineral data need to be used with caution, but seemingly in this case specific garnet-type analysis can be used to help determine sediment provenance.

The abundance of Fe-minerals in the estuary

One of the aims of this research was to study the presence and amounts of Fe minerals in the modern sedimentary system in order to develop a fundamental understanding of Fe-rich clay minerals in ancient and deeply buried sandstone oil and gas reservoirs. The northern and southern arms of the estuary contain the following Fe-bearing minerals: chlorite, biotite, pyrite, Fe oxide and siderite (Fig. 6, Table 1). The sandstone- and andesite-sourced sediments in the northern arm contain about 5.4% of Fe-bearing minerals including dominant chlorite (4.3%) and lesser amounts of biotite, pyrite and siderite. The granite-sourced sediments in the southern arm contain half the quantity of Fe-bearing minerals (2.8%), although these are also dominated by chlorite (2.4%). SEM-analysis of the sediment revealed that the chlorite is in the silt-fraction and is probably detrital. The trace quantities of biotite are also probably detrital. In the northern Irt arm of the estuary, both chlorite and biotite are probably derived from the andesites of the Borrowdale Volcanic Group andesite as they have not been reported in the Triassic sandstones. Chlorite has been reported as an alteration product of primary Fe-rich minerals in the granite hinterland supplying sediment to the southern Esk arm of the estuary. The intermediate igneous volcanic rocks seem to have supplied more detrital chlorite than the acid igneous plutonic rocks.

Fe oxides represent a small fraction of the sediment in both arms of the estuary. The Fe-oxides identified by QEMSCAN potentially include hematite and less-crystalline phases such as goethite and limonite. There is slightly more Fe oxide in the southern Esk arm of the estuary than the northern Irt arm (0.13 v. 0.08%; Fig. 6, Table 1). The Fe oxides probably include the weathering products of Fe silicates in the hinterland and flocculation of Fe complexes transported by the rivers into the estuary. The slightly higher quantity of Fe oxide in the Irt core is not surprising since the Irt drains the New Red Sandstone as well as the Fe-rich andesites of the Borrowdale Volcanic Group.

The sand-dominated, quartz-rich Esk core contains almost no pyrite in contrast to the less-clean sediment in the Irt core (Fig. 6, Table 1). Pyrite, in modern sediments, is a product of bacterial sulphate reduction; organic matter in the sediment is the reducing agent for the aqueous sulphate that is derived from the twice daily inundation of the estuary with seawater. In the Esk sediment, the redox boundary was not observed in the 1 m core. The Irt sediment has pyrite as the dominant Fe mineral (2.4%) at a depth of about 70 cm, suggesting that bacterial sulphate reduction has led to the local conversion of detrital Fe-minerals into pyrite. It is noteworthy that, despite samples at this depth being more clay-rich than sediment above and below, it contains relatively less chlorite. This seems to suggest that the Fe-bearing chlorite, as well as less volumetrically important minerals such as biotite, Fe oxides and heavy minerals, have locally supplied the iron to create pyrite.

The presence of siderite was not detected using XRD as it occurs in small amounts and its diagnostic peaks tend to be obscured by diffraction peaks from other minerals. QEMSCAN identified small amounts of siderite throughout the cores in both arms of the estuary. Siderite is present in roughly uniform amounts. There is slightly more siderite in the pyrite-free Esk core than the pyrite-bearing Irt core. Calcite in the estuarine sediment is likely to be derived from indigenous shell fragments with increasing amounts of calcite signifying the degree of marine influence. The amount of calcite is approximately uniform in Esk and Irt cores with most occurring in the upper parts of the cores (Fig. 6) suggesting that the most recent sediment has experienced a bigger marine influence than older sediment. In contrast, siderite abundance is decoupled from calcite, indicating a different origin. In the absence of dissolved sulphide (resulting from bacterial sulphate reduction), the available iron is able to form iron carbonate (siderite) if aqueous carbonate is present. The dissolved carbonate could be supplied from dissolved atmospheric CO_2, microbial metabolisis, or other biological processes.

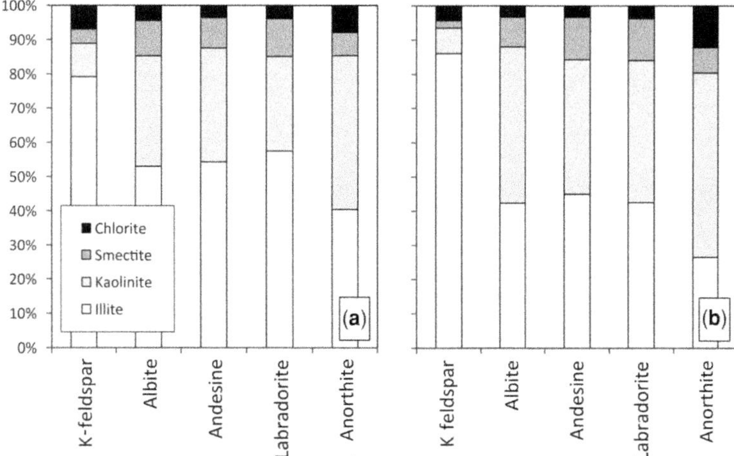

Fig. 8. Mineral association (mineral contact) between clay minerals and feldspars including K-feldspar, albite ($Na_{1.0-0.9}Ca_{0.0-0.1}$), andesine ($Na_{0.5-0.7}Ca_{0.5-0.3}$), labradorite ($Na_{0.3-0.5}Ca_{0.7-0.5}$) and anorthite ($Na_{0.0-0.1}Ca_{1.0-0.9}$). There is a broad association rate between K-feldspar with illite and plagioclase with kaolinite, however, small variation is also noted. (**a**) Irt estuarine sediments showing a strong contact between kaolinite and plagioclase and significant association between K-feldspar and illite and illite-dominated illite–chlorite. In this chart, andesite and labradorite show significant contact with illite-dominated illite–chlorite as well. (**b**) Esk estuary sediment suggesting strong association between kaolinite and plagioclase and K-feldspar and illite.

Clay mineral-coated grains in the estuary

Throughout the estuary, sand grains are variably coated with clay minerals (Fig. 4). Illite and kaolinite are the main clay minerals that occur as grain coatings in the estuarine sand, with a much smaller amount of chlorite found as a grain-coating phase in the grain mount samples (Fig. 4).

Illite and kaolinite are found in approximately equal proportions in the sediment (Table 1). The QEMSCAN data have been processed to reveal which clay minerals preferentially coat the feldspars. Chlorite, in this modern sediment, rarely occurs as a coating on feldspars (Fig. 8) despite being a common mineral in the sediment (Fig. 6). K-feldspar preferentially occurs in coatings with illite in both the Irt and Esk estuary sediments (Fig. 8). The plagioclase family of minerals tend to have a more equal proportion of illite and kaolinite, although anorthite has a very high proportion of grain coating kaolinite. The data displayed in Figure 8 suggest that different clay minerals preferentially coat different feldspar types. It seems unlikely that this could be a physical phenomenon; there seems to be no way that different clay minerals could selectively attach to different feldspar types. Instead, the patterns in Figure 8 suggest that different feldspar minerals have undergone relatively rapid alteration, since deposition after the last glaciations at 17 ka, to different types of clay minerals. K-feldspar that is preferentially coated by illite suggests that illite is a direct alteration product of K-feldspar. Anorthite that is preferentially coated by kaolinite suggests that kaolinite is a direct alteration product of anorthite. Three simple end-member alteration reactions that explain these textures can be written:

$$\underset{\text{K-feldspar}}{2.5KAlSi_3O_8} + 2H^+ \Rightarrow \underset{\text{illite}}{K_{0.5}Al_{2.5}Si_{3.5}O_{10}(OH)_2} + \underset{\text{silica}}{4SiO_2} + 2K^+ \qquad (R1)$$

$$\underset{\text{albite-plagioclase}}{2NaAlSi_3O_8} + H_2O + 2H^+ \Rightarrow \underset{\text{kaolinite}}{Al_2Si_2O_5(OH)_4} + \underset{\text{silica}}{4SiO_2} + 2Na^+ \qquad (R2)$$

$$\underset{\text{anorthite-plagioclase}}{2CaAl_2Si_2O_8} + 2H_2O + 4H^+ \Rightarrow \underset{\text{kaolinite}}{2Al_2Si_2O_5(OH)_4} + 2Ca^+ \qquad (R3)$$

The clay mineral associations with specific feldspar minerals thus suggest that localized silicate alteration reactions occur relatively rapidly (i.e. in less than the 17 kyr since the last glaciations when estuarine sediment began to accumulate). Figure 8 reveals that specific feldspars alter to different types of clay minerals. The absence of chlorite, or other Fe–Mg-bearing clay minerals, as a grain-coating minerals on feldspars may thus be a direct consequence of feldspars not containing the requisite Fe and Mg. The chlorite present in the sediment may be a detrital mineral, in contrast to illite and kaolinite, which seem to be partly the result of continued *in situ* alteration of the primary sediment. Further work is needed to assess whether the chlorite is partly a result of detrital Fe–Mg mineral alteration, although the siderite and pyrite occurrence patterns (Fig. 6) suggest that Fe mineral occurrence is partly controlled by the supply of sulphide via bacterial sulphate reduction and carbonate (possibly via organic decay processes).

Conclusions

(1) The main minerals throughout the post-glacial sediments of the Ravenglass Estuary in Cumbria (NW England, UK), which are defined by BSE and QEMSCAN analyses, are: quartz, K-feldspar, plagioclase, muscovite, biotite, illite, chlorite, kaolinite, smectite, pyrite, calcite, dolomite and siderite.

(2) The arkosic sandstone- and andesite-sourced sediments of the northern arm of the estuary, drained by the River Irt, are quartz rich with an average Q/QF index of 81 $\pm$ 6% and a Q/(QFL) index of 72 $\pm$ 11%.

(3) The granite-sourced sediments of the southern arm of the estuary, drained by the River Esk, are even more quartz-dominated with an average Q/QF index of 88 $\pm$ 2% and a Q/(QFL) index of 84 $\pm$ 4%.

(4) Feldspars are marginally dominated by K-feldspar, representing about 56% of the total feldspars in both arms of the estuary, irrespective of provenance.

(5) The dominant form of plagioclase in both arms of the estuary is albite but andesine, labradorite and anorthite are also found; albite represents about 80% of the plagioclase in both the Esk and Irt arms of the estuary.

(6) Heavy minerals identified by QEMSCAN analysis include amphibole, garnet, apatite, rutile, zircon, ilmenite and epidote. The garnet in the granite-sourced sediments is dominated by almandine, while the sandstone- and andesite-sourced sediment can be differentiated from the granite-sourced sediment by the presence of Ca-rich grossular.

(7) The sandstone- and andesite-sourced sediments in the northern arm contain about 5.4% Fe-bearing minerals, including

dominant chlorite (4.3%) and lesser amounts of biotite, pyrite and siderite. The granite-sourced sediments in the southern arm contain half the quantity of Fe-bearing minerals (2.8%), although these are also dominated by chlorite (2.4%).

(8) Throughout the estuary sand grains are variably coated with clay minerals, particularly illite and kaolinite.

(9) Throughout the estuary detrital K-feldspar grains are preferentially associated with illite-dominated clay mineral coats, whereas plagioclase mineral grains are preferentially associated with kaolinite-dominated clay mineral coats. This difference in association hints at localized early diagenesis in the sediment with K-feldspar locally converted to illite and plagioclase locally converted to kaolinite.

(10) Chlorite rarely seems to be associated with grain coatings on feldspar grains, possibly because it is predominantly a detrital mineral, in contrast to illite and kaolinite, which seem to be partly the result of continued *in situ* alteration of the primary sediment.

Authors would like to thank Richard Windmill and Peter Wellsbury at CGG Robertson (UK) for undertaking the QEMSCAN analysis and for their support.

References

AAGAARD, P., EGEBERG, P.K., SAIGAL, G.C., MORAD, S. & BJORLYKKE, K. 1990. Diagenetic albitization of detrital K feldspars in Jurassic, lower Cretaceous and Tertiary clastic reservoir rocks from offshore Norway, II. Formation water chemistry and kinetic considerations. *Journal of Sedimentary Petrology*, **60**, 575–581.

AAGAARD, P., JAHREN, J., HARSTAD, A.O., NILSEN, O. & RAMM, M. 2000. Formation of grain coating chlorite in sandstones: laboratory synthesized v. natural occurrences. *Clay Minerals*, **35**, 261–269.

ALLEN, J.L., JOHNSON, C., HEUMANN, M., GOOLEY, J. & GALLIN, W. 2012. New technology and methodology for assessing sandstone composition: a preliminary case study using quantitative electron microscope scanner (QEMSCAN®). *Geological Society of America Special Papers*, **487**, 177–194.

ALLER, R. & ALLER, J. 1998. The effect of biogenic irrigation intensity and solute exchange on diagenetic reaction rates in marine sediments. *Journal of Marine Research*, **56**, 905–936.

BAKER, J.C., HAVORD, P.J., MARTIN, K.R. & GHORI, K.A.R. 2000. Diagenesis and petrophysics of the early Permian Moogooloo sandstone, Southern Carnarvon Basin, Western Australia. *AAPG Bulletin*, **84**, 250–265.

BARNES, R., AMBROSE, K., HOLLIDAY, D. & JONES, N. 1994. Lithostratigraphical subdivision of the Triassic Sherwood sandstone group in west Cumbria. *Proceedings of the Yorkshire Geological Society*, **50**, 51–60.

BERNER, R. 1980. *Early Diagenesis: A Theoretical Approach*. Princeton University Press, Princeton.

BHUIYAN, M., RAHMAN, J., DAMPARE, S. & SUZUKI, S. 2011. Provenance, tectonics and source weathering of modal fluvial sediments of the Brahmaputra-Jamuna River, Bangladesh: inference from geochemistry. *Journal of Geochemical Exploration*, **111**, 113–137.

BJØLYKKE, K. 1998. Clay mineral diagenesis in sedimentary basins-a key to the prediction of rock properties. Examples from the North Sea Basin. *Clay Minerals*, **33**, 15–34.

BOKUNIEWICZ, H. 1995. Sedimentary systems of coastal-plain estuaries. *In*: PERILLO, G.M. (ed.) *Developments in Sedimentology*, **53**, 49–67, https://doi.org/10.1016/S0070-4571(05)80023-8

BOLES, J.R. 1982. Active albitization of plagioclase, Gulf coast Tertiary. *American Journal of Science*, **282**, 165–180.

BOUSHER, A. 1999. *Ravenglass Estuary: Basic Characteristics & Evaluation of Restoration Options*. RESTRAT-WP 1.4, Restrad-Td 12, Draft issue 4. Westlakes Scientific Consulting Ltd, Moor Row.

BOYLE, E., COLLIER, R., DENGLER, A., EDMOND, J., NG, A. & STALLARD, R. 1974. On the chemical mass-balance in estuaries. *Geochimica et Cosmochimica Acta*, **38**, 1719–1728.

BOYLE, E., EDMOND, J. & SHOLKOVITZ, E. 1977. The mechanism of iron removal in estuaries. *Geochimica et Cosmochimica Acta*, **41**, 1313–1324.

BROWN, P., SOPER, N. & MILLER, J. 1964. Age of the principal intrusions of the Lake District. *Proceedings of the Yorkshire Geological Society*, **34**, 331–342.

BURLEY, S. & WORDEN, R.H. 2003. Sandstone diagenesis: the evolution of sand to stone. In BURLEY, S.D. & WORDEN, R.H. (eds) *Sandstone Diagenesis, Recent and Ancient*. International Association of Sedimentologists, **4**. Blackwell Scientific, Oxford, 3–44.

CARACCIOLO, L., TOLOSANA-DELGADO, R., LE PERA, E., VON EYNATTEN, H., ARRIBAS, J. & TARQUINI, S. 2012. Influence of granitoid textural parameters on sediment composition: implications for sediment generation. *Sedimentary Geology*, **280**, 93–107.

COLTER, V. & EBBERN, J. 1978. The petrography and reservoir properties of some Triassic sandstones of the Northern Irish Sea Basin. *Journal of Geological Society, London*, **135**, 57–62.

DALRYMPLE, R.W., ZAITLIN, B.A. & BOYD, R. 1992. Estuarine facies models – conceptual models and stratigraphic implications. *Journal of Sedimentary Petrology*, **62**, 1130–1146.

DANESHVAR, E. 2012. *Clay Minerals and Provenance Study in Modern Sedimentary Environments*. Lambert Academic Publishing, Berlin.

DEER, W.A., HOWIE, R.A. & ZUSSMAN, J. 2013. *An Introduction to the Rock-Forming Minerals*. 3rd edn. Mineralogical Society Publication, London.

DREVER, J. & HURCOMB, D. 1986. Neutralization of atmospheric acidity by chemical weathering in an alpine drainage basin in the North Cascade Mountains. *Geology*, **14**, 221–224.

DREVER, J. & ZOBRIST, J. 1992. Chemical weathering of silicate rocks as a function of elevation in the southern Swiss Alps. *Geochimica et Cosmochimica Acta*, **56**, 3209–3216.

ENTWISLE, D.C., HOBBS, P.R.N., JONES, L.D., GUNN, D. & RAINES, M.G. 2005. The relationships between effective porosity, uniaxial compressive strength and Sonic Velocity of Intact Borrowdale Volcanic Group Core samples from Sellafield. *Geotechnical & Geological Engineering*, **23**, 793–809.

FAN, D., NEUSER, R., SUN, X., YANG, Z., GUO, Z. & ZHAI, S. 2008. Authigenic iron oxide formation in the estuarine mixing zone of the Yangtze River. *Geo-Marine Letters*, **28**, 7–14.

FITTON, G.J. 1972. The genetic significance of almandine-pyrope phenocrysts in the calc-alkaline Borrowdale Volcanic Group, Northern England. *Contributions to Mineralogy and Petrology*, **36**, 231–248.

GRAUCH, R.I., EBERL, D., BUTCHER, A.R. & BOTHA, P. 2008. Quantitative mineralogy of fine-grained sedimentary rocks: a preliminary look at QEMSCAN. *Microscopy and Microanalysis*, **14**, 532–533.

HABERLAH, D., STRONG, C., PIRRIE, D., ROLLINSON, G., GOTTLIEB, P., BOTHA, P. & BUTCHER, A.R. 2011. Automated petrography applications in Quaternary sciences. *Quaternary Australasia*, **28**, 3–12.

HAUGHTON, P.D., TODD, S.P. & MORTON, A.C. 1991. Sedimentary provenance studies. *In*: HAUGHTON, P.D., TODD, S.P. & MORTON, A.C. (eds) *Development in Sedimentary Provenance Studies*. Geological Society, London, Special Publications, **57**, 1–11.

HILLIER, S. 2003. Quantitative analysis of clay and other minerals in sandstones by X-ray powder diffraction (XRPD). *In*: WORDEN, R.H. & MORAD, S. (eds) *Clay Mineral Cements in Sandstones*. International Association of Sedimentologists, Special Publications, **34**, 213–251.

HILLIER, S. & VELDE, B. 1992. Chlorite interstratified with a 7 Å mineral: an example from offshore Norway and possible implications for the interpretation of the composition of diagenetic chlorites. *Clay Minerals*, **27**, 475–486.

HUANG, W. 1993. Stability and kinetics of kaolinite to boehmite conversion under hydrothermal conditions. *Chemical Geology*, **105**, 197–214.

MCDOUGALL, D.A. 2001. The geomorphological impact of Loch Lomond (Younger Dryas) Stadial plateau icefields in the central Lake District, northwest England. *Journal of Quaternary Science*, **16**, 531–543.

MCILROY, D., WORDEN, R.H. & NEEDHAM, S. 2003. Faeces, clay minerals and reservoir potential. *Journal of the Geological Society, London*, **160**, 489–493.

MERRITT, J.W. & AUTON, C.A. 2000. An outline of the lithostratigraphy and depositional history of Quaternary deposits in the Sellafield district, west Cumbria. *Yorkshire Geological Society*, **53**, 129–154.

MILDOWSKI, A.E., GILLESPIE, M.R., NADEN, J., FORTEY, N.J., SHEPHERD, T.J., PEARCE, J.M. & METCALFE, R. 1998. The petrology and paragenesis of fracture mineralization in the Sellafield area, west Cumbria. *Proceedings of the Yorkshire Geological Society*, **52**, 215–241.

MOORE, D. & REYNOLDS, R. 1997. *X-Ray Diffraction and the Identification and Analysis of Clay Minerals*. 2nd edn. Oxford University Press, Oxford.

MORAD, S. 1988. Albitized microcline grains of post-depositional and probable detrital origins in Brøttum Formation sandstones (Upper Proterozoic), Sparagmite region of southern Norway. *Geological Magazine*, **125**, 229–239.

MORAD, S., BERGAN, M., KNARUD, R. & NYSTUEN, P. 1990. Albitization of detrital plagioclase in Triassic reservoir sandstones from the Snorre field, Norwegian North Sea. *Journal of Sedimentary Petrology*, **60**, 411–425.

MORTON, A.C. 1987. Influence of provenance and diagenesis on detrital garnet suites in the Paleocene Forties sandstone, central North Sea. *Journal of the Sedimentary Petrology*, **57**, 1027–1032.

MOSELEY, F. (ed.) 1978. *The Geology of the Lake District*. Yorkshire Geological Society Occasional Publication, Leeds, **3**, 41–44.

PARSONS, I., THOMPSON, P., LEE, M.R. & CAYZER, N. 2005. Alkali feldspar microtextures as provenance indicators in siliciclastic rocks and their role in feldspar dissolution during transport and diagenesis. *Journal of Sedimentary Research*, **75**, 921–942.

PIRRIE, D., POWER, M.R., ROLINSON, G., CAMM, S., HUGHES, S., BUTCHER, A.R. & HUGHES, P. 2003. The spatial distribution and source of arsenic, copper, tin and zinc within the surface sediments of the Fal Estaury, Cornwall, UK. *Sedimentology*, **50**, 579–595.

RAMM, M. 2000. Reservoir quality and its relationship to facies and provenance in Middle to Upper Jurassic sequences, northeastern North Sea. *Clay Minerals*, **35**, 77–94.

RODDAZ, M., VIERS, J., BRUSSET, S., BABY, P., BOUCAYR, & HERAIL, G. 2006. Controls on weathering and provenance in the Amazonian foreland basin: insight from major and trace element geochemistry of Neogene Amazonian sediments. *Chemical Geology*, **226**, 31–65.

RUNDLE, C. 1979. Ordovician intrusions in the English Lake District. *Journal of the Geological Society, London*, **136**, 29–38.

SIMPSON, B. 1934. The petrology of the Eskdale (Cumberland) granite. *Proceedings of the Geologists Association*, **45**, 17–34.

SIONNEAU, T., BOUT-ROUMAZEILLES, V., BISCAYE, P., VAN VLIET-LANOE, B. & BORY, A. 2008. Clay mineral distributions in and around the Mississippi River watershed and Northern Gulf of Mexico: sources and transport patterns. *Quaternary Science Reviews*, **27**, 1740–1751.

SOPER, N. 1987. The Ordovician batholith of the English Lake District. *Geological Magazine*, **124**, 481–482.

STARKEY, H.C., BLACKMON, P.D. & HAUFF, P.F. 1984. *The Routine Mineralogical Analysis of Clay-Bearing Samples*. US Geological Survey Bulletin, 1563.

STRONG, G., MILODOWSKI, A.E., PEARCE, J.M., KEMP, S.J., PRIOR, S.V. & MORTON, A.C. 1994. The petrology and diagenesis of Permo-Triassic rocks of the Sellafield area, Cumbria. *Proceedings of the Yorkshire Geological Society*, **50**, 77–89.

THIRY, M. 2000. Palaeoclimatic interpretation of clay minerals in marine deposits: an outlook from the continental origin. *Earth-Science Reviews*, **49**, 201–221.

TUCKER, M.E. 2001. *Sedimentary Petrology: An Introduction to the Origin of Sedimentary Rocks*. Blackwell, Oxford.

TYRRELL, S., LELEU, S., SOUDERS, A.K., HAUGHTON, P.D. & DALY, J.S. 2011. K-feldspar sand-grain provenance in the Triassic, west of Shetland: distinguishing first-cycle and recycled sediment sources? *Geological Journal*, **44**, 692–710.

VELDE, B. 1985. *Clay Minerals: A Physico-Chemical Explanation of their Occurrence*. Developments in Sedimentology, **40**. Elsevier, Amsterdam.

VELDE, B. & CHURCH, T. 1999. Rapid clay transformations in Delaware salt marshes. *Applied Geochemistry*, **14**, 559–568.

WORDEN, R.H. & BURLEY, S.D. 2003. *Sandstone Diagenesis: Recent and Ancient*. Reprint series volume 4, International Association of Sedimentologists, Oxford.

WORDEN, R.H. & MORAD, S. 2003. Clay minerals in sandstones: controls on formation, distribution and evolution. *In*: WORDEN, R.H. & MORAD, S. (eds) *Clay Mineral Cements in Sandstones*. International Association of Sedimentologists, Special Publications, **34**, 3–41.

ZHANG, J., WEN HUANG, W. & CHONG SHI, M. 1990. Huanghe (Yellow River) and its estuary: sediment origin, transport and deposition. *Journal of Hydrology*, **120**, 203–223.

Evolution of small-scale flow barriers in German Rotliegend siliciclastics

BENJAMIN BUSCH[1]*, REBECCA WINKLER[1], KEYVAN OSIVANDI[2], GEORG NOVER[3], ALEXANDRA AMANN-HILDENBRAND[4] & CHRISTOPH HILGERS[1]

[1]*Institute of Reservoir-Petrology, EMR – Energy and Mineral Resources Group, RWTH Aachen University, Wuellnerstraße 2, 52062 Aachen, Germany*

[2]*RWE DEA AG, Hamburg, Ueberseering 40, 22297 Hamburg, Germany*

[3]*Steinmann-Institute of Geology, Mineralogy und Palaeontology, University Bonn, Meckenheimer Allee 169, 53115 Bonn, Germany*

[4]*Institute of Geology and Geochemistry of Petroleum and Coal, EMR – Energy and Mineral Resources Group, RWTH Aachen University, Lochnerstraße 4-20 (House B), 52062 Aachen, Germany*

**Correspondence: Benjamin.Busch@emr.rwth-aachen.de*

Abstract: Many siliciclastic reservoirs contain millimetre-scale diagenetic and structural phenomena affecting fluid flow. We identified three major types of small-scale flow barriers in a clastic Rotliegend hydrocarbon reservoir: cataclastic deformation bands; dissolution seams; and bedding-parallel cementation. Deformation bands of various orientations were analysed on resistivity image logs and in core material. They are mainly conjugates, and can be used to validate seismically observable faults and infer subseismic faults. Bedding-parallel dissolution seams are related to compaction and post-date at least one set of deformation bands. Bedding-parallel cementation is accumulated in coarser-grained layers and depends on the amount of clay coatings.

Apparent permeability data related to petrographical image interpretation visualizes the impact of flow barriers on reservoir heterogeneity. Transmissibility multiplier calculations indicate the small efficiency of the studied deformation bands on flow properties in the reservoir. Deformation bands reduce the host-rock permeability by a maximum of two orders of magnitude. However, host-rock anisotropies are inferred to reduce the permeability by a maximum of four orders of magnitude. The relative timing of these flow barriers, as well as the assessment of reservoir heterogeneities, are the basis for state-of-the-art reservoir prediction modelling.

Deformation bands are zones of localized deformation in granular media, and are frequently reported from siliciclastic rocks and limestones (e.g. Antonellini & Aydin 1995; Fossen *et al.* 2007; Legler & Marchel 2008; Wennberg *et al.* 2013). Generally, three end members of deformation bands are kinematically classified as shear bands, compaction bands and dilation bands (Fossen *et al.* 2007). The orientation of the localization plane with respect to the principal stress orientation and the deformation band type differs for these three end members (Bésuelle & Rudnicki 2004, fig. 5.16). Shear localization forms at low effective stresses, compaction at higher effective stresses and dilation is linked to decreasing effective pressures (Wong *et al.* 1997; Bésuelle & Rudnicki 2004). Low porosity and small grain size exert an influence on the physical process of strain localization and will result in an increased magnitude of the compactive yield strength (Wong *et al.* 1997; David *et al.* 2001; Schultz *et al.* 2010).

The wide range of varying microstructures in deformation bands is reflected by the diverse terminology, addressing the mineralogical composition (e.g. the incorporation of clay minerals into the deformation band by shearing results in phyllosilicate bands), the kinematic or physical processes of formation (e.g. dilational deformation bands, cataclastic deformation bands and compaction bands) or hybrids of different mechanisms (e.g. shear-enhanced compaction bands). Cataclastic deformation bands are frequently observed around larger normal faults in soft sediment and weakly lithified rocks, and display the incipient stage of faulting and strain hardening (Antonellini & Aydin 1995; Fossen 2010; Ballas *et al.* 2012; Soliva *et al.* 2013). They

From: Armitage, P. J., Butcher, A. R., Churchill, J. M., Csoma, A. E., Hollis, C., Lander, R. H., Omma, J. E. & Worden, R. H. (eds) 2018. *Reservoir Quality of Clastic and Carbonate Rocks: Analysis, Modelling and Prediction.* Geological Society, London, Special Publications, **435**, 141–160.
First published online November 18, 2015, https://doi.org/10.1144/SP435.3

are reported to have no or only a very little impact on reservoir properties due to their spatially limited geometry and the limited interconnectivity in three dimensions (e.g. Fossen & Bale 2007), while their clustering may act as a permeability barrier that reduces fluid flow by up to four orders of magnitude (e.g. Saillet & Wibberley 2013).

The geometry, evolution and different scaling relationships of deformation bands have been studied in many outcrops (e.g. Antonellini & Aydin 1994; Fisher & Knipe 2001; Ogilvie & Glover 2001; Schultz & Fossen 2002; Davatzes & Aydin 2003; Eichhubl *et al.* 2004; Olsson *et al.* 2004; Parry *et al.* 2004; Eichhubl & Flodin 2005; Sternlof 2006; Ahmadov *et al.* 2007; Rotevatn *et al.* 2007; Johansen & Fossen 2008; Schultz *et al.* 2008; Guo *et al.* 2009; Kolyukhin *et al.* 2009; Fossen 2010; Fossen *et al.* 2011; Ballas *et al.* 2012; Chemenda *et al.* 2012; Exner *et al.* 2013; Nicol *et al.* 2013; Saillet & Wibberley 2013; Schueller *et al.* 2013; Schultz *et al.* 2013; Soliva *et al.* 2013; Awdal *et al.* 2014; Torabi 2014). While deformation bands often strike parallel to extension faults, local stress perturbations around faults, fault splays and in relay ramps may result in multiple sets of deformation bands (e.g. Antonellini & Aydin 1995; Johansen *et al.* 2005; Rotevatn *et al.* 2007).

In this study, we analyse the apparent permeability anisotropy of moderately to very-well-sorted Rotliegend sandstone cores, and the influence of cataclastic deformation bands and bedding anisotropies on fluid flow. From microstructural analyses, we derive a scaling relationship that considers the width of the cataclastic deformation bands and the grain size of the undisturbed host rock, which is correlated with the apparent permeability data. Data mining tools are used to better display variations of deformation bands in a reservoir. Finally, transmissibility calculations in relation to deformation-band frequency and the assessment of deformation bands from core material are critically discussed.

Geological setting

The Rotliegend gas-producing reservoir is situated in a part of the north–south-orientated fan-shaped Graben setting in the North German Basin in the area between Bremen, Hamburg and the Elbe river (Kayser 2006). It is part of the Southern Permian Basin, and accumulated thick successions of Upper Rotliegend continental siliciclastics and evaporites. The Permian horst-and-graben structure (Fig. 1) was affected by Triassic–Cretaceous

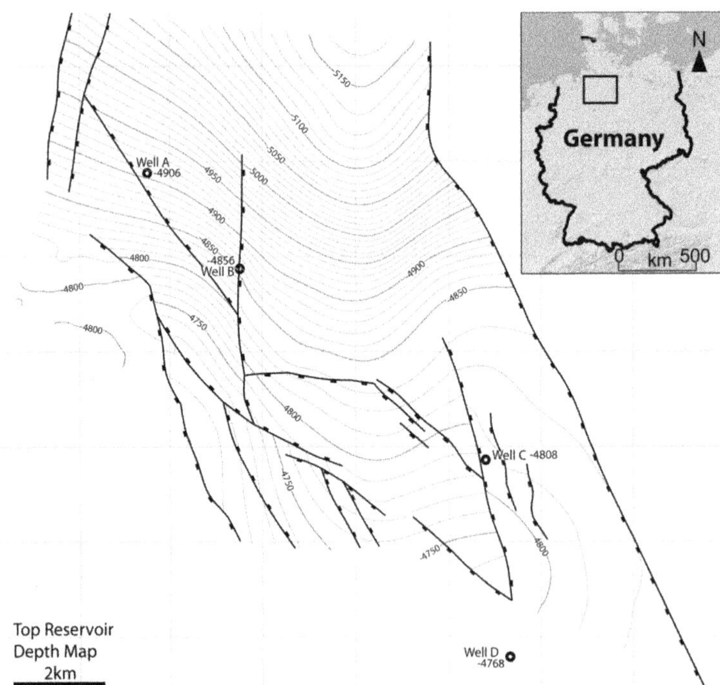

Fig. 1. Location of the study area in northern Germany (right), and the structure and depth map of the top of the reservoir unit including well locations (left).

extension and the formation of Zechstein salt structures (Vackiner 2011). The geological evolution was followed by Upper Cretaceous north–south-orientated compression, concurrent inversion and further anticlockwise stress rotation to the recent stress field in the Late Eocene–Miocene (Lohr *et al.* 2007; Kley *et al.* 2008; Legler & Marchel 2008). The sediments have mainly been deposited in an arid–semi-arid desert environment and the red–grey sandstones fall into the field of lithic subarkoses–subarkoses (Legler & Marchel 2008). The main reservoir target for exploration and production is the Upper Rotliegend Havel Subgroup (266–262 Ma: Menning 1995) (Fig. 2), which has a thickness of 300–400 m in the studied wells. It overlies Lower Rotliegend volcanics at the base. The present lithotypes, as characterized on resistivity image logs, are divided into dune, sandsheet and alluvial-fan deposits.

Materials and methods

The database comprises resistivity image logs, gamma-ray logs and lithological logs of four wells, including core material. The available samples were 14 plugs from core material for permeability measurements and 32 petrographical thin sections of four wells. Overall, plugs were taken from dune and sandsheet deposits at a reservoir depth of between 4768 and 5251 m.

Plugs, 30 mm in diameter and 40 mm in length, were used for permeability measurements, conducted in a high-pressure flow cell at well-defined pressure (P) and monitored temperature conditions ($P_{conf} = 6.5$ and 32 MPa, $P_{fluid} = 0.1$–2 MPa, $T = 25 \pm 3$°C). In order to prevent bypass of the permeating fluid, the samples were tightly sealed with a rubber sleeve before being placed in the water-filled autoclave. Argon was used as permeant for the pressure-pulse decay measurements. Fluid pressures on both sides of the sample were continuously monitored, with the upstream pressure (P_1) held constantly at 1.8 MPa. The downstream pressure (P_2) within a previously calibrated reservoir ($V_2 = 1.02 \times 10^{-5}$ m^3) was allowed to equilibrate, starting from atmospheric conditions (Freund & Nover 1995; Fowler *et al.* 2005). Apparent gas permeability coefficients were derived numerically by least-square fitting to the measured pressure decay data. Calculation was based on Darcy's law for compressible media (e.g. Hildenbrand *et al.* 2002):

$$\frac{dn}{dt} = \frac{V_2\,dP_2}{dt} = -\frac{k_{gas}A(P_2^2 - P_1^2)}{\eta 2x}. \qquad (1)$$

Here, the amount of mass (dn [mol]) moving through the sample is calculated from the pressure change with time (dP_2/dt [Pa s^{-1}]) in reservoir V_2. A [m^2] is the cross-sectional area of the plug, η is the dynamic viscosity [Pa s], x the sample length [m] and k_{gas} [m^2] is the apparent gas permeability. A reliable Klinkenberg correction could only be performed for seven samples. In order to obtain values for all samples, we additionally calculated Klinkenberg-corrected permeability values (k_∞) according to a trend given by Jones & Owens (1980), which was based on more than 100 tight sandstones samples. The trend indicates larger slip factors for rocks with lower permeability. Both methods yielded largely consistent k_∞ values. In general, Klinkenberg-corrected permeability coefficients are 12% (±8%) lower than measured apparent gas permeabilities.

However, in the following, we will concentrate on the interpretation of the apparent gas permeabilities. This is to avoid the usage of (probably inaccurate) extrapolation. As all measurements were conducted under the same experimental conditions, apparent permeability values of the present sample set can be used for heterogeneity analysis (comparison between samples).

All permeability experiments were performed parallel to bedding. To investigate the influence of

Group		Subgroup	Formation	Age
				258
			Hannover	
		Elbe		260
Rotliegend	Upper Rotliegend		Dethlingen	
				262
			Mirow	
		Havel		264
			Parchim	
				266
		Müritz		
	Lower Rotliegend			296

Fig. 2. Stratigraphic subdivision of the Rotliegend in northern Germany (after Schöner 2006; stratigraphic ages from Menning 1995).

deformation bands on bedding-parallel apparent permeability, different pairs of sample plugs were prepared (undisturbed host rock v. samples with cataclastic deformation bands). The corresponding sample pairs were taken directly adjacent to one another from core material.

Two samples were measured twice at a 6.5 MPa confining pressure before and after peak pressures of 30 MPa. The permeability was determined at the given pressure steps.

In addition, data on Klinkenberg-corrected permeability was provided for low confining pressures (1.8 MPa) using air as the permeant. Plugs parallel to, as well as subnormal to, bedding allow access to the impact of bedding horizons on flow. To reduce the influence of possible microfractures that formed as a response to external stress release after coring (Holt 1994), permeabilities of plugs were measured stepwise at increasing confining pressures up to 32 MPa.

Based on this dataset, transmissibility multipliers were calculated according to Manzocchi *et al.* (1999) for use in reservoir models:

$$TM = \left[1 + \frac{t_{def}}{L}\left(\frac{k_{host} - k_{def}}{k_{def}}\right)\right]^{-1} \quad (2)$$

where TM is the transmissibility multiplier, t_{def} is the cumulative thickness of deformation bands in the calculated interval [m], L is the length of the calculated interval [m], k_{host} is the apparent host-rock permeability [m^2] and k_{def} is the the apparent deformation band permeability [m^2]. For all calculations and assessments of deformation band occurrence around faults, a damage zone width of 20 m has been assumed (Guo *et al.* 2009). This width was also adapted to assess transmissibility multipliers of deformation bands in a damage zone.

Petrographical methods include standard transmitted light microscopy, as well as cathodoluminescence (CL) microscopy performed with a HC3-LM, operated at an acceleration voltage of 13.5 kV and a beam current density of 0.6–0.8 mA mm^{-2}. Of the 32 petrographical thin sections, five were prepared without a cover slip for CL microscopy, which is the main tool to outline quartz-cemented cataclastic quartz grains. Optical porosity was measured from thin sections stained with a low-viscosity fluorescent dye. The representative elementary area (REA) was established by incrementally increasing the evaluated area on a photomosaic from 10 × 10 pixels up to the maximum area of the photomosaic (max. 4000 × 4000 pixels on a 12 000 × 4000 pixel photomosaic within one layer). It was determined that the REA had been reached when the measured optical porosity stopped changing. For the correlation of grain size and deformation band width, the average grain size was determined by averaging grain sizes measured along a grid in the host rock on either side of the deformation band. These data were taken adjacent to the site where deformation band width was measured. Orientations of deformation bands were plotted on lower-hemisphere Schmidt nets.

The analysis and data mining of revised WellCAD data (resistivity image log, gamma-ray log and lithological log) was performed with the graphical user interface 'Rattle' (Williams 2011), which is based on 'R', an environment and programming language for statistical computing. The 'latticist' plug-in allows the study of changes in orientation with increasing depth on marginal plots, which show a distribution for defined intervals (margins). The depth resolution of orientation data (dip direction, dip), derived from resistivity image logs embedded in WellCAD, can be adjusted by changing the number of created margins for the target variable depth. For this study, the number of margins was set to 4 to allow a significant number of deformation bands per interval ($\geq$40).

Results

Petrography

Most samples are similar in mineralogical composition and consist mostly of quartz (40–55%) with alkali feldspar and plagioclase (6–18% combined), lithoclasts (12–23%), and iron oxide grains. Authigenic cements include quartz (0–15%), calcite (0–6%) and feldspar (0–5%). Accessory phases are always hematite and illite coatings and pore-lining chlorite.

The studied cataclastic deformation bands from core material all show a simple-shear displacement between 0.4 and 15 mm. All samples contain discrete variations in grain size, be it in the form of foresets of grainfall or grainflow laminae in dune sandstones or deflation layers in sheetsand deposits.

The undisturbed host rock for dune sandstones is marked by a characteristic pinstripe lamination (Fig. 3a, c–e), which is due to grain-size changes during deposition. Sand sheets (Fig. 3b) are characterized by planar bedding and occasional deflation lags, with an overall maximum of clay content of 1%.

Heterogeneities in the mineral distribution of different lithotypes or within rocks of the same lithotype were not observed. Quartz overgrowth cementation is only observed in rocks with an incomplete coat coverage. Of the discussed flow barriers, only dissolution seams result in a significant heterogeneity of the mineral distribution by enriching clay minerals within the dissolution seam.

Petrographical analyses show that almost all studied deformation bands contain cataclasis of different degrees and cementation (Fig. 3a–d). Cataclasis in the studied samples is followed by subsequent quartz cementation. Thus, cataclasis is often invisible in transmitted light microscopy but shows up well in hot CL microscopy, which displays intragranular fractures of detrital quartz cemented by quartz of different luminescence (Fig. 4c).

Thicknesses of deformation bands vary and do not necessarily depend on the amount of displacement along the deformation band. Generally, thinner bands (0.2–1 mm thickness) occur in tightly cemented, densely packed and generally finer-grained host rocks. Thicker bands, up to 10 mm wide, occur in less cemented layers and rather coarse-grained intervals (Fig. 3b, c). A correlation between the average grain size in the host rock and the deformation band width exists for the studied cataclastic deformation bands, with a coefficient of determination of $R^2 = 0.74$ (Fig. 5a). All studied samples of the host rock show a wide spread in grain diameters from <0.1 to 0.9 mm, occasionally including grains as large as 1.1 mm. The skewness of the distribution, as well as the Trask sorting, only poorly correlate with the width of cataclastic deformation bands (Fig. 5b). The median grain size resulted in the best correlation with deformation band width, only differing slightly from the average grain size in most cases (Fig. 5b).

Incipient deformation bands show grain–grain contacts with minor chemical compaction along quartz grain boundaries and quartz cements in adjacent pore space. They are orientated oblique to bedding, forming a load-bearing framework. Incipient deformation bands are occasionally aligned in linear arrays at a high angle to bedding (Fig. 3d). These arrays of grain–grain contacts and linear quartz overgrowth cements are also observed in the transition zone between the deformation band core and the undisturbed host rock (Fig. 4a). In areas of two cataclastic bands merging into one band, no increase in cataclasis is observed, only an increase in chemical compaction (Fig. 3e & f). Localized bedding-parallel cementation occurs predominantly in fine-grained layers (e.g. Fig. 3e).

Next to the 29 cataclastic deformation bands, only a few other types of deformation bands or subvertical flow barriers were observed, which include two cementation bands (no shear component, continuous cementation with calcite at a medium to high angle to bedding: Fig. 2a), one disaggregation band (characterized by simple shear, disaggregation and dissolution) and one vein (characterized by simple shear, dilation and calcite cementation).

The visible porosity inside the band was optically assessed by fluorescent staining and compared to the host rock (Fig. 6). Porosity inside a deformation band was analysed in one sample using digital image analysis and was found to be $2.1 \pm 0.2\%$. Host-rock porosity ranges from $10.8 \pm 0.7\%$ in the fine-grained layers to $15.5 \pm 0.5\%$ in coarser-grained layers of that same sample. In another sample, the porosity inside the deformation band was much lower, at $0.18 \pm 0.05\%$. Host-rock values in this sample range from 1.2 ± 0.08 to $2.9 \pm 0.3\%$ for two coarse-grained bedding intervals of the host rock, and from 0.4 to 1.3% for one fine-grained layer.

Two dissolution seams in the studied samples are enriched in illitic clay minerals, which are present as grain-coating clays in most parts of the undisturbed samples. They developed parallel to bedding planes and foresets in dune layers, which suggests formation during chemical compaction (also frequently termed 'pressure dissolution'). However, chemical compaction post-dates at least one set of cataclastic deformation bands, as derived from cross-cutting relationships (Fig. 3b). In most samples, dissolution seams are concentrated in the fine-grained intervals. In the case of non-densely packed finer-grained layers, the area surrounding the dissolution seam is intensely cemented.

Data mining

Resistivity image logs were used to analyse the orientation and depth of the flow barriers. The conductive medium in the horizons is mainly the porewater. Flow barriers on resistivity image logs appear as more resistive features due to the loss or lack of porosity, and are, in the case of cataclastic deformation bands, orientated oblique to bedding and show a slight offset of the bedding planes. Bedding-parallel compaction is most likely to be masked by the bimodal sorting of the sediment. The lithotype association was applied consistently throughout the four different wells.

The dependency of deformation band density on lithotypes. The pre-picked lithologies in lithological log files were analysed to determine the number of deformation bands per metre (DB/m). No consistent trends concerning the different lithological units (dunes, sand sheets, alluvial fans) could be established (Fig. 7). The number of deformation bands per metre in well B range from 0 to 5.7 DB/m. The 50% data accumulation between the first (25%, Q1) and the third (75%, Q3) quartile shows the widest spread for the dune facies lithotype. All lithotypes exhibit a median of 0 DB/m for well B. In well D, the deformation band frequency ranges from 0 to 7.75 DB/m. The median for all lithotypes

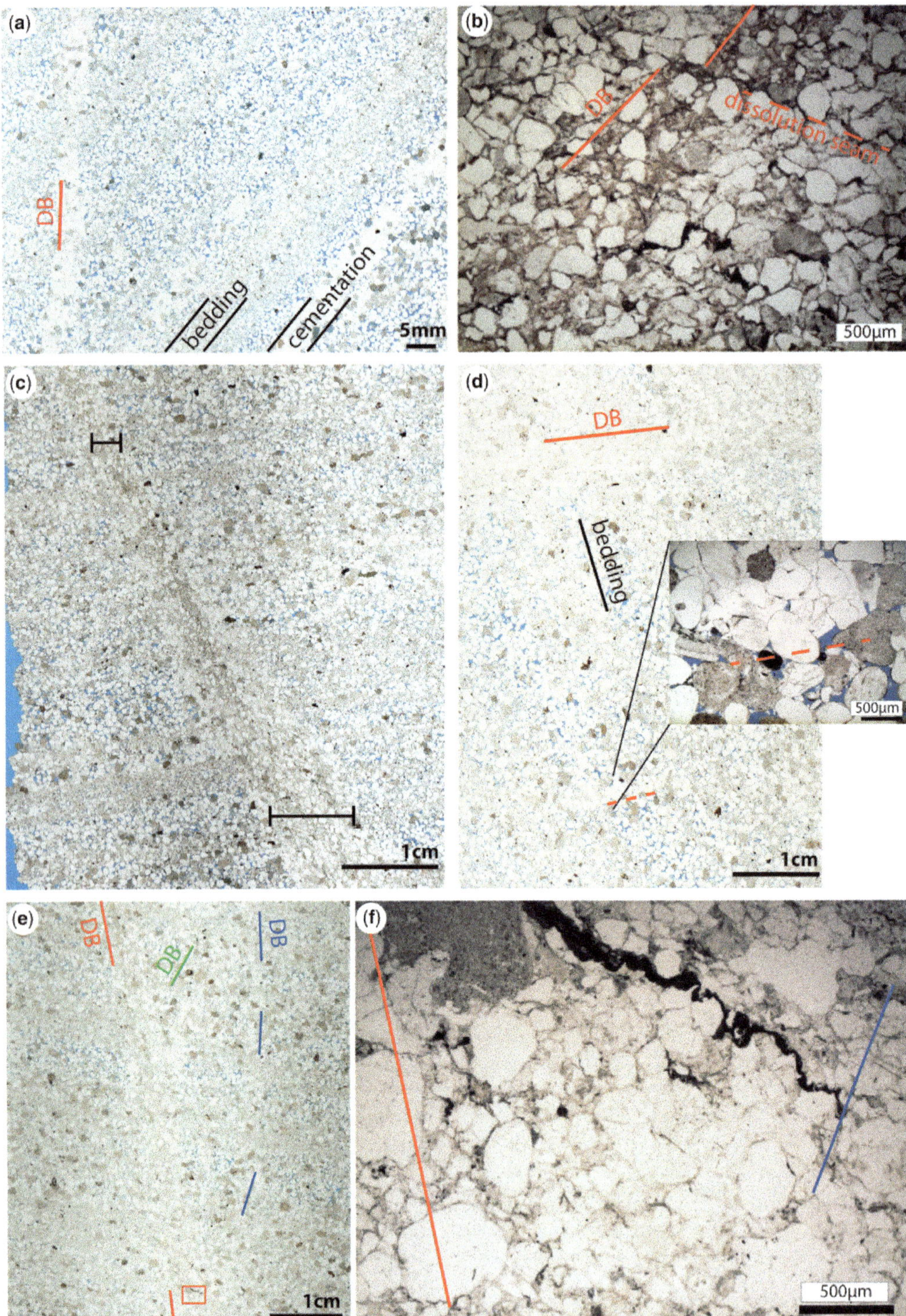

Fig. 3. (**a**) Cataclastic deformation band at a high angle to bedding marks a zone of low porosity. The host rock shows differences in grain size in the bedding and bedding-parallel cementation. (**b**) Compaction-related dissolution seam parallel to bedding, which sets off a deformation band, indicates early formation of a deformation band prior

is between 0 and 1 DB/m, but does not exhibit a clear accumulation within one lithotype.

The dependency of deformation band density on faults. Fault dependency was studied qualitatively from resistivity image logs and by calculating correlation coefficients of fault displacements v. number of deformation bands in the vicinity around faults. The lithotype dependency has been neglected due to the minor mineralogical differences of often closely spaced lithotypes. The maximum of observed displacements on faults was 30 cm. Well log data show no clear accumulation of deformation bands around small-scale normal faults. A correlation could not be established either to the occurrence or to the displacement of small-scale faults interpreted from resistivity image logs (Fig. 8). In comparison to the other wells, well D (crosses in Fig. 8) exhibits a larger number of deformation bands within the studied intervals, even at small observed displacements.

Marginal plots: orientation data. The depth-dependent orientation trends were visualized by marginal plots, which split the dataset into four depth intervals (margins). Each depth interval in one well contains the same number of deformation bands. The intervals therefore can cover different lengths of the well section. The plots for the four studied wells are presented in Figures 9 and 10. The first row in each plot visualizes the distribution of deformation bands along the well log interval. The second row shows the dip direction and number of deformation bands in each depth interval. The third row shows the dip and number of deformation bands in each depth interval with the according dip angle.

In wells B and D, the deformation bands are accumulated in the second and third depth interval (blue and green line) (Fig. 9a, b). These depth intervals cover parts of the Rotliegend stratigraphy of the Havel subgroup in both wells. The dip directions of resistive features in well B show a somewhat consistent bimodal distribution with a minor maximum at around 270° for all depth intervals and a second major maximum at about 90°. This reflects the consistent conjugate alignment of deformation bands along the well. The lowermost interval in well B (red line) comprises resistive (veins) and conductive features (fractures) picked in the underlying Rotliegend volcanics interval of the log section, which explains the deviation from the consistent trend. The very minor maximum of dip directions towards about 180° in the third depth interval (green line) points to the accumulation of a few resistive features deviating from the consistent trend. The unimodal distribution of dip values ranges around 60° and reflects a conjugate extensional system.

Well D also shows a bimodal distribution of deformation band dip directions (Fig. 9b). However, their strike does not match with those of well B, indicating a different local stress regime around well D during deformation band formation. Maxima at approximately 40° and 210° dip direction suggest a conjugate arrangement in the uppermost (purple line) depth interval. Dip directions also suggest a clockwise rotation of about 20° with depth, as indicated by aligning the peaks of the following two deeper depth intervals. At greater depth (blue to red line intervals) deformation bands dipping towards 300° become increasingly more prominent, which indicates a similar clockwise rotation in strike. The dip values show an accumulation of shallower dips in the interval between 5079 and 5150 m (green line). In contrast to well B, dips are dominantly bimodal, with one maximum at a more shallow dip of 50° and a second maximum at a steeper dip. There appears to be an increasingly steeper dip with increasing depth from approximately 70° to 80°.

In well A, the number of deformation bands increases with depth (Fig. 10a). Dip directions of deformation bands show a bimodal distribution with maxima at about 90° and 270°, indicating a conjugate set of deformation bands. Similar to well D, steepest dips of around 80° are observed at the deepest depth intervals. An anticlockwise rotation in dip direction is indicated by arranging the peaks of the top three intervals (purple, blue and green line) around 90°. A second more prominent azimuth of about 200° is present in the lowermost depth interval. The lowermost interval (5199–5424 m) for this well only comprises fractures of the Rotliegend volcanics, causing the deviation from the consistent trend.

In well C, the deformation bands are accumulated in the second and third depth interval (blue and green line) (Fig. 10b). The distribution of dip directions has four somewhat consistent maxima

Fig. 3. (*Continued*) to chemical compaction. (**c**) Cataclastic deformation band thickness (black lines) increases with an increase in average host-rock porosity and grain size. (**d**) Cemented area parallel to the deformation band (DB, red line) forming a load-bearing structure in a relatively porous host rock. (**e**) Thin section with three linking cataclastic deformation bands (DB), the red box covers the area of f). (**f**) In the concourse area of two linking cataclastic deformation bands, an increase in chemical compaction but no increase in cataclasis is observed. Porosity is coloured in blue.

Fig. 4. (**a**) Cemented margin of a cataclastic deformation band interpreted to be the remains of an incipient cataclastic band. The pore space (blue colour) away from the deformation band is uncemented (upper-left and right-hand corner). (**b**) Mainly chloritic grain coatings (light green colour) on coarse-grained quartz grains preserve porosity in deeply buried sandstone samples. (**c**) Comparison of a cataclastic band in cathodoluminescence (CL, top image: bright blue colours are illite and feldspar; dark blue colours are quartz; orange colours are carbonates) and in transmitted light (bottom, plane polarized light (PPL)). Occasionally, one can infer the cataclasis in ppl images, but often the broken and healed grains of different quartz generations are only visible in CL microscopy.

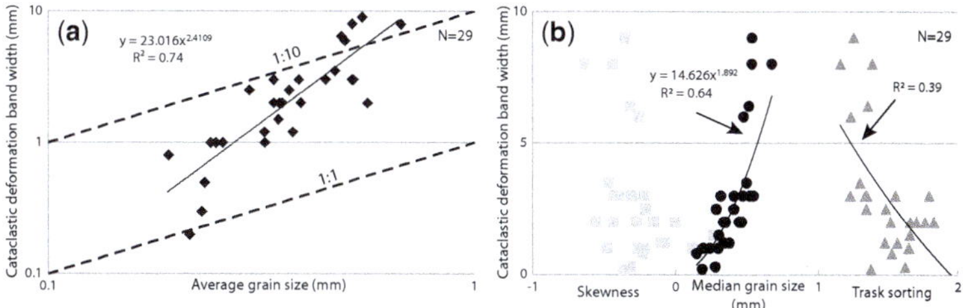

Fig. 5. (**a**) Correlation of average grain size in the undisturbed host rock v. deformation band width. The trend indicates that a larger grain size of the host rock results in wider deformation bands. Dashed lines indicate a 1:1 and 1:10 scaling relationship. (**b**) Correlation of skewness of the grain-size distribution, median grain size (in mm) and sorting v. the width of cataclastic deformation bands. The median grain size shows the best correlation.

at approximately 290°. The second maximum is located between 70° and 120°. The 70° dip direction corresponds well with the dip direction extracted from seismic line interpretation and is present in the lowermost interval, as well as subordinately in the shallowest interval (red and purple lines). The dominant dip values steepen with depth (blue to red line intervals) from about 55° to 70°. However, steepest dips of 80° are also dominant in the shallowest depth interval (4888–5073 m, purple line).

The strike of deformation bands present in resistivity image logs locally differs from seismically observable fault strike (Fig. 11). In the case of well A, the resistivity image log, covering all of the Havel subgroup, shows no small-scale resistive features indicating faults of the given strike (145°), as is noted in the structural map (Fig. 1). The fault that is noted on the structural map was derived only from production data (Fig. 11, ellipse around well A). This fault strike derived from production data can only be seen in small-scale faults, fractures and resistive features (veins) of the underlying Rotliegend volcanics. However, it is not reflected by resistivity anisotropies (deformation bands, faults, veins) in the reservoir unit. In the case of well D, the resistivity log data (strike of deformation bands, faults, veins) suggests that only one orientation of faults interpreted from seismic sections matches one of the two main orientations of the small-scale resistive features. In addition, the second orientation is accumulated in a certain depth interval (green line Fig. 9b).

Permeability

The apparent permeability at increasing confining pressures decreases non-linearly. Measured data were fitted by exponential regressions, which are given in Figure 12a–d. Generally, the decrease in apparent permeability from 6.5 to 32 MPa confining pressure was less than one order of magnitude (Fig. 12a–c).

The stress dependency of measured apparent permeability can well be described by an exponential function (cf. David *et al.* 1994). The stress-dependency factor, γ, of our samples ranges between 0.005 and 0.079 MPa^{-1}.

The apparent permeability of the host-rock samples (Fig. 12, squares) are larger than those of adjacent samples containing a deformation band (Fig. 12, diamonds). At higher confining pressures, this difference remains almost constant for each

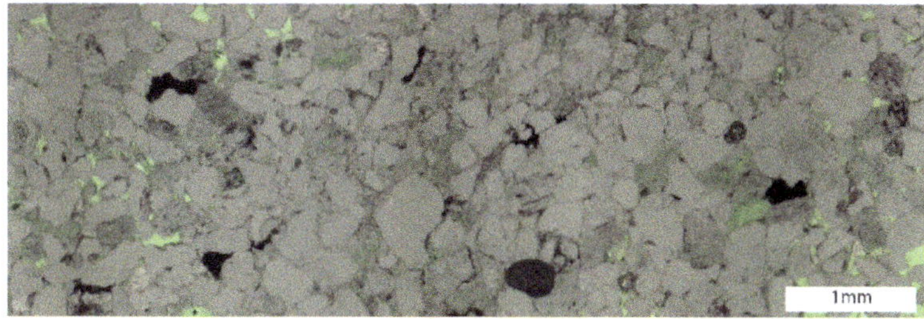

Fig. 6. Thin section stained with fluorescent dye to highlight porosity (green color) in comminuted cataclastic deformation band. Dyed porosity inside the deformation band (central part of image) can clearly be differentiated from the porosity of the host rock (left and right side of image). Digital image interpretation and thresholding of the green fluorescent dye resulted in 2.1 ± 0.2% visible porosity in the deformation band.

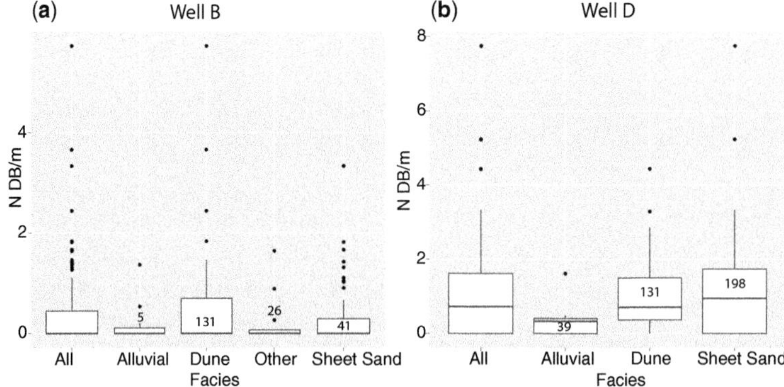

Fig. 7. Box-whisker plots for two wells B (**a**) and D (**b**) present the number of deformation bands per metre (DB/m) v. lithotype. Numbers next to the boxes indicate the number of deformation bands analysed in this lithotype. No clear accumulation of deformation bands in one lithotype as characterized from well logs and resistivity image logs can be determined. The dependency of deformation bands to fault throw, grain size and distance to faults had to be neglected owing to the restricted 1D database.

sample set with less than one to more than one order of magnitude (Fig. 12a, b). Only sample set 4 shows a continuous increase in apparent permeability reduction from one to nearly two orders of magnitude during increasing confining pressure (Fig. 12c).

The apparent permeability reduction of sample 7 is smaller than in other more porous samples at 30 MPa confining pressure (Fig. 12a, squares), which is due to specific microstructual differences. Here, the deformation band is located in a fine-grained, well-cemented and densely packed host rock. The thin nature of the deformation band in this fine-grained host rock only reduces the apparent permeability by 0.7 orders of magnitude with respect to the undisturbed sample at 30 MPa confining pressure (Fig. 12a). Coarser-grained host rocks favour the development of wider deformation bands, which result in a larger apparent permeability reduction (Fig. 12b, c).

The Klinkenberg-corrected permeability across bedding-parallel cementation and dissolution seams could only be inferred from low confining pressure data using air as the permeant (grey crosses and box in Fig. 12d). These data indicate that the permeability measured perpendicular to bedding

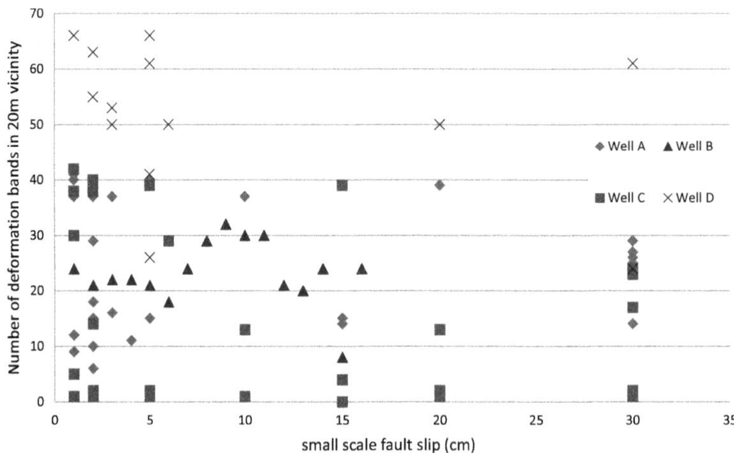

Fig. 8. Plot of small-scale fault slip taken from image logs v. the number of deformation bands in a 20 m vicinity around the fault. No clear trend concerning the dependency of deformation band density in a certain interval v. fault slip can be established from the 1D image logs. A possible lithotype dependency has been neglected.

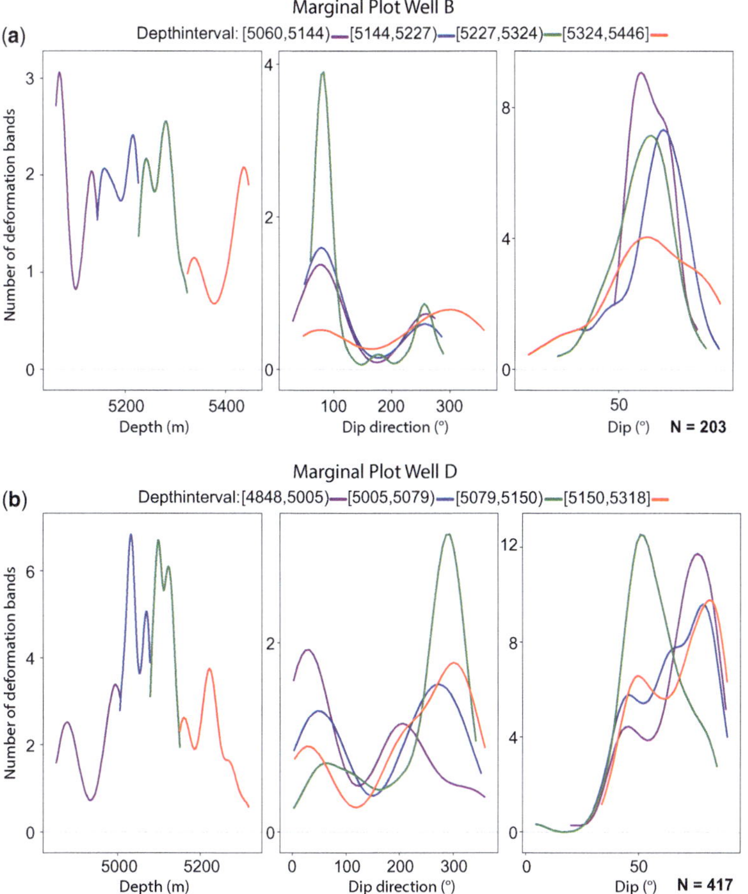

Fig. 9. Marginal plots for wells B (**a**) and D (**b**) for different depth intervals. The total number of deformation bands is given as N. The four depth intervals (margins) from top to bottom are a purple, blue, green and a red line. The corresponding depth intervals are given at the top of each plot. (First row) Distribution of deformation bands along the log interval. (Second row) Azimuth of deformation bands in each depth interval; note the accumulation in well D between 5079 and 5150 m (green interval). (Third row) Dip of the deformation bands in each interval; note the inhomogeneous distribution in well D in comparison to well B.

(grey crosses, Fig. 12d) is up to four orders of magnitude lower than the corresponding permeability parallel to bedding (grey box in Fig. 12d).

Two experiments on host-rock samples (3B and 7B) after a subsequent stress reduction from 30 MPa down to 6.5 MPa resulted in a lower apparent permeability of 10 and 15%, respectively.

During the measurements, several samples, which are not presented, fractured at low, as well as at high, confining pressures. One sample completely disintegrated to sand, others partly disaggregated.

Transmissibility multipliers were calculated for peak confining pressures of 30 MPa using the given apparent permeability data calculated for the width of the deformation band. These transmissibility multipliers (equation 2) were calculated for deformation band frequencies that were determined from the actual resistivity image logs (sample series A). A worst-case scenario (sample series B) was assumed for 100 DB/m and a frequency taken from a published field example (sample series C) from Johansen & Fossen (2008) (Table 1). Thresholds for the sealing efficiency are based on transmissibility multiplier calculations by Shipton *et al.* (2005), who suggested that deformation bands are sealing at $TM < 0.0005$ on geological timescales and $TM < 0.001$ on production timescales. Transmissibility multipliers show that the sampled deformation bands do not completely seal-off fluid flow in the

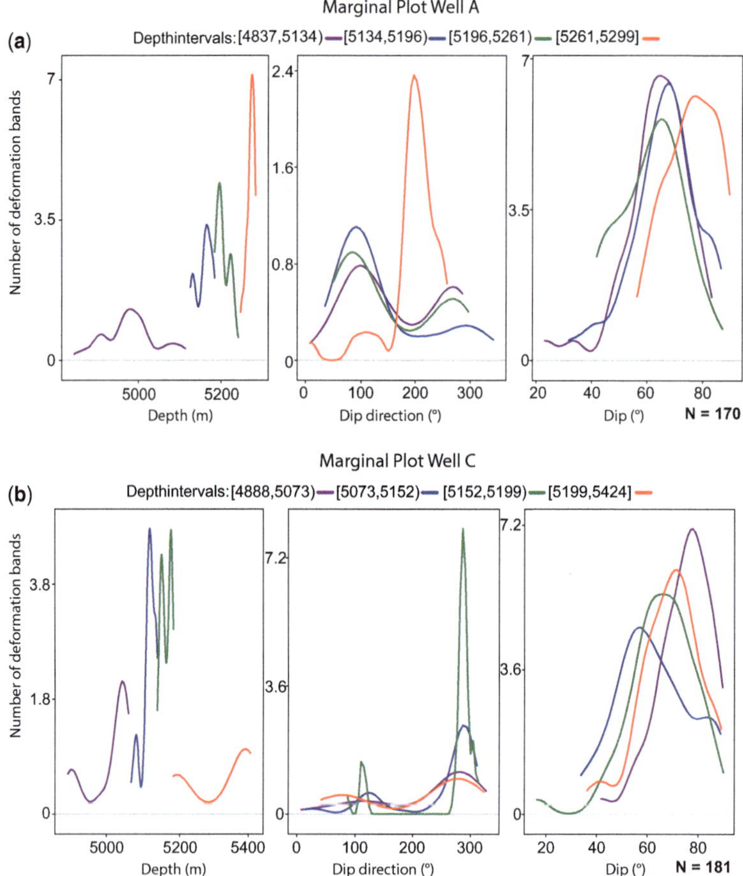

Fig. 10. Marginal plots for wells A (**a**) and C (**b**) for four different depth intervals. (First row) Distribution of deformation bands along the log interval. (Second row) Azimuth of deformation bands in each interval. (Third row) Dip of deformation bands in each interval. Well C: lowermost interval (5199–5424 m) represents fractures and veins in the volcanic units.

reservoir under the given assumptions, not even at reservoir confining pressure conditions (Table 1). Contrary to this, production data have shown that flow barriers do stop flow parallel to bedding.

Discussion

Petrography

The optical evaluation of deformation bands from thin sections, coupled with apparent permeability measurements, better constrains the impact of deformation bands on fluid flow in the reservoir. The absence of grain coatings and the high reactivity of fractured quartz grains enhance the cementation of cataclastic grains (Fisher *et al.* 2000; Fossen *et al.* 2007). In the studied samples, such masking of cataclasis in deformation bands is visualized by CL microscopy, which is recommended to verify the presence of brittle deformation in grains.

A correlation of deformation band thickness and grain size was described by Wennberg *et al.* (2013, fig. 12 and references therein) for carbonates and sandstones. Our results for sandstones are in accordance with their findings, with a good correlation of cataclastic deformation band thickness v. the host-rock grain size adjacent to the band (Fig. 5). Such correlation also links to our petrophysical measurements, which show an increase in the apparent permeability reduction with increasing deformation band width. It allows a first apparent permeability prediction and thus points to the role of deformation band width on fluid flow in a reservoir (Fig. 13). A greater number of data should be generated to further support the overall trend.

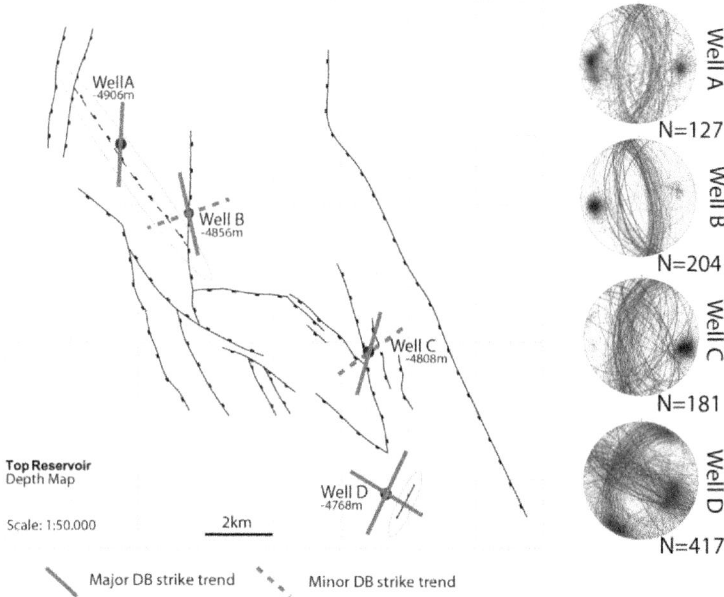

Fig. 11. Structural map of the top of the reservoir. Black lines are normal faults; grey lines on the well locations represent the strike of small-scale flow barriers as interpreted on resistivity image logs; dashed lines represent minor strike sets deduced from resistivity image logs. Circles indicate areas that might be reassessed by small-scale structural data with a depth resolution. The denoted depths represent the depth of the top of the reservoir unit. Only resistive features (deformation bands) oblique to bedding are visualized in the Schmidt net plots.

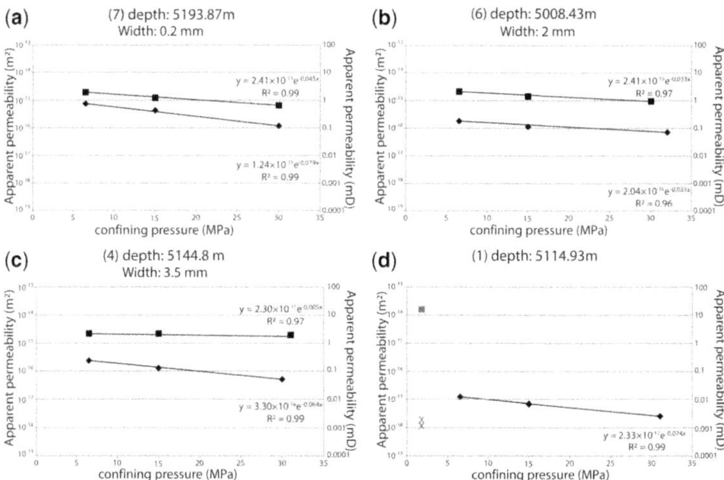

Fig. 12. Apparent permeability results for four pairs of samples (1, 4, 6 and 7). Diamonds are measurements parallel to the bedding of samples that include a deformation band; boxes are samples without a deformation band parallel to bedding. Crosses are measurements perpendicular to bedding. (**a**)–(**c**) Apparent permeability decreases non-linearly. Thin deformation bands (sample 7, a) develop less of an apparent permeability anisotropy than wider deformation bands (sample 4, c). The decrease in apparent permeability correlates with the deformation band width in the three samples where a correlation was possible. (**d**) The plot for sample 1 visualizes the permeability anisotropy of the regular host rock with samples measured perpendicular to bedding (crosses) as opposed to a measurement parallel to bedding (box). The displayed samples were collected within 10 cm of the sampling location of sample 1 (black diamonds). The Klinkenberg-corrected datapoints (grey boxes and crosses) were measured at a confining pressure of 1.8 MPa and using air as the permeant.

Table 1. *Compilation of transmissibility multiplier (TM) calculations for samples 4, 6 and 7 with different deformation band densities (DB/m)*

	Sample No.	k_{def} (m^2)	k_{host} (m^2)	P_{conf} (MPa)	DB/m (*N*)	Thickness (*t*) (m)	*TM*	Leak/seal	
								Geological timescale	Production timescale
A	4	4.53×10^{-18}	1.95×10^{-15}	30	1	3.50×10^{-3}	0.002	Leak	Leak
	6	2.74×10^{-18}	2.40×10^{-15}	32	0.5	1.50×10^{-3}	0.005	Leak	Leak
	7	9.08×10^{-19}	6.23×10^{-16}	30	0.5	2.00×10^{-4}	0.001	Leak	Leak
B	4	4.53×10^{-18}	1.95×10^{-15}	30	100	3.50×10^{-3}	0.002	Leak	Leak
	6	2.74×10^{-18}	2.40×10^{-15}	32	100	1.50×10^{-3}	0.005	Leak	Leak
	7	9.08×10^{-19}	6.23×10^{-16}	30	100	2.00×10^{-4}	0.001	Leak	Leak
C	4	4.53×10^{-18}	1.95×10^{-15}	30	17.25	3.50×10^{-3}	0.002	Leak	Leak
	6	2.74×10^{-18}	2.40×10^{-15}	32	17.25	1.50×10^{-3}	0.005	Leak	Leak
	7	9.08×10^{-19}	6.23×10^{-16}	30	17.25	2.00×10^{-4}	0.001	Leak	Leak

Input parameters are apparent permeability calculated for the deformation band width and thickness. The length of the reference interval is 20 m. Confining pressures at which apparent permeability has been determined are given.

The experimental observations of Cheung *et al.* (2012) demonstrate the important role of grain-size distribution on the physical properties during the formation of a deformation band. In their experimental set-up, the authors created compaction bands, which are end members of deformation bands without any considerable shearing. Sandstones that have a comparably narrow grain-size distribution (spread below 300 μm) formed localized compaction bands, whereas comparably wide distributions (spread of 700 μm) resulted in distributed cataclastic flow (Cheung *et al.* 2012). Our localized cataclastic deformation bands, however, formed in a host rock with a wide spread of grain-size distributions (>800 μm) and with shear displacement. Thus, experimental results cannot be transferred to our data with respect to localization. No consistent trend could be established from the width of the grain-size distribution to the width of the cataclastic deformation bands. Furthermore, no correlation is evident for the deformation band width v. the skewness of grain-size distributions and the sorting of the host rock (Fig. 5b).

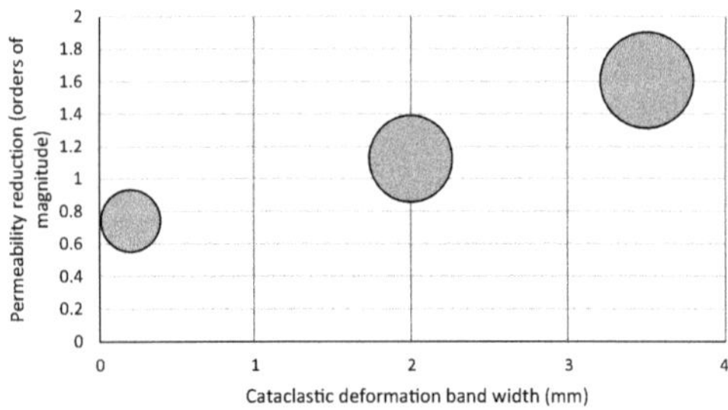

Fig. 13. Apparent permeability across cataclastic deformation bands inferred from deformation band width, based on a combination of grain size v. cataclastic deformation band width correlation plot, and cataclastic deformation band width and apparent permeability reduction plot. The point size corresponds with the grain size. This correlation allows the reservoir quality to be predicted if deformation bands are to be expected in a reservoir unit.

The evolution of deformation bands can well be explained by the localization of stress at load-bearing frameworks, which form during initial mechanical and chemical compaction. Small cemented arrays at a high angle to bedding and crossing porous beds may represent such a load-bearing framework, with solution along grain boundaries and precipitation in the adjacent pore space (see Figs 2d & 14). These frameworks may evolve towards a discrete cataclastic deformation band (Fig. 14) due to the overall cementation and strain hardening (Antonellini & Aydin 1995; Fossen 2010; Ballas *et al.* 2012; Soliva *et al.* 2013).

The bedding-parallel cementation is mainly concentrated in the finer-grained parts of the sandstones, although locally coarser-grained layers are also affected (Fig. 3a, d, e). Samples point to a smaller coat coverage in fine-grained beds, which results in larger reactive quartz surfaces and causing higher cementation (cf. Bloch *et al.* 2002).

Data mining

Data mining, especially marginal plots, proved to be a good method to add a depth resolution to spatial orientation data. The visualization of changes in dip direction and dip with increasing depth cannot be made on a Schmidt net plot.

Given that deformation bands form sub-parallel and prior to faults, deformation bands and their orientation can be used to infer the orientation of

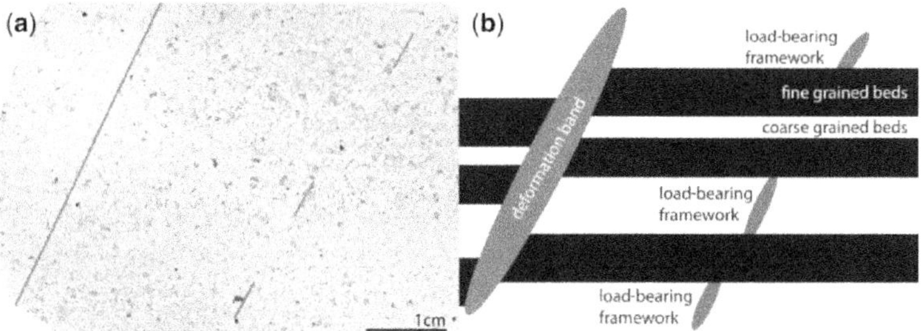

Fig. 14. (**a**) Thin section of a cataclastic deformation band (grey line) and load-bearing structures (dashed line) in the high-porosity beds aligned parallel to the deformation band. The image on the right is the same as in Figure 2d. (**b**) Conceptual model of the formation of cataclastic deformation bands in high-porosity sandstones. The formation of a load-bearing framework during chemical compaction acts as a precursor from which a cataclastic deformation band develops. The load-bearing framework is developed sub-parallel to deformation bands, which supports the hypothesis. Porosity is higher in the coarse-grained intervals.

larger-scale faults. Thus, faults depicted on seismic lines and inferred from production data may be validated by faults and deformation bands from the resistivity image logs, and vice versa.

The change in strike and steepening of deformation bands with depth in well D may indicate the influence of a flower structure (Woodcock & Schubert 1994), although the geometry of the fault planes cannot be visualized with the limited dataset provided for this work. The accumulation of a deformation band strike in a certain depth interval in well D that is not reflected by seismically observable faults may indicate a subseismically observable fault trend (Fig. 11, circle at well D). The local variations in strike and dip of the deformation bands are ascribed to changes in the local stress field and can be compared to the interaction of subseismically observable faults with seismically observable faults (cf. Johansen *et al.* 2005, fig. 4). The deviation of deformation band strike from seismically observable fault strike in well C is attributed to interactions between closely spaced normal faults and is similar to the setting presented in Antonellini & Aydin (1995, fig. 7).

The resistivity image logs of well A show an absence of subseismically observable faults, deformation bands and veins matching the strike of a large seismically observable fault on the structural reservoir map. However, several field studies show the occurrence of deformation bands around normal faults (e.g. Antonellini & Aydin 1995; Johansen & Fossen 2008; Fossen & Rotevatn 2012) and their use as a proxy of nearby faults. Such a proxy may, thus, verify reservoir-scale faults interpreted from seismic sections (in our case subsalt) or production data. Although deformation bands generally predate fault formation (Antonellini & Aydin 1995) and are thus not sufficient criteria to infer faults, the mismatch of resistivity and reservoir map data led to a successful revision of the structural map.

Dependencies of deformation band densities in different siliciclastic lithologies could not be established, which may be ascribed to similar lithotypes. However, several field studies clearly demonstrate the accumulation of deformation bands in more competent beds (Schultz & Fossen 2002). Furthermore, additional sets of deformation bands related to additional faults in the vicinity of the well affect any regular distribution. An additional factor affecting deformation band density is the proximity to faults (Antonellini & Aydin 1995), which is shown to be consistent if the host-rock grain size at a similar distance to a fault is the same (Griffiths *et al.*, this volume, in press). The observations by Cheung *et al.* (2012) also point towards a grain-size distribution control on the occurrence of compaction bands that might also be applicable to deformation bands with a simple-shear component. The number and displacements of observed faults derived from resistivity images could not be linked to different lithotypes due to the one-dimensional (1D) data in wells.

Published data from field exposures clearly show an increased number of deformation bands around faults and a correlation with fault displacement (e.g. Antonellini & Aydin 1995; Rotevatn & Fossen 2011; Griffiths *et al.*, this volume, in press). Since a well represents an arbitrarily orientated scanline across a fault and surrounding lithotypes in the reservoir, and the lithotypes are similar, a correlation might be expected. However, well log data show no clear or consistent accumulation of deformation bands either in their occurrence or in the displacement of small-scale faults interpreted from resistivity image logs (Fig. 8).

Permeability

Apparent permeability data of samples, including a cataclastic deformation band and associated undisturbed host-rock samples, show that the studied cataclastic deformation bands have a negative impact on production.

The non-linear decrease in apparent permeability with increasing confining pressure is attributed to the closure of microfractures, as well as to granular compaction and inelastic pore collapse (cf. Bernabe 1987; Holt 1994; Fortin *et al.* 2005; Saillet & Wibberley 2013). The decrease in permeability with increasing confining pressure as marked by the stress dependency factor, γ, which is between 0.005 and 0.079 MPa^{-1}, is within or slightly above that given in literature for sandstones and tight sandstones (Yale 1984 in David *et al.* 1994; Saillet & Wibberley 2013).

The maximum reduction of two orders of magnitude for a measured cataclastic deformation band is in the range of previously measured samples (e.g. Fossen & Bale 2007; Tueckmantel *et al.* 2012). Experimentally created compaction bands within sandstones led to a permeability reduction of two–three orders of magnitude (Vajdova *et al.* 2004; Baud *et al.* 2012; Deng *et al.* 2015). Brittle shear faulting in an experimental set-up by Zhu & Wong (1997) resulted in a permeability decrease of less than one–two orders of magnitude compared to the undisturbed rock. Effects of chemical compaction and dissolution are also considered to have a diverse impact on fluid flow, with stylolites in carbonates showing an unimpeded flow vertically and enhanced flow laterally (Heap *et al.* 2013). In all our samples, dissolution seams contain clay and are orientated parallel to bedding.

While all our apparent permeability data were measured along bedding and were lower across sheared cataclastic deformation bands, the permeability reduction is higher across cemented and bimodally distributed bedding and bedding-parallel dissolution seams. This corresponds with results of Baud *et al.* (2012), who reported that compaction bands parallel to bedding would form the most efficient flow barriers. The application to a field scale, however, needs to include an assessment of the connectivity of all deformation band types in three dimensions to form extensive barriers to fluid flow (e.g. Fossen & Bale 2007).

The provided Klinkenberg-corrected permeabilities are not considered for further correlation with the deformation band samples. However, the three samples were taken within 10 cm of the apparent permeability sample 1, and are presented to display the anisotropy between fluid-flow parallel compared to perpendicular to bedding within the same lithotype (Fig. 12d).

The permeability of a sample containing a deformation band is mainly controlled by the lower permeability of the deformation band. Considering the entire sample length results in larger permeability coefficients compared to the thin deformation band itself (cf. Main *et al.* 2000; Baud *et al.* 2012). As the deformation band will be the controlling factor in the permeability difference between samples with and without deformation bands (cf. Main *et al.* 2000; Baud *et al.* 2012), the permeability measurements were also evaluated with regard to the width of the deformation band. The apparent permeability of the deformation band itself is one (wide cataclastic deformation bands) to two orders (narrow cataclastic deformation bands) of magnitude lower than the bulk apparent permeability of such host-rock samples containing a deformation band.

The assessment of the transmissibility multipliers, calculated with the apparent permeability coefficients of the deformation band, will only significantly change to become sealing if the permeability anisotropy exceeds more than three orders of magnitude, irrespective of the number of deformation bands per interval.

A relative permeability reduction after a phase of stress release from 30 to 6.5 MPa was attributed to irreversible granular compaction, which results in permanent deformation (cf. Saillet & Wibberley 2013) and inelastic pore collapse (Fortin *et al.* 2005). Although no repeated stress cycles were performed, the permeability stress hysteresis is highly probable. If a less confined host-rock sample after peak confining pressures results in lower permeabilities than the same sample at high confining pressures, permeability stress hysteresis as measured in other siliciclastic rocks (e.g. Faulkner & Rutter 2000) can strongly be inferred.

As several plug samples were (partly) disintegrated after application of confining pressures of up to 32 MPa, we suggest that the critical pressure marking the onset of grain crushing and inelastic pore collapse (e.g. Wong *et al.* 1997; Tembe *et al.* 2008) has been exceeded during these measurements. We suggest that reservoir pore fluid pressures were very high to maintain host-rock stability at the current burial depth. Thus, a significant effective stress increase during production could lead to depletion-induced compaction failures (cf. Schutjens *et al.* 2004). Maintaining the initial pore pressures might be essential in preventing reservoir integrity failures.

Orientation-dependent permeability anisotropies, parallel and perpendicular to bedding, can be attributed to pore-shape anisotropies, grain-size differences (clay size to sand size), sorting, sedimentary fabric, compaction, cementation, microcracking and chemical compaction (Benson *et al.* 2005; Louis *et al.* 2005; Armitage *et al.* 2011). The measured reduction corresponds to observed bedding-parallel cementation by either quartz or calcite, chemical compaction and bimodal grain-size distributions on a microscopic scale. Sedimentary fabric-induced permeability anisotropies of sandstones of up to four orders of magnitude are in accordance with the findings of Armitage *et al.* (2011).

Conclusion

Almost all deformation bands are cataclastic, with the fractures cemented by quartz. Cemented cataclastic fractures are not visible in transmitted light microscopy but can be seen using cathodoluminescence (CL) microscopy. Intersections of dissolution seams and deformation bands indicate that the deformation bands formed prior to and during chemical compaction.

Marginal plots of resistivity log data highlight the rotation of deformation bands and fractures with depth, which may point to flower structures. Data mining from image logs was successfully used to re-evaluate structural reservoir maps and seismic-scale faults. It may further indicate subseismic fault trends at specific depth intervals. Lithology-dependent variations of deformation bands in the reservoir units could not be established from image logs and cores owing to the similar mineralogy of the beds.

A scaling relationship for cataclastic deformation bands correlates an increase in deformation band width with an increase in average host-rock grain size.

Well-cemented and densely packed fine-grained host rocks contain thinner deformation bands with a lower apparent permeability contrast of less than one order of magnitude. More porous and coarser-grained host rocks reach apparent permeability contrasts of up to two orders of magnitude along bedding. However, the impact of bedding-parallel dissolution seams and cementation parallel to fine-grained bedding has a larger impact on fluid flow by up to four orders of magnitude.

The failure of some samples at or below 30 MPa confining pressure may be indicative of formation intervals at high pore fluid pressures that, during reservoir depletion, may lead to depletion-induced compaction failure, and hence to reduced flow.

The authors would like to thank RWE DEA AG, Hamburg, for provision of data and the permission to publish the results of this study, as well as for support during the study. We thank Keyvan Osivandi, Nils Buurman, Holger Rieke, Philipp Berger and Robert Lippmann for stimulating discussions; Bernhard Krooss and Reinhard Fink are thanked for their contributions to the permeability measurements. The authors would like to acknowledge the constructive and thorough reviews of Dan Faulkner, an anonymous reviewer and editor Peter Armitage, which significantly improved this manuscript.

References

Ahmadov, R., Aydin, A., Karimi-Fard, M. & Durlofsky, L. J. 2007. Permeability upscaling of fault zones in the Aztec Sandstone, Valley of Fire State Park, Nevada, with a focus on slip surfaces and slip bands. *Hydrogeology Journal*, **15**, 1239–1250.

Antonellini, M. & Aydin, A. 1994. Effect of faulting on fluid flow in porous sandstones: petrophysical properties. *American Association of Petroleum Geologists Bulletin*, **78**, 355–377.

Antonellini, M. & Aydin, A. 1995. Effect of faulting on fluid flow in porous sandstones: geometry and spatial distribution. *American Association of Petroleum Geologists Bulletin*, **79**, 642–671.

Armitage, P. J., Faulkner, D. R., Worden, R. H., Aplin, A. C., Butcher, A. R. & Iliffe, J. 2011. Experimental measurement of, and controls on, permeability and permeability anisotropy of caprocks from the CO2 storage project at the Krechba Field, Algeria. *Journal of Geophysical Research*, **116**, B12208.

Awdal, A., Healy, D. & Alsop, G. I. 2014. Geometrical analysis of deformation band lozenges and their scaling relationships to fault lenses. *Journal of Structural Geology*, **66**, 11–23.

Ballas, G., Soliva, R., Sizun, J.-P., Benedicto, A., Cavailhes, T. & Raynaud, S. 2012. The importance of the degree of cataclasis in shear bands for fluid flow in porous sandstone, Provence, France. *American Association of Petroleum Geologists Bulletin*, **96**, 2167–2186.

Baud, P., Meredith, P. & Townend, E. 2012. Permeability evolution during triaxial compaction of an anisotropic porous sandstone. *Journal of Geophysical Research*, **117**, B05203.

Benson, P. M., Meredith, P. G., Platzman, E. S. & White, R. E. 2005. Pore fabric shape anisotropy in porous sandstones and its relation to elastic wave velocity and permeability anisotropy under hydrostatic pressure. *International Journal of Rock Mechanics and Mining Sciences*, **42**, 890–899.

Bernabe, Y. 1987. The effective pressure law for permeability during pore pressure and confining pressure cycling of several crystalline rocks. *Journal of Geophysical Research*, **92**, 649.

Bésuelle, P. & Rudnicki, J. W. 2004. Localization: shear bands and compaction bands. *In*: Gueguen, Y. & Boutéca, M. (eds) *Mechanics of Fluid-Saturated Rocks*. Elsevier Academic Press, Amsterdam, 219–321.

Bloch, S., Lander, R. H. & Bonnell, L. 2002. Anomalously high porosity and permeability in deeply buried sandstone reservoirs: origin and predictability. *American Association of Petroleum Geologists Bulletin*, **86**, 301–328.

Chemenda, A. I., Wibberley, C. & Saillet, E. 2012. Evolution of compactive shear deformation bands: numerical models and geological data. *Tectonophysics*, **526–529**, 56–66.

Cheung, C. S. N., Baud, P. & Wong, T.-f. 2012. Effect of grain size distribution on the development of compaction localization in porous sandstone. *Geophysical Research Letters*, **39**, L21302.

Davatzes, N. C. & Aydin, A. 2003. Overprinting faulting mechanisms in high porosity sandstones of SE Utah. *Journal of Structural Geology*, **25**, 1795–1813.

David, C., Wong, T.-f., Zhu, W. & Zhang, J. 1994. Laboratory measurement of compaction-induced permeability change in porous rocks: implications for the generation and maintenance of pore pressure excess in the crust. *Pure and Applied Geophysics*, **143**, 425–456.

David, C., Menendez, B., Zhu, W. & Wong, T.-f. 2001. Mechanical compaction, microstructures and permeability evolution in sandstones. *Physics and Chemistry of the Earth, Part A: Solid Earth and Geodesy*, **26**, 45–51.

Deng, S., Zuo, L., Aydin, A., Dvorkin, J. & Mukerji, T. 2015. Permeability characterization of natural compaction bands using core flooding experiments and three-dimensional image-based analysis: comparing and contrasting the results from two different methods. *American Association of Petroleum Geologists Bulletin*, **99**, 27–49.

Eichhubl, P. & Flodin, E. 2005. Brittle deformation, fluid flow, and diagenesis in sandstones at Valley of Fire State park, Nevada. *In*: Pederson, J. & Dehler, C. M. (eds) *Interior Western United States*. Geological Society of America, Field Guide, **6**, 151–167.

Eichhubl, P., Taylor, W. L., Pollard, D. D. & Aydin, A. 2004. Paleo-fluid flow and deformation in the Aztec Sandstone at the Valley of Fire, Nevada – Evidence for the coupling of hydrogeologic, diagenetic, and tectonic processes. *Geological Society of America Bulletin*, **116**, 1120–1131.

Exner, U., Kaiser, J. & Gier, S. 2013. Deformation bands evolving from dilation to cementation bands in a hydrocarbon reservoir (Vienna Basin, Austria). *Marine and Petroleum Geology*, **43**, 504–515.

Faulkner, D. R. & Rutter, E. H. 2000. Comparisons of water and argon permeability in natural clay-bearing fault gouge under high pressure at 20 °C. *Journal of Geophysical Research*, **105**, 16 415–16 426.

Fisher, Q. J. & Knipe, R. 2001. Permeability of faults within siliciclastic petroleum reservoirs of the North Sea and Norwegian continental shelf. *Marine and Petroleum Geology*, **18**, 1063–1081.

Fisher, Q. J., Knipe, R. J. & Worden, R. H. 2000. Microstructures of deformed and non-deformed sandstones from the North Sea: implications for the origins of quartz cement in sandstones. *In*: Worden, R. H. & Morad, S. (eds) *Quartz Cementation in Sandstones*. Blackwell Science, Oxford, 129–146.

Fortin, J., Schubnel, A. & Guéguen, Y. 2005. Elastic wave velocities and permeability evolution during compaction of Bleurswiller sandstone. *International Journal of Rock Mechanics and Mining Sciences*, **42**, 873–889.

Fossen, H. 2010. Deformation bands formed during soft-sediment deformation: observations from SE Utah. *Marine and Petroleum Geology*, **27**, 215–222.

Fossen, H. & Bale, A. 2007. Deformation bands and their influence on fluid flow. *American Association of Petroleum Geologists Bulletin*, **91**, 1685–1700.

Fossen, H. & Rotevatn, A. 2012. Characterization of deformation bands associated with normal and reverse stress states in the Navajo Sandstone, Utah: discussion. *American Association of Petroleum Geologists Bulletin*, **96**, 869–876.

Fossen, H., Schultz, R. A., Shipton, Z. K. & Mair, K. 2007. Deformation bands in sandstone: a review. *Journal of the Geological Society, London*, **164**, 1–15.

Fossen, H., Schultz, R. A. & Torabi, A. 2011. Conditions and implications for compaction band formation in the Navajo Sandstone, Utah. *Journal of Structural Geology*, **33**, 1477–1490.

Fowler, C. M., Stead, D., Pandit, B. I., Janser, B. W., Nisbet, E. G. & Nover, G. 2005. A database of physical properties of rocks from the Trans-Hudson Orogen, Canada. *Canadian Journal of Earth Sciences*, **42**, 555–572.

Freund, D. & Nover, G. 1995. Hydrostatic pressure tests for the permeability-formation factor relation on crystalline rocks from KTB drilling project. *Surveys in Geophysics*, **16**, 47–62.

Griffiths, J., Worden, R. H., Faulkner, D. R. & Edwards, A. In press. Deformation band development as a function of intrinsic host-rock properties in Triassic Sherwood Sandstone. *In*: Armitage, P. J., Butcher, A. R. et al. (eds) *Reservoir Quality of Clastic and Carbonate Rocks: Analysis, Modelling and Prediction*. Geological Society, London, Special Publications, **435**, https://doi.org/10.1144/SP435.11

Guo, J., McCaffrey, K., Jones, R. & Holdsworth, R. 2009. The spatial heterogeneity of structures in high porosity sandstones: variations and granularity effects in orientation data. *Journal of Structural Geology*, **31**, 628–636.

Heap, M. J., Baud, P., Reuschle, T. & Meredith, P. G. 2013. Stylolites in limestones: barriers to fluid flow? *Geology*, **42**, 51–54.

Hildenbrand, A., Schlömer, S. & Krooss, B. M. 2002. Gas breakthrough experiments on fine-grained sedimentary rocks. *Geofluids*, **2**, 3–23.

Holt, R. M. 1994. Effects of coring on petrophysical measurements. *In*: *International Symposium of the Society of Core Analysts*, Stavanger, Norway. Society of Core Analysts, Fredericton, Canada, 77–86.

Johansen, T. E. S. & Fossen, H. 2008. Internal geometry of fault damage zones in interbedded siliciclastic sediments. *In*: Wibberley, C. A. J., Kurz, W., Imber, J., Holdsworth, R. E. & Collettini, C. (eds) *The Internal Structure of Fault Zones: Implications for Mechanical and Fluid-Flow Properties*. Geological Society, London, Special Publications, **299**, 35–56, https://doi.org/10.1144/SP299.3

Johansen, T. E. S., Fossen, H. & Kluge, R. 2005. The impact of syn-faulting porosity reduction on damage zone architecture in porous sandstone: an outcrop example from the Moab Fault, Utah. *Journal of Structural Geology*, **27**, 1469–1485.

Jones, F. O. & Owens, W. W. 1980. A laboratory study of low-permeability gas sands. *Journal of Petroleum Technology*, **32**, 1631–1640.

Kayser, A. 2006. *Herkunft, Auftreten und Visualisierung von Permeabilitätsbarrieren in einer Gaslagerstätte in Sandsteinen des Rotliegenden (Südliches Permbecken, Deutschland)*. PhD thesis, Philipps-Universität Marburg, Germany.

Kley, J., Franzke, H.-J. et al. 2008. Strain and stress. *In*: Littke, R., Bayer, U., Gajewski, D. & Nelskamp, S.

(eds) *Dynamics of Complex Intracontinental Basins – The Central European Basin System*. Springer, Berlin, 97–124.

KOLYUKHIN, D., SCHUELLER, S., ESPEDAL, M. S. & FOSSEN, H. 2009. Deformation band populations in fault damage zone – impact on fluid flow. *Computational Geosciences*, **14**, 231–248.

LEGLER, B. & MARCHEL, C. 2008. Deformation bands in Permian reservoir sandstones (Rotliegend, Northwest Germany). *Freiberger Forschungshefte*, **C528**, 125–138.

LOHR, T., KRAWCZYK, C. M. *ET AL*. 2007. Strain partitioning due to salt: insights from interpretation of a 3D seismic data set in the NW German Basin. *Basin Research*, **19**, 579–597.

LOUIS, L., DAVID, C., METZ, V., ROBION, P., MENÉNDEZ, B. & KISSEL, C. 2005. Microstructural control on the anisotropy of elastic and transport properties in undeformed sandstones. *International Journal of Rock Mechanics and Mining Sciences*, **42**, 911–923.

MAIN, I. G., KWON, O., NGWENYA, B. T. & ELPHICK, S. C. 2000. Fault sealing during deformation-band growth in porous sandstone. *Geology*, **28**, 1131–1134.

MANZOCCHI, T., WALSH, J. J., NELL, P. & YIELDING, G. 1999. Fault transmissibility multipliers for flow simulation models. *Petroleum Geoscience*, **5**, 53–63, https://doi.org/10.1144/petgeo.5.1.53

MENNING, M. 1995. A numerical time scale for the Permian and Triassic periods: an integrated time analysis. *In*: SCHOLLE, P. A., PERYT, T. M. & ULMER-SCHOLLE, D. S. (eds) *The Permian of Northern Pangea, Volume 1: Paleogeography, Paleoclimates, Stratigraphy*. Springer, Berlin, 77–97.

NICOL, A., CHILDS, C., WALSH, J. J. & SCHAFER, K. W. 2013. A geometric model for the formation of deformation band clusters. *Journal of Structural Geology*, **55**, 21–33.

OGILVIE, S. R. & GLOVER, P. W. J. 2001. The petrophysical properties of deformation bands in relation to their microstructure. *Earth and Planetary Science Letters*, **193**, 123–142.

OLSSON, W. A., LORENZ, J. C. & COOPER, S. P. 2004. A mechanical model for multiply-oriented conjugate deformation bands. *Journal of Structural Geology*, **26**, 325–338.

PARRY, W. T., CHAN, M. A. & BEITLER, B. 2004. Chemical bleaching indicates episodes of fluid flow in deformation bands in sandstone. *American Association of Petroleum Geologists Bulletin*, **88**, 175–191.

ROTEVATN, A. & FOSSEN, H. 2011. Simulating the effect of subseismic fault tails and process zones in a siliciclastic reservoir analogue: implications for aquifer support and trap definition. *Marine and Petroleum Geology*, **28**, 1648–1662.

ROTEVATN, A., FOSSEN, H., HESTHAMMER, J., AAS, T. E. & HOWELL, J. A. 2007. Are relay ramps conduits for fluid flow? Structural analysis of a relay ramp in Arches National Park, Utah. *In*: LONERGAN, L., JOLLY, R. J. H., RAWNSLEY, K. & SANDERSON, D. J. (eds) *Fractured Reservoirs*. Geological Society, London, Special Publications, **270**, 55–71, https://doi.org/10.1144/GSL.SP.2007.270.01.04

SAILLET, E. & WIBBERLEY, C. A. J. 2013. Permeability and flow impact of faults and deformation bands in high-porosity sand reservoirs: southeast Basin, France, analog. *American Association of Petroleum Geologists Bulletin*, **97**, 437–464.

SCHÖNER, R. 2006. *Comparison of Rotliegend sandstone diagenesis from the northern and southern margin of the North German Basin, and implications for the importance of organic maturation and migration*. PhD thesis, Friedrich-Schiller-Universität Jena, Germany.

SCHUELLER, S., BRAATHEN, A., FOSSEN, H. & TVERANGER, J. 2013. Spatial distribution of deformation bands in damage zones of extensional faults in porous sandstones: statistical analysis of field data. *Journal of Structural Geology*, **52**, 148–162.

SCHULTZ, R. A. & FOSSEN, H. 2002. Displacement–length scaling in three dimensions: the importance of aspect ratio and application to deformation bands. *Journal of Structural Geology*, **24**, 1389–1411.

SCHULTZ, R. A., SOLIVA, R., FOSSEN, H., OKUBO, C. H. & REEVES, D. M. 2008. Dependence of displacement–length scaling relations for fractures and deformation bands on the volumetric changes across them. *Journal of Structural Geology*, **30**, 1405–1411.

SCHULTZ, R. A., OKUBO, C. H. & FOSSEN, H. 2010. Porosity and grain size controls on compaction band formation in Jurassic Navajo Sandstone. *Geophysical Research Letters*, **37**, 1–5.

SCHULTZ, R. A., KLIMCZAK, C., FOSSEN, H., OLSON, J. E., EXNER, U., REEVES, D. M. & SOLIVA, R. 2013. Statistical tests of scaling relationships for geologic structures. *Journal of Structural Geology*, **48**, 85–94.

SCHUTJENS, P. M. T. M., HANSSEN, T. H., HETTEMA, M. H. H., MEROUR, J., DE BREE, P., COREMANS, J. W. A. & HELLIESEN, G. 2004. Compaction-induced porosity/permeability reduction in sandstone reservoirs: data and model for elasticity-dominated deformation. *SPE Reservoir Evaluation & Engineering*, **7**, 202–216.

SHIPTON, Z. K., EVANS, J. P. & THOMPSON, L. B. 2005. The geometry and thickness of deformation-band fault core and its influence on sealing characteristics of deformation-band fault zones. *In*: SORKHABI, R. & TSUJI, Y. (eds) *Faults, Fluid Flow, and Petroleum Traps*. American Association of Petroleum Geologists, Memoirs, **85**, 181–195.

SOLIVA, R., SCHULTZ, R. A., BALLAS, G., TABOADA, A., WIBBERLEY, C., SAILLET, E. & BENEDICTO, A. 2013. A model of strain localization in porous sandstone as a function of tectonic setting, burial and material properties; new insight from Provence (southern France). *Journal of Structural Geology*, **49**, 50–63.

STERNLOF, K. R. 2006. *Structural geology, propagation mechanics and hydraulic effects of compaction bands in sandstone*. PhD thesis, Stanford University, CA.

TEMBE, S., BAUD, P. & WONG, T.-F. 2008. Stress conditions for the propagation of discrete compaction bands in porous sandstone. *Journal of Geophysical Research*, **113**, 1–16.

TORABI, A. 2014. Cataclastic bands in immature and poorly lithified sandstone, examples from Corsica, France. *Tectonophysics*, **630**, 91–102.

Tueckmantel, C., Fisher, Q. J., Grattoni, C. A. & Aplin, A. C. 2012. Single- and two-phase fluid flow properties of cataclastic fault rocks in porous sandstone. *Marine and Petroleum Geology*, **29**, 129–142.

Vackiner, A. A. 2011. *Sedimentary facies reconstruction and kinematic restoration of an Upper Permian tight gas field, north-western Germany*. PhD thesis, RWTH Aachen University, Germany.

Vajdova, V., Baud, P. & Wong, T.-f. 2004. Permeability evolution during localized deformation in Bentheim sandstone. *Journal of Geophysical Research*, **109**, 1–15.

Wennberg, O. P., Casini, G., Jahanpanah, A., Lapponi, F., Ineson, J., Wall, B. G. & Gillespie, P. 2013. Deformation bands in chalk, examples from the Shetland Group of the Oseberg Field, North Sea, Norway. *Journal of Structural Geology*, **56**, 103–117.

Williams, G. J. 2011. *Data Mining with R and Rattle: The Art of Excavating Data for Knowledge Discovery*. Springer, New York.

Wong, T.-f., David, C. & Zhu, W. 1997. The transition from brittle faulting to cataclastic flow in porous sandstones: mechanical deformation. *Journal of Geophysical Research*, **102**, 3009–3025.

Woodcock, N. H. & Schubert, C. 1994. Continental strike-slip tectonics. *In*: Hancock, P. (ed.) *Continental Deformation*. Pergamon Press, Oxford, 251–263.

Zhu, C. & Wong, T.-f. 1997. The transition from brittle faulting to cataclastic flow: permeability evolution. *Journal of Geophysical Research*, **102**, 3027–3041.

Deformation band development as a function of intrinsic host-rock properties in Triassic Sherwood Sandstone

JOSHUA GRIFFITHS[1]*, DANIEL R. FAULKNER[1], ALEXANDER P. EDWARDS[2] & RICHARD H. WORDEN[1]

[1]*School of Earth, Ocean and Ecological Sciences, University of Liverpool, Liverpool L69 3GP, UK*

[2]*Ikon Science, Teddington, Middlesex TW11 0JR, UK*

**Correspondence: j.griffiths1@liverpool.ac.uk*

Abstract: Deformation bands significantly alter the local petrophysical properties of sandstone reservoirs, although it is not known how the intrinsically variable characteristics of sandstones (e.g. grain size, sorting and mineralogy) influence the nature and distribution of deformation bands. To address this, cataclastic deformation bands within fine- and coarse-grained Triassic Sherwood Sandstone at Thurstaston, UK were analysed, for the first time, using a suite of petrographical techniques, outcrop studies, helium porosimetry and image analysis. Deformation bands are more abundant in the coarse-grained sandstone than in the underlying fine-grained sandstone. North- and south-dipping conjugate sets of cataclastic bands in the coarse-grained sandstone broadly increase in density (defined by number/m^2) when approaching faults. Microstructural analysis revealed that primary grain size controls deformation band density. Deformation bands in both coarse and fine sandstones led to significantly reduced porosity, and so can represent barriers or baffles to lateral fluid flow. Microstructural data show preferential cataclasis of K-feldspar grains within the host rock and deformation band. The study is of direct relevance to the prediction of reservoir quality in several petroleum-bearing Lower Triassic reservoirs in the near offshore, as deformation band development occurred prior to Carboniferous source-rock maturation and petroleum migration.

The oil and gas industry has expressed a growing interest in deformation bands because they are subseismic, tabular zones of strain localization that can cause large changes to a reservoir's petrophysical properties (Ballas *et al.* 2013). Examples of permeability alteration include the Clair Field, west of Shetland, UK, where deformation bands provide a conduit to lateral fluid flow during the early stages of deformation band formation that initially increased porosity (Baron *et al.* 2008). At the Anschutz Ranch East Field, Wyoming, USA, deformation bands separate clean sandstones and bitumen-stained sandstones, implying a strong impact on oil and gas movement (Solum *et al.* 2010). At the Arroyo Grande Field, California, USA, steam conductivity parallel to deformation bands is reported to be nine times higher than conductivity perpendicular to deformation bands, with tar deposits present on only one side of the deformation bands (Solum *et al.* 2010). Data presented in this study could potentially maximize near-term production targets in the Morecambe, Hamilton, Douglas and Lennox oil and gas fields within the neighbouring East Irish Sea Basin.

Millimetres to centimetres in width, with lengths of several metres or more (Schultz & Soliva 2012), deformation bands have been kinematically classified as one of three end members, namely: dilation bands (pore volume increase); shear bands (pore volume increase, decrease or no change); or compaction bands (pore volume decrease) (Aydin *et al.* 2006; Torabi 2014). Deformation bands in this study will be classified by the predominant deformation mechanism: disaggregation; phyllosilicate smear; cataclasis; and solution and cementation (Fossen *et al.* 2007).

Disaggregation bands form due to shear-induced disaggregation of grains by grain rolling, grain-boundary sliding and the breakage of cements bonding grains, but show little or no evidence of grain crushing (Schultz *et al.* 2010). Phyllosilicate bands form in sandstones which contain >10–15% platy minerals, and 'deformation bands with clay smearing' form in sandstones which have a clay content >40% (Fisher & Knipe 2001; Cerveny *et al.* 2004; Fossen *et al.* 2007). Mechanical grain fracturing is the dominant process in cataclastic bands, where compaction and reorganization of broken

From: Armitage, P. J., Butcher, A. R., Churchill, J. M., Csoma, A. E., Hollis, C., Lander, R. H., Omma, J. E. & Worden, R. H. (eds) 2018. *Reservoir Quality of Clastic and Carbonate Rocks: Analysis, Modelling and Prediction.* Geological Society, London, Special Publications, **435**, 161–176.
First published online January 19, 2016, https://doi.org/10.1144/SP435.11

grains significantly reduces porosity (Cerveny *et al.* 2004; Fossen 2010). Solution bands are produced when chemical compaction, or pressure solution, is the dominant process; they commonly form at shallow depths and contain minimal cataclasis (Fossen *et al.* 2007). Fresh mineral surfaces exposed by grain-boundary sliding and/or grain crushing provide preferential sites for cementation, thus creating cementation bands (Fossen *et al.* 2007).

The primary aim of this paper is to establish how intrinsic host-rock properties (grain size, grain size distribution, porosity and mineralogy) control the nature and distribution of deformation bands. Using a range of petrographical techniques, image analysing software and field techniques, this paper will address the following specific questions, using examples of deformation bands for the first time from Thurstaston, Wirral, UK (Lower Triassic Sherwood Sandstone Group):

- What types of deformation band are present at Thurstaston?
- What is the spatial relationship between deformation band density and fault proximity?
- What is the relationship between the nature of deformation bands and the instrinsic host-rock properties?
- What is the potential impact of the presence of deformation bands in nearby reservoirs in the same lithology?

Geological setting

Early Permian rifting formed a predominantly north–south-orientated asymmetrical half-graben, deepening towards the east (Mikkelsen & Floodpage 1997). The Cheshire Basin (Fig. 1a) formed in the hanging wall of the Wem–Red Rock Fault (Knott 1994; Beach *et al.* 1997; Mikkelsen & Floodpage 1997), a northerly continuation of the Permo-Triassic rift system (Rowe & Burley 1997) that extended from the Wessex Basin to the Scottish Inner Hebrides. Thermal subsidence prolonged rifting until the mid-Triassic, with normal faulting during the early Triassic and Jurassic modifying the basin morphology. Tertiary contraction generated uplift of up to 1500 m (Knott 1994; Beach *et al.* 1997; Ware & Turner 2002), with intra-Triassic uplift resulting in 700–900 m of erosion (Mikkelsen & Floodpage 1997; Rowley & White 1998; Ware & Turner 2002).

Potential organic-rich Carboniferous source-rock sediments (Mikkelsen & Floodpage 1997) in the Cheshire Basin are overlain unconformably by the Permian Collyhurst Sandstone, Manchester Marl and the Kinnerton Sandstone (Rowe & Burley 1997). The Sherwood Sandstone Group (the focus of this study) overlies the Permian sediments, and is a 1500 m-thick succession composed of the Cheshire Pebble Beds Formation, the Wilmslow Sandstone Formation and the Helsby Sandstone Formation (Rowe & Burley 1997). The UK migrated from approximately 10° to 30° N of the equator during the Permo-Triassic (Tellam & Barker 2006).

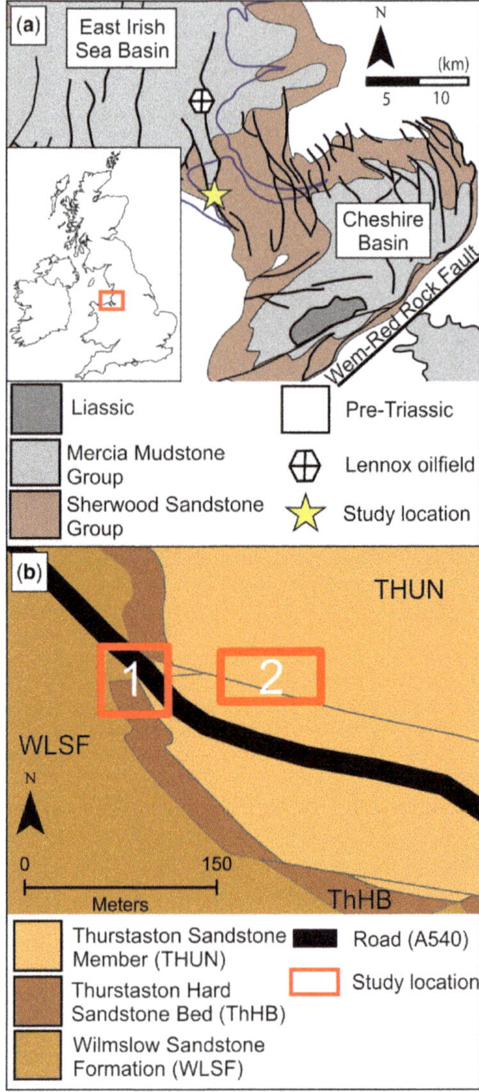

Fig. 1. (**a**) A simplified geological map of the Cheshire and East Irish Sea basins (edited from Meadows 2006). (**b**) The geology of Thurstaston and the locations used in this study: Telegraph Road (1) and Thurstaston Common (2). Maps adapted from Lexicon of Named Rock Units [XLS geospatial data], Scale 1:50000, Tiles: GB, Version 2011, British Geological Survey, UK. Using: EDINA Geology Digimap Service, http://digimap.edina.ac.uk, downloaded April 2013.

The depositional environment of the Sherwood Sandstone Group in the Cheshire Basin was mixed aeolian and fluvial. The Triassic river systems flowed NW from the Cheshire Basin into the East Irish Sea Basin (Meadows 2006). Post-Triassic successions have been removed across most of the Cheshire Basin following Cretaceous and Tertiary uplift; the youngest deposits (Pre-Quaternary) within the Cheshire Basin are middle Liassic (Rowe & Burley 1997). Meadows (2006) provided a synthesis of existing stratigraphic nomenclature applied to the Early and Middle Triassic Sherwood Sandstone Group in NW England, including East Irish Sea Basin equivalent units based on well correlations that contain various oil and gas discoveries (Knott 1994; Meadows 2006).

Methods

Data were collected from the Cheshire Basin within the Sherwood Sandstone Group at Thurstaston (Fig. 1), on the western side of the Wirral peninsula, 7 km SW of Birkenhead. The presence of deformation bands at Thurstaston has been documented (Knott 1994; Beach *et al.* 1997); however; this paper provides the first detailed analysis of deformation band distribution, as well as a petrographical description and interpretation.

Field data

The outcrop of Lower Triassic Wilmslow Sandstone Formation at Thurstaston (Beach *et al.* 1997), composed of aeolian dune and interdune strata, provides world-class examples of sandstone deformation bands. The Thurstaston Sandstone Member is incorporated within the Wilmslow Sandstone Formation in this study following the most recent pronouncement on the Sherwood Sandstone from the British Geological Society (see Meadows 2006 and references therein). Spatial relationships between Lower Triassic aeolian dune and interdune facies have been well documented within the Cheshire Basin, UK (Mountney & Thompson 2002; Mountney 2012). Two outcrop locations (Fig. 1b) provide a three-dimensional view of the deformation bands:

- Location 1 – Telegraph Road (Figs 2a & 3a), contains conjugate sets of deformation bands within the damage zone of three slip surfaces. The orientation, density and thickness of deformation bands were recorded with proximity to three slip surfaces (over a 5 m linear scanline perpendicular to faulting), within fine-grained (mean grain size of *c.* 170 μm) and overlying coarse-grained sandstone (mean grain size of *c.* 540 μm).
- Location 2 – Thurstaston Common (Figs 2b & 4), allowed for a more extensive study of the relationship between slip surface proximity and deformation band density within coarse-grained sandstone.

Two 30 m linear north–south transects (approximately perpendicular to the strike of the major slip surface) allowed for measurements of fault and deformation band density, spacing and orientation. Using a collection of field photographs covering approximately 1 m^2 of exposure subparallel to bedding, the anastomosing geometry of the deformation bands was captured in detail.

Host-rock grain size and grain-size distribution data were collected using a Beckman Coulter LS13 320 Laser Diffraction Particle Size Analyser (LPSA) for five undeformed coarse- and fine-grained sandstone samples. Owing to the friable nature of both the fine- and coarse-grained sandstones, samples required only gentle disaggregation by hand, and this was analysed under an optical microscope to ensure full disaggregation. As histograms are sensitive to bin selection, sorting was defined by the gradient of cumulative frequency curves (Cheung *et al.* 2012). Grain-size range was calculated by $D_{90}-D_{10}$. D_{90} is the grain size at the upper bound of the 90% fraction, whereas D_{10} is the grain size at the upper bound of the finest 10% fraction. X-ray diffractograms generated from PANalytical X'pert Pro MPD X-ray diffractometer (XRD) quantified mineralogy of both the host rock and the deformation bands within the fine- and coarse-grained sandstones layers.

Microstructural characteristics and petrophysical properties

Orientated samples were sectioned along a north–south plane in order to reveal depositional and diagenetic features, prior to vacuum impregnation with blue epoxy resin to reduce friability and to highlight porosity. A Meiji 9000 optical microscope fitted with an Infinity 1.5 camera with Infinity Analyser software was used to carry out an initial reconnaissance of polished thin sections. Secondary electron images (SE) were collected using a Philips XL30 SEM equipped with an Oxford Instruments Secondary X-ray detector from gold–palladium-coated deformation bands and host rock. Backscattered electron images (BSE) were collected using a Hitachi (TM3000) scanning electron microscope (SEM) and Philips XL30 SEM. Using a Philips XL30 SEM equipped with a K.E. Developments Ltd cathodoluminescence (CL) detector (D308122), SEM-CL images were obtained at 10 kV and spot size 7. SEM-CL images took up to 25 min to collect

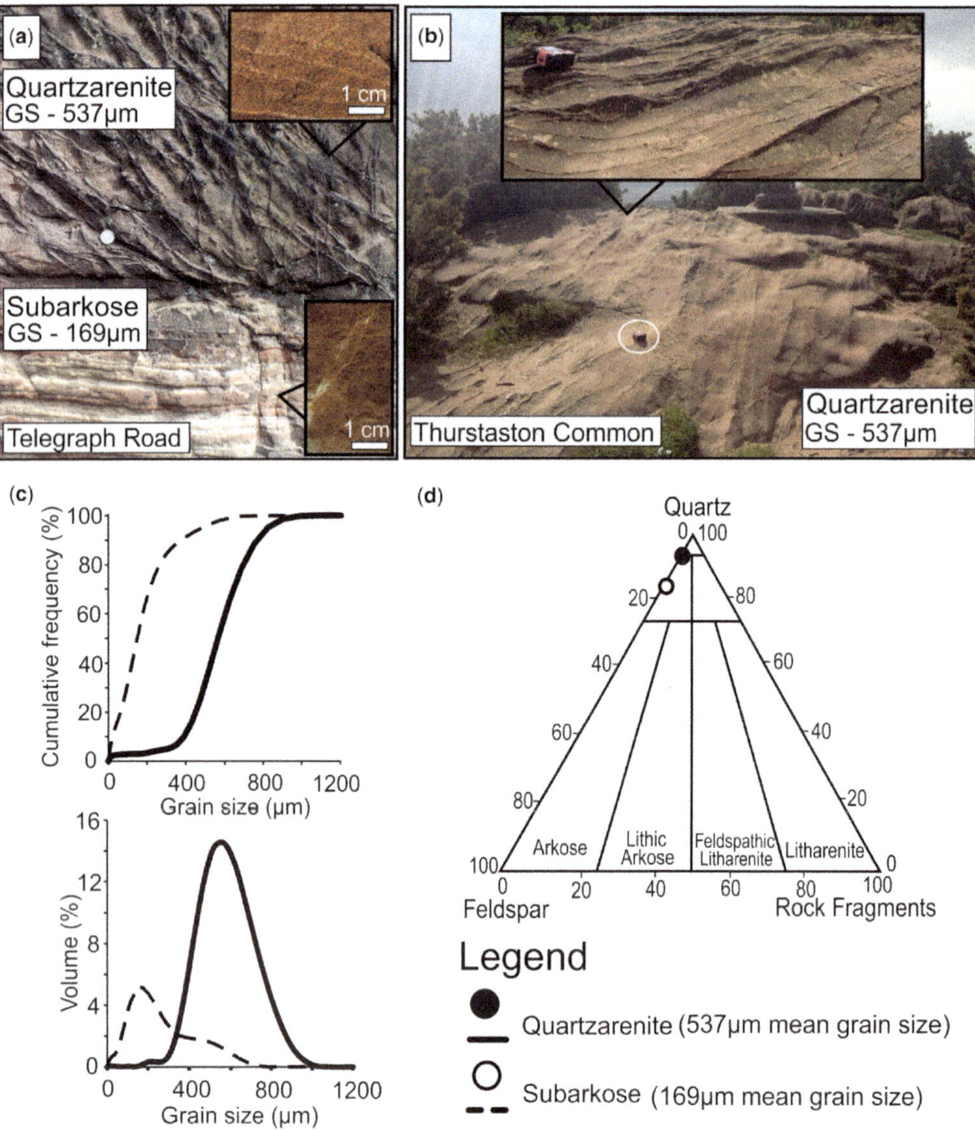

Fig. 2. (**a**) Deformation bands confined to the coarse-grained sandstone at Telegraph Road. (**b**) Deformation bands at Thurstaston Common showing a positive relief. (**c**) Grain-size analysis of both the fine- and coarse-grained sandstone at Telegraph Road. (**d**) XRD-determined mineralogy of both the fine- and coarse-grained sandstone at Telegraph Road.

and were gathered by integrating the signal of 16 frames using a slow scanning raster.

Helium porosimetry was used to calculate porosity within the undeformed host rock of both the fine- and coarse-grained sandstone (two core plugs per sandstone). Owing to the friable nature of both sandstones, the core plugs used for the helium porosimetry were not perfectly cylindrical, resulting in a porosity error margin of 4%. Porosity heterogeneity at a deformation band scale (typically <1 mm) cannot be captured on the scale of a core plug (*c.* 25 mm in diameter) and, instead, a petrographical image analysis is typically used (Antonellini *et al.* 1994). It should also be noted that porosity values documented within the literature using helium porosimetry are typically higher than those calculated using digital image analysis (Anselmetti *et al.* 1998; Ogilvie *et al.* 2001). In

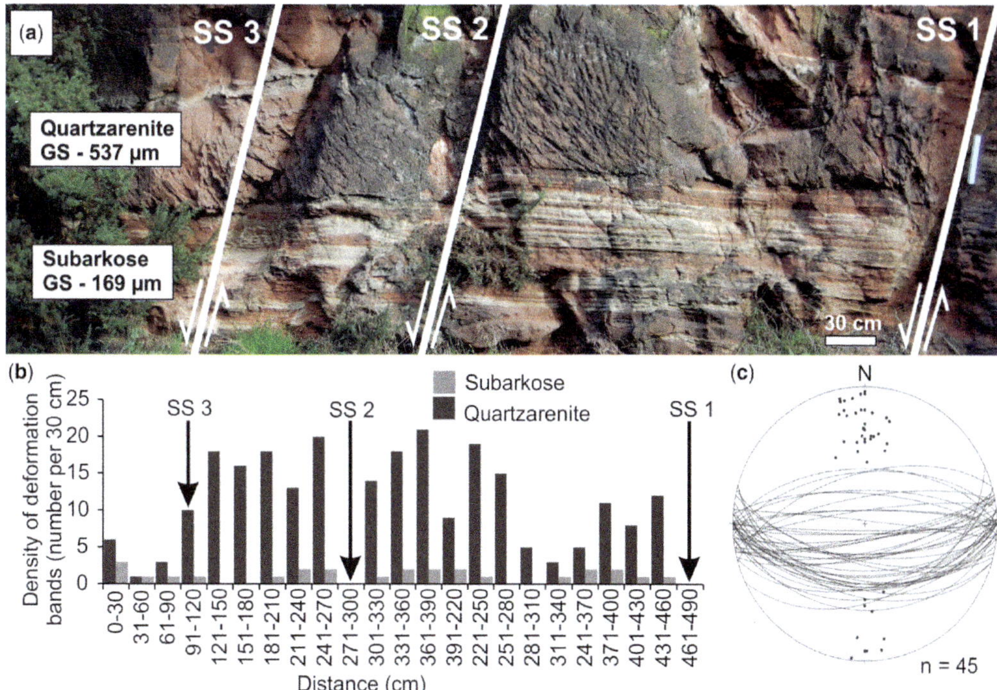

Fig. 3. Outcrop at Telegraph Road. (**a**) White lines illustrate the position of the normal faults. The majority of deformation bands end abruptly at the boundary between the overlying coarse-grained quartz arenite and the underlying fine-grained subarkose sandstone. (**b**) Density of deformation bands with proximity to slip surfaces (SS). (**c**) Stereonet representation of the orientation of deformation bands (*n* refers to the number of measurements).

order to calculate porosity within the deformation band and host rock, BSE images (converted to an 8-bit format) have been digitized in ImageJ Analyser (Schneider *et al.* 2012), creating an array of pixels that are assigned a grey-level intensity. Pixel segmentation was then undertaken using a thresholding formula in which black pixels (porosity) and grey pixels (host rock) were differentiated, allowing for the quantification of total optical porosity. Fifteen images (with varying fields of view ranging from *c.* 0.25 to 1.00 mm^2) have been analysed for both fine- and coarse-grained sandstone (i.e. for both the undeformed host rock and deformation band) to ensure accurate results (Ehrlich *et al.* 1991).

Results

Field data

Host-rock properties. Laser particle size analyser and X-ray diffraction data are displayed in Figure 2. Both the coarse-grained sandstone (mean grain size of *c.* 540 μm) and the underlying fine-grained sandstone (mean grain size of *c.* 170 μm) are moderately sorted, with a grain size range of approximately 325 and 340 μm, respectively (Fig. 2c). Minor fining-upward sequences (medium to coarse grained) are present within the overlying cross-stratified foresets of the coarse-grained sandstone (aeolian facies). However, the majority of any pre-existing primary depositional structures are no longer recognizable owing to intense deformation. The underlying fine-grained sandstone (interdune facies) exhibits subhorizontal, centimetre-scale, wavy sandstone laminae with negligible change in grain size both temporally and spatially. X-ray diffraction analysis of the coarse-grained sandstone identified a dominance of quartz (96%), a small quantity of K-feldspar (4–5%) and a trace of illite. X-ray diffraction analysis of the fine-grained sandstone produced a slightly lower percentage of quartz (83%), and an increase in K-feldspar (11%) and illite (6–7%). In all tested samples, there is a negligible difference in the mineralogy (and mineral abundances) of the deformation band and the host rock. The coarse-grained sandstone is classified as a quartz arenite and the fine-grained sandstone is classified as a subarkosic sandstone (Fig. 2d) according the QFR classification (Folk *et al.* 1970).

Fault kinematics and deformation band distribution. The Wilmslow Sandstone Formation (Thurstaston Sandstone Member) at Telegraph Road (Fig. 3a) is faulted by three WNW-trending, high-angle (>80°), north-dipping, normal faults (with respect to bedding), with striations suggesting a minor component of right-oblique slip. Slip surface 1 (SS 1) has an offset of 64 cm, slip surface 2 (SS 2) of 19.5 cm and slip surface 3 (SS 3) of 7 cm: all are subperpendicular to the main NE-trending Formby Point Fault (Fig. 1a) that extends many kilometres northwards, forming a bounding fault to the Lennox oilfield (Yaliz & Chapman 2003). A north- and south-dipping conjugate set (an acute angle of *c.* 55°) of deformation bands display an east–west orientation at both Telegraph Road and Thurstaston Common, parallel to faulting (Figs 3c & 4d). Deformation bands are sporadic within the underlying fine-grained subarkosic sandstone beds and form in swarms within the overlying coarse-grained quartz-arenite sandstone (Fig. 3). Deformation bands are largely confined to the overlying coarse-grained sandstone, and commonly end abruptly at the fine-grained sandstone boundary (Fig. 2a). Deformation bands range from 0.05 mm to 1.2 cm in width, displaying mm-scale offset. The relationship between deformation band density and fault proximity at Telegraph Road is displayed in Figure 3b. Deformation bands broadly increase in density with proximity to the faults. There is no obvious correlation between deformation band density and the magnitude of fault offset.

Slip surfaces observed at Thurstaston Common (Fig. 4) are a continuation of the high-angle WNW-trending faults that can be observed in cross-section at Telegraph Road (Figs 1b & 3). Where slip planes have initiated and offset has occurred, deformation bands tend to localize and orientate in broad zones in proximity to the slip planes. The anastomosing map pattern of the deformation bands (Fig. 4a), as well as the linkage structures (Fig. 4c), can be recognized at the mm- to cm-scale.

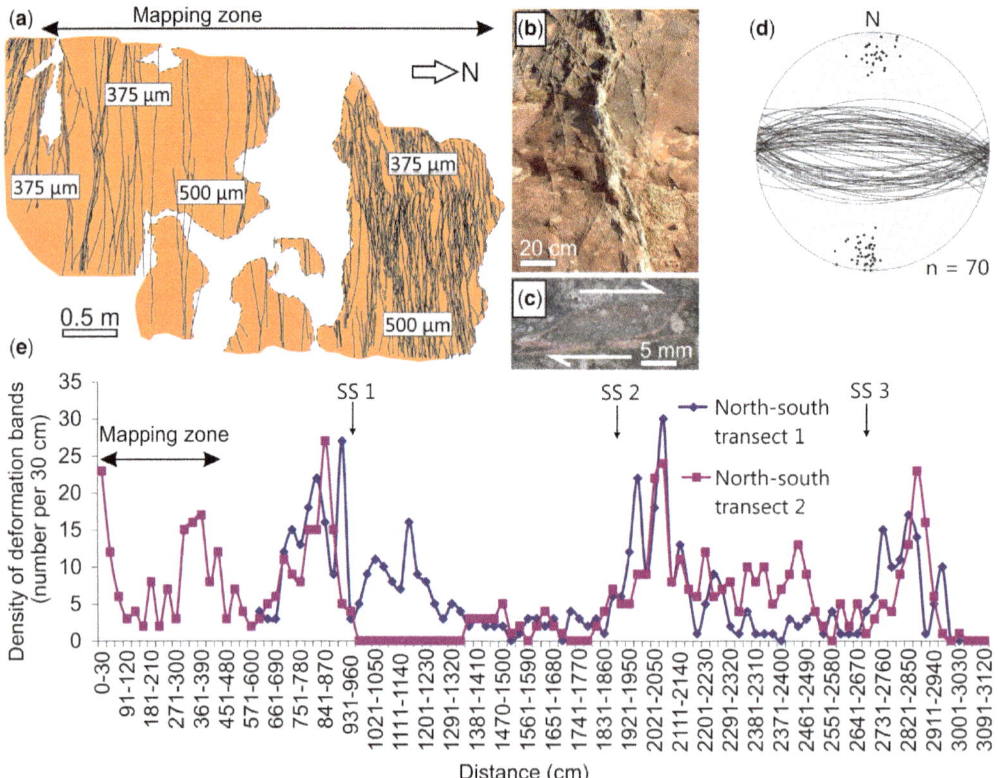

Fig. 4. Outcrop at Thurstaston Common. (**a**) Map of the zones of deformation bands. (**b**) Plan view of deformation bands surrounding a slip surface. (**c**) Linkage structure indicating a strong shear component. (**d**) Stereonet representation of the orientation of deformation bands (*n* refers to the number of measurements). (**e**) Density of deformation bands with proximity to slip surfaces (SS).

Microstructures and petrophysical properties

At both study sites (Telegraph Road and Thurstaston Common), deformation bands are classified as cataclastic bands, with the mechanical fracturing of grains being the predominant deformation mechanism. Secondary electron images (Fig. 5a–c) highlight the friable nature of weakly quartz-cemented host-rock grains and intense localized cataclasis, limited to the deformation band core, within the overlying coarse-grained sandstone. Cathodoluminescence images (Fig. 5e) reveal Hertzian grain–grain interaction, with the deformation

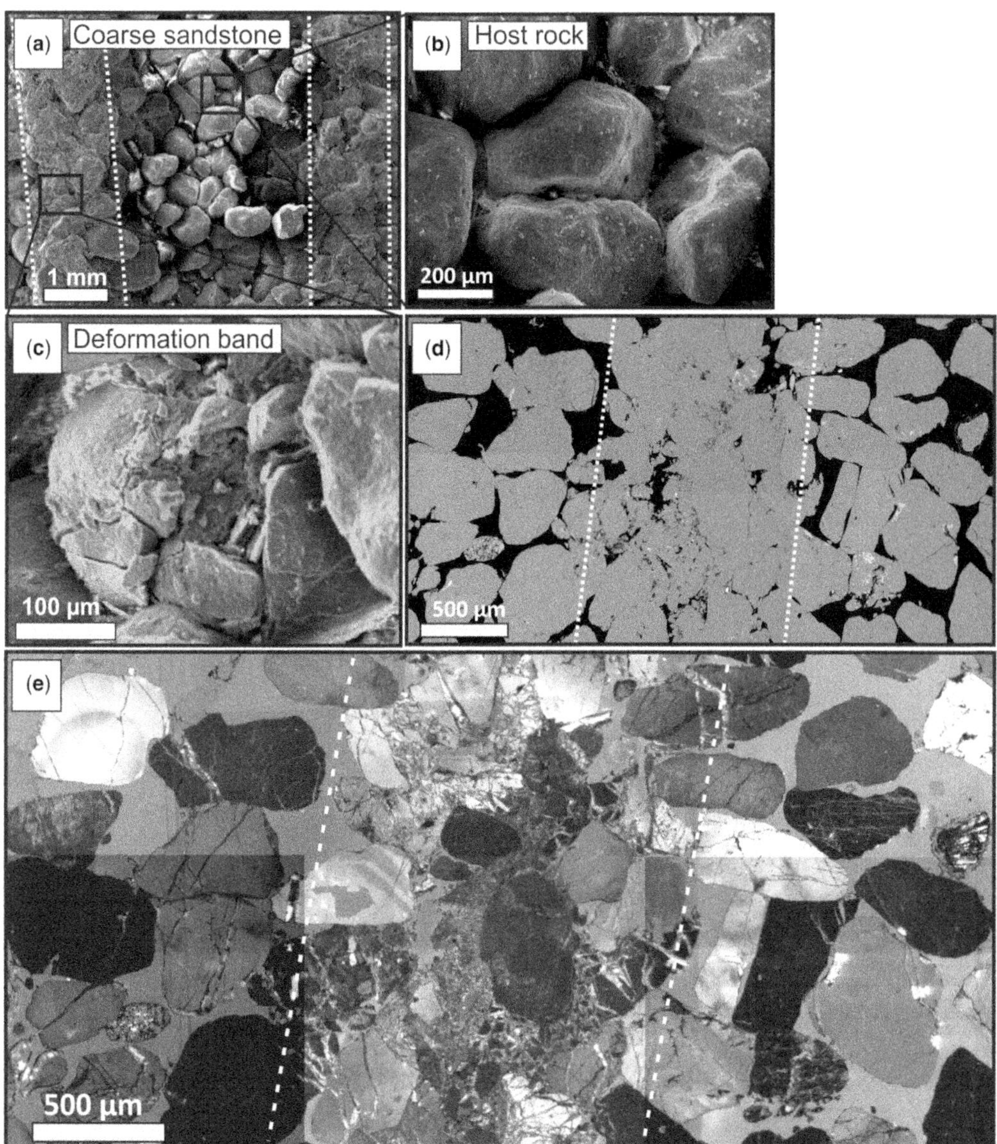

Fig. 5. Deformation bands within the coarse-grained quartzarenite (bandwidth is inferred by the dashed white lines). (**a**)–(**c**) Secondary electron images illustrating the strain localization, poorly cemented host rock and intense cataclasis within the deformation band core. (**d**) BSE image of the deformation band and host rock. (**e**) Collated CL images revealing quartz cementation of the deformation band core and Hertzian fractures.

band core composed of interlocking, fragmented quartz grains cemented by quartz.

The fine-grained host rock is moderately cemented (Fig. 6b), with pore-filling quartz reducing friability; localized comminution of grains produces a deformation band core composed of interlocking detrital clast fragments (Fig. 6c). K-feldspar is preferentially fractured (Fig. 6d–f) within both the deformation band core and the proximal host rock, indicating a strong shear component with a K-feldspar grain being entrained into the deformation band (Fig. 6f). The undeformed host-rock porosity for coarse- and fine-grained sandstone using helium porosimetry is 32 and 15%, respectively. Mean porosity data calculated using image analysis are as follows: the

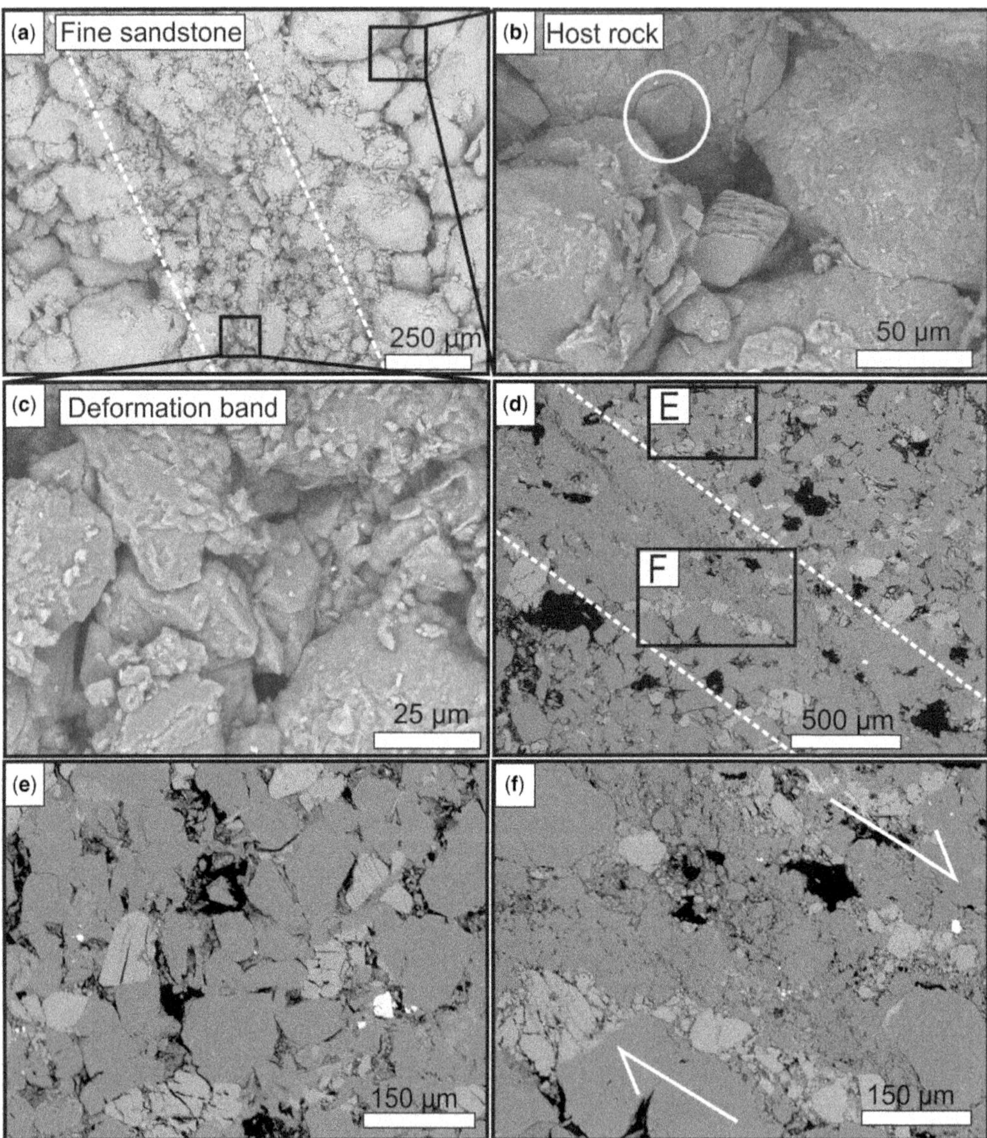

Fig. 6. Deformation bands within the fine-grained subarkosic sandstone (band width is inferred by the dashed white lines). (**a**)–(**c**) Secondary electron images depicting strain localization and grain comminution within the deformation band core, surrounded by a moderately cemented host rock (circled in white). (**d**)–(**f**) BSE images showing preferential fracturing of K-feldspar within the host rock and deformation band core. A strong shear component is indicated by the entrainment of K-feldspar into the deformation band core.

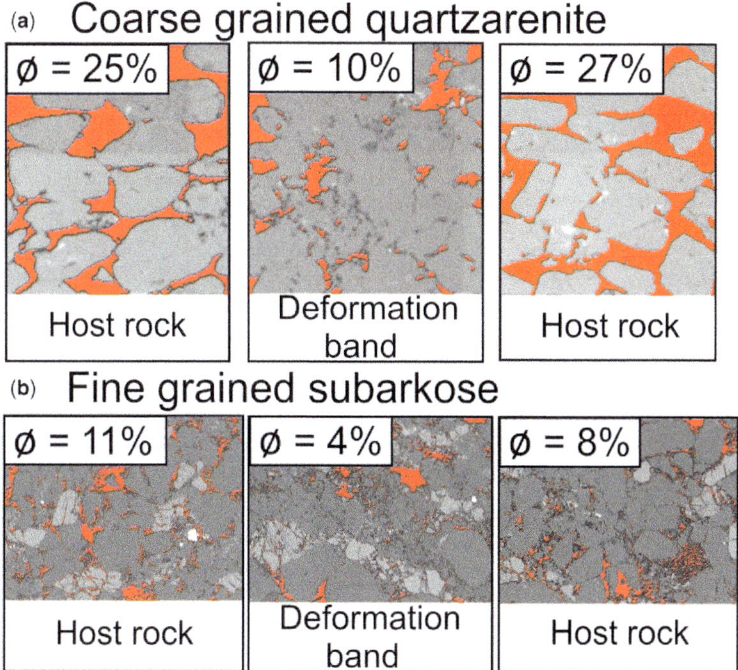

Fig. 7. Porosity calculations using ImageJ analysis software for both the deformation band core and host rock: (**a**) coarse-grained quartzarenite; (**b**) fine-grained subarkosic sandstone.

undeformed host-rock porosity of coarse-grained sandstone (Fig. 7a) is 26%; porosity has been reduced to 10% within the deformation band core in the coarse sandstones (Fig. 7a); undeformed host-rock porosity for the fine-grained sandstone (Fig. 7b) is 10%; porosity has been reduced to 4% within the deformation band core in the fine sandstones (Fig. 7b).

Discussion

Deformation band distribution and fault proximity

Fault-zone architecture is well documented both in the field and in experimental studies (Antonellini *et al.* 1994; Antonellini & Pollard 1995; Caine *et al.* 1996; Faulkner *et al.* 2010). A typical fault zone comprises of a fault core surrounded by a damage zone (Faulkner *et al.* 2010; Schueller *et al.* 2013). The fault core is an area of localized strain that accommodates the majority of displacement (Faulkner *et al.* 2010; Schueller *et al.* 2013). Damage zones in porous sandstones form by growth of deformation bands prior to the initiation of a slip surface (Schueller *et al.* 2013). Shear strain, state of stress, rock type and microstructural deformation mechanisms are key controls in fault-zone architecture (Ngwenya *et al.* 2003). Hydraulic properties of faults and intrinsic properties of host rocks evolve spatially and temporarily, producing heterogeneous permeability.

At both Telegraph Road and Thurstaston Common, deformation bands broadly increase in density with fault proximity (Figs 3 & 4), consistent with 106 outcrop scanlines recording predominantly cataclastic band density in porous sandstones surrounding extensional faults documented by Schueller *et al.* (2013). Before the initiation of a slip surface, it is evident that deformation band density reaches a maximum of around 20–25 bands per 30 cm section independent of fault displacement, analogous to critical microfracture density recorded within low-porosity granodiorite by Mitchell & Faulkner (2009). The trace of isolated deformation bands in outcrops at Thurstaston Common tend to be straight: however, zones of deformation have an anastomising profile showing linkage structures between neighbouring segments (Antonellini *et al.* 1994), similar to the duplex structure described by Cruikshank *et al.* (1991*a*, *b*). The presence of linkage structures suggests a sense of shear, as they resemble miniature restraining bends (Davis 1999). Deformation band lozenges (defined as the rock volumes between deformation bands) at

Thurstaston Common (Figs 2b & 4b) closely compare to those documented within Goblin Valley, Utah, USA (Awdal *et al.* 2014). Early studies explained the development of closely spaced cataclastic bands (Aydin 1978; Aydin & Johnson 1978) in proximity to low-displacement faults (<10 m throw) by the strain-hardening model, showing an increase in deformation band density with fault displacement (Nicol *et al.* 2013). Cataclastic deformation bands have been suggested to strengthen during formation, thus leading to subsequent band formation within relatively weaker wall rock, adjacent to the earlier-formed bands (Nicol *et al.* 2013). Density counts, the positive relief of deformation bands, linkage structures and microstructural analysis (porosity reduction, interlocking quartz fragments, intense grain comminution increasing grain angularity, shear compaction and preferential quartz cementation) at Thurstaston all support a strain-hardening model, resulting in an increase in deformation band density with fault proximity (Nicol *et al.* 2013). Anomalous results, such as the spike in deformation band density within the mapping zone at Thurstatson Common (Fig. 4e), may be explained by an alteration in host-rock cohesion by neighbouring slip surfaces. Deformation band development explained by a geometric model (see Nicol *et al.* 2013 and references therein) infers that deformation bands are strain weakened and form clusters at geometric complexities or irregularities on faults. Further three-dimensional analysis of the fault geometry and a better understanding of the relative timings of slip-surface formation would be required to apply a geometric model at Thurstaston.

Distribution-localization of deformation bands as a function of intrinsic host-rock properties

Mineralogy. The mineralogy of the host rock is an important controlling factor, with different minerals having varying chemical stability, shape, strength and vulnerability to cleavage fractures (Aydin *et al.* 2006). Mineralogically mature, coarse-grained quartzarenite samples (Fig. 2) display highly localized cataclasis within the deformation band (Fig. 5), with little host-rock fracturing in comparison to the underlying subarkosic sandstone (Fig. 6). In addition to intense cataclasis within the deformation band core, feldspathic subarkosic samples in this study show a higher degree of grain fracturing within the host rock (Fig. 6). Because feldspar fractures at lower differential stress than quartz grains (Rawling & Goodwin 2003), an increase in host-rock deformation may be a result of a selective grain-size reduction of weak grains (Fig. 6e). Preferential feldspar grain-size reduction has also been documented within conjugate sets of deformation bands within poorly consolidated arkosic sands of the Vienna Basin, Austria, by Exner & Tschegg (2012). Intense cataclasis creating angular grains and broadening the grain-size distribution considerably lowers porosity as a result of more efficient grain packing (Main *et al.* 2001; Ogilvie & Glover 2001; Tueckmantel *et al.* 2012).

Porosity, grain size and sorting. Cataclastic deformation bands are common in high-porosity (*c.* 10–35%) sands and sandstones deformed at low confining pressures of <40 MPa at shallow depths of <3 km (Nicol *et al.* 2013). Samples with high porosity have lower rock strength than low-porosity samples as pore spaces coalesce, thus increasing the likelihood of pore collapse and so promoting volumetric reduction deformation (Aydin *et al.* 2006). The critical minimum porosity for deformation band development and propagation is lowered by the addition of shear to compaction (Fossen *et al.* 2011). In order to advance the understanding of fluid migration into subsurface reservoirs, it is important to note that the distribution of porosity (and permeability) in deformed high-porosity sandstones can be markedly anisotropic (Farrell *et al.* 2014). Whilst mapping of a thin section using image analysis yields a more detailed microscale (mm-scale) porosity profile, porosity values may also be dependent upon the scale of the measurement and the thin-section orientation with respect to the orientation of the pores (Ogilvie *et al.* 2001). As expected, helium porosity values are slightly higher than those calculated using image analysis as image analysis does not include microporosity. In addition to mineralogy and porosity, factors such as grain size (Zhang *et al.* 1990; Yin *et al.* 1993; Lothe *et al.* 2002) and sorting (Cheung *et al.* 2012) significantly alter the probability of deformation band development and propagation.

It is well documented that larger grain sizes deform under lower effective stresses than finer-grain material (Zhang *et al.* 1990; Yin *et al.* 1993; Lothe *et al.* 2002; Schultz & Siddharthan 2005; Schultz *et al.* 2010; Tueckmantel *et al.* 2012). Since sandstones at Thurstaston have a similar sorting and fall within the porosity range that allows for deformation band development, it is assumed that host-rock grain size is the principal control on deformation band density. Coarser grains have few contact points, which leads to a larger stress concentration and promotes grain-size reduction (Zhang *et al.* 1990; Yin *et al.* 1993; Lothe *et al.* 2002) in the form of Hertzian grain–grain interaction (Fig. 8a). Hertzian fractures are explained by a complex stress field that is set up when a spherical indenter is pressed onto the surface of an isotropic material. The stresses under and around the indenter contact are compressive; however, outside the

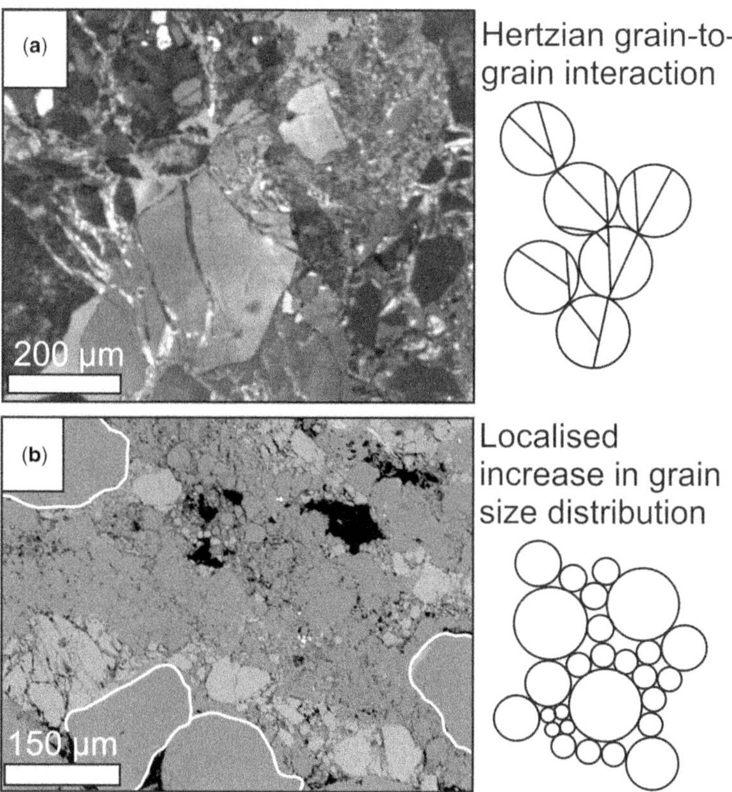

Fig. 8. (**a**) Hertzian grain-contact fracture, creating force chains of fractures propagating into neighbouring grains, promoted by the coarser grain size. The schematic illustration is adapted from Soliva *et al.* (2013). (**b**) Localized increase in grain-size distribution within the fine-grained sandstone, allowing smaller grains to distribute the load over larger particles and so reducing the tensile stress.

contact circle, a radially directed tensile stress is created (Frank & Lawn 1967; Master 2012). Results are consistent with deformation band development within Navajo Sandstone sequences with varying grain size and porosity values at Buckskin Gulsch, Utah, USA (Schultz *et al.* 2010). The corresponding yield envelopes for layers within the Navajo Sandstone are documented to be largest for the fine-grained, less porous sandstones, and smallest for the largest values of porosity and average grain sizes (Schultz *et al.* 2010). The fact that there is a higher density of deformation bands within the overlying coarse-grained sandstone compared to the fine-grained sandstone suggests strain incompatibility between the layers. However, although the density of localized deformation may be different, more strain may have been accommodated through distributed deformation via porosity loss (without fracture) in the finer-grain-sized unit.

In considering mineralogy and grain-size distribution, it is possible to surmise that an increase in K-feldspar content within sandstones will produce a wider grain-size distribution, since K-feldspar has been shown to fracture under lower differential stress than quartz (Rawling & Goodwin 2003; Exner & Tschegg 2012). Thus, a high K-feldspar content may potentially inhibit the development of deformation bands within more feldspathic sandstones, as a non-uniform grain-size distribution allows smaller grains to distribute the load over large particles, and so reduces stress concentrations between grains (Sammis & Ben-Zion 2008; Cheung *et al.* 2012). Petrographical evidence (Fig. 8b) within fine-grained sandstones support the 'constrained comminution' model proposed by Sammis *et al.* (1987), with a localized increase in grain-size distribution within the deformation band core allowing for survivor (or relict) grains.

Implication for sandstone reservoirs

During the appraisal and development of oil and gas fields, analogue studies are helpful for predicting the potential impact on subsurface fluid-flow

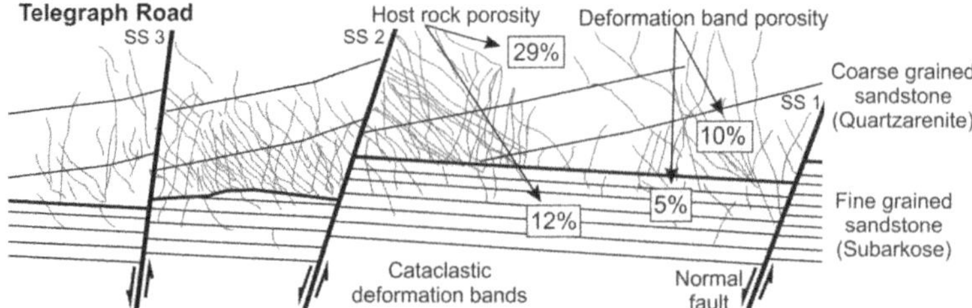

Fig. 9. Schematic synthesis illustration. Deformation bands broadly increase with proximity to faulting. Deformation bands are predominantly restricted to the coarse-grained sandstone. Deformation bands in this study would lower the reservoir quality and potentially compartmentalize the sandstone reservoir.

behaviour. Unfortunately, as the classification and petrophysical measurements of cataclastic deformation bands are not systematic in the literature, it is very difficult to yield a meaningful comparison of results from different study areas (Saillet & Wibberley 2013). However, by combining this study with other analogue studies and experimental datasets, it is likely that reservoir quality predictions will be greatly improved. Cataclastic deformation bands in the literature commonly display lower permeability than the host rock, maximum reductions being of the order of five–six magnitudes and average reductions being around two–three orders (Saillet & Wibberley 2013). Clusters of cataclastic bands have been shown to be as efficient seals as fault cores, withholding up to about a 1 m column of oil and CO_2 (Torabi 2014). The porosity reduction documented in this study would (locally, at least) greatly reduce the reservoir quality, acting as a baffle to fluid flow. For deformation bands to affect well performance, bands must extend over typical well drainage areas: 0.5–1 km^2 for onshore and shallow offshore wells; and 5 km^2 in deep-water wells (Brandenburg *et al.* 2012). Although deformation bands are commonly confined within the damage zones of faults, examples of deformation bands extending over such a large scale have been documented: for example, deformation bands extend approximately 7.5 km^2 at the Valley of Fire, Nevada, USA (Brandenburg *et al.* 2012). In addition to vertical and horizontal continuity and intrinsic host-rock properties, the reservoir-scale impact will also depend on their permeability, orientation, connectivity and abundance (Sternlof *et al.* 2004; Brandenburg *et al.* 2012). The addition of quartz cement, lowering the porosity within the deformation band core, further increases the likelihood of reservoir compartmentalization. Unless accompanied by quartz cement, deformation bands in North Sea reservoirs have not proved to be problematic to oil and gas production (Solum *et al.* 2012). Figure 9 provides a schematic synthesis of the likely distribution of deformation bands and resulting porosity loss associated with conjugate sets of deformation bands within two sandstones with varying intrinsic host-rock properties. If encountered within core, reservoir geologists may use a combination of analogue studies in order to predict the extent of subseismic deformation bands, and the impact on petrophysical properties and reservoir performance. From another point of view, subseismic fault-development mechanisms may be understood by the intrinsic geometry of damage zones connected to the processes of fault growth (Schueller *et al.* 2013).

Specific importance to the East Irish Sea Basin

The Wilmslow Sandstone Formation, part of the Sherwood Sandstone Group, continues north and west into the East Irish Sea Basin, where it is locally known as the St Bees Sandstone Formation (Meadows 2006). The Sherwood Sandstone Group is

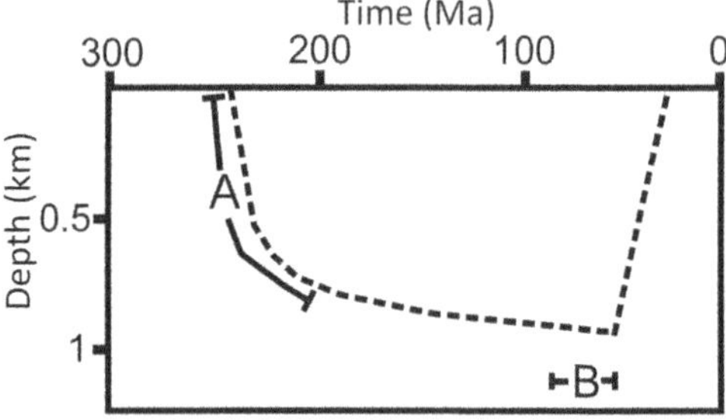

Fig. 10. Schematic illustration of the burial history of the Sherwood Sandstone Group and the timing of deformation band development at Thurstaston. 'A' is the Permo-Triassic rifting forming north–south-trending faults and both the Cheshire and East Irish Sea basins. Development of WNW-trending transfer faults and deformation bands within the Cheshire Basin and, possibly, the East Irish Sea Basin. 'B' is the timing of the underlying Carboniferous source-rock maturation and migration of hydrocarbons into the neighbouring East Irish Sea Basin.

a significant petroleum reservoir within the East Irish Sea Petroleum Province (Duncan *et al.* 1998). A schematic burial history of the outcrop at Thurstaston, including the possible timing of deformation band development (Fig. 10), has been developed based on burial curves constructed by Rowley & White (1998) and the timing of WNW-trending faults suggested by Chadwick (1997). Apatite fission-track analysis from an outcrop 5 km NW of Thurstaston Common seemed to suggest a maximum palaeo-temperature, prior to early Tertiary uplift and cooling, of 90–100°C (Green *et al.* 1997). However, in the undeformed matrix, the high intergranular volume implies limited burial and compaction, and negligible quartz cement (Fig. 6b) implies a maximum temperature much less than 80°C based on depth v. host-rock quartz cement relationships (Worden *et al.* 2000).

Carboniferous source rocks matured during the Late Cretaceous–Early Tertiary, and then oil and gas migrated into the Lower Triassic Sherwood Sandstone (Duncan *et al.* 1998). The timing of the development of WNW-trending faults within both the Cheshire and East Irish Sea basins is poorly constrained. Knott (1994) suggested that WNW-trending faults within the Cheshire Basin formed under a NW–SE-trending maximum horizontal compressive stress regime, present since the Paleocene. It is unlikely that the WNW-trending faults at Thurstaston have formed under a compressive regime, since faults clearly display extensional offsets. Instead, it is possible that these faults, which have formed subperpendicular to, and cross-cut, the main northô–south-trending faults are transfer fault (Chadwick 1997). Adding to the complexity, the extensional direction may not have remained constant throughout the evolution of the Cheshire Basin (Chadwick 1997). Despite some uncertainty on the timing of deformation band development within the damage zones of the WNW-trending faults, faulting occurred prior to Tertiary uplift, source-rock maturation and hydrocarbon migration (Fig. 10). Oil and gas migration may have, therefore, been affected by deformation bands in the oil- and gas-bearing offshore equivalent outcrop. As the deformation bands are locally quartz-cemented (thus, further reducing porosity and permeability lower than achieved by simple comminution), careful analysis of cores and borehole image logs for deformation band occurrence and their stratigraphic constraints should be undertaken during field appraisal and the development of oil- and gas-bearing structures in the basin centres.

Conclusions

- Deformation bands in the Triassic sandstone exposed at Thurstaston are cataclastic with a strong component of shear and porosity reduction.
- Deformation bands at Thurstaston broadly increase in density (number/m^2) with proximity to faulting over a scale of several metres.
- Deformation band distribution at Thurstaston is predominantly controlled by grain size. Deformation bands are more abundant within the overlying coarse-grained quartzarenite, and sporadic within the underlying fine-grained subarkose. K-feldspar is preferentially fractured in comparison to quartz grains.
- Deformation bands in Triassic sandstones from Thurstaston have significantly altered the petrophysical properties of the intact rock. Porosity is substantially reduced relative to the matrix due to intense cataclasis and localized quartz cementation. The potential impact of deformation bands in nearby reservoirs in the same lithology could have a detrimental effect on reservoir quality and well performance.

We would like to acknowledge Oliver Hinds for his guidance on statistical grain-size analysis. We are also grateful to John Bedford for assistance during the operation of the helium porosimeter and James Utley for XRD analysis. This manuscript has benefited considerably from the comments of Dr D. Healey and an anonymous reviewer.

References

ANSELMETTI, F. S., LUTHI, S. & EBERLI, G. P. 1998. Quantitative characterization of carbonate pore systems by digital image analysis. *American Association of Petroleum Geologists Bulletin*, **82**, 1815–1836.

ANTONELLINI, M. A. & POLLARD, D. D. 1995. Distinct element modelling of deformation bands in sandstone. *Journal of Structural Geology*, **17**, 1165–1182.

ANTONELLINI, M. A., AYDIN, A. & POLLARD, D. D. 1994. Microstructure of deformation bands in porous sandstones at Arches National Park, Utah. *Journal of Structural Geology*, **16**, 941–959.

AWDAL, A., HEALY, D. & ALSOP, G. I. 2014. Geometrical analysis of deformation band lozenges and their scaling relationships to fault lenses. *Journal of Structural Geology*, **66**, 11–23.

AYDIN, A. 1978. Small faults formed as deformation bands in sandstone. *Pure and Applied Geophysics*, **116**, 913–930.

AYDIN, A. & JOHNSON, A. M. 1978. Development of faults as zones of deformation bands and as slip surfaces in sandstones. *Pure and Applied Geophysics*, **116**, 931–942.

AYDIN, A., BORJA, R. I. & EICHHUBL, P. 2006. Geological and mathematical framework for failure modes in granular rock. *Journal of Structural Geology*, **28**, 83–98.

BALLAS, G., SOLIVA, R., SIZUN, J.-P., FOSSEN, H., BENEDICTO, A. & SKURTVEIT, A. 2013. Shear-enhanced compaction bands formed at shallow burial conditions; implications for fluid flow (Provence, France). *Journal of Structural Geology*, **47**, 3–15.

Baron, M., Parnell, J., Darren, M., Carr, A., Przyjalgowski, M. & Feely, M. 2008. Evolution of hydrocarbon migration style in a fractured reservoir deduced from fluid inclusion data, Clair Field, west of Shetland, UK. *Marine and Petroleum Geology*, **25**, 153–172.

Beach, A., Brown, J. L., Welbon, A. I., McCallum, J. E., Brockbank, P. & Knott, S. 1997. Characteristics of fault zones in sandstones from NW England: application to fault transmissibility. *In*: Meadows, N. S., Trueblood, S. R., Hardman, M. & Cowan, G. (eds) *Petroleum Geology of the Irish Sea and Adjacent Areas*. Geological Society, London, Special Publications, **124**, 315–324, https://doi.org/10.1144/GSL.SP.1997.124.01.19

Caine, J. S., Evans, J. P. & Forster, C. B. 1996. Fault zone architecture and permeability structure. *Geology*, **11**, 1025–1028.

Brandenburg, J. P., Alpak, F. O., Solum, J. G. & Naruk, A. 2012. A kinematic trishear model to predict deformation bands in a fault-propagation fold, East Kaibab monocline, Utah. *American Association of Petroleum Geologists Bulletin*, **96**, 109–132.

Cerveny, K., Davies, R., Fox, G. D. R., Kaufman, P., Knipe, R. & Krantz, R. 2004. Reducing uncertainty with fault-seal analysis. *Oilfield Review*, **16**, 38–51.

Chadwick, R. A. 1997. Fault analysis of the Cheshire Basin, NW England. *In*: Meadows, N. S., Trueblood, S. R., Hardman, M. & Cowan, G. (eds) *Petroleum Geology of the Irish Sea and Adjacent Areas*. Geological Society, London, Special Publications, **124**, 297–313, https://doi.org/10.1144/GSL.SP.1997.124.01.18

Cheung, C. S., Baud, P. & Wong, T. F. 2012. Effect of grain size distribution on the development of compaction localization in porous sandstone. *Geophysical Research Letters*, **39**, L21302.

Cruikshank, K. M., Zhao, G. & Johnson, A. M. 1991*a*. Analysis of minor fractures associated with joints and faulted joints. *Journal of Structural Geology*, **13**, 865–886.

Cruikshank, K. M., Zhao, G. & Johnson, A. M. 1991*b*. Duplex structures connecting fault segments in Entrada sandstone. *Journal of Structural Geology*, **13**, 1185–1196.

Davis, G. H. 1999. *Structural Geology of the Colorado Plateau Region of Southern Utah, with Special Emphasis on Deformation Bands*. Geological Society of America, Special Papers, **342**.

Duncan, W. I., Green, P. F. & Duddy, I. R. 1998. Source Rock burial history and seal effectiveness: key facts to understanding hydrocarbon exploration potential in the East and Central Irish Sea Basins. *American Association of Petroleum Geologists Bulletin*, **82**, 1401–1415.

Ehrlich, R., Crabtree, S. J., Horkowitz, K. O. & Horkowitz, J. P. 1991. Petrography and reservoir physics I: objective classification of reservoir porosity (1). *American Association of Petroleum Geologists Bulletin*, **75**, 1547–1562.

Exner, U. & Tschegg, C. 2012. Preferential cataclastic grain size reduction of feldspar in deformation bands in poorly consolidated arkosic sands. *Journal of Structural Geology*, **43**, 63–72.

Farrell, N. J. C., Healy, D. & Taylor, C. W. 2014. Anisotropy of permeability in faulted porous sandstones. *Journal of Structural Geology*, **63**, 50–67.

Faulkner, D. R., Jackson, C. A. L., Lunn, R. J., Schlische, R. W., Shipton, Z. K., Wibberley, C. A. J. & Withjack, M. O. 2010. A review of recent developments concerning the structure, mechanics and fluid flow properties of fault zones. *Journal of Structural Geology*, **32**, 1557–1575.

Fisher, Q. & Knipe, R. 2001. The permeability of faults within siliciclastic petroleum reservoirs of the North Sea and Norwegian Continental Shelf. *Marine and Petroleum Geology*, **18**, 1063–1081.

Folk, R. L., Andrews, P. B. & Lewis, D. W. 1970. Detrital sedimentary rock classification and nomenclature for use in New Zealand, New Zealand. *Journal of Geology and Geophysics*, **13**, 937–968.

Fossen, H. 2010. Deformation bands formed during soft-sediment deformation: observations from SE Utah. *Marine and Petroleum Geology*, **27**, 215–222.

Fossen, H., Schultz, R. A., Shipton, Z. K. & Mair, K. 2007. Deformation bands in sandstone; a review. *Journal of the Geological Society, London*, **164**, 755–769, https://doi.org/10.1144/0016-76492006-036

Fossen, H., Schultz, R. A. & Torabi, A. 2011. Conditions and implications for compaction band formation in the Navajo Sandstone, Utah. *Journal of Structural Geology*, **33**, 1477–1490.

Frank, F. C. & Lawn, B. R. 1967. On the theory of Hertzian fracture. *Proceedings of the Royal Society of London, Series A. Mathematical and Physical Sciences*, **299**, 291–306.

Green, P. F., Duddy, I. R. & Bray, R. J. 1997. Variation in thermal history styles around the Irish Sea and adjacent areas: implications for hydrocarbon occurrence and tectonic evolution. *In*: Meadows, N. S., Trueblood, S. R., Hardman, M. & Cowan, G. (eds) *Petroleum Geology of the Irish Sea and Adjacent Areas*. Geological Society, London, Special Publications, **124**, 73–93.

Knott, S. D. 1994. Fault zone thickness v. displacement in the Permo-Triassic sandstones of NW England. *Journal of the Geological Society, London*, **151**, 17–25, https://doi.org/10.1144/gsjgs.151.1.0017

Lothe, A. E., Gabrielsen, R. H., Hagen, N. B. & Larsen, B. T. 2002. An experimental study of the texture of deformation bands: effects on the porosity and permeability of sandstones. *Petroleum Geoscience*, **8**, 195–207, https://doi.org/10.1144/petgeo.8.3.195

Main, I., Mair, K., Kwon, O., Elphick, S. & Ngwenya, B. 2001. Experimental constraints on the mechanical and hydraulic properties of deformation bands in porous sandstones; a review. *In*: Holdsworth, R. E., Strachan, R. A., Magloughlin, J. F. & Knipe, R. J. (eds) *The Nature and Tectonic Significance of Fault Zone Weakening*. Geological Society, London, Special Publications, **186**, 43–63, https://doi.org/10.1144/GSL.SP.2001.186.01.04

Master, S. 2012. Hertzian fractures in the sub-dwyka Nooitgedacht striated pavement, and implications for the former thickness of Karoo strata near Kimberley, South Africa. *South African Journal of Geology*, **115**, 561–576.

MEADOWS, S. N. 2006. The correlation and sequence architecture of the Ormskirk Sandstone Formation in the Triassic Sherwood Sandstone Group of the East Irish Sea Basin, NW England. *Geological Journal*, **41**, 93–122.

MIKKELSEN, P. W. & FLOODPAGE, J. B. 1997. The hydrocarbon potential of the Cheshire Basin. *In*: MEADOWS, N. S., TRUEBLOOD, S. R., HARDMAN, M. & COWAN, G. (eds) *Petroleum Geology of the Irish Sea and Adjacent Areas*. Geological Society, London, Special Publications, **124**, 161–183, https://doi.org/10.1144/GSL.SP.1997.124.01.10

MITCHELL, T. M. & FAULKNER, D. R. 2009. The nature and origin of off-fault damage surrounding strike-slip fault zones with a wide range of displacements: a field study from the Atacama fault system, northern Chile. *Journal of Structural Geology*, **31**, 802–816.

MOUNTNEY, N. P. 2012. A stratigraphic model to account for complexity in aeolian dune and interdune successions. *Sedimentology*, **59**, 964–989.

MOUNTNEY, N. P. & THOMPSON, D. B. 2002. Stratigraphic evolution and preservation of aeolian dune and damp/wet interdune strata: an example from the Triassic Helsby Sandstone Formation, Cheshire Basin, UK. *Sedimentology*, **49**, 805–833.

OGILVIE, S. R. & GLOVER, P. W. J. 2001. The petrophysical properties of deformation bands in relation to their microstructure. *Earth and Planetary Science Letters*, **193**, 129–142.

OGILVIE, S. R., ORRIBO, J. M. & GLOVER, P. W. J. 2001. The influence of deformation bands upon fluid flow using profile permeametry and positron emission tomography. *Geophysical Research Letters*, **28**, 61–64.

NGWENYA, B. T., KWON, O., ELPHICK, S. C. & MAIN, I. G. 2003. Permeability evolution during progressive development of deformation bands in porous sandstones. *Journal of Geophysical Research: Solid Earth*, **108**, 2343.

NICOL, A., CHILDS, C., WALSH, J. J. & SCHAFER, K. W. 2013. A geometric model for the formation of deformation band clusters. *Journal of Structural Geology*, **55**, 21–33.

RAWLING, G. C. & GOODWIN, L. B. 2003. Cataclasis and particulate flow in faulted, poorly lithified sediments. *Journal of Structural Geology*, **25**, 317–331.

ROWE, J. & BURLEY, S. D. 1997. Faulting and porosity modification in the Sherwood Sandstone at Alderley Edge, northeastern Cheshire: an exhumed example of fault-related diagenesis. *In*: MEADOWS, N. S., TRUEBLOOD, S. R., HARDMAN, M. & COWAN, G. (eds) *Petroleum Geology of the Irish Sea and Adjacent Areas*. Geological Society, London, Special Publications, **124**, 325–352, https://doi.org/10.1144/GSL.SP.1997.124.01.20

ROWLEY, E. & WHITE, N. 1998. Inverse modelling of extension and denudation in the East Irish Sea and surrounding areas. *Earth and Planetary Science Letters*, **161**, 57–71.

SAILLET, E. & WIBBERLEY, C. A. 2013. Permeability and flow impact of faults and deformation bands in high-porosity sand reservoirs: Southeast Basin, France, analog. *American Association of Petroleum Geologists Bulletin*, **97**, 437–464.

SAMMIS, C., KING, G. & BIEGEL, R. 1987. The kinematics of gouge deformation. *Pure and Applied Geophysics*, **125**, 777–812.

SAMMIS, C. G. & BEN-ZION, Y. 2008. Mechanics of grain-size reduction in fault zones. *Journal of Geophysical Research*, **113**, B02306.

SCHNEIDER, C. A., RASBAND, W. S. & ELICEIRI, K. W. 2012. NIH Image to ImageJ: 25 years of image analysis. *Nature Methods*, **9**, 671–675.

SCHUELLER, S., BRAATHEN, A., FOSSEN, H. & TVERANGER, J. 2013. Spatial distribution of deformation bands in damage zones of extensional faults in porous sandstones: statistical analysis of field data. *Journal of Structural Geology*, **52**, 148–162.

SCHULTZ, R. A. & SIDDHARTHAN, R. 2005. A general framework for the occurrence and faulting of deformation bands in porous granular rocks. *Tectonophysics*, **411**, 1–18.

SCHULTZ, R. A. & SOLIVA, R. 2012. Propagation energies inferred from deformation bands in sandstone. *International Journal of Fracture*, **176**, 135–149.

SCHULTZ, R. A., OKUBO, C. H. & FOSSEN, H. 2010. Porosity and grain size controls on compaction band formation in Jurassic Navajo Sandstone. *Geophysical Research Letters*, **37**, 1–5.

SOLIVA, R., SCHULTZ, R. A., BALLAS, G., TABOADA, A., WIBBERLEY, C., SAILLET, E. & BENEDICTO, A. 2013. A model of strain localization in porous sandstone as a function of tectonic setting, burial and material properties; new insight from Provence (southern France). *Journal of Structural Geology*, **49**, 50–63.

SOLUM, J. G., BRANDENBURG, J. P., NARUK, S. J., KOSTENK, O. V., WILKINS, S. J. & SCHULTZ, R. A. 2010. Characterization of deformation bands associated with normal and reverse stress states in the Navajo Sandstone, Utah. *American Association of Petroleum Geologists Bulletin*, **94**, 1453–1475.

SOLUM, J. G., BRANDENBURG, J. P. & NARUK, S. J. 2012. Characterization of deformation bands associated with normal and reverse stress states in the Navajo Sandstone, Utah: reply. *American Association of Petroleum Geologists Bulletin*, **96**, 877–890.

STERNLOF, K. R., CHAPIN, J. R., POLLARD, D. D. & DURLOFSKY, L. J. 2004. Permeability effects of deformation band arrays in sandstone. *American Association of Petroleum Geologists Bulletin*, **88**, 1315–1329.

TELLAM, J. H. & BARKER, R. D. 2006. Towards prediction of saturated-zone pollutant movement in groundwaters in fractured permeable-matrix aquifers: the case of the UK Permo-Triassic sandstones. *In*: BARKER, R. D. & TELLAM, J. H. (eds) *Fluid Flow and Solute Movement in Sandstones: The Onshore UK Permo-Triassic Red Bed Sequence*. Geological Society, London, Special Publications, **263**, 1–48, https://doi.org/10.1144/GSL.SP.2006.263.01.01

TORABI, A. 2014. Cataclastic bands in immature and poorly lithified sandstone, examples from Corsica, France. *Tectonophysics*, **630**, 91–102.

TUECKMANTEL, C., FISHER, Q. J., GRATTONI, C. A. & APLIN, A. C. 2012. Single-and two-phase fluid flow properties of cataclastic fault rocks in porous sandstone. *Marine and Petroleum Geology*, **29**, 129–142.

WARE, P. D. & TURNER, J. P. 2002. Sonic velocity analysis of the Tertiary denudation of the Irish Sea Basin. *In*: DORÉ, A. G., CARTWRIGHT, J. A., STOKER, M. S., TURNER, J. P. & WHITE, N. (eds) *Exhumation of the North Atlantic Margin: Timing, Mechanisms and Implications for Petroleum Exploration.* Geological Society, London, Special Publications, **196**, 355–370, https://doi.org/10.1144/GSL.SP.2002.196.01.19

WORDEN, R. H., MAYALL, M. & EVANS, I. J. 2000. The effect of ductile-lithic sand grains and quartz cement on porosity and permeability in Oligocene and lower Miocene clastics, South China Sea: prediction of reservoir quality. *American Association of Petroleum Geologists Bulletin*, **84**, 345–359.

YALIZ, A. & CHAPMAN, T. 2003. The Lennox Oil and Gas Field, Block 110/15, East Irish Sea. *In*: GLUYAS, J. G. & HICHENS, H. M. (eds) *United Kingdom Oil and Gas Fields: Commemorative Millennium Volume*. Geological Society, London, Memoirs, **20**, 87–96, https://doi.org/10.1144/GSL.MEM.2003.020.01.07

YIN, H., MARVKO, G. & NUR, A. 1993. Grain size effects on porosity, permeability and acoustic velocities in granular materials. *Eos, Transactions of the American Geophysical Union*, **74**, 568.

ZHANG, J., WONG, T. F. & DAVIS, D. M. 1990. Micromechanics of pressure-induced grain crushing in porous rocks. *Journal of Geophysical Research: Solid Earth*, **95**, 341–352.

Diagenetic controls on the location of reservoir sweet spots relative to palaeotopographical and structural highs

JESSICA E. POTEET[1,2], ROBERT H. GOLDSTEIN[1]* & EVAN K. FRANSEEN[1,2]

[1]*Department of Geology, Kansas Interdisciplinary Carbonates Consortium (KICC), University of Kansas, 1475 Jayhawk Boulevard, Lawrence, KS 66045, USA*

[2]*Kansas Geological Survey, 1930 Constant Avenue, Lawrence, KS 66047, USA*

**Correspondence: gold@ku.edu*

Abstract: Many carbonate reservoirs are located on top, or down the flanks, of extant structural highs or syndepositional palaeo-highs. This study examines diagenesis in Pennsylvanian oolitic reservoirs close to the crest and down the flank of a long-lived anticline. It illustrates that the position of the best reservoir quality shifted back and forth during successive diagenetic events. Cement stratigraphy shows that early diagenesis did not enhance reservoir character significantly. Most oomoldic porosity formed penecontemporaneously with compaction. Fluid-inclusion and stable isotope data indicate that late cements precipitated during burial conditions by refluxing brines and later hydrothermal fluids. After initial burial, greater permeability existed downdip, where smaller amounts of early meteoric cement allowed for compaction. Subsequent reflux cementation initially degraded downdip reservoirs preferentially and then progressed updip, resulting in relatively uniform reservoir porosity. Later hydrothermal events are most important in affecting the distribution of the highest quality present-day reservoir. Highest porosity is preserved in wells down the flanks of the structure, where hydrothermal cements are not as prevalent. Understanding the effect of diagenesis on location of the best reservoir in relation to palaeotopographical and structural highs allows for the prediction of reservoir quality using seismic and mapping data typically available in the subsurface.

Predictability of the location of the best porosity and permeability in carbonate reservoirs is among the greatest challenges in oil and gas exploration. Commonly, the problem is due to post-depositional diagenetic processes that overprint depositional and stratigraphic relationships to such an extent that the diagenesis dominates the location of the best reservoir. As many carbonate reservoirs are discovered on the basis of stratigraphic or structural relief, this study compares an area that was a palaeo-high, and is a current structural high, to an adjacent area that was a palaeo-low and is down the flank of an extant anticlinal structure. To make this study broadly applicable, the comparison of reservoir quality on the high v. the low is carried through each major diagenetic step to examine the impact of commonly observed diagenetic processes, such as early subaerial exposure and meteoric diagenesis, burial processes, brine reflux, structural rejuvenation, and hydrothermal fluid flow.

There is broad-based evidence that ties diagenetic differences, and resultant reservoir quality, to a location relative to high or low positions. Although not an exhaustive review, the following summarizes a range of carbonate reservoirs around the world where position relative to a high has had a diagenetic impact. In some examples, early meteoric diagenesis has affected reservoir porosity where surfaces of subaerial exposure have interrupted sedimentation. Porosity creation or reduction from meteoric waters varies laterally, depending on whether the reservoir was in a palaeo-high v. palaeo-low position. This has been observed in reservoirs such as the Pennsylvanian–Permian of the Permian Basin and the Midcontinent, Tengiz of Kazakhstan, offshore Angola, Jurassic Smackover, Lisburne Group of Alaska, and Arab D (Moore & Druckman 1981; Saller *et al.* 1994, 1999; Eichenseer *et al.* 1999; Carlson *et al.* 2003; Kenter *et al.* 2006*a*, *b*; Buijs & Goldstein 2012). Typically, duration of subaerial exposure is longer on palaeo-highs and shorter in palaeo-lows, and thus the effects of meteoric diagenesis varies from high to low.

Lateral variation in degree of early meteoric diagenesis impacts burial porosity formation and compaction. Although the impact of mesogenetic porosity creation has recently been questioned by Ehrenberg *et al.* (2012), there is strong evidence of its impact on reservoir rocks. Our study provides strong support for burial and hydrothermal porosity modification. Others, such as the Jurassic Smackover reservoirs (e.g. Moore & Druckman 1981; Heydari 2000), provide strong evidence of burial dissolution and compaction, and that the effects

From: Armitage, P. J., Butcher, A. R., Churchill, J. M., Csoma, A. E., Hollis, C., Lander, R. H., Omma, J. E. & Worden, R. H. (eds) 2018. *Reservoir Quality of Clastic and Carbonate Rocks: Analysis, Modelling and Prediction.* Geological Society, London, Special Publications, **435**, 177–215.
First published online December 21, 2016, https://doi.org/10.1144/SP435.18

of these processes on reservoir character were impacted by laterally variable earlier diagenesis.

Similarly, brine reflux diagenesis (Adams & Rhodes 1960) follows updip to downdip flow paths related to surface sources of brines and fluid density differences. Downdip flow leads to vertical and lateral variability of dolomitization, calcite cementation, and anhydrite precipitation. Brine reflux is among the most important processes in low-temperature carbonate diagenesis, and had an important impact on reservoir character of a large number of reservoirs including: Permian reservoirs of the Permian Basin; Pennsylvanian–Permian reservoirs of the Midcontinent; Arab D; and, possibly, basin margin Cretaceous reservoirs of Mexico (Enos 1977; Moore *et al.* 1988; Goldstein *et al.* 1991; Enos & Stephens 1993; Amthor *et al.* 1994; Saller & Henderson 1998; Luczaj & Goldstein 2000; Cantrell *et al.* 2004; Al-Helal *et al.* 2012).

The flow of warm fluids in the subsurface, including hydrothermal fluids, typically is impacted by structure, stratigraphic discontinuities and oil–water contacts. Warm fluids may flow upwards until they cool or reach a stratigraphic or fluid discontinuity. Such fluids may be pumped by faulting, valved by fracturing or driven by local thermal sources, and regional head differences (e.g. Oliver 1986; Garven 1995; Jones & Xiao 2006; Davies & Smith 2006; Smith & Davies 2006; Hiemstra & Goldstein 2015; King & Goldstein 2016). This leads to gradients in porosity, mineralization, and a thermal history in oil and gas reservoirs and Mississippi Valley Type (MVT) ore deposits. These structurally controlled gradients in reservoirs and MVTs include: cementation below oil–water contacts in the Abu Dhabi and Amu Darya Basin reservoirs (Neilson *et al.* 1998); MVT ores on the flanks of large structures (Leach *et al.* 2005; Shelton *et al.* 2011); porosity enhancement in reservoirs related to fluid mixing (Salas *et al.* 2007); porosity enhancement below baffles to fluid flow in the Lisburne Group and USA Midcontinent (Jameson 1994; King & Goldstein 2016); cementation in the Outer Moray Firth from upward flow of fluids and CO_2 outgassing (Hendry & Poulsom 2006); structural control on the formation of high-temperature dolomite in the Isle of Man (Hendry *et al.* 2015); fracture control on diagenesis in Khuff reservoirs (Esrafili-Dizaji & Rahimpour-Bonab 2009; Ameen *et al.* 2010); porosity enhancement and CO_2 and H_2S increases in Zechstein reservoirs (Biehl *et al.* 2016); both porosity enhancement and reduction in Jurassic reservoirs in the western desert of Egypt (Rossi *et al.* 2002); regional fluid flow and modification of Devonian reservoirs in western Canada (Mountjoy *et al.* 1999); and modification of the downslope part of Tengiz in Kazahkstan, and Indian Basin in New Mexico (Jones & Xiao 2006; Hiemstra & Goldstein 2015).

This project evaluates the relationship between structural position (high or low) and reservoir quality for: (1) early subaerial exposure; (2) burial processes; (3) brine reflux; (4) structural rejuvenation; and (5) hydrothermal fluid flow. It examines all of these processes in one reservoir system, so the impact of earlier stages on later stages can be assessed in determining the evolution of location of reservoir sweet spots. To decrease the number of variables and broaden applicability of the system studied, this project focuses on oolitic reservoirs because of their original relative uniformity and because they are common reservoirs in the geological record (e.g. Moore 1986; Nurmi & Neuberger 1987; Ehrenberg *et al.* 2007; Shabestari *et al.* 2009; Al-Muraikhi *et al.* 2012; Rong *et al.* 2012; Adam & Abdullatif 2013). This study focuses on one such oomoldic limestone that has been interpreted to be regressive in nature, deposited during a relative fall in sea level down the flank of a Pennsylvanian topographical high, the Central Kansas Uplift (CKU). This unit is the Raytown Limestone Member of the Iola Limestone (Kansas City Group) (Fig. 1) and is colloquially known as the G-zone in the subsurface, where it is a producing unit in oil fields throughout Kansas (Watney 1980; Callewaert 1987; Byrnes *et al.* 2003; Raef *et al.* 2005).

Geological setting

Pennsylvanian cyclothemic strata in Kansas were deposited in an epeiric sea. The Raytown Limestone in the subsurface of central Kansas was deposited as part of a broad, shallow shelf interrupted by palaeotopographical highs, such as the CKU (Fig. 2a). The CKU area received Pennsylvanian sediments that were thinner than adjacent areas (Jewett 1951; Walters 1958). Extensive uplift and erosion during the Late Mississippian and Early Pennsylvanian (Carboniferous) occurred atop the CKU and vicinity (Newell *et al.* 1987). Glacioeustatic sea-level fluctuations during the Pennsylvanian resulted in cyclothems that lap onto and cover the CKU, indicating that the CKU was a positive topographical feature during Pennsylvanian deposition. The oolites in these units are interpreted to have formed in en echelon shoals in response to high-frequency sea-level oscillations across this positive feature (Watney 1980; Byrnes *et al.* 2002) responding to glacioeustacy with durations and amplitudes comparable to late Pleistocene glacioeustatic cycles (Rasbury *et al.* 1998).

Cyclothems in the Lansing and Kansas City groups mark times when the seas transgressed and regressed over the Kansas shelf, depositing cycles of sandstone, shale and limestone. According to Heckel (1977), a transgressive–regressive succession of a

Series	Stage	Group	Sub-group	Formation	Member
Upper Pennsylvanian	*Missourian*	Kansas City	Zarah	Lane Shale	Bonner Springs Shale
					Farley Limestone
					Island Creek Shale
				Wyandotte Limestone	Argentine Limestone
					Quindaro Shale
					Frisbie Limestone
				Liberty Memorial Shale	
				Iola Limestone	**Raytown Limestone**
					Muncie Creek Shale
					Paola Limestone
			Linn	Chanute Shale	Cottage Grove Sandstone
					Thayer Coal Bed
					Noxie Sandstone
				Dewey Limestone	Cement City Limestone
					Quivira Shale
				Nellie Bly Formation	
				Cherryvale Formation	Westerville (Drum) Limestone
					Wea Shale
					Block Limestone
					Fontana Shale

Fig. 1. Stratigraphic position of the Raytown Limestone Member of the Iola Limestone (cyclothem), and the overlying Argentine Limestone Member of the Wyandotte Limestone (indicated by highlighted lettering) (modified from Heckel & Watney, 2002).

typical Pennsylvanian Kansas cyclothem consists of: (1) a relatively thick, nearshore to terrestrial outside shale, and localized coal; (2) a thin, transgressive middle limestone; (3) a thin, offshore, non-sandy dark grey to black core shale; and (4) a thicker regressive upper limestone (Fig. 3) (Heckel 1977). Oolite can be found either at the base of the transgressive middle limestone or capping the regressive upper limestone of a classic Kansas cyclothem (Heckel 2002). The Raytown Limestone is the regressive 'upper' limestone of the Iola Cyclothem, forming during shoal water conditions (Heckel 2002). In the subsurface, this reservoir unit averages 6–12 m (Morgan 1952; Parkhurst 1959), and chiefly consists of oolitic grainstone and packstone facies.

Stratigraphically, regional Pennsylvanian and Early Permian cyclothems give way to an evaporite-rich section in the Permian that provided a source of high-salinity fluids. Additionally, strata in the studied area lie north of deep Pennsylvanian basins, the Arkoma and Anadarko basins (Fig. 2a), which were foreland systems of the Ouachita Fold and Thrust Belt (Ham & Wilson 1967; Houseknecht 1986). These basins provided a deep source of high-temperature fluids. The Mississippian–Early Permian Ouachita Orogeny was responsible for significant deformation. Another event of deformation in the area included reactivation during the Cretaceous–Palaeogene Laramide Orogeny.

Diagenetic setting

Previous research has shown the effect of diagenesis on reservoir quality in Pennsylvanian regressive oolites (Walton *et al.* 1995; Byrnes *et al.* 2002, 2003). Pennsylvanian regressive limestones (originally aragonite; i.e. Sandberg 1985) typically are altered early by subaerial exposure associated with relative sea-level fall, including palaeosols (e.g. Watney 1983, 1984, 1985), and dissolution and cementation (Heckel 1983; Prather 1984; Watney 1984, 1985; Phares 1991; Lebeau 1997). As duration of subaerial exposure increases on palaeotopographical highs, such as the crest of the CKU,

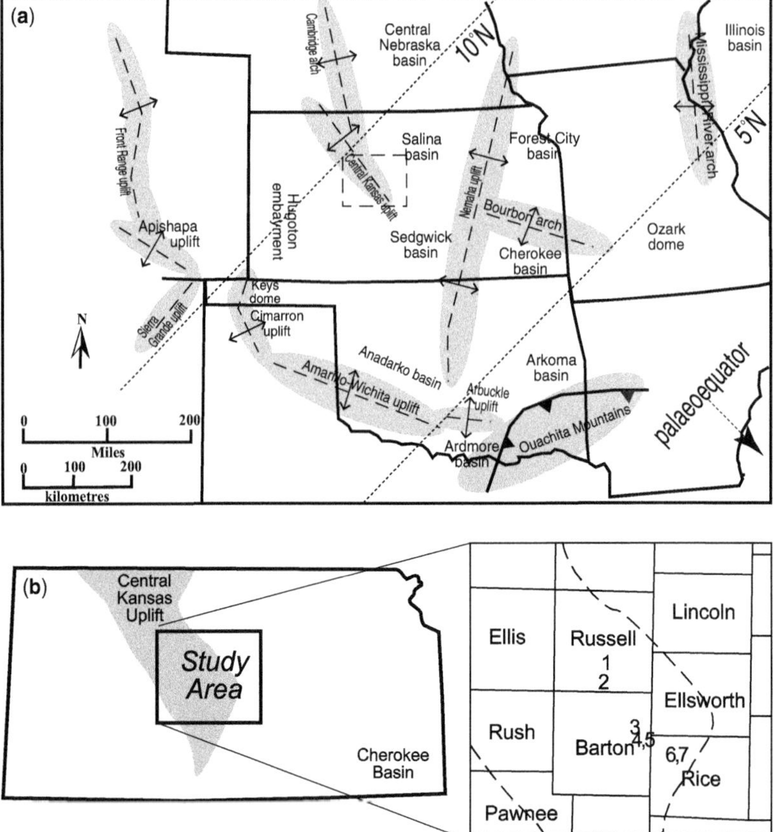

Fig. 2. (**a**) Interpreted paleostructure and paleolatitude map of the Midcontinent USA for the Pennsylvanian Period (modified from Wanless 1969; Heckel 1977; Johnson 2004). Dashed line box shows location of study area; (**b**) Location of cores used for this study and their relation to the Central Kansas Uplift (CKU), indicated by dashed outline on county map. From north to south – Well 1: Carter-Colliver CO2 Injection well, Hall Gurney Field, central Russell County. Well 2: Michealis #1, Trapp Field, south central Russell County. Well 3: Drews #A-1, Ames Northwest Field, eastern Barton County. Well 4 and 5: H. Rader #1 and Rader #2, Rick Field, eastern Barton County. Well 6 and 7: Hafferman #1 and E.E. Tobias #1, Chase-Silica Field, northwestern Rice County. These two wells indicated by the southern-most locations are on the margin of the CKU, and thus hypothesized to be palaeotopographically lower than the other wells in the study area.

it is hypothesized that diagenetic impact would be greatest in updip positions rather than in downdip positions (e.g Buijs & Goldstein 2012; Li *et al.* 2014). This variability in early diagenesis would likely have an impact on burial processes such as compaction.

Multiple projects have demonstrated regional Permian events of low-temperature brine reflux that had a diagenetic impact on underlying Pennsylvanian carbonates (Wojcik *et al.* 1994; Luczaj & Goldstein 2000). As the Pennsylvanian CKU had an impact on the stratigraphy in the overlying Pennsylvanian and Permian sections, it would be expected that this impact might vary from palaeohigh to flank of the CKU.

Further studies in the Midcontinent USA have demonstrated the impact of late-stage diagenetic processes on pore space evolution (e.g. Phares 1991; LeBeau 1997). In particular, regional and local flow of hydrothermal fluids has deposited MVT ores, enhanced porosity and thermal maturity, and precipitated baroque dolomite, calcite and quartz (see King & Goldstein 2016).

Methods

Seven cores from central Kansas were used for this study. They were taken from the Rick and Ames NW fields of eastern Barton County, the Chase-Silica

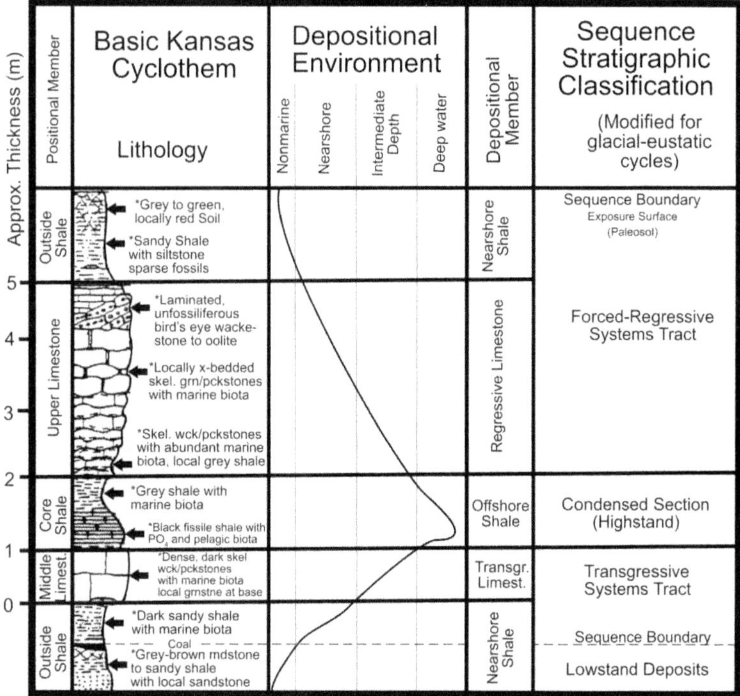

Fig. 3. Diagram illustrating lithologies, fossil content, sea-level position, depositional environments, depositional member nomenclature and sequence stratigraphic classification of a typical Kansas cyclothem (modified from Heckel 2002). The Raytown Limestone is an 'Upper Limestone' in the Iola Cyclothem, recording shoal water conditions.

Field of NW Rice County, and the Hall-Gurney and Trapp fields of southern Russell County (Fig. 2b). Cores were chosen so that various topographical levels along the flank of the CKU were represented to evaluate palaeotopographical controls on diagenesis. The cores chosen were also picked because they already had porosity and permeability data for the Raytown Limestone's oolitic grainstone facies, which were measured from 1.0 × 2.0 inch core plugs (2.54 × 7.62 cm) (Byrnes *et al.* 2002, 2003). The porosity and percentage of each major diagenetic phase were also determined in thin section using the visual comparison chart of Terry & Chilingar (1955). Given the scope of this study and the complexity of diagenetic alteration, this approach was deemed sufficient to show major shifts in porosity through time. Tests of the visual estimate technique show no bias towards area percent or average standard deviation. With accuracy estimated at 10% (Dennison & Shea 1966), the approach is deemed sufficient for evaluating major shifts in location of the best porosity through time, and the controlling mechanisms for those shifts. Improved quantification is viewed as a next step, incorporating image analysis and a larger number of thin sections.

Ninety thin sections were prepared for transmitted light and cathodoluminescence petrography. Carbonate cement types were characterized in thin section by petrographical examination in transmitted light and cathodoluminescence. Thin sections were dual-stained with Alizarin red-S and potassium ferricyanide following Lindholm & Finkelman (1972). Cathodoluminescence petrography (Meyers 1991) was performed using a Cambridge Image Technology Ltd Clmk4 system mounted on a Leitz SM-LUX-POL microscope. Operating conditions for luminescence were a 14 kV acceleration potential and an approximately 0.5 mA gun current.

Ten thick sections were prepared for fluid-inclusion work following the techniques of Goldstein & Reynolds (1994). Microthermometric analyses were performed using a Linkam THMSG 600 stage. After microthermometric analysis, cathodoluminescence and transmitted-light photomicrographs were overlain in Adobe Photoshop to determine the luminescence character of cement associated with each measured fluid inclusion.

Table 1. *Detailed summary data of facies observed*

Lithofacies	Grain composition	Sedimentary structures	Texture	EOD interpretation*
Oolitic grainstone**	Ooids most common; bioclasts common but not abundant (crinoids, brachiopods, bryozoans); peloids and pellets present but sparse	Diverse array of sedimentary structures (see Table 2)	Oolitic to oomoldic; well sorted, grain supported (0.3–0.9 mm diameter); bioclasts are abraded (locally moldic, 0.5–3.0 mm length); prevalent stylolitization	Linked assemblage of marine environments, from low-energy shallow subtidal to intertidal to beach to supratidal; 0–10 m water depth (see Table 2 for a detailed interpretation)
Fossiliferous oolitic packstone	Ooids most common; bioclasts very common (more diverse than grainstone facies, crinoids, brachiopods, bivalves, bryozoans, gastropods and foraminifera); peloids present	Lack of physical sedimentary structures due to pervasive bioturbation; rarely, but notable, individual burrows are discernable	Oolitic to oomoldic; less sorted; grain (0.25–0.7 mm) support with micritic matrix; bioclasts (1–10 mm long; minimal abrasion; locally moldic); prevalent stylolitization	Facies is always associated with the grainstone facies, occurring at the base of depositional cycles. Deeper, more seawards (relative to the grainstone facies) subtidal marine environment; 4–25 m water depth
Clay-rich shale	Diverse assemblage of fauna; bioclasts (foraminifers including fusulinids, crinoids, gastropods, brachiopods and bryozoans); no non-biological coarse grains discernable	Dominated by fine laminations	Clay-rich shale with carbonate lenses; lenses contain majority of bioclasts; prevalent stylolitization	Facies is found atop the oolitic facies. Very thin; poorly developed. Deep, normal marine, normal salinity and oxygenation, low-energy environment, 50–100 m depth
Wackestone–packstone	Bioclasts are diverse and common (benthic foraminifera, crinoids, brachiopods, bryozoans, gastropods, bivalves, and calcareous and siliceous sponges and sponge spicules); minor ooids; sparse peloids or pellets	Mottled fabrics and chaotically orientated bioclasts are common, and are interpreted to result from pervasive bioturbation; physical sedimentary structures are not apparent	Bioclasts in grain support in a micrite matrix, with patches of bioclasts supported by micrite; bioclasts abraded, except forams; prevalent stylolitization, with solution-enhanced fractures and vugs	Facies is found atop the shale facies. Below fair weather wave base, epeiric sea environment, 10–50 m water depth

*EOD, environments of deposition.
**Subfacies of oolitic grainstone are presented in Table 2.

Table 2. *Detailed summary data of subfacies of oolitic grainstone observed*

Subfacies	Structures and clasts	EOD interpretation*
Mottled oolitic grainstone	Prevalent interpreted bioturbation and more bioclasts than other subfacies; bioclasts are, on average, less abraded than bioclasts in other subfacies; high diversity and abundance of typical normal marine fauna, including crinoids and bryozoans	Low-energy, shallow to somewhat deeper water marine (1–10 m water depths) environment
High-angle cross-stratification oolitic grainstone	High-angle, tabular cross-beds with abraded bioclasts and sparse peloids; more sparse bioclasts associated with this subfacies	Most common subfacies seen in core (*c.* 40%); shallow subtidal to intertidal setting, in a depth of less than about 5 m
Low-angle cross-stratification and planar oolitic grainstone	Low-angle, tabular cross-beds to planar, parallel beds with abraded bioclasts and sparse peloids; moderate abundance of bioclasts	Shallow subtidal to beach setting, in a water depth of approximately 2 m or less
Fenestral oolitic grainstone	Fenestrae, vugs and rhizomolds within the low-angle cross-stratification subfacies	Intertidal to supratidal settings and experienced subaerial exposure; found in the top 0.5 m of Raytown Limestone

*EOD, environments of deposition.

Isotopic samples were drilled from polished billets. Despite careful microsampling, some of the carbonate cement samples were unavoidably contaminated with some of the adjacent cements. Isotopic analysis, as well as most sample preparation, was conducted at the University of Kansas's Keck Paleoenvironmental and Environmental Stable Isotope Laboratory. Samples were analysed using a Finnigan-MAT 253 IRMS with a Kiel III automated carbonate reaction device. Carbonate reference standards NBS-18 and NBS-19 were used during analyses. All results are reported in per mil relative to V-PDB. Samples with a standard deviation of greater than 0.07‰ were not used in analyses. Average standard deviation for $\delta^{18}O$ samples is 0.03‰, and average standard deviation for $\delta^{13}C$ samples is 0.05‰.

Lithofacies and depositional environment

Four lithofacies and five subfacies were defined for this study. The facies are defined on the basis of grain composition, texture and sedimentary structures. Subfacies classification relied solely on sedimentary structures. Detailed facies and subfacies descriptions are presented in Tables 1 & 2 and conceptualized in Figure 4. A more detailed examination of the porosity and its evolution are in the 'Diagenesis' and 'Porosity evolution v. structural and palaeotopographical position' sections. The following summarizes interpreted depositional environments for facies. Details are discussed in Poteet (2007).

The oolitic grainstone facies (Figs 5–7), comprising several subfacies, makes up the bulk of the rock studied. Depositional environments range from moderate-energy, shallow- to moderate- water depth (1–10 m) marine settings, to higher-energy, shallow subtidal to intertidal settings (<5 m), to intertidal to supratidal settings with subaerial exposure. Fenestrae and rhizomolds occur in the top 0.5 m of the Raytown Limestone, within oolitic grainstone facies. Settings are interpreted as shoal to beach marine systems and subaerial exposure. The fossiliferous oolitic packstone facies (Figs 5 & 6) is interpreted to have been deposited seawards of the oolitic grainstone facies, in a lower-energy, normal marine environment (4–25 m water depth). It is found adjacent to and interfingering with the oolitic grainstone facies. The clay-rich shale facies (Fig. 7) overlies the oolitic grainstone and fossiliferous oolitic packstone facies. Characteristics (Table 1) suggest it was deposited in a fully oxygenated, normal-salinity, low-energy marine environment (50–100 m water depth). The fossiliferous wackestone–packstone facies (Figs 7 & 8) overlies the clay-rich shale facies and is interpreted to have been deposited in a subtidal, open-marine environment below fair-weather wave base (10–50 m water depth).

Stratigraphy

Data and interpretations discussed in this stratigraphy section come from the Carter–Colliver CO_2 Injection core description, and plug data and thin section descriptions from all cores (Fig. 4). The

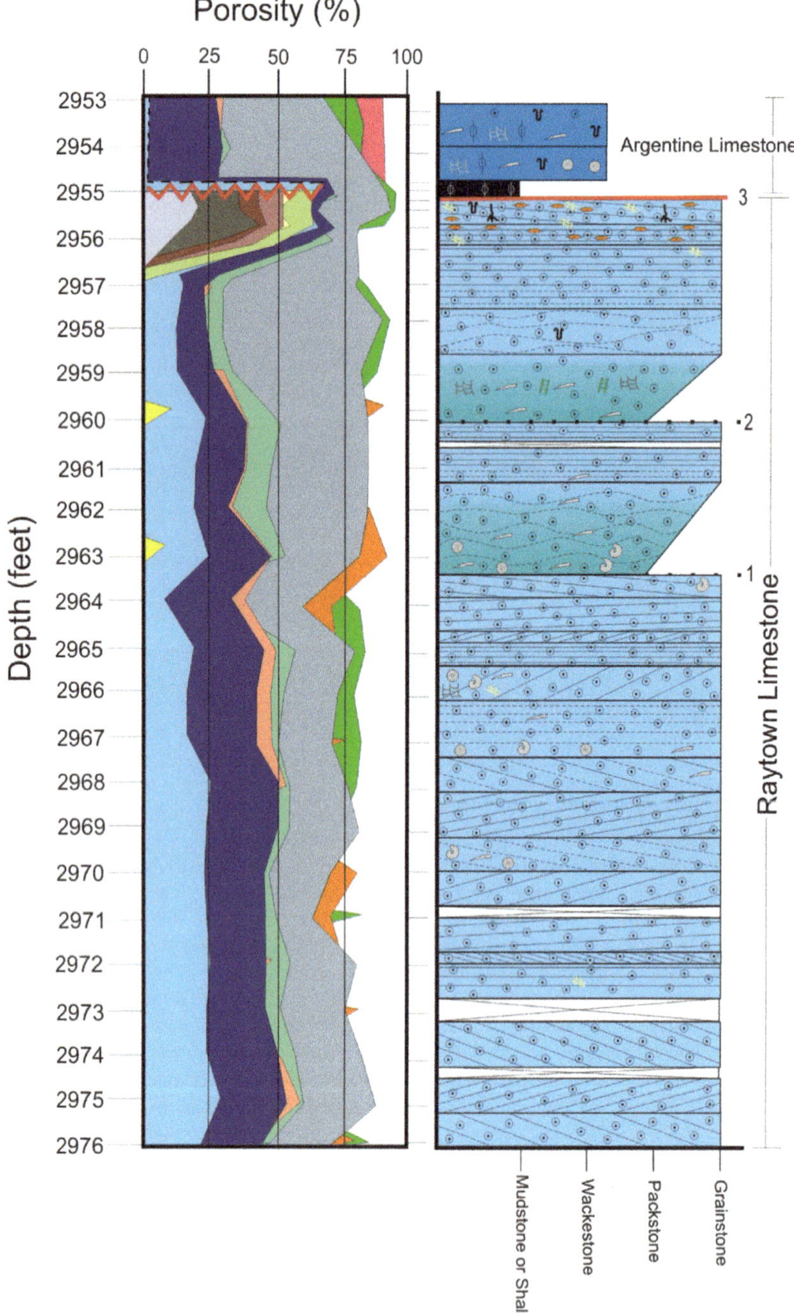

Fig. 4. (Left) Vertical porosity graph depicting porosity occlusion by each cement for a given depth. Extant porosity data were measured from core plugs from the Carter–Colliver CO_2 Injection core, and porosity occlusion per each cement was estimated from thin sections taken from the same core. The left vertical axis is depth (in ft) and the horizontal axis is porosity reduction (in %). The grey lines on the right vertical axis of the porosity graph indicate locations where porosity data were collected in the core. Porosity values between these data locations are interpreted. (Right) Stratigraphic column of the Carter–Colliver CO_2 Injection core. Cycle numbers are indicated on the right, next to the top of the respective cycle boundary. A key of the symbols and colours for the vertical porosity graph and stratigraphic column is provided.

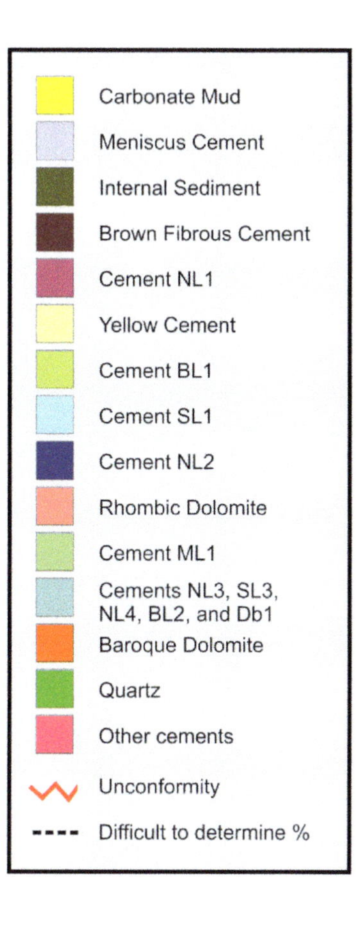

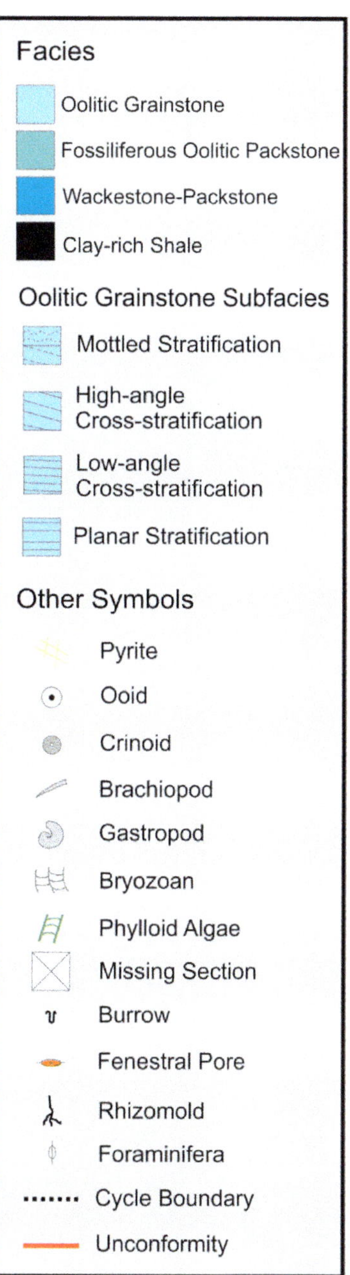

Fig. 4. *Continued.*

Raytown Limestone is capped by clay-rich shale, which in turn is overlain by fossiliferous packstone and wackestone. Facies above the Raytown Limestone of the Iola Cyclothem are associated with the overlying Wyandotte Cyclothem (Fig. 1).

Three cycles (1–>3.5 m, or 3.5–>12 ft, in thickness) were delineated in the Raytown Limestone for this study (Fig. 4). The lowermost, cycle 1, is not complete in the Carter–Colliver core, and is completely oolitic grainstone. Cycles 2 and 3 are interpreted as shoaling-upwards packages of thin oolitic packstone with abundant bioclasts at the base, grading upwards into oolitic grainstone. Upward trends within cycles 2 and 3 include decreasing

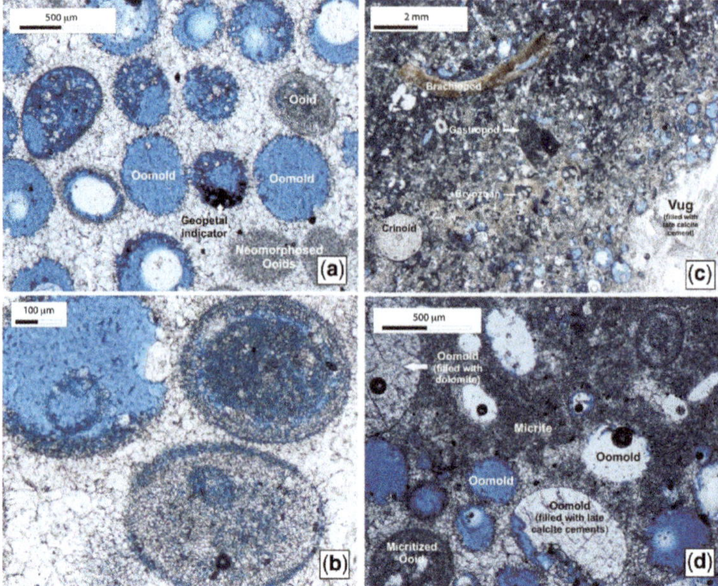

Fig. 5. Transmitted light photomicrographs showing typical oolitic grainstone (left) and fossiliferous oolitic packstone (right). (**a**) Oomolds and neomorphosed ooids, some with geopetal indicators. Material surrounding grains are later calcite cements. (**b**) Three neomorphosed ooids with varying degrees of dissolution. (**c**) Near the base of the Raytown Limestone, with large bioclasts (labelled) and smaller bioclasts. (**d**) Close-up of oolitic packstone with cement-reduced oomolds.

mud matrix and bioturbation, increasing ooid size, and increasing dominance of cross-bedding. In the grainstone facies within upper parts of these cycles, bedding style transitions upwards, from high-angle cross-stratified beds to low-angle cross-stratified beds or planar stratified beds. From these observations, each cycle within the Raytown Limestone is interpreted to record some shallowing. The features indicative of subaerial exposure in the uppermost centimetres of the Raytown Limestone (Fig. 7) are superimposed on intertidal and subtidal facies of this cycle. The subaerial exposure of subtidal facies necessitates a relative fall in sea level at the top of the Raytown Limestone (Enos & Perkins 1979; Goldhammer *et al.* 1987; Goldstein *et al.* 1991). A relative sea-level fall at the top of the Raytown

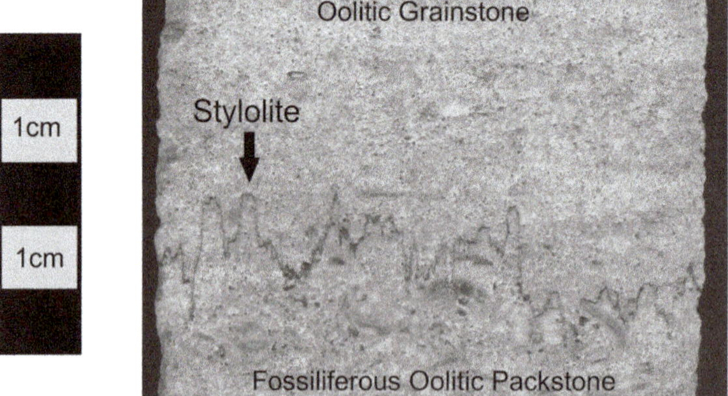

Fig. 6. Oolitic packstone below the stylolite, and oolitic grainstone facies above. Oolitic packstone differs from oolitic grainstone in two key factors: the presence of micrite within interparticle pore space; and, on average, larger bioclasts. Several large bioclasts can be seen directly below and truncated by the stylolite.

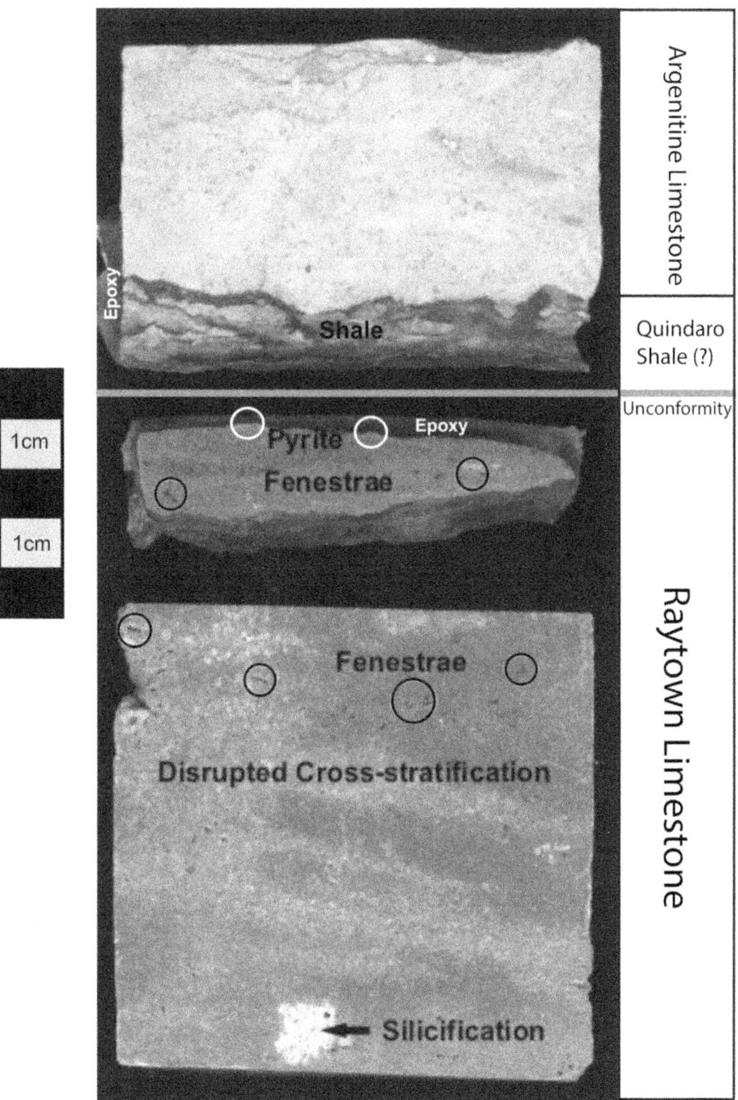

Fig. 7. Below the grey line is cross-stratification and fenestral porosity at the top of the Raytown Limestone. Black circles indicate fenestral pores, and white circles indicate pyrite. A patch of silicified limestone is marked with a black arrow. Above the grey line is the Argentine Limestone and thin, clay-rich shale with carbonate lenses. Here, the Argentine Limestone is a fully bioturbated packstone. Surrounding the core is clear epoxy, used to secure fragile pieces of rock during thin section preparation.

Limestone is also supported by the sea-level curve of Heckel (1989), produced from his investigations of Pennsylvanian cyclothems in the Midcontinent.

The clay-rich shale unit with carbonate lenses and marine fossils (50–100 m water depth) represents transgression and maximum marine flooding on top of the Raytown Limestone. It is interpreted as the deeper-water 'core shale', or Quindaro Shale, of the Wyandotte Cyclothem. The fossiliferous wackestone–packstone directly above the shale (10–50 m water depth) indicates deposition during a regression, and is thus interpreted to be the Argentine Limestone (see Heckel 2002 for outcrop equivalents).

These observations compare favourably to those of Leonard (2006), who studied outcrops of the Iola Limestone in Kansas and Iowa, and found as many as two smaller-scale parasequences in the Raytown

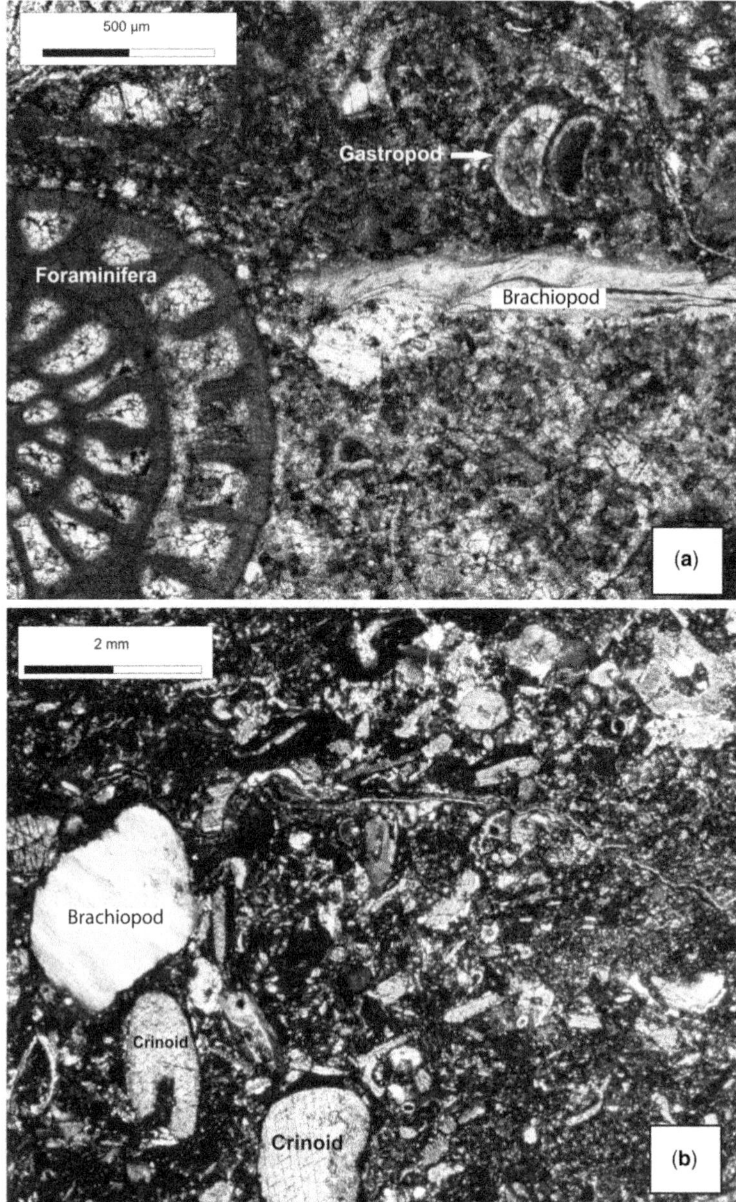

Fig. 8. Photomicrographs of typical fossiliferous wackestone–packstone facies in the overlying Argentine Limestone. (**a**) Packstone with a fusulinid (with intraparticle porosity occluded with late diagenetic cements) and gastropod adjacent to a brachiopod fragment and other smaller indistinguishable bioclasts in a carbonate mud matrix. (**b**) Packstone with recognizable brachiopod and crinoid fragments, and other unidentifiable bioclasts in a carbonate mud matrix.

Limestone. No evidence of subaerial exposure was noted, and this may be because outcrops of the Iola Cyclothem are in eastern and SE Kansas, away from the palaeotopographical high of the CKU (Wright & Leonard 2007).

Diagenesis

The paragenesis of the Raytown Limestone in the study area consists of 29 separate events. The relative timing of each diagenetic event is summarized

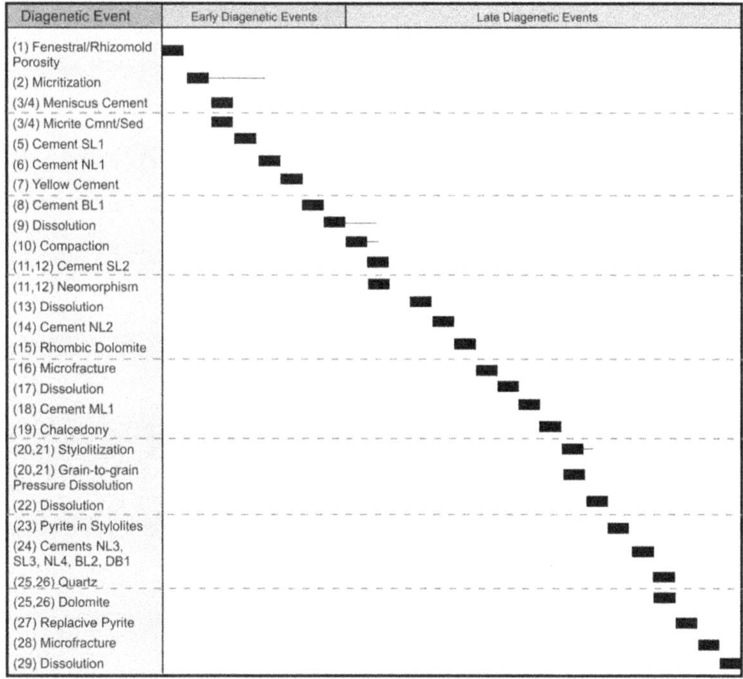

Fig. 9. The paragenetic sequence for the Raytown Limestone. Early diagenetic events occurred prior to the deposition of the overlying Argentine Limestone. All cements forming during the early stages are found only in the upper 0.5 m of the Raytown Limestone. Late-stage diagenetic events occur after the deposition of the overlying unit, and are found throughout the entire Raytown Limestone, as well as the Argentine Limestone. The length of the boxes is not considered to be a quantitative representation of absolute duration. All conclusions made for the paragenetic sequence are based on microscopic observations incorporating cross-cutting and superpositional relationships.

in Figure 9. The paragenesis is subdivided into early and late stages. Early-stage events (1–8) are present only in the upper 0.5 m of the Raytown Limestone and are absent in the overlying Argentine Limestone, indicating that they occurred prior to deposition of that unit (Figs 9 & 10). Petrographical characteristics of early-stage cements are summarized in Table 3. Late-stage events (9–29) occurred after the deposition of the overlying limestone, as they are found in both the Raytown Limestone and Argentine Limestone (Figs 9 &11). Petrographical characteristics of late-stage cements are summarized in Table 4. Some events are listed jointly with other events (e.g. precipitation of quartz is either Event 25 or 26 and precipitation of dolomite is either Event 25 or 26). This was done because no petrographical evidence was found to order them relative to one another. The mineralogy of each cement in the paragenesis is calcite, unless otherwise noted in the description.

What follows is an abbreviated description of the features associated with each event in the paragenesis, along with estimates of its impact on minus-cement porosity. Further detailed descriptions of each diagenetic event are found in Poteet (2007), including description of each cement under cathodoluminesence. Minus-cement porosity includes the percent volume of void space in the rock, added to the volume of cements and other post-depositional pore-filling material. As events 1–8 are found only in the upper 0.5 m of the Raytown Limestone, percentage porosity is given for both the upper 0.5 m, as well as for the Raytown Limestone as a whole. For events 9–29, percentage porosity is given for the Raytown Limestone as a whole. Data and calculations are available at http://hdl.handle.net/1808/21744.

Early-stage events

Event 1: Fenestral and rhizomoldic pores. After deposition of grains (ooids, bioclasts), the first diagenetic event is the creation of fenestral and rhizomoldic porosity in the top 0.5 m of the Raytown Limestone (Fig. 12). Fenestrae manifest themselves as horizontal, multigranular-roofed, millimetre- to

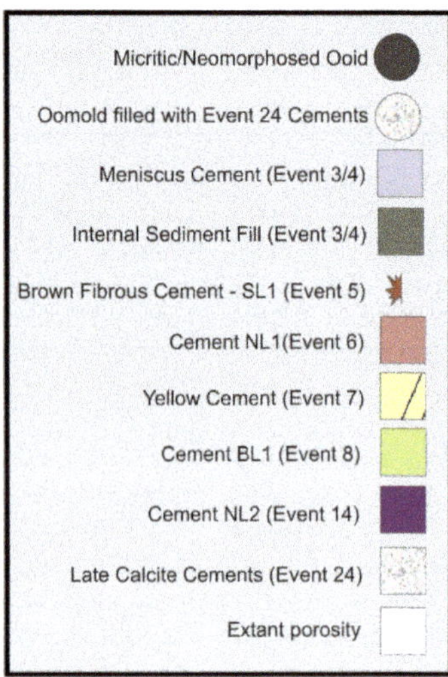

Fig. 10. Cartoon illustrating the typical relationships between cementation events within the upper 0.5 m of the Raytown Limestone. Light purple denotes meniscus cements that asymmetrically coat micritic ooids. Some ooids dissolve later and become filled with Event 24 late calcite cements. Sediment covers the meniscus cements and, in places, directly covers ooids. This is the reasoning for listing the meniscus cement and sediment penecontemporaneously within the paragenetic sequence. The next cement in the paragenesis is the brown cement, SL1, which occurs as fibrous or globular bundles of calcite. The pink cement denotes cement NL1, which in-fills around the crystal bundles of cement SL1. The next cement precipitated is the very localized yellow cement (YC), seen here in yellow with banding. The yellow cement is only seen in one thin section, in several fenestral pores and in adjacent interparticle pores. The last early cementation event is BL1, shown in light green. These cements cover all the previous cements, from the internal sediment fill, to NL1, to the yellow cement, where applicable. Cement NL2, a later cement in the paragenesis, is shown here occluding interparticle porosity. Finally, the remaining pore space, as well as the later-created oomoldic pore space, is filled with Event 24 late calcite cements.

centimetre-sized, cement-reduced and solution-enlarged pores. They are lined with early cements (events 3–8) and subsequently filled with later calcite cements (Event 24). Root molds are common, and occur as semi-cylindrical, cement-filled molds. Many rhizomolds contain alveolar textures (Fig. 12c, d) (Esteban 1974; Esteban & Klappa 1983). The rhizomolds are primarily filled with Event 24

Table 3. *Summary of the cements of the early stage of paragenesis (events 1–8)*

Name of cement (Event No.)	Appearance in transmitted light	Appearance in cathodoluminescence	Size	Porosity type (moldic, etc.)	Porosity (top 0.5 m %/total Raytown Limestone %)	Evidence for timing
Meniscus cements (MC) (Event 3–4)	Colour: clear • Asymmetrical coating of grains • Bladed cement morphology	Non-luminescent to slightly luminescent	60–90 μm	Fenestral and interparticle porosity	10–60%/1.5%	First cement to coat grains penecontemporaneous with sediment
Internal sediment (Event 3–4)	Colour: dark brown to black • Distinctive three layers • Thickest in fenestrae	Moderately to brightly luminescent	10–130 μm	Fenestral and interparticle porosity	15–70%/1.5%	Locally coats grains before meniscus cement
SL1 (Event 5)	Colour: brown • Fibrous or equidimensional patches • Asymmetrically coats grains	Slightly luminescent	25–150 μm	Fenestral and interparticle porosity	5–15%/0.7%	Crystal clusters precipitated on meniscus cement and sediment
NL1 (Event 6)	Colour: clear to slightly cloudy • Associated with cement SL1 • Equant crystal morphology	Non-luminescent	30–200 μm	Fenestral and interparticle porosity	5–15%/0.7%	Coats cement SL1, but not YC
Yellow cement (YC) (Event 7)	Colour: yellow to light brown • Isopachous • Fibrous • Very localized	Non- to moderately luminescent	150 μm	Fenestral and interparticle porosity	2–3%/0.1%	Coats cement NL1 when present
BL1 (Event 8)	Colour: clear • Equant crystal morphology	Brightly luminescent	10–20 μm	Moldic, rhizomoldic and interparticle porosity	5%/0.2%	Coats cement NL1 primarily; and YC, where found

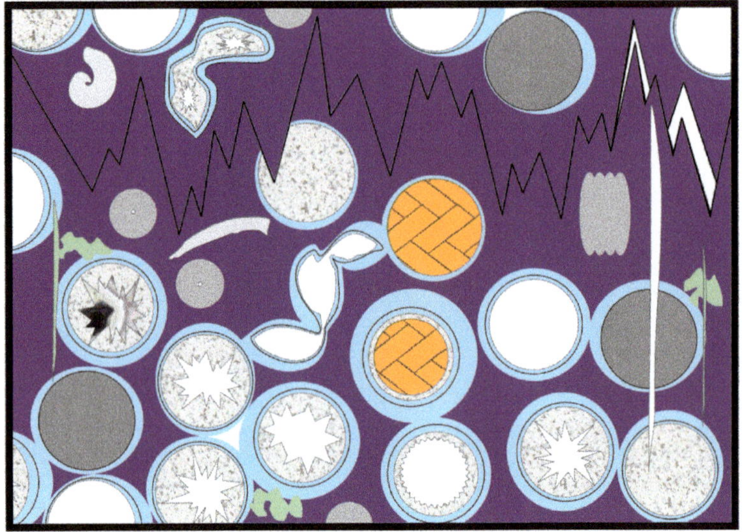

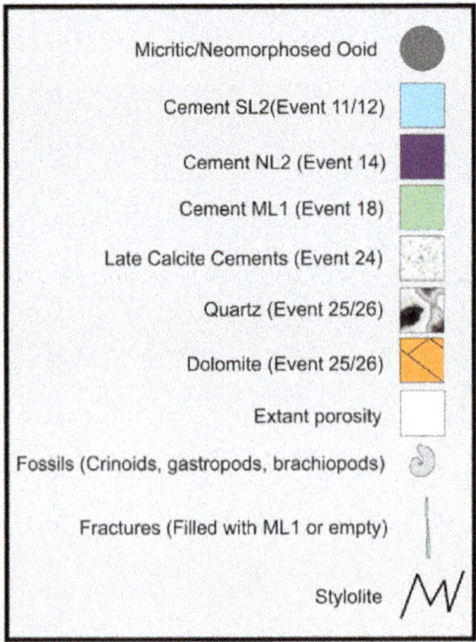

Fig. 11. Cartoon illustrating the paragenesis in the majority of the Raytown Limestone. Compaction creates spastoliths. Cement SL2 (light blue) isopachously to asymmetrically coats the inside and outside of oomolds and spastoliths, as well as the outside only of micritic ooids. Cement NL2 (purple) fills in the majority of interparticle pore space, conforming to the morphology of the newly created spastoliths, where found. Cement ML1 (green) replaces cements SL2 and NL2 in localized patches, and fills in fractures that cross-cut cements SL2 and NL2. Stylolites (Event 20–21) cross-cut ooids, creating oomoldic porosity, and stylolites experience solution enhancement. Event 24 cements, SL3, NL3, SL4, BL2 and DB1 (white and grey stippled pattern), partially or completely occlude oomoldic porosity. Quartz (white to black marbled pattern) and baroque dolomite (orange brick pattern) occlude a small percentage of oomolds after precipitation of Event 24 cements. Solution-enlarged fractures with no cements cross-cut all cements. Not shown is the rhombic dolomite because it is not common.

calcite cements. Earlier cements may not have precipitated within the rhizomolds because the molds may have been occupied by roots for some period of time.

Event 2: Micritization. Micritic envelopes around fossil and ooid grains are observed in almost every thin section in the study. Burrow fill is also micritized.

Event 3 or 4: Meniscus cementation (MC). Meniscus cements coat grains asymmetrically (Fig. 12b, c). The precipitation of meniscus cements is penecontemporaneous with internal sediment. Some pores have meniscus cement coatings before internal sedimentation, whereas other pores have internal sedimentation before meniscus cementation. The cement reduced the primary interparticle porosity of the upper 0.5 m by 20–60% and fenestral porosity, on average, by less than 10%. However, it reduced the porosity of the Raytown Limestone reservoir as a whole by only 2%.

Event 3 or 4: Micritic cement and sedimentation. This feature has characteristics of both a sediment and a cement. It coats pores primarily, but, locally, acts as a geopetal fill (Figs 5a & 12b, c). Micrite coats meniscus cement dominantly, but in several locations it is the first coating of grains and pores, and the meniscus cement coats sediment. The sediment coating is thus labelled coincident with the meniscus cement. It is composed primarily of depositional micrite. Within the upper 1 m, the sediment reduced 15–30% of the interparticle porosity and 30–70% (average of 40%) of the fenestral porosity. The sediment's reduction of the porosity of the whole reservoir is only 2%.

Event 5: Brown fibrous cementation (SL1). This brown-hued cement lines meniscus cement and internal sediment (events 2 or 3) (Fig. 12c, d). It is irregularly distributed in fenestral and interparticle pores, and unevenly coats grains. This geometry forms an asymmetrical crust that does not respond to gravity. Cement morphology is fibrous to equidimensional patches. In the top 0.5 m of the Raytown Limestone, where cement SL1 occurs in cement-reduced interparticle pores, it reduces approximately 5% of that pore space. It reduces 5–15% of cement-reduced fenestral pores and approximately 1% of the total porosity of the Raytown Limestone.

Event 6: Precipitation of NL1. This cement occurs mostly in fenestral pores and rarely in interparticle pores, filling in around and coating cement SL1 (Fig. 12c). NL1 occludes only 5–15% of cement-reduced fenestral porosity, and 5% of cement-reduced interparticle porosity. This cement reduces total Raytown Limestone porosity by 1%.

Event 7: Precipitation of yellow cement (YC). Cement YC is rare, and is only seen in one thin section. It occurs as isopachous bands of fibrous cement, lining pores, and overlying cements SL1 and NL1 (Fig. 12c). The cement reduces 20–80% of the fenestral and interparticle pores in which it occurs, with 30% as the average; however, because it only occurs in a small number of pores, the actual percentage of all cement-reduced fenestral and interparticle porosity reduced in the thin section is only 2–3%. Cement YC has a negligible impact on the total porosity (less than 0.5%).

Event 8: Precipitation of BL1. BL1 is clear and equant in transmitted light; in cathodoluminescence, it consists of a series of very thin cement zones of varying degrees of bright luminescence (BL1), found on cements MC, SL1, NL1 and YC within fenestral, rhizomoldic and interparticle pore space. The most recognizable of these cements is cement BL1. BL1 and its associated cements reduce total porosity (fenestral + rhizomoldic + interparticle) of the upper 0.5 m by 5%, and reduce total Raytown Limestone porosity by 1%.

Late-stage events

Event 9: Initiation of ooid and bioclast dissolution. The timing for initial ooid and bioclast dissolution is constrained by which cements are found in oomolds or biomolds. Cements MC, SL1, NL1, YC and BL1 are not found in moldic porosity, and all cements that follow in the paragenesis are found in moldic porosity. Ooid dissolution, however, appears to have been a long-lived process, and, perhaps, occurred episodically as discrete events throughout late parts of the paragenesis (see events 13, 17 and 28).

Event 10: Initiation of compaction. The timing of initial compaction is constrained by the presence of later cements coating spastoliths. A spastolith, as used in this study, is defined as an ooid or coated grain that has been partially dissolved and crushed, generally resulting in shearing of the concentric laminations and the separation of cortical layers away from their nuclei (Tucker 1988; Scholle & Ulmer-Scholle 2003). SL2 is the initial cement internally and externally coating crushed fragments in spastoliths (Event 11–12) (Fig. 13c). The moldic dissolution needed to allow crushing (Event 9) and coating of cement SL2 (Event 11–12) constrains timing. Later cements, including cement NL2 of Event 14, fill interparticle space that was deformed by the creation of spastoliths, which constrains timing further.

Event 11 or 12: Precipitation of SL2. Cement SL2 is the first cement to partially fill interparticle pores within the bulk of the Raytown Limestone below the upper 0.5 m. It is also the first cement in the paragenesis to be present in the overlying Argentine Limestone. SL2 is the first cement precipitated on the outside and inside of spastoliths and non-deformed oomolds (Fig. 13c, d), which constrains its timing as post-inception of compaction. In the two most easterly wells, Hafferman #1 and E.E.

Table 4. *Summary of the cements of the late stage of paragenesis (events 9–29)*

Name of cement (Event No.)	Appearance in transmitted light	Appearance in cathodoluminescence	Size	Porosity types (moldic, etc.)	Porosity % total Raytown	Evidence for timing
SL2 (Event 11–12)	Colour: clear to cloudy • Asymmetrical coating of grains • Equant to bladed crystal morphology	Slightly luminescent	20–70 μm	Moldic and interparticle porosity	10–15%	Locally, coats BL1 (Event 8). First cement to coat oomolds; coats spastoliths, therefore precipitation occurs after compaction
NL2 (Event 14)	Colour: clear to slightly cloudy • Most widespread cement • Medium equant crystals	Non-luminescent with brightly luminescent hairline fringes	75–125 μm	Moldic and interparticle porosity	25–35%	Cement fills in interparticle and moldic pore space after cement SL2
Rhombic dolomite (Event 15)	Colour: clear • Fine grained • Unimodal planar rhomboids	Non-luminescent	25–80 μm	Interparticle porosity	1%	Fills in extant pore space left over after precipitation of NL2. Occurs before ML1, since ML1 recrystallizes whole patches of interparticle space filled with NL2 and dolomite
ML1 (Event 18)	Colour: cloudy • Recrystallization of earlier cements	Moderately luminescent with slightly luminescent bands on either side	Undeterminable	Interparticle porosity and fracture fill	3%	Fills fractures that cross-cut earlier cements, recrystallizes earlier cements

Late calcite cements: NL3, SL3, NL4, BL2 and DB1 (Event 24)	Colour: clear • Fine- to medium-grained, equant crystals	NL3: non-luminescent SL3: slightly luminescent NL4: non-luminescent BL2: brightly luminescent DB1: dull banded – sector zoned	30–100 μm average	Interparticle and moldic porosity; solution-enlarged fracture and stylolite-related porosity; fenestral and rhizomoldic	15–30%	Fills fractures that cross-cut previous cements (SL2, NL2 and ML1); precipitates in oomolds on cement SL2; fills interparticle porosity after NL2
Quartz (Event 25–26)	Colour: clear • Fine to coarse, equant crystals • Both micro- and megaquartz present	Non-luminescent	10–250 μm	Interparticle and moldic porosity; solution-enlarged stylolite-related porosity	3–5%	Fills oomolds internally, after precipitation of Event 24 cements
Baroque dolomite (25–26)	Colour: cloudy • Unimodal, non-planar • Highly fractured	Non-luminescent	Variable; relative to oomold diameter	Moldic porosity; solution-enlarged fracture- and stylolite-related porosity	3–5%	Coincident with quartz; fills fractures, stylolites

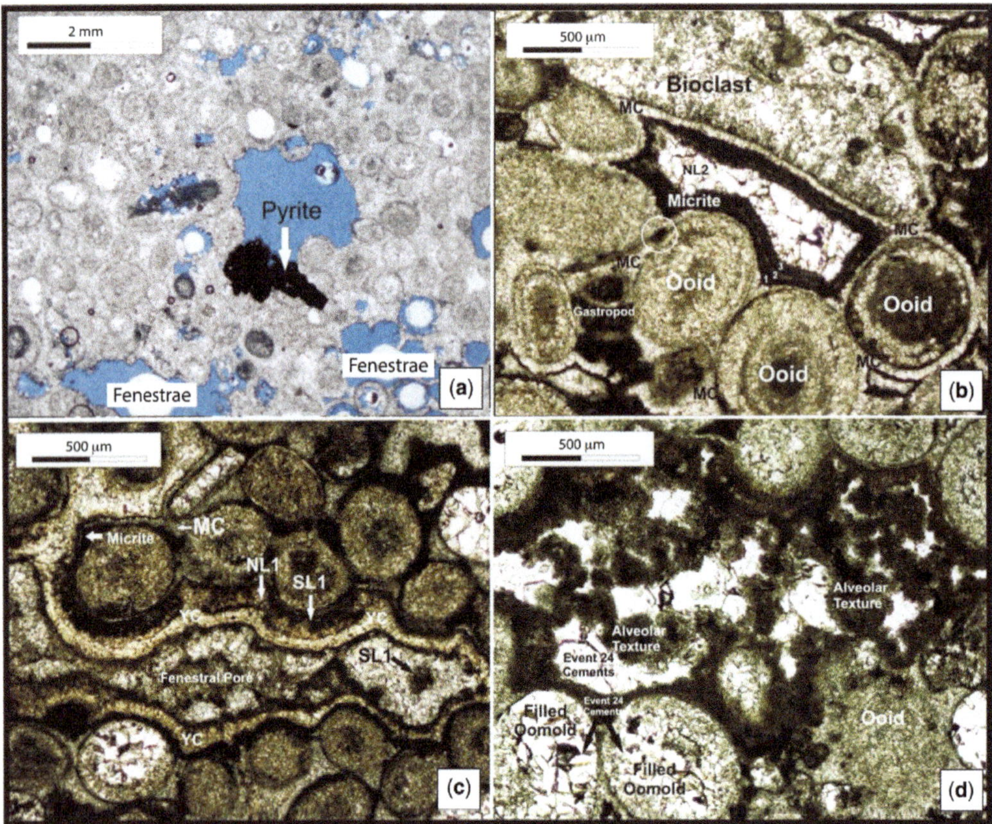

Fig. 12. Photomicrographs of early diagenetic events. (**a**) Euhedral pyrite (Event 27) in the upper 0.5 m of the Raytown Limestone, next to a solution-enlarged fenestral pore (Event 1), and replacing grains and cements. (**b**) Pore with three layers of internal sediment. Internal sediment and meniscus cements (MC) occur penecontemporaneously. Evidence for this timing is shown in the white circle, where sediment lines grains before the meniscus cements. In the rest of the field of view, the meniscus cements line grains first. A micritized gastropod in the bottom left is filled with internal sediment. (**c**) Fenestral and interparticle porosity with meniscus cements, and internal sediment coating grains penecontemporaneously. SL1 (brown fibrous crystal bundles), NL1 (dark orange and brown cement overlying SL1) and the yellow calcite cement (YC) (yellow isopachous cement overlying NL1), also found in the fenestral pore and adjacent interparticle porosity, overlie MC and sediment, thereby making cements SL1, NL1 and YC younger than MC and the sediment. (**d**) Alveolar texture filling rhizomoldic porosity. Late calcite cements (Event 24) fill oomolds and rhizomolds.

Tobias #1, cement SL2 manifests itself as thick, equigranular, continuous coatings on grains. In other wells, cement SL2 forms discontinuous coatings with variably sized equant crystals. SL2 occludes 10–25% of oomoldic porosity, by lining oomold or spastolith walls. SL2 does not fully occlude cement-reduced oomoldic or interparticle pores. Thickness is highly variable in interparticle pores, reducing primary porosity by 5–75%. On average, though, primary interparticle porosity is reduced by 10–20%, with this value gradually increasing downhole (Fig. 4). The reduction of porosity by SL2 for the Raytown Limestone as a whole is 10–15%.

Event 11 or 12: Local neomorphism of micritized grains. Most grains that had been micritized subsequently underwent neomorphism to produce microspar and pseudospar (Fig. 5b). This neomorphism is likely to be coincident with precipitation of cement SL2, as the neomorphic spar produced has similar cathodoluminescence to cement SL2. If cement SL2 is, indeed, the same cement seen in neomorphic textures, this constrains timing to coincide with the precipitation of cement SL2, and it has the same evidence for timing as SL2. Ooids are the predominant grains that underwent neomorphism. Many neomorphosed ooids preserve relict ooid coatings, with differing amounts of microporosity.

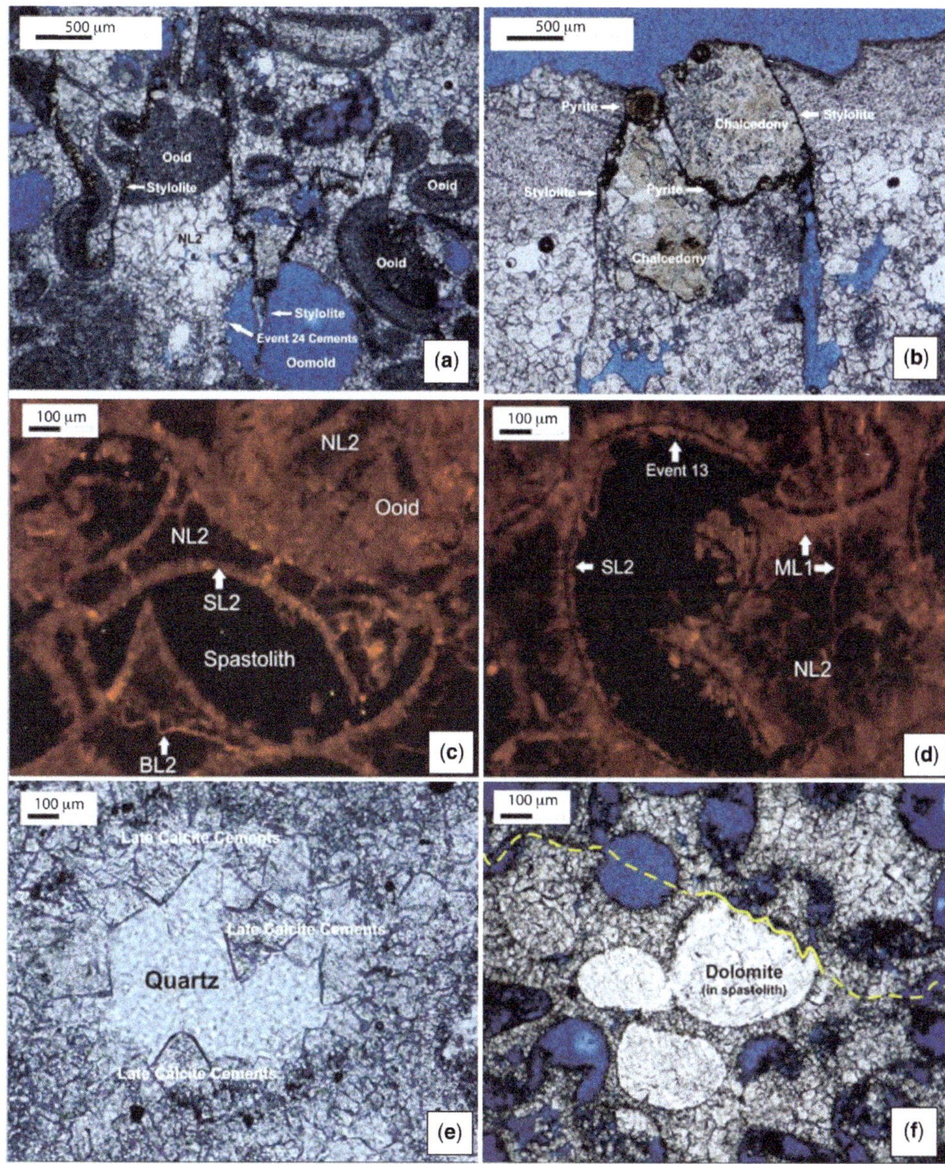

Fig. 13. Photomicrographs of late diagenetic events. (**a**) Photomicrograph of a stylolite truncating and offsetting grains. The stylolite projects into oomold and preserves a neomorphosed ooid, indicating that the stylolite formed prior to ooid dissolution. Late calcite cements (Event 24) along the internal oomold perimeter indicate that they came after stylolitization. (**b**) Photomicrograph of chalcedony (Event 19) being truncated by stylolite. Stylolite exhibits solution-enhanced secondary porosity and pyrite precipitate. (**c**) Cathodoluminescence photomicrograph depicting a spastolith coated with SL2, cement NL2 fills primary interparticle porosity, BL2 (Event 24) cement fills microfracture and the ooid has been neomorphosed. (**d**) Cathodoluminescence photomicrograph illustrating cement SL2 coating the inside and outside of crushed oomolds (spastoliths), and cement NL2 forming around a spastolith geometry in the interparticle pore space. ML1 is replacing cement SL2 and NL2, and filling fractures. Dissolution from Event 13 is evident in this thin section as truncated terminations on SL2 cement. (**e**) Late calcite cements (Event 24) precipitated along the oomold perimeter, with megaquartz occluding the rest of the moldic porosity, thereby indicating that the quartz post-dates the late calcite cements. (f) Baroque dolomite precipitated within a secondary solution-enhanced moldic pore associated with stylolitization (Event 20–21), indicating that dolomite came after stylolitization. The path of the stylolite is marked with a solid yellow line where it intersects the dolomite-filled spastolith, and with a dashed yellow line where it intersects unfilled spastoliths.

Event 13: Dissolution. The only evidence for this minor dissolution event is the truncation of cement SL2 (Fig. 12d). These truncations are seen in most thin sections from the study, but occur locally, and not in every oomold. No discernable or relevant porosity increase is detected.

Event 14: Precipitation of NL2. This pervasive cement is seen in every thin section in the study, including samples from the overlying limestone. It reduces the majority of primary interparticle pore space within the Raytown unit, and only rarely occurs in oomolds (Fig. 13). Where present in oomolds, this cement tends to occlude all remaining primary interparticle porosity or moldic porosity. In the upper 0.5 m of the Raytown Limestone, this cement completely or partially fills several of the largest fenestral and primary interparticle pores. It reduces pore space in the upper 0.5 m of the Raytown Limestone by 5–50%, with an average of 15%. Cement NL2 reduces the total oomoldic porosity by 5% because it is rarely found in cement-reduced oomolds. Conversely, this cement diminishes primary cement-reduced interparticle porosity by 25–95%. NL2 reduces the total Raytown Limestone porosity by 25–35%.

Event 15: Precipitation of rhombic dolomite. Very fine-grained rhombs of unimodal, planar-e (euhedral) dolomite are found in intercrystalline pores of the NL2 cement. It is a difficult cement to distinguish, as it is similar to NL2 in transmitted and CL light. Staining (Dickson 1965) reveals the dolomite's distribution. It is a volumetrically insignificant cement, reducing less than 1% of the total Raytown Limestone porosity.

Event 16: Microfracturing, brecciation and microfaulting. Thin, but widespread, microfractures cut across SL2, NL2 and the rhombic dolomite, from events 11 to 15 (Fig. 13d). In places, the microfracturing is so substantial, that the interparticle cements appear brecciated. No palaeotopographical or spatial control is apparent in their distribution. They appear throughout the sampled area. Also evident are micro-scale faults. Most of the ooids that are intersected by microfaults are completely dissolved (Event 17). These microfractures and microfaults are solution enlarged, and contain later cements, including Event 18 calcite cement, Event 24 late calcite cements and Event 25–26 dolomite.

Event 17: Dissolution. This event of dissolution manifests itself as an enhancement of the porosity created by the microfractures and microfaults of Event 16. These microfractures and microfaults are filled with cement ML1 (Event 18) and Event 24 calcite cements, but not earlier cements (SL2, NL2 and rhombic dolomite).

Event 18: Precipitation of and replacement with ML1. Cement ML1 occurs inside microfractures (Event 16) and replaces earlier cements, mostly SL2 (Fig. 13d). Timing of replacement can be evaluated by comparing the cathodoluminescence characteristics of non-altered parts of cements SL2 and NL2 within interparticle pore space with replaced patches that cross-cut these original cements. This cement commonly occludes fracture porosity and replaces earlier cements. On the whole, this cement reduces the total porosity of the Raytown Limestone by 3%.

Event 19: Silicification – Chalcedony replacement. Chalcedony replacement is most readily seen in micritic areas, such as burrows and micritized ooids (Fig. 7). The replacement fabric cross-cuts cement in interparticle pores and ooid grains, preserving relict ooid laminations. Chalcedony commonly occurs near and is truncated by stylolites (Fig. 13b). Truncation constrains the timing of chalcedony replacement to before stylolitization, but after the cementation of the cements it cross-cuts, including NL2 and ML1.

Event 20 or 21: Stylolitization. Stylolites (Figs 6 & 13) are common. Timing of stylolitization is constrained by the types of cements found in the solution-enhanced porosity associated with the dissolution of the stylolitized areas. Earlier cements (SL2, NL2 and ML1) are not seen in this type of secondary porosity, and stylolites cross-cut these cements. All precipitation events listed after stylolitization occur in the solution-enhanced porosity. Stylolitization is listed as concurrent with grain-to-grain pressure dissolution because no location was found where one chemical compaction feature cross-cut another.

Event 20 or 21: Grain-to-grain pressure dissolution. This type of compaction feature is rare. Grain-to-grain suturing occurs below the upper 0.5 m of the Raytown Limestone, where early cements (events 1–8) are not present. These chemical compaction features cross-cut intermediate-aged cements (SL2, NL2 and ML1), but deformed grains have later cements precipitated along the deformation contact, constraining the time of grain-to-grain pressure dissolution.

Event 22: Dissolution. Almost all stylolites observed are followed by noticeable dissolution, with most of the dissolutional porosity occurring at the apices of the stylolites (Fig. 13b). These secondary pores cross-cut grains and earlier cements (SL2, NL2 and ML1). In one thin section, a stylolite projects into an oomold that has late calcite cement (Event 24) precipitated along the perimeter, which indicates that the ooid was cross-cut by the stylolites and then dissolved (Fig. 13a).

Event 23: Precipitation of pyrite in stylolites. Irregularly shaped pyrite concentrations occur within solution-enlarged stylolites (Fig. 13b). Later cement (Event 24) precipitated on the pyrite.

Event 24: Precipitation of late calcite cements (NL3, SL3, NL4, BL2 and DB1). These cements are widespread, and fill remaining pore space in cement-reduced fenestral pores, reduce and/or occlude cement-reduced interparticle pores of the Raytown Limestone, reduce solution-enlarged pores that formed after stylolitization, reduce solution-enlarged fractures, and reduce and/or occlude cement-reduced oomolds in both the Raytown and overlying limestone (Figs 5c, d, 12d & 13a, c, e). In the upper 0.5 m of the Raytown Limestone, these cements are the only cements that fill oomolds, occluding approximately 90% of the oomoldic porosity. Event 24 cements also occlude the rest of the fenestral porosity by, on average, 20% of the porosity. It is key to note that these later cements are not in every pore in the upper 0.5 m, as early cements (events 2–8) filled some pores entirely. Below the upper 0.5 m of the Raytown Limestone, in most of the unit, Event 24 cements precipitate in microfractures and in other solution-enhanced pores, interparticle pores and oomoldic pores. Event 24 cements, on average, reduce oomoldic porosity by 25%. It commonly precipitated as large crystals along the perimeter of oomolds, although, locally, large crystals also occlude the oomold entirely, or form as syntaxial overgrowths on crinoid fragments that served as ooid nuclei. Cement NL2 (Event 14) occluded the majority of the interparticle porosity, but Event 24 calcite cements are still volumetrically significant by reducing around 25% of the total primary interparticle porosity. Event 24 cements reduced the overall porosity of the entire Raytown Limestone by 15–30%.

Event 25 or 26: Precipitation of baroque dolomite. Nonplanar, unimodal baroque dolomite occurs in spastoliths, secondary porosity associated with stylolites, and oomolds adjacent to fractures (Figs 5d & 13f). Some oomolds are lined by Event 24 cements, and are then filled with baroque dolomite, indicating that the dolomite occurred after mechanical and chemical compaction, stylolitization, at least one fracture event (Event 16), and precipitation of Event 24 cements. Staining of samples reveals that the baroque dolomite consists of two precipitation events, a non-ferroan phase and a ferroan phase. Overall, the porosity reduction of the Raytown Limestone associated with the baroque dolomite is 3–5%.

Event 25 or 26: Precipitation of micro- and megaquartz in oomolds. Micro- and megaquartz are primarily found in oomolds. Locally, megaquartz crystals precipitate within interparticle pores. Oomolds containing these quartz phases also have Event 24 late calcite cements precipitated along the perimeter (Fig. 13e), indicating that the calcite cements grew first along the edge of the oomold, then quartz filled the rest of the mold. Quartz is also found in secondary porosity associated with stylolitization. This event is placed coincident with late baroque dolomite precipitation as no cross-cutting relationships were found. Total porosity reduction of the Raytown Limestone by micro- and megaquartz is 3–5%.

Event 27: Pyrite replacement. Coarse, euhedral crystals of pyrite are seen in the Raytown and overlying limestones, and cross-cut all cements and grains (Fig. 12a). No examples of the pyrite truncating late fractures could be found, which is why this event is inferred to occur before late fracturing. Locally, pyrite is most evident at the boundary between the Raytown Limestone and the overlying clay-rich shale, where the shale appears truncated by or deformed around the pyrite. This pyrite event is distinctive from the earlier Event 23 pyrite because Event 23 pyrite is anhedral in morphology, and does not cross-cut Event 24 late calcite cements.

Event 28: Fracturing. Fractures associated with this event remain open. The fractures cross-cut all lithofacies, cements and grains, therefore placing this event later than any cementation event within the paragenesis.

Event 29: Dissolution. Late fractures (Event 28) are solution-enlarged locally, and not reduced by any cement. Porosity is enhanced by as much as 5% locally.

Integration of petrography, stable isotopes and fluid inclusions

In this section, petrographical and geochemical data are integrated to yield interpretations of origin of cements. Fluid-inclusion homogenization temperature and final melting temperature of ice data are presented in Figure 14a, b. Primary fluid inclusions are identified on the basis of distribution along concentric growth zones or orientation in the direction of crystal growth. Data from primary fluid inclusions are reported for cements SL2, NL2, late calcite cements and baroque dolomite. Distribution of fluid-inclusion data in fluid-inclusion assemblages (FIAs) is recorded to evaluate variability of data and quality of fluid-inclusion data (i.e. Goldstein

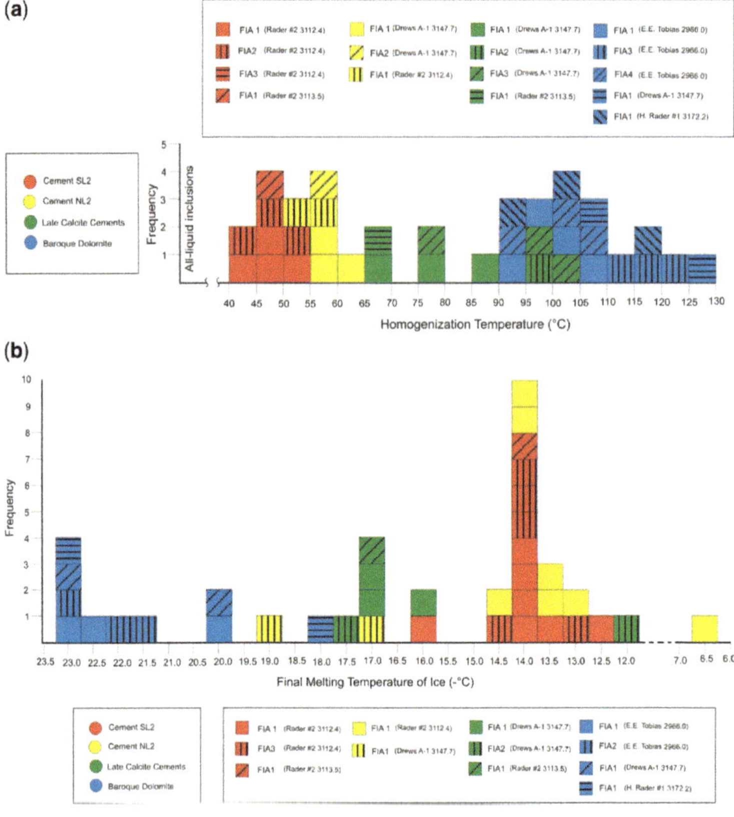

Fig. 14. (**a**) Histogram displaying homogenization temperatures (T_h) for various FIAs within specific cements. Each cement is labelled with a specific colour, and each FIA is given a specific pattern. (**b**) Histogram displaying the final melting temperature of ice ($T_{m_{ice}}$) for various FIAs within specific cements. Colours and patterns are the same as in (a).

& Reynolds 1994). Stable isotope data, $\delta^{18}O$ and $\delta^{13}C$ values, are presented in Figure 15 for depositional micrite, early cements (events 3–7), SL2 (Event 10–11), NL2 (Event 14), late cement (Event 24) and baroque dolomite (Event 25–26). Data and calculations are available at http://hdl.handle.net/1808/21744.

Origin of early cements (events 3–7)

Petrographical observations place early cements at the beginning of the paragenesis (events 3–7), prior to chemical and mechanical compaction and before deposition of the overlying strata. Petrographical observations are indicative of vadose diagenesis and a close association with other indicators of subaerial exposure (fenestral fabrics, rhizoliths). One grouping of the stable isotope data has negative $\delta^{13}C$ values. These samples come from the uppermost 3 cm of the Raytown Limestone. The negative $\delta^{13}C$ values offer strong support for meteoric diagenesis (Allan & Matthews 1982) due to ^{12}C-enrichment associated with soil-gas CO_2 from the oxidation of organic matter. Three other samples of the same early cements were taken 15 cm below the top of the Raytown Limestone, slightly deeper than the other samples. These samples yield positive $\delta^{13}C$ values that are similar to those of depositional micrite, an indication that carbon is marine-rock-dominated during meteoric diagenesis, just 15 cm below the top of the Raytown Limestone. Oxygen isotopic composition of these samples are −5.5 to −6‰, several per mil more negative than least altered Pennsylvanian marine calcite (Mii *et al.* 1999), and consistent with other Pennsylvanian meteoric calcite (Goldstein 1991). Higher in the Raytown Limestone (upper 3 cm), the cements yield less negative $\delta^{18}O$ values, in the range of −3.5 to −5‰. These values are more negative than least altered marine calcite values (Mii *et al.* 1999), and support the interpretation of meteoric cementation. The several per mil enrichment in comparison to samples taken from slightly deeper in the Raytown Limestone is consistent with the fractionation

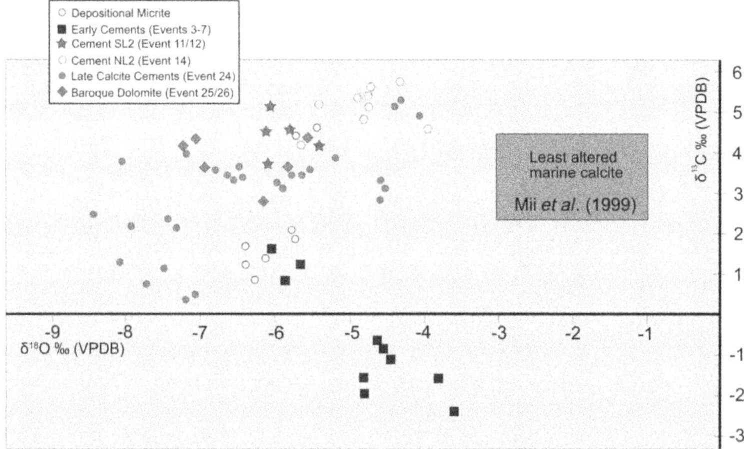

Fig. 15. Distribution of the oxygen and carbon isotopic data from all samples. Dark grey box represents the oxygen and carbon isotopic compositions of least-altered calcite derived from surface-temperature Late Pennsylvanian seawater from Mii *et al.* (1999).

associated with evaporation proximal to a subaerial surface (e.g. Allan & Matthews 1982). The presence of all-liquid, primary fluid inclusions corroborates the low-temperature interpretation for events 3–7 cements.

Origin of cement SL2 (Event 11 or 12)

Petrographical observations indicate that cement SL2 was precipitated after compaction, indicating a late origin. All-liquid, primary fluid inclusions indicate that the temperature of entrapment was below about 50°C, and two-phase primary fluid inclusions yield homogenization temperatures of 43–52°C (Fig. 14a). Associated gas-rich fluid inclusions indicate the heterogeneous entrapment of both gas and aqueous liquid. This eliminates the need for a pressure correction (Goldstein & Reynolds 1994). Thus, SL2 precipitated at temperatures from less than about 50°C to just above 50°C. These temperatures are consistent with normal burial temperatures during the Permian, assuming a geothermal gradient of 35°C km^{-1} (Stavnes 1982) and a mean annual surface temperature of 25–27°C (Benison & Goldstein 1999). This calculation implies burial temperatures consistent with burial beneath 0.5–1.0 km of overburden. The mid-Permian– Raytown section in this area is 0.5–1.0 km in thickness. Thus, fluid-inclusion temperatures are consistent with cement precipitation during the Early to mid-Permian. $T_{m_{ice}}$ (final melting temperature of ice) data (Fig. 14b) indicate precipitation from a brine of 12.6–19.5 wt% NaCl equivalent. Low eutectic temperatures near −54°C suggest the system may have had a $NaCl–CaCl_2$ composition (Goldstein & Reynolds 1994). All observations point to Permian reflux as the origin of fluids accountable for the precipitation of SL2. The idea of density-driven reflux is supported by abundant anhydrite and halite in the Permian section (Zeller 1968; Benison 1997). Permian density-driven reflux (Adams & Rhodes 1960) has been interpreted by others for this area (Anderson 1989; Luczaj & Goldstein 2000; Buijs 2005). Substantial evaporite deposition did not begin until the Leonardian, therefore fluids that precipitated cement SL2 are likely to have originated during the Leonardian or later.

Using the fluid-inclusion temperatures of 43–52°C, and measured $\delta^{18}O$ values of −5.4 to −6.1‰, the isotopic compositions of pore fluids ranged from −0.4 to +1.9‰ VSMOW (O'Neil *et al.* 1969; Friedman & O'Neil 1977) (Fig. 15). These pore fluid values are more positive than those interpreted for Permian seawater (−3 to −1‰) (VSMOW) (Veizer *et al.* 1999; Korte *et al.* 2003), supporting the interpretation that pore fluids originated from the evaporation of seawater or other fluids at the surface during the Permian.

Origin of cement NL2 (Event 14)

The petrographical observations indicate that cement NL2 was precipitated after compaction, indicating a late origin. Fluid-inclusion observations yield primary all-liquid and two-phase inclusions, as well as gas-rich inclusions. Homogenization temperatures range from 45 to 61°C (Fig. 14a), consistent with burial temperatures for the 1 km of burial achieved during Permian. $T_{m_{ice}}$ data (Fig. 14b) indicate precipitation from a brine of 9.9–21.8 wt% NaCl equivalent. Low eutectic temperatures near −52°C suggest the system may have had a NaCl–$CaCl_2$ composition (Goldstein & Reynolds 1994). Citing the same reasons as cement SL2, cement NL2 is also interpreted to originate from Permian refluxing brines.

Using the 45–61°C temperature range, and measured $\delta^{18}O$ of −3.9 to −5.6‰, the isotopic composition of pore fluids ranged from +0.5 to +4.9‰ VSMOW (O'Neil *et al.* 1969; Friedman & O'Neil 1977) (Fig. 15). These pore fluid values are distinctly more positive than those interpreted for Permian seawater (−3 to −1‰) (VSMOW) (Veizer *et al.* 1999; Korte *et al.* 2003), supporting the interpretation that pore fluids originated from the evaporation of seawater or other fluids at the surface during the Permian.

Origin of cements NL3, SL3, NL4, BL3 and DB1 (Event 24)

Petrographical observations indicate that Event 24 cements were precipitated after compaction (Event 10), stylolitization (Event 20/21) and at least one pervasive microfracture event (Event 16). The timing of cement precipitation can be evaluated in the context of regional events. There are two orogenic events, which occurred after the deposition of the Raytown Limestone that could produce pervasive microfractures: (1) Pennsylvanian–Early Permian Ouachita Orogeny; and (2) Late Cretaceous–Palaeogene Laramide Orogeny (Merriam 1963; English & Johnston 2004; Watney *et al.* 2008). Earlier cements SL2 and NL2 are interpreted to have precipitated from refluxing brines originating from evaporation of mid-Permian seawater. Since the Ouachita event ended before the mid-Permian reflux, it is not likely that Ouachita deformation was the cause of Event 16 fracturing. That leaves the Laramide Orogeny as the alternative structural event for producing the observed pervasive microfracturing event(s), and, in turn, indicates that Event 24 cements precipitated during or after the Laramide deformation. Thus, when calculating a burial temperature for Event 24, the calculation should use overburden and surface temperatures found on the CKU during the Late Cretaceous and Early Tertiary, essentially the time of maximum burial. Given a geothermal gradient of 35°C km^{-1} (Stavnes 1982), mean annual surface temperature in the Midcontinent from the Cretaceous of 24–30°C (27°C average: Barron *et al.* 1995; Ufnar *et al.* 2002) and maximum burial depths of 1.4 km at the end of the Cretaceous for the top of the Missourian Stage rocks (Newell 1997), the maximum burial temperature would have been 76°C. Current subsurface formation temperatures in the Raytown Limestone are only 32–35°C.

Primary fluid inclusions yield homogenization temperatures of 65–102°C (Fig. 14a), temperatures at and well above maximum burial temperatures. As earlier cements have low homogenization temperatures and preserve all-liquid fluid inclusions, these values do not result from thermal re-equilibration of the inclusions. The variable T_h (homogenization temperature) data in individual FIAs and among FIAs is thus an indication of variable temperature during cement precipitation. Therefore, a combination of two processes can be considered for the origin of Event 24 calcite cement by incorporating the fluid-inclusion and stable isotopic data.

One process explains why some FIAs in Event 24 cements have some fluid-inclusion and isotopic data similar to data generated for Events 11 or 12, and 14. Events 11 or 12, and 14 are interpreted to represent normal burial temperatures and brine reflux. The low-end homogenization temperatures measured for Event 24 cements could be produced by normal burial temperatures, only slightly higher than those for SL2 and NL2. $T_{m_{ice}}$ data (Fig. 14b) indicate precipitation from a brine of 15.3–20.5 wt% NaCl equivalent. This highly saline brine is comparable to fluid inclusions in cements SL2 and NL2, and may have originated as a refluxing fluid. The most isotopically positive stable isotope data (Fig. 15) are in the same range as earlier reflux cements ($\delta^{18}O_{calcite} > -6‰$; $\delta^{13}C_{calcite} > 3.5‰$). Using a range of temperatures of 65–102°C, and measured $\delta^{18}O_{calcite}$ of −4.1 to −8.4‰, the isotopic compositions of pore fluids calculated for these late calcite cements range from +0.8 to +9.8 ‰ V-SMOW (O'Neil *et al.* 1969; Friedman & O'Neil 1977). These pore fluid values are distinctly more positive than those interpreted for Permian seawater (−3 to −1‰) (V-SMOW) (Veizer *et al.* 1999; Korte *et al.* 2003), supporting the interpretation that pore fluids originated from the evaporation of seawater or other fluids at the surface during the Permian. Therefore, it is realistic to hypothesize that these cements were produced by fluids similar to those that produced the earlier cements SL2 and NL2, although at a slightly higher temperature consistent with Cretaceous burial depths.

Some data, however, are dissimilar to data produced by SL2 and NL2, and indicate that an additional process must have been active during Event 24. Some homogenization temperatures are much higher than the maximum burial temperature of 76°C, and both $\delta^{18}O$ values and $\delta^{13}C$ values are more negative than the earlier cements. These data argue for hydrothermal injection. Fluid-inclusion data indicate that fluid salinity and temperature fluctuated through time. T_h data indicate that single FIAs record variable temperatures; and T_h of successive FIAs rise and fall. These data indicate that warm hydrothermal fluid must have been repeatedly injected into cooler rock during precipitation of Event 24 cements. Stable isotope data support this hydrothermal hypothesis as well. Some data have more negative $\delta^{18}O$ values than previous reflux-produced cements.

If hydrothermal fluids precipitated some of the Event 24 cements, this places the initiation of hydrothermal injection during the Late Cretaceous–Early Tertiary. Several researchers have found evidence for injection of hot fluids in Kansas during the Late Cretaceous or Early Tertiary. Coveney *et al.* (2000), using radiometrically dated calcite and fluid-inclusion data, established an influx of hydrothermal fluids in the southern Midcontinent at 67 Ma. Luczaj & Goldstein (2000) showed that hydrothermal activity must have been younger than a Permian U–Pb date produced from dolomite. Blackburn *et al.* (2008) indicated that kimberlites near the study area were emplaced in Kansas around 85–110 Ma and were subjected to local hydrothermal reheating at approximately 65 Ma. This bolsters the idea of hydrothermal activity in the region, and places this activity within a plausible geological time frame.

Ascribing the pervasive microfracturing in central Kansas to Laramide tectonics is not common among previous researchers in the area, but stress trajectory of faults in NE Kansas suggests potential activity during the Laramide Orogeny (Ohlmacher & Berendsen 2005). There is also significant eastwards tilting of strata in Kansas that took place in the late Mesozoic and Cenozoic, which can be attributed to the Laramide Orogeny, and could have caused major regional fracturing (Merriam 1963). If microfracturing is, indeed, a regional phenomenon, then the implications for central Kansas's reservoirs are far-reaching. A regional network of interconnected fractures would provide excellent pathways for hydrocarbon migration, as well as conduits for later fluids associated with late cementation and dissolution. Hypotheses for timing of oil migration into Pennsylvanian rocks on the CKU centre around oil generation in the Anadarko Basin, with subsequent migration into Lansing and Kansas City group reservoirs in Pennsylvanian–Permian time (Walters 1958; Gerhard 2004). Recent data argue for much later timing of migration (Newell 1997; Sorenson 2003). This study supports the idea of a relatively late timing for oil migration into Pennsylvanian reservoirs.

The Event 24 data, therefore, are best explained as forming during maximum burial, either late in the Cretaceous or just after, as hydrothermal injection pulsed on and off, and, therefore, hydrothermal conditions alternated with normal burial conditions to precipitate cements.

Origin of baroque dolomite (Event 25 or 26)

Petrographical observations indicate that the baroque dolomite precipitated very late in the paragenetic sequence, after most events. Fluid-inclusion and stable isotope data are consistent with this interpretation.

Primary two-phase fluid inclusions yield homogenization temperatures of 90–130°C (Fig. 14a). Based on the preservation of low-temperature FIAs in earlier cements, it is unlikely that these inclusions were altered by thermal re-equilibration. These values are well above normal burial temperatures at the 1.4 km of maximum burial (Stavnes 1982; Barron *et al.* 1995; Newell 1997; Ufnar *et al.* 2002). $T_{m_{ice}}$ data (Fig. 14b) indicate precipitation from a brine of greater than 22.5 wt% NaCl equivalent. The variability of T_h data within and among FIAs indicates pulsed hydrothermal fluid flow. Other researchers have suggested a hydrothermal origin for baroque dolomites in the Midcontinent as well, but many interpreted the hydrothermal fluids to be related to Pennsylvanian–Permian Ouachita–Arkoma tectonics, not Cretaceous–Tertiary Laramide-related events (Bethke & Marshak 1990; Wojcik 1991; Walton *et al.* 1995; Goldstein & King 2014; Ramaker *et al.* 2014; King & Goldstein 2016).

Using a range of temperatures of 90–130°C, and measured $\delta^{18}O$ of −5.5 to −7.2‰ (Fig. 15), the isotopic composition of pore fluids would have ranged from +0.8 to +6.6‰ VSMOW (Anderson & Arthur 1983; Land 1985). These calculations reveal that pore fluids were markedly more positive than Pennsylvanian or Permian seawater (Mii *et al.* 1999; Veizer *et al.* 1999). This is consistent with baroque dolomite forming from hydrothermal basinal brines.

Porosity evolution v. structural and palaeotopographical position

A major component of this study is to determine conceptually how diagenetic processes controlled evolution of porosity and permeability. The goal is to postulate predictive models for locations of reservoirs in relation to synsedimentary and later structural features, similar to the CKU. Conceptual models for understanding the location of the best reservoirs through time are most easily constrained by integrating the paragenesis with the evolution of porosity, and comparing wells in different locations. Today, the best reservoir quality is on the flank of the CKU rather than closer to its crest (see data immediately below; Fig. 16), but this has not

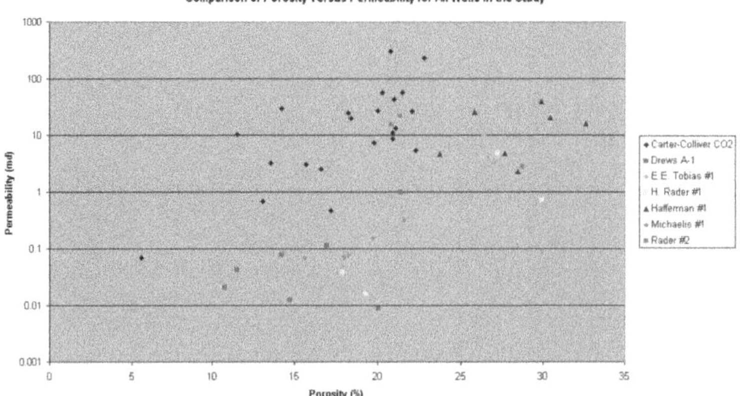

Fig. 16. Graph comparing the porosity and permeability of the oolitic grainstone facies of the Raytown Limestone for all wells in the study. Carter–Colliver, Michealis #1, Drews #A-1, Rader #1 and Rader #2 are considered updip wells. Hafferman #1 and E.E. Tobias #1 are considered downdip wells.

been the case through geological time. The position of the highest reservoir quality has shifted through time as diagenetic conditions varied spatially across this structural relief.

Conceptual models for the location of reservoir quality through time require integrating the diagenetic history with the evolution of porosity. The diagenetic history of the Raytown Limestone is most easily summarized by seven major steps that affected porosity, shown graphically in Figure 17: (1) deposition; (2) meteoric diagenesis; (3) compaction and dissolution; (4) initiation of reflux; (5) extensive reflux; (6) microfracturing, stylolitization and dissolution; and (7) late cements. For each step

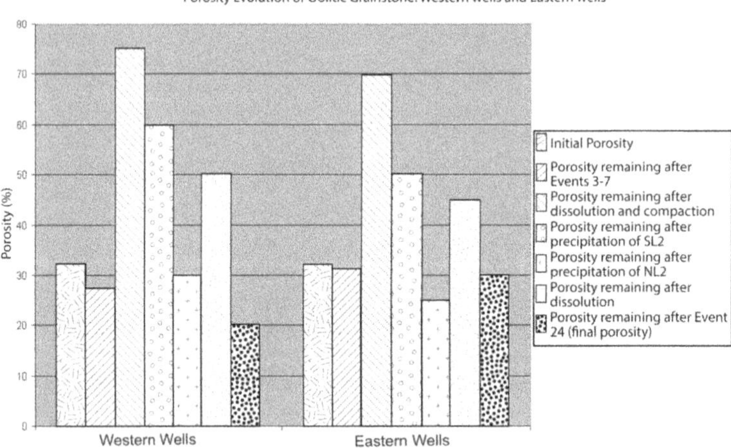

Fig. 17. Graph comparing the porosity evolution for the Raytown Limestone in eastern and western wells in the study area. Initial porosity is assumed to be 32%. The second major step in the evolution is the precipitation of early meteoric cements (after Events 3–7). The next step estimates porosity after ooid dissolution and compaction. Eastern wells experienced more compaction to yield lower porosity. The fourth step shows porosity remaining after the precipitation of cement SL2, a Permian reflux cement (Event 11). More SL2 cement was precipitated in eastern wells than western wells. The next step is the porosity remaining after precipitation of cement NL2, another Permian reflux cement (Event 14). There is less cement NL2 in eastern wells than western wells. The sixth step represents dissolution events 17 and 22, after stylolitization and fracturing. The final step shows the porosity remaining after precipitation of Event 24 late calcite cements (NL3, SL3, NL4, BL2, and DB1). Eastern wells have less of these late calcite cements than western wells, ultimately yielding higher reservoir porosity.

listed, porosity data and hypotheses for possible mechanisms of porosity enhancement or reduction are discussed. This is followed by postulating a conceptual model for each (Fig. 18), which discusses where the highest reservoir quality would be found, updip or downdip, at that step in the evolution of the diagenesis. Some of the listed steps include discussions of implications for porosity in the Raytown Limestone and other oomoldic reservoirs.

Extant porosity and permeability in relation to the structural position on the CKU

Extant porosity and permeability for the seven cored wells in the study area are presented in Figure 16. Core plugs from oolitic grainstone (Byrnes *et al.* 2002, 2003) yield porosities from 5.6 to 32.6%, with 85% of the samples measured having porosities of between 13 and 29%. Permeabilities are from 0.02 to 297 md. Thin-section porosity for eight samples of the fossiliferous oolitic packstone range from 5 to 20%, with an average of 12.5%. Extant porosity types for the fossiliferous oolitic packstone facies are dominantly oomoldic, with cement reduction. Today, the two easternmost wells, Hafferman #1 and E.E. Tobias #1, have, on average, the highest extant porosities (mean of 28% in the study), and also have relatively high permeabilities (mean of 17 md). These two eastern wells sample the Raytown in lower palaeotopographical and structural positions than the other wells to the west (Fig. 2), as they sit on the flank of the CKU. In contrast, average porosities and permeabilities from the updip wells to the west are lower (mean of 20% and 5 md). Data and calculations are available at http://hdl.handle.net/1808/21744.

Step 1: deposition of the Raytown Limestone

Step 1 is the original deposition of the unit. Depending on the packing configuration of the ooids, the original porosity for a perfectly sorted oolite would be 26–48% (Graton & Fraser 1935). Based on theoretical calculations, Graton & Fraser (1935) predicted that the porosity for randomly packed, perfectly sorted spheres is 35%. This estimate is too high for the Raytown Limestone because the ooids are not perfectly sorted and have variable packing. Given these variables, 32% porosity is a reasonable assumption for the initial porosity in the Raytown oolite.

Step 2: early meteoric diagenesis

Location of reservoir quality. The second major step in the paragenesis is precipitation of early meteoric cement, events 3–7. Where the top of the Raytown is preserved, the early cements reduce 45–100% of the porosity in the upper 0.5 m, with an average of 80%. Below this upper 0.5 m, the early cements are only minor components. On average, early cements (Table 3) reduce only 1% of the Raytown Limestone's porosity in the eastern wells and 5% in the most complete core of the western wells. Cores in the eastern area do not preserve samples of the top of the Raytown Limestone, and, thus, the 1% value is likely to underestimate the actual total porosity reduction. However, it is clear that immediately after early meteoric diagenesis, total porosity had not been altered much from the original depositional porosity. As the major effect of subaerial diagenesis was cementation, not dissolution, one might hypothesize that reservoirs from updip areas experienced longer times of subaerial exposure and cementation than downdip reservoirs (e.g. Saller *et al.* 1999; Buijs & Goldstein 2012). The data support this hypothesis; however, this may be due to a lack of core preserved at the top of the Raytown Limestone from eastern (downdip) wells. It can be inferred, therefore, that after early meteoric cementation, palaeotopographically lower wells would preserve higher porosity and permeability because pore systems had not been as extensively reduced during meteoric diagenesis (Figs 17 & 18a).

Discussion of meteoric diagenesis and reservoir quality. Meteoric cementation was not a major factor in extant porosity in the reservoirs. As meteoric cements are not precipitated within oomolds, initial dissolution occurred after meteoric cementation. Both findings contradict common perceptions that reservoir oomoldic porosity is created during initial subaerial exposure, or that early meteoric cements play a pivotal role in porosity occlusion (e.g. Watney 1984; Byrnes *et al.* 2000; Wilke & Carr 2001; Esrafili-Dizaji & Rahimpour-Bonab 2009). Petrographical data indicate that the first phase of ooid dissolution (Event 9) possibly was a long-lived phenomenon, occurring penecontemporaneously with burial and compaction.

Other investigations of shoaling-upwards oolites and oolitic reservoirs indicate that patterns of early meteoric cementation can exert a great control on porosity. Dominant diagenetic patterns in the Upper Jurassic Smackover Formation of the southern USA are interpreted as being related to early meteoric diagenesis (Moore & Druckman 1981; Humphrey *et al.* 1985; Llinas 2002). Many reservoirs in the Smackover have precompaction oomoldic porosity, as well as decreased permeabilities from precompactional cementation (Moore & Druckman 1981). The Mississippian Brofiscin Oolite of South Wales (Hird & Tucker 1988) and the Upper Jurassic Arab Formation oolite (Alsharhan & Whittle 1995) both have prevailing interpretations of cementation during subaerial exposure. The Brofiscin Oolite

(a)

Step 2 - Early Meteoric Cementation.
More meteoric cement (indicated by darker yellow shading) from subaerial exposure occurs upslope, thus downslope settings see less porosity occlusion, and are better reservoir candidates (indicated by red star) after the end of Step 2.

(b)

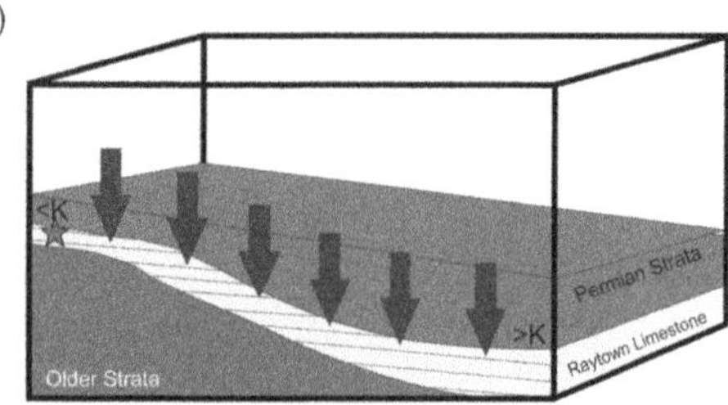

Step 3 - Compaction.
More meteoric cement was precipitated upslope, coating grains more. Less coating of grains downslope allowed for more compaction, creating less porosity, but greater permeability. Thus, upslope settings are better reservoir candidates after the end of Step 3.

(c)

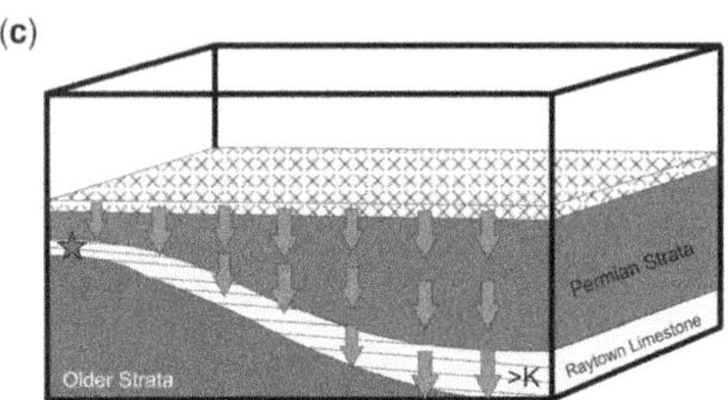

Step 4 - Initiation of Permian Reflux.
The increased permeability downslope after compaction allowed for more precipitation of cement SL2 by refluxing brines. More cement means greater occlusion of porosity downslope, thus, upslope settings are better reservoir candidates after the end of Step 4.

(d)

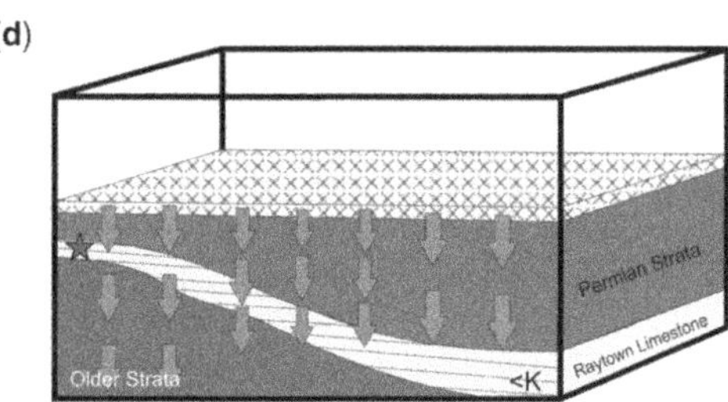

Step 5 - Extensive Permian Reflux.
The decrease in porosity and permeability downslope after precipitation of SL2 caused a lateral shift in the locus of cementation during further reflux diagenesis. Even though porosity is still greater in upslope settings, reservoir distinction decreases by the end of Step 5.

(e)

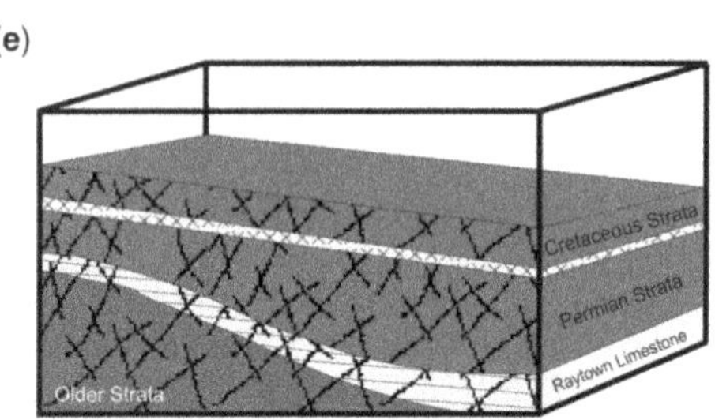

Step 6 - Fracturing and Dissolution.
Extensive fracturing increases and equalizes permeability both upslope and downslope. Subsequent dissolution occurs equally throughout the unit. Upslope settings are still better reservoir candidates, but only by a margin of several per cent porosity after Step 6.

(f)

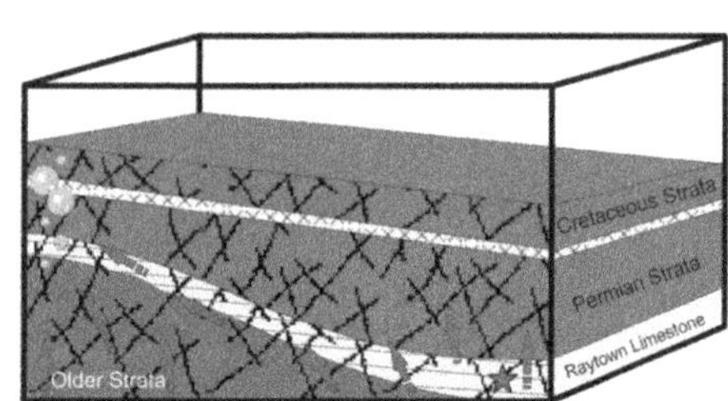

Step 7 - Outgassing of Hydrothermal Fluids.
Event 24 late calcite cements precipitate from both refluxing brines and hydrothermal fluids. More cement is precipitated upslope because of outgassing or updip convection. This makes downslope settings the best reservoir candidates at the end of diagenesis.

Fig. 18. Six schematic cartoons (**a**)–(**f**) illustrating the main processes of porosity and permeability creation or destruction for steps 2–7 of the conceptual model for porosity evolution. Predicted best reservoir is indicated by a red star.

(Hird & Tucker 1988) is dominated by a first-generation calcite spar, interpreted as meteoric vadose and phreatic in origin, which occluded much of the porosity and decreased permeability considerably. Alsharhan & Whittle (1995) hypothesized that sabkha-related cementation created a stable framework of grains responsible for the retention of oomoldic porosity. Other studies of shoaling-upwards carbonates show intense early meteoric cementation updip and less cementation downdip

(Goldstein 1988; Dickson & Saller 1995; Carlson *et al.* 2003). The Raytown Limestone shows a similar pattern, but the overall percentage of early meteoric cementation appears to be lower than in many other Carboniferous examples, which commonly cite 32–60% of the porosity reduced by early meteoric cementation (Goldstein 1988; Carlson *et al.* 2003). A recent study by Buijs & Goldstein (2012) showed only 5% average porosity reduced by meteoric calcite in upper Palaeozoic cyclothemic reservoir rocks of western Kansas. The reason for the small amount of meteoric cement in those settings may be related mostly to a sequence architecture with a hydrogeology that prevents groundwater recharge and discharge.

The findings of this study, of the Raytown Limestone reservoir in central Kansas, imply that the paradigm of subaerial diagenesis dominating the origin of oomoldic porosity may not apply to all oolitic reservoirs.

Step 3: compaction and dissolution

Location of reservoir quality. The third major step within the paragenesis is dissolution and compaction. Dissolution was after meteoric cementation but before mid-Permian reflux cementation. It appears to have formed directly before and penecontemporaneously with compaction experienced during burial. Through back-calculations, starting at the final measured porosity and subtracting the amount of porosity reduced from later cements, it is estimated that the dissolution event or events created a significant amount of oomoldic porosity, increasing porosity by 35% or greater in both eastern and western areas. Compaction, though, decreased some of this pore space, notably by 7% more in the eastern wells compared to western wells. It is clear that this led to a shift in the position of the highest porosity to updip areas. As much of this porosity is oomoldic, it is speculated that the increased compaction and crushing of rims around molds downdip may have led to higher permeabilities in downdip locations at this stage in the paragenesis. Similar fabrics clearly lead to higher permeabilites in moldic reservoirs (e.g. Byrnes *et al.* 2003; Lucia 2007).

There are two possible hypotheses to explain this differential compaction between localities. The first is that there was substantially greater overburden above the eastern wells prior to the mid-Permian, leading to greater compaction. An isopachous map showing the convergence between the top of the Lansing Group and the top of the Permian Stone Corral Formation indicates a 500 ft (150 m) difference in overburden between deposits atop the CKU and deposits on the flank (Merriam 1963). This minor difference in Permian burial between eastern and western areas is unlikely to account for the greater compaction in the eastern area. An alternative hypothesis is that the greater early meteoric cementation updip stabilized the mechanical properties of the rock in western areas (updip) and led to decreased compaction updip in comparison to downdip. Thus, for oolitic reservoirs that had experienced only early meteoric cementation, followed by subsequent dissolution and compaction during burial, updip oolitic reservoirs would retain the best porosity because compaction would have been inhibited by increased meteoric cementation updip (Figs 17 & 18b).

Discussion of reservoir implications of compaction and dissolution. On the basis of the observations from this study, it is hypothesized that palaeotopography has an effect on early meteoric cementation. This study hypothesizes that updip areas acquired slightly more meteoric cement than their downdip counterparts due to an assumed longer exposure time updip. Although this small amount of cement does little to affect the porosity in the earliest stages, it could stabilize the framework of grains, protecting them somewhat from compaction. In areas lacking the early cement, compaction was more prevalent in modifying fabrics. This downdip-localized compaction would have degraded porosity, and could have enhanced permeability, developing better connections between oomoldic pores. This postulated increase in permeability could control the locus of cementation for subsequent cementation events. Thus, the location of early cementation may have an effect on where later diagenesis occurs within the reservoir system.

Timing of the early phase of oomoldic pore formation has been constrained to just after initial subaerial exposure at the top of the Raytown Limestone, but before mid-Permian reflux cementation. The origin of the fluids that were responsible for this dissolution remains enigmatic. Speculatively, these fluids could have been meteoric waters from repeated subaerial exposure events after deposition of the overlying Pennsylvanian strata, connate marine fluids or brines that were refluxing throughout the system prior to mid-Permian cementation.

Step 4: initiation of reflux diagenesis

Location of reservoir quality. At Step 4 (cement SL2), the earliest phase of precipitation of Permian reflux cement, the eastern (downdip) wells received greater amounts of SL2 cement (Event 11) in comparison to the western (updip) wells. In downdip wells, SL2 coats grains continuously and reduces 4% more porosity than in updip wells. This results in 7% less porosity in the eastern wells than in the western wells after precipitation of SL2, clear evidence that the highest reservoir quality shifts updip after precipitation of this calcite cement (Figs 17 & 18c).

Discussion of reservoir implications of initial reflux cementation. Two hypotheses can account for the greater amount of cement SL2 being precipitated downdip. One hypothesis is that greater permeability in the eastern wells, generated by the more extensive earlier compaction in the downdip systems, allowed for greater fluid flow during early stages of reflux, and thus more precipitation of SL2 cements. The second hypothesis is that downdip localities encountered a greater flow of reflux fluids because they were subject to fluid migration not only directly downwards, but also laterally from updip areas of the CKU. In Step 5 (discussed below), it appears that later reflux cements are more abundant updip, and this suggests that the second hypothesis, of increased reflux on the flank, is unlikely. Therefore, a conceptual model can be proposed for reservoirs with regressive oolitic deposits that have experienced early meteoric diagenesis, creation of oomoldic porosity, compaction and one event of reflux diagenesis. The model predicts better reservoirs in updip positions as opposed to downdip positions. The explanation for this prediction ultimately centres on an assumption of improved permeability downdip because of a greater degree of earlier compaction, allowing for greater precipitation from cementing fluids downdip.

Step 5: extensive reflux diagenesis

Location of reservoir quality. In Step 5, cement NL2 reduced significant amounts of pore space during continued reflux, decreasing porosity in the eastern area (downdip) by 25% and in the western area (updip) by 28%. Increased cementation updip at this stage largely closed the gap between porosity updip v. downdip, with 28% remaining updip as opposed to 25% downdip.

One could hypothesize that the large amount of SL2 (Step 4) in the downdip wells decreased permeability, shifting the locus of more rapid cementation laterally into updip areas (Event 12), during Step 5. At this stage in the paragenesis, after extensive reflux diagenesis, updip sites are still the better reservoirs, but the dissimilarity between updip and downdip settings is reduced by almost equalizing porosity. This may be because earlier stages of cementation, which were more abundant downdip, decreased permeability, shifting cementation laterally to areas of higher permeability updip (Figs 17 & 18d).

Discussion of reservoir implications of Permian reflux cementation. After dissolution and compaction, mid-Permian reflux diagenesis began, creating cements during steps 4 and 5. Salinity-driven fluid flow is not a new interpretation for cementation within Phanerozoic sedimentary rocks (Carpenter 1978; Spencer 1987), and is a proposed mechanism for the origin of late diagenetic cements in Pennsylvanian strata in SW Kansas (Buijs 2005), NW Kansas (Anderson 1989) and SE Kansas (Wojcik 1991). An intriguing aspect of this reflux cementation is the shift in locus of cementation as permeability of the reservoir evolves. The data from this study suggest that reflux fluids precipitated most cement where permeability may have been greatest. Pore-lining cements, such as those of SL2, can reduce permeability by thousands of millidarcies (Budd 2002). Initially, greater cementation by cement SL2 occurred downdip because permeability was greater in the downdip reservoir rock. Later, greater cementation by cement NL2 may have shifted updip because permeability was preferentially reduced downdip after SL2 precipitated. By the end of reflux diagenesis, extant porosity between updip and downdip settings was within 5% of one another. After extensive reflux diagenesis, reservoir heterogeneity was minimized because of this process of shifting the locus of cementation.

Step 6: microfracturing, stylolitization and dissolution

Location of reservoir quality. The sixth step is solution enhancement during events 17 and 22 after events of through-going microfracturing (Event 16) and stylolitization (Event 20–21). As microfractures appear to have no location control, they affected both updip and downdip areas more or less equally, and regionally increased permeability. One could argue that microfracturing provided the major permeability conduits for corrosive fluids to migrate through the Raytown Limestone, causing solution enhancement of porosity. This dissolution is interpreted to have formed well after burial, subsequent to active reflux diagenesis, but prior to initiation of hydrothermal mineral precipitation. In both areas, porosity was enhanced by 20% from dissolution, yielding higher porosity in western areas because of their higher porosity earlier after Step 5. This indicates that regressive oolitic deposits that had undergone meteoric cementation, dissolution and compaction, extensive reflux cementation, and subsequent dissolution via pervasive microfracture permeability would produce reservoirs approximately equal in quality whether updip or downdip (Figs 17 & 18e). It is also arguable that this stage of the paragenesis is when the Raytown Limestone experienced the best reservoir conditions. Fluid flow was no longer constrained by cement-reduced pore throats and spastoliths, but was enhanced by a laterally and vertically widespread fracture network that had been solution enlarged.

Discussion of reservoir implications of microfracturing and dissolution. A major implication of porosity evolution associated with Step 6 is the creation of oomoldic porosity late within the paragenesis. Regional microfracturing may have provided conduits for corrosive fluids. The source of these corrosive fluids is speculative, but possibilities include fluids from initial pulses of hydrothermal fluid flow or mixing between hydrothermal fluids and existing formation waters.

Step 7: precipitation of late calcite cements

Location of reservoir quality. The final step is the precipitation of Event 24 cements (NL3, SL3, NL4, BL2 and DB1). Event 24 cements decrease porosity in all wells, but more so in updip western wells. The final extant porosity in the reservoirs is controlled by how much pore space is reduced by these cements; 16% porosity reduction in eastern wells v. an average of 28% reduction in western wells. Although palaeotopographically higher (western) wells had slightly higher porosity immediately before this stage, it appears that the amount of porosity reduction from late cementation dominates the extant reservoir porosity. As Event 24 cements were precipitated, at least in part by hydrothermal fluids, several hypotheses could be proposed to explain the increased cementation updip. Lower pressures updip could have resulted in outgassing of a phase enriched in CO_2 (e.g. Leach *et al.* 1997; Hendry & Poulsom 2006), causing precipitation of cements preferentially in updip areas. Also, preferred migration updip, from density-driven convection (see King & Goldstein 2016), may be a reasonable interpretation as well. Finally, it is possible that local intensity of fault pumping (e.g. Davies & Smith 2006) related to deformation of the CKU area had an impact. Therefore, a conceptual model can be proposed for locating the best reservoir relative to structure, in areas that have undergone hydrothermal cementation. The model would indicate that downdip deposits might preserve more porosity than updip deposits, just as is observed in the extant reservoirs studied here. Furthermore, the implications of observations at this final stage are that late processes dominate in the location of the best reservoir quality. Late-stage fracturing and late-stage dissolution (Step 5) were essential for making much of the reservoir porosity and permeability. Later-stage hydrothermal cementation had a negative effect on the reservoir, affecting updip systems more than downdip systems, and shifting high reservoir quality to downdip settings (Figs 17 & 18f).

Discussion of reservoir implications of late cementation. The implications of porosity reduction by late calcite cements (Event 24), as well as dolomite (Event 25–26) are possibly the most important of this study. Data reveal that final extant porosity is controlled by the amount of cementation by these latest cements. In this study, final extant porosity in the grainstones is not a function of original depositional texture (Byrnes *et al.* 2002), nor is it related to initial subaerial exposure directly after deposition. Instead, final porosity is controlled by the percentage of late porosity reduced by late, primarily calcite, cements. Reduction of porosity is greatest in updip areas, hypothetically, by means of outgassing or density-driven convection. These data and observations should influence predictive models of reservoir location, as hydrothermally related cementation could be the ultimate control on the remaining porosity. Research by Davies & Smith (2006) supports the findings of this study's data, demonstrating that hydrothermal dolomite (HTD) facies have a structural control.

Additionally, since this reduction of late porosity may not have occurred until during or after Laramide microfracturing, models of porosity location should incorporate a model of structural control. Laramide microfracturing and hydrothermal fluid flow were related in our study. It is envisioned that there were multiple events of fracturing, each followed by or synchronous with pulses of hydrothermal fluids through the newly created microfracture network, followed by quiescence and a return to normal burial conditions until the next episode of fracturing commenced. Overall, structural control and hydrothermal fluid flow may prove to be the best predictive model for locating porosity in reservoir rocks in similar settings.

Conclusions

This study integrated porosity and permeability core-plug data, core-based sedimentology and stratigraphy, transmitted-light and cathodoluminescence petrography, fluid-inclusion microthermometry, and stable isotopic analyses to create conceptual models for porosity and permeability evolution by studying the Pennsylvanian Raytown Limestone oomoldic reservoirs of central Kansas. Cores in the study were chosen to reflect the spectrum of palaeotopography and structural position updip and downdip relative to the Central Kansas Uplift (CKU). The two easternmost cores are considered palaeotopographically and structurally low wells (downdip), and the other five cores in the study are considered palaeotopographically and structurally high wells (updip).

The oolitic reservoirs of the Raytown Limestone consist of two–three shoaling-upwards cycles. The top cycle is capped by a surface of subaerial exposure that can be traced down the flank of the CKU. Thus, this unit is a good model for other regressive grainstone reservoirs that drape palaeotopography.

The study shows that the structural setting can be used predictively to determine the location of areas with the best reservoir quality:

- Early Meteoric cement reduces less than 5% of the total Raytown Limestone porosity, and has a minimal effect on final porosity and permeability values. Palaeotopographically higher wells preserve less porosity early within the paragenesis due to interpreted longer subaerial exposure and, therefore, more meteoric cementation.
- Oomoldic porosity was created during several discrete time intervals of dissolution, all occurring well after initial subaerial exposure. The first significant event of dissolution occurred after meteoric cementation, but before mid-Permian reflux cementation, directly before and coincident with compaction.
- After this initial period of burial, dissolution and compaction, updip oolitic reservoirs retain the highest porosity. This is because compaction was inhibited in updip wells by the greater meteoric cementation updip. In downdip wells, compaction occurred preferentially because of the near absence of meteoric calcite cements. This may have led to greater permeability in downdip wells because of the crushing of oomoldic coatings and increased connections between oomolds.
- Intermediate-stage calcite cements precipitated from mid-Permian refluxing brines. The locus of reflux cementation may have been controlled by permeability at the time of reflux, with greater cementation initially occurring in downdip wells. After initial preferred cementation from refluxing fluids in downdip wells, the highest reservoir quality would have been in updip wells.
- The next stage of reflux cementation preferentially cemented updip areas, decreasing reservoir quality from what it had been. It is postulated that the earlier phase of reflux cementation preferentially decreased permeabilities downdip, shifting the locus of preferred reflux fluid flow updip. This updip shift in cementation decreased the differences between downdip and updip reservoir quality, so that, after this stage of cementation, little difference existed between updip and downdip settings.
- After the reflux cementation, there was microfracturing throughout the area in both updip and downdip areas. This is interpreted to result from Cretaceous–Tertiary deformation associated with the Laramide Orogeny. This microfracturing provided conduits for corrosive fluids to migrate through the Raytown Limestone, causing solution enhancement of porosity.
- The porosity enhancement was followed by precipitation of late calcite and dolomite cements, some of which showed evidence for mixing of normal temperature formation brines with hydrothermal fluids, and others that were from hydrothermal fluids. Updip localities experienced twice as much late cementation as downdip localities. This may be due to updip outgassing of hydrothermal fluids, convection or differing intensities of fault pumping of hydrothermal fluids, any of which could have led to increased hydrothermal cementation in updip areas. These late events lead to a shift of highest reservoir quality away from updip localities and to downdip localities in the current reservoirs.

Structural or palaeotopographical settings can be integrated with diagenetic conceptual models developed here to predict the location of the highest reservoir quality, which is expected to shift position through time. Predicting the location of high reservoir quality from the structural setting depends on knowing the history of diagenesis of the extant reservoir. Depending on the final diagenetic stage of the reservoir, the highest reservoir quality may be either updip or downdip.

This project is a distillation of much of the Master of Science thesis of the senior author, Jessica Poteet. Funding for this project was provided by the sponsors of the Kansas Interdisciplinary Carbonates Consortium (KICC), Chevron, The Kansas Geological Survey, The University of Kansas Department of Geology, The Kansas Geological Foundation, and the Desk and Derrick Association. This paper benefitted from careful reviews by James Hendry, Anita Csoma, and an anonymous reviewer. Their comments are greatly appreciated, as the final paper was improved because of their hard work.

References

Adam, A. & Abdullatif, O. 2013. Diagenetic evolution and subsequent implications on reservoir qualities, Khartam Member, Permo-Triassic Khuff Formation, central Saudi Arabia. *Geophysical Research Abstracts*, **15**, EGU General Assembly 2013, 9416.

Adams, J.E. & Rhodes, M.L. 1960. Dolomitization by Seepage Refluxion. *American Association of Petroleum Geologists Bulletin*, **44**, 1912–1920.

Al-Helal, A.B., Whitaker, F.F. & Xiao, Y. 2012. Reactive transport modeling of Brine Reflux: dolomitization, anhydrite precipitation, and porosity evolution. *Journal of Sedimentary Research*, **82**, 196–215.

Allan, J.R. & Matthews, R.K. 1982. Isotope signatures associated with early meteoric diagenesis. *Sedimentology*, **29**, 797–818.

Al-Muraikhi, H.R., Dutta, D. *et al.* 2012. Distribution of the reservoir properties within the Minagish Oolite and Ratawi Limestone reservoirs (Kuwait): a complex interplay of sedimentation, depositional architecture and diagenesis. *Paper presented at the Fourth Arabian Plate Geology Workshop, Late Jurassic/Early Cretaceous Evaporite–Carbonate–Siliciclastic Systems of the Arabian Plate*, 9–12 December 2012, Abu Dhabi, United Arab Emirates.

Alsharhan, A.S. & Whittle, G.L. 1995. Carbonate-evaporite sequences of the Late Jurassic southern and southwestern Arabian Gulf. *American Association of Petroleum Geologists Bulletin*, **79**, 1608–1630.

Ameen, M.S., Buhidma, I.M. & Rahim, Z. 2010. The function of fractures and in-situ stresses in the Khuff reservoir performance, onshore fields, Saudi Arabia. *American Association of Petroleum Geologists Bulletin*, **94**, 27–60.

Amthor, J.E., Mountjoy, E.W. & Machel, H.G. 1994. Regional-scale porosity and permeability variations in Upper Devonian Leduc buildups: implications for reservoir development and prediction in carbonates. *American Association of Petroleum Geologists Bulletin*, **78**, 1541–1559.

Anderson, J.E. 1989. *Diagenesis of the Lansing and Kansas City Groups (Upper Pennsylvanian), Northwestern Kansas and Southwestern Nebraska*. Master's thesis, University of Kansas, Lawrence, KS.

Anderson, T.F. & Arthur, M.A. 1983. Stable isotopes of oxygen and carbon and their application to sedimentologic and paleoenvironmental problems. *In*: Arthur, M.A., Anderson, T.F., Kaplan, I.R., Veizer, J. & Land, L.S. (eds) *Stable Isotopes in Sedimentary Geology*. Society of Economic Paleontologists and Mineralogists (SEPM), Short Course Notes, **10**, 1.1–1.151.

Barron, E.J., Fawcett, P.J., Peterson, W.H., Pollard, D. & Thompson, S.L. 1995. A 'simulation' of middle Cretaceous climate. *Paleoceanography*, **10**, 953–962.

Benison, K.C. 1997. *Field Descriptions of Sedimentary and Diagenetic Features in Red Beds and Evaporites of the Nippewalla Group (Middle Permian), Kansas and Oklahoma*. Kansas Geological Survey, Open-File Report **97-21**.

Benison, K.C. & Goldstein, R.H. 1999. Permian paleoclimate data from fluid inclusions in halite. *Chemical Geology*, **154**, 113–132.

Bethke, C.M. & Marshak, S. 1990. Brine migrations across North America: the plate tectonics of ground water. *Annual Review of Earth and Planetary Sciences*, **18**, 287–317.

Biehl, B.C., Reuning, L., Schoenherr, J., Lüders, V. & Kukla, P.A. 2016. Impacts of hydrothermal dolomitization and thermochemical sulfate reduction on secondary porosity creation in deeply buried carbonates: A case study from the Lower Saxony Basin, northwest Germany. *American Association of Petroleum Geologists Bulletin*, **100**, 597–621.

Blackburn, T., Stockli, D.F., Carlson, R.W. & Berendsen, P. 2008. (U–Th)/He dating of kimberlites – a case study from north-eastern Kansas. *Earth and Planetary Science Letters*, **275**, 111–120.

Budd, D.A. 2002. The relative roles of compaction and early cementation in the destruction of permeability in carbonate grainstones: a case study from the Paleogene of West-Central Florida, U.S.A. *Journal of Sedimentary Research*, **72**, 116–128.

Buijs, G.J.A. 2005. *Carbonate diagenesis of Pennsylvanian and Permian cyclic strata from the Hugoton embayment, western Kansas, United States of America*. PhD dissertation, University of Kansas.

Buijs, G. & Goldstein, R.H. 2012. Sequence architecture and palaeoclimate controls on diagenesis related to subaerial exposure of icehouse cyclic Pennsylvanian and Permian carbonates. *In*: Ketzer, M. & Morad, S. (eds) *Linking Diagenesis to Sequence Stratigraphy*. International Association of Sedimentologists, Special Publications, **45**, 55–80.

Byrnes, A.P., Watney, W.L., Guy, W.J. & Gerlach, P. 2000. Oomoldic reservoirs of central Kansas: controls on porosity, permeability, capillary pressure and architecture. *American Association of Petroleum Geologists, Annual Meeting Abstracts*, 90914.

Byrnes, A.P., Watney, W.L., Guy, W.J., Buijs, G. & Gerlach, P.M. 2002. *Oomoldic Reservoirs of Central Kansas, Controls on Porosity, Permeability, Capillary Pressure, and Architecture*. Kansas Geological Survey, Open-File Report **2002-48**.

Byrnes, A.P., Franseen, E.K., Watney, W.L. & Dubois, M.K. 2003. *The Role of Moldic Porosity in Paleozoic Kansas Reservoirs and the Association of Original Depositional Facies and Early Diagenesis with Reservoir Properties*. Kansas Geological Survey, Open-File Report **2003-32**.

Callewaert, D.L. 1987. *Hydrocarbon production potential of the Iola limestone (Kansas City Group, Upper Pennsylvanian) in parts of Stafford and Pawnee Counties, Kansas*. Master's thesis, Wichita State University.

Cantrell, D., Swart, P. & Hagerty, R. 2004. Genesis and characterization of dolomite Arab-D reservoir, Ghawar Field, Saudi Arabia. *GeoArabia*, **9**, 11–36.

Carlson, R.C., Goldstein, R.H. & Enos, P. 2003. Effects of subaerial exposure on porosity evolution in the Carboniferous Lisburne Group, northeastern Brooks Range, Alaska, U.S.A. *In*: Ahr, W.M., Harris, P.M., Morgan, W.A. & Somerville, I.D. (eds) *Permo-Carboniferous Carbonate Platforms and Reefs*. American Association of Petroleum Geologists, Memoirs, **83**, 269–290.

Carpenter, A.B. 1978. Origin and chemical evolution of brines in sedimentary basins. *In*: Johnson, K.S. & Russell, J.A. (eds) *Thirteenth Annual forum on the Geology of Industrial Minerals*. Oklahoma Geological Survey Circular, **79**, 60–77.

Coveney, R.M., Ragan, V.M. & Brannon, J.C. 2000. Temporal benchmarks for modeling Phanerozoic flow of basinal brines and hydrocarbons in the southern Midcontinent based on radiometrically dated calcite. *Geology*, **28**, 795–798.

Davies, G.R. & Smith, L.B. 2006. Structurally controlled hydrothermal dolomite reservoir facies: an overview. *American Association of Petroleum Geologists Bulletin*, **90**, 1641–1690.

Dennison, J.M. & Shea, J.H. 1966. Reliability of visual estimates of grain abundance. *Journal of Sedimentary Petrology*, **36**, 81–89.

Dickson, J.A.D. 1965. A modified staining technique for carbonates in thin section. *Nature*, **205**, 587.

Dickson, J.A.D. & Saller, A.H. 1995. Identification of subaerial exposure surfaces and porosity preservation in Pennsylvanian and Lower Permian shelf limestones, eastern Central Basin platform, Texas. *In*: Budd, D.A., Saller, A.H. & Harris, P.M. (eds) *Unconformities and Porosity in Carbonate Strata*. American Association of Petroleum Geologists, Memoirs, **63**, 239–257.

Ehrenberg, S.N., Nadeau, P.H. & Aqrawi, A.A.M. 2007. A comparison of Khuff and Arab reservoir potential throughout the Middle East. *American Association of Petroleum Geologists Bulletin*, **91**, 275–286.

Ehrenberg, S.N., Walderhaug, O. & Bjørlykke, K. 2012. Carbonate porosity creation by mesogenetic dissolution: reality or illusion? *American Association of Petroleum Geologists Bulletin*, **96**, 217–233.

Eichenseer, H.Th., Walgenwitz, F.R. & Biondi, P.J. 1999. Stratigraphic control on facies and diagenesis of dolomitized oolitic siliciclastic ramp sequences (Pinda Group, Albian, offshore Angola). *American Association of Petroleum Geologists Bulletin*, **83**, 1729–1758.

English, J.M. & Johnston, S.T. 2004. The Laramide Orogeny: what were the driving forces. *International Geology Review*, **46**, 833–838.

Enos, P. 1977. Tamabra limestone of the Poza Rica trend, Cretaceous, Mexico. *In*: Cook, H.E. & Enos, P. (eds) *Deep-Water Carbonate Environments*. Society of Economic Paleontologists and Mineralogists (SEPM), Special Publications, **25**, 273–314.

Enos, P. & Perkins, R.D. 1979. Evolution of Florida Bay from island stratigraphy. *Geological Society of America Bulletin*, **90**, 59–83.

Enos, P. & Stephens, B.P. 1993. Mid-Cretaceous basin margin carbonates, east-central Mexico. *Sedimentology*, **40**, 539–556, https://doi.org/10.1111/j.1365-3091.1993.tb01349.x

Esrafili-Dizaji, B. & Rahimpour-Bonab, H. 2009. Effects of depositional and diagenetic characteristics on carbonate reservoir quality: a case study from the South Pars gas field in the Persian Gulf. *Petroleum Geoscience*, **15**, 325–344, https://doi.org/10.1144/1354-079309-817

Esteban, M. 1974. Caliche textures and *Microcodium*. *Society Geological Italiana Bulletin*, **92**, (Suppl.), 105–125.

Esteban, M. & Klappa, C.F. 1983. Subaerial exposure. *In*: Scholle, P.A., Bebout, D.G. & Moore, C.H. (eds) *Carbonate Depositional Environments*. American Association of Petroleum Geologists, Memoirs, **33**, 25–37.

Friedman, I. & O'Neil, J.R. 1977. Compilation of stable isotope fractionational factors of geochemical interest. *In*: Fleischer, M. (ed.) *Data of Geochemistry*. 6th edn. United States Geological Survey, Professional Papers, **440-KK**, 1–117.

Garven, G. 1995. Continental-scale groundwater flow and geologic processes. *Annual Review of Earth and Planetary Sciences*, **23**, 89–117.

Gerhard, L.C. 2004. *A New Look at an Old Petroleum Province. Kansas Geological Survey, Current Research in Earth Sciences Bulletin*, **250**, Part 1, 1–27.

Goldhammer, R.K., Dunn, P.A. & Hardie, L.A. 1987. High frequency glacio-eustatic seal-level oscillations with Milankovitch characteristics recorded in Middle Triassic platform carbonates in northern Italy. *American Journal of Science*, **287**, 853–892.

Goldstein, R.H. 1988. Cement stratigraphy of Pennsylvanian Holder Formation, Sacremento Mountains, New Mexico. *American Association of Petroleum Geologists Bulletin*, **72**, 425–438.

Goldstein, R.H. 1991. Stable isotope signatures associated with paleosols, Pennsylvanian Holder Formation, New Mexico. *Sedimentology*, **38**, 67–77.

Goldstein, R.H. & King, B.D. 2014. Impact of hydrothermal fluid flow on Mississippian reservoir properties, southern Midcontinent. *Paper presented at the Unconventional Resources Technology Conference (URTeC) Unconventional Resources Technology Conference Denver*, 25–27 August 2014, Denver, Colorado, USA.

Goldstein, R.H. & Reynolds, T.J. 1994. *Systematics of Fluid Inclusions in Diagenetic Minerals*. Society of Economic Paleontologists and Mineralogists (SEPM), Short Course Notes, **31**.

Goldstein, R.H., Anderson, J.E. & Bowman, M.W. 1991. Diagenetic responses to sea-level change: integration of field, stable-isotope, paleosol, paleokarst, fluid-inclusion, and cement-stratigraphy research to determine history and magnitude of sea-level fluctuation. *In*: Franseen, E.K. & Watney, W.L. (eds) *Sedimentary Modeling: Computer Simulations and Methods for Improved Parameter Definition*. Kansas Geological Survey Bulletin, **233**, 139–162.

Graton, L.C. & Fraser, H.J. 1935. Systematic packing of spheres – with particular relationship to porosity and permeability. *American Association of Petroleum Geologists Bulletin*, **43**, 785–909.

Ham, W.E. & Wilson, J.L. 1967. Paleozoic epeirogeny and orogeny in the central United States. *America Journal of Science*, **265**, 332–407.

Heckel, P.H. 1977. Origin of phosphatic black shale facies in Pennsylvanian cyclothems of Mid-continent North America. *American Association of Petroleum Geologists Bulletin*, **61**, 1045–1068.

Heckel, P.H. 1983. Diagenetic model for carbonate rocks in Midcontinent Pennsylvanian eustatic cyclothems. *Journal of Sedimentary Petrology*, **53**, 733–759.

Heckel, P.H. 1989. Updated Middle–Upper Pennsylvanian eustatic sea-level curve for midcontinent North America and preliminary biostratigraphic characterization. *Compte Rendu XI Congres International de Stratigraphie et de Geologie du Carbonifere*, **4**, 160–185.

Heckel, P.H. 2002. Genetic stratigraphy and conodont biostratigraphy of Upper Desmoinesian-Missourian (Pennsylvanian) cyclothem succession in Midcontinent North America. *In*: Henderson, C.M. & Bamber, E.W. (eds) *Carboniferous and Permian of the World*. Canadian Society of Petroleum Geologists, Memoirs, **19**, 99–119.

Heckel, P.H. & Watney, W.L. 2002. *Revision of Stratigraphic Nomenclature and Classification of the Pleasanton, Kansas City, Lansing, and Lower Part of the Douglas Groups (Lower Upper Pennsylvanian, Missourian) in Kansas*. Kansas Geological Survey Bulletin, **246**.

Hendry, J.P. & Poulsom, A.J. 2006. Sandstone-hosted concretions record evidence for syn-lithification seismicity, cavitation processes, and Palaeocene rapid burial of Lower Cretaceous deep marine sandstones

(Outer Moray Firth, UK North Sea). *Journal of the Geological Society, London*, **163**, 447–460, https://doi.org/10.1144/0016-764905-033

HENDRY, J.P., GREGG, J.M., SHELTON, K.L., SOMERVILLE, I.D. & CROWLEY, S.F. 2015. Origin, characteristics and distribution of fault-related and fracture-related dolomitization: insights from Mississippian carbonates, Isle of Man. *Sedimentology*, **62**, 717–752.

HEYDARI, E. 2000. Porosity loss, fluid flow and mass transfer in limestone reservoirs: application to the upper Jurassic Smackover formation, Mississippi. *American Association of Petroleum Geologists Bulletin*, **84**, 100–118.

HIEMSTRA, E.J. & GOLDSTEIN, R.H. 2015. Repeated injection of hydrothermal fluids into downdip carbonates: a diagenetic and stratigraphic mechanism for localization of reservoir porosity, Indian Basin Field, New Mexico, USA. *In*: AGAR, S.M. & GEIGER, S. (eds) *Fundamental Controls on Fluid Flow in Carbonates: Current Workflows to Emerging Technologies*. Geological Society, London, Special Publications, **406**, 141–177, https://doi.org/10.1144/SP406.1

HIRD, K. & TUCKER, M.E. 1988. Contrasting diagenesis of two Carboniferous Oolites from South Wales: a tale of climatic influence. *Sedimentology*, **35**, 587.

HOUSEKNECHT, D.W. 1986. Evolution from passive margin to foreland basin – the Atoka Formation of the Arkoma basin, south-central U.S.A. *In*: ALLEN, P.A. & HOMEWOOD, P. (eds) *Foreland Basins*. International Association of Sedimentologists, Special Publications, **8**, 327–345.

HUMPHREY, J.D., RANSOM, K.L. & MATTHEWS, R.K. 1985. Early meteoric diagenetic controls of Upper Smackover Production, Oaks Field, Louisiana. *American Association of Petroleum Geologists Bulletin*, **70**, 70–85.

JAMESON, J. 1994. Models of porosity formation and their impact on reservoir description, Lisburne Field, Prudhoe Bay, Alaska. *American Association of Petroleum Geologists Bulletin*, **78**, 1651–1678.

JEWETT, J.M. 1951. *Geologic Structures in Kansas*. Kansas Geological Survey Bulletin, **90**, Part 6.

JOHNSON, T.A. 2004. *Stratigraphy, Depositional Environments, and Coalbed Gas Potential of Middle Pennsylvanian (Desmoinesian Stage) Coals – Bourbon Arch Region, Eastern Kansas*. Kansas Geological Survey, Open-File Report **2004-38**.

JONES, G.D. & XIAO, Y. 2006. Geothermal convection in the Tengiz carbonate platform, Kazakhstan: reactive transport models of diagenesis and reservoir quality. *American Association of Petroleum Geologists Bulletin*, **90**, 1251–1272.

KENTER, J.A.M., HARRIS, P.M., COLLINS, J.F., WEBER, L.J., KUANYSHEVA, G. & FISCHER, D.J. 2006*a*. Late Visean to Bashkirian platform cyclicity in the central Tengiz buildup, Precaspian Basin, Kazakhstan: depositional evolution and reservoir development. *In*: HARRIS, P.M. & WEBER, L.J. (eds) *Giant Hydrocarbon Reservoirs of the World: From Rocks to Reservoir Characterization and Modeling*. American Association of Petroleum, Memoirs, **88**, 7–54.

KENTER, J.A.M., HARRIS, P.M., KUANYSHEVA, G., FISCHER, D.J. & STEFFEN, K.L. 2006*b*. Facies and reservoir quality variations in the Late Visean to Bashkirian outer platform, rim, and flank of the Tengiz buildup, Pricaspian Basin, Kazakhstan. *In*: HARRIS, P.M. & WEBER, L.J. (eds) *Giant Hydrocarbon Reservoirs of the World: From Rocks to Reservoir Characterization and Modeling*. American Association of Petroleum, Memoirs, **88**, 55–95.

KING, B.D. & GOLDSTEIN, R.H. 2016. History of hydrothermal fluid flow in the midcontinent, USA: the relationship between inverted thermal structure, unconformities, and porosity distribution. *In*: ARMITAGE, P.J., BUTCHER, A.R. *ET AL*. (eds) *Reservoir Quality of Clastic and Carbonate Rocks: Analysis, Modelling and Prediction*. Geological Society, London, Special Publications, **435**. First published online August 17, 2016, https://doi.org/10.1144/SP435.16

KORTE, C., KOZUR, H.W., JOACHIMSKI, M.M. & VEIZER, J. 2003. Strontium, oxygen and carbon isotope records of Permian seawater. *Geophysical Research Abstracts*, **5**, 5.

LAND, L.S. 1985. The origin of massive dolomite. *Journal of Geological Education*, **33**, 112–125.

LEACH, D.L., APODACA, L.E., REPETSKI, J.E., POWELL, J.W. & ROWAN, E.L. 1997. *Evidence for Hot Mississippi Valley-Type Brines in the Reelfoot Rift Complex, South-Central United States, in Late Pennsylvanian–Early Permian*. United States Geological Survey, Professional Paper, **1577**.

LEACH, D.L., SANGSTER, D.F. *ET AL*. 2005. Sediment-hosted lead–zinc deposits: A global perspective. *Economic Geology*, **100**, 561–607.

LEBEAU, J.A. 1997. *Geologic controls on porosity and permeability in the Bethany Falls Limestone, Collier Flats Field, Comanche County, Kansas*. Master's thesis, University of Kansas, Lawrence, KS.

LEONARD, K.W. 2006. Correlating and characterizing small-scale flooding surfaces within a high-frequency sequence using conodont distribution patterns: Iola Limestone (Upper Pennsylvanian), Midcontinent USA. *Geological Society of America Abstracts with Programs*, **38**, 4.

LI, Z., GOLDSTEIN, R.H. & FRANSEEN, E.K. 2014. Climate, duration and mineralogy controls on meteoric diagenesis. *Interpretation*, **2**, (3), SF111–SF123.

LINDHOLM, R.C. & FINKELMAN, R.B. 1972. Calcite staining: semiquantitative determination of ferrous iron. *Journal of Sedimentary Petrology*, **42**, 239–245.

LLINAS, J.C. 2002. Diagenetic history of the Upper Jurassic Smackover Formation and its effects on reservoir properties: Manila Sub-basin, eastern Gulf Coastal Plain. *Gulf Coast Association of Geological Societies Transactions*, **52**, 631–644.

LUCIA, F.J. 2007. *Carbonate Reservoir Characterization: An Integrated Approach*. Springer, Berlin.

LUCZAJ, J.A. & GOLDSTEIN, R.H. 2000. Diagenesis of the lower Permian Krider Member, southwest Kansas: fluid inclusion, U–Pb, and fission-track evidence for reflux dolomitization during latest Permian time. *Journal of Sedimentary Research*, **70**, 762–773.

MERRIAM, D.F. 1963. *The Geologic History of Kansas*. Kansas Geological Survey Bulletin, **162**.

MEYERS, W.J. 1991. Calcite cement stratigraphy: an overview. *In*: BARKER, C.E. & KOPP, O.C. (eds) *Luminescence Microscopy: Quantitative and Qualitative Aspects*. Society of Economic Paleontologists and

Mineralogists (SEPM), Short Course Notes, **25**, 133–146.

Mii, H.-S., Grossman, E.L. & Yancey, T.E. 1999. Carboniferous isotope stratigraphies of North America: implications for Carboniferous paleoceanography and Mississippian glaciation. *Geological Society of America Bulletin*, **111**, 960–973.

Moore, C.H. 1986. Regional Jurassic Smackover dolomitization; importance, origin, and controls. *Geophysics*, **51**, 1321–1322.

Moore, C.H. & Druckman, Y. 1981. Burial diagenesis and porosity evolution, upper Jurassic Smackover, Arkansas and Louisiana. *American Association of Petroleum Geologists Bulletin*, **65**, 597–628.

Moore, C.H., Chowdhury, A. & Chan, L. 1988. Upper Jurassic Smackover Platform Dolomitization Northwestern Gulf of Mexico a Tale of Two Waters. *In*: Shukla, V. & Baker, P. (eds) *Sedimentology and Geochemistry of Dolostones*. Society of Economic Paleontologists and Mineralogists (SEPM), Special Publications, **43**, 175–190.

Morgan, J.V. 1952. Correlation of radioactive logs of the Lansing and Kansas City Groups in central Kansas. *Journal of Petroleum Technology*, **4**, 111–118.

Mountjoy, E.W., Machel, H.G., Green, D., Duggan, J. & Williams-Jones, A.E. 1999. Devonian matrix dolomites and deep burial carbonate cements: a comparison between the Rimbey Meadowbrook reef trend and the deep basin of west-central Alberta. *Bulletin of Canadian Petroleum Geology*, **47**, 487–509.

Neilson, J.E., Oxtoby, N.H., Simmons, M.D., Simpson, I.R. & Fortunatova, N.A. 1998. The relationship between petroleum emplacement and carbonate reservoir quality: examples from Abu Dhabi and the Amu Darya Basin. *Marine and Petroleum Geology*, **15**, 57–72.

Newell, D.K. 1997. Comparison of maturation data and fluid-inclusion homogenization temperatures to simple thermal models: implications for thermal history and fluid flow in the Midcontinent. *Kansas Geological Survey, Current Research in Earth Sciences Bulletin*, **240**, 13–27.

Newell, D.K., Watney, W.L., Cheng, S.W.L. & Brownrigg, R.L. 1987. *Stratigraphic and Spatial Distribution of Oil and Gas Production in Kansas*. Kansas Geological Survey, Subsurface Geology Series, **9**, 1–86.

Nurmi, R. & Neuberger, D. 1987. Diagenetic evolution and petrophysical characteristics of oomoldic facies in United States and Middle East reservoirs. *American Association of Petroleum Geologists Bulletin*, **71**, 995.

Ohlmacher, G.C. & Berendsen, P. 2005. Kinematics, mechanics, and potential earthquake hazards for faults in Pottawatomie County, Kansas, USA. *Tectonophysics*, **396**, 227–244.

Oliver, J. 1986. Fluids expelled tectonically from orogenic belts: their role in hydrocarbon migration and other geologic phenomena. *Geology*, **14**, 99–102.

O'Neil, J.R., Clayton, R.N. & Mayeda, T.K. 1969. Oxygen isotope fractionation in divalent metal carbonates. *Journal of Chemical Physics*, **51**, 5547–5558.

Parkhurst, R.W. 1959. *Surface to subsurface correlations and oil entrapment in the Lansing and Kansas City Groups (Pennsylvanian) in northwestern Kansas*. MS thesis, University of Kansas, Lawrence, KS.

Phares, R.A. 1991. *Characterization and reservoir performances of the Lansing-Kansas City 'I' and 'J' Zones (Upper Pennsylvanian) in the Pen Oil Field, Graham County, Kansas*. MS thesis, University of Kansas, Lawrence, KS.

Poteet, J.E. 2007. *Porosity and permeability evolution of the Raytown Limestone oolite, central Kansas*. MS thesis, University of Kansas, Lawrence, KS.

Prather, B.E. 1984. Deposition and diagenesis of an Upper Pennsylvanian cyclothem from the Lansing–Kansas City groups, Hitchcock County, Nebraska. *In*: Hyne, N.J. (ed.) *Limestones of the Midcontinent*. Tulsa Geological Society, Special Publications, **2**, 393–419.

Raef, A.E., Miller, R.D., Franseen, E.K., Byrnes, A.P. & Watney, W.L. 2005. 4D seismic to image a thin carbonate reservoir during a miscible CO_2 flood: Hall-Gurney Field, Kansas, USA. *The Leading Edge*, **24**, 521–526.

Ramaker, E.M., Goldstein, R.H., Franseen, E.K. & Watney, W.L. 2014. What controls porosity in cherty fine-grained carbonate reservoir rocks? Impact of stratigraphy, unconformities, structural setting and hydrothermal fluid flow: Mississippian, SE Kansas. *In*: Agar, S. & Geiger, S. (eds) *Fundamental Controls on Fluid Flow in Carbonates: Current Workflows to Emerging Technologies*. Geological Society, London, Special Publications, **406**, 179–208, https://doi.org/10.1144/SP406.2.

Rasbury, E.T., Hanson, G.N., Meyers, W.J., Holt, W.E., Goldstein, R.H. & Saller, A.H. 1998. U–Pb dates of paleosols: constraints on late Paleozoic cycle durations and boundary ages. *Geology*, **26**, 403–406.

Rong, H., Yangquan, J. et al. 2012. Relationship between heterogeneity and seismic velocities of the Yudongzi Triassic oolitic reservoirs in the Erlangmiao area, northwest Sichuan Basin, China. *Journal of Petroleum Science and Engineering*, **100**, 81–98.

Rossi, C., Goldstein, R.H., Ceriani, A. & Marfil, R. 2002. Fluid inclusions record thermal and fluid evolution in reservoir sandstones, Khatatba Formation, Western Desert, Egypt: a case for fluid injection. *American Association of Petroleum Geologists Bulletin*, **86**, 1773–1799.

Salas, J., Taberner, D., Esteban, M. & Ayora, C. 2007. Hydrothermal dolomitization, mixing corrosion and deep burial porosity formation: numerical results from 1-D reactive transport models. *Geofluids*, **7**, 99–111.

Saller, A.H. & Henderson, N. 1998. Distribution of Porosity and Permeability in Platform Dolomites. Insight from the Permian of West Texas. *American Association of Petroleum Geologists Bulletin*, **82**, 1528–1550.

Saller, A.H., Dickson, J.A.D. & Boyd, S.A. 1994. Cycle stratigraphy and porosity in Pennsylvanian and Lower Permian shelf limestones, eastern Central Basin Platform, Texas. *American Association of Petroleum Geologists Bulletin*, **78**, 1820–1842.

Saller, A.H., Dickson, J.A.D. & Matsuda, F. 1999. Evolution and distribution of porosity associated with subaerial exposure in upper Paleozoic platform limestones, West Texas. *American Association of Petroleum Geologists Bulletin*, **83**, 1835–1854.

SANDBERG, P.A. 1985. Aragonite cements and their occurrence in ancient limestones. *In*: SCHNEIDERMANN, N. & HARRIS, P.M. (eds) *Carbonate Cements*. Society of Economic Paleontologists and Mineralogists (SEPM), Special Publications, **36**, 33–57.

SCHOLLE, P.A. & ULMER-SCHOLLE, D.S. 2003. *A Color Guide to the Petrography of Carbonate Rocks: Grains, Textures, Porosity, Diagenesis*. American Association of Petroleum Geologists, Memoirs, **77**.

SHABESTARI, G.M., WORDEN, R.H. & MARSHALL, J.D. 2009. Source to cement in the Great Oolite Reservoir, Storrington Oil Field, Weald Basin, south of England. *Journal of Sciences, Islamic Republic of Iran*, **20**, 41–45.

SHELTON, K.L., BEASLEY, J.M., GREGG, J.M., APPOLD, M.S., CROWLEY, S.F., HENDRY, J.P. & SOMERVILLE, I.D. 2011. Evolution of a Carboniferous carbonate-hosted sphalerite breccia deposit, Isle of Man. *Mineralium Deposita*, **46**, 859–880.

SMITH, L.B. & DAVIES, G.R. 2006. Structurally controlled hydrothermal alteration of carbonate reservoirs: Introduction. *American Association of Petroleum Geologists Bulletin*, **90**, 1635–1640.

SORENSON, R.P. 2003. A dynamic model for the Permian Panhandle and Hugoton gas fields, Western Anadarko Basin. *AAPG Search and Discovery article 20015, from poster session presented at the 2003 AAPG Mid-Continent Section Meeting*, October 12–14, 2003, Tulsa, Oklahoma, USA.

SPENCER, R.J. 1987. Origin of Ca–Cl brines in Devonian formations, western Canada sedimentary basin. *Applied Geochemistry*, **2**, 373–384.

STAVNES, S.A. 1982. *A Preliminary Study of the Subsurface Temperature Distribution in Kansas and Its Relationship to the Geology*. Kansas Geological Survey, Open-File Report **82-10**.

TERRY, R.D. & CHILINGAR, G.V. 1955. Summary of 'Concerning some additional aids in studying sedimentary formations' by M. S. Shvetsov. *Journal of Sedimentary Petrology*, **25**, 229–234.

TUCKER, M.E. (ed.) 1988. *Techniques in Sedimentology*. Blackwell Scientific, Oxford.

UFNAR, D.F., GONZALEZ, L.A., LUDVIGSON, G.A., BRENNER, R.L. & WITZKE, B.J. 2002. The mid-Cretaceous water bearer: isotope mass balance quantification of the Albian hydrologic cycle. *Palaeogeography, Palaeoclimatology, Palaeoecology*, **188**, 51–71.

VEIZER, J., ALA, D. *ET AL*. 1999. $^{87}Sr/^{86}Sr$, $\delta^{18}O$, $\delta^{13}C$ evolution of Phanerozoic seawater. *Chemical Geology*, **161**, 59–88.

WALTERS, R.F. 1958. Differential entrapment of oil and gas in Arbuckle dolomite of central Kansas. *American Association of Petroleum Geologists Bulletin*, **42**, 133–132.

WALTON, A.W., WOJCIK, K.M., GOLDSTEIN, R.H. & BARKER, C.E. 1995. Diagenesis of Upper Carboniferous rocks in the Ouachita foreland shelf in mid-continent USA: an overview of widespread effects of a Variscan-equivalent orogeny. *Geologische Rundschau*, **84**, 535–551.

WANLESS, H.R. 1969. Marine and non-marine facies of the Upper Carboniferous of North America. *In*: *Compte Rendu de Congress International Stratigraphy Carboniferous, Sheffield, Volume 1*, 293–336.

WATNEY, W.L. 1980. *Cyclic sedimentation of the Lansing–Kansas City Groups in Northwestern Kansas and Southwestern Nebraska*. Kansas Geological Survey Bulletin, **220**.

WATNEY, W.L. 1983. Carbonate dominated shelf cycles in the Late Pennsylvanian of midcontinent-intrabasinal and extrabasinal controls on sedimentation and early diagenesis. *American Association of Petroleum Geologists Bulletin*, **67**, 567.

WATNEY, W.L. 1984. Recognition of favorable reservoirs trends in Upper Pennsylvanian cyclic carbonates in western Kansas. *In*: HYNE, N.J. (ed.) *Limestones of the Midcontinent*. Tulsa Geological Society, Special Publications, **2**, 201–245.

WATNEY, W.L. 1985. Evaluation of the significance of tectonic, sedimentary control v. eustatic control of Upper Pennsylvanian cyclothems in the western midcontinent. *In*: *Recent Interpretation of Late Paleozoic Cyclothems: Proceedings of the Third Annual Meeting and Field Trip Conference, Society of Economic Paleontologists and Mineralogists Midcontinent Section*. Kansas Geological Survey, Lawrence, Kansas, 105–140.

WATNEY, W.L., FRANSEEN, E.K., BYRNES, A.P. & NISSEN, S. 2008. Evaluating structural controls on the formation and properties of Carboniferous carbonate reservoirs in the northern midcontinent, U.S.A. *In*: LUKASIK, J. & SIMO, J.A. (eds) *Controls on Carbonate Platform and Reef Development*. Society of Economic Paleontologists and Mineralogists (SEPM), Special Publications, **89**, 125–146.

WILKE, N.A. & CARR, T.R. 2001. Sequence stratigraphy of oomoldic reservoir analogs in the Swope Formation (Missourian Series, Pennsylvanian System) in eastern Kansas and western Missouri. *American Association of Petroleum Geologists and Society of Economic Paleontologists and Mineralogists, Annual Meeting, Abstracts*, **10**, 90906.

WOJCIK, K.M. 1991. *Diagenesis of Pennsylvanian sandstones and limestones, Cherokee basin, southeastern Kansas: importance of regional fluid flow*. PhD dissertation, University of Kansas.

WOJCIK, K.M., GOLDSTEIN, R.H. & WALTON, A.W. 1994. History of diagenetic fluids in a distant foreland area, Middle and Upper Pennsylvanian, Cherokee basin, Kansas: fluid inclusion evidence. *Geochimica et Cosmochimica Acta*, **58**, 1175–1192.

WRIGHT, N. & LEONARD, K.W. 2007. A tale of two cyclothems: a comparison of two sections of the Iola high frequency sequence using sequence stratigraphy and conodont distribution patterns (Upper Pennsylvanian, Iowa and Kansas). *Geological Society of America Abstracts with Programs*, **39**, 57.

ZELLER, D.E. (ed.) 1968. *The Stratigraphic Succession in Kansas*. Kansas Geological Survey Bulletin, **189**.

Evaluation of porosity change during chemo-mechanical compaction in flooding experiments on Liège outcrop chalk

WENXIA WANG[1,2], MERETE V. MADLAND[1,2], UDO ZIMMERMANN[1,2*], ANDERS NERMOEN[2,4], REIDAR I. KORSNES[1,2], SILVANA R. A. BERTOLINO[3] & TANIA HILDEBRAND-HABEL[2,4]

[1]*Department of Petroleum Technology, University of Stavanger, Norway*

[2]*The National IOR Centre of Norway, University of Stavanger, Ullandhaug, 4036 Stavanger, Norway*

[3]*FaMAF, Universidad Nacional de Córdoba, Medina Allende s/n, Ciudad Universitaria, Córdoba, Argentina*

[4]*IRIS AS, International Research Institute of Stavanger, Ullandhaug, 4036 Stavanger, Norway*

**Correspondence: udo.zimmermann@uis.no*

Abstract: The mechanical strength, porosity and permeability of chalk are affected by chemical and mineralogical changes induced by fluids that are chemically out of equilibrium with the host rock. Here, two high-porosity Upper Cretaceous chalk cores from Liège were tested at effective stresses beyond yield at 130°C during flooding with $MgCl_2$ and NaCl brines. Core L1 (flooded by $MgCl_2$ brine) deformed more than L2 (flooded with NaCl brine), with volumetric strains of 9.4% and 5.1%, respectively. The porosity losses estimated from strain measurements alone are 5.82% for L1 and 3.01% for L2. However, this approach does not account for dissolution and precipitation reactions. Porosity calculations that are based on strain measurements in combination with (i) the weight difference between saturated and dry cores and (ii) the solid density measurement before and after flooding show an average porosity reduction of 3.69% between the two methods for L1. This discrepancy was not observed for core L2 (with the NaCl brine). The rock and effluent chemistry show that Ca^{2+} dissolved and Mg^{2+} is retained within the core for the L1 experiment. Therefore, accurate porosity calculations in chalk cores that are flooded by non-equilibrium brines (e.g. $MgCl_2$) require both the volumetric strain and chemical alteration to be considered.

Chalk strength, which is associated with the elastic moduli, the onset of yield and the rates of deformation, has been the focus of several research communities since the 1980s, when the subsidence in the Ekofisk field (Norwegian North Sea) was discovered (Wiborg & Jewhurst 1986). Seawater injection was initiated in 1987 to improve the oil production and to stop or reduce subsidence. Oil production improved significantly, but reservoir compaction in the water-saturated regions continued, in contrast to the regions with no water breakthrough. This observation indicates the existence of a water weakening effect on the chalk's mechanical properties. Numerous studies have since been initiated to improve the understanding of: (i) the relation between the pore fluid composition and mechanical properties (Newman 1983; Madland 2005; Korsnes *et al.* 2006*a*, *b*); (ii) the effect of mechanical strength as temperature decreases during injection (Korsnes *et al.* 2008; Madland *et al.* 2008); and (iii) the relation between the changed brine chemistry and deformation rates under constant stress conditions (e.g. Hellmann *et al.* 2002; Korsnes 2007; Madland *et al.* 2011; Nermoen *et al.* 2015).

During the continuous injection of brines such as synthetic seawater (SSW) that are not in chemical equilibrium with the host rock, equilibrium calculations at the pressures and temperatures relevant to some reservoirs indicate that the brines are supersaturated in terms of Mg^{2+}, leading to the precipitation of new mineral phases (Hiorth *et al.* 2008). Consequently, the fluids become undersaturated in other elements such that additional secondary dissolution is triggered (Madland *et al.* 2011). In contrast, chalk flooding experiments with simpler brines, such as $MgCl_2$, which are not in equilibrium with the host

From: Armitage, P. J., Butcher, A. R., Churchill, J. M., Csoma, A. E., Hollis, C., Lander, R. H., Omma, J. E. & Worden, R. H. (eds) 2018. *Reservoir Quality of Clastic and Carbonate Rocks: Analysis, Modelling and Prediction.* Geological Society, London, Special Publications, **435**, 217–234.
First published online October 26, 2016, https://doi.org/10.1144/SP435.10

rock, show that the dissolution of minerals may lead to the precipitation of Mg-bearing carbonate minerals (Madland *et al.* 2013).

The chalk reservoir at Ekofisk field (North Sea) is currently undergoing compaction during seawater injection, despite the reduction in effective stress. An increase in pore pressure will usually reduce or even stop compaction when no chemical interaction between the rock and brine is taking place. Reservoir compaction by pore volume reduction is an important driving mechanism that forces pore fluids towards the production wells. At the same time, the reduction in pore volume changes the flow properties (i.e. by reducing the permeability) of the reservoir. To support qualified reservoir management decisions and EOR implementation it is therefore important to have accurate estimates of the pore space evolution during fluid injection.

The objective of this article is to show that the non-equilibrium nature of the rock–fluid interface induces mineralogical changes and to demonstrate how those mineralogical changes affect the mechanical properties of the rock. These results lead to a more accurate way of estimating porosity compared with using bulk volumetric strain alone. This approach is necessary because significant changes in the mass and density, and hence, the solid volume, occur during the injection of fluids that are chemically out of equilibrium with the host chalk rocks. This effect also may also be relevant for other reservoir rock types where chemo-mechanical interactions occur.

Methods and experimental procedures

Core material and flooding fluids

Large sample blocks were taken from the quarry close to Hallembaye (Cimenterie Belge Reunié (CBR) Lixhe quarry; Slimani 2001) which is situated in the Liège region (Belgium) at 50°44′ 51.16″N and 05°38′52.17″E. The petrology, mineralogical and rheological characteristics of the chalk have been studied in detail (Felder 1975; Molenaar & Zijlstra 1997; Slimani 2001; Strand *et al.* 2007; Hjuler & Fabricius 2009). The porosity varies between 40% and 45% and the chalk is composed of *c.* 97% carbonate minerals.

In this study, 0.219 M $MgCl_2$ and 0.657 M NaCl were used as the injection fluids (Table 1).

Experimental preparation and procedure

Two samples were drilled from a large chalk block and machined into cylindrical shapes with initial diameters of 37.05 and 38.15 mm and lengths of 68.88 and 72.65 mm for tests L1 and L2, respectively. Because both cores have a length to diameter ratio that is close to two their strengths should be insensitive to differences in their diameters (Fjær *et al.* 2008). After preparation, both cores were dried for 12 h in a heating chamber at 100°C to remove more than 99% of the water in the core before the dry mass was determined. The calcite crystals are unaffected by the drying conditions. Then the cores were evacuated by vacuum before being saturated with distilled water and weighed to obtain the saturated mass. The difference between the saturated and dry masses is attributed to the mass of the pore water, which is then used to estimate the pore volume fraction (i.e. porosity). The same experimental procedure was used on both cores. The saturated cores were mounted into two identical triaxial cells (see Fig. 1 for a description of the set-up). The cells were equipped with a heating jacket and a regulating system (Omron E5CN) with precise proportional integral derivative (PID) temperature control ($\pm$0.1°C). Three Gilson (model 307 HPLC) pressure pumps were connected to each triaxial cell enabling the independent control of the piston pressure, the confining pressure (radial stress) and the injected flow rate (i.e. pumps 1–3 in Fig. 1). The confining pressure balances the piston such that the axial stress has to be calculated from the confining pressure plus the piston pressure, (see equation 1 in Nermoen *et al.* 2015). A back-pressure regulator was placed downstream of the core to control the pore pressure and allow for continuous sampling of the effluent fluids. To isolate the cores from the confining oil, a heat shrinkage sleeve (1–1/2″ FEP with a diameter of 33–43 mm and a wall thickness of 0.5 mm) was installed between the core and

Table 1. Injected brine composition for tests L1 and L2, flow rate and test duration

Test	Flow rate (ml min^{-1})	Test duration (days)	Cl^- (mol l^{-1})	Mg^{2+} (mol l^{-1})	Na^+ (mol l^{-1})	Ion strength (mol l^{-1})	TDS (g l^{-1})
L1	0.022	65	0.438	0.219	0	0.657	20.84
L2	0.025	90	0.657	0	0.657	0.657	38.40

TDS, total dissolved solids.

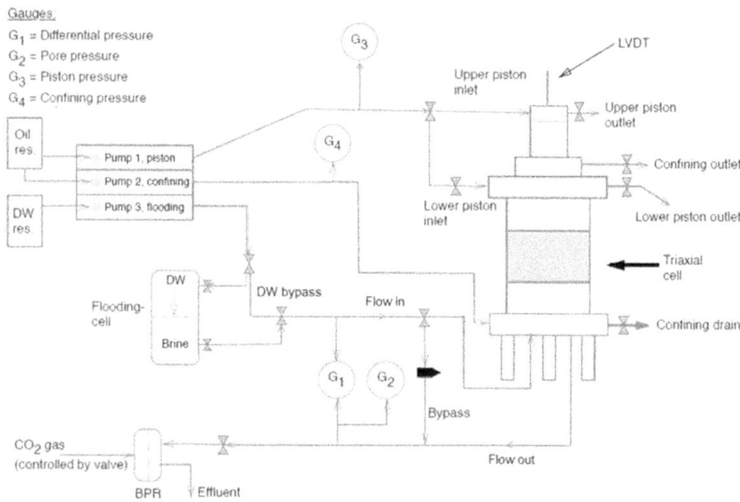

Fig. 1. The triaxial cell used in the flooding tests. Three pumps (the piston, confining and flooding pumps) were connected to the cell. The flooding cell contained DW (upper chamber) and $MgCl_2/NaCl$ (lower chamber). During the flooding period the cell was refilled as soon as the fluid ran out. This whole set-up was connected to the computer and complete data, including the confining, pore and piston pressures, flooding rate and time, were logged automatically.

the confining oil. After core mounting, the confining pressure was increased to 0.5 MPa. Thereafter the experiments were conducted according to the following stages:

(1) The samples were flooded with three pore volumes (PVs) of distilled water (DW) overnight at ambient temperature to remove any salt precipitates that could affect the flooding test. This flooding procedure does not significantly alter the geochemical measurements of the non-salt minerals in the core.
(2) Flooding was then switched to an $MgCl_2$ brine for L1 and an NaCl brine for L2, by attaching the piston cell into the flow loop (see Fig. 1). The ion compositions of the injected brines are shown in Table 1. Throughout the rest of the test, the flow rate was set to one initial PV per day, which corresponds to 0.022 and 0.025 ml min^{-1} for L1 and L2, respectively (Table 1).
(3) The confining pressure and pore pressure were increased to 1.2 and 0.7 MPa, respectively, before the temperature was raised to 130°C. The pore pressure and temperature were subsequently kept constant throughout the test.
(4) The confining pressure was increased from 1.2 MPa by injecting hydraulic oil into the confining chamber at a constant flow rate using pump 2. During pressurization the piston pressure was set to 0.5 MPa to slightly exceed friction of the piston (0.3–0.4 MPa). This increased the axial stress to a value of 0.1–0.2 MPa above the radial stress. Because the additional axial stress is small compared with the radial stress, the stress condition can be considered near-hydrostatic. The axial strain was measured by an external axial linear variable displacement transducer (LVDT) placed on top of the piston to monitor the sample length with time. The stress–strain behaviour was monitored as the confining pressure was increased (see Fig. 2). When the rock began to deform plastically (Fig. 3), i.e. when the stress–strain behaviour became non-linear, the stress at the onset of yield was noted (7.5 and 8.5 MPa for L1 and L2, respectively). The targets for the confining pressure were selected to be 2 MPa above the yield stress, i.e. 9.5 and 10.5 MPa for cores L1 and L2.
(5) The axial deformation at constant temperature, stress and pressure conditions (termed creep) was monitored for 65 and 90 days during continuous flooding with $MgCl_2$ and NaCl brines for tests L1 and L2, respectively. The creep magnitude through time is shown in Figure 3. The pore pressure and confining pressure varied within 0.1 MPa such that the effective stresses were stable throughout the test period.

The pore pressure, hydraulic pressure difference, confining pressure, piston pressure, sample length (axial strain) and flooding time were logged

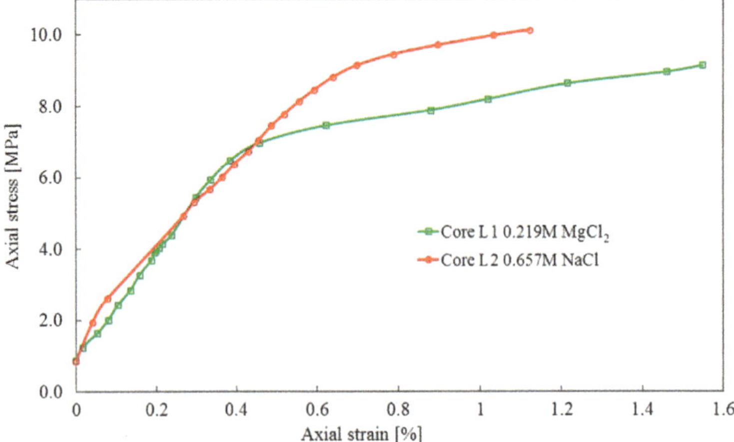

Fig. 2. Axial stress v. axial strain for the L1 and L2 cores flooded with $MgCl_2$ and NaCl, respectively, at 130°C. Note that L2 has higher yield point (8.5 MPa) than the core L1 (7.5 MPa).

continuously via a LabView programme. Before dismantling the core, the sample was cleaned by injecting three PVs of distilled water to avoid the precipitation of salts from the $MgCl_2$ and NaCl brines during drying. The saturated weight was measured immediately after the cell was dismantled. The core was then placed in the drying cabinet at 100°C and weighed several times until the mass stayed constant. Both experiments have been performed in the same way so the brine dependent behaviour can be identified. The basic core properties before and after flooding are presented in Table 2. Then the cores were cut into seven slices using a Struers Discotom-5 cutting machine. The slices were labelled FL1-1 to FL1-7 for flooded L1 and FL2-1 to FL2-7 for flooded L2 along the core axis from the inlet to the outlet (see Fig. 4 for the sectioning scheme). The slices were used for geochemical analyses of the rock to identify the mineralogical changes along the core.

Ion chromatography

During the experiment, the effluents of the samples were collected for chemical comparison between the produced and injected water. The ionic

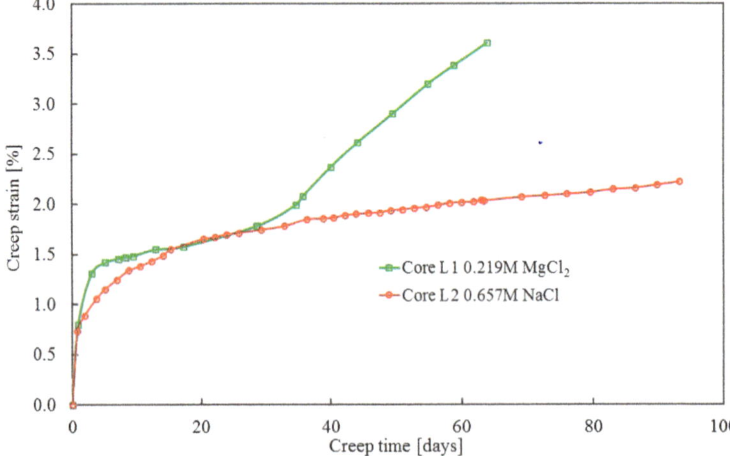

Fig. 3. Axial creep strain v. creep time for L1 and L2 under different creep stress at 130°C. The different creep stresses for L1 and L2 are 9.5 and 10.5 MPa respectively. Note that the L1 test accumulates more creep strain over 65 days than the L2 test, which was flooded for 90 days.

Table 2. *Basic core measurements before and after the tests performed at 130°C for L1 and L2, respectively*

Test	Initial core values										
	Length (mm)	Diameter (mm)	Dry mass (g)	Saturated mass (g)	Bulk volume (cm^3)	Solid volume (saturation) (cm^3)	Pore volume (saturation) (cm^3)	Porosity (saturation) (%)	Density (pycnometry) ($g\ cm^{-3}$)	Mineral density (saturation) ($g\ cm^{-3}$)	Solid volume (pycnometry) (cm^3)
L1 ($MgCl_2$ at 9.5 Mpa)	68.88	37.05	112.38	144.65	74.26	41.99	32.27	43.45	2.68	2.68	41.93
L2 (NaCl at 10.5 Mpa)	72.65	38.15	126.05	162.12	83.05	46.98	36.07	43.43	2.69	2.68	46.86

Test (three sets of FL1)	Flooded core material after the test										
	Vol. strain (%)	Dry mass (g)	Saturated mass (g)	Bulk volume (truncated wedge) (cm^3)	Density (pycnometry) ($g\ cm^{-3}$)	Solid volume (saturation) (cm^3)	Pore volume (saturation) (cm^3)	Solid volume (pycnometry) (cm^3)	Porosity (method 1) (%)	Porosity (method 2) (%)	Porosity (method 3) (%)
FL1	9.4	110.28	137.38	67.27	2.70	40.17	27.10	40.84	37.63	40.29	39.28
FL2	5.1	125.51	157.85	78.77	2.69	46.43	32.34	46.66	40.42	41.06	40.77

Comparison of the dry and DW-saturated weights with the solid volume, pore volume and porosity (ϕ) estimates based on both saturation and the pycnometer results from Table 3. In addition, volumetric strains and porosities are estimated using the methods 1–3 as explained in the 'Porosity calculation' section. Indices s1 to s7 relate to the position along the axis of the sample from inlet to outlet (see Fig. 4). 'Test' indicates three sets of tests for FL 1.

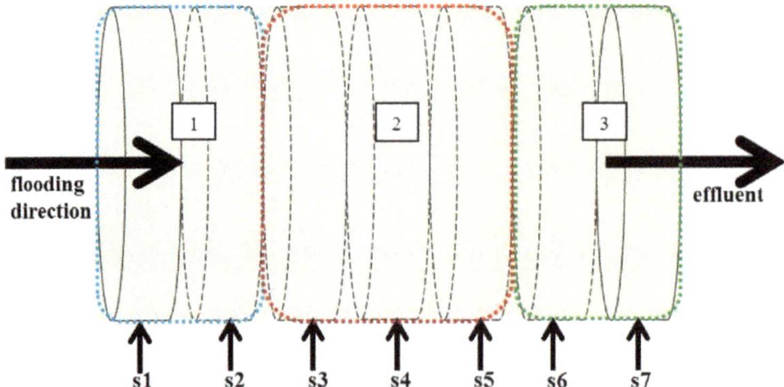

Fig. 4. Schematic of the sectioning of the core after the experiment. The subsamples were analysed using different methods. The inlet is at s1 and the outlet at s7. The samples were grouped as indicated (1–3) for pycnometry measurements.

concentrations were analysed with a Dionex ICS-3000 ion-exchange chromatograph. The analyses were performed with an ICS 3000 CD Conductivity Detector. IonPac AS16 and IonPac CS12A were used as the anion and cation exchange columns, respectively. The sampled effluents were diluted (Gilson, GX-271) to stay in the linear region of the calibration curve and the ionic concentrations were calculated on the basis of an external standard (SMOW; standard mean ocean water). The following ions have been quantified and plotted through time in Figure 5: Mg^{2+}, Na^{+}, Cl^{-} and Ca^{2+}.

Pycnometry

The solid volume was measured with an AccuPyc II 1340 gas pycnometer, which measures the amount of displaced helium gas within a porous rock sample. The dry sample was placed into the volume chamber and helium gas molecules rapidly filled the pores. A chamber of 35 cm^3 was selected to provide the best fit with the samples. Estimates of the average mineral densities were obtained based on the solid volume and dry mass (see Table 3). The pycnometry measurements represent a complementary way of estimating the porosity, in addition to the porosity estimate from the difference between the dry and DW-saturated cores.

Field emission gun scanning electron microscopy (FEG-SEM)

The fresh surfaces from each slice (Fig. 4) were analysed using a Zeiss Supra 35VP field emission SEM in high vacuum mode with an accelerating voltage of 12 kV, an aperture size of 30 μm and a working distance of 10–12 mm. EDAX Genesis energy-dispersive X-ray spectroscopy (EDS) was used to determine the mineralogical and elemental compositions of the chalk samples. Because the core dominantly consisted of calcite, the EDS system was calibrated using an Iceland spar calcite crystal before the samples were analysed.

X-Ray diffraction

X-Ray diffraction patterns were obtained from bulk samples of the different slices (in Fig. 4) in random mounts using a Bruker D5005 diffractometer, Cu–Kα radiation, 0.02° per step at 1 s per step. The measurements were repeated in certain samples to evaluate the reproducibility of the results; particularly for cases involving newly formed trace minerals. The available amount for measurements was around 1 g. The XRD analyses were performed at the Institute of Earth Sciences Jaume Almera (ICTJA, Barcelona, Spain).

Whole-rock geochemistry

Representative sample material from each slice (see Fig. 4) was carefully separated from the core and milled to a very fine mesh in a clean agate mill. Geochemical data were obtained by inductively coupled plasma mass spectrometry (ICP-MS) analysis at Acme laboratory (Vancouver, Canada). Details of the analytical method and processing technique can be found at http://acmelab.com but are summarized here. The samples were first ground in an agate mill. The milled sample was then mixed with an $LiBO_2/Li_2B_4O_7$ flux in crucibles and fused in a furnace. The cooled bead was dissolved in ACS grade nitric acid and analysed by ICP-MS. Loss on ignition (LOI) was determined by igniting a sample split then measuring the weight loss. A 1 g sample was weighed into a tarred crucible and ignited to 1000°C for 1 h before being cooled and weighed again. The loss in weight is the LOI of the sample.

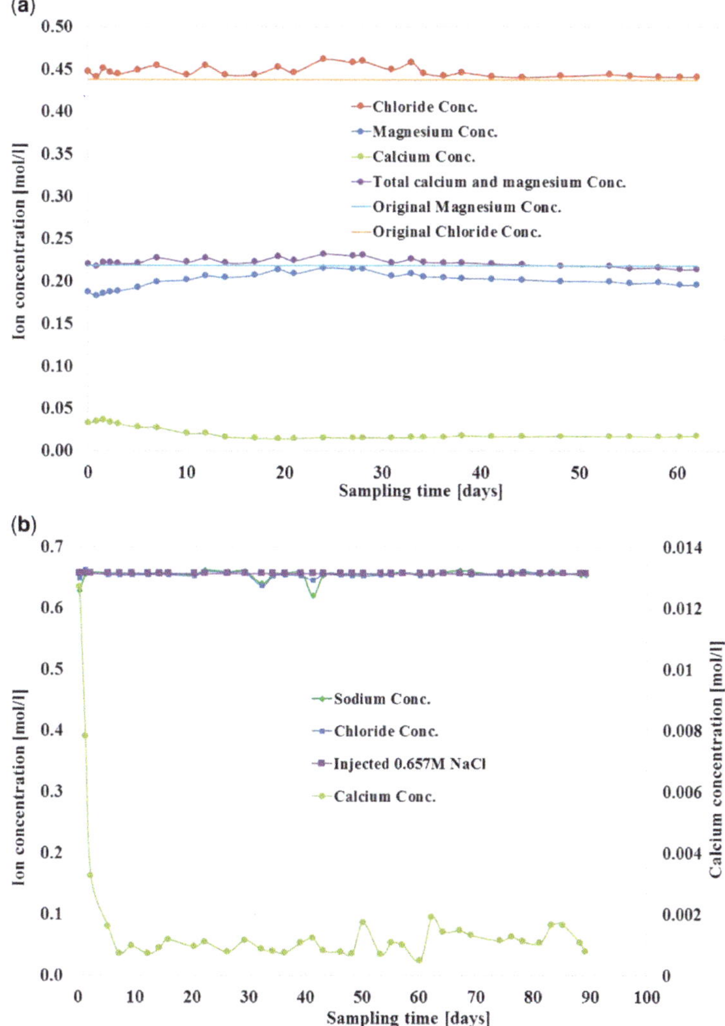

Fig. 5 (**a**) Mg^{2+}, Ca^{2+} and Cl^- concentrations in the sampled effluent from core L1 flooded with a 0.219 M $MgCl_2$ brine. (**b**) Na^+, Ca^{2+} and Cl^- concentrations in the sampled effluent from core L2 flooded with 0.657 M NaCl. The test temperature was set to 130°C. Sampling started after the hydrostatic loading was completed.

Table 3. *Densities obtained from pycnometer measurements of flooded core samples*

Pycnometry results		
Test	Slices	Density (g cm^{-3})
FL1	s1 + s2	2.71
	s3 + s4 + s5	2.70
	s6 + s7	2.70
FL2	s1 + s2	2.69
	s3 + s4 + s5	2.69
	s6 + s7	2.69

Total carbon and sulphur were determined by the Leco method. Here, an induction flux was added to the prepared sample then ignited in an induction furnace. A carrier gas sweeps up the released carbon to be measured by adsorption in an infrared spectrometric cell. The results are the total concentrations attributed to the presence of carbon and sulphur in all of the components. An additional 14 elements were measured after dilution in aqua regia. The prepared sample was digested with a modified aqua regia solution with equal parts concentrated HCl, HNO_3 and deionized water for one hour in a heating block or hot water bath. The sample volume was

Table 4A. *Data for the FL1 and FL2 cores and untested end pieces of the L1 and L2 cores from Liège: (A) Geochemical isotope concentrations. Orig., unflooded chalk sample; avg., average; A, unflooded sample adjacent to the inlet (s1 in Fig. 4); B, unflooded sample adjacent to the outlet (s7 in Fig. 4)*

	SiO_2 %	Al_2O_3 %	Fe_2O_3 %	MgO %	CaO %	Na_2O %	K_2O %	TiO_2 %	P_2O_5 %	MnO %	LOI %	Sum %
Orig. L1 – A	2.21	0.38	0.19	0.29	53.98	0.05	0.1	0.02	0.07	0.02	42.6	99.86
Orig. L1 – B	2.15	0.37	0.18	0.27	54.1	0.05	0.09	0.02	0.07	0.02	42.6	99.87
Orig. L1 (avg.)	**2.18**	**0.38**	**0.19**	**0.28**	**54.04**	**0.05**	**0.10**	**0.02**	**0.07**	**0.02**	**42.6**	**99.87**
FL1 (avg.)	**2.00**	**0.37**	**0.13**	**1.85**	**52.32**	**0.04**	**0.05**	**0.02**	**0.12**	**0.02**	**42.8**	**99.84**
Orig. L2 – A	1.92	0.32	0.24	0.27	54.21	0.04	0.07	0.02	0.14	0.02	42.6	99.85
Orig. L2 – B	1.89	0.32	0.20	0.27	54.14	0.04	0.07	0.02	0.13	0.02	42.8	99.85
Orig. L2 (avg.)	**1.91**	**0.32**	**0.22**	**0.27**	**54.18**	**0.04**	**0.07**	**0.02**	**0.14**	**0.02**	**42.7**	**99.85**
FL2 (avg.)	**1.85**	**0.34**	**0.13**	**0.28**	**54.62**	**0.06**	**0.05**	**0.02**	**0.10**	**0.02**	**42.3**	**99.86**
Slice 1 to 7 of the $MgCl_2$ – flooded core (Test ID: L1)												
FL1-1	1.70	0.33	0.12	1.72	52.84	0.03	0.04	0.02	0.10	0.03	42.9	99.83
FL1-2	2.00	0.39	0.13	2.34	51.51	0.03	0.05	0.02	0.11	0.02	43.2	99.83
FL1-3	2.09	0.39	0.15	3.21	50.76	0.04	0.06	0.02	0.18	0.03	42.8	99.82
FL1-4	2.09	0.40	0.12	2.07	51.94	0.04	0.06	0.02	0.16	0.02	42.8	99.84
FL1-5	2.06	0.37	0.16	1.38	52.60	0.04	0.06	0.02	0.12	0.02	42.9	99.85
FL1-6	2.09	0.37	0.13	1.16	53.24	0.04	0.06	0.02	0.08	0.02	42.5	99.85
FL1-7	1.95	0.35	0.12	1.04	53.37	0.04	0.05	0.02	0.11	0.02	42.7	99.85
Slice 1 to 7 of the NaCl – flooded core (Test ID: L2)												
FL2-1	1.44	0.27	0.10	0.30	54.41	0.04	0.04	0.02	0.05	0.02	43.1	99.85
FL2-2	1.80	0.35	0.15	0.26	55.46	0.04	0.05	0.02	0.09	0.02	41.6	99.85
FL2-3	1.76	0.33	0.18	0.27	55.68	0.05	0.05	0.02	0.07	0.02	41.4	99.85
FL2-4	1.96	0.35	0.15	0.28	54.72	0.05	0.06	0.02	0.10	0.02	42.0	99.86
FL2-5	1.97	0.37	0.10	0.29	54.13	0.06	0.06	0.02	0.13	0.02	42.6	99.87
FL2-6	1.97	0.36	0.11	0.29	54.16	0.07	0.06	0.02	0.13	0.02	42.6	99.87
FL2-7	2.05	0.38	0.12	0.28	53.77	0.08	0.06	0.02	0.11	0.02	42.9	99.87

increased with dilute HCl solutions and splits of 0.5 g were analysed. All of the measured concentrations fall within the standard range (see Table 4), and accuracy and precision are between 2% and 3%.

Constitutive relations and porosity calculations

At any time the bulk volume of a porous material (V_b) is given by the volume of the pores (V_p) plus the volume of the solids (V_s),

$$V_b = V_s + V_p \tag{1}$$

Porosity (ϕ, expressed as a percentage) is calculated from the ratio

$$\phi = \frac{V_p}{V_b} \tag{2}$$

The initial porosity (ϕ_0) is calculated by the difference between the saturated and dry weights ($M_{sat,0}$ and $M_{dry,0}$) divided by the density of distilled water (ρ_{DW}) multiplied by the initial bulk volume ($V_{b,0}$),

$$\phi_0 = \frac{M_{sat,0} - M_{dry,0}}{\rho_{DW} V_{b,0}} \tag{3}$$

After the experiment was finished and dismantled, the diameter (D_i) was measured at intervals (L_i) along the core to provide accurate estimates of the total bulk volume from the sum of the truncated wedges $i = 1, 2, \ldots, 7$ along the core,

$$V_{b,end} = \sum_i \frac{L_i \pi}{3}\left(\left(\frac{D_i}{2}\right)^2 + \left(\frac{D_i}{2}\right)\left(\frac{D_{i+1}}{2}\right) + \left(\frac{D_{i+1}}{2}\right)^2\right) \tag{4}$$

This procedure was used because the volumetric deformation was not uniform along the sample, i.e. radial deformation was greater near the centre of the core than the ends. The overall change in the bulk volume before and after testing was used to estimate the total volumetric strain ε_{vol}, expressed as a percentage according to

$$\varepsilon_{vol} = \frac{V_{b,0} - V_{b,end}}{V_{b,0}} \tag{5}$$

It is standard procedure in triaxial experiments to define the positive strain when the volume or length of the core plug is reduced. Given a porous material, as defined in equation (1), any change in

Table 4B. *Data for the FL1 and FL2 cores and untested end pieces of the L1 and L2 cores from Liège: (B) isotope geochemistry and selected trace element whole-rock geochemistry. Orig., unflooded chalk sample; avg., average; A, unflooded sample adjacent to the inlet (s1 in Fig. 4); B, unflooded sample adjacent to the outlet (s7 in Fig. 4)*

	Ba (ppm)	Rb (ppm)	Sr (ppm)	Zr (ppm)	Y (ppm)	ΣREE (ppm)	Total carbon	$\delta^{13}C_{SMOW}$ (%)	$\delta^{18}O_{SMOW}$ (‰)	Temperature estimate (‰)
Orig. L1 – A	30.0	2.2	1008	5.3	7.4	21.8	11.9	1.495	−1.985	20.2
Orig. L1 – B	32.0	2.2	982	5.0	7.2	22.5	11.9	1.609	−1.670	18.8
Orig. L1 (avg.)	**31.0**	**2.2**	**995**	**5.2**	**7.3**	**22.1**	**11.9**	**1.552**	**− 1.828**	**19.5**
FL1 (avg.)	**12.1**	**2.7**	**961**	**9.5**	**8.1**	**25.6**	**12.4**	**1.592**	**− 2.049**	**20.5**
Orig. L2 – A	31.0	3.6	908	7.5	9.0	26.3	11.7	1.599	−1.911	19.9
Orig. L2 – B	35.0	3.6	912	8.8	8.2	23.9	12.0	1.803	−1.607	18.6
Orig. L2 (avg.)	**33.0**	**3.6**	**910**	**8.2**	**8.6**	**25.1**	**11.8**	**1.701**	**− 1.759**	**19.2**
FL2 (avg.)	**17.9**	**2.3**	**1007**	**7.7**	**7.5**	**24.1**	**12.6**	**1.652**	**− 1.887**	**19.8**
Slices 1–7 of the $MgCl_2$ – flooded core (Test L1)										
FL1-1	14.0	2.5	1007	7.2	7.1	23.1	11.9	1.549	−2.028	20.4
								1.591	−1.955	20.1
FL1-2	13.0	2.6	958	7.5	8.2	24.1	12.1	1.638	−1.991	20.2
								1.651	−2.017	20.3
FL1-3	12.0	2.7	891	8.9	9.2	26.8	12.5	1.498	−2.469	22.4
								1.542	−2.322	21.7
FL1-4	10.0	3.1	922	10.0	8.9	28.7	12.4	1.432	−2.312	21.7
								1.571	−2.115	20.8
FL1-5	12.0	2.9	963	11.5	8.3	25.3	12.7	1.613	−1.961	20.1
								1.674	−1.930	20.0
FL1-6	11.0	2.7	1008	11.0	7.0	25.7	12.6	1.697	−1.732	19.1
								1.630	−1.907	19.9
FL1-7	13.0	2.6	978	10.4	8.2	25.3	12.4	1.592	−1.995	20.2
								1.605	−1.947	20.0
Slices 1–7 of the NaCl – flooded core (Test L2)										
FL2-1	13.0	1.8	1043	8.1	7.1	23.4	12.5	1.686	−1.863	19.7
								1.627	−1.913	19.9
FL2-2	16.0	2.1	1039	5.7	7.4	22.5	12.7	1.661	−1.909	19.9
								1.647	−1.885	19.8
FL2-3	15.0	1.9	1034	5.6	7.0	21.9	12.7	1.663	−1.810	19.4
								1.662	−1.847	19.6
FL2-4	24.0	2.3	1020	7.8	8.2	24.7	12.8	1.614	−1.903	19.8
								1.625	−1.980	20.2
FL2-5	18.0	2.7	947	9.3	7.0	26.7	12.5	1.722	−1.852	19.6
								1.627	−1.903	19.8
FL2-6	19.0	2.5	968	8.6	8.2	23.3	12.6	1.637	−1.880	19.7
								1.672	−1.846	19.6
FL2-7	20.0	2.8	997	9.0	7.8	26.5	12.8	1.642	−1.898	19.8
								1.641	−1.933	20.0

*Palaeo- and flooding temperatures have been calculated after Anderson & Arthur (1983) with salinity values taken from Wright (1987). Highlighted in grey are the significant chemical changes discussed in the text. **Bold**, average values for the different samples.

the bulk volume (ΔV_b) is given by the change in the solid volume (ΔV_s) plus the change in the pore volume (ΔV_p),

$$\Delta V_b = \Delta V_s + \Delta V_p \qquad (6)$$

If all of the deformation is accommodated by the reduction in pore volume and the solid volume remains constant ($\Delta V_s = 0$) (either there is no rock–fluid interaction or the mineral density and mass alteration cancel out) then,

$$\Delta V_b = \Delta V_p \qquad (7)$$

If this assumption is valid then the porosity may be determined by combining equations (2, 4, 5 & 6)

$$\phi_{end,1} = \frac{V_p + \Delta V_b}{V_b + \Delta V_b} = \frac{\phi_0 - \varepsilon_{vol}}{1 - \varepsilon_{vol}} \qquad (8)$$

Equation (8) is referred to as method 1 in the porosity calculations.

The second way of estimating the porosity after the test (method 2) relies on the difference between the saturated and dry masses after the test in combination with equation (3),

$$\phi_{\text{end},2} = \frac{(M_{\text{sat,end}} - M_{\text{dry,end}})}{\rho_{\text{dw}} V_{\text{b,end}}} \tag{9}$$

The third way of quantifying the porosity is by using pycnometry measurements. After the core was dismantled from the triaxial cell and the dry and saturated masses were obtained, it was cut into 7 slices. To provide more accurate solid volume measurements in the pycnometer the slices were grouped into 3 sets (s1 + s2, s3 + s4 + s5 and s6 + s7, see Fig. 4). Using the solid volume and mass we estimated the average mineral density of each set ($\rho_{s,i}$) from,

$$\rho_{s,i} = \frac{M_{\text{dry},i}}{V_{s,i}} \tag{10}$$

where $i = 1, 2, 3$. The average solid mineral density of the whole core after flooding was calculated from the weighted average (by volume),

$$\rho_{\text{s,end}} = \frac{V_{s,1}\rho_1 + V_{s,2}\rho_{s,2} + V_{s,3}\rho_{s,3}}{\sum_i^3 V_{s,i}} \tag{11}$$

With these data we could estimate the porosity by combining the average solid density ($\rho_{\text{s,end}}$), the dry mass ($M_{\text{dry,end}}$) and bulk volume ($V_{\text{b,end}}$) after the experiment

$$\phi_{\text{end},3} = \frac{V_p}{V_b} = 1 - \frac{V_s}{V_b} = 1 - \frac{M_{\text{dry,end}}}{V_{\text{b,end}}\rho_{\text{s,end}}} \tag{12}$$

Equation (12) is referred to as method 3 for estimating the porosity after the test. By comparison we see that methods 2 and 3 take changes in mass and density into account, while method 1 only considers the volumetric strain changes.

Results

Porosity calculations before and after the experiment

The porosity of the tested chalk before the experiment based on the pore volume and the bulk volume (according to equation 3) is 43.45% and 43.43% for L1 and L2, respectively (Table 2). After flooding tests, ε_{vol} for L1 (9.4%) is larger than for L2 (5.1%) although both the total creep times and the effective stresses were higher for the L2 test than for compared with the L1 test (the L1 $MgCl_2$-flooded core was tested for 65 days at an effective stress of 9.5 MPa while the L2 NaCl-flooded core was tested for 90 days at 10.5 MPa). The effective stress during creep was set to be 2 MPa above the yield stress for both experiments (7.5 and 8.5 MPa for L1 and L2, respectively. See Fig. 2). The time evolution in the observed creep for the L1 test in Figure 3 displays an intriguing effect: at *c.* 35–40 days the axial creep rate increased. A similar acceleration in the creep rate induced by $MgCl_2$ flow was observed by Madland *et al.* (2011). This accelerated creep behaviour is not observed when the core is flooded by more inert brines such as NaCl. With time, the brine composition is the dominant factor, together with the effective stress, in determining the total observed volumetric deformation.

After the experiments, the dry mass changed for both cores. In the $MgCl_2$-flooded core (L1) 2.1 g was lost, which corresponds to 1.87% over 65 days. The NaCl-flooded core (L2) lost a mass of 0.54 g, corresponding to 0.43% (see Table 2) over 90 days. Hence, the $MgCl_2$ brine is more reactive towards the chalk than the NaCl brine. Given a specific density of approximate 2.70 g cm^{-3}, the mass losses for L1 and L2 correspond to solid volume losses of 1.09 and 0.20 cm^3, respectively. The weight difference between the DW-saturated cores and the dried cores was used to calculate the pore volume. For L1 the pore volume changed from 32.27 cm^3 before the test to 27.10 cm^3 after the test, while for L2 the pore volume changed from 36.07 cm^3 to 32.34 cm^3. At the same time the bulk volume changed from 74.26 cm^3 to 67.27 cm^3 for L1 and from 83.05 cm^3 to 78.77 cm^3 for L2. The average density of L1 increased from 2.68 g cm^{-3} to 2.70 g cm^{-3}, while there is no change in density for L2 after the experiment (Table 2). The resulting porosities that were calculated according to the three methods (in equation 8, 9 & 12) are shown in Table 2. For L1, method 1 (from the volumetric strain) results in a porosity reduction from 43.45% to 37.63%. However when using methods 2 and 3, which include the mass and density alterations in addition to the volume after the test, the porosity calculation revealed a smaller reduction than for method 1. The porosity changed from 43.45% to 40.29% and 39.28% for each method. The porosity estimates from method 1, 2 and 3 for L2 are 40.42%, 41.06% and 40.77%, respectively, from the original porosity of 43.43%. The L2 core shows a smaller variation in the porosity values from the three methods compared to the L1 core. By including the effect of chemical alteration in the porosity calculation, the difference in the final porosity between the $MgCl_2$- and NaCl-flooded cores is less significant, even though total volumetric strain is close to factor of two higher for the $MgCl_2$-flooded core (Table 2).

Effluent chemistry

The chemical analysis of the effluent during injection of $MgCl_2$ and NaCl confirms that chemical reactions occur inside the cores, as shown by the difference in both the density and the mass after flooding. Throughout the experiment the effluent from the $MgCl_2$-flooded core (L1) is depleted in Mg^{2+} ions (see Fig. 5a). The average Mg^{2+} ion concentration of the effluent is 0.202 mol l^{-1} in comparison with the 0.219 mol l^{-1} concentration of the injected brine, i.e. it is depleted by −0.017 mol l^{-1}. On the other hand, the average calculated Ca^{2+} concentration of the effluent water is 0.021 mol l^{-1}, corresponding to a net gain in the core material. The sum of the magnesium and calcium ion concentrations in the effluent (*c.* 0.222 mol l^{-1}) is, however, reasonably close to the injected $MgCl_2$ concentration (0.219 mol l^{-1}; Table 1), such that stoichiometry is almost preserved, indicating that a one-to-one ion exchange reaction occurs. Hence, the tested core experiences a significant magnesium gain and calcium loss. At the same time, the injected chloride reaches its original concentration (0.438 mol l^{-1}; Table 1) after only one test day (i.e. one PV see Fig. 5a). It should be noted that the number of test days includes the loading phase.

As can be seen in the lower plot in Figure 5b from the NaCl-flooded core (L2), the Na^+ and Cl^- ions reached their original concentrations (0.657 mol l^{-1}) after only one test day. Thereafter, the concentrations of Cl^- (0.653 mol l^{-1}) and Na^+ (0.654 mol l^{-1}) stayed constant during the whole experiment (see Fig. 5b). The calcium concentration of the effluent increased to a maximum value of 0.013 mol l^{-1} before decreasing to constant value around 0.001 mol l^{-1} after *c.* 7–8 days. These fluid flux and compositional data indicate that nearly 0.4 g of calcium carbonate was dissolved from the core, which agrees well with the loss in mass in Table 2.

FEG-SEM

The chalk from Liège is a relatively pure coccolithic mudstone; EDS analyses indicate more than 51 wt% CaO. Although the chalk contains only small quantities of non-calcite minerals, SEM micrographs document the widespread occurrence of submicrometre-sized quartz as well as clay minerals and other phyllosilicates (see Fig. 6 for an SEM micrograph of untested Liège chalk). Qualitative EDS measurements show that the dominant silicate phase is SiO_2 (*c.* 4.5 wt%), while other oxides occur in very low amounts: 0.9 wt% Al_2O_3, 0.3 wt% MgO and 0.02 wt% Na_2O.

No alteration is observed using SEM-EDS in the L2 core with NaCl injection. Neither imaging nor qualitative EDS measurements indicate any compositional changes induced by the injection of the brine. In contrast, the core flooded with the $MgCl_2$ solution shows significant changes in the mineralogical composition (Fig. 7). After the injection of the $MgCl_2$ brine into Liège chalk, the amount of magnesium is strongly increased (Table 4), with MgO contents of up to 3.2 wt% after 65 days of flooding. SEM imaging shows that highest magnesium contents are encountered in aggregates of newly formed submicrometre-sized crystallites scattered in the chalk. These crystallites are most commonly

Fig. 6. SEM micrograph of Liège chalk before flooding. Scale bar, 2 μm.

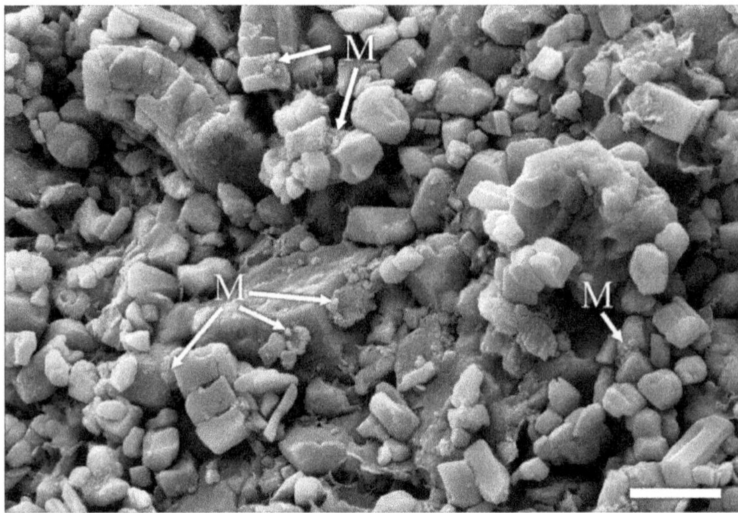

Fig. 7. SEM micrograph of Liège chalk after flooding with $MgCl_2$. Scale bar, 2 μm. M, Mg-rich carbonates or magnesite formed on top of pre-existing calcite grains.

observed close to the inlet and decrease in abundance along the core. Because of a relatively large sample interaction volume, the composition of grains smaller than *c.* 1 μm cannot unequivocally be determined using EDS as the signals from the new mineral phase and the primary calcite are mixed. However, the EDS measurements indicate that magnesium-rich carbonate minerals are being formed, in agreement with results from an experiment lasting 516 days with continuous flooding of an $MgCl_2$ solution (Zimmermann *et al.* 2015), and a 1072-day chemo-mechanical flow test using the same $MgCl_2$ solution (Nermoen *et al.* 2015), where a complete chemical re-working was documented.

XRD

The chalk from Liège appears to be homogeneous in XRD studies. Besides calcite, only very scarce quartz and traces of feldspar can be identified (see the 1L-d and 2L-d in Figs 8 & 9). The samples from the NaCl-flooded core (L2 in Fig. 9) contain slightly more quartz and feldspar compared to the L1 core (in Fig. 8). In the L1 experiment ($MgCl_2$ brine), the XRD analyses identified larger amount of Mg-bearing minerals in slice FL1-3 than slice FL1-1 (Fig. 8). Slice FL1-3 (Fig. 8) shows scarce magnesite and traces of a phase that could correspond to norsethite (peaks marked with an asterisk in Fig. 8), while traces of high-Mg calcite (Brown 1980) have been found in slice FL1-4. In L2 (NaCl-brine) no significant mineralogical changes are observed on the XRD patterns (Fig. 9).

Whole-rock geochemistry and isotope geochemistry

The chalk from Liège is also relatively homogeneous in terms of its geochemistry, reflecting the homogeneous mineralogy documented in Table 5. Table 4 shows the variation of four selected samples taken from the same block, which is compared with an average value of chalk from the same exposure ($n = 10$; unpublished data by Zimmermann). CaO concentrations are around 54.04 wt% and MgO abundances are as low as 0.27–0.29 wt%. SiO_2 varies only slightly along the core. Traces of clastic materials have been introduced to the chalk as observed from the sum of all rare earth elements (ΣREE), which increases faintly in between the samples (Table 4B). Although the samples have been taken from the same block, $\delta^{13}C_{SMOW}$ (where SMOW indicates Vienna Standard Mean Ocean Water) isotope values vary between 1.495‰ and 1.803‰ and $\delta^{18}O_{SMOW}$ between −2.049‰ and −1.607‰ (Table 4B), which therefore reflects the natural variation in stable isotopes for this rock material.

After flooding with NaCl (L2), the chalk does not show any significant geochemical changes (Table 4). $\delta^{13}C_{SMOW}$ values are similar to the unflooded samples of the chalk and only $\delta^{18}O_{SMOW}$ values are slightly more negative (Table 4). When the chalk was flooded with $MgCl_2$ significant changes in the MgO and CaO contents are observed. The amount of MgO increased from 0.28 wt% to 1.85 wt% (i.e. from 0.31 g to 2.05 g; highlighted in grey in Table 4), while the CaO concentration

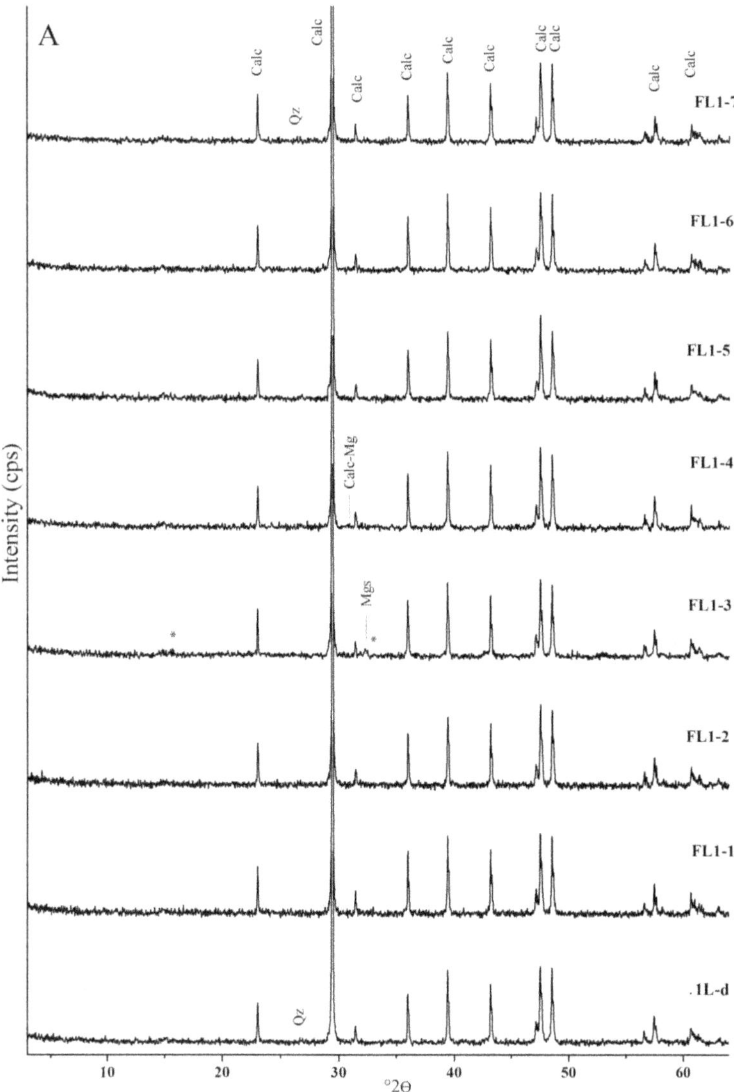

Fig. 8. XRD patterns of unflooded and $MgCl_2$-flooded Liège chalk (test L1). FL1-1 to FL1-7 are the seven slices that were acquired along the core axis (see Fig. 4) and 1L-d is the unflooded chalk. Calc, calcite; Mgs, magnesite; Calc-Mg, for Mg calcite; Feld, feldspars; Qz, quartz; the star symbols (*) mark peaks matching those of norsethite.

decreased from 54.04 wt% to 52.32 wt%, which corresponds to a net weight of 60.73 g before and 57.70 g after the experiment. However, the absolute weight per cent increase of MgO is almost exactly equal to the decrease in CaO at around 1.6 wt% (Table 4), indicating a correlation. Interestingly, it can also be observed that slice 3 reveals a higher MgO composition than slice 1. The reasons for the observed spatial dynamics along the core are complex and deserve further study. The alteration profile also is probably influenced by the interaction between the fluid flux and reaction kinetics. The only other element with significant changes in flooded L1 is Sr, which decreased after flooding.

Discussion

Our experiments indicate that geochemical reactions between the injected pore fluids and the rock mass may impact both the porosity and the

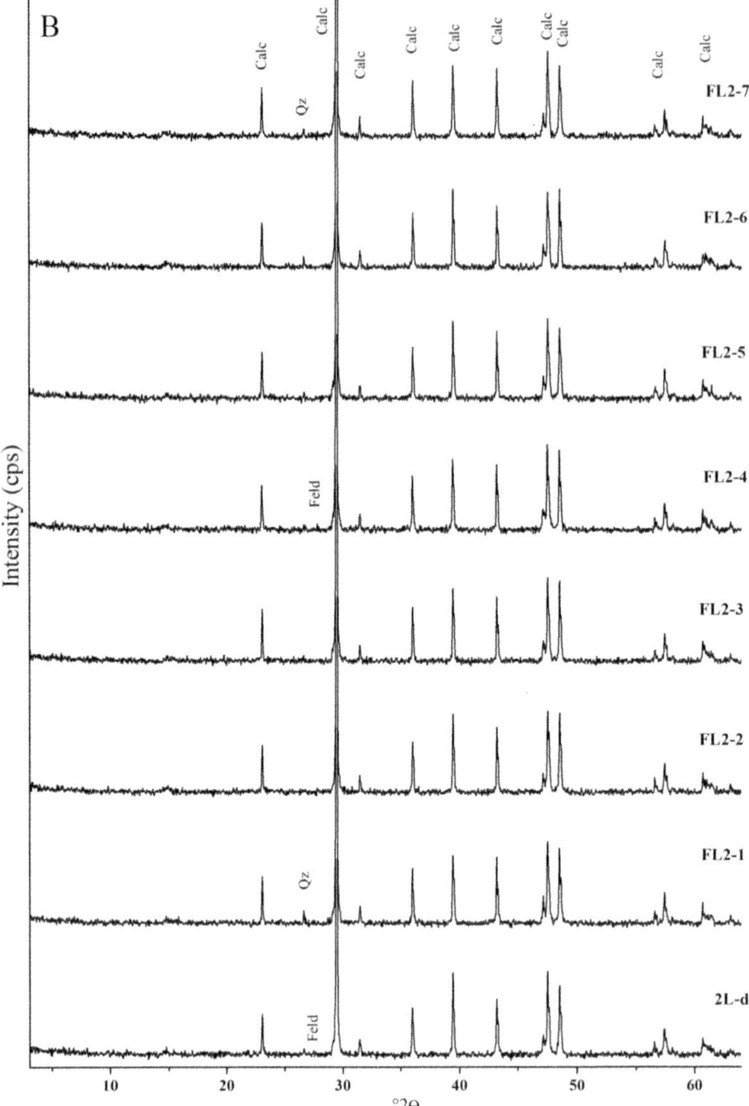

Fig. 9. XRD patterns of unflooded and NaCl flooded Liège chalk (test L2). FL2-1 to FL2-7 are the seven slices that were acquired along the axis of the core (see Fig. 4) and 2L-d is the untested Liège chalk. Calc, calcite; Feld, feldspars; Qz, quartz.

geomechanical behaviour during loading and creep. In particular, flooding of the chalk sample by $MgCl_2$ brines (L1) resulted in a reduction in the solid mass as well as an increase in the average solid density as calcite dissolved and Mg-bearing phases precipitated. A control experiment (L2) involving the flooding of a comparable sample by an NaCl brine, in contrast, showed much smaller geochemical alterations than the $MgCl_2$-flooded core. The L1 experiment (with geochemical alteration) shows greater volumetric strain compared with the control experiment, indicating that chemical reactions affect the overall volumetric creep.

Detailed SEM-EDS analyses revealed that minerals with a significant amount of MgO are abundant in the inlet of the core and slightly decrease through the sample (see the SEM images of unflooded chalk in Fig. 6 and flooded chalk in Fig. 7). This is confirmed by quantitative whole-rock geochemistry, which points to a 2 g increase in MgO

Table 5. *Comparison of the isotope geochemistry and geochemistry of the unflooded samples presented in this study (§) and the lower end, upper end and average values (#) for the Liège samples from Wang* et al. *(2013)*

	SiO_2 (%)	Al_2O_3 (%)	Fe_2O_3 (%)	MgO (%)	CaO (%)	Na_2O (%)	K_2O (%)	TiO_2 (%)	P_2O_5 (%)	MnO (%)	LOI (%)	Sum (%)	Ba (ppm)	Rb (ppm)	Sr (ppm)	Zr (ppm)	Y (ppm)	ΣREE (ppm)	Total carbon (%)	$\delta_{13}C_{SMOW}$ (‰)	$\delta_{18}O_{SMOW}$ (‰)	Temperature estimate (°C)
3L-A	2.01	0.35	0.13	0.27	54.14	0.05	0.08	0.02	0.12	0.02	42.7	99.86	36.0	3.4	945.6	6.9	8.1	26.3	12.19	1.550	−1.574	18.4
3L-B	2.00	0.35	0.19	0.26	53.86	0.05	0.08	0.02	0.13	0.02	42.9	99.85	41.0	3.7	961.8	7.7	8.5	24.3	12.02			
4L-A	1.84	0.32	0.12	0.27	54.20	0.04	0.07	0.02	0.14	0.02	42.8	99.87	32.0	3.0	887.8	7.5	8.5	23.0	12.15	1.578	−1.633	18.7
4L-B	1.83	0.32	0.14	0.27	54.34	0.05	0.07	0.02	0.14	0.02	42.7	99.86	35.0	3.0	915.5	9.2	8.7	25.0	12.05			
5L-A	1.87	0.33	0.12	0.26	54.59	0.04	0.07	0.02	0.10	0.02	42.4	99.86	35.0	3.1	959.9	6.5	7.7	22.9	12.15	1.502	−1.751	19.2
5L-B	1.83	0.32	0.13	0.26	54.39	0.04	0.08	0.02	0.10	0.02	42.6	99.84	38.0	3.1	958.6	6.5	7.7	23.3	12.05			
6L-A	1.99	0.32	0.16	0.25	54.13	0.04	0.07	0.02	0.08	0.02	42.7	99.84	38.0	3.2	973.0	8.2	6.9	22.7	11.97	1.511	−1.676	18.9
6L-B	2.03	0.32	0.14	0.25	54.34	0.04	0.06	0.02	0.09	0.02	42.5	99.85	39.0	2.8	968.5	7.5	7.0	24.0	12.17			
7L-A	2.03	0.36	0.13	0.26	54.19	0.04	0.08	0.02	0.08	0.02	42.6	99.85	37.0	3.9	932.6	8.0	7.9	22.2	12.25	1.618	−1.709	19.0
7L-B	2.06	0.37	0.14	0.26	54.04	0.04	0.08	0.02	0.09	0.02	42.7	99.85	35.0	3.7	952.3	7.0	8.0	24.6	12.13			
# (average)	1.95	0.34	0.14	0.26	54.22	0.04	0.07	0.02	0.11	0.02	42.66	99.85	36.6	3.3	945.6	7.5	7.9	23.8	12.11	1.552	−1.669	18.8
§ (average)	2.04	0.35	0.20	0.28	54.11	0.05	0.08	0.02	0.10	0.02	42.70	99.86	32.00	2.9	953.0	6.7	8.0	23.6	11.90	1.626	−1.793	19.4

$n = 10$	SiO_2 %	Al_2O_3 %	Fe_2O_3 %	MgO %	CaO %	Na_2O %	K_2O %	TiO_2 %	P_2O_5 %	MnO %	LOI %	SUM %
Lower	1.83	0.32	0.12	0.25	53.86	0.04	0.06	0.02	0.08	0.02	42.40	99.84
Upper	2.06	0.37	0.19	0.27	54.59	0.05	0.08	0.02	0.14	0.02	42.90	99.87
Average	1.95	0.34	0.14	0.26	54.22	0.04	0.07	0.02	0.11	0.02	42.66	99.85
§ (average)	2.04	0.35	0.20	0.28	54.11	0.05	0.08	0.02	0.10	0.02	42.70	99.86

$n = 10$	Ba (ppm)	Rb (ppm)	Sr (ppm)	Zr (ppm)	Y (ppm)	ΣREE PPM	Total C (%)	$\delta^{13}C_{SMOW}$ (‰)	$\delta^{18}O_{SMOW}$ (‰)	Temperature estimate (°C)
Lower	32.0	2.8	887.8	6.5	6.9	22.2	12.0			
Upper	41.0	3.9	973.0	9.2	8.7	26.3	12.3			
Average	36.6	3.3	945.6	7.5	7.9	23.8	12.1	1.566	−1.778	19.3
§ (average)	32.0	2.9	953.0	6.7	8.0	23.6	11.9	1.626	−1.793	19.4

$n = 10$ for geochemistry.

(*c.* 5–10× the original amount) in the flooded chalk (Table 4A). Values are not constant through the core and do not decrease from the inlet into the core, but vary from one slice to the next, surprisingly, the highest value for MgO is found in slice FL1-3. This trend is not yet understood, but may be related to differing microfacies or complex patterns of fluid flow through the chalk sample. However, the exact progress of the fluid–rock interplay can only be analysed when the involved chemical and mineralogical components are determined and quantified and coupled to the flow field within the sample. In addition, the alteration pattern is likely to reflect both the kinetics of the reaction and the fluid flux rates (Jones & Xiao 2005). The increase in MgO is probably associated with the formation of magnesite and Mg-bearing calcite. Other alteration products associated with the flooding process include norsethite, a Ba-rich carbonate associated with the dissolution of other carbonate phases (Gamsjäger *et al.* 1998). Zimmermann *et al.* (2013, 2015) propose that the lack of chemical equilibrium between the brine and the core triggers the dissolution of calcium carbonate and leads to the precipitation of Mg-bearing minerals. In long-term tests, new growth of magnesium carbonate could be verified with SEM-EDS studies, XRD analyses, nanoSIMS (nano secondary ion mass spectrometry) observations and oxygen isotopic data (Zimmermann *et al.* 2013). The solid volume could be affected by the chemical reaction between the rock and the fluid (Ruiz-Agudo *et al.* 2014; Nermoen *et al.* 2015). In the L1 experiment ($MgCl_2$ flooding test for 65 days), the bulk volume and solid volume decreased (Table 2) and the density increased (Table 3) because of calcium dissolution and the precipitation of new minerals. The solid volume is given by ratio of the solid mass and the average mineral density. Consequently, the geochemical alterations change the volume of the solids. Changes in both the bulk volume and the solid volume therefore need to be incorporated into porosity calculations when chemo-mechanical flow experiments are being interpreted.

When calculating the resulting porosity for the $MgCl_2$-flooded core (L1) with method 1, the volumetric strain only results in a porosity of 37.63%, indicating an absolute decrease of 5.82% (Table 2). When applying method 2 the calculated porosity is only 40.29%, i.e. an absolute reduction of less than 3.16%. By using the solid density measurement (pycnometry, method 3), the porosity reduces by about 4.17% (see Table 2). Methods 2 and 3 provide more accurate porosity values because they take into account the chemical alteration of the solid volume in addition to the volumetric strain, whereas method 1 only considers only the bulk volumetric strain. Therefore, coupling the volumetric strain directly to porosity is inadequate when chemical reactions occur. In the L1 experiment Mg–Ca exchange led to the formation of magnesite and high-Mg calcite and the dissolution of calcite.

In contrast, the results from the L2 experiment (the NaCl-flooded core), the effluent concentration changes are quite modest during the test, given that the sodium and chloride concentrations are similar to the injected ion concentrations (Fig. 5b). In addition, the SEM-EDS analyses do not reveal any new mineral growth and the whole-rock geochemistry shows no significant changes for the samples before and after flooding (Table 4A, B). In this experiment, the porosity calculations are rather similar for all three methods without the systematic differences that are seen in the L1 experiment. This lack of chemical alteration is consistent with geochemical, SEM and XRD data. However, the NaCl experiment (L2) may have dissolved some of the chalk (it lost *c.* 0.54 g over 90 days). It is therefore valid to use method 1 to estimate the porosity development as the reaction rate is slow for L2 when compared with the total volume of the injected fluid and the duration of the experiment.

There are always uncertainties associated with experimental work. Here, we discuss how experimental error may affect our conclusions regarding the porosity estimate. The porosity is estimated using three methods. For method 2 the porosity estimates originate from the saturated and dry weight difference according to equation (9). Here, the errors are attributed to the weight measurements and bulk volume estimates (from the truncated wedge in equation 4). Repeated bulk volume estimates display a natural variability on the order of 0.1–0.2 ml. During the DW saturation procedure only partial saturation could occur, which would yield an underestimate of the saturated weight and thus an underestimate of the porosity using this technique. However, the pycnometry tests (see Table 3) can be used to measure the solid density, which can be used to estimate the initial solid volume. If the bulk volume is known, the pore volume can be estimated from equation (1). The pore volume measurement from the pycnometry measurement and from the DW saturation and drying indicates an initial saturation of more than 99.5% for both tests. The dry weight of the plugs was measured after the plug had been placed in a heating cabinet until the weight is stable (at least 24 h). The repeatability in the porosity estimate from method 2 originates from errors in the bulk volume as well as in the saturated and dry weights and adds up to less than 1%. In order to gain certainty in the estimates the pycnometry measurements using method 3 in equation (12) are applied. Method 3 is partially independent of method 2 and may therefore provide a complementary estimate to method 1. Here, the largest error source is attributed to the solid volume estimate from the helium gas pycnometry. On the basis of repeatability tests the solid density is estimated within ± 0.02 g cm^{-3}. Errors of this magnitude in the solid density measurements yield porosity errors on the order of $<0.5\%$. In method 1, the porosity is estimated from the volumetric strain measurements from the bulk volume measurements before and after the test (see equation 5). This estimate is mathematically correct under the assumption that the total solid volume is conserved during compaction. The source of error is found in the bulk volume measurements after the test. Before the test the diameters along the plug vary by less than 0.01 mm. However, after hydrostatic compactionthis is not the case and the varying diameters are accounted for by calculating the volumes of the truncated wedge.

The resulting changes in the produced effluent concentrations, and the post-experimental analysis of the tested chalk sample, demonstrate that mineral reactions occur that change the volume. These changes are linked to the dissolution of Ca-bearing minerals, and the precipitation of Mg-bearing minerals. These mineralogical changes, exemplified by dissolution of e.g. calcium carbonate (density 2.69 g cm^{-3}) and the precipitation of magnesite (density of 3.00 g cm^{-3}), lead to measurable volume changes. It should be noted that the solid volume changes may be explained by both differences in the solid mass and the solid density in equation (12). In the L1 test (flooded with $MgCl_2$) a mass loss of 2.1 g and a density increase of 0.03 g cm^{-3} were both observed in slice 1 and the change in the solid volume is 1.09 cm^3, while for L2 the solid volume is changed by only 0.20 cm^3 (see Table 2).

Conclusions

Chemo-mechanical experiments on the Upper Cretaceous onshore chalk from Liège with 0.657 M NaCl and 0.219 M $MgCl_2$ brines at Ekofisk temperatures have revealed that both mechanical compaction and dissolution/precipitation reactions play a role in the porosity development. By contrast, mechanical compaction alone is the primary cause of porosity loss in a companion experiment involving an NaCl brine. A chalk sample that has been flooded with $MgCl_2$ shows significant changes in the rock chemistry. Total MgO in the core increases from 0.3 g to 2.05 g and a similar decrease is observed for CaO. This observation was confirmed by whole-rock geochemistry, effluent chemistry and SEM-EDS observations. The latter provides evidence for the precipitation of magnesium calcite or even magnesite. XRD analyses show scarce magnesite and minute amounts of high-Mg calcite. The geochemical alteration associated with $MgCl_2$ flooding changed the geomechanical properties of the chalk by accelerating the strain rate.

An additional effect of the geochemical alteration is that the strain alone is not a sufficient basis for accurately determining the pore volume change. Porosity estimations for geochemically altered reservoirs must account for the effects of the reactions on the solid mass, the volume and the density (Nermoen *et al.* 2015). These geochemical effects result in more complex reservoir behaviour compared with mechanical compaction alone for non-reactive rock–fluid combinations. Such effects may markedly influence the outcome of EOR/IOR (improved oil recovery) operations in fields where injection fluids react with the reservoir rock. Precise porosity estimates are of crucial importance for accurate resource potential estimates and to predict the flow properties of reservoirs that are undergoing chemo-mechanical compaction processes.

The authors thank the Faculty of Science and Technology for the PhD grant for W. Wang. The authors thank COREC for financial support for this research study and all authors acknowledge the Research Council of Norway and the industry partners, ConocoPhillips Skandinavia AS, BP Norge AS, Det Norske Oljeselskap AS, Eni Norge AS, Maersk Oil Norway AS, DONG Energy A/S, Denmark, Statoil Petroleum AS, ENGIE E&P NORGE AS, Lundin Norway AS, Halliburton AS, Schlumberger Norge AS and Wintershall Norge AS of The National IOR Centre of Norway for support. The research presented is integral part of the PhD thesis of W. Wang at UiS. We also thank two anonymous reviewers and Peter Armitage as well as Ron Lander for their thoughtful comments, which improved the original manuscript. We also appreciate the handling by the editors.

References

Anderson, T.F. & Arthur, M.A. 1983. Stable isotopes of oxygen and carbon and their application to sedimentologic and environmental problems. *In*: Arthur, M.A., Anderson, T.F., Kaplan, I.R., Veizer, J. & Land, L.S. (eds) *Stable Isotopes in Sedimentary Geology*. SEPM Short Course Notes, **10**, 1–151.

Brown, G. 1980. Associated minerals. Chapter 6. *In*: Brindley, G.W. & Brown, G. (eds) *Crystal Structures of Clay Minerals and Their X-ray Identification*. Mineralogical Society Monographs, **5**, 361–410.

Felder, W.M. 1975. Lithostratigrafie van het Boven-Krijt en het Dano-Montien in Zuid-Limburg en het aangrenzende gebied. ('Lithostratigraphy of the Upper Cretaceous and the Dano-Montien in southern Limburg and the neighboring region.') *In*: Zagwijn, W.H. & Van Staalduinen, C.J. (eds) *Toelichting bij de geologische overzichtskaarten van Nederland*. Rijks Geologische Dienst, Haarlem, 63–72.

Fjær, E., Holt, R.M., Horsrud, P., Raaen, A.M. & Risnes, R. 2008. *Petroleum Related Rock Mechanics*. 2nd edn. Elsevier, Amsterdam.

Gamsjäger, H., Königsberger, E. & Preis, W. 1998. Solubilities of metal carbonates. *Pure & Applied Chemistry*, **70**, 1913–1920.

Hellmann, R., Renders, P.J.N., Gratier, J.P. & Guiguet, R. 2002. Experimental pressure solution compaction of chalk in aqueous solutions. Part 1. Deformation behavior and chemistry. *In*: Hellmann, R. & Wood, S.A. (eds) *Water Rock Interactions, Ore Deposits, and Environmental Geochemistry: A tribute to David A. Crerar*. The Geochemical Society, Special Publications, **7**, 129–152.

Hiorth, A., Cathles, L.A., Kolnes, J., Vikane, O., Lohne, A., Korsnes, R.I. & Madland, M.V. 2008. A chemical model for the seawater-CO_2-carbonate system – aqueous and surface chemistry. Paper presented at the International Symposium of the Society of Core Analysts, Abu Dhabi, United Arab Emirates, 1–12.

Hjuler, M.L. & Fabricius, I.L. 2009. Engineering properties of chalk related to diagenetic variations of Upper Cretaceous onshore and offshore chalk in the North Sea area. *Journal of Petroleum Science and Engineering*, **68**, 151–170.

Jones, G.D. & Xiao, Y.T. 2005. Dolomitization, anhydrite cementation, and porosity evolution in a reflux system: Insights from reaction transport models. *AAPG Bulletin*, **89**, 577–601.

Korsnes, R.I. 2007. *Chemical induced water weakening of chalk by fluid-rock interaction, A mechanistic study*. PhD thesis, University of Stavanger.

Korsnes, R.I., Strand, S., Hoff, Ø., Pedersen, T., Madland, M.V. & Austad, T. 2006*a*. Does the chemical interaction between seawater and chalk affect the mechanical properties of chalk?. *In*: Cottheium, A.V., Charlier, R., Thimus, J.F. & Tshibangu, J.P. (eds) *Eurock 2006. Multiphysics Coupling and Long Term Behaviour in Rock Mechanics*. The International Symposium of the International Society for Rock Mechanics, 9–12 May 2016, Liège, Belgium, 427–434.

Korsnes, R.I., Madland, M.V. & Austad, T. 2006*b*. Impact of brine composition on the mechanical strength of chalk at high temperature. *In*: Cottheium, A.V., Charlier, R., Thimus, J.F. & Tshibangu, J.P. (eds) *Eurock 2006. Multiphysics Coupling and Long Term Behaviour in Rock Mechanics*. The International Symposium of the International Society for Rock Mechanics, 9–12 May 2016, Liège, Belgium, 133–140.

Korsnes, R.I., Madland, M.V., Austad, T., Haver, S. & Røsland, G. 2008. The effects of temperature on the water weakening of chalk by seawater. *Journal of Petroleum Science & Engineering*, **60**, 183–193.

Madland, M.V. 2005. *Water weakening of chalk: A mechanistic study*. PhD thesis, University of Stavanger.

Madland, M.V., Midtgarden, K., Manafov, R., Korsnes, R.I., Kristiansen, T.G. & Hiorth, A. 2008. The effect of temperature and brine composition on the mechanical strength of Kansas chalk. Paper presented at the International Symposium of the Society of Core Analysts, Abu Dhabi, United Arab Emirates, 29 October–2 November 2008, 1–6.

Madland, M.V., Hoirth, A. *et al.* 2011. Chemical alterations induced by rock–fluid interactions when injecting brines in high porosity chalks. *Transport in Porous Media*, **87**, 679–702.

Madland, M.V., Zimmermann, U. *et al.* 2013. Neoformed dolomite in flooded chalk for EOR processes. Paper presented at the 75th EAGE Conference & Exhibition incorporating SPE EUROPEC, 10–13 June 2013, London, UK.

Molenaar, N. & Zijlstra, J.J.P. 1997. Differential early diagenetic low-Mg calcite cementation and rhythmic hardground development in Campanian-Maastrichtian chalk. *Sedimentary Geology*, **109**, 261–281.

Nermoen, A., Korsnes, R.I., Hiorth, A. & Madland, M.V. 2015. Porosity and permeability development in compacting chalks during flooding of nonequilibrium brines: insights from long-term experiment. *Journal of Geophysical Research*, **120**, 2935–2960.

Newman, G. 1983. The effect of water chemistry on the laboratory compression and permeability characteristics of some north sea chalks. *Journal of Petroleum Technology*, **35**, 976–980.

Ruiz-Agudo, E., Putnis, C.V. & Putnis, A. 2014. Coupled dissolution and precipitation at mineral-fluid interfaces. *Chemical Geology*, **383**, 132–146.

Slimani, H. 2001. New species of dinoflagellate cysts from the Campanian–Danian chalks at Hallembaye and Turnhout (Belgium) and at Beutenaken (the Netherlands). *Journal of Micropalaeontology*, **20**, 1–11.

Strand, S., Hjuler, M.L., Torsvik, R., Pederse, J.I., Madland, M.V. & Austad, T. 2007. Wettability of chalk impact of silica, clay content and mechanical properties. *Petroleum Geoscience*, **13**, 69–80.

Wang, W., Zimmermann, U., Madland, M.V., Bertolino, S.R.A., Lie, B.V. & Morosova, V. 2013. The search of on-shore equivalents for reservoir chalk in the North Sea: Rare Earth Elements, $\delta^{13}C$ and $\delta^{18}O$ isotopes as indicators? Paper presented at Norsk Geologisk Forening Vinterkonferansen, Oslo, 8–10 January 2013, 112–113.

Wiborg, R. & Jewhurst, J. 1986. Ekofisk subsidence detailed and solutions accessed. *Oil and Gas Journal*, **2**, 47–51.

Wright, E.K. 1987. Stratification and paleocirculation of the Late Cretaceous Western Interior Seaway of North America. *Geological Society of America Bulletin*, **99**, 480–490.

Zimmermann, U., Madland, M.V., Bertolino, S.A.R., Hildebrand-Habel, T., Hiorth, A. & Korsnes, R.I. 2013. Tracing fluid flow in flooded chalk under long term test conditions. Paper presented at the 75th EAGE Conference & Exhibition incorporating SPE EUROPEC, 10–13 June 2013, London, UK.

Zimmermann, U., Madland, M.V. *et al.* 2015. Evaluation of the compositional changes during flooding of reactive fluids using SEM, Nano-SIMS, XRD and whole-rock geochemistry. *AAPG Bulletin*, **99**, 791–805.

Hydrothermal dolomitization: simulation by reaction transport modelling

ALBERTO CONSONNI*, ALFREDO FRIXA & CHIARA MARAGLIULO

Eni Exploration & Production, via Maritano 26, 20097 San Donato Milanese, Italia

**Correspondence: alberto.consonni@eni.com*

Abstract: Reaction transport modelling (RTM) has been successfully used to simulate dolomitization and to predict the lateral extension of dolomitized bodies as potential hydrocarbon reservoirs in the subsurface as well as to give insights into the dolomitization process itself. Geological configurations that have been tested include reflux dolomitization, thermal convection and dolomitization in compactional burial settings. In this study the hydrothermal dolomitization model has been tested with a RTM approach where a limestone reservoir in an intermediate burial setting is fluxed in the presence of deep feeding faults. Hydrothermal fluids migrate upwards from the deep basin areas through extensional fault conduits reaching more permeable levels in the reservoir. The fluids enter the primary interparticle porosity and dolomitize the limestone creating moldic and vuggy porosity. The simulations demonstrate the importance of faulting in this dolomitization process. An effective control on dolomitization is thus exerted by the permeability distribution. Moreover, the type and reactivity of the fluids entering the limestone reservoir exert important controls on the extent of the dolomite bodies.

Dolomitization of carbonate layers may occur at different times during their geological evolution, from synsedimentary environments to deep burial settings. Interpretation of the dolomitization process is key for making predictions about the resulting geometry of dolomitized bodies. This is of great importance for the oil industry because dolomitized bodies often have greater porosity than the original limestone. Different models of dolomitization have been proposed: early dolomitization models, occurring in synsedimentary conditions with the influx of seawater-derived fluids; and burial dolomitization models, occurring at higher depth and temperature conditions through the action of subsurface fluids. For each proposed model the geometry and characteristics of the dolomitized bodies are peculiar and dependent on the fluid flow pattern and the chemistry of the flowing fluids (Warren 2000; Machel 2004). Numerical modelling, and RTM in particular, (Whitaker *et al.* 2004; Xiao & Jones 2006) is a powerful tool to validate conceptual models, to verify that mass balance, kinetic and thermodynamic constraints are honoured and to give relative importance to the physical parameters affecting the dolomitization process.

RTM has been successfully used (Jones & Xiao 2005; Consonni *et al.* 2010; Whitaker & Xiao 2007, 2010) to interpret and predict the formation and distribution of dolomite bodies, which may form potential hydrocarbon reservoirs in the subsurface, and to give insights into the dolomitization process itself. Simulation of various early (near-surface) diagenetic dolomitization models has demonstrated that RTM can be used for reflux dolomitization and thermal convection.

Jones & Xiao (2005) investigated the temporal and spatial distribution of dolomitization in a reflux system, with a sensitivity to different parameters including brine concentration, temperature, flow rate and porosity–permeability heterogeneity. Simulations proved that the rates of dolomitization are sensitive to the flow rate and the brine chemistry. Moreover, the authors verified that the models were able to predict the general spatial and temporal trends in dolomitization and porosity observed in major dolomite reservoirs.

Whitaker & Xiao (2010) investigated early burial dolomitization in an isolated platform driven by geothermal (Kohout) convection of seawater. Simulations showed that sediment permeability and reactive surface area, which are commonly inversely related, are key controls.

The reproduction and evaluation of dolomitization from compactional fluid flow in burial settings has been tested in Consonni *et al.* (2010). In this study simulations were run to evaluate the origin and evolution of the dolomitizing fluids and to provide insights regarding the distribution of the dolomitized bodies in a Jurassic carbonate reservoir in the Po Plain, Italy. A basin simulation was used to create a 3D model of the compacting system and to constrain the flow rate reaching the reservoir. Simulations showed that dolomitization was an efficient process even at low temperatures. Fluid

From: Armitage, P. J., Butcher, A. R., Churchill, J. M., Csoma, A. E., Hollis, C., Lander, R. H., Omma, J. E. & Worden, R. H. (eds) 2018. *Reservoir Quality of Clastic and Carbonate Rocks: Analysis, Modelling and Prediction.* Geological Society, London, Special Publications, **435**, 235–244.
First published online June 14, 2016, https://doi.org/10.1144/SP435.13

composition is one of the main constraints, seawater-derived compaction fluids have proven to be efficient for dolomitization due to the relatively high Mg content. Simulations also confirmed that permeability is the most important factor that influences fluid flow and, consequently, the dolomite distribution in the formation.

The importance of faults and fractures that act as conduits for fluid flow in sedimentary basins and drive diagenesis has been increasingly recognized in recent years (Laubach *et al.* 2010). Faults are able to transmit fluids formed in the deep basin or basement into structurally higher and cooler strata, where they can interact with a mineral assemblage that may be in disequilibrium with them, causing dissolution, recrystallization or combination of these processes. Dolomitization of limestone with the replacement of pre-existing calcite may be caused by fault-related fluid flow when fluids with an appropriate Mg content may flow from hotter and deeper parts of a basin into a cooler reservoir. These hydrothermal dolomites may become potential reservoirs; conceptual models have been proposed to predict their characteristics from observations at wells in area far apart wells (Davies & Smith 2006). In contrast to other dolomitization processes, it is difficult to constrain important parameters such as fluid sources, flow rate and chemistry, timing of dolomitization and temperature.

In this study the hydrothermal dolomitization model has been tested with an RTM approach on a West African lacustrine carbonate reservoir of the Toca Fm (Barremian age). An integrated diagenetic study allowed the identification of a dolomite phase replacing previous limestone. Different investigative techniques allowed us to hypothesize a mechanism of dolomitization by hypersaline high-temperature fluids interacting with crustal rocks.

Two question have been addressed: can the proposed model of hydrothermal dolomitization justify the amount of dolomite encountered and the extension of dolomitized body, considering mass balance, fluid chemistry, chemical thermodynamics and kinetic constraints? If yes, which are the important parameters influencing the process? To answer those questions dolomitization by hydrothermal waters coming from deep faults has been simulated in an intermediate burial context: high-temperature fluids migrating through extensional faults that originate in deeper parts of the basin react with permeable limestone beds and dolomitize them.

Differently from other dolomitization models, the volume and type of fluids entering the system are difficult to evaluate. The RTM approach has then been used as a numerical laboratory to test and quantify the formation of dolomite bodies as a response to variable fluid flow rate and reactivity. Results will be shown for a seawater composition and a highly concentrated brine. Moreover, other parameters of the model (temperature, fluid flow rate and duration) have been considered as uncertain, and thus different values have been used for the thermal regime of the basin, the fluid flow rate and the fluid flow duration. Each dolomitization case has been analysed to determine the dolomite content, and the dolomitized volume in a specified region of the reservoir has been calculated.

Background geology

The Kambala oil and gas field is located in the Lower Congo Basin (Cabinda Offshore), approximately 8 km west of the present coast line, and in shallow water depths (approximately 15 m). The field is a NW–SE trending structural–stratigraphic trap (Scheevel *et al.* 2004). Its width and length are 2.1 km and 5.4 km, respectively, whereas the productive intervals reside at depths greater than 3000 m (Scheevel *et al.* 2004).

The Barremian Toca Fm is the main reservoir rock of the field. It consists of lacustrine carbonates deposited on and around the basement highs. The reservoir is sealed by the Aptian salt (Loeme Formation). The Toca Fm is divided in two units, named Lower and Upper, separated by the Middle Bucomazi and laterally grading to the Middle Upper Bucomazi (Wasson *et al.* 2013; Frixa *et al.* 2014).

According to McHargue (1990), Wasson *et al.* (2013), Frixa *et al.* (2014), the Lower Toca is dominated by finely dolomitized, ooid/oncoid grainstones to wackestones and microbialites. During its deposition, the sagging of basinal areas and the subsidence of fault crests were insufficient to flood all remaining hinterland, thus lacustrine shelf areas above the wave base are narrow. Moving offshore, siltstones and shales with occasional microbial growth are deposited in the deeper, calmer waters of the Bucomazi. In the two analysed wells the thickness of the Lower Toca ranges from 74–81 m.

The Middle Bucomazi is a continuous siltstone and shale section with thinly interbedded carbonates as dark grey oncoidal packstones. In the examined section the thickness is approximately 10–18 m. The anoxic, organic-rich laminated Middle Bucomazi shales are the main source rock of the field.

The Upper Toca is the best reservoir unit, being dominated by fossiliferous grainstones to packstones (mollusc coquinas). These lithotypes were later strongly hydrothermally dolomitized and/or silicified close to major faults. In the analysed wells, the Upper Toca thickness ranges from 46–51 m. In this period the lake level was thought to be higher than the lower Toca.

Among the analysed core samples, the best reservoir facies are represented by coarse mollusc coquina limestone and hydrothermal dolomite, which is the subject of this study. Two types of dolomite have been recognized. A finely crystalline, probably early diagenetic dolomite that shows low reservoir properties and a coarsely crystalline, probably late diagenetic zebra dolomite, with very good reservoir properties.

Dolomitization is associated with strong silica precipitation in the Upper Toca Fm, as indicated by mineralogical and chemical analyses that also document the presence of detrital palygorskite, typical of lacustrine environments under arid conditions and, in the Lower Toca Fm, of talc replacing ooids that suggest strong hydrothermal conditions.

The integration of different investigative techniques allowed us to hypothesize a mechanism of dolomitization by hyper saline fluids that probably mixed with modified hypersaline and fresh waters, interacting with crustal rocks, as indicated by strontium isotope analysis. High homogenization temperatures registered in fluid inclusions (100–140°C) could be due to crustal thinning, whereas the hyper saline fluids (20–25 NaCl eq. %) could have migrated from the Aptian Loeme Fm salt or be originally present in the deep lake sediments.

Methods

RTM was performed using the TOUGHREACT code (Xu *et al.* 2004). The TOUGHREACT simulator is a non-isothermal RTM capable of simulating multiphase fluid flow, heat and solute transport. It can be applied to 1D, 2D or 3D domains with physical and chemical heterogeneities. Grid techniques used by the simulators allow the description of complex geometries, as for geological structures. For water–rock interactions the code can simulate the reactivity of minerals using a thermodynamic equilibrium approach or a kinetic approach.

A geological section (Fig. 1a) has been chosen to represent the geometry of a transect from a West African lacustrine carbonate reservoir of the Toca Fm (Barremian age) at the time when hydrothermal dolomitization is supposed to have happened (Albian). It contains from bottom to top: a deep shaly layer; the limestone lower reservoir; an intercalated marly layer, less porous and permeable than the reservoirs; the limestone upper reservoir; a permeable sandstone layer; a shaly caprock; and a shallow shaly layer. The depth of reservoirs reaches around 3000 m in the left part of the section. The formation is cut from the bottom to the reservoir layers by faults that may act as conduits for the flow of hydrothermal fluids.

The section has been discretized with a regular horizontal step of 200 m, and the main petrophysical parameters assigned. The main parameters of the rock layers used in the simulations are summarized in Table 1 and the distribution of the input porosity is plotted in Figure 1b. Input porosity varies between 5% and 10% for the limestone reservoirs and the sandstone layer above them, it is below 5% for the intercalated marly layer and caprock. The mineralogical composition of the reservoirs is mainly calcite (98%), with a low initial amount of dolomite (1%) to start the dolomite reaction. The marly intercalated layer has the same mineralogy as the reservoirs.

The temperature was assigned using a constant surface temperature of 25°C and a geothermal gradient of $0.03°C\ m^{-1}$. The temperature rises from 80–100°C in the zones closest to the faults for the lower reservoir and from 70–100°C for the upper reservoir. A second thermal gradient of $2.5°C\ m^{-1}$ is used to test the sensitivity to thermal regime.

The section is considered to be impermeable at the bottom, impermeable at the vertical left and right side, open at the top. Five grid elements have

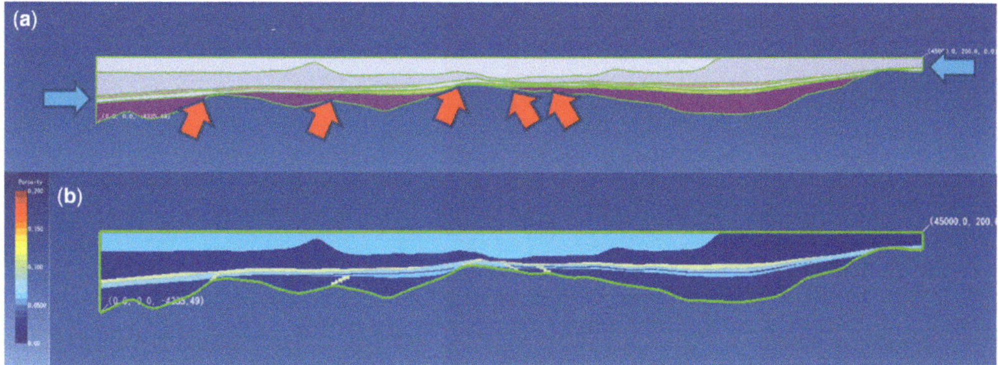

Fig. 1. **(a)** Simulated section. The positions of the faults are indicated by red arrows and the reservoir layers are indicated by blue arrows. The length of the section is 45 km, with a maximum depth of 4.5 km. Colours in the section are used to graphically differentiate the geological layers. **(b)** Input porosity.

Table 1. *Rock parameters of simulated layers*

	Porosity	Permeability (md)	Calcite	Dolomite
Shale	5	1	1	1
Shale	2	0,1	1	1
Sandstone	10	50	1	1
Upper res	7	30	98	1
Limestone	3	1	98	1
Lower res	5	10	98	1
Shale	3	0,1	1	1

Input porosity varies between 5% and 10% for the limestone reservoir and the sandstone layers above them, it is below 5% for intercalated marly layer and caprock. The mineralogical composition of the reservoirs is mainly calcite (98%), with a low initial amount of dolomite (1%) to start the dolomite reaction. The marly intercalated layer has the same mineralogy as the reservoirs. Layers are ordered from top to bottom in the geological section. Calcite and dolomite abundances are in percent of rock volume. The remaining fraction of rock volume closing to 100% is considered as a non-reactive material.

Table 2. *Composition for seawater and brine*

	Seawater concentration ($mol\ l^{-1}$)	Brine concentration ($mol\ l^{-1}$)
Ca	0.01829	0.5489
Cl	0.6307	3.949
pH	7.5	5.8
C	0.001628	0.004113
K	0.01003	0.06905
Mg	0.07764	0.1852
Na	0.4684	2.349
Si	7.12E-05	1.00E-04
S	0.0115	2.81E-04

been used to input flow (one for each fault). The flow rate is poorly constrained for this type of flow, no data are available to fix the precise values. In Consonni *et al.* (2010) fluid flow entering the reservoir was the result of the channelization of compactional flows in the permeable reservoir limestone, hence the flow rate was constrained by a basin modeling application and a value of 0.03 $m\ a^{-1}$ was calculated. The rate used here (0.79 $m\ a^{-1}$) is around 30 times larger, which can be justified due to the concentration of flow in faults that are very small conduits from a geometrical point of view, and thus able to efficiently channelize the flow. A value of 1.58 $m\ a^{-1}$ will be considered for the fluid flow rate in the sensitivity analysis. The duration of flow is probably the repetition of multiple events of fluid flow and periods with no flow. In this context the duration used in the simulations has to be considered as the sum of multiple activity periods; values of 500 ka and 1000 ka have been used.

As a starting hypothesis for water composition, seawater has been used. To test the effect of water variation a brine from a similar reservoir in the South Atlantic has been chosen. Values of dissolved species for the two water compositions are reported in Table 2. They substantially differ in total dissolved species (seawater less concentrated than brine), pH (7.5 v. 5.8) and Mg/Ca ratio (4.2 v. 0.34).

Two preliminary modelling runs were done: an analysis of the dolomitization power of the waters and a run for fluid flow with no reactions. An analysis of the dolomitization power of the waters has been done to test if seawater and brine are able to dolomitize an initial calcitic rock, and to assess the importance of fluid composition on dolomitization process. This has been executed with one-cell simulations. A set-up has been produced to give a quick assessment about the ability of the water to replace calcite by dolomite. The mineral assemblage has been set with an initial 1% both calcite and dolomite (98% non-reactive material) with porosity of 10%. The water composition was seawater (Table 2) with the temperature fixed at 75°C. A one-cell simulation has been executed in order to evolve the system towards equilibrium, the dolomitizing power of seawater is defined by the equilibrium content of calcite and dolomite. The procedure has been repeated for a brine composition (Table 2).

The run for fluid flow alone is used to analyze the flow patterns on the geological section, this is key to understanding the subsequent process of dolomitization, this being substantially driven by fluid flow. The simulation was executed on the whole geological section to model the fluid flow without the interactions between the fluid chemistry and the rocks, this is possible with the TOUGHREACT simulator.

Transport and reaction runs were then executed to test the hydrothermal dolomitization model. Five simulations have been set (Table 3): a base case (run 'a') and four simulations varying four parameters (run 'b–e'). Run 'a' tested the reliability of the model and allowed an estimation of the amount of dolomite encountered and the extension of the dolomitized bodies. After that four simulations have been executed to vary the input flow rate in the faults, the water composition, the thermal regime and the duration of flow in a systematic way in order to evaluate the effect of each parameter on dolomitization and to define which are the important parameters influencing the process.

Results

One-cell simulations for the determination of dolomitizing power have been executed for seawater (Fig. 2a) and brine (Fig. 2b). Calcite dissolves and dolomite precipitates until an equilibrium is reached within the fluid. The brine is more

Table 3. *Simulations parameters*

Model run	Simulated time (ka)	Flow ($m\ a^{-1}$)	Water	Thermal regime (°C per 100 m)
a	500	0.79	Seawater	3
b	500	1.58	Seawater	3
c	500	0.79	Brine	3
d	500	0.79	Seawater	2.5
e	1000	0.79	Seawater	3

Five simulations have been set: a base case (run 'a'), four simulations varying flow rate, water composition, thermal regime and flow duration (runs 'b–e'). Run 'a' was used to test the reliability of the model and to estimate the amount of dolomite encountered and the extension of the dolomitized body. The other four simulation have been executed to evaluate the effect of each parameter on dolomitization and to define which are the important parameters influencing the process.

efficient in causing the dissolution of calcite and precipitation of dolomite. This happens despite the Mg/Ca ratio for seawater being higher than brine. This means that the Mg/Ca ratio is not the only parameter of water composition that is important for causing dolomitization – the total amount of Mg (0.1852 mol l^{-1} for brine, 0.07764 mol l^{-1} for seawater) seems to be more important. In fact, increasing the Mg/Ca ratio favours the stability of dolomite with respect to calcite, whereas the volume of calcite dissolved and precipitated as dolomite directly depends on the amount of Mg expressed (in mol) in a given volume of water.

A model run with a non-reacting fluid flow has been executed on the geological section. Fluids flow along the faults (Fig. 3a), enter the reservoir layers and flow within them up to the zone in the basin where they are able escape at the top. The marly intercalated layer acts as a separator to flow between the lower and upper reservoir layers. The sandstone layer above the reservoirs also plays a role; due to its relatively high permeability it is able to capture part of the fluid flow. Flow analysis shows that after leaving the faults hot fluids follow the porosity and permeability patterns of the rocks. This result is a direct consequence of the permeability contrast between the carbonate reservoir layers, the intercalated marl, the caprock and the sandstone layer. Hence, this seems to be a useful guide for the interpretation of dolomitization results – the process will be effective in the zones reached by the dolomitizing fluids.

The base case simulation has been executed, values have been assumed for flow rate, flow duration, water composition and thermal regime of the basin using the parameters of Table 3, run 'a'. The amount of dolomite at the end of the simulation is plotted in Figure 3b, porosity in Figure 3c. Five dolomitized bodies are evident both in Figures 3b and 3c, located where calcite has been substituted by dolomite. The final porosity is enhanced from 5–7% up to 20%.

The replacement of calcite by dolomite proceeds from the entry point of hot fluids into the limestone layers; dolomite bodies are present where the faults cut the reservoir layers. Dolomitization is more developed in the reservoir layers (Fig. 4a), both lower and upper, and seems more extensive in the upper reservoir. The upper reservoir has a higher permeability than the lower, hence dolomitizing fluids enter this layer more easily. The marly layer between the two reservoirs is also dolomitized, because some of the fluids are able to enter this layer. Permeability contrasts, not temperature, seem to be the most important driver for dolomitization.

Dolomitized bodies, independent of the layer in which they are located, are the objects in the geological section with highest porosity and permeability, reaching values that make them good potential reservoirs (20% porosity). Dolomitized bodies have significant horizontal dimensions depending on the initial permeability, from 1 km in the marly intercalated layer to 2 km in the upper reservoir. Hence, model run 'a' indicates that this combination of the input parameters is effective in producing dolomite bodies in the limestone layers. Dolomitized bodies develop where faults reach the reservoir layers with an extension dependent on the initial permeability distribution.

To evaluate the robustness of these observations, and to give a relative importance to the input parameters, model runs 'b', 'c' and 'd' have been executed (Table 3). To test the dependence of the results on the fluid flow rate in the faults the rate has been doubled (run 'b'). To test a different water composition, a brine from a similar reservoir in the South Atlantic (Table 2) has been considered (run 'c'). To test the importance of the basin thermal regime the value of the geothermal gradient has been lowered from 0.03°C m^{-1} to 0.025°C m^{-1} (run 'd'). Results are presented for the central part of the geological section (Fig. 4), where important potential reservoirs are present. Doubling the flow rate results in more extensive dolomite bodies (Fig. 4b), whereas their positions near the fault intersections with the reservoirs are the same as run 'a'. In run 'b' the marly layer is also partly affected by dolomitization. This result seem reasonable, considering that doubling the flow rate doubles the volume of dolomitizing fluid entering the system. Run 'c' (with the brine) gives results very similar to run 'b' (Fig. 4c). In this case the dolomitizing power has been doubled (Fig. 2), hence the dolomitized bodies are extended. Run 'd' did not show obvious differences from run 'a', thus temperature does not seem to be a primary

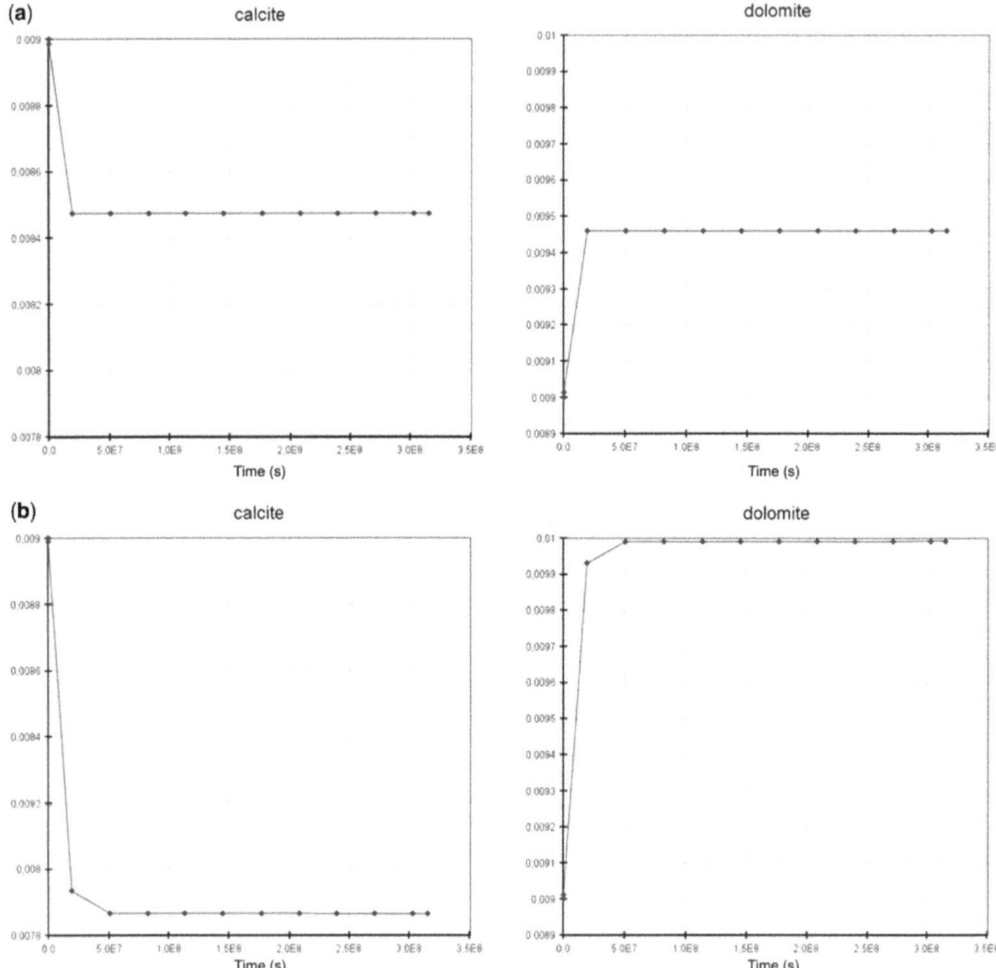

Fig. 2. Dolomitizing power of seawater **(a)** and brine **(b)**. y axes, volume fraction.

control in the dolomitization process (Fig. 4d). In fact temperature may influence the reaction rate, but the amount of calcite dissolving and dolomite precipitating is controlled by the fluid volume and the Mg content entering the system.

Runs 'b' and 'c' show that dolomitized bodies extend further into the reservoir layers and coalesce, forming bodies that extend for some kilometres horizontally. In the upper reservoir a large dolomite body forms with horizontal dimension of more than 5 km. This could be important for petroleum exploration – the resulting bodies may have high porosities (up to 20%) that extend continuously for kilometres.

To quantify the differences between model runs the volume of rock with complete dolomitization has been calculated for one dolomitized body (Fig. 4, the body on the left). Limestone complete dolomitization volumes are reported in Table 4. The volume of the dolomitized body is directly proportional to the fluid flow rate and the Mg content but is insensitive to temperature. Fluid flow and water composition thus seem to be the most important parameters controlling dolomitization, whereas temperature seems to be less important.

The duration of the flow has been kept constant for runs 'a–d' at 500 ka. Results are dependent from flow duration, all other parameters being constant the flow volume entering the system (and hence the Mg amount) is directly proportional to the flow duration. Run 'e' has been executed using a flow duration of 1000 ka. The resulting dolomitization distribution is similar to that obtained with an enhanced flow rate and brine (Fig. 4e) and the volume of the dolomitized body is proportional to the duration of the process (Table 4).

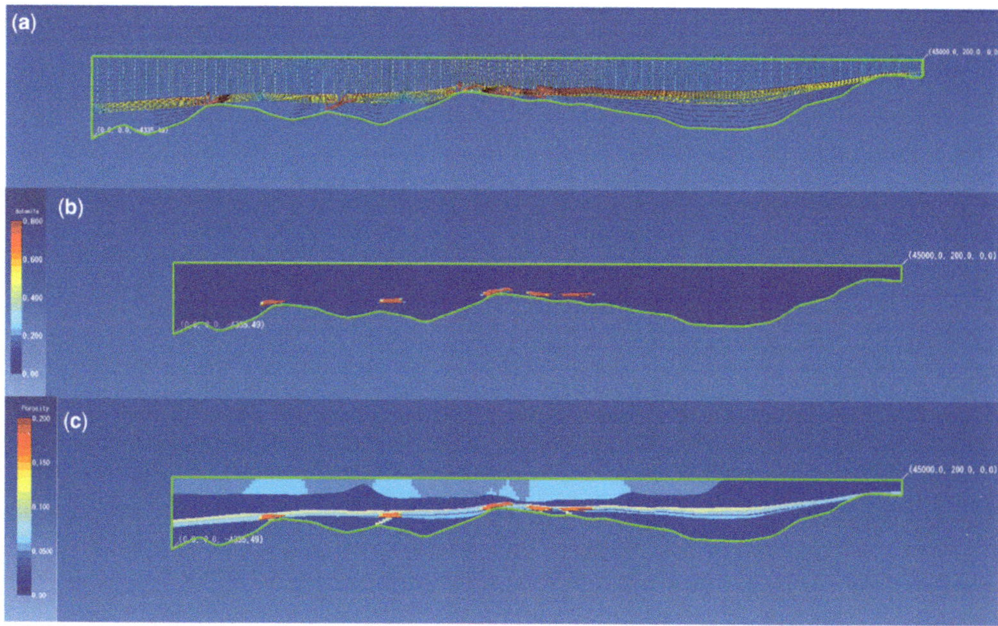

Fig. 3. (**a**) Simulated flow measured in volume fraction units. The fluid regime is considered to be constant during the simulations. (**b**) Simulated dolomite distribution after 500 ka of flow. (**c**) Simulated porosity. The dolomitization process enhances the porosity from an initial 5–7% up to 20% in the reservoir dolomitized bodies.

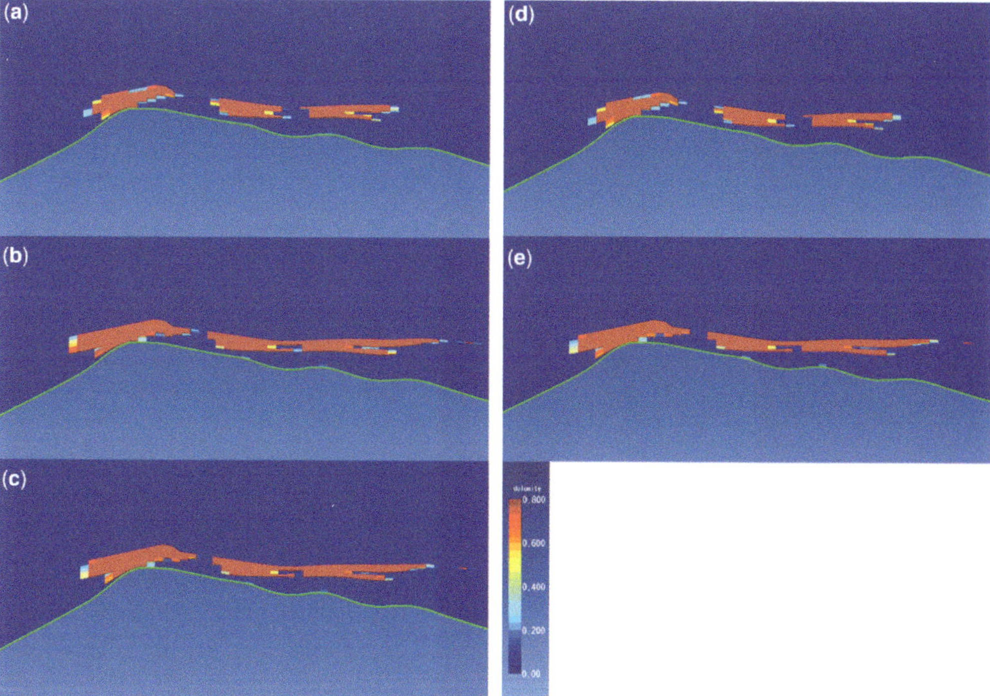

Fig. 4. Dolomite volume for all simulation runs (**a–e**) in the central part of the section measured in volume fraction units. Zoom has been used for the volume calculations; the body considered is the first on the left.

Table 4. *Dolomitized volumes*

Model run	Dolomitized volume (millions of m^3)
a	75
b	132.9
c	123.6
d	75
e	126.8

Discussion

The main question of this work is whether the proposed model of hydrothermal dolomitization can account for the amount of dolomite encountered and the extension of dolomitized bodies, considering mass balance, fluid chemistry, chemical thermodynamics and kinetic constraints. A set of simulations of hydrothermal dolomitization has been executed, varying some of the input parameters. All of the simulations indicate that the model is effective in producing dolomitized bodies in the limestone reservoirs, located at the intersections of the faults with the reservoirs and extending into the reservoir layers for hundreds of metres to kilometres.

The bodies are located at the positions where the faults that drive hot reactive fluids cut the reservoirs; the location is independent of the input parameters used in the simulations (Fig. 4). The input parameters affect the lateral extension of the bodies into the reservoirs and hence their volumes. This confirms the idea that faults intersecting reservoirs may be considered as discontinuity points (Laubach *et al.* 2010) where fluids in disequilibrium with the mineral assemblage of the reservoir interact causing calcite dissolution and the precipitation of dolomite.

Fluids enter the reservoir layers at fault intersections, flow along the limestone reservoirs following higher permeability layers and dolomitize them (Fig. 3). The lateral extension of the dolomitized bodies is higher in the upper and lower reservoirs than in the marly intercalated layer (Fig. 4) due to higher permeability in the reservoir layers. Moreover, the lateral extension is higher in upper than lower reservoir layer due to increased permeability in the upper layer relative to the lower reservoir layer. As is found for other dolomitization models (Consonni *et al.* 2010), the distribution of permeability is the most important parameter that determines the position of the dolomitized bodies and one of the more effective in predicting their extension.

The second question involves the relative importance of the parameters that influence the process. Four different parameters have been varied to assess their impact on the volume of the dolomitized bodies, flow duration, flow rate, water composition and thermal regime. Flow duration and flow rate both have a direct proportional effect on the volume of the dolomitized bodies, the fluid volume entering the system (and hence the Mg content) being directly proportional to the flow duration and flow rate (Fig. 4e, b and Table 4). In fact the multiplication of the two parameters gives the total amount of fluid (and hence Mg) entering the limestone reservoirs. In the simulations, the duration of flow has been considered as continuous whereas in nature multiple fluid flow events alternate with periods of no flow, and hence final dolomitization is probably the sum of multiple dolomitization events. The duration used in the simulations has to be considered as the sum of multiple periods of activity.

Temperature seems to be less important for the dolomitization process (Fig. 4d). Fluids entering the system react with the limestone, dissolve calcite and precipitate dolomite using the Mg contained in them until equilibrium is reached. The fluid then continues to flow along the reservoir without reacting with the rock as it is then in equilibrium with it. The process continues until calcite is totally dissolved at the entry point, then the process moves on in the direction of flow, causing dolomitization away from the entry point. Given sufficient time, the dolomitization process can affect the limestone reservoir layer far from the entry point. This process is not dependent on temperature, but only on the fluid amount and Mg content entering the system.

The role of water composition has not been assessed in a systematic way: only two water compositions have been tested. Nevertheless, some conclusions can be drawn. The brine from the South Atlantic reservoir results in double the amount of dolomitization of the seawater, despite a lower Mg/Ca ratio as it has a high Mg content. This second factor seems to be a determinant for the amount of dolomitization – the amount of Mg is directly proportional to the amount of calcite that can be reprecipitated as dolomite. The impact of water composition on dolomitization deserves a systematic analysis; waters from different sources and with different compositions should be analysed.

This hydrothermal dolomitization model has some peculiar aspects compared with the other dolomitization models when simulated with an RTM approach. In the reflux and thermal convection processes fluid flow is regulated by the geometry of the carbonate platform (Jones & Xiao 2005). The compaction case presents more uncertainty, but an estimate of the fluid flow rates can be done by looking at the compacting volume that is producing the fluids (Consonni *et al.* 2010). In hydrothermal dolomitization the flow along faults is difficult to estimate and thus only poorly constrained estimations may be done. Fluid composition can be estimated to be marine for thermal convection and hyper concentrated marine due to evaporation for reflux processes (Jones & Xiao 2005; Whitaker & Xiao 2010). The compaction case is more uncertain and no straightforward assumption can be made (Consonni *et al.* 2010), even if assumptions can be made using the mineral composition of the host rocks the fluids are coming from. Hydrothermal dolomitization presents large uncertainties with respect to the fluid type and only two trials have been done. Therefore rough estimations have been made for the input parameters and RTM has been used as a numerical laboratory to quantify the influence of these parameters and to check the feasibility of the dolomitization process.

Conclusions

Hydrothermal dolomitization has been simulated for lacustrine limestones from West Africa. Hydrothermal fluids entering carbonate layers from deep-seated faults dissolve calcite and precipitate dolomite-forming bodies. Five simulations have been executed to test the importance of flow duration, flow rate, temperature and water composition. The main conclusions can be summarized as follows:

(1) Hydrothermal dolomitization is an important diagenetic process that potentially affects the petrophysical characteristics of carbonate reservoir rocks and is a viable process in forming dolomite bodies.
(2) Dolomite bodies are located at the intersections of the faults that drive reactive fluids in the limestone reservoir layers and extend into the reservoir following the permeability distribution of the rock.
(3) Lateral extension of the bodies and hence their volumes are determined by fluid composition, flow duration and flow rate.
(4) Dolomitization is directly proportional to the flow rate and duration whereas temperature seems to be less important. The sensitivity analysis shows that water composition is very important for the process, although the dependence of dolomitization on water composition has to be analysed more deeply.
(5) There are large uncertainties in the parameters that affect hydrothermal dolomitization, hence RTM seems to be a powerful tool to quantify the influence of these parameters and to check the feasibility of the dolomitization process.

References

Consonni, A., Ronchi, P., Geloni, C., Battistelli, A., Grigo, D., Biagi, S., Gherardi, F. & Gianelli, G. 2010. Application of numerical modelling to a case of compaction driven dolomitization: a Jurassic palaeohigh in the Po Plain, Italy. *Sedimentology*, **57**, 209–231.

Davies, G.R. & Smith, L.B. 2006. Structurally controlled hydrothermal dolomite reservoir facies: an overview. *American Association of Petroleum Geologists Bulletin*, **90**, 1641–1690.

Frixa, A., Maragliulo, C., Consonni, A. & Ortenzi, A. 2014. *Dolomitization of the lacustrine carbonates of the Toca Fm. (Kambala Field, offshore Cabinda) and quantitative diagenetic modeling*. Paper SPE 171960-MS presented at the Abu Dhabi International Petroleum Exhibition and Conference, 10–13 November 2014, Abu Dhabi, UAE.

Jones, G.D. & Xiao, Y. 2005. Dolomitization, anhydrite cementation and porosity evolution in a reflux system: insights from reactive transport models. *American Association of Petroleum Geologists Bulletin*, **89**, 577–601.

Laubach, S.E., Eichhubl, P., Hilgers, C. & Lander, R.H. 2010. Structural diagenesis. *Journal of Structural Geology*, **32**, 1866–1872.

Machel, H.G. 2004. Concepts and models of dolomitization: a critical reappraisal. *In*: Braithwaite, C.J.R., Rizzi, G. & Darke, G. (eds) *The Geometry and Petrogenesis of Dolomite Hydrocarbon Reservoirs*. Geological Society, London, Special Publications, **235**, 7–63.

McHargue, T.R. 1990. Stratigraphic development of Proto-South Atlantic rifting in Cabinda, Angola – a Petroliferous Lake Basin. *In*: Katz, B.J. (ed.) *Lacustrine Basin Exploration: Case Studies and Modern Analogs*. AAPG Memoirs, Tulsa, OK, **50**, 307–326.

Scheevel, J., Domingos, F., Nogueira, J., Fernandes, M., Skander, L., Costa, L., Lomando, A. & Kienast, V. 2004. *Analysis and Modeling of Fracture-enhanced Production in Lacustrine Carbonate Reservoirs at Kambala Field, Cabinda Province, Angola, West Africa*. AAPG Hedberg Research Conference, 8–11 February 2004, Austin, Texas.

Warren, J. 2000. Dolomite: occurrence, evolution and economically important associations. *Earth-Science Reviews*, **52**, 1–81.

Wasson, M.S., Saller, A. & Self, D. 2013. *Controls on reservoir development in the Toca formation of Block 0, offshore Cabinda, Angola*. AAPG Annual Convention & Exhibition, 19–22 May 2013, Pittsburgh, Pennsylvania.

Whitaker, F. & Xiao, Y. 2007. *Reaction transport modelling of geothermal convection: a viable mechanism for early dolomitization platform? Abstract presented at the Bathurst Meeting*, 16–18 July 2007, Norwich.

Whitaker, F. & Xiao, Y. 2010. Reactive transport modeling of early burial dolomitization of carbonate platforms by geothermal convection. *AAPG Bulletin*, **94**, 889–917.

Whitaker, F., Smart, P.L. & Jones, G.D. 2004. Dolomitization: from conceptual to numerical models. *In*: Braithwaite, C., Rizzi, G. & Darke, G. (eds) *The Geometry and Petrogenesis of Dolomite Hydrocarbon*

Reservoirs. Geological Society, London, Special Publications, **235**, 99–139.

XIAO, Y. & JONES, D. 2006. *Reactive Transport Modeling of Carbonate and Siliciclastic Diagenesis and Reservoir Quality Prediction*. Society of Petroleum Engineers Report **101669**.

XU, T., SONNENTHAL, E.L., SPYCHER, N. & PRUESS, K. 2004. *TOUGHREACT User's Guide: A Simulation Program for Non-Isothermal Multiphase Reactive Geochemical Transport in Variably Saturated Geologic Media*. Lawrence Berkeley National Laboratory Report LBNL-**55460**.

Reservoir quality prediction of deep-water Oligocene sandstones from the west Niger Delta by integrated petrological, petrophysical and basin modelling

O. K. CHUDI*, HELEN LEWIS, D. A. V. STOW & J. O. BUCKMAN

Institute of Petroleum Engineering, Heriot-Watt University, EH14 4AS Edinburgh, UK

**Correspondence: obinna.chudi@pet.hw.ac.uk*

Abstract: Petroleum exploration and production in the region of the Niger Delta to date has mainly focused on the onshore, deltaic and offshore deep-water Miocene successions. Although Miocene turbidites have been the principal deep-water target, deeper-lying Oligocene sandstones are now being considered for exploration. This study targets an area beneath the Niger Basin slope at a present-day water depth of 800–1500 m. Within this study area, the Miocene to Recent sands above a burial depth of 3600 m show very good reservoir quality with porosities as high as 35% and permeabilities in the Darcy range. The aim of this study is to predict the reservoir quality and properties of the Oligocene sandstones below 3800 m using basin modelling to predict conditions where quartz cementation will take place and quartz cementation models to predict the amount of cementation and hence the potential porosity loss. Modelling results show that the Oligocene sandstones have been exposed to conditions favourable for quartz precipitation, but that less than 14% of the original porosity will have been occluded by quartz cement. These results are in agreement with elemental analysis from both petrophysical and petrological observation of thin sections. Although the deeper-lying Oligocene sandstones are likely to have reduced reservoir quality due to the presence of quartz overgrowth cementation, it appears likely that the volume of cement is relatively low and the Oligocene succession should be considered a viable play.

The Miocene clastic succession of the prolific Niger Delta province has been a major target for hydrocarbon exploration since the 1950s and is responsible for hosting most of the discoveries in both the onshore and offshore parts of the delta system (Doust & Omatsola 1990; Saugy & Eyer 2003; Reijers 2011). In part, this is due to the excellent porosity and permeability characteristics of the poorly consolidated reservoir sands (Weber 1971). Porosities of up to 35% and permeabilities of more than 3 D have been reported from well logs and cores across producing fields in the delta region. There has been no or minimal cementation and very little precipitation of authigenic minerals, which are dominated by quartz overgrowths, within the primary porosity. The reservoir quality of the Miocene interval has only undergone a low level of degradation.

As the quest for further hydrocarbon reserves increases, exploration and production is gradually moving from the shallow and more easily identifiable reservoirs of the Miocene to deeper-lying Oligocene plays that have been exposed to higher temperatures and hence physical and chemical conditions that are more conducive to quartz precipitation. The aim of this study is to estimate the degree of quartz cementation in unexplored, deeply buried, Oligocene deep-water clastic sediments of the basin slope of the Niger Delta region. We apply an integration and cross-comparison of basin modelling, cementation modelling, petrological analysis and petrophysical multimineral elemental analysis.

The Niger Delta has been the centre of attraction for hydrocarbon exploration and production on the continental margin of West Africa for over four decades (Haack *et al.* 2000) and has consequently has thoroughly studied (Short & Stauble 1967; Weber 1971; Whiteman 1982; Ejedawe *et al.* 1984; Damuth 1994; Haack *et al.* 2000). The delta is situated in the Gulf of Guinea and covers an area of approximately 140 000 km^2 with a maximum clastic sediment thickness of about 12 km at the basin centre (Damuth 1994). The study area is located in the translational tectonic province in the mid-slope of the Niger Delta deep-water system (Fig. 1). The proven reservoir intervals are Miocene channel sands with porosity values greater than 30% and permeability in the Darcy range. Of over 30 wells drilled in the study area only one well penetrated the top of the Oligocene succession, which is the primary interval of interest for this study.

Generally reservoir rock quality is controlled by variables such as the grain size, initial depositional porosity, mechanical compaction, chemical compaction, mineralogy and volume of pore filling

From: Armitage, P. J., Butcher, A. R., Churchill, J. M., Csoma, A. E., Hollis, C., Lander, R. H., Omma, J. E. & Worden, R. H. (eds) 2018. *Reservoir Quality of Clastic and Carbonate Rocks: Analysis, Modelling and Prediction.* Geological Society, London, Special Publications, **435**, 245–264.
First published online December 14, 2016, https://doi.org/10.1144/SP435.8

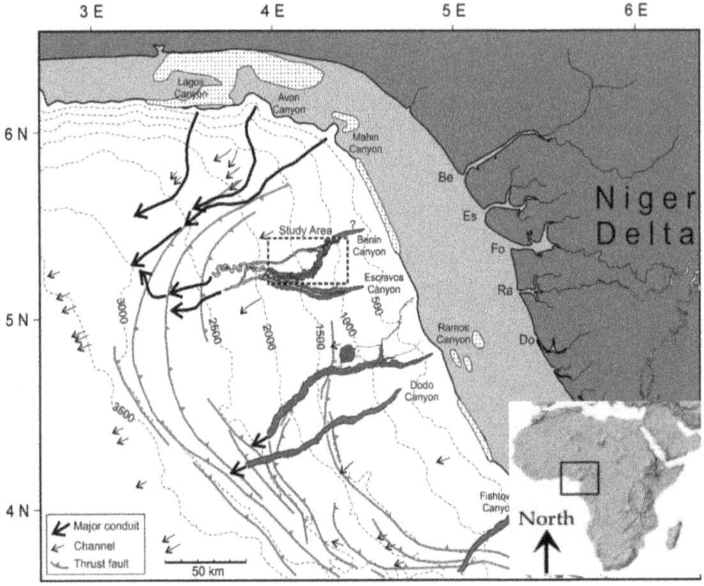

Fig. 1. Regional location map showing the western Niger Delta and study area. Adapted from (Deptuck *et al.* 2007).

cement (Worden & Morad 2009). The diagenetic changes that influence reservoir properties, such as porosity and permeability, are affected by an interplay of these factors under conditions of increased effective stress, temperature and burial (Taylor *et al.* 2010). The regional geothermal gradient of the whole Niger Delta region ranges from 1.3 to 1.8°C per 100 m in the centre of the basin and increases up-dip and northwards to about 2.7–5.5°C per 100 m (Nwachukwu 1976; Doust & Omatsola 1990). These controlling variables have influenced the strategy and the technique adopted herein for petrophysical evaluation across fields within the delta system.

The accurate prediction of sandstone reservoir quality is a key challenge for hydrocarbon exploration and production. A good understanding of reservoir quality is required throughout the entire lifecycle of a reservoir; from exploration (where it is required for the estimation of the initial hydrocarbon reserves) to the appraisal and development stages where the porosity and permeability distribution in a reservoir are necessary for well-placement optimization and to estimate economic cut-off limits that control estimates of hydrocarbon pore volumes, recoverable reserves and production rates (Sneider 1990). A fundamental control on the quality of a sandstone reservoir, especially for siliciclastic systems, is quartz cementation. This is controlled by the temperature, pressure (effective stress) and surface area of the quartz substrate, which is itself a function of the grain size, abundance of grain coatings and quartz clast abundance (Walderhaug 2000; Worden & Morad 2009).

We present a synthesis of petrological, petrophysical and basin modelling studies to characterize the poorly understood potential Oligocene reservoirs of the mid-slope, deep-water part of the Niger Delta Basin. The focus of this work is to better understand and investigate the reservoir potential of the Oligocene interval buried below 3800 m, and most importantly its diagenetic history, using a novel integrated methodology. The burial, temperature and pressure histories have been considered, in addition to their key controls on quartz cementation. Petrological analysis of thin sections and polished block samples of the hydrocarbon-charged shallower Miocene samples were studied and inferences made to predict the mineralogy of the uncored Oligocene interval that was penetrated by a single well (A1). The petrological study was integrated into multimineral petrophysical study to predict the composition of the sandstones from well log responses. We further consider the applicability of this approach and these findings to analogous subsurface systems around the world.

Geological setting

Rifting and break-up of Africa from South America in this region initiated in the Early Cretaceous, with a triple junction located approximately beneath the region that is now the Niger Delta. By the Late Cretaceous, spreading of the Atlantic was underway along the southern and western arms of this tri-rift systems, whereas the NE arm, the Benue Trough, became a failed rift system. By the Palaeocene, the Niger Delta region had begun to receive significant sediment input from the land. Through the Palaeocene and Eocene there were probably at least three principle depocentres linked to marked sediment progradation over the subsiding continental–oceanic lithospheric transition zone. By the Oligocene, these three depocentres had merged into the single major arcuate depocentre that is now the Niger Delta. The increased sediment load on a cooling and subsiding basement further enhanced subsidence in the region. Since this time the delta has prograded in a generally SW direction creating a series of major depositional belts that represent the most active portion of the delta at each stage of development. The principal sediment provenance for the western delta region has been the Benue–Niger drainage system, whereas the eastern delta region was sourced from the Cross River system. The present day Niger Delta complex takes the form of a constructive arcuate delta, with mud diapirism beneath the slope playing a major role in controlling sediment distribution (Whiteman 1982).

Lithostratigraphy

The Tertiary Niger Delta complex comprises a tripartite lithostratigraphic sequence: the Benin, Agbada and Akata formations (Fig. 2). Each of these formations represents deposition in one of three main sedimentary environments: continental, transitional and marine, respectively. These depositional environments are arranged laterally from proximal to distal and, as a result of marked progradation, they also stack up vertically with the continental Benin Formation overlying the transitional (marginal marine) Agbada Formation, which in turn overlies the marine/deep-marine Akata Formation. We describe each briefly below from proximal to distal.

The proximal continental environments comprise sediments of the upper delta plain and fluvial feeder system and are represented by predominantly sandy lithologies. The sediments are poorly sorted, granular and pebbly, coarse to very fine grained. They are assigned to the Benin Formation, which occurs wholly onshore at the present day, but also extends a short distance offshore in the subsurface (Short & Stauble 1967). The known age of the Benin Formation is from Oligocene to Recent although, presumably, an equivalent lithostratigraphic unit also acted as sediment source for the older delta deposits. The Benin Formation generally exhibits poor reservoir quality and so is not an effective reservoir target.

The transitional, marginal marine environment comprises interbedded sandstones and shales assigned to the Agbada Formation. The shale units are more prominent with depth, becoming progressively thicker than the intercalated sandstone units; this reflects the seaward advance of the delta system through time (Whiteman 1982). The sandstones constitute the principle reservoir units in the Niger Delta Basin, exhibiting excellent reservoir quality. They were most likely charged through large-scale open growth faults that were connected to adjacent kitchen areas. The Agbada Formation extends throughout the Niger Delta region and represents deposition in both lower delta plain and marginal marine to continental shelf environments.

The interplay between subsidence and sediment supply, and the relative change in sea-level, has driven transgressive and regressive cycles that are responsible for alternating sandstones and shale sequences (Doust & Omatsola 1990). The known age of the Agbada Formation ranges from Eocene to Recent.

The distal and fully marine environments are represented by a more mud-rich succession known as the Akata Formation. The marine shales of this sequence range from Palaeocene to Recent in age, and are mainly characterized by hemipelagic slope sediments intercalated with turbidite sandstones. Regional studies and experimental analysis suggest that the Akata Formation provides the principal hydrocarbon source rock for the Niger Delta province (Evamy *et al.* 1978; Lambert-Aikhionbare & Ibe 1984; Tuttle *et al.* 1999). The Akata shales are extensive across the whole region, whereas the turbidite sands are more localized as deep-water channel and lobe deposits.

These channel-lobe turbidites provide the principal deepwater reservoirs of Miocene age as well as the proposed Oligocene reservoir targets. Similar features are recognized through the Pliocene section as well as across the present day Niger Delta slope. Certainly there is no evidence either from seismic interpretation or from borehole studies to suggest any significant change in sediment provenance or depositional processes between the Oligocene and Miocene systems. We are therefore confident in using what we know of the Miocene reservoirs as a template for understanding the Oligocene.

Tectonics and structure

The end of Atlantic rifting in the Late Cretaceous gave way to gravity-driven tectonism as the primary deformational process affecting delta development. Owing to a rapidly prograding delta, low-permeability, fine-grained, pro-delta muds became

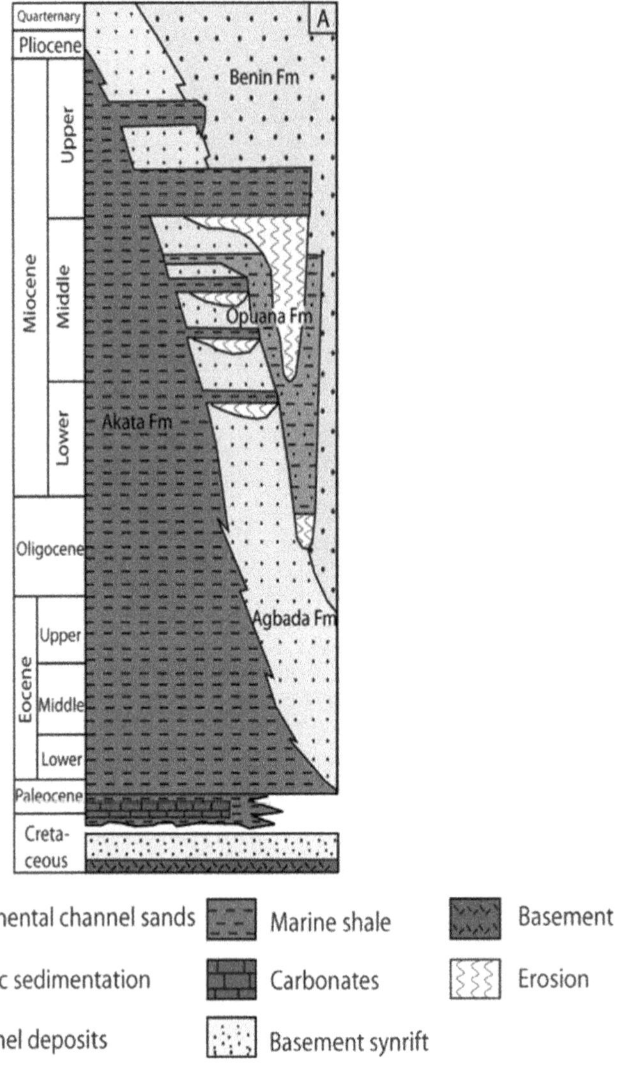

Fig. 2. Stratigraphy of the Niger Delta (after Corredor *et al.* 2005).

overpressured and formed mud diapirs as a response (Whiteman 1982). Diapirism began in the Miocene and is still taking place today both at the delta front and extending further into the deep-water basin. The growth of the diapirs resulted in the development of growth faults, which offset the shallower parts of the section and flatten onto a décollement surface at the top of the hemipelagic muds. The overall structural trend is oriented in a NW–SE direction (Haack *et al.* 2000). Three structural provinces generally define the delta (Fig. 3): (a) an extensional province that extends from onshore to the outer shelf region, marked by large-scale listric growth faults; (b) a translational province across the upper and middle slope, dominated by mud diapirs and folds; and (c) a compressional province in the lower slope region, triggered by tectonic-scale downslope movement of sediment due to gravity gliding, which leads to toe thrust features and a marked fold and thrust belt (Bilotti & Shaw 2005; Corredor *et al.* 2005; Deptuck *et al.* 2007).

Methods

Petrography

Petrological analyses were conducted on 17 samples acquired from four wells in the study area.

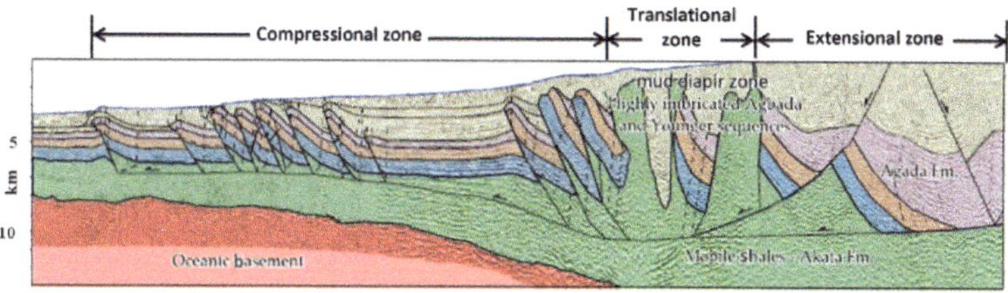

Fig. 3. Regional cross-section showing the three structural provinces of the Niger Delta (modified from Corredor *et al.* 2005).

The cored samples were taken from the Miocene interval; from the Middle Miocene at a depth of approximately 2470 m and from the lower Miocene at 3450 m. Samples taken from depths shallower than 3000 m were unconsolidated sands. All samples were made into thin sections and polished blocks as appropriate. The samples were dried at 38°C and vacuum-impregnated with epoxy resin. Ultraviolet glue was further applied to mount cover slips. For the purpose of SEM studies, samples were also made into polished blocks by drying at 38°C and placing in a 25–30 mm diameter mould. The granular samples were filled with epoxy resin, mixed by stirring and placed under vacuum to remove air bubbles.

Techniques used included light optical microscopy and scanning electron microscopy (SEM) using back-scattered electron imaging (BSE) and cathodoluminescence (CL) analysis. Samples were examined uncoated and in low-vacuum mode, using a Quanta 650 FEG SEM and XL30 LaB6 ESEM.

Optical microscopy was used for mineralogical identification, and where the optical identification of quartz overgrowths was difficult SEM was used. CL analyses were carried out on the SEM with a Centaurus CL detector. A combined BSE and CL analysis was undertaken using a similar method to that outlined by Evans *et al.* (1994), which involves the analysis of pairs of BSE and CL images from polished block samples. A set of 13 samples were selected for this analysis. CL is particularly useful in distinguishing detrital quartz grains from syntaxial quartz overgrowths.

Petrological analysis was carried out over the Middle and Lower Miocene intervals to investigate the probable presence of microcrystalline quartz or chlorite rims, as they can impede the nucleation of quartz overgrowth if present (Bloch *et al.* 2002; Marchand *et al.* 2002; Taylor *et al.* 2010).

Petrophysical analysis

Wireline log data from four wells within the study area were analysed to characterize the porosity and permeability of the reservoir zones penetrated. Of the four wells studied only one well penetrated the Oligocene interval. The density log was used to calculate porosity and these values were calibrated against core porosity where available. Permeability was estimated using the neural network technique, where sets of input logs (gamma ray, sonic, density and neutron logs) were trained to recognize the core-derived, stress-corrected air permeability, acquired from routine core analysis (Sonde *et al.* 2011).

A multimineral model based on petrophysical elemental analysis (ELAN) of open-hole logs was used to compute the volume of mineralogical components within the intervals of interest. ELAN uses log curves and the response parameters of the tools to compute the volumetric constituents of the minerals and fluid in the formation. This method derives the relative quantities, or relative volumes, of the mineral components that would most probably produce the set of measurements recorded by the logging instruments. The three-way relationship between tools (T), response parameters (R) and formation component volume (V) is shown in Figure 4.

Given the data represented by any two corners of the triangle (Fig. 4), the third can be determined. In this study, T and R are used to compute V. For quality control, forward-modelling of R and V are used to reconstruct T – the input logs. The reconstructed logs are compared against the input data to determine the quality of the volume results. Four principal minerals (quartz, feldspar, zircon and kaolinite) were modelled in ELAN based on their abundances determined in the petrographic study.

Basin modelling and diagenetic evolution

Since the early 1980s basin modelling has been used to study the burial and thermal evolution of a basin; particularly in relation to petroleum generation, expulsion, migration, accumulation and preservation (Welte & Yalcin 1988; Wygrala 1988; Talçin 1991; Hermanrud 1993; Underdown & Redfern 2008). Basin modelling has also been used by some authors to study diagenetic evolution and to infer its

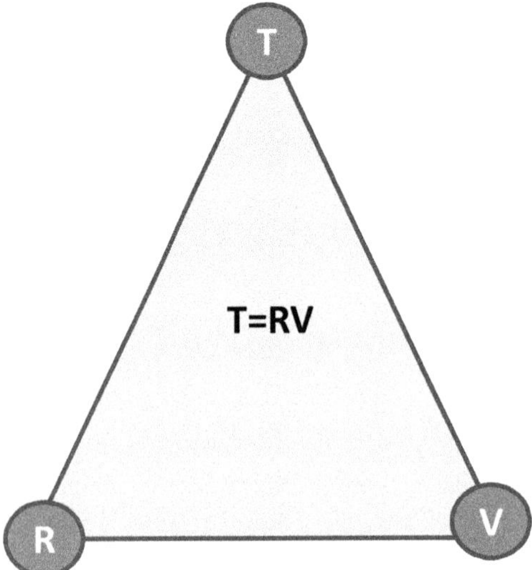

Fig. 4. Schematic illustration of ELAN. T represents the tool vector, which in this case is the input log data (gamma ray, resistivity, neutron, density and calculated porosity logs) and R is the response matrix, which is a pre-defined value for the reading each tool would give for 100% of each formation component. R values were determined on the basis of the known mineralogical responses to the physics of the different tools as defined using the Schlumberger Techlog software package. V is the volume vector, which is the volume of each of the formation components.

impact on reservoir quality, particularly on porosity (Sombra & Chang 1997; Walderhaug 2000) by attaching a calculation of the appropriate chemical reactions to the states generated by the basin model at different burial positions. Siever (1983) first associated chemical reaction calculations with fluid type and conditions, with stress and fluid pressure states and with temperature, permitting an estimate of the transformation of organic matter to petroleum and also of quartz precipitation from a silica-rich fluid. Before this study by Siever (1983), the compaction models proposed by Athy (1930) were used to calculate porosity evolution using the degree of sediment compaction, but these relationships were sufficient to calculate the effect of porosity changes due to cementation or dissolution. Starting with Siever (1983), models can relate the calculated (or otherwise derived) basin history to temperature and to appropriate diagenetic reactions. However basin-modelling studies are certainly not the only method available to estimate palaeothermal and palaeochemical states. For example, palaeotemperatures have been regularly inferred from the petrographic study of fluid inclusions, and particularly from chemical composition and isotopic signatures. Sombra & Chang (1997) follow this approach, using time–depth indices (TDI) to quantify the influence of burial history on the evolution of sandstone porosity.

Basin modelling. The 2D basin model uses the transect shown in Figure 3, oriented NE–SW and extending over 160 km across the extensional, translational and compressional parts of the Niger Delta Basin (Bilotti & Shaw 2005; Corredor *et al.* 2005; Deptuck *et al.* 2007; Fig. 5). A total of seven chronostratigraphic horizons, the Seafloor, Lower Pliocene, Upper Miocene, Mid Miocene, Lower Miocene, Oligocene and Basement were digitized and used in this study. Each of the resulting defined units was assigned suitable lithologies. Where no well control was available, each lithology was initially assumed to be laterally continuous. Other basic requirements for modelling include boundary conditions, calibration data (corrected bottom hole temperature (BHT), vitrinite reflectance, pressure, porosity and present-day hydrocarbon accumulations) and source rock properties, particularly the kinetic parameters that are responsible for the type and amount of generated petroleum. Interpreted faults on the 2D transect were also digitized, with priority given to the faults that are likely to have a significant impact on hydrocarbon migration. The fault selection method used adopted similar criteria to those chosen by Derks *et al.* (2012).

Quartz cement modelling. In this study the basin modelling tool used is PetroMod V. 2013, which has an additional quartz cementation calculation module based on the Walderhaug quartz cementation model (Walderhaug 2000). This module was well optimized with measured data (corrected BHT, vitrinite reflectance, pressure and porosity). The amount of quartz cementation over the time–depth range represented in the basin model was then calculated. However, Walderhaug (2000) expressed the effect of cementation as a function of the change in quartz surface area whereas PetroMod models cementation as a function of the rate of porosity loss. Equation (1) shows the relationship expressed by the Walderhaug quartz cementation model:

$$A = \frac{(1 - C)6fV\phi}{D\phi_0} \quad (1)$$

where A is the change in quartz surface area, ϕ is the porosity, ϕ_0 is the original porosity, C is the fraction of the quartz grain surface coated by clay or other substances, f is the volume fraction of quartz clasts, V is the sample volume and D is the average quartz grain diameter. PetroMod uses the cementation equation shown in equation (2). Both cementation modelling algorithms vary with assigned detrital mineralogy, quartz grain size, quartz grain coating abundance, temperature and A. The porosity loss

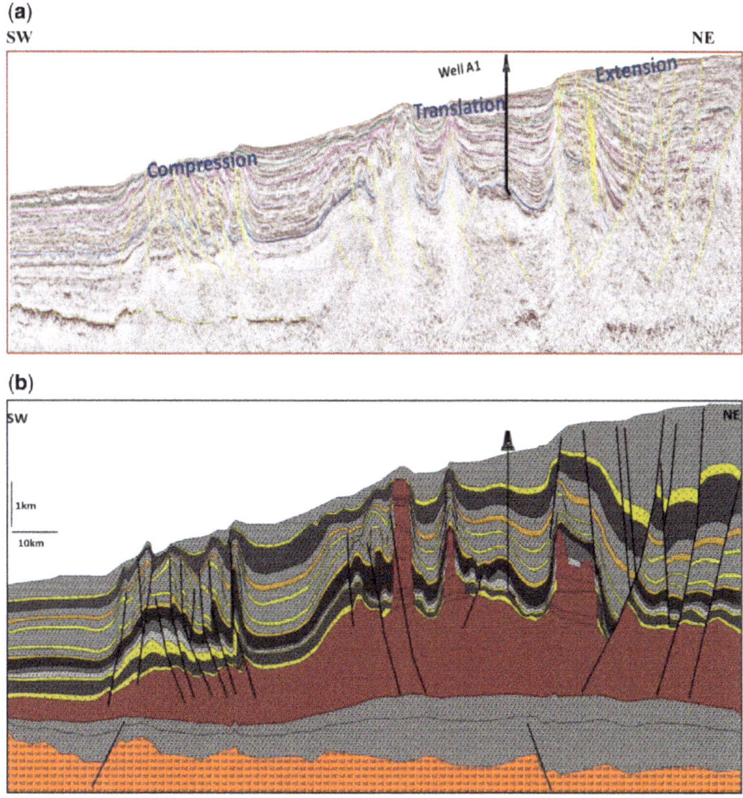

Fig. 5. Interpreted regional 2D transect across the 3 structural provinces (**a**) and digitized 2D transect (**b**).

rate due to quartz cementation that has been adopted in PetroMod is therefore expressed as:

$$\frac{\delta\phi cc}{\delta t} = \frac{m}{\rho}\frac{(1-C)6f}{D}\frac{\phi}{\phi_o}Ae^{-E_o/RT} \quad (2)$$

where C is the quartz grain coating factor, f is the quartz grain volume fraction, D is the average quartz grain diameter, ϕ is the porosity, ϕ_o is the original porosity, ϕcc is the porosity loss, t is time. A and E_o are respectively the frequency factor and activation energy of quartz precipitation. M represents the mole-mass and ρ the density of quartz (Walderhaug 1994).

The average grainsize and quartz grain volume fraction used in the quartz cementation calculations were based on the results of the petrological study. A dominant grain size range of 125–500 μm is seen in the Miocene. No core has yet been taken from the Oligocene interval. So, in the absence of any grain size measurements, it was assumed that the Oligocene reservoirs in the study area, which are below 3800 m, would have a similar grain size as the Lower Miocene reservoir at 3450 m, the deepest cored section available to this study.

The PetroMod calculation assumes that quartz is sourced by stylolitization. In addition, it is also assumed that the precipitation phase of the overall quartz dissolution, transportation and precipitation process is the slowest and hence acts as the rate-controlling step in the entire process (Walderhaug 2000).

By implementing the Walderhaug quartz cementation model as part of the simulation, the percentage of the pore space that is occluded by cement was estimated. The results show that cement volume tends to increase both with depth and with the landward and up-dip distance along the regional transect (Fig. 6) as water column thickness decreases and mean overburden increases. Little or no cement is predicted above 3000 m but cementation starts to develop from about 3450 m depth within the Lower Miocene reservoirs with less than 5% of the pore space now containing quartz cement. The cement volume increases steadily with depth into the potential Oligocene reservoirs with values of close to 14% of pre-cementation pore space. The minor quartz cement observed from the petrographic study for the Lower Miocene reservoir and the absence of quartz

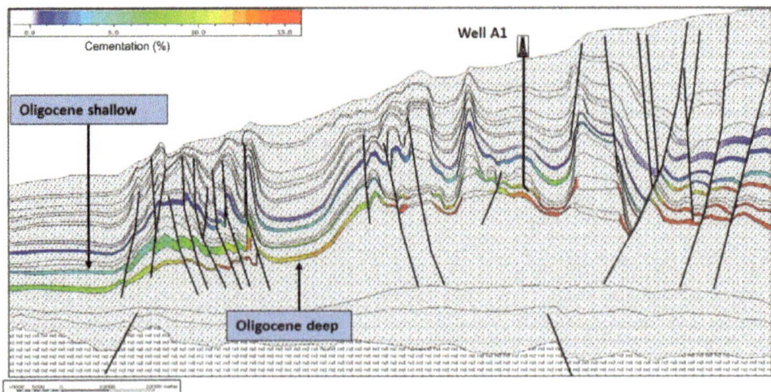

Fig. 6. Overlay of predicted quartz cement on a 2D transect. In the vicinity of well A1 the cement volume increases with depth from Miocene to Oligocene.

in Middle Miocene agrees with this modelled result. The modelled result is also displayed as a 1D extraction at the well location from the 2D model with cement volume plotted against depth (Fig. 7).

Results

Miocene sediments

The Miocene reservoirs in the study area are typically composed of loosely consolidated sands that are moderately sorted and display a dominant grain size of medium to fine, although coarser grain sizes also exist. The sands are classed as quartz arenites (according to the classification of McBride 1963) as they are largely composed of quartz with less than 15% feldspar and a minor lithic component (<2%). The excellent quality of the reservoir sands is largely attributed to the unconsolidated nature of the sands with little or no cementation present.

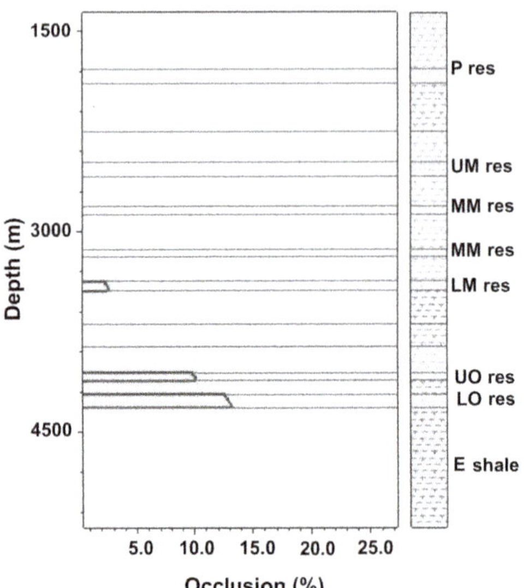

Fig. 7. Fraction of the pore space occluded by quartz cement plotted against depth: a 1D extract from the 2D model in Figure 14. Res, reservoir; P, Pliocene; UM, Upper Miocene; MM, Middle Miocene; LM, Lower Miocene; UO, Upper Oligocene; LO, Lower Oligocene; E, Eocene.

A total of 15.1 m of cores was studied; 12 m across the Middle Miocene and 3.1 m across the Lower Miocene sequence. The Middle Miocene reservoir facies comprise massive, unconsolidated sands that are typical of the Bouma Ta facies, with minor normal grading seen in the sands. High-density turbidity current flow is the most likely depositional mechanism. Two facies have been noted inthe Lower Miocene: (1) light grey, very fine consolidated massive sands that are moderately to well-sorted, representing deposition by high-density turbidity currents, and (2) interlaminated very fine sands and silts that are characterized by millimetre-scale parallel and convolute lamination. Some dark laminations are also noted and are most likely to be mud or organic material. These facies probably represent deposition by dilute turbidity currents. Figure 8 is a schematic of the sedimentological log of the core section studied, with representative core photographs of facies where samples for petrographic analysis were taken.

Petrological analysis

Detrital mineralogy. The Miocene samples from the three wells examined by optical microscopy are composed dominantly of quartz, comprising up to 85% monocrystalline grains with minor polycrystalline quartz, 10% feldspar and 0–5% lithic fragments of possible igneous and metamorphic origin (Fig. 9).

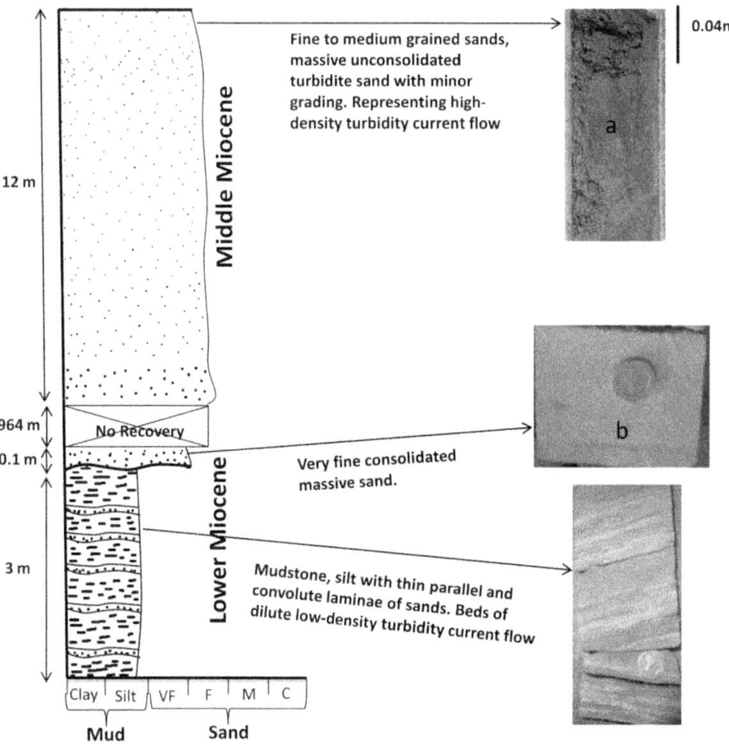

Fig. 8. Sedimentological log of core from well A1 across the Middle Miocene and Lower Miocene sections. Core photos (right) represent typical examples of the lithologies described; petrological samples were taken from a and b.

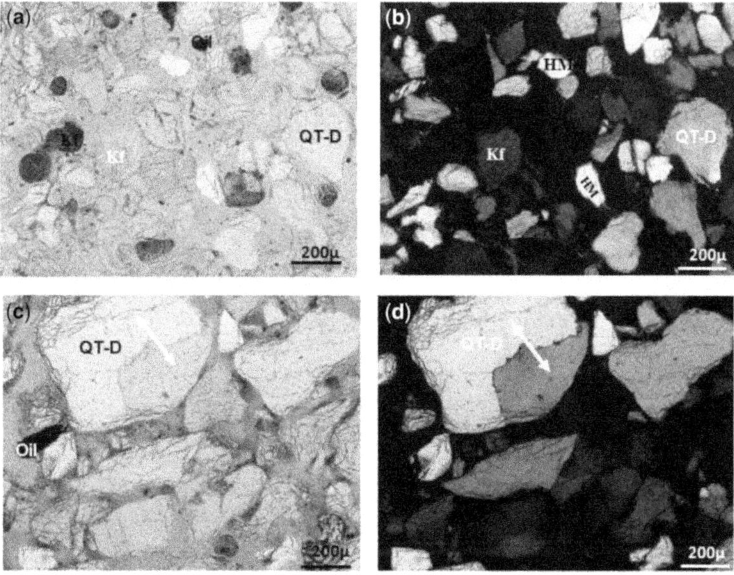

Fig. 9. Photomicrographs in both plane-polarized light and cross-polarized light of Middle Miocene (**a, b**) and Lower Miocene reservoirs (**c, d**). QT-D, detrital quartz; Kf, K-feldspar; arrows identify detrital quartz.

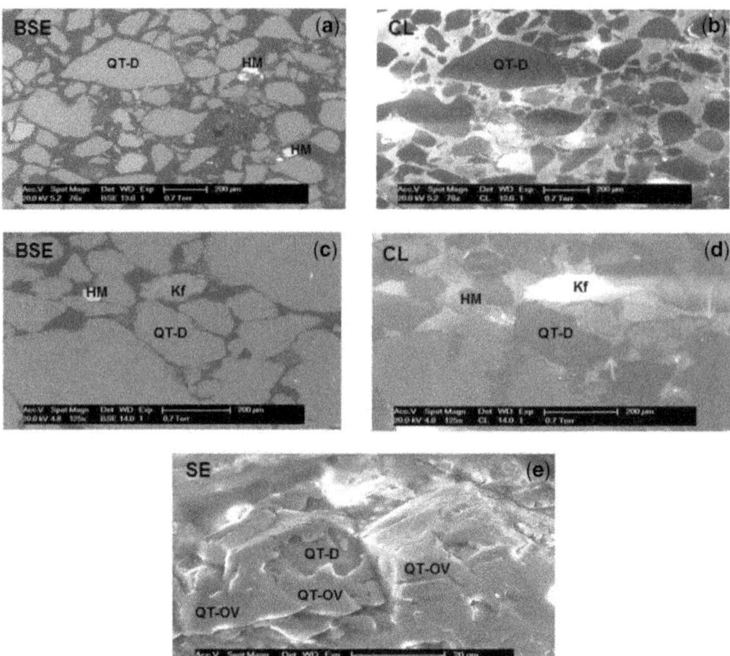

Fig. 10. Upper Miocene polished block samples with no quartz cement seen from both BSE and CL plates (**a, b**), while overgrowths (indicated by arrows) are present in the Lower Miocene samples (**c, d** and **e**), detrital quartz (QT-D), quartz overgrowth (QT-OV), K-feldspar (Kf), heavy mineral (HM).

Muscovite and heavy minerals (probably zircon) occur in trace amounts. Heavy minerals are easily recognizable using BSE micrographs as highly luminescent grains, as depicted in Figure 10a, c. Grain size ranges between fine and medium sand (125–500 μm), with sorting from poor to well-sorted and grain roundness from subangular to angular.

Diagenetic mineralogy

Quartz cement. Quartz cements are rarely observed in thin sections in the Middle Miocene samples located at depths of less than 3000 m. However, thin sections of samples retrieved from the Lower Miocene reservoirs at 3450 m do show some quartz overgrowths. In some cases they occur as euhedral overgrowths protruding into primary pores. The overgrowths range in size from individual small euhedral crystal growths of 3–20 μm to grain-rimming cements of less than 30 μm in thickness. Overgrowths are not widespread but are locally developed around quartz grains. CL and BSE analyses allowed distinction between authigenic and detrital quartz. Authigenic quartz rims are typically less luminescent than their detrital host (Fig. 10d). Topographic secondary electron images also reveal quartz overgrowths with well-defined crystal faces over the surface of the detrital host (Fig. 10e). BSE analyses of the same sample showed no distinction as both detrital and authigenic quartz display the same false colour grey (Fig. 10c).

Grain coating. The SEM analysis of the samples does not reveal the presence of any grain-coating minerals of early origin. The Upper Miocene samples are coated with thick 'dead'/bituminous oil (Fig. 11). The presence of this oil is indicative of biodegradation due to exposure of the Middle Miocene reservoir to temperatures lower than 60°C (Peters *et al.* 1996).

Clays: SEM examination revealed the presence of kaolinite-type clay in both the Upper Miocene and the Lower Miocene successions. These clays occur as stacked pseudo-hexagonal crystals of euhedral booklets (Fig. 12a–c). In the Lower Miocene succession kaolinite is observed blocking primary intergranular porosity, suggesting that the kaolinite clay developed before the main phase of mechanical compaction (Gier *et al.* 2008 Fig. 12c).

Petrophysics

The porosity calculated from wireline log analysis (for example, Fig. 13) decreases with depth as

Fig. 11. BSE image of Upper Miocene samples showing quartz grains surrounded by oil.

seen from the porosity depth plot of four wells in the study (Fig. 14). Figure 14 also reveals that only the estimated porosities from Well A1 (represented by a triangle), which is the only well that encountered the Oligocene reservoirs, have average porosity values of 20%. The porosity for the Miocene samples generally ranges from 21% to 37% across all of the wells (see Fig. 14).

The results of the multimineral analysis are displayed in Figure 15. Tracks three to seven of Figure 15a show the input logs while the last track shows the multimineral model derived from the open-hole logs. The modelled quartz volume is seen to increase with depth as observed from the cross plot of quartz volume as a fraction of the total reservoir component (mineral, fluid and rock) plotted against depth (Fig. 15b) with two trend lines seen. However, it is not possible to verify the proportions of detrital and authigenic quartz from the multimineral model. The first trend line (a) indicates a steady increase in the quartz volume of close to 60% with depth from the Upper Miocene to the Middle Miocene reservoirs. An offset is seen in the second trend (b), representing the quartz volume for the Lower Miocene down to the Oligocene with an estimated quartz volume of close to 80% in the Oligocene reservoir intervals.

Basin modelling

Temperature and pressure history. The thermal history calculated by the 2D basin model shows that the Middle Miocene and younger reservoirs in the study area have never been exposed to

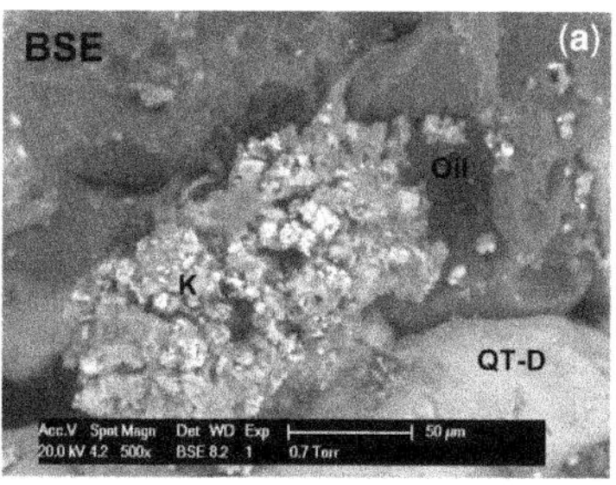

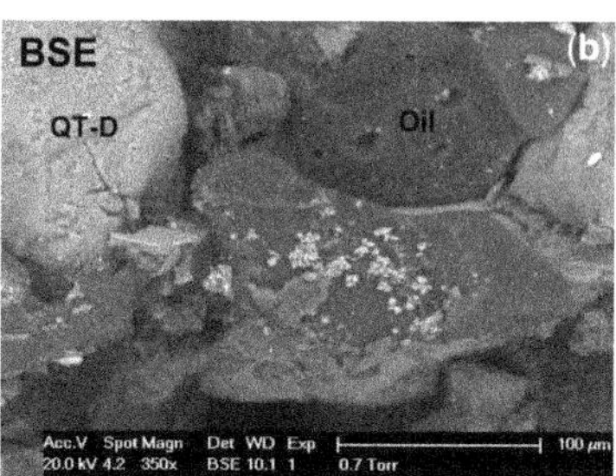

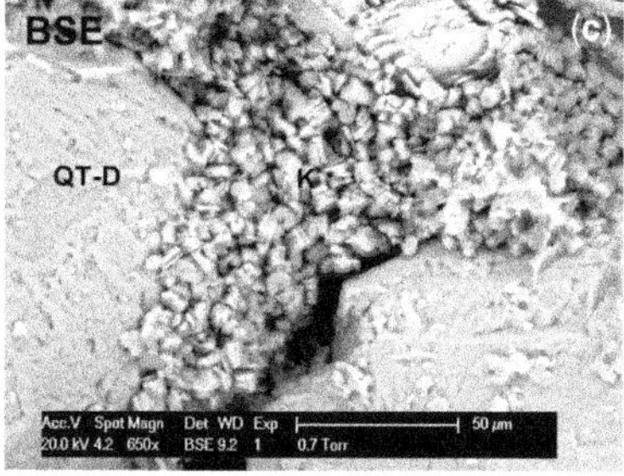

Fig. 12. BSE images of Upper Miocene and Lower Miocene sandstones. Images show authigenic kaolinite surrounded by oil (**a, b**) and pore-filling kaolinite crystals (**c**). QT-D, detrital quartz; QT-OV, quartz overgrowth; K, kaolinite.

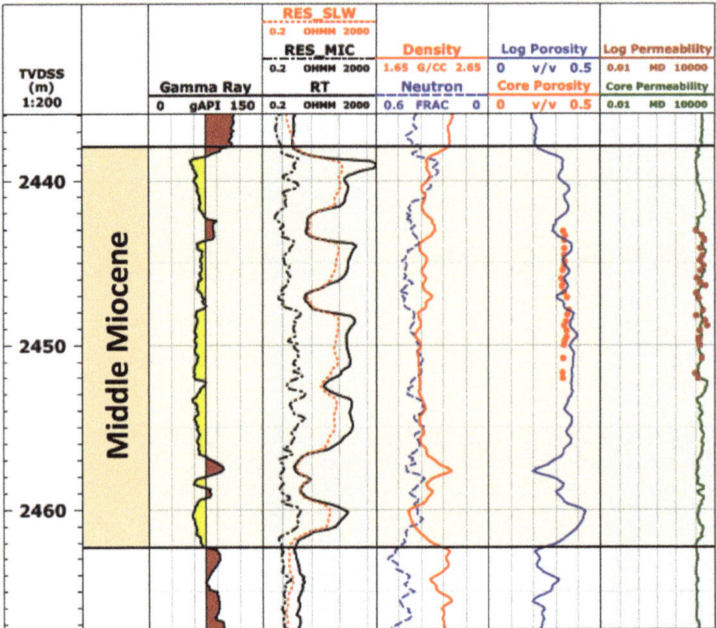

Fig. 13. Petrophysical log over part of the cored reservoir interval. Column 1 shows true vertical depth in metres. Column 3 gives gamma ray trace in API units. Column 4 gives wellbore log values of resistivity in ohm-metres as solid lines and column 5 shows the wellbore log derived density and neutron values, in grams per cubic centimetre, both as solid lines. Column 6 shows measured porosity from core as dots and the calculated porosity is given as a solid line. Column 7 shows permeability measured from core as dots and the equivalent calculated permeability as a solid line.

temperatures above 60°C (Fig. 16a), which is below the 70–135°C initiation of effective quartz precipitation temperature range (Walderhaug 1994; Bjørkum *et al.* 1998). On the other hand both the Lower Miocene and the Oligocene intervals are calculated to have reached temperatures above

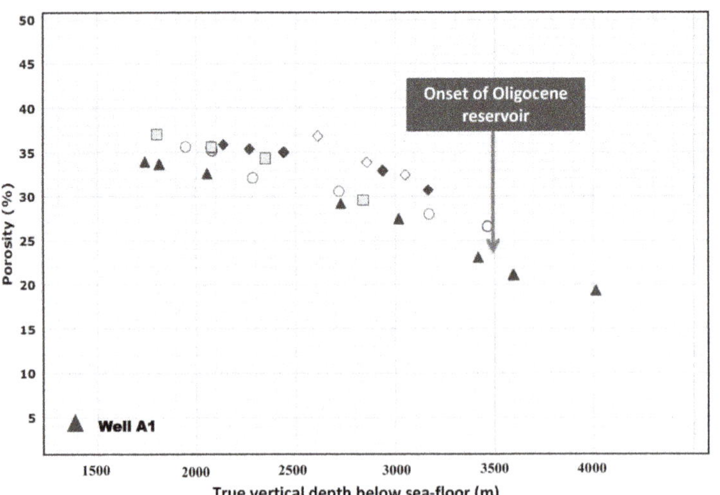

Fig. 14. Reservoir porosity plotted against true vertical depth for a series of wells that penetrated the Miocene section plus Well A1 that also penetrated the Oligocene section, positioned as is shown by the arrow. Well A1, the deepest well, is shown by triangles. Figure shows a reduction in porosity with increasing depth for all wells.

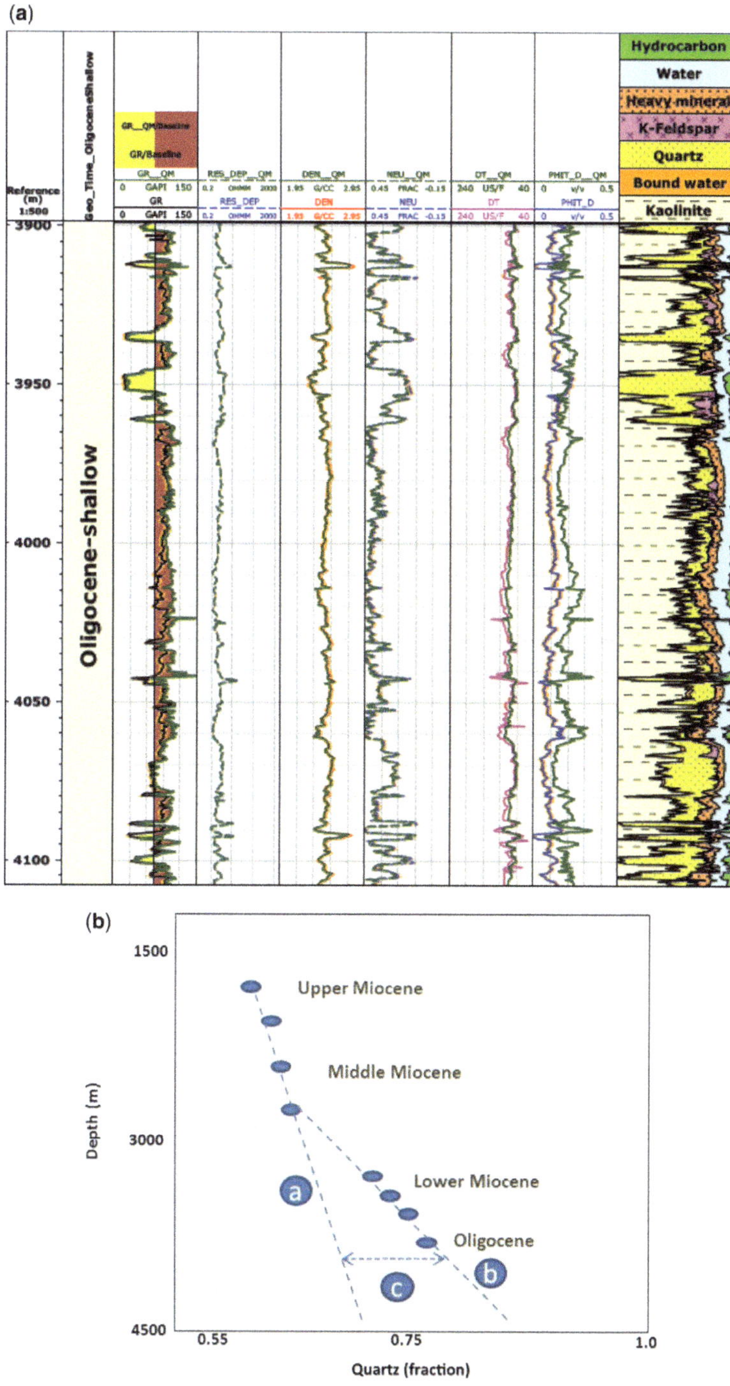

Fig. 15. (**a**) Modelled mineral volumes are shown in the far right column (quartz, yellow; feldspar, pink; mica, black; kaolinite, cream with dash). Column 1 shows true vertical depth subsea in metres. Column 3 gives gamma ray traces. Column 4 gives wellbore log values of resistivity, column 5 shows the wellbore log derived density and column 6 shows neutron values. Units are as given in Fig. 10. Column 3 upwards show wellbore electric log values in the normal manner and also a log trace with the suffix QM. Traces suffixed QM display inverse-calculated values that use the modelled mineral volumes rather than wellbore electric log data. (**b**) Cross plot of modelled quartz volume against depth.

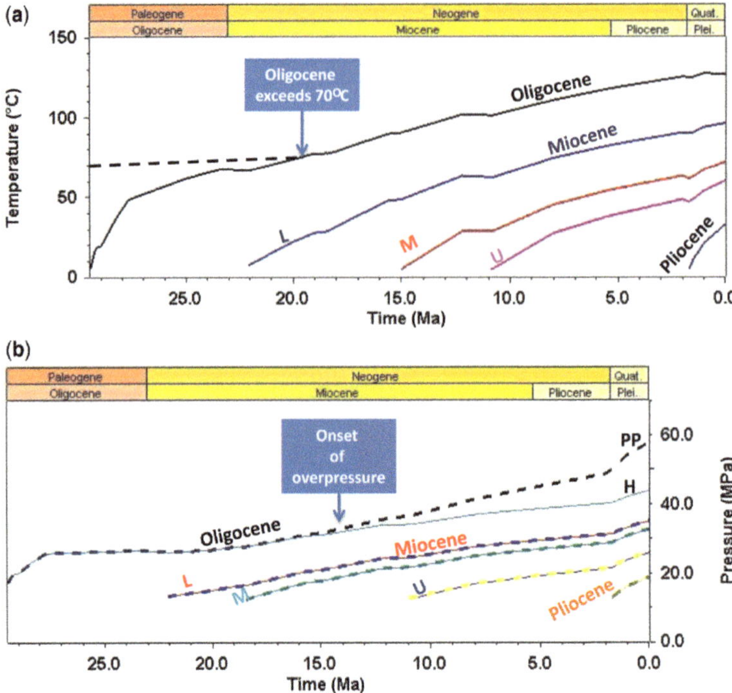

Fig. 16. Modelled thermal (**a**) and pressure (**b**) history. The pressure histories (b) are plots for both hydrostatic pressure (solid line) and pore pressure (dashed lines). L, Lower Miocene; M, Middle Miocene; U, Upper Miocene.

70°C. The Oligocene has been exposed to temperatures above 70°C from the Early Miocene (19 Ma) to the present (reaching 90°C in the Middle Miocene and 120°C bythe Pleistocene, Fig. 16a). In contrast, the Lower Miocene reservoirs encountered temperatures of 70°C and above from the Early Pliocene, reaching about 80°C at the present day, which can explain the very minor quartz overgrowths observed in samples of this age. It is well established that overpressured reservoirs retain porosity because mechanical compaction is inhibited (Taylor *et al.* 2010).

The simulated pressure history of the reservoirs indicates that the Miocene reservoirs have not developed overpressure, but that the Oligocene rocks did encounter overpressure from the Middle Miocene, the overpressure value having continued to increase to the present day having a value of about 12 MPa, as seen from the separation of the modelled pore pressure (PP) and hydrostatic pressure (H) lines (Fig. 16b). Notwithstanding, the overpressured Oligocene sandstones are not likely to have avoided all quartz cementation because they were exposed to temperatures high enough for cementation from the Early Miocene (19 Ma), which is before the significant cementation in the Late Miocene (10 Ma): the temperature for quartz cementation was reached 9 Ma before overpressure developed.

Quartz cement modelling. By implementing the Walderhaug quartz cementation model as part of the simulation, the percentage pore space occluded by cement was estimated. The results show that cement volume tends to increase both with depth and up-dip (Fig. 6) as water column thickness decreases up-dip and mean overburden increases. Little or no cement is predicted above 3000 m in the area of Well 1A but cementation starts to develop from about 3450 m depth within the Lower Miocene reservoirs. In sandstones at this depth, less than 5% of the pore space has been occluded by quartz cement. The cement volume increases steadily with depth into the potential Oligocene reservoirs where values of close to 14% occlusion of the pore space are reached. The minor quartz cement observed from the petrological study for the Lower Miocene reservoir and the absence of quartz in the Middle Miocene reservoir agrees with this simulated result. The simulated result is also displayed as a 1D extraction at the well location from the 2D model with cement volume plotted against depth (Fig. 7).

The trend observed in the quartz model also closely resembles the trend from the multimineral analysis of Figure 15b. A corresponding decrease in modelled porosity from Miocene to Oligocene reservoirs is seen when the modelled porosity is plotted against depth. A good match is seen between the modelled porosity and the average calculated porosity from the well log (Fig. 17).

A 1D column that was extracted from a 2D basin model near the deepest well shows the volume fraction of pore space filled with quartz cement plotted against depth (see Fig. 7). The figure shows an increase in cement with depth as is expected; a similar trend is predicted from multimineral petrophysical evaluation, where major rock mineral volumes have been estimated from well log responses. A good match is seen between the original input log and the reverse calculated versions of the logs with a suffix of ELAN attached to the name of the curve (Fig. 15a). This is a good quality control for the estimated volumes of quartz, feldspar, zircon and kaolinite.

A cross plot of the fraction of quartz volume against depth shows an increase in quartz with depth. There is a clear demarcation of two trends, a higher gradient down to the Middle Miocene (trend a) and a gentler gradient from Lower Miocene down to the Oligocene (trend b). It is not possible to differentiate between detrital and authigenic quartz as contributors to the total quartz as indicated on this plot. However, if we assume a stable sediment supply yielding a constant quartz volume input, then trend a can be interpreted as a function of physical compaction due to the weight of the overburden, such that the quartz volume increases slowly and steadily as a percentage of the total rock volume. Whereas the second trend b shows a more rapid increase of quartz volume with depth, which can be interpreted as a function of both compaction and quartz volume increase due to cementation. The separation of trends a and b, designated line c, can be assumed to be additional quartz volume of about 10% due to quartz cement. This assumption is supported by: (1) the modelled quartz cement that is based on the Walderhaug algorithm, which shows an onset of quartz cementation from the Lower Miocene, increasing volumetrically in the Oligocene to a quartz cement volume of less than 14% (Fig. 7); and (2) the presence of quartz overgrowths noted from the petrographic inspection of the Lower Miocene samples (Fig. 10d, e). No quartz cement was seen in samples from shallower or younger reservoirs.

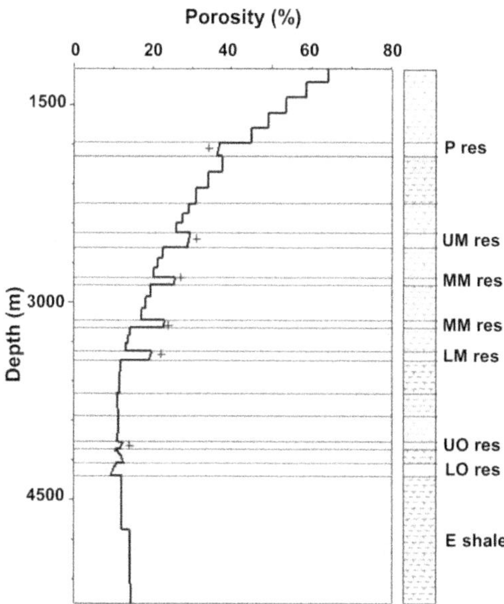

Fig. 17. Modelled porosity (black line) v. depth calibrated against the calculated porosity from well logs (black crosses). Res, reservoir; P, Pliocene; UM, Upper Miocene; MM, Middle Miocene; LM, Lower Miocene; UO, Upper Oligocene; LO, Lower Oligocene; E, Eocene.

Discussion

The focus of this work is to better understand the reservoir potential of the Oligocene interval buried below 3800 m – most importantly how, and to what extent, quartz cementation has reduced porosity below that attributable to the depositional character and mechanical compaction with burial. The burial-related mechanical compaction, temperature and pressure histories have been calculated by basin modelling and have then been used to calculate the probable amounts of quartz cementation throughout the burial history, given sediment properties such as quartz content and grain size, which are known to control cementation rate.

There is direct evidence of some quartz cementation at the lower Miocene. CL images of samples from the deepest cored Lower Miocene interval at 3450 m reveal the presence of syntaxial quartz overgrowths (Fig. 10d, e). However the overgrowth is less than 30 μm thick, suggesting that the effect on porosity here is likely to be minimal. This expectation is supported by the porosity calculated from density logs for this interval, which shows an average value of about 25% (Fig. 14). This value is consistent with only mechanical compaction. The presence of quartz overgrowths in the Lower Miocene is taken as indicative of an increase in cementation with depth and consequently the potential for a significant degree of quartz cementation within the Oligocene reservoir. There is, however, no direct evidence from the Oligocene, so quartz cementation calculations require a critical assessment of the assumed parameters.

The main parameters that control quartz cementation rate as calculated here are mineralogy, grain size and the degree of grain coating, together with temperature, pore fluid pressure (particularly its deviation from the hydrostatic pore pressure) and potentially the presence of petroleum. These are discussed below for the Oligocene, with recognition of the level to which the values are known.

Mineralogy

The Oligocene has not been sampled directly so we must make an estimate of the likely mineralogy on the basis of an assessment of the overlying Miocene succession. There is no reason to assume that sediment supply to the deep water varied significantly during this time interval, as the provenance and supply routes have not changed, so we can use values taken from the Miocene. This is true throughout the Miocene section, as well as into the Pliocene. The sandstone turbidite reservoirs of Miocene age are all quartz arenites (i.e. >90% quartz by weight). Assuming a depositional porosity of 40% by volume, we have calculated an initial volume percent of 54% quartz, which we have used in the quartz cementation calculations. There is a general absence of calcite and relatively low percentages of clay minerals, so we would not expect marked cementation or pore-clogging by other authigenic minerals.

Grain size

Following a similar reasoning and based on an average of the range of mean grain-size values found in Miocene reservoirs, we have assumed a value for the Oligocene sands of 0.12 mm, i.e. fine-grained sandstone. It is interesting to compare this with other turbidite reservoirs of similar age. In the Lower Congo Basin, for example, field data indicate that the petroleum-bearing Oligocene reservoirs have grain sizes that vary from very fine to very coarse (0.125 –2 mm) (Dessus & Abreu 2002; Pelleau *et al.* 2002; Delattre *et al.* 2004). The Oligocene reservoirs of the Campos Basin are mostly medium- to coarse-grained (0.25–1 mm) (Souza *et al.* 1989; Pinto *et al.* 2001; Bruhn *et al.* 2003). In our model calculations, different input grain sizes have a significant effect on calculated quartz cementation amount. A grain size of 0.05 mm (coarser silt) produced 19% of the pore volume filled with cement whereas a grain size of 2 mm (very coarse sand) produced a cement volume of 1.8%. It is possible that the Oligocene sands may have been deposited in a marginally more distal slope setting. Indeed, there is evidence from our interpretation of depositional environments that the Oligocene reservoirs represent deposition in ponded mid-slope lobes whereas the Miocene reservoirs are inferred slope channel deposits (Chudi 2015). However, it is unlikely that the very slightly finer grain size of the Oligocene sandstones would make more than a 1–2% difference in the occlusion of porosity by quartz cementation.

Grain coating

In this study, the Miocene samples observed in SEM and under the optical microscope did not show any evidence of grain coating; this was true for all samples studied including those from the earliest Miocene. However, in order to assess the sensitivity to coating, we systematically increased the grain coating value in our model from 0 to 100%, and as coating percentage increased the calculated volume of quartz cement decreased steadily. Full (100%) grain coating (perhaps unsurprisingly) completely inhibits cementation, yielding no cement, whereas a complete absence of grain coating favours cementation, and yielded up to 14% occlusion of the pore space by cement. The evidence we have indicates that the latter figure is more likely to be representative of the Oligocene section.

Temperature

Temperature is calculated directly in the basin model as is pore pressure, together with its deviation from the hydrostatic pressure gradient. These values have been used to calculate the degree of quartz cementation. Sensitivity studies have also been performed around these values to accommodate reasonable variations either side of the basin simulation results. The heat flow history within this passive margin was modelled using the McKenzie heat flow model, with peak heat flow of 110 mW m^{-2} reached during rifting (at 100 Ma) and gradually dropping to a present day 53 mW m^{-2}. This choice of heat flux over the basin evolution resulted in a cement volume of just below 14% of the available pore space. When synrift heat flow values between 65 and 110 mW m^{-2} (Allen and Allen 1990) were used in the basin model, the calculated cement volume remained effectively unchanged, presumably because the conditions for cementation had not been reached in the reservoir rocks. However, when the heat flow was altered during the Miocene, the presumed time of quartz cementation onset, the calculated cement pore volume was significantly affected. Using a Miocene heat flow 35 mW m^{-2} produces low cementation (<8%) whereas a higher heat flow of 65 mW m^{-2} yielded relatively higher cementation (up to 17%). This is to be expected as effectively the time of onset of cementation is being changed.

Neither a variable peak heat flow nor a Miocene variation is justified geologically and basin models that incorporate these heat flows do not produce values that match the measured present-day temperature or vitrinite data. Nevertheless, these calculations are a good illustration of the significant impact of heat flow on quartz diagenesis where a change in heat flow value is justified.

Onset temperature for quartz cementation

The onset temperature of quartz cementation has been reported to be between 70°C and 135°C (Walderhaug 1994; Walderhaug 2000; Worden & Morad 2009; Taylor *et al.* 2010). The bulk of the literature suggests values closer to the lower end of this range for significant onset. Many of the models also have no lower cut-off temperature, but simply infer that reaction rates will be so low as to have no noticeable effect. For this study, we have taken the lower end of this range, i.e. 70°C, as the temperature for the onset of quartz precipitation in order to model the maximum likely adverse effect on potential reservoir properties. The Upper and Middle Miocene has no evidence of quartz cementation, which is commensurate with its low formation temperature throughout its burial history. The Lower Miocene section shows a low volume of quartz cement (average of less than 5% of the pore space) in the study area, which can be attributed to the shallow burial depth of the sediments (800–1300 m) with a temperature that has just entered the window for the onset of quartz cementation. However, the Oligocene section is buried at depths greater than 3800 m and has therefore experienced temperatures exceeding 70°C for a longer time period. Our models show up to 14% of the pore space has suffered quartz cementation. We therefore suggest that our results, both observation and modelling, support the case for the onset of quartz cementation at the low end of the published temperature range.

Pressure

The Middle and Lower Miocene reservoirs do not develop overpressure in the 2D basin model but the Oligocene reservoirs do develop overpressure starting in the Middle Miocene (14 Ma) and continuing to increase up to the present day. Overpressure is typically expected to result in reduced cementation as compared with the hydrostatic equivalent. An in-depth investigation of the effect of overpressure development on cementation is beyond the scope of this study, but generally overpressure can be thought of as a fluid pressure state that is representative of a deeper location but at a lower temperature than would normally be associated with that deeper location. Here the model shows that the Oligocene reservoir becomes overpressured at 14 Ma, which is about 6 Ma after the conditions for cementation have been reached. If correct, earlier cementation would not have been affected by overpressure but later cementation may well have been impeded.

Hydrocarbons

Oil charge is another potential influence on cementation, although the effectiveness of petroleum in reducing the cementation rate, or stopping it completely, is widely debated (Marchand *et al.* 2002; Wilkinson *et al.* 2004; Taylor *et al.* 2010). Hydrocarbon coating agents were observed during SEM analysis of the Middle Miocene samples and dead oil was observed coating grains (Fig. 11). Considering that early oil emplacement can halt or reduce quartz cement precipitation (Taylor *et al.* 2010), it is therefore possible that the presence of this oil seen only in the Middle Miocene samples may have helped in preserving the porosity.

Comparative Oligocene reservoirs. Here we briefly compare the potential Oligocene reservoirs of the Niger Delta that we have studied with two other Oligocene turbidite systems along the South Atlantic Margin: (1) the Lower Congo Basin, offshore Angola, and (2) the Campos Basin, offshore SE Brazil. The quality of the reservoirs and the corresponding key controlling factors are of primary importance.

The Lower Congo Basin is one of several different sub-basins developed along the West African passive margin, and is one of the most prolific, with major discoveries made within both Oligocene and Miocene turbidite systems of the Malembo Formation (Broucke *et al.* 2004). The most significant discoveries have been made in the Oligocene section, most notably the Girassol field, which has oil in place estimated at 1550 mmbbl and recoverable reserves placed at 725 mmbbl trapped within Oligocene turbidite channel–levee complexes. The field is located within a water depth of 1350–1450 m (Pelleau *et al.* 2002).

Whereas the Oligocene sediments of the Niger Delta slope system are found at depths greater than 3800 m below sea-level (up to 3000 m below the seafloor), the Oligocene reservoirs in Girassol field are located at a depth of 2400 m below sea-level (around 1000 m below the seafloor). The lower Congo Oligocene reservoirs have therefore undergone less burial and compaction. This is reflected in the wholly unconsolidated nature of the fine to very coarse sands, which exhibit excellent reservoir quality, having porosities of up to 40% and permeabilities greater than 5 D. The Congo Basin reservoirs are also normally pressured with present-day temperatures of 58–69°C (Dessus & Abreu 2002; Pelleau *et al.* 2002; Delattre *et al.* 2004). The regional geology of the area does not suggest that the Malembo Formation has at any time been exposed to any physiochemical conditions that would have led to the precipitation of any significant porosity-occluding diagenetic minerals, such as early quartz or calcite cements (Broucke *et al.* 2004; Gay *et al.* 2006).

Conjugate to the basins in West Africa is the Campos Basin of the Brazilian margin that lies beneath the coastal plain, continental shelf and slope along the western South Atlantic Ocean. The Campos Basin is the most prolific of the twelve east Brazilian offshore basins. (Bruhn *et al.* 2003). Turbidites constitute the main reservoir play in the Campos Basin, representing 37 different oilfields, including the Marlim Complex super giant field, which encompasses the Marlim Field itself and the surrounding fields of East Marlim, West Marlim and South Marlim (Souza *et al.* 1989; Bruhn *et al.* 2003). The total estimated oil in place for the Marlim Complex is about 13.9 billion barrels of oil, with the Marlim field itself accounting for over 57% of the total volume of hydrocarbon (Souza *et al.* 1989). The principal reservoir of the Marlin field is the Oligocene Carapebus Member of the Campos Formation. The Marlim field was discovered by the first exploratory well drilled at a water depth of about 850 m. The Oligocene reservoirs, located at a depth of about 2700 m below sea-level (i.e. <2000 m below the seafloor), are known for three outstanding qualities: (1) predictability from seismic data, (2) good hydraulic connectivity, and (3) excellent petrophysical properties. Reservoir average porosities and permeabilities typically range between 27% and 30%, and 1000–2000 mD respectively (Pinto *et al.* 2001).

The sand-rich reservoirs of the Marlim field comprise turbidite lobes, which accumulated in wide intraslope depressions. The reservoir facies comprise mostly amalgamated graded beds of medium to very fine grained sandstones that are poorly consolidated, poorly sorted and have a low volume of silt, clay and diagenetic minerals. The low degree of cementation in the Oligocene reservoirs is attributed to the following factors (Moraes 1989; Soldan *et al.* 1995): (1) reservoir temperatures of about 80°C today and hence short residence times within the quartz-cementation window, (2) early oil emplacement that has inhibited cementation, (3) early carbonate cementation that persisted to depth and inhibited pressure dissolution, and (4) late-stage carbonate dissolution and creation of secondary porosity.

Conclusions

(1) The potential Niger Delta Oligocene reservoir of the study area is interpreted to have been subjected to temperatures at or above 70°C from the Early Miocene to the present day. Our use of this temperature as the starting temperature for the onset of quartz precipitation is validated by the observation and modelling results presented.

(2) Petrological and petrophysical analyses of quartz in the Lower Miocene suggest an increase in quartz overgrowth volume with depth, indicating that the Oligocene sediments are likely to be cemented.

(3) The possible volume of pore space occluded by cement in the Oligocene has been predicted to be (just) less than 14% of the available pore space.

(4) Despite both modelled results and inferences suggesting the possibility of quartz cementation in the potential Oligocene reservoir, the quartz cement volume predicted from 2D simulations is not sufficient to prevent the Oligocene interval of the Agbada Formation from being a viable hydrocarbon reservoir.

(5) The properties of comparable Oligocene reservoirs from the Lower Congo Basin and Campos Basin are more similar to the reservoir properties of the Miocene turbidites of the Niger Delta that have been buried to a depth range of 2400–3000 m than those of the Oligocene system. It is very likely that one of the principal factors controlling reservoir quality is the depth of burial and the associated physiochemical conditions, since the best quality reservoirs are seen at depths of less than 3000 m below sea-level in all three basins.

(6) Temperature-induced physicochemical factors that favour quartz cementation are prevalent in sandstones that are buried deeper than 3000 m within the Niger Delta deepwater system. The potential Oligocene reservoirs are currently located at depths >3800 m below sea-level.

(7) Modelling carried out in this study suggest that reservoir properties of the Oligocene turbidite sands of the Niger Delta slope system are only slightly reduced from those of the Miocene (up to 14% porosity occlusion). They therefore represent a very viable future exploration target.

(8) We suggest that this integrated approach using petrological, petrophysical and basin modelling is an important and novel methodology with widespread applicability to more accurate prediction of sandstone reservoir quality in the subsurface.

We wish to express our appreciation to Shell in Nigeria and The Hague for providing the data and particularly to Daniel Agbaire of Shell Nigeria and Andy Bell of Shell Global Solutions International, The Hague, for their expert advice. Also special thanks to Thomas Hantschel, Daniel Palmowski and Nour Koronful of Schlumberger Aachen for allowing the use of PetroMod software and for all the technical discussions, which have considerably enhanced this work. The authors are extremely grateful to the reviewers of this article for their copious efforts. This project is part of Obinna Chudi's PhD, which has been generously sponsored by the Petroleum Technology Development Fund (PTDF), Nigeria.

References

Allen, P.A. & Allen, J.R. 1990. *Basin Analysis: Principles and Applications*. Blackwell Scientific Publications, Oxford.

Athy, L.F. 1930. Density, porosity, and compaction of sedimentary rocks. *Bulletin of the American Association of Petroleum Geologists*, **14**, 1–24.

Bilotti, F. & Shaw, J.H. 2005. Deep-water Niger Delta fold and thrust belt modeled as a critical-taper wedge: the influence of elevated basal fluid pressure on structural styles. *AAPG Bulletin*, **89**, 1475–1491.

Bjørkum, P.A., Oelkers, E.H., Nadeau, P.H., Walderhaug, O. & Murphy, W.M. 1998. Porosity prediction in quartzose sandstones as a function of time, temperature, depth, stylolite frequency, and hydrocarbon saturation. *AAPG Bulletin*, **82**, 637.

Bloch, S., Lander, R.H. & Bonnell, L. 2002. Anomalously high porosity and permeability in deeply buried sandstone reservoirs: origin and predictability. *AAPG Bulletin*, **86**, 301–328.

Broucke, O., Temple, F., Rouby, D., Robin, C., Calassou, S., Nalpas, T. & Guillocheau, F. 2004. The role of deformation processes on the geometry of mud-dominated turbiditic systems, Oligocene and Lower–Middle Miocene of the Lower Congo basin (West African Margin). *Marine and Petroleum Geology*, **21**, 327–348.

Bruhn, C.H., Gomes, J.A.T., Lucchese, J.R.C. & Johann, P.R. 2003. Campos Basin: reservoir characterization & management–Historical overview & future challenges. Paper presented at the *Offshore Technology Conference*, 5–8 May 2003, Houston, 5–8.

Chudi, O.K. 2015. *Predicting Oligocene reservoir potential in the deep-water Western Niger Delta: an integrated basin modelling and diagenetic study*. PhD thesis, Heriot-Watt University.

Corredor, F., Shaw, J.H. & Bilotti, F. 2005. Structural styles in the deep-water fold and thrust belts of the Niger Delta. *AAPG Bulletin*, **89**, 753–780.

Damuth, J.E. 1994. Neogene gravity tectonics and depositional processes on the deep Niger Delta continental margin. *Marine and Petroleum Geology*, **11**, 320–346.

Delattre, E., Authier, J.F., Rodot, F., Petit, G. & Alfenore, J. 2004. Review of sand control results and performance on a deep water development-A case study from the Girassol Field Angola. *Paper presented at the SPE Annual Technical Conference and Exhibition*, 26–29 September 2004, Houston, TX.

Deptuck, M.E., Sylvester, Z., Pirmez, C. & O'byrne, C. 2007. Migration–aggradation history and 3-D seismic geomorphology of submarine channels in the Pleistocene Benin-major Canyon, western Niger Delta slope. *Marine and Petroleum Geology*, **24**, 406–433.

Derks, J., Swientek, O., Fuchs, T., Kauerauf, A., Al-Quattan, M. & Al-Saeed, M. 2012. *Three-Dimensional Basin and Petroleum System Model of the Cretaceous Burgan Formation, Kuwait: Model-in-Model, High-Resolution Charge Modeling. Basin Modeling*. New Horizons in Research and Applications, AAPG Hedberg Series, **4**, 159.

Dessus, J.L. & Abreu, J. 2002. Girassol: drilling and completion experience gained through first 12 wells. *Offshore Technology Conference*, 6–9 May, Houston, Texas, **14168**, 6–9.

Doust, H. & Omatsola, E. 1990. Niger Delta. *In*: Edwards, J.D. & Santogrossi, P.A. (eds), *Divergent/Passive Margin Basins*. American Association of Petroleum Geologists, **48**, 201–238.

Ejedawe, J., Coker, S., Lambert-Aikhionbare, D., Alofe, K. & Adoh, F. 1984. Evolution of oil-generative window and oil and gas occurrence in Tertiary Niger Delta Basin. *Bulletin of the American Association of Petroleum Geologists*, **68**, 1744–1751.

Evamy, B.D., Haremboure, J., Kamerling, P., Knaap, W.A., Molly, F.A. & Rowlands, P.H. 1978. Hydrocarbon habitat of Tertiary Niger Delta. *American Association of Petroleum Geologists Bulletin*, **62**, 1–39.

Evans, J., Hogg, A.J., Hopkins, M.S. & Howarth, R.J. 1994. Quantification of quartz cements using combined SEM, CL, and image analysis. *Journal of Sedimentary Research*, **64**.

Gay, A., Lopez, M., Cochonat, P., Séranne, M., Levaché, D. & Sermondadaz, G. 2006. Isolated seafloor pockmarks linked to BSRs, fluid chimneys, polygonal faults and stacked Oligocene–Miocene turbiditic palaeochannels in the Lower Congo Basin. *Marine Geology*, **226**, 25–40.

Gier, S., Worden, R.H., Johns, W.D. & Kurzweil, H. 2008. Diagenesis and reservoir quality of Miocene sandstones in the Vienna Basin, Austria. *Marine and Petroleum Geology*, **25**, 681–695.

Haack, R.C., Sundararaman, P., Diedjomahor, J.O., Xiao, H., Gant, N.J., May, E.D. & Kelsch, K. 2000. Chapter 16: Niger delta petroleum systems, Nigeria. *In*: Mello, M.R. & Katz, B.J. (eds) *Petroleum Systems of South Atlantic Margins*. American Association of Petroleum Geologists, Memoir, **73**, 213–231.

Hermanrud, C. 1993. Basin modelling techniques-an overview. In *Basin Modelling: Advances and Applications*. Norwegian Petroleum Society, Special Publications, **3**, 1–34.

Lambert-Aikhionbare, D. & Ibe, A. 1984. Petroleum source-bed evaluation of Tertiary Niger delta: discussion. *Bulletin of the American Association of Petroleum Geologists*, **68**, 387–389.

Marchand, A.M., Smalley, P.C., Haszeldine, R.S. & Fallick, A.E. 2002. Note on the importance of hydrocarbon fill for reservoir quality prediction in sandstones. *AAPG Bulletin*, **86**, 1561–1572.

Mcbride, E.F. 1963. A classification of common sandstones. *Journal of Sedimentary Research*, **33**.

MORAES, M.A. 1989. Diagenetic evolution of Cretaceous-Tertiary turbidite reservoirs, Campos Basin, Brazil. *AAPG Bulletin*, **73**, 598–612.

NWACHUKWU, S. 1976. Approximate geothermal gradients in Niger Delta Sedimentary Basin. *AAPG Bulletin*, **60**, 1073–1077.

PELLEAU, R., SERCEAU, A. & TOTAL FINA ELF. 2002. The Girassol development: project challenges.

PETERS, K.E., MOLDOWAN, J.M., MCCAFFREY, M.A. & FAGO, F.J. 1996. Selective biodegradation of extended hopanes to 25-norhopanes in petroleum reservoirs. Insights from molecular mechanics. *Organic Geochemistry*, **24**, 765–783.

PINTO, C., ANTONIO, C., GUEDES, S.S., BRUHN, C.H., GOMES, J.A.T. & NETTO, J. 2001. Marlim complex development: a reservoir engineering overview. *Paper presented at the SPE Latin American & Caribbean Petroleum Engineering Conference*, Buenos Aires, Argentina, 25–28 March 2001.

REIJERS, T. 2011. Stratigraphy and sedimentology of the Niger Delta. *Geologos*, **17**.

SAUGY, L. & EYER, J.A. 2003. Fifty years of exploration in the Niger Delta (West Africa). *In*: HALBOUTY, M.T. (ed.) *Giant Oil and Gas Fields of the Decade 1990–1999*. AAPG Memoirs, **78**, 211–226.

SIEVER, R. 1983. Burial history and diagenetic reaction kinetics. *AAPG Bulletin*, **67**, 684–691.

SHORT, K. & STAUBLE, A. 1967. Outline of geology of Niger Delta. *Bulletin of the American Association of Petroleum Geologists*, **51**, 761–779.

SNEIDER, R.M. 1990. *Reservoir Description of Sandstones. Sandstone Petroleum Reservoirs*. Springer, New York.

SOLDAN, A., CERQUEIRA, J., FERREIRA, J., TRINDADE, L., SCARTON, J. & CORÁ, C. 1995. Giant Deep Water Oil Fields in Campos Basin, Brazil: a Geochemical Approach. *Revista Latino-Americana de Geoquimica Organica*, **1**, 14–27.

SOMBRA, C.L. & CHANG, H.K. 1997. Burial history & porosity evolution of Brazilian Upper Jurassic to Tertiary sandstone reservoirs. *In*: KUPECZ, J.A., GLUYAS, J. & BLOCH, S. (eds) *Reservoir Quality Prediction in Sandstones and Carbonates*. AAPG Memoir, **69**, 79–89.

SONDE, A., OMOBUDE, O., CHUDI, O., OGHENE, U. & COKER, T. 2011. Integrated 3D modelling in a structurally complex brown field: a foundation for improved reservoir management and optimisation of further development. *Nigeria Annual International Conference and Exhibition*, 2011. Society of Petroleum Engineers.

SOUZA, J., SCARTON, J., CANDIDO, A., CRUZ, C. & CORA, C. 1989. The Marlim & Albacora Fields: geophysical geological & reservoir aspects. Paper presented at the *Offshore Technology Conference*, Houston TX, 1–4 May 1989.

TALÇIN, M.N. 1991. Basin modeling and hydrocarbon exploration. *Journal of Petroleum Science and Engineering*, **5**, 379–398.

TAYLOR, T.R., GILES, M.R. *ET AL*. 2010. Sandstone diagenesis and reservoir quality prediction: models, myths, and reality. *AAPG Bulletin*, **94**, 1093–1132.

TUTTLE, M.L., CHARPENTIER, R.R. & BROWNFIELD, M.E. 1999. *The Niger Delta Petroleum System: Niger Delta Province, Nigeria, Cameroon, and Equatorial Guinea, Africa*. US Department of the Interior, US Geological Survey Open-File Report **99-50-H**.

UNDERDOWN, R. & REDFERN, J. 2008. Petroleum generation and migration in the Ghadames Basin, north Africa: a two-dimensional basin-modeling study. *AAPG Bulletin*, **92**, 53–76.

WALDERHAUG, O. 1994. Precipitation rates for quartz cement in sandstones determined by fluid-inclusion microthermometry and temperature-history modeling. *Journal of Sedimentary Research*, **64**, 324–333.

WALDERHAUG, O. 2000. Modeling quartz cementation and porosity in Middle Jurassic Brent Group sandstones of the Kvitebjørn field, northern North Sea. *AAPG Bulletin*, **84**, 1325–1339.

WEBER, K. 1971. Sedimentological aspects of oil fields in the Niger Delta. *Geologie en Mijnbouw*, **50**, 559–576.

WELTE, D. & YALCIN, M. 1988. Basin modelling – a new comprehensive method in petroleum geology. *Organic Geochemistry*, **13**, 141–151.

WHITEMAN, A. 1982. *Nigeria: Its Petroleum Geology, Resources & Potential*. Vol. **1**. Springer, Netherlands.

WILKINSON, M., STUART HASZELDINE, R., ELLAM, R.M. & FALLICK, A. 2004. Hydrocarbon filling history from diagenetic evidence: brent Group, UK North Sea. *Marine and Petroleum Geology*, **21**, 443–455.

WORDEN, R. & MORAD, S. 2009. Quartz cementation in oil field sandstones: a review of the key controversies. *In*: WORDEN, R. & MORAD, S. (eds) *Quartz Cementation in Sandstones*. Blackwell Publishing Ltd, Oxford, **29**, 1–20.

WYGRALA, B. 1988. Integrated computer-aided basin modeling applied to analysis of hydrocarbon generation history in a Northern Italian oil field. *Organic Geochemistry*, **13**, 187–197.

Diagenesis, plagioclase dissolution and preservation of porosity in Eocene and Oligocene sandstones at the Greeley oil field, southern San Joaquin basin, California, USA

D. T. NGUYEN, R. A. HORTON, JR* & A. B. KAESS

California State University, Bakersfield, Department of Geological Sciences, 9001 Stockdale Highway, Bakersfield, CA 93311, USA

**Correspondence: rhorton@csub.edu*

Abstract: The Vedder (Oligocene) and Kreyenhagen (Eocene) sandstones at the Greeley oil field consist of arkosic to subarkosic arenites and wackes deposited in shallow marine environments. Burial depths of the Vedder sandstones exceed 3150 m and the reservoir temperature is 124°C. The Kreyenhagen sandstones are buried to greater than 3920 m and the reservoir temperature is estimated to be *c.* 135°C. These sandstones are currently at or very near their deepest burial depths.

The textural relationships of the diagenetic minerals suggest syndepositional formation of glauconite, phosphate and pyrite, followed by early precipitation of pore-lining clay coatings and carbonate cements along with framework-grain fracturing and possibly dissolution. With increasing burial, dissolution of the framework grains continued, accompanied by the albitization of feldspars, the formation of K-feldspar and quartz overgrowths, the precipitation of kaolinite and other clays and possibly the precipitation of late carbonate cements. Finally, hydrocarbon migration and the formation of pyrite occurred during late diagenesis.

Porosity preservation and reservoir quality are primarily the result of plagioclase dissolution occurring as the strata approached their current burial depths. Mass balance calculations indicate the significant export of aluminium out of the sands. Thus secondary porosity produced by plagioclase dissolution has replaced the primary porosity destroyed by compaction, and now accounts for the majority of the porosity in these rocks.

Predicting reservoir quality in sandstones is a valuable tool for hydrocarbon exploration and production. This requires knowledge of the processes that may have affected porosity development within the target strata. In the southern San Joaquin basin (Fig. 1), hydrocarbon production mainly comes from sandstone reservoirs. As these sandstones were buried, they underwent diagenesis, including compaction, deformation, dissolution, cementation, replacement and recrystallization of minerals. These processes can significantly alter the reservoir quality by reducing or increasing porosity and permeability (Schmidt & McDonald 1979; Ehrenberg 1990). Understanding the diagenetic processes of a target reservoir may increase the possibility of exploration and production success (Galloway 1979; Hayes 1979). One issue that has long been controversial is the role of framework-grain dissolution on the overall development of porosity in sandstones and whether aluminium is conserved or exported (Hayes 1979; Bjorlykke 1984; Boles 1984; MacGowan & Surdam 1988; Harris 1989; Hayes & Boles 1992; Wilkinson *et al.* 2001; Horton *et al.* 2009).

California's San Joaquin basin has been a major petroleum-producing area for over a century. In mature oil provinces such as this continued exploration often focuses on more deeply buried strata as shallow reservoirs become depleted. In order to improve the chances of successful exploration, knowledge of the factors that control reservoir quality at depth is critical. The Greeley oil field is located *c.* 16 km NW of Bakersfield, California (Fig. 1) near the axis of the San Joaquin basin. It has been producing hydrocarbons since 1936. Production comes mainly from sandstones in the Vedder Fm and the Stevens sandstones within the Monterey Fm (California DOGGR 1998). Although sandstones in the Kreyenhagen Fm have not been exploited at Greeley, they have been productive at the nearby Wasco oil field. Burial depths of the Vedder Fm exceed 3155 m and the Kreyenhagen sandstones are buried at depths greater than 3920 m. Petrological aspects of the Vedder Fm along the eastern margin of the basin and of Kreyenhagen sandstones on the western margin of the basin have been well documented (e.g. Clarke 1973; Olson *et al.* 1986; Bent 1988; Hayes & Boles 1992; Taylor 2007), however, there have been no detailed published studies of the Vedder and Kreyenhagen sandstones near the axis of the basin.

As Greeley is among the deepest productive oil fields in the San Joaquin basin, knowledge of the

From: ARMITAGE, P. J., BUTCHER, A. R., CHURCHILL, J. M., CSOMA, A. E., HOLLIS, C., LANDER, R. H., OMMA, J. E. & WORDEN, R. H. (eds) 2018. *Reservoir Quality of Clastic and Carbonate Rocks: Analysis, Modelling and Prediction.* Geological Society, London, Special Publications, **435**, 265–282.
First published online June 28, 2016, https://doi.org/10.1144/SP435.14

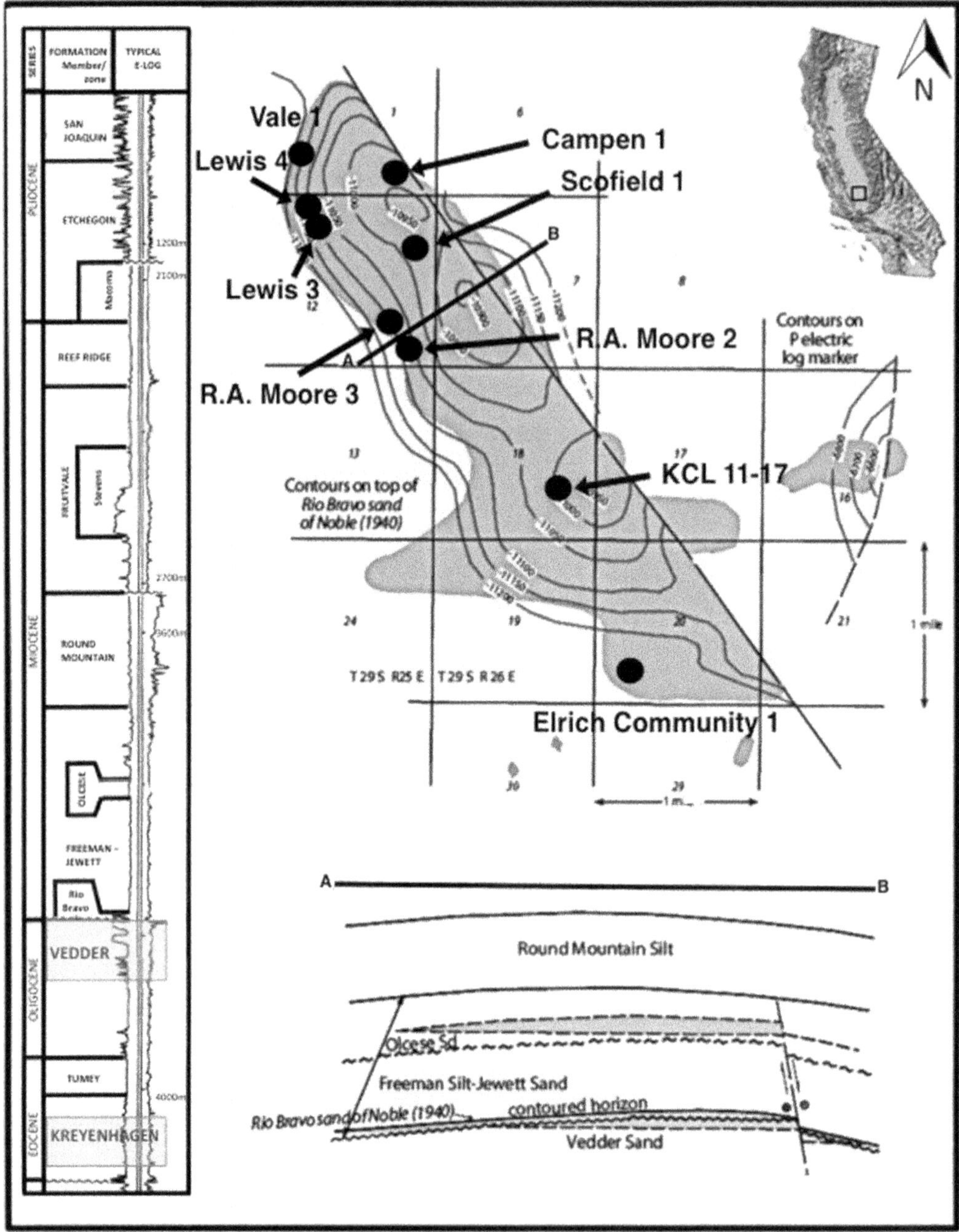

Fig. 1. The location of Campen 1, KCL 11–17, Lewis 3 and 4, R.A. Moore 2 and 3, Scofield and Vale 1 and Elrich Community 1 wells at the Greeley oil field, located in the southern San Joaquin Basin, and a stratigraphic column of the Vedder and Kreyenhagen sandstones. Modified from Gautier & Scheirer (2007).

processes that control reservoir quality there will be helpful to future exploration in the basin. In addition, knowledge of the controls on reservoir quality at Greeley oil field may be useful in other mature petroleum provinces as exploration focuses on more deeply buried strata. This study examines diagenetic processes in the sandstones of the Vedder and Kreyenhagen Fms at the Greeley oil field, including the roles of feldspar dissolution and export of aluminium in the evolution of porosity and overall development of reservoir quality, thus contributing to the ongoing discussion on this important topic.

Geologic setting

The San Joaquin basin is bounded by the Sierra Nevada to the east, the Temblor Range to the west and the San Emigdio and Tehachapi Mountains to the south. It is part of the Great Valley forearc basin that formed in the Late Jurassic (Bartow 1991). It comprises approximately 9 km of marine and non-marine sediments deposited during the Late Mesozoic and Cenozoic (Bandy & Arnal 1969; Bartow 1991).

In the southern San Joaquin basin, the strata on the eastern margin consist of continental facies whereas strata on the western side consist of shallow to deep marine facies (Bartow 1991; Wagoner 2009). Structurally, the eastern margin of the basin has undergone little deformation compared with the western margin of the basin. En echelon fold sets on the west side are associated with the migration of the Medocino triple junction and the formation of the San Andreas transform fault in the Oligocene (Bartow 1991), as well as more recent transpression (Niemi & Hall 1992; Grove & Niemi 2005). Active tectonics along the western margin of the North American plate resulted in periods of uplift and subsidence within the basin. The marine Vedder and Kreyenhagen Fms are transgressive units deposited in the Oligocene and Eocene (Bartow & McDougall 1984; Bloch & Olson 1990; Bartow 1991).

The Vedder Fm underlies the Rio Bravo sandstones of the Freeman-Jewett Fm (Fig. 1) and is separated from the Kreyenhagen Fm by the Tumey Fm (Gautier & Scheirer 2007). The Vedder Fm covers a large portion of the San Joaquin basin and ranges up to 610 m in thickness close to the southern basin boundary (Richardson 1966). It is *c.* 381 m thick at the Greeley oil field (Sullivan & Weddle 1960; Bloch & Olson 1990). The upper portion of the formation consists of well-sorted, very fine to coarse grey sandstones. The lower portion of the formation consists of dark brown shale; it is separated from the upper portion by thin, tight, grey sandstones. These sandstones were derived from the Sierra Nevada and deposited in a shallow marine setting (Bent 1985, 1988; Olson 1988; Bloch & Olson 1990).

The late-Middle to Late Eocene Kreyenhagen Fm is composed of mostly deep marine shales with minor sandstones and siltstones. The formation is thickest to the west and thins to the north and east (Isaacson & Blueford 1984). The upper portion of the formation contains sandstones that are correlative to the productive 'Mushrush' sandstone (Reid 1988) at the nearby (24 km along depositional strike) Wasco oil field. The Kreyenhagen Fm covers most of the San Joaquin basin and was derived from source areas to the south, west and east of the basin (Clarke 1973). It is one of the basin's major hydrocarbon source rocks (Isaacson & Blueford 1984; Lillis & Magoon 2007).

Structure

The Greeley oil field is located on the north flank, just off the crest, of the Bakersfield Arch – a NE–SW trending anticline that separates the Buttonwillow sub-basin to the north from the Maricopa (aka Tejon) sub-basin to the south and along which many important San Joaquin basin oil fields are located. The Greeley oil field is 6.5 km long and ranges from 1.5–3 km wide (Fig. 1). The structure of the field is a NW–SE trending anticlinal fold, part of a series of en echelon fold sets that host oil reservoirs both on and off the Bakersfield Arch. On the eastern margin of the field lies the Greeley fault, with the downthrown side to the east. The fault has a vertical displacement of *c.* 76 m (Sullivan & Weddle 1960).

Methods

Forty-nine thin sections of Vedder sandstones from eight wells (Campen 1, KCL 11–17, Lewis 3 and 4, R.A. Moore 2 and 3 and Scofield and Vale 1) and 14 thin sections of Kreyenhagen sandstones from one well (Elrich Community 1) (Fig. 1) were analysed using transmitted-light optical petrography. The depths of the Vedder samples are 3444–3510 m and the depths of the Kreyenhagen samples are 4098–4118 m. Each thin section was point-counted (300 points per thin section) using a petrographic microscope to quantify mineralogy and porosity (all porosity values reported herein refer to thin-section porosity unless otherwise noted). Point-count data are reported as volume per cent (Stauffer 1966; Halley 1978; Byers 1990). Inter- and intragranular porosity was recorded. In addition, where pore space could reasonably be interpreted as secondary and the dissolved mineral could be identified, this information was also recorded (Harris 1989). Grain size (Wentworth 1922), roundness and sorting were estimated using standard diagrams (Scholle 1979). Diagenetic features such as cementation, alteration, dissolution and compaction were also recorded.

The data collected from the petrographic analyses are plotted on ternary diagrams following the methodologies of Dickinson (1970), Dott (1964) as modified by Pettijohn *et al.* (1987) and Harris (1989). Quantitative analysis of the compaction of these sandstones was documented using the contact index (CI) and tight-packing index (TPI) methodologies of Wilson & McBride (1988). Selected thin sections were analysed using a Hitachi scanning electron microscope (SEM) equipped with an Oxford INCA energy-dispersive X-Ray spectrometer (EDS) and a Gatan cathodoluminescence (CL) system. The SEM-EDS analytical system provided the chemical components of the minerals whereas the SEM-CL system provided the textural relationships between the detrital grains, cements and fractures.

The point-count results were used to calculate the mass transfer of aluminium by comparing the volumes of authigenic kaolinite and secondary porosity formed by feldspar dissolution (Hayes & Boles 1992). The calculations assumed that dissolution of feldspars released aluminium ions that precipitated as kaolinite. The composition of $Na_{0.7}Ca_{0.3}Al_{1.3}Si_{2.7}O_8$ (An_{30}) was chosen for detrital plagioclase based on Boles & Ramseyer's (1988) study on albitization in the San Joaquin basin as well as detrital plagioclase compositions determined during this study. For each mole of plagioclase dissolved, 0.65 mol of kaolinite could precipitate:

$$\underset{\text{plagioclase}}{Na_{0.7}Ca_{0.3}Al_{1.3}Si_{2.7}O_8} + H^+ + 3.75H_2O$$
$$\rightarrow \underset{\text{kaolinite}}{0.65Al_2Si_2O_5(OH)_4} + 1.4H_4SiO_4 + 0.7Na^+$$
$$+ 0.3Ca^{2+} + 0.3OH^-$$

For each mole of K-feldspar dissolved, 0.5 mol of kaolinite could precipitate:

$$\underset{\text{K-feldspar}}{KAlSi_3O_8} + H^+ + 4.5H_2O$$
$$\rightarrow \underset{\text{kaolinite}}{0.5Al_2Si_2O_5(OH)_4} + 2H_4SiO_4 + K^+$$

Other sources of aluminium ions are volumetrically minor and have been excluded from the mass-balance calculations. Also, the expected volumes of kaolinite do not take into account that portion of secondary porosity for which the dissolved minerals could not be determined; thus they are conservative.

Results

Detrital composition and texture

The Vedder sandstones range in grain size from fine to medium sand (Fig. 2a, b). The Kreyenhagen sandstones range from fine to coarse sand (Fig. 2c, d). Framework grains vary from angular to round and the sandstones vary from poorly to well sorted. In general, the sandstones are moderately sorted with subangular grains. Poor sorting in some samples is due to smaller fragments created by the fracturing of the framework grains. The sandstones are composed of quartz (mainly monocrystalline grains), plagioclase and K-feldspars (microcline and orthoclase) and rock fragments (metamorphics, volcanics, cherts and granitic microphanerites). The mean value for the Vedder sandstones is $Q_{34}F_{52}Rf_{14}$ and for the Kreyenhagen sandstones is $Q_{69}F_{24}Rf_7$. Accessory detrital minerals include biotite, muscovite, chlorite, epidote, hornblende, titanite and zircon. The Vedder Fm sandstones consist of arkosic arenites and wackes (Fig. 3a, Pettijohn *et al.* 1987). The Kreyenhagen sandstones consist of arkosic to subarkosic arenites and wacke (Fig. 3a). The Vedder sandstones plot in the basement uplift field on a QFL diagram (Dickinson 1970) and Kreyenhagen sandstones plot in the transitional continental field (Fig. 3b).

The feldspar content decreases with depth (Figs 4a & 5a), with plagioclase (Fig. 4b) decreasing more than K-feldspars (Fig. 4c). Plagioclase averages 20% by volume in the Vedder sandstones and 7% by volume in the Kreyenhagen sandstones. SEM-EDS analyses indicate detrital plagioclase compositions in the Vedder Fm range from An_{55} to An_2 (Fig. 5b). Detrital plagioclase ranges from An_{38} to An_8 in the Kreyenhagen sandstones, but only two analyses exceeded An_{30}. Some plagioclase grains also contain up to 1.0 wt% of Sr and/or 0.1 wt% of Ba; there is a significant decrease in the Sr content of plagioclase in the Kreyenhagen sandstones compared with the Vedder sandstones. Detrital plagioclase luminesces pink (Fig. 6a, b), pinkish-purple and blue.

Both the Vedder and Kreyenhagen sandstones contain an average of 12.5% K-feldspars (Fig. 4c). The Vedder sandstones consist of 7.5% of microcline and 5.1% of orthoclase. In the Kreyenhagen sandstones, 8.1% is microcline and 4.0% is orthoclase. Most detrital K-feldspars contain some Na (Fig. 5b) and some are Ba-rich (Fig. 6). K-feldspars luminesce pink (Fig. 6) and blue.

Compaction

A quantitative analysis of the compaction in 12 arenite samples of the Vedder sandstones and 11 of the Kreyenhagen sandstones was undertaken following the methodologies of Wilson & McBride (1988). The CI is the average of the total number of contacts between each point-counted framework grain and the surrounding grains, and the TPI is the average number of long-, concavo-convex- and sutured contacts between each point-counted grain and the surrounding grains. The CI for the Vedder sandstones ranges from 2.9–6.1 and the TPI ranges

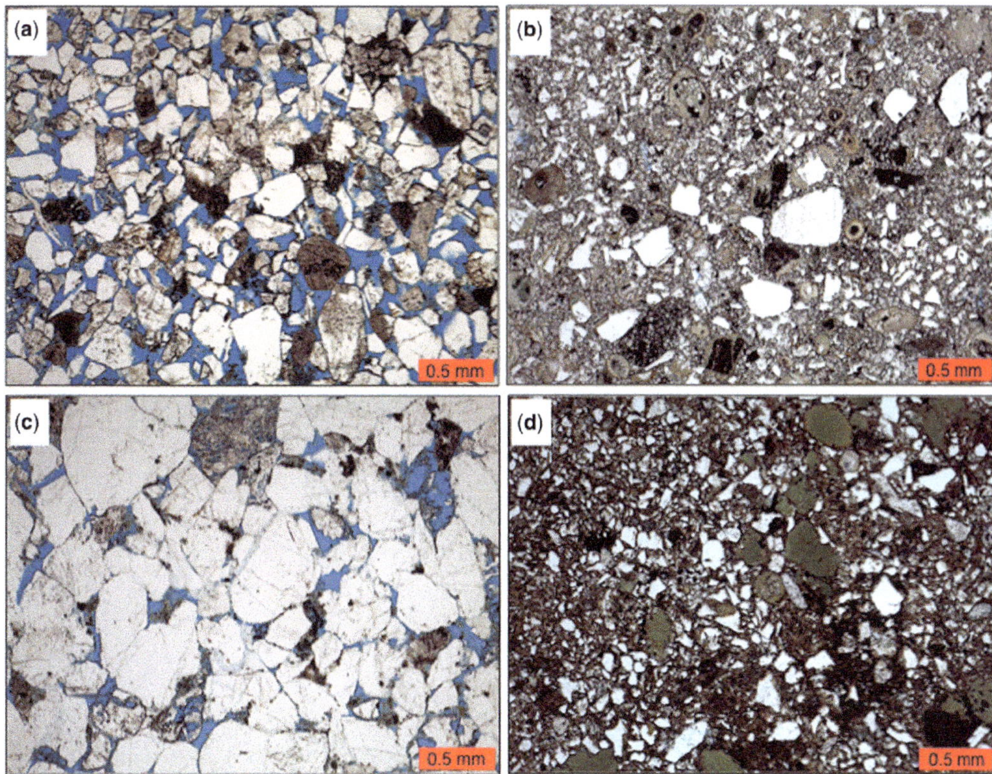

Fig. 2. Plane-polarized images of the Vedder and Kreyenhagen sandstones. Porosity is filled with blue epoxy. (**a**, **b**) Typical and atypical textures of the Vedder sandstones. (**c**, **d**) Typical and atypical textures of the Kreyenhagen sandstones. The typical texture of the sandstones is arenite and the atypical texture is wacke or significantly cemented.

from 2.3–6.0. For the Kreyenhagen sandstones, the CI ranges from 4.9–7.0 and the TPI ranges from 3.5–6.8. The CI values of the Vedder and Kreyenhagen sandstones show a minor increase with depth, but there is no significant change in the TPI. Floating grains are not common in arenites, but are more common in those samples with significant cement (phosphate or carbonate) volumes.

Detrital feldspars, volcanics, glauconite, biotite and muscovite are commonly deformed. Some volcanics, feldspars and biotite have been altered and squeezed into adjacent pore spaces to form pseudomatrix. Early compaction between the framework grains created point and tangential contacts. Late compaction between the framework grains caused pressured-induced dissolution, forming long, concavo-convex, and sutured contacts.

Fractures

Fracturing of the feldspars (Fig. 6a, b), quartz (Fig. 7a, b) and micas is common. Fracture growth occurred along grain boundaries and/or cleavage planes. Fracturing within detrital grains (Fig. 6a, b) and cements is common at shallower depths. Through-going fractures are more common in deeper samples (Fig. 7a, b). Fracturing created various sizes of angular fragments, thus increasing the angularity of grains and decreasing the degree of sorting.

Authigenic minerals

Glauconite

Glauconite is present in both the Vedder and Kreyenhagen sandstones. Glauconite grains range from subround to round (Fig. 7c). It replaced quartz (Fig. 7d) and feldspars (Fig. 8a) along the fractures and grain boundaries. Glauconite is most common in medium grained sandstones with abundant matrix.

Phosphate

Phosphate is present in several samples in the Vedder sandstones. It occurs as individual grains, coatings on framework grains and cement. The

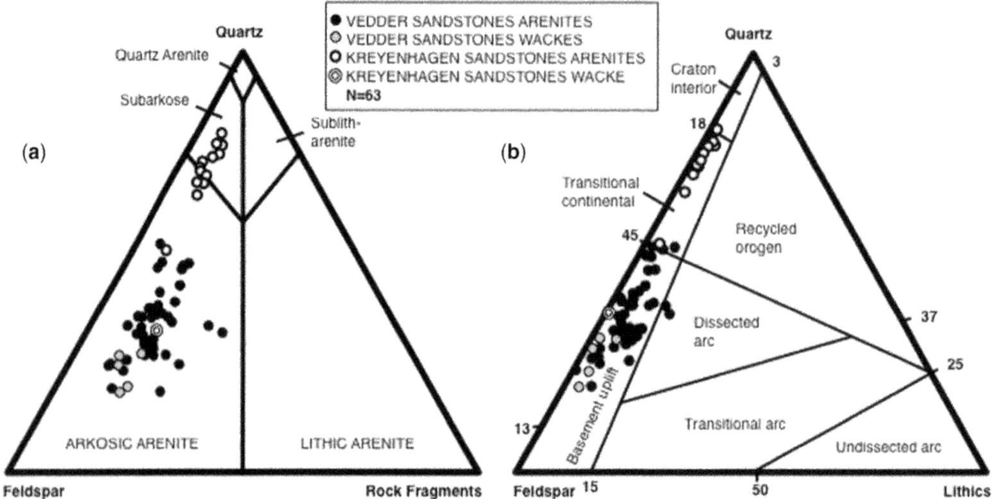

Fig. 3. (**a**) QFRf ternary diagram (Pettijohn *et al.* 1987) for the Oligocene Vedder and Eocene Kreyenhagen sandstones. The results suggest that the deeper samples (Kreyenhagen sandstones) are more quartz-rich than shallower samples (Vedder sandstones). (**b**) QFL ternary diagram (Dickinson 1970) for the Vedder and Kreyenhagen sandstones. The results indicate that most of the samples are derived from the basement uplift and transitional continental fields.

phosphate coatings occur as light-, medium- and dark-brown zones (Figs 7c & 8b, c). The outer zones are darker than the inner zones. Phosphate replaced glauconite (Fig. 7c), quartz (Fig. 8b) and feldspars (Fig. 8c). Phosphate replacement often occurred along the grain boundaries and sometimes along fractures. Phosphate cement filled the intergranular pore space. Some phosphate cement

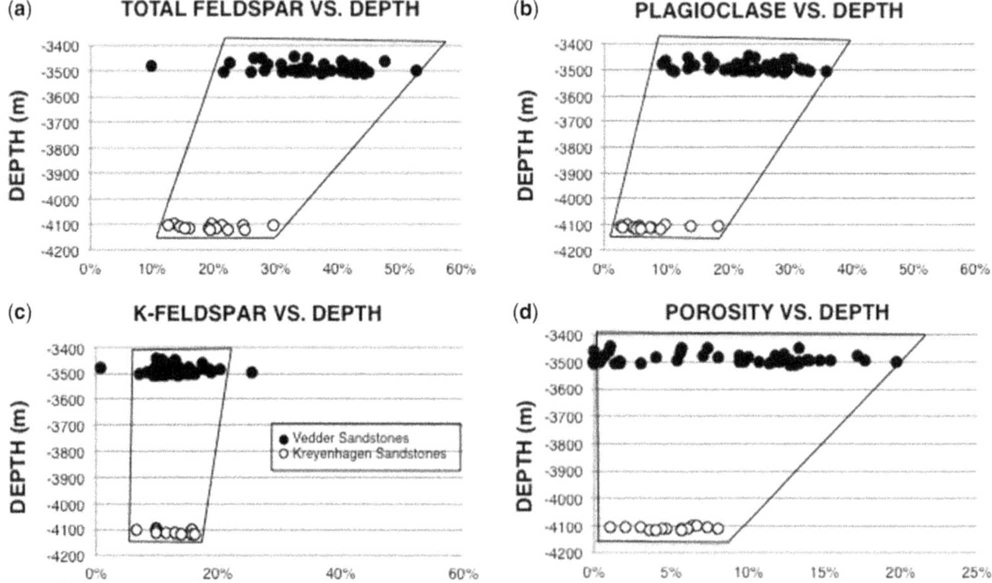

Fig. 4. (**a–c**) Percentages of feldspars decrease with increasing depth. (**d**) Thin-section porosity of the Vedder and Kreyenhagen sandstones. Data are derived from point counts. $N = 63$.

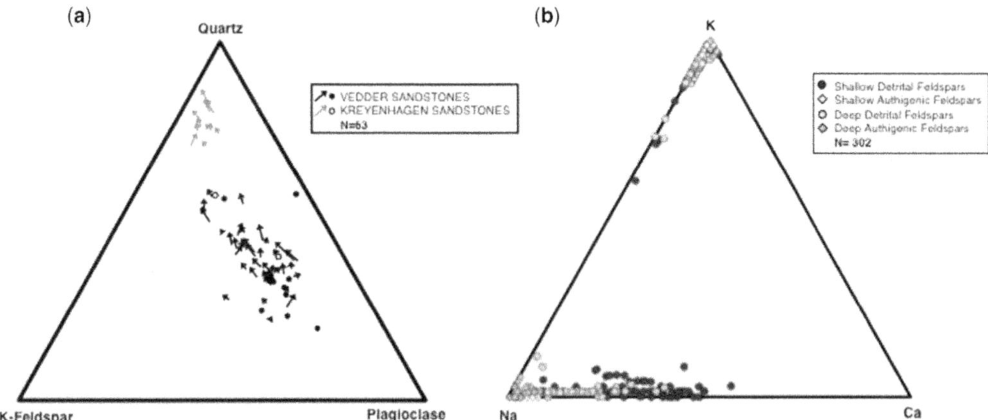

Fig. 5. (**a**) QKP ternary diagram plotted for the Vedder and Kreyenhagen sandstones following the methodology of Harris (1989). The compositions are adjusted for dissolution. The arrows point to the present composition, they originate at the reconstructed compositions. The dots indicate no change in composition. (**b**) K–Na–Ca ternary diagram of detrital and authigenic feldspars.

contains fragments of feldspars and quartz. The cement occurs as light- and dark-brown zones. Optically the outer zones are darker than the inner zones.

Pyrite

Authigenic pyrite crystals occur throughout the sandstones, ranging from 0–5% of the rock. Pyrite formed along the biotite cleavage planes, within hydrocarbon-filled pores (Fig. 9a, b), between feldspars and their phosphate coatings and as isolated crystals or groups of crystals within pore spaces. The crystals shapes range from subhedral to euhedral and are up to 0.04 mm long.

Clays

Authigenic clays occur as coatings on detrital grains (Fig. 9c) and as pore-lining and pore-filling cements (Fig. 9c).The authigenic clays were tentatively identified using the petrographic microscope and confirmed using SEM-EDS analyses. They include chlorite, kaolinite (Fig. 9c) and a highly birefringent clay whose main cation is potassium. This is consistent with illite, but contains small amounts of Na, Ca, Mg and Fe, suggesting the presence of small amounts of smectite or chlorite Most samples contain all of these clays; they range from 0–18% of the rock volume with an average volume of 2%. Clays in these sandstones are mostly products of altered feldspars and rock fragments.

Illitic clays occur as pore linings or grain coatings and are commonly found near or intergrown with pore-filling kaolinite. These clays range from 0% to 18% in volume. In some cases illitic coatings occur between detrital quartz grains and quartz overgrowths.

Chlorite is present in very small amounts, typically less than 1%. It most commonly occurs as a product of biotite alteration, in which patches of chlorite are found along biotite cleavage planes.

Kaolinite is present in most samples as pore-filling cement, ranging from traces to 9% in volume (average 2%). It occurs as booklets, up to 20 μm thick, partially to completely filling the pore space. Kaolinite often formed adjacent to dissolved feldspars or enclosed within remnant clay coatings from partially dissolved feldspars (Fig. 9c). In the Lewis 3 well, kaolinite has completely replaced precursor grains. In volcanic rock fragments feldspars have partially altered to kaolinite while the rest of the grains remain unaltered.

Carbonate

Carbonate cements occur in most of the samples, ranging from traces to 57%. These include dolomite (Figs 8d & 10a) and calcite (Figs 9d & 10b), with calcite being the most abundant. Calcite occurs as a poikilotopic cement filling the pore spaces and replacing quartz (Fig. 10b), plagioclase (Fig. 10b), K-feldspars and volcanics. In the early stages of calcite replacement, large portions of the grains remained unaltered and the minerals can be identified. In the later stages of calcite replacement the grains are completely replaced and cannot be identified. As a result, the porosity is reduced to negligible in several samples and the framework grains appeared to be floating.

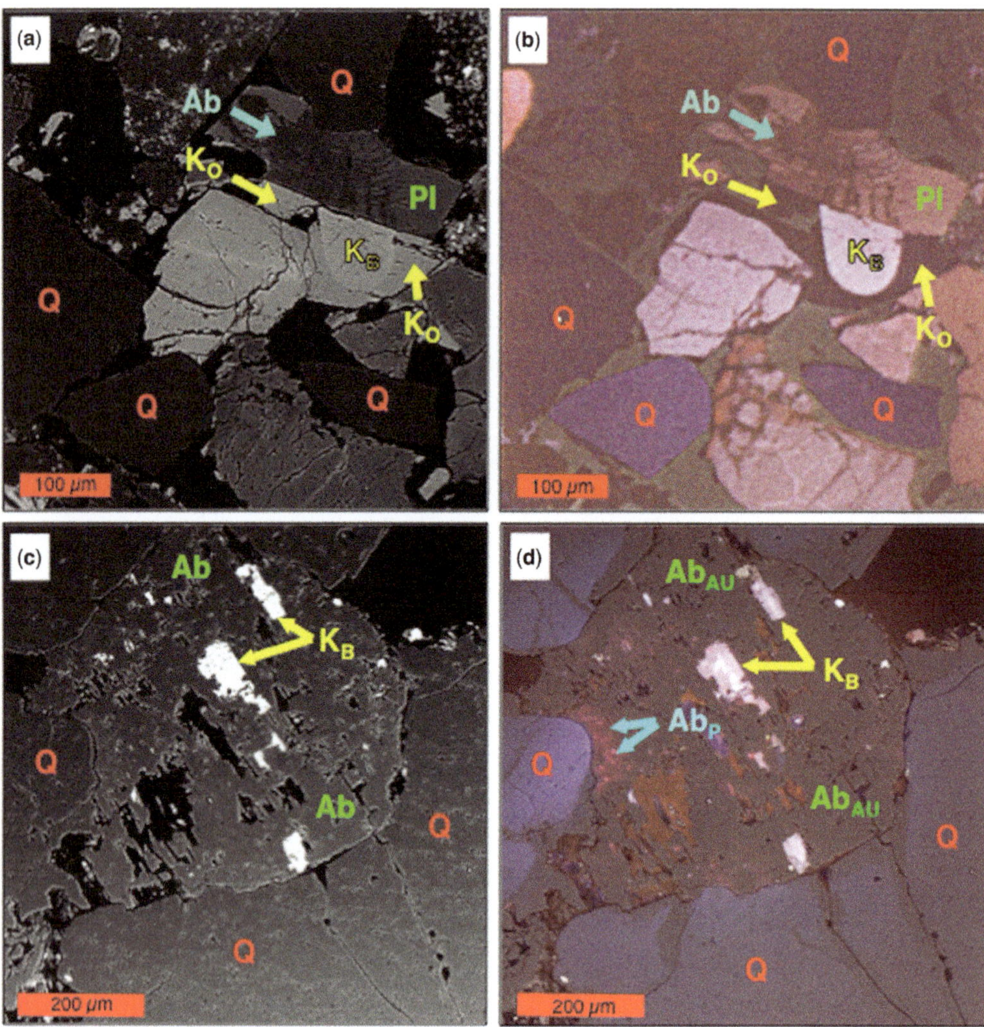

Fig. 6. SEM-BSE (**a**) and SEM-CL (**b**) images of albitized (Ab) plagioclase (Pl), Ba-rich K-feldspar (K_B), K-feldspar overgrowths (K_O), and quartz (Q). Albitization occured along the concentric fractures in plagioclase. K-feldspar overgrowths formed along plagioclase and Ba-rich K-feldspar grain boundaries. Plagioclase luminesces pink, Ba-rich K-feldspar luminesces bright pink, albite and K-feldspar overgrowths are non-luminescent, and quartz luminesces blue. Well: R.A. Moore 2, depth: 3499 m. (**c**) SEM-BSE image of almost completely albitized Ba-rich K-feldspar surrounded by quartz. (**d**) Composite image of SEM-BSE and SEM-CL images of authigenic albite (Ab_{AU}) replacing Ba-rich K-feldspar, but not albite from perthite (Ab_P). K-feldspar and albite in perthite luminesce pink, detrital quartz luminesces blue, and authigenic albite and quartz are non-luminescent. Well: ELR, depth: 4104 m.

Dolomite is present as blocky cement (Figs 8d & 10a) and as subhedral to euhedral rhombic crystals. It occurs in pore spaces, between biotite cleavage planes and replacing quartz (Fig. 8d), feldspars (Fig. 8d), phosphate and calcite. Some crystals contain Fe-rich zones. In some instances, clay coatings (Fig. 8d) within crystals suggest at least two episodes of dolomite cement.

Albite

Authigenic albite replaced plagioclase (Figs 6a, b, 10 & 11) and K-feldspars (Figs 6c, d & 10a). Albitization of detrital feldspars occurred along fractures (Figs 6a, 10 & 11) and/or original grain boundaries (Fig. 6c). Some detrital feldspars in deeper samples (Fig. 6c) are more albitized compared with detrital

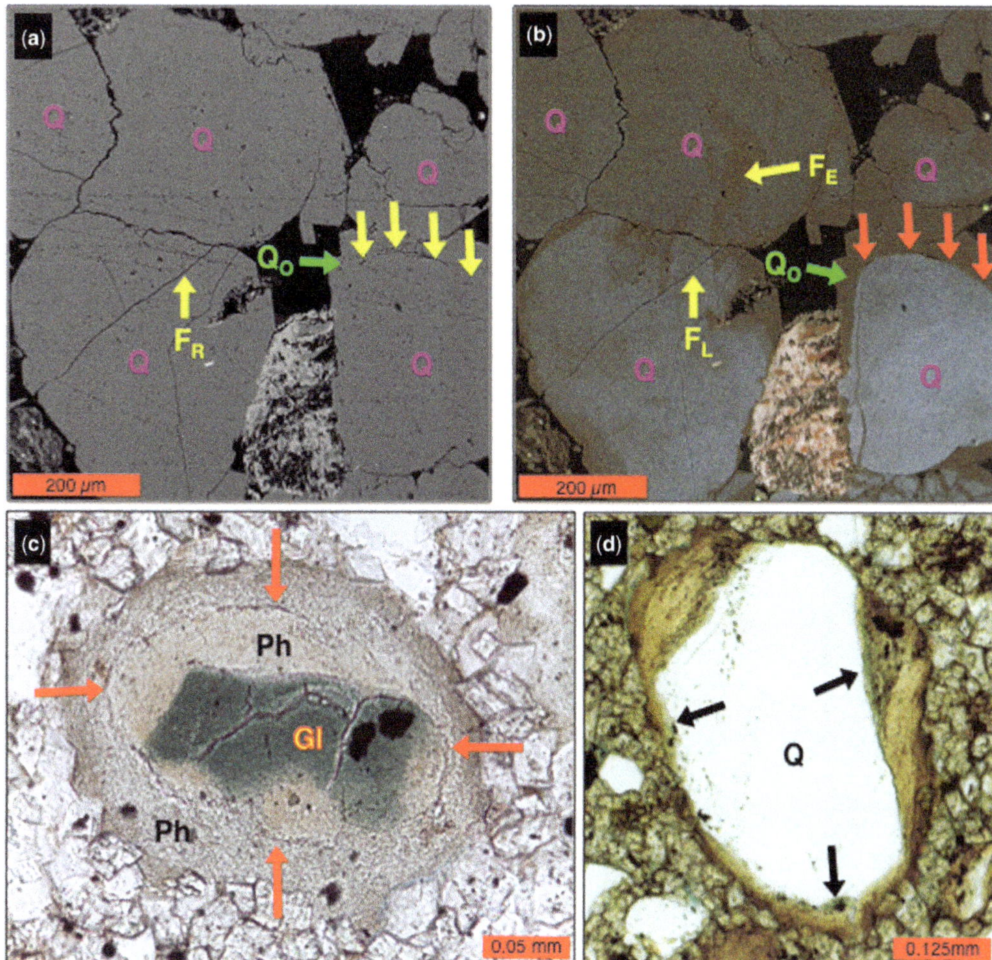

Fig. 7. (**a**) SEM-BSE image of fractured (F_R) quartz (Q) and quartz overgrowth (Q_O) on quartz grains. The yellow arrows point to the original grain boundary between detrital and authigenic quartz. (**b**) Composite image of SEM-BSE and SEM-CL images revealing quartz overgrowth on a rounded quartz grain (red arrows) and multiple episodes of fracturing. Early fractures (F_E) in quartz were healed by secondary quartz whereas late fractures (F_L) have not. Detrital quartz luminesces blue, authigenic quartz is non-luminescent. Well: ELR, depth: 4104 m. (**c**) Zoned phosphate (P) replacing glauconite (Gl). The red arrows point to the original grain boundary. Well: R.A. Moore 2, depth: 3477 m. Plane-polarized light. (**d**) Glauconite (arrows) replacing quartz (Q). Well: R. A. Moore 2, depth: 3477 m. Plane-polarized light.

feldspars in shallower samples (Figs 6a, 10a & 11). The average composition for albite is $Ab_{97.3}An_{2.2}Or_{0.5}$ in shallower samples and $Ab_{98.5}An_{1.0}Or_{0.5}$ in deeper samples. SEM-CL imaging revealed authigenic albite is non-luminescent (Fig. 6) in these sandstones.

K-feldspar

Authigenic K-feldspar occurs as overgrowths and within fractures in plagioclase (Figs 6a, b, 10a & 11) and K-feldspar (Figs 6a, b & 10a). It is most abundant in shallower samples, especially in the KCL 11–17 well. The overgrowths usually exhibit optical continuity between the detrital grains and they are non-luminescent (Figs 6a, b). In some cases K-feldspar overgrowths replace dolomite cement (Fig. 10a). Partial dissolution of the K-feldspar overgrowths occurred in deeper samples. SEM-EDS analyses indicate that the K-feldspar overgrowths are almost pure (Fig. 5b) with average compositions of $Or_{98.9}Ab_{0.9}An_{0.2}$ in Vedder

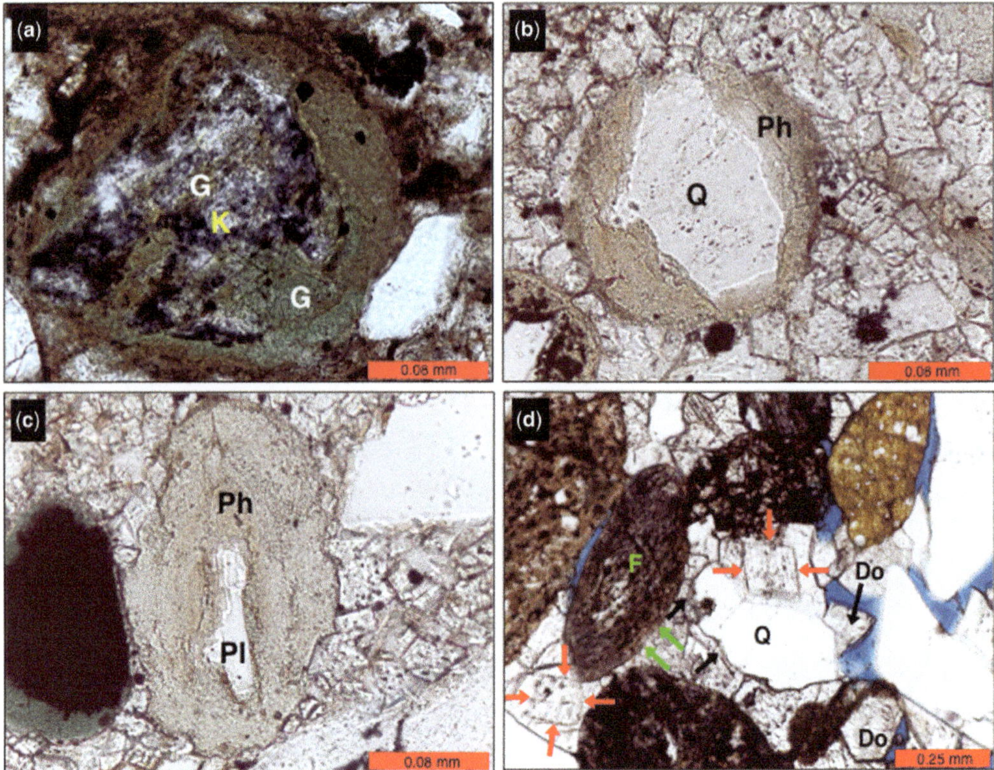

Fig. 8. (**a**) Glauconite (G) replacing K-feldspar (K). Well: ELR, depth: 4104 m. Cross-polarized light. (**b**, **c**) Phosphate (Ph) replacing quartz (Q; **b**) and plagioclase (Pl; **c**). Well: R. A. Moore 2, depth: 3477 m. Plane-polarized light. (**d**) Pore-filling dolomite (Do) cement that has replaced feldspar (F) and quartz. Black and green arrows point to the original grain boundaries of the quartz and feldspar. Clay coatings (red arrows) on dolomite rhombs within the blocky cement suggest at least two episodes of dolomite precipitation. Well: Campen 1, depth: 3466 m. Plane-polarized light.

sandstones and $Or_{99.5}Ab_{0.4}An_{0.1}$ in Kreyenhagen sandstones.

Quartz

Authigenic quartz occurs as overgrowths (Figs 7a, b & 9d) and fills fractures (Fig. 7b) within detrital quartz grains. Authigenic quartz is usually optically continuous with the detrital grains and often cannot be identified under the petrographic microscope. Some detrital grains have clay coatings, separating them from the overgrowths. In other cases euhedral crystal facies (Figs 7a & 9d) indicate the presence of overgrowths. Petrographic and SEM-BSE (Fig. 7a) images only show unhealed fractures within detrital quartz grains, but SEM-CL images reveal non-luminescent secondary quartz that has healed fractures in detrital quartz grains (Fig. 7b). Continuing compaction resulted in the subsequent formation of fractures that have not been healed by authigenic quartz.

Pseudomatrix

Pseudomatrix is found scattered throughout the rock, ranging from 0–2% of the rock volume. It is most common in samples with small amounts of volcanic rock fragments. Pseudomatrix formed from squeezing of micas and rock fragments into the adjacent pore throats. Some partially dissolved feldspars formed pseudomatrix as well.

Rutile, Barite and Anhydrite

Trace amounts of authigenic rutile, barite and anhydrite are present in some samples. Subhedral to euhedral rutile crystals (up to 20 μm long and 10 μm wide) occur in association with pyrite crystals and clays. Poikilotopic barite cement fills the intergranular pores in tightly packed sandstones and possibly replaces feldspars. Anhydrite occurs as coatings on carbonate cements and between biotite cleavage planes. The actual amount of anhydrite

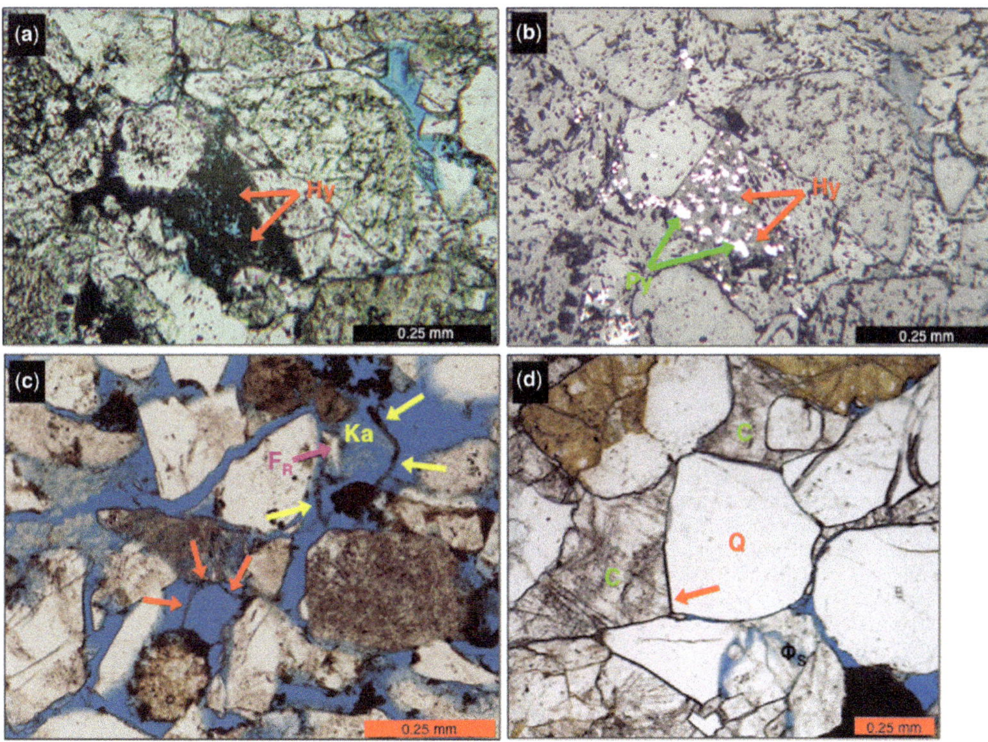

Fig. 9. Pyrite (Py) formed within hydrocarbon (Hy) in secondary pore created by framework-grain dissolution under (**a**) transmitted light and (**b**) reflected light. Well KCL 11–17, depth: 3496 m. (**c**) Plane-polarized light image of kaolinite (Ka) formed adjacent to remnants of feldspar (F_R) within a remnant clay coating (yellow arrows). Red arrows show the remnant clay coatings of completely dissolved grains. Well: R.A. Moore 3, depth: 3497 m. (**d**) Plane-polarized light image of sharp contact between calcite (C) cement and quartz (Q) overgrowth (red arrow) suggests that the quartz overgrowth formed prior to calcite precipitation. Secondary porosity (Φ_S) is present in feldspar grain. Well: Campen 1, depth: 3477 m.

is unknown as some probably dissolved during thin-section manufacture.

Hydrocarbons

Although most hydrocarbons were removed during thin-section manufacture, patches of hydrocarbons occur within pore spaces in most samples (Figs 9a, b). Volatile loss from these old cores is likely to have changed the nature of these hydrocarbons, so no attempt was made to further characterize them.

Porosity

Thin-section porosity within these sandstones ranges from traces to 20% (Fig. 4d). It varies between samples collected within the same depth intervals and only a few feet apart, but in general porosity decreases with depth. Intragranular porosity averages 0.5% and intergranular porosity averages 7% of the rock volume. Intragranular porosity is most likely of secondary origin whereas intergranular porosity can be of primary or secondary origin. Following the methodologies of Schmidt & McDonald (1979) and Shanmugam (1985), secondary porosity is estimated to make up from 50% to nearly 100% of the intergranular porosity in these sandstones. Porosity within wackes is significantly less than within arenites due to the presence of matrix. Cementation by carbonates, K-feldspar and quartz has significantly reduced porosity in some samples. The formation of pseudomatrix resulted in a minor porosity reduction.

Secondary porosity commonly resulted from the dissolution of detrital feldspars (Figs 9c, d & 10a), particularly plagioclase. However, carbonate and K-feldspar cements, as well as detrital quartz, micas and volcanics, were also affected by dissolution. Figure 12 is a ternary diagram adjusted for grain dissolution (Harris 1989) indicating shifts towards more quartz-rich compositions in the arenites but little change in the wackes.

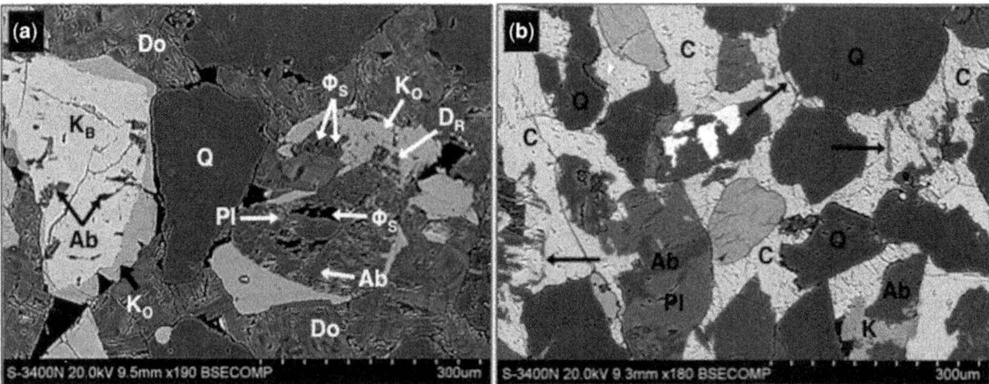

Fig. 10. (**a**) SEM-BSE image of zoned dolomite (Do) cemented sandstone. Albitized (Ab) plagioclase (Pl) with remnants of original plagioclase and albitized Ba-rich K-feldspar (K_B) grains exhibit K-feldspar overgrowths (K_O), but quartz (Q) does not have quartz overgrowths. Remnants of dolomite (D_R) within K-feldspar overgrowths suggest that the overgrowths have replaced dolomite. Secondary porosity (Φ_S) within the plagioclase and dolomite indicate a later dissolution event that did not affect the authigenic K-feldspar. Also notice the cross-cutting relationship of K-feldspar overgrowth around euhedral dolomite, indicating that K-feldspar formed after dolomite. Well: KCL 11–17, depth: 3496 m. (**b**) SEM-BSE image of calcite (C), quartz and albitized (Ab) plagioclase (Pl) and K-feldspar (K). Calcite has replace quartz and plagioclase; the red and green arrows show the original plagioclase and quartz edges respectively. Well: ELR, depth: 4084 m.

Mass balance

The potential for mass transfer of aluminium into or out of the sandstones was investigated. The amounts of kaolinite cement present in the samples were compared with the amounts of kaolinite cement that would have been produced if all of the aluminium released during documented feldspar dissolution was incorporated into kaolinite. The per cent imbalance of kaolinite suggests export or import of aluminium during diagenesis.

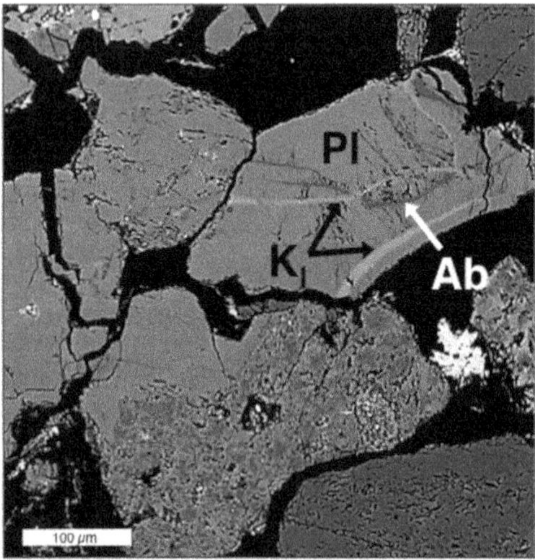

Fig. 11. SEM-BSE image of albitized (Ab) plagioclase (Pl) exhibit K-feldspar infilled (K_I) fractures. The cross-cutting relationship between albite and K-feldspar infilled fractures indicates that albitization occurred before authigenic K-feldspar. Well: KCL 11–17, depth: 3452 m.

The amounts of kaolinite cement precipitated from dissolved plagioclase and K-feldspar assumes that there is 0% microporosity in the kaolinite. The microporosity of kaolinite in these sandstones was examined by using the SEM-BSE images from 650× to 1300× magnification; the results indicate that kaolinite has an average of 40% microporosity. The amount of kaolinite cement, adjusted for 40% microporosity, that could precipitate from one volume of dissolved plagioclase is 1.08 and one volume of dissolved K-feldspar is 0.76.

Results of the mass balance calculations are shown in Figure 13. The dashed lines in Figure 13 represent a balance between estimated and observed kaolinite; points that plot above the line indicate aluminium import whereas points below indicate aluminium export, at least on the scale of a thin section. If no adjustment is made for microporosity within kaolinite cements the number of points indicating import or export are roughly equal. However, when the results are adjusted for the measured average of 40% microporosity, the results suggest aluminium export in 57% of the samples, aluminium import in 19% and neither export nor import of aluminium in 24%, These calculations are conservative in that completely dissolved (and thus unidentified)

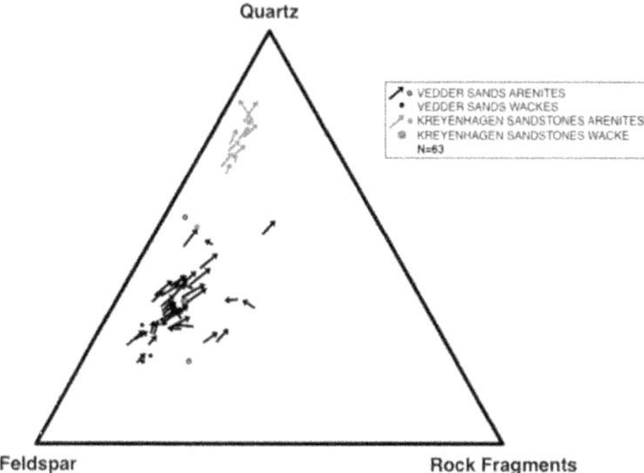

Fig. 12. QFRf ternary diagram for the Vedder and Kreyenhagen sandstones adjusted for grain dissolution. The arrows point to the present composition, they originate at the reconstructed compositions (Harris 1989). The dots indicate no change in composition.

grains were not included, nor was any adjustment made for secondary porosity that may have been destroyed by subsequent compaction.

Discussion

Various diagenetic processes have affected reservoir quality in the Vedder and Kreyenhagen sandstones including compaction, deformation, dissolution, cementation, alteration and recrystallization of minerals. These processes are controlled by the pore-water chemistry, depositional environment, temperature and pressure (Hayes & Boles 1992). The paragenetic sequence (Fig. 14) is established based on the textural relationships of the diagenetic minerals. These events started at the time of deposition and some continue to the present day.

The syndepositional formation of glauconite and phosphate occurred in early diagenesis. The presence of coeval glauconite and phosphate suggests these sandstones were deposited in submarine conditions between 100 m and 300 m depth (Odin & Letolle 1980; Jimenez-Millan *et al.* 1998; Stonecipher 1999, 2000), although Stonecipher (1999) concluded that glauconite can be reworked and deposited in any depositional sequence. Glauconite grains can form by alteration of clays (Stonecipher 1999), micas (Tapper & Fanning 1968) fossil-rich sediments and fecal pellets (Harder 1980) in very early diagenesis. Phosphate has replaced glauconite (Fig. 7c), and glauconite and phosphate have replaced feldspars (Fig. 8a, c) and quartz (Figs 7d, 8b). Replacement of silicate grains suggests ongoing formation of glauconite and phosphate at deeper depths during early diagenesis.

Some pyrite formed between cleavage planes and expanded the fabric of detrital biotite grains. It also formed between detrital feldspars and phosphate coatings. These suggest pyrite precipitation during early diagenesis.

Illitic clays often occur as coatings on detrital grains, and sometimes separate the detrital grains from overgrowths. This suggests that the clay coatings formed during early diagenesis. They are more abundant in shallower samples than deeper samples.

Calcite cement formed during early diagenesis. It precipitated into pore spaces and replaced framework grains. The framework grains can sometimes be identified by unreplaced remnants of the grains within the cement. In other instances, poikilotopic calcite cement completely replaced framework grains and their former presence is only recognizable through remnant clay coatings within the calcite.

Authigenic albite in these sandstones is almost pure (Fig. 5b). Most of the feldspars in the deeper samples (Fig. 6c) are more albitized compared with feldspars in shallower samples (Figs 6a, 10a & 11). Albitization replaced detrital plagioclase (Figs 6a, b, 10 & 11) and K-feldspar (Figs 6c, d &

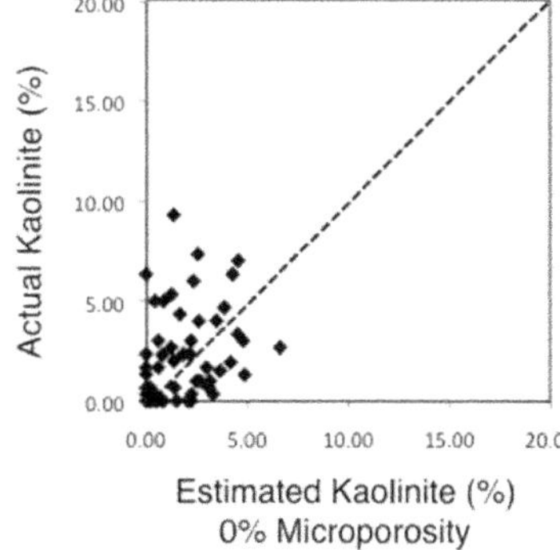

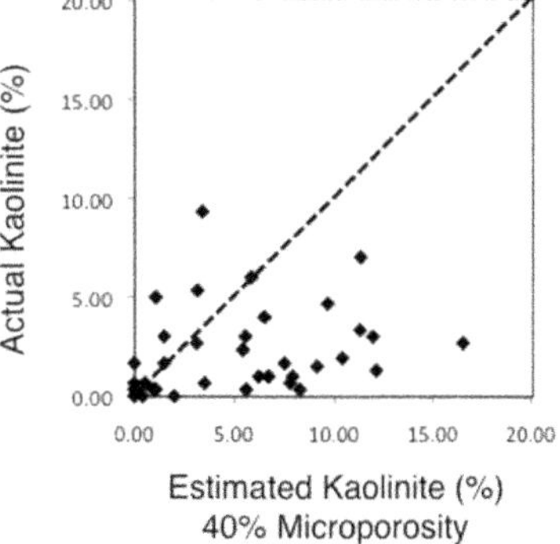

Fig. 13. Estimated v. actual kaolinite. The dashed lines indicate estimated kaolinite equals actual kaolinite. Points above the line indicate aluminium import, those below the lines indicate aluminium export.

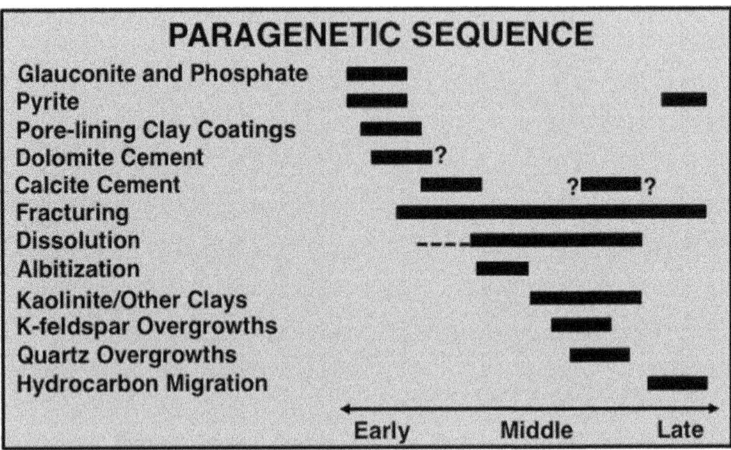

Fig. 14. Paragenetic sequence for the Vedder and Kreyenhagen sandstones. Dashed line indicates minor dissolution at that time.

10a), but not detrital albite (Fig. 6c, d). Albitization occurs in a wide (Gold 1987; Boles & Ramseyer 1988) or narrow (Boles 1982; Morad *et al.* 1990) range of temperature in middle of diagenesis. Albitization may occur between 75°C (Morad *et al.* 1990) and 124°C, which equates to depths of 1.8–3.4 km at the Greeley oil field.

Biotite has commonly been altered to chlorite and clay. Alteration typically occurred along the cleavage planes, which suggests the event occurred during late diagenesis after significant compaction.

Authigenic quartz (Figs 7a, b & 9d) and K-feldspar (Figs 6a, b, 10a & 11) are present in more than half of the samples. The absence of quartz and K-feldspar overgrowths in some calcite-cemented samples indicate calcite formed during early diagenesis before the quartz and K-feldspar overgrowths. In addition, the presence of euhedral quartz overgrowths in sharp contact with calcite cement (Fig. 9d) suggests that authigenic quartz formed prior to a second period of calcite precipitation during late diagenesis.

The cross-cutting relationship between the albite and K-feldspar-infilled fractures indicates that albitization occurred before precipitation of authigenic K-feldspar (Fig. 11). K-feldspar overgrowths replaced dolomite cement and penetrated into the albitized plagioclase (Fig. 10a) indicating that the overgrowths formed after precipitation of dolomite and dissolution of plagioclase.

Compaction of the framework grains started after deposition. It rearranged the grains creating point and tangential contacts during early diagenesis. As compaction increased, long, concavo-convex, and sutured grain-to-grain contacts formed by pressure-induced dissolution. Micas and volcanic-rock fragments were often deformed to create pseudomatrix that filled the adjacent pore space. As a result porosity and permeability were reduced. The detrital grains within these sandstones contain an average of at least three contacts with adjacent grains. Although other sandstones in the western San Joaquin basin show an increase of CI and TPI with increasing depth (Taylor 2007; Horton *et al.* 2009), there is no significant difference between the shallower and deeper units at Greeley oil field. This may be due to the difference in composition between the Vedder and Kreyenhagen sandstones (Fig. 3).

Compaction created fractures within individual grains and cements. Fractures occurred in feldspars, micas, quartz and dolomite-cement. Fracturing created various sizes of angular fragments that, with increasing burial, rotated relative to each other, changing the textures of the sandstones. Fracturing also allowed the exchange of ions between the grains' interiors and pore fluids resulting in the formation of dolomite crystals between biotite cleavage planes as well as albitization (Figs 6, 10 & 11) and dissolution (Figs 9c & 10a) of feldspars and - to a lesser extent – micas, volcanics, quartz and carbonates.

Dissolution is most common in feldspars. Preferential dissolution of feldspars shifted the composition of sandstones towards the quartz corner on the ternary diagrams as shown in Figures 5a and 12. Plagioclase (Fig. 4b) was affected by dissolution more than K-feldspar (Fig. 4c). Plagioclase in wackes is more calcic than in the arenites, though the data are limited. Figure 5b shows a significant shift towards more sodic compositions in deeper (Kreyenhagen) samples at Greeley. The normal range of Sierran plagioclase is An_{15}–An_{50} (Boles & Ramseyer 1988). At the nearby Wasco oil field

(24 km to the NW) correlative sandstones in the Kreyenhagen Formation contain abundant plagioclase in the An_{30}–An_{50} range (Olabisi 2015) and it is probable that such plagioclase was originally present at Greeley. This suggests that calcic-plagioclase was more susceptible to dissolution (Boles 1984; Milliken *et al.* 1989). The loss of the most calcic-plagioclase (An_{30}–An_{50}) at Greeley was accompanied by a decrease in the Sr contents of plagioclase, suggesting Sr substitution for Ca (Schultz *et al.* 1989) in the plagioclase. The significant loss of Ca-rich plagioclase between 3444 m and 4118 m suggests a pronounced period of dissolution at these depths.

Authigenic kaolinite precipitated in the pore spaces, adjacent to partially and completely feldspar dissolved grains. This suggests Al^{3+} for kaolinite precipitation was derived from dissolved feldspar. Kaolinite cement formed after significant compaction and dissolution of the framework grains, suggesting that it formed during late diagenesis.

Hydrocarbon migration occurred during late diagenesis after significant dissolution of detrital and authigenic minerals. Continued hydrocarbon maturation within the reservoir resulted in formation of pyrite (Fig. 9a, b) as Fe^{2+} in the pore waters reacted with sulfur liberated from the hydrocarbons.

Porosity was reduced in these sandstones by compaction, cementation (Figs 6a, b, 7a–c, 8b–d, 9d, 10 & 11), precipitation of clays (Figs 8d & 9c) and formation of pseudomatrix; it was increased by the dissolution of the framework grains (Figs 9c & 10a) and cements (Fig. 10a). As dissolution progressed, continued compaction further reduced primary porosity as well as secondary porosity and caused secondary intergranular porosity to resemble primary porosity (Schmidt & McDonald 1979; Horton *et al.* 2009). As result, more than 50% of intergranular porosity in these sandstones is likely of secondary origin.

Mass balance calculations suggest, on the scale of a thin section, an overall export of aluminium (Fig. 13). These calculations do not take into account albitization of plagioclase, dissolution of other alumino-silicates, secondary porosity for which the precursor mineral could not be determined or any secondary porosity that may have been destroyed by continued compaction – all of which would have released additional aluminium into the pore fluids. Thus, an overall export of aluminium from these sandstones is likely, particularly in the Kreyenhagen Fm where a large compositional shift due to the loss of plagioclase is indicated (Figs 4b & 5). It also suggests that porosity formed by feldspar dissolution had a net positive effect on porosity by replacing pore spaces that were destroyed by compaction, thereby maintaining an open pore system. Thus, it was a major factor in the overall development of reservoir quality at this oil field.

Laumontite cement has been widely documented in the San Joaquin basin (e.g. Bloch *et al.* 1993; Noh & Boles 1993; Taylor 2007), including in the Vedder and Kreyenhagen sandstones at the nearby Wasco oil field (Olabisi 2015), but no laumontite was observed during the present study. Laumontite can precipitate as result of dissolution or albitization of plagioclase (Boles & Coombs 1977; Boles 1982; Helmold & van de Kamp 1984) and the formation of laumontite can occur at different temperatures and depths (McCulloh *et al.* 1978). The reported temperature of the Vedder sandstones at the Greeley field is 124°C, but the temperature of the Kreyenhagen sandstones is unknown (California DOGGR 1998). At the Wasco oil field, the measured temperatures reported for the Standard Oil Mushrush 5 well (log on file with California DOGGR (2014)) were reported as 135°C for the Vedder Formation and 145°C for the Kreyenhagen sandstones. As the Kreyenhagen sandstones at Greeley are buried to approximately the same depth as the Vedder Formation at Wasco, an estimated temperature of *c.* 135°C for the Kreyenhagen sandstones at the Greeley field is reasonable. At the Wasco oil field laumontite cement is common in the Kreyenhagen sandstones and present in small amounts in the Vedder sandstones, and there is no evidence of laumontite dissolution in those strata (Olabisi 2015). Dissolution of plagioclase (Figs 5 & 10a), which could provide Ca^{+2} ions for laumontite precipitation, is widespread in the Kreyenhagen sandstones at the Greeley oil field. Thus, the lack of laumontite at the Greeley oil field suggests that the temperature of the Vedder sandstones at the Greeley field is too low for laumontite to form and the temperature of the Kreyenhagen sandstones is at the boundary for laumontite precipitation. It also suggests that these sandstones are currently at or very near their deepest burial depths.

This has significant implications for the structural history of this portion of the San Joaquin basin. The Greely oil field sits near the crest of the Bakersfield Arch, and the Kreyenhagen sandstones occur at an approximately 500 m shallower depth than the Kreyenhagen sandstones at Wasco. If the Kreyenhagen sandstones at Greeley are currently at their maximum burial depth as indicated by the lack of laumontite, then the Bakersfield Arch must have formed before the deposition of the uppermost 500 m sediments in the overlying strata. At Wasco the top of the Pliocene Etchegoin Fm is approximately 1325 m thick with its top at a depth of approximately 1340 m (California DOGGR 1998). At Greeley the Etchegoin Formation is only 1190 m thick with its top at a depth of 1050 m (California DOGGR 1998). As the difference in elevation between the two oil fields is less than 20 m, the differences in the burial depths and the thicknesses of what are generally interpreted as shallow marine sandstones (Scheirer & Magoon 2007) are best explained by growth of the Bakersfield Arch during deposition of the Etchegoin Fm in the Early to Middle Pliocene and continuing into the Late Pliocene or early Pleistocene.

Gold (1987) suggested that plagioclase dissolution occurs mainly at temperatures between 100°C and 130°C. This is the approximate temperature range to which the Kreyenhagen sandstones were exposed during the onset and growth of the Bakersfield Arch. As growth of the Arch dramatically slowed the rate of burial, the Kreyenhagen sandstones were exposed to these temperatures for a prolonged period, resulting in significant plagioclase removal. The Vedder Sandstones did not reach these temperatures until much later, so the amount of plagioclase dissolution, although substantial (Fig. 5), was less than in the Kreyenhagen Fm. Therefore, a primary control on the development of reservoir quality in these rocks appears to be the length of time that the rocks remained within the 100–130°C temperature window. This might be useful as a predictive tool in other basins where the growth of anticlines during burial may have prolonged the exposure of target strata to this temperature interval.

Summary

The Vedder and Kreyenhagen sandstones at the Greeley oil field are arkosic to subarkosic arenites and wackes (Fig. 3). The sandstones contain abundant quartz, plagioclase and K-feldspars, and minor rock fragments. Common accessory minerals include biotite, muscovite, chlorite, epidote, glauconite, hornblende, pyrite and zircon. Cements are generally minor and include calcite, dolomite, kaolinite and mixed-layer illite–smectite or illite–chlorite. Other diagenetic features include the replacement of quartz and feldspars by glauconite, phosphate and carbonates, albitization of plagioclase and K-feldspars and precipitation of K-feldspar and quartz both as overgrowths and within fractures.

Reservoir quality was controlled primarily by the compaction and the dissolution of feldspars. Secondary porosity due to feldspar dissolution is well known as an important component of overall reservoir quality (Hayes 1979), but whether this process enhances porosity or simply redistributes it has been the subject of considerable debate (Hayes 1979; Bjorlykke 1984; Boles 1984; McBride 1987: MacGowan & Surdam 1988; Harris 1989; Hayes & Boles 1992; Wilkinson *et al.* 2001; Horton *et al.* 2009). At the Greeley oil field the deeper Kreyenhagen samples are more quartz-rich than shallower Vedder samples due to the preferential dissolution of feldspars. Dissolution affected plagioclase more than K-feldspar and preferentially removed those with highest calcium contents, resulting in higher K-feldspar to plagioclase ratios and more Na-rich plagioclase in the deeper strata. The removal of plagioclase also significantly altered QFRf and QFL ratios.

The amount of plagioclase dissolution appears to be controlled by the length of time the sandstones resided at temperatures between 100°C and 130°C (Gold 1987) as the rate of burial decreased due the growth of the Bakersfield Arch. Mass balance calculations indicate the net export of aluminium from most of the samples studied, at least on the scale of a thin section. Compaction reduced both primary and secondary porosity through fracturing and readjustment of detrital grains and remnants of partially dissolved grains. This created secondary intergranular porosity that now makes up the majority of porosity in these reservoirs. Thus, dissolution of feldspar had a net positive effect on reservoir quality by adding new pore space to offset the loss of porosity due to compaction.

This research is funded by: California State University at Bakersfield (CSUB) National Science Foundation Center for Research Excellence in Science and Technology (award #1137774), CSUB Weddle Scholarship, CSUB Graduate Student-Faculty Collaborative Initiative, CSUB Student Research Scholarship, Pacific Section – American Association of Petroleum Geologists, and San Joaquin Geological Society. Special thanks to Charles James and Eric Hierling at the California Well Sample Repository, and Elizabeth Powers and Stephanie Caffee at the CSUB Department of Geological Sciences.

References

Bandy, O.L. & Arnal, R.E. 1969. Middle tertiary basin development, San Joaquin Valley, California. *Geological Society America Bulletin*, **80**, 783–820.

Bartow, J.A. 1991. *The Cenozoic evolution of the San Joaquin Valley, California*. U.S. Geological Survey Professional Paper 1501.

Bartow, J.A. & McDougall, K. 1984. *Tertiary stratigraphy of the Southeastern San Joaquin Valley, California*. U.S. Geological Survey Bulletin, 1529-J.

Bent, J.V. 1985. *Petrographic reconnaissance of the Upper Oligocene–Middle Miocene sandstones of the San Joaquin Basin, California*. MS thesis, Stanford University.

Bent, J.V. 1988. Paleotectonics and provenance of Tertiary sandstones of the San Joaquin Basin, California. *In*: Graham, S.A. (eds) *Studies of the Geology of the San Joaquin Basin*. Pacific Section SEPM Publications, **60**, 109–127.

Bjorlykke, K. 1984. Formation of secondary porosity: how important is it? *American Association of Petroleum Geologists Memoir*, **37**, 277–286.

BLOCH, R.B. & OLSON, H.C. 1990. Stratigraphy and structural history of the Lower and Middle Miocene section, eastside San Joaquin Valley. *In*: KUESPERT, J.G. & REID, S.A. (eds) *Structure, Stratigraphy and Hydrocarbon Occurrences of the San Joaquin Basin, California*. Pacific Section SEPM Publications, **64**, 287–291.

BLOCH, S., KELLEY, S.A., CORRIGAN, J.D. & KELTON, M.J. 1993. Predictable distribution pattern of laumontite in Paleogene sandstones of the Tejon Block, southern San Joaquin Basin (California). *In*: *AAPG and SEPM, Annual Meeting Program with Abstracts*, April 25–28, 1993, New Orleans, Louisiana, USA, 77.

BOLES, J.R. 1982, Active albitization of plagioclase, Gulf Coast Tertiary. *American Journal of Science*, **282**, 165–180.

BOLES, J.R. 1984. Secondary porosity reactions in the Stevens sandstone, San Joaquin Valley, California. *In*: MCDONALD, D.A. & SURDAM, R.C. (eds) *Clastic Diagenesis*. AAPG Memoirs, **37**, 217–224.

BOLES, J.R. & COOMBS, D.S. 1977. Zeolite facies alteration of sandstones in the Southland Syncline, New Zealand. *American Journal of Science*, **277**, 982–1012.

BOLES, J.R. & RAMSEYER, L. 1988. Albitization of plagioclase and vitrinite reflectance as paleothermal indicators, San Joaquin basin. *In*: GRAHAM, S.A. (eds) *Studies of the Geology of the San Joaquin Basin*. Pacific Section SEPM Publications, **60**, 129–139.

BYERS, F.M., JR. 1990. *Procedure for Determination of Volume Constituents in Thin Sections of Rocks*. Los Alomos National Laboratory TWS-EES-DP-102, R2, 8 p. http://pbadupws.nrc.gov/docs/ML0317/ML03 1760791.pdf

CALIFORNIA DOGGR 1998. Oil and Gas Fields in California: Sacramento, California, USA, **CD-1**, 192–193.

CALIFORNIA DOGGR 2014. Mushrush 5 well (UWI 02909487). http://owr.conservation.ca.gov/WellRe cord/029/02909487/02909487_1949-06-18_1Electri c%20Log.pdf

CLARKE, S.H., JR. 1973. *The Eocene Point of Rocks Sandstone: Provenance, mode of deposition and implications for the history of offSET ALong the San Andreas fault in central California*. PhD thesis, University of California, Berkeley.

DICKINSON, W.R. 1970. Interpreting detrital modes of wacke and arkose. *Journal of Sedimentary Petrology*, **40**, 695–707.

DOTT, R.H., JR. 1964. Wacke, greywacke and matrix – what approach to immature sandstone classification? *Journal of Sedimentary Petrology*, **34**, 625–632.

EHRENBERG, S.N. 1990. Relationship between diagenesis and reservoir quality in sandstones of the Garn Formation, Haltenbanken, Mid-Norwegian continental shelf. *AAPG Bulletin*, **74**, 1538–1558.

GALLOWAY, W.E. 1979. Diagenetic control of reservoir quality in arc-derived sandstones: implications for petroleum exploration. *In*: SCHOLLE, P.A. & SCHLUGER, P.R. (eds) *Aspects of Diagenesis*. SEPM Special Publications, **26**, 251–262.

GAUTIER, D.L. & SCHEIRER, A.H. 2007. *Miocene Total Petroleum System – Lower Bakersfield Arch Assessment Unit of the San Joaquin Basin Province*. U.S. Geological Survey Professional Papers, **1713**, chapter 14.

GOLD, P.B. 1987. Textures and geochemistry of authigenic albite from Miocene sandstones, Louisiana Gulf Coast. *Journal of Sedimentary Petrology*, **57**, 353–362.

GROVE, K. & NIEMI, T.M. 2005. Late Quaternary deformation and slip rates in the northern San Andreas fault zone at Olema Valley, Marin County, California. *Tectonophysics*, **401**, 231–250.

HALLEY, R.B. 1978. Estimating pore and cement volumes in thin section. *Journal of Sedimentary Petrology*, **48**, 642–650.

HARDER, H. 1980. Syntheses of glauconite at surface temperatures. *Clays and Clay Minerals*, **28**, 217–222.

HARRIS, N.B. 1989. Diagenetic quartzarenite and destruction of secondary porosity: an example from the Middle Jurassic Brent sandstone of northwest Europe. *Geology*, **17**, 361–364.

HAYES, J.B. 1979. Sandstone diagenesis: the hole truth. *In*: SCHOLLE, P.A. & SCHLUGER, P.R. (eds) *Aspects of Diagenesis*. SEPM Special Publications, **26**, 127–139.

HAYES, M.J. & BOLES, J.R. 1992. Volumetric relations between dissolved plagioclase and kaolinite in sandstones: implications for aluminium mass transfer in the San Joaquin basin, California. *In*: HOUSEKNECHT, D.W. & PITTMAN, E.D. (eds) *Origin, Diagenesis, and Petrophysics of Clay Minerals in Sandstones*. SEPM Special Publications, **47**, 111–123.

HELMOLD, K.P. & VAN DE KAMP, P.C. 1984. Diagenetic mineralogy and controls on albitization and laumontite formation in Paleogene arkoses, Santa Ynez Mountains, California. *In*: MCDONALD, D.A. & SURDAM, R.C. (eds) *Clastic Diagenesis*. AAPG Memoirs, **37**, 239–276.

HORTON, R.A., JR, MCCULLOUGH, P.T., & HANSON, D.M. 2009. Grain dissolution, compaction, and development of secondary intergranular porosity, western San Joaquin Basin, California. *In*: KNAUER, L.C. & BRITTON, A. (eds) *Contributions to the Geology of the San Joaquin basin, California*. Pacific Section AAPG Special Publications, **MP48**, 141–165.

ISAACSON, K.A. & BLUEFORD, J.R. 1984. Kreyenhagen Formation and related rocks – a history. *In*: BLUEFORD, J.R. (ed.) *Kreyenhagen and Related Rocks*. Pacific Section SEPM Publications, **37**, 1–7.

JIMENEZ-MILLAN, J., MOLINA, J.M., NIETO, F., NIETO, L. & RUIZ-ORTIZ, P.A. 1998. Glauconite and phosphate peloids in Mesozoic carbonate sediments (Eastern Subbetic Zone, Betic Cordilleras, SE Spain). *Clay Minerals*, **33**, 547–559.

LILLIS, P.G. & MAGOON, L.B. 2007. Petroleum systems of the San Joaquin Basin Province, California–Geochemical characteristics of oil types. *In*: SCHEIRER, A.H. (eds) *Petroleum Systems and Geologic Assessment of Oil and Gas in the San Joaquin Basin Province, California*. U.S. Geological Survey Professional Papers, Denver, Colorado, USA, **1713**, chapter 9.

MACGOWAN, D.B. & SURDAM, R.C. 1988. Difunctional carboxylic acid anions in oil-field waters. *Organic Geochemistry*, **12**, 245–259.

MCBRIDE, E.F. 1987. Diagenesis of the Maxon Sandstone (Early Cretaceous), Marathon region, Texas: a diagenetic quartzarenite. *Journal of Sedimentary Petrology*, **57**, 98–107.

McCulloh, T.H., Cashman, S.M. & Stewart, R.J. 1978. Diagenetic baselines for interpretive reconstructions of maximum burial depths and paleotemperatures in clastic sedimentary rocks. *In*: Oltz, D.F. (eds) *A Symposium in Geochemistry: Low Temperature Metamorphism of Kerogen and Clay Minerals*. Pacific Section SEPM Publications, 65–96.

Milliken, K.L., McBride, E.F. & Land, L.S. 1989. Numerical assessment of dissolution v. replacement in the subsurface destruction of detrital feldspars, Oligocene, south Texas. *Journal of Sedimentary Petrology*, **59**, 740–757.

Morad, S., Bergan, M., Knarad, R. & Nystuen, J.P. 1990. Albitization of detrital plagioclase in Triassic reservoir sandstones from the Snorre Field, Norwegian North Sea. *Journal of Sedimentary Petrology*, **60**, 411–425.

Niemi, T.M. & Hall, N.T. 1992. Late Holocene slip rate and recurrence of great earthquakes on the San Andreas fault in northern California. *Geology*, **20**, 195–198.

Noh, J.H. & Boles, J.R. 1993. Origin of zeolite cements in the Miocene sandstones, North Tejon oil fields, California. *Journal of Sedimentary Petrology*, **63**, 248–260.

Odin, G.S. & Letolle, R. 1980. Glauconiztion and phosphatization environments: a tentative comparison. *In*: Bentor, Y.K. (ed.) *Marine Phosporites–Geochemistry, Occurence, Genesis*. SEPM Special Publications, **29**, 227–237.

Olabisi, O.E. 2015. *Petrology and porosity development in Oligocene and Eocene sandstones of the Wasco oil field, San Joaquin basin, California, USA*. MS thesis, California State University, Bakersfield.

Olson, H.C. 1988. Oligocene-middle Miocene depositional systems north of Bakersfield, California: eastern basin equivalents of the Temblor Formation. *In*: Graham, S.A. (ed.) *Studies of the Geology of the San Joaquin Basin*. Pacific Section SEPM Publications, **60**, 189–205.

Olson, H.C., Miller, G.E. & Bartow, J.A. 1986. Stratigraphy, paleoenvironment and depositional setting of Tertiary sediments, southeastern San Joaquin Basin. *In*: Bell, P. (ed.) *Structure and Stratigraphy of the East Side, San Joaquin Valley*. Pacific Section AAPG, April 18–19, Bakersfield, California, 18–54.

Pettijohn, F.J., Potter, P.E. & Siever, R. 1987. *Sand and Sandstone*. 2nd edn. Springer, New York.

Reid, S.A. 1988. Late Cretaceous and Paleogene sedimentation along the east side of the San Joaquin basin. *In*: Graham, S.A. (ed.) *Studies of the Geology of the San Joaquin Basin*. Pacific Section SEPM Publications, **60**, 157–171.

Richardson, E.E. 1966. *Structure Contours on Top of the Vedder Sand, Southeastern San Joaquin Valley, California*. U.S. Geological Survey.

Scheirer, A.H. & Magoon, L.B. 2007. *Neogene Gas Total Petroleum System–Neogene Nonassociated Gas Assessment unit of the San Joaquin Basin Province*. U. S Geological Survey Professional Paper, 1713, chapter 22.

Schmidt, V. & McDonald, D.A. 1979. Texture and recognition of secondary porosity in sandstones. *In*: Scholle, P.A. & Schluger, P.R. (eds) *Aspects of Diagenesis*. SEPM, Special Publications, **26**, 209–225.

Scholle, P.A. 1979. *A Color Illustrated Guide to Constituents, Textures, Cements, and Porosities of Sandstones and Associated Rocks*. AAPG Memoirs, 28.

Schultz, J., Boles, J.R. & Tilton, G. 1989. Tracking calcium in the San Joaquin basin, California: a strontium isotopic study of carbonate cements at North Coles Levee. *Geochimica et Cosmochimica Acta*, **53**, 1991–1999.

Shanmugam, G. 1985. Significance of secondary porosity in interpreting sandstone composition. *AAPG Bulletin*, **69**, 378–384.

Stonecipher, S.A. 1999. Genetic characteristics of glauconite and siderite: implications for the origin of ambiguous isolated marine sand bodies. *In*: Bergman, K.M. & Snedden, J.W. (eds) *Isolated Shallow Marine Sand Bodies: Sequence Stratigraphic Analysis and Sedimentologic Interpretation*. SEPM Special Publications, **64**, 191–204.

Stonecipher, S.A. 2000. Applied sandstone diagenesis – practical petrographic solutions for a variety of common exploration, development and production problems. SEPM Short Course Notes, **50**.

Stauffer, P.H. 1966. Thin-section analysis: a further note. *Sedimentology*, **7**, 261–263.

Sullivan, J.C. & Weddle, J.R. 1960. Rio Bravo oil field. *In*: Department of Natural Resources Division of Oil and Gas (eds) *Summary of Operations, California Oil Fields*. Annual Report of the State Oil and Gas Supervisor, Sacramento, California, USA, **46**, 27–38.

Tapper, M. & Fanning, D.S. 1968. Glauconite pellets: similar X-Ray patterns from individual pellets of lobate and vermiform morphology. *Clays and Clay Minerals*, **16**, 275–283.

Taylor, B.L. 2007. *Petrography and diagenesis of the Eocene Point of Rocks Sandstone, McKittrich oil field, Kern County, California*. MS thesis, California State University, Bakersfield.

Wagoner, J. 2009. *3D Geologic Modeling of the Southern San Joaquin Basin for the Westcarb Kimberlina Demonstration Project – A Status Report*. Lawrence Livermore National Laboratory, LLTNL-TR-410813.

Wentworth, C.K. 1922. A scale of grade and class terms for clastic sediments. *Journal of Geology*, **30**, 377–392.

Wilkinson, M., Milliken, K.L. & Haszeldine, R.S. 2001. Systematic destruction of K-feldspar in deeply buried rift and passive margin sanstones. *Journal of the Geological Society*, **158**, 675–683.

Wilson, J.C. & McBride, E.F. 1988. Compaction and porosity evolution of Pliocene Sandstones, Ventura Basin, California. *AAPG Bulletin*, **72**, 664–681.

History of hydrothermal fluid flow in the midcontinent, USA: the relationship between inverted thermal structure, unconformities and porosity distribution

BRADLEY D. KING[1,2]* & ROBERT H. GOLDSTEIN[1]

[1]*Department of Geology, Kansas Interdisciplinary Carbonates Consortium, The University of Kansas, 1475 Jayhawk Boulevard, Lawrence, KS 66045, USA*

[2]*Present address: ConocoPhillips Company, 550 Westlake Park Boulevard, 12099 Two Westlake, Houston, TX 77079, USA*

**Correspondence: Brad.D.King@ConocoPhillips.com*

Abstract: A comprehensive study of the Cambrian–Ordovician Arbuckle Group suggests that multiple fluid migration events have affected reservoir porosity via fractures and preferred stratigraphic horizons. Fluid inclusion homogenization temperatures from late-stage precipitates yield temperatures higher than can be explained by burial conditions or an elevated geothermal gradient. Fluid inclusion melting temperatures yield salinity values that indicate multiple fluids evolving through time. Hydrocarbon fluid inclusions in late-stage baroque dolomite suggest oil migration concurrent with hydrothermal fluid flow. Depleted $\delta^{13}C$ and $\delta^{18}O$ values provide evidence for a high-temperature basinal fluid source as well as for the preferential flow of hydrothermal fluids through permeable zones in the Mississippian and Arbuckle Group, where pore systems related to paleokarst are overlain by less permeable units. Radiogenic strontium isotopic data support fluid–rock interaction with siliciclastic material or basement rock at some point during the fluid migration history. Variable $^{87}Sr/^{86}Sr$ values suggest multiple sources for the fluids responsible for the cements and a transition from an advective fluid flow system to a vertical fluid flow system. The ancient aquifer system was vertically connected during migration of hydrothermal fluids, and a temperature-controlled vertical density gradient appears to have played an important role in late-stage porosity evolution, focusing the hottest fluids in the upper sections of permeable layers.

The Cambrian–Ordovician Arbuckle Group is a deep saline aquifer within the Ozark Plateau aquifer system (OPAS) of south-central Kansas (Figs 1 & 2). The Arbuckle Group and age-equivalent units have been noted as having considerable porosity, which is believed to have been modified by dolomitization, karsting, fracturing and hydrothermal alteration (Kerans 1988; Gao & Land 1991; Kupecz & Land 1991; Montanez & Read 1992; Gao *et al.* 1995; Simo & Smith 1997; Franseen 2000; Franseen *et al.* 2004). The expulsion of fluids from the Anadarko basin, after extensive fluid–rock interaction, may be the source of the hydrothermal brines that are believed to be responsible for late-stage mineral precipitation (Fig. 1; Gao 1990; Gao & Land 1991; Musgrove & Banner 1993).

The mineral assemblage observed in the Arbuckle Group and overlying units is similar to that commonly observed in Mississippi Valley-type (MVT) deposits (Oliver 1986; Sverjensky 1986; Bethke & Marshak 1990; Garven 1993; Leach & Sangster 1995). Previous studies that correlated regional MVT deposits with basinal-sourced and orogenically influenced fluid flow models were used to help establish the timing, extent and control of hydrothermal fluid flow within the Arbuckle Group (Table 1; Sharp 1978; Cathles & Smith 1983; Garven & Freeze 1984*a*, *b*; Leach & Rowan 1986; Oliver 1986; Sverjensky 1986; Garven 1993; Leach & Sangster 1995; Coveney *et al.* 2000). The tectonic activity associated with the Ouachita (Late Paleozoic) and Laramide (Late Mesozoic–Early Cenozoic) orogenies, as well as the resultant structures, is hypothesized to drive topographic or tectonically controlled fluid flow, affecting strata in the study area (Figs 1 & 2).

The Wellington 1-32 and Vulcan cores, drilled in south-central Kansas (Fig. 1), provided the opportunity for core observations and petrographic analysis of the Arbuckle Group and overlying units. Petrographic methods include study of the core in hand sample, transmitted-light, reflected light, cathodoluminescence (CL) and UV epifluorescence petrography and back-scattered electron microscopy (BSEM). Petrographic observations have been combined with fluid inclusion analysis and carbon,

From: Armitage, P. J., Butcher, A. R., Churchill, J. M., Csoma, A. E., Hollis, C., Lander, R. H., Omma, J. E. & Worden, R. H. (eds) 2018. *Reservoir Quality of Clastic and Carbonate Rocks: Analysis, Modelling and Prediction.* Geological Society, London, Special Publications, **435**, 283–320.
First published online August 17, 2016, https://doi.org/10.1144/SP435.16

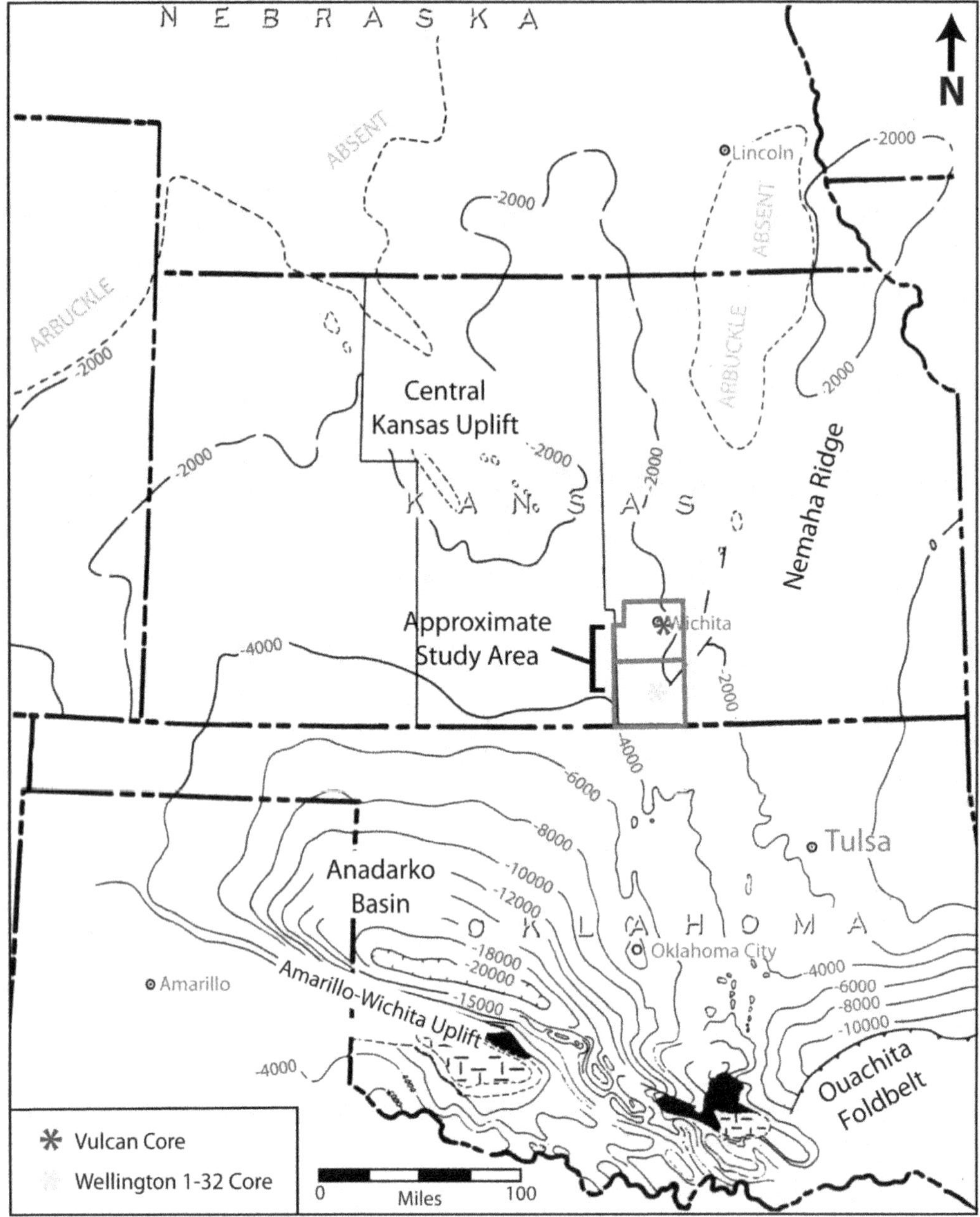

Fig. 1. Regional structural map on the top of the Arbuckle Group displaying the location of the Anadarko basin and the surrounding structural features relative to the study area. Modified from Franseen *et al.* (2004).

oxygen and strontium isotope analyses to aid in understanding the history and controls on fluid flow in the Arbuckle Group and overlying units. The integration of those data allow for the determination of the structure of the hydrothermal fluids, where warm fluids appear to have risen to the top of stratigraphic units during regionally advective fluid flow . A similar fluid and thermal structure may give rise to increased hydrothermal alteration of the stratigraphic intervals immediately below unconformities and other stratigraphic discontinuities in many areas around the world.

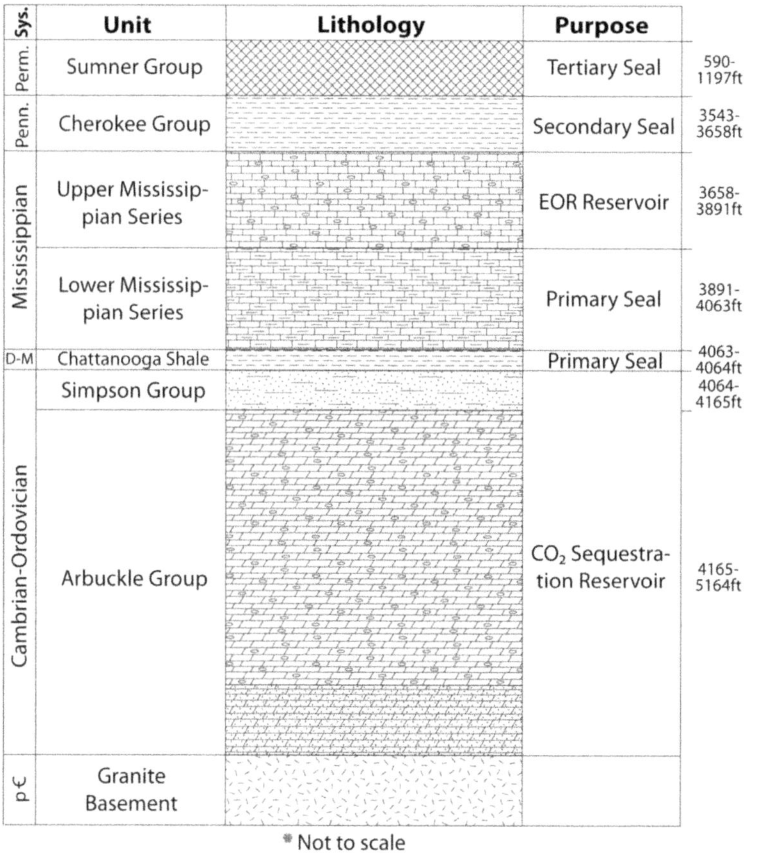

Sys.	Unit	Lithology	Purpose	
Perm.	Sumner Group		Tertiary Seal	590-1197ft
Penn.	Cherokee Group		Secondary Seal	3543-3658ft
Mississippian	Upper Mississippian Series		EOR Reservoir	3658-3891ft
Mississippian	Lower Mississippian Series		Primary Seal	3891-4063ft
D-M	Chattanooga Shale		Primary Seal	4063-4064ft
Cambrian-Ordovician	Simpson Group			4064-4165ft
Cambrian-Ordovician	Arbuckle Group		CO_2 Sequestration Reservoir	4165-5164ft
p€	Granite Basement			

* Not to scale

Fig. 2. Stratigraphic section of Precambrian through Permian strata, displaying units of note in this study and the larger scale CO_2 sequestration project. Depth markers correlate with depths observed in the Wellington 1-32 core (based on Scheffer 2012).

The evolution of porosity and late fluid history of the Arbuckle Group is the focus of this study. Interpreting the complex relationship between diagenesis, fluid flow and structural deformation enables the identification and characterization of fluid migration within the Arbuckle Group, as well as the analysis of potential connectivity with overlying units. Establishing a correlation between diagenetic events and the potential controls that produce the events can enhance understanding of the timing, control and distribution of porosity evolution in the Arbuckle Group, as well as similar reservoirs around the world.

Geological background

The Cambrian–Ordovician Arbuckle Group is present in the subsurface throughout the majority of Kansas, with the exception of the northeastern and northwestern portions of the state that are associated with ancient uplifts (Franseen *et al.* 2004; Fig. 1). The unit unconformably rests on either Precambrian or Cambrian strata and is topped by a major regional unconformity that is overlain by Middle Ordovician, Mississippian or Pennsylvanian strata (Fig. 2; Franseen *et al.* 2004). In general, the unit gently dips to the south, with a westerly dip on the east side of the Nemaha anticline in the SE part of the state (Fig. 1; Franseen *et al.* 2004). The unit thickens to the south, with the thickest portions occurring in SE Kansas (Franseen *et al.* 2004).

The Arbuckle Group was deposited as part of the Sauk Sequence and is a major component of the Great American Bank. Franseen *et al.* (2004) described the Arbuckle Group as shallow-shelf carbonate strata, consisting primarily of coarse-grained skeletal, intraclastic, oolitic, peloidal, dolograinstones/packstones, finer-grained mudstones, thrombolites and stromatolites, intraclastic conglomerate and breccia and minor shale and siltstone. Following a fall in sea-level in the Early Ordovician the unit

Table 1. *Summary of regional-scale hydrothermal systems where similar mineral assemblages have been observed in the study area (south Kansas) and elsewhere in the USA, including direct dates of late-stage cements*

Region of study	Hydrothermal fluid flow mechanism	Specific driving forces	Interpreted age of hydrothermal deposit	Direct dates	Reference
SE Kansas	Late Palaeozoic Ouachita orogeny	Progressively deeper thrust faulting	Late Pennsylvanian–Early Permian		Young (2010)
	Late Palaeozoic Ouachita orogeny		Late Pennsylvanian		Wojcik *et al.* (1992, 1994, 1997)
Tri-State	Late Palaeozoic–Late Mesozoic tectonic activity	Expelled pore fluids or gravity-driven flow	Pennsylvanian–Late Cretaceous	251 ± 11 Ma; 137 ± 3 Ma; 66 ± 2 Ma	Coveney *et al.* (2000)
	Late Palaeozoic Alleghenian–Ouachita orogeny	Gravity-driven flow	Late Permian–Early Triassic	251 ± 11 Ma	Brannon *et al.* (1996*b*)
	Late Palaeozoic Ouachita orogeny		>Late Jurassic	>183 Ma	Arne (1992), Leach *et al.* (2001)
SE Missouri	Late Palaeozoic Alleghenian–Ouachita orogeny		Late Pennsylvanian–Early Permian	286 ± 20 Ma	Wisniowiecki *et al.* (1983), Leach *et al.* (2001)
	Late Palaeozoic Alleghenian–Ouachita orogeny		Early Permian	273 ± 10 Ma	Symons *et al.* (1998*a*), Leach *et al.* (2001)
			Late Permian–Eocene	241 ± 10 Ma; 203 ± 10 Ma; 100–50 Ma	Arne *et al.* (1990), Leach *et al.* (2001)
			Devonian	392 ± 21 Ma	Lange *et al.* (1983), Leach *et al.* (2001)
North Arkansas	Late Palaeozoic Alleghenian–Ouachita orogeny		Permian	265 ± 20	Pan *et al.* (1990), Leach *et al.* (2001)
Central Missouri	Late Palaeozoic Alleghenian orogeny		Pennsylvanian	303 ± 17 Ma	Symons & Sangster (1991), Leach *et al.* (2001)
Central Tennessee	Late Palaeozoic Alleghenian orogeny		Late Permian–Early Triassic	245 ± 10 Ma	Lewchuck & Symons (1996)
	Late Palaeozoic Alleghenian orogeny	Gravity-driven flow	Late Permian	260 ± 42 Ma	Brannon *et al.* (1996*a*)
East Tennessee	Late Palaeozoic Alleghenian orogeny		Late Pennsylvanian–Early Permian	286 ± 20 Ma	Bachtadse *et al.* (1987), Leach *et al.* (2001)

(*Continued*)

Table 1. *Continued*

Region of study	Hydrothermal fluid flow mechanism	Specific driving forces	Interpreted age of hydrothermal deposit	Direct dates	Reference
	Late Palaeozoic Alleghenian orogeny		Pennsylvanian	316 ± 8 Ma	Symons & Stratakos (2000)
	Middle Palaeozoic Acadian orogeny	Thrust faulting	Devonian	377 ± 29 Ma	Nakai *et al.* (1990), Leach *et al.* (2001)
	Middle Palaeozoic Acadian orogeny	Thrust faulting	Mississippian	347 ± 20 Ma	Nakai *et al.* (1990, 1993), Leach *et al.* (2001)

was subaerially exposed, allowing for deep weathering and erosion to produce the regionally extensive Arbuckle–Ellenburger–Knox–Prairie du Chien–Beekmantown–St. George karst plain; this extensive karsting generated caves, joint-controlled solution features and collapse breccias throughout the Arbuckle Group and age-equivalent units (Kerans 1988; Kupecz & Land 1991; Montanez & Read 1992; Simo & Smith 1997; Franseen *et al.* 2004). The unit was subsequently buried and fractured by structural deformation probably associated with the Ouachita and possibly the Laramide orogenies (Franseen *et al.* 2004). Previous studies have found evidence for warm, basin-derived brines modifying the porosity of the Arbuckle Group relatively late in the established paragenesis (Gao *et al.* 1995). Evidence for regional-scale precipitation of similar mineral assemblages, related to hydrothermal activity and potential basin dewatering, has been observed in areas affected by MVT deposition (Table 1; Cathles & Smith 1983; Garven & Freeze 1984*a*, *b*; Leach & Rowan 1986; Oliver 1986; Sverjensky 1986; Bethke & Marshak 1990; Wojcik *et al.* 1992; Garven 1993; Leach & Sangster 1995; Young 2010).

Paragenesis

The paragenetic sequence of the Arbuckle Group can be separated into 23 major diagenetic events. The events have been divided into an early and late diagenetic stage, relative to the onset of burial. The timing and effects on porosity have been summarized in Figure 3. For the purpose of this article, the focus will be on events 14–23, which are

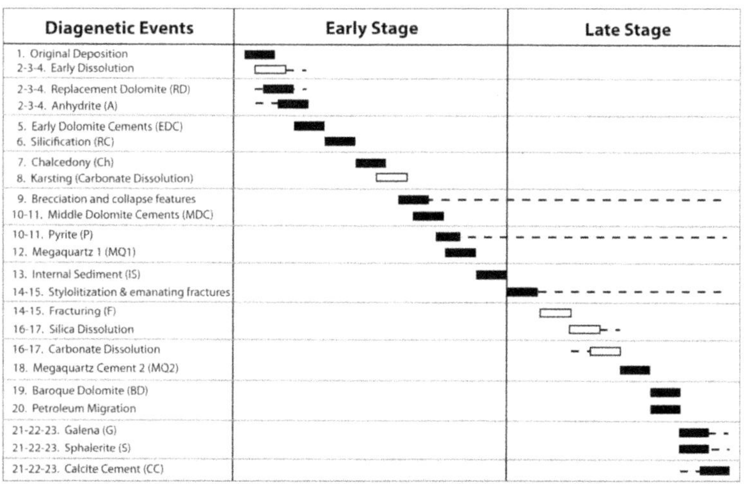

Fig. 3. Paragenesis of the Arbuckle Group with events separated into early and late stages. Porosity evolution is illustrated with solid boxes indicating a decrease in porosity and empty boxes indicating an increase in porosity. Dashed lines indicate a level of uncertainty regarding the duration of some events.

grouped into the late diagenetic stage because they formed after burial and compaction. The late diagenetic stage includes evidence for hydrothermal alteration, with mineral assemblages similar to MVT deposits. Pore types are classified using Choquette & Pray (1970). Folk (1965) was used to describe calcite textures and Gregg & Sibley (1984) was used to describe dolomite textures. Any percentages provided are visual estimates from the core and should be viewed with uncertainty.

Early diagenetic stage

The early diagenetic stage consists of 13 events that affected porosity, all occurring prior to burial. Primary porosity has been preserved within the section in some instances, mostly in the form of fenestral and interparticle porosity (e.g. Franseen *et al.* 2004). Early replacement dolomite (RD; events 2–4) appears to have decreased overall porosity, replacing existing carbonate sediment and reducing interparticle pores. Early dissolution events, probably due to intraformational sea-level oscillations, temporarily enhanced porosity and created vugs and caverns (Ross & Ross 1988; Franseen *et al.* 2004). These early dissolution features appear to have created preferential fluid flow conduits for the fluids responsible for subsequent precipitation of cements, and are reduced by early and late diagenetic cements. At the culmination of Arbuckle Group deposition, a regionally extensive subaerial exposure event during the Middle Ordovician generated substantial porosity – particularly in the upper portions of the unit – through extensive erosion, karsting and collapse breccias (Kerans 1988; Gao & Land 1991; Kupecz & Land 1991; Montanez & Read 1992; Gao *et al.* 1995; Simo & Smith 1997; Franseen *et al.* 2004). Pores associated with this event are commonly reduced by cements and probably acted as enhanced fluid flow conduits for the fluids that precipitated those cements.

Late diagenetic stage

Events 14–15

Stylolitization and emanating fractures. The burial of the Arbuckle Group led to compaction and stylolitization that has continued to develop over time. Stylolites are present throughout the entire unit. Nearly all of the stylolites observed display fractures that emanate from peaks. Fractures and stylolites cross-cut early cements (events 10–11) and can be found occluded by baroque dolomite (event 19) (Fig. 4c). Silica dissolution associated with events 16–17 can be observed along fractures emanating from stylolites (Fig. 4d). The requirement for burial and the presence of late cements filling fractures tentatively places the onset of stylolitization as the first late-stage diagenetic event, with the development of pressure solution features as sediment load increased and time progressed (Machel 1999; Mountjoy *et al.* 1999; Heimstra 2003). Stylolitization probably reduced overall porosity, whereas emanating fractures locally enhanced it. Precipitation of late cements in fractures has destroyed the majority of the porosity created.

Fracturing. Vertical and sub-vertical fractures (F) can be observed, and are typically partially to fully occluded with later cements (events 18–23); fractures are solution enhanced in some examples (Fig. 5a–d). Fractures are typically isolated, relatively short and bedding-constrained. Fractures cross-cut middle dolomite cements (MDC) and megaquartz 1 (MQ1) and are reduced with only late cements, allowing for an approximate placement in the late diagenetic history of the Arbuckle Group (Fig. 5a–d). Fractures associated with this event may correlate with deformation caused by the Late Palaeozoic Ouachita orogeny. However, the lack of petrographic observations suggesting a later fracturing event reveals the possibility for a genetic link between at least some of the fractures and the Late Mesozoic–Early Cenozoic Laramide orogeny.

Events 16–17

Silica dissolution. The dissolution of chert and chalcedony along fractures – some of which appear to be associated with stylolites (events 14–15) – provides evidence for a dissolution event increasing porosity by <5% in chert and chalcedony. The porosity resulting from this event can be observed throughout the core. Tripolitic chert is seen in both cores; the dissolution of chalcedony can be observed along fractures that emanate from stylolites (Figs 4d & 6a, b). The fact that dissolution is observed propagating from fractures associated with stylolites requires that at least some of the dissolution occurred after burial and compaction, making low-temperature meteoric diagenesis unlikely for at least some of the dissolution observed. Silica dissolution has been proposed by others as possibly being genetically linked with the onset of hydrothermal activity in the study area (Young 2010; Ramaker *et al.* 2014).

Carbonate dissolution. Relatively late dissolution of carbonate material is indicated by large (>10 mm) vugs that lack early cements but display partial to full occlusion with late cements (events 18–23) (Fig. 6c, d). Porosity resulting from this event can be observed throughout the cores and has enhanced porosity by <1%. The lack of early-stage cements, more specifically cements that are observed in porosity believed to be created by the dissolution associated with the major regional

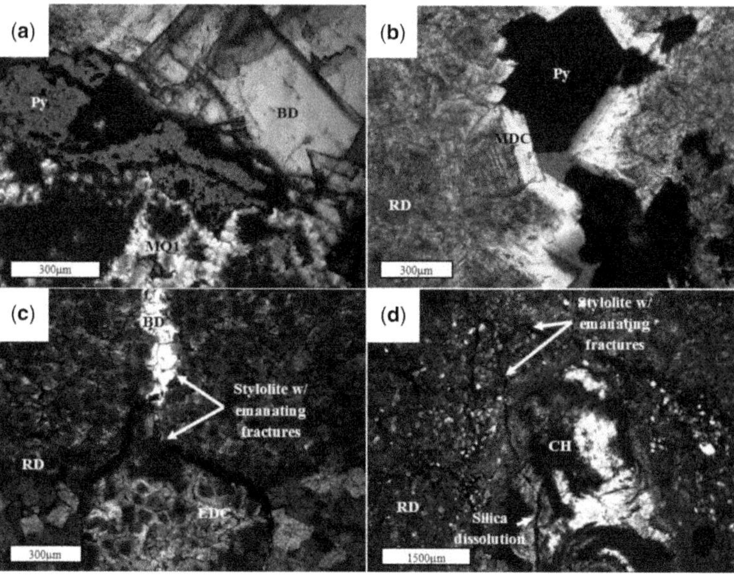

Fig. 4. (**a**) Combination of bright-field and crossed-polarized light photomicrograph of pyrite (Py; events 10–11) after MQ1 (event 12), followed by precipitation of BD (event 19) (sample 4977.7 from Wellington 1-32); (**b**) Transmitted light photomicrograph of vug likely associated with event 8 dissolution, lined with MDC (events 10–11) and then pyrite (events 10–11) (sample 5070.6 from Wellington 1-32); (**c**) Transmitted light photomicrograph of RD (events 2–4) and early dolomite cements (EDC; event 5) truncated by stylolite and emanating fractures (events 14–15), fractures are subsequently filled with BD (event 19) (sample 4460.7 from Wellington 1-32); (**d**) Transmitted light photomicrograph of a stylolite cross-cutting RD (event 2–4) material, an emanating fracture cross-cuts chalcedony (Ch; event 7) with silica dissolution (events 16–17) occurring along the fracture (sample 5070.6 from Wellington 1-32).

subaerial exposure event (event 8), suggests that this porosity must have occurred after burial and during late-stage diagenesis. The close association of this event with the precipitation of megaquartz cement 2 (event 18) and baroque dolomite (event 19) suggests that carbonate dissolution may be linked to the introduction of the fluids responsible for these events.

Event 18

Megaquartz cement. Megaquartz cement 2 (MQ2) can be observed in samples throughout the cores, decreasing porosity by $<1\%$. MQ2 is present as euhedral quartz crystals lining fractures and vugs (Figs 5b, 6b, c & 7a). MQ2 is clear in transmitted light and has small (<5 µm) all-liquid and two-phase fluid inclusions that feather out in the direction of crystal growth (see ‘Fluid Inclusion’ section). MQ2 is non-luminescent under CL. MQ2 precipitation occurred in porosity associated with events 16–17 and can be observed precipitating into pore space after infiltration of internal sediment (Fig. 7a), positioning it late in the paragenesis. The presence of baroque dolomite (event 19) after MQ1 brackets this event between events 16–17 and event 19 (Fig. 4b & 7b).

Event 19

Baroque dolomite cement. Baroque (saddle) dolomite (BD) is the most common late-diagenetic cement observed in the Arbuckle Group and can be observed in samples throughout the unit in the Wellington 1-32 core and in the upper Arbuckle Group recovered in the Vulcan core. This event reduces porosity by $<5\%$. BD can be found partially to fully occluding fractures and vugs occurring early in the paragenesis and as late as events 16–17 (Fig. 5a–d). Replacement of material not previously dolomitized appears to be common in areas that fractured later in the paragenesis (events 14–15). In thin section, BD displays a xenotopic-C texture with pore- and fracture-lining saddle-shaped crystals characterized by curved crystal terminations and sweeping extinction under crossed polars (Figs 5b & 7d). BD crystals are clear under transmitted light, with cloudy cores and growth bands with a high abundance of fluid inclusions (Fig. 13a, b). BD is the only cement in the Arbuckle Group with recognizable hydrocarbon fluid inclusions (Fig. 13c, d). BD is non-fluorescent under UV illumination and very dully concentrically banded under CL with a thin double band of bright orange luminescence in some examples (Fig. 8a–d). When immersed in an

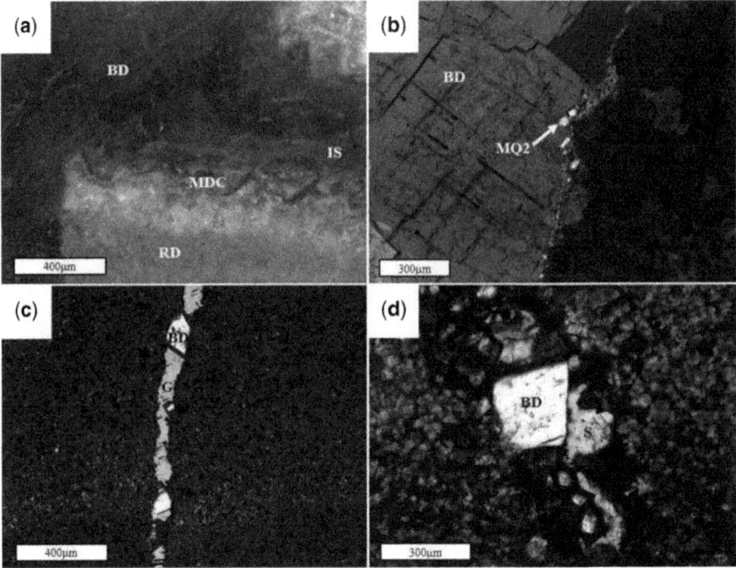

Fig. 5. (**a**) UV epifluorescence photomicrograph of an early vug lined with MDC (events 10–11) overlain by internal sediment (IS; event 13). The vug is subsequently fractured (events 14–15) and porosity is occluded with BD (event 19) (sample 4984.4B from Wellington 1-32); (**b**) Crossed-polarized light photomicrograph of dolomitized material fractured (events 14–15) and then lined with MQ2 (event 18) and then BD (event 19) (sample 4470.9A from Wellington 1-32); (**c**) Bright-field photomicrograph of carbonate material that had been cut by a fracture (events 14–15) and filled by BD (event 19) followed by galena (G; events 21–23) (sample 4324.9 from Wellington 1-32); (**d**) Transmitted light photomicrograph of dolomitized material fractured (event 14–15) and subsequently filled with BD (event 19) and then sphalerite (S; events 21–23) (sample 4325.3 from Wellington 1-32).

Alizarin Red S:Potassium Ferrocyanide (ARS:PF) solution, BD from the Vulcan core stains blue (ferroan), whereas BD from the Wellington core does not take a stain (non-ferroan). BD can be observed in pore space following MQ2 (Fig. 5b & 7b) and prior to calcite, sphalerite and galena (Fig. 5c, d & 7c, d). The textures and morphologies of the BD crystals are indicators of high temperatures (60–150°C) at the time of precipitation (Radke & Mathis 1980; Gregg & Sibley 1984; Davies & Smith 2006).

Event 20

Petroleum migration. Petroleum fluid inclusions are observed within BD (event 19), and staining along fractures associated with events 14–15 also fluoresces under UV illumination. Petroleum fluid inclusions are present in pseudosecondary or secondary, as well as primary, fluid inclusion assemblages (FIAs) in BD from the Arbuckle Group (Fig. 13c, d), suggesting migration during BD precipitation and possibly for some time after.

Events 21–23

Galena. Galena is also a relatively rare precipitate within the Arbuckle Group and was observed in only two samples near the top of the unit (158 ft from the top of unit in Wellington 1-32 and 12 ft from the top of the unit in Vulcan), decreasing porosity by <0.5%. Galena exists as large cubic crystals that are present in open pore space and fractures (Fig. 5c). Galena can be observed precipitating into an open fracture (events 14–15) following the precipitation of BD (Fig. 5c), but no paragenetic relationships were observed between galena, sphalerite or calcite.

Sphalerite. Sphalerite is also a relatively rare precipitate within the Arbuckle Group and was only observed in two samples in the upper portion of the Wellington 1-32 core (159 and 255 ft from the top of the unit), decreasing porosity by <0.5%. Sphalerite exists as coarse crystals that appear yellow-brown in the core and are yellow in thin section (Fig. 7d). A fine-grained ore with small dolomite rhombs precipitates along with the sphalerite cement (Fig. 7d). Sphalerite can be observed precipitating after BD (Figs 5d & 7d) into open fractures but no paragenetic relationships were observed between sphalerite, galena or calcite. The absence of petroleum fluid inclusions in sphalerite constrains the migration of petroleum as occurring prior to sphalerite precipitation.

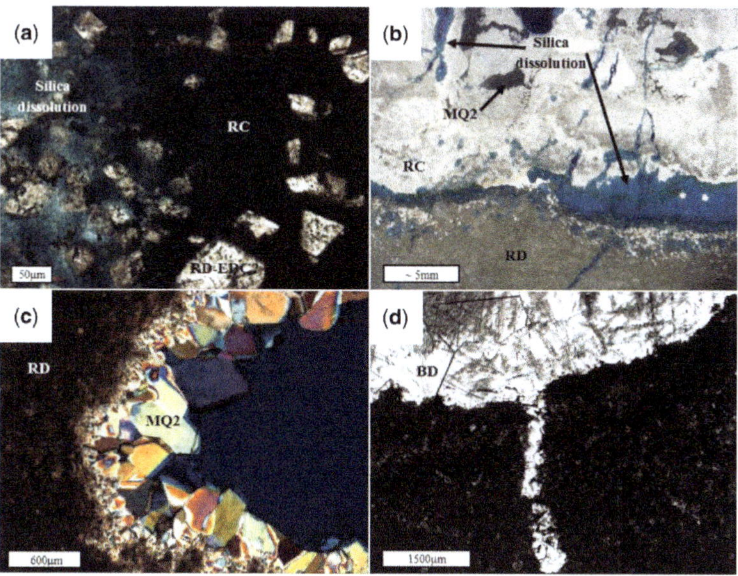

Fig. 6. (**a**) Transmitted light photomicrograph of carbonate material that appears to have been partially dolomitized (RD; event 2–4) or EDC (event 5)) then silicified (RC; event 6) and finally affected by silica dissolution (events 16–17) that produced tripolitic chert (sample 3-17 (3967.3 ft) from the Vulcan core); (**b**) Reflected light photomicrograph of RD (events 2–4) that was completely silicified (RC; event 6) in the upper portion of photo, subsequent fracturing (events 14–15) allowed for silica dissolution (events 16–17) and then precipitation of MQ2 (event 18) (sample 4401.7 from Wellington 1-32); (**c**) Crossed-polarized light photomicrograph of RD (events 2–4) subjected to dissolution associated with events 16–17. The vug is then lined with MQ2 (event 18) (sample 3-5A (4005.5 ft) from the Vulcan core); (**d**) Transmitted light photomicrograph of dolomitized material that is fractured (events 14–15), followed by late dissolution probably associated with events 16–17 and subsequently filled with BD (event 19) (sample 5081.9 from Wellington 1-32).

Calcite cement. Calcite cement (CC) is a minor precipitate in the Arbuckle Group and was only observed in four samples (3 in the Wellington 1-32 core and 1 in the Vulcan core), decreasing porosity by $<1\%$. The samples span from the base to the top of the unit. CC precipitates as very coarse, isolated crystals of equant spar (Fig. 7c). CC is clear in thin section with deformation twins and partially healed microfractures common. CC is a light yellow luminescent in CL, but shows no zoning or banding (Fig. 8c, d). This cement can be observed filling pore space after BD (Fig. 7c) but no relationships were observed between calcite, galena or sphalerite. Work by other authors has demonstrated that calcite typically occurs as the last stage of cement precipitation in late-stage deposits related to hydrothermal activity (Hagni & Grawe 1964; Voss *et al.* 1989; Garven 1993; Brannon *et al.* 1996*a*; Coveney *et al.* 2000), tentatively placing calcite as the last diagenetic event in the paragenesis. The absence of petroleum fluid inclusions in calcite cement also helps to bracket the migration of petroleum as occurring prior to calcite precipitation.

BSEM analysis

BSEM analysis was performed on four BD samples, with at least one sample obtained from the bottom, middle and top of the Arbuckle Group. This was performed to support or refute the potential for BD recrystallization and to illustrate the impact recrystallization may have had on microthermometric data obtained from fluid inclusions. The images collected reveal at least four separate dolomite growth zones with rare patches of calcite, as well as extensive fracturing and subsequent recrystallization associated with what appears to be the latest stage of crystal growth (Fig. 9). Additional growth zones may be present, but these were difficult to discern with certainty throughout the samples.

The four distinct zones consist of an early dark-grey zone (EDZ), a middle light-grey zone (MLZ), a middle medium-grey zone (MMZ) and a late dark-grey zone (LDZ). The LDZ is a thin growth band that is approximately 40 µm thick; this zone appears to be the last stage of BD crystal growth. Because this growth band was so thin, it was only recognized in one sample. The dark-grey BSE of this zone

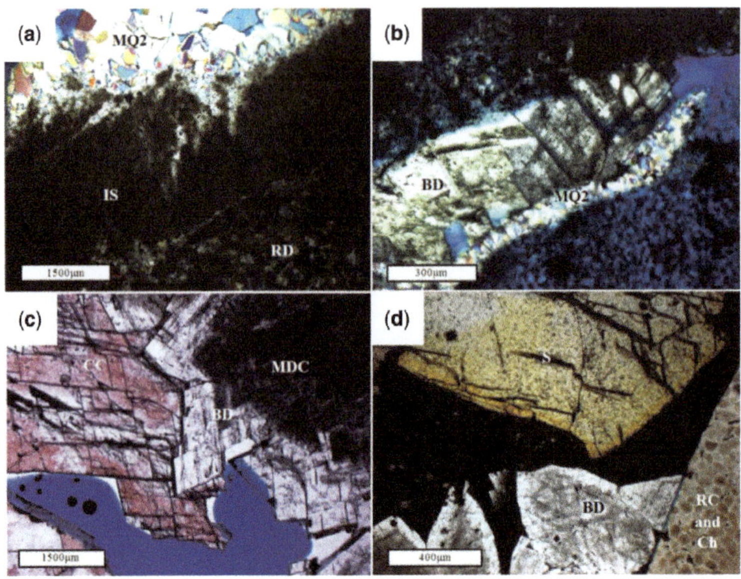

Fig. 7. (**a**) Crossed-polarized light photomicrograph of RD (events 2-3-4) affected by dissolution that is likely associated with event 8. Porosity is subsequently filled with IS (event 13) and then precipitation of MQ2 (event 18) (sample 4806.0 from Wellington 1-32); (**b**) Crossed-polarized light photomicrograph of silica dissolution (events 16–17) enhancing porosity, porosity is occluded with MQ2 (event 18) and then BD (event 19) (sample 4966.5 from Wellington 1-32); (**c**) Transmitted light photomicrograph of brecciated material that is lined with MDC (events 10–11) and then BD (event 19), with CC (events 21–23) that has taken an ARS stain reducing porosity last (sample 5061.5A from Wellington 1-32); (**d**) Transmitted light photomicrograph of silicified material (replacement chert (RC; event 6) and chalcedony (Ch; event 7)) that is fractured (events 14–15) and then filled with BD (event 19), sphalerite (S; events 21–23) and fine-grained ore (opaque) associated with sphalerite appearing to precipitate into the fracture following BD (sample 4421.1B from Wellington 1-32).

appears to correspond with the dark-grey BSE of the recrystallized fabric associated with microfractures, suggesting recrystallization of earlier zones occurred during precipitation of this zone as a cement; if this is the case, then the LDZ is actually present in all samples. Fluid inclusions are most abundant in this zone and are concentrated in recrystallized lineaments along microfractures that appear to be associated with this last stage of dolomite precipitation and recrystallization (Fig. 9).

Fluid inclusion analysis

The following section discusses fluid inclusion petrography and microthermometry of late-stage cements, including MQ2, BD and CC. The methodology and terminology proposed by Goldstein & Reynolds (1994) is used for fluid inclusion analysis. The term FIA is used throughout and is defined as the most finely discriminated, petrographically distinguishable group of inclusions within a given mineral phase (Goldstein & Reynolds 1994; Goldstein 2003). Goldstein & Reynolds (1994) define a consistent FIA as one in which $\geq$90% of homogenization temperature (T_h) data fall within a range of 10–15°C; and an inconsistent FIA as one that yields more variable T_h data. Consistent T_h data are interpreted to mean that thermal reequilibration of fluid inclusions has not occurred (Goldstein & Reynolds 1994). The microthermometric data comprise T_h and final melting temperatures of ice (Tm_{ice}). Tm_{ice} values are translated to salinity in wt% NaCl equivalent (wt% NaCl eq.) using the AqSoVir program from Bakker (2008) and the equation from Bodnar (1993):

$$\text{Salinity} = 0.00 + 1.78\theta - 0.0442\theta^2 + 0.000557\theta^3 \quad (1)$$

where θ equals the depression of the freezing point in °C (Bodnar 1993). Tm_{ice} data were only collected from fluid inclusions with gas bubbles present before final melting. T_h data were preferentially measured from fluid inclusions that appeared to result from homogeneous entrapment or were likely the liquid-rich end member of an FIA with evidence for heterogeneous entrapment (Goldstein & Reynolds 1994). The first melting temperature of several

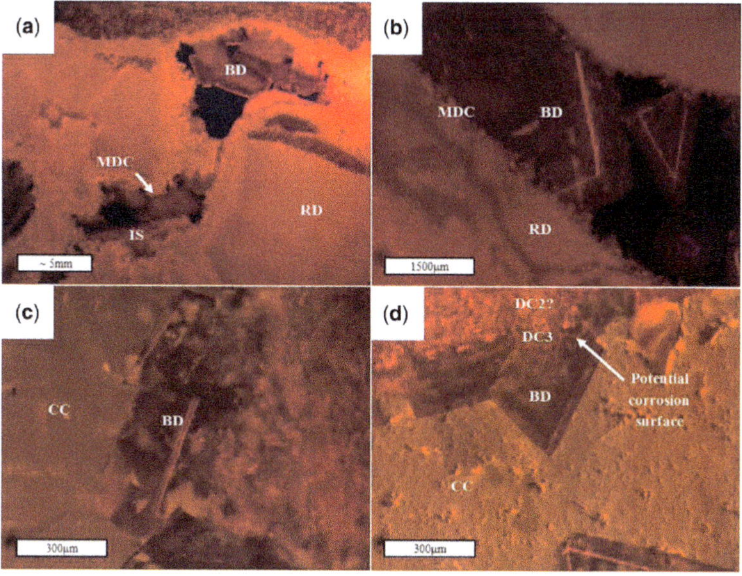

Fig. 8. All images are CL photomicrographs: (**a**) RD (events 2–4) was affected by dissolution associated with event 8, porosity was then lined with MDC (events 10–11), then infiltrated with IS (event 13) and subsequently lined with BD (event 19) displaying thick dull-red bands and the double red-orange band as the last stage of crystal growth (sample 4524.8 from Wellington 1-32); (**b**) Brecciated material (event 9) lined with MDC (events 10–11) and then filled with BD (event 19) displaying dull-red bands and only one band from the double red-orange band (sample 4755.4 from Wellington 1-32); (**c**) Pore lined with BD (event 19) displaying a double red-orange band, followed by the dull-yellow luminescence of CC (events 21–23) (sample 3-14 (3979.1 ft) from the Vulcan core); (**d**) Porosity lined with early dolomite cement 2 (DC2; event 5), early dolomite cement 3 (DC3; events 10–11) and BD (event 5), with a potential corrosion surface between DC3 and BD; CC (events 21–23) subsequently fills porosity (sample 3-11 (3983.2 ft) from the Vulcan core).

samples was approximated in some instances and is considered to have some degree of uncertainty due to petrographic limitations. An interpretation of fluid inclusion data is provided for each diagenetic event.

MQ2

Petrography. MQ2 is clear under transmitted light and rarely contains primary FIAs distributed in feather-like patterns, oriented in the direction of crystal growth (Fig. 10a, b). Fluid inclusions are typically small (<5 μm) and rounded, with slight elongation in the direction of crystal growth (Fig. 10a, b). Both two-phase inclusions (aqueous liquid and vapour) and small, all-liquid aqueous inclusions are present (Fig. 10b). The small all-liquid inclusions are considered to represent metastability (failure to nucleate a vapour bubble due to the small inclusion size) rather than low temperatures of entrapment. Liquid:gas ratios among inclusions in a single FIA appear inconsistent and there are some gas-dominated inclusions (Fig. 10a, b). Petrographic pairing is not observed in fluid inclusions, suggesting that necking-down after a phase change has not caused the variability in the liquid:gas ratio (Goldstein & Reynolds 1994).

Microthermometry. The general paucity of fluid inclusions limited the number of microthermometric measurements. The relatively small size (<5 μm) of inclusions led to petrographic limitations and it was only possible to collect T_h data from a total of nine inconsistent primary FIAs. The T_h measurements varied from 87.0–157.0°C (Fig. 12).

Tm_{ice} data were collected from five inconsistent primary FIAs. Tm_{ice} occurred between −1.8 and −3.7°C (Fig. 11), corresponding with salinities ranging from 3.1 to 6.0 wt% NaCl eq.

Interpretation. If the small, all-liquid inclusions are considered to reflect metastability, then the T_h data from the two-phase inclusions within inconsistent FIAs (87.0–157.0°C) suggest that megaquartz precipitation is most likely dominated by high-temperature conditions. Inconsistent liquid:gas ratios in FIAs with gas-dominated inclusions and no evidence for petrographic pairing strongly suggest entrapment during heterogeneous conditions. Because the fluid inclusions within FIAs contain

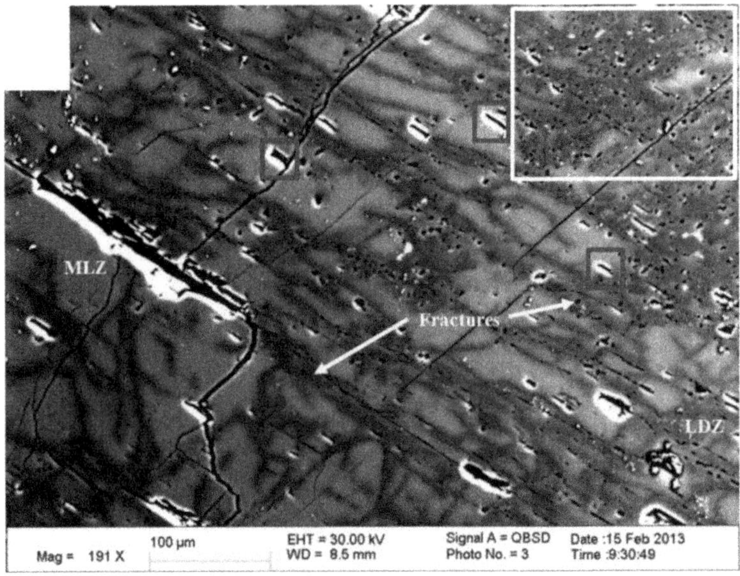

Fig. 9. BSEM high-magnification photomicrograph of the MLZ showing extensive fracturing and subsequent recrystallization by the LDZ, earlier fluid inclusions are larger and present in earlier MLZ (highlighted with black (print)/red (online) squares), while later, smaller fluid inclusions are present along fractures and latest stage of recrystallization (white square).

liquid-rich and gas-rich end members of the heterogeneous system, liquid-rich end members would provide an estimate of temperatures during entrapment (Goldstein & Reynolds 1994); in an attempt to collect T_h data from liquid-rich end members (two-phase fluid inclusions with the smallest gas bubbles) it is inevitable that data collection from inclusions that formed with miniscule amounts of a separate gas phase would have occurred. If this is the case, the lowest measured temperatures (*c.* 90°C) would represent the most realistic estimate of entrapment temperature, whereas the higher temperature range would represent heterogeneous entrapment of gas along with liquid.

Salinities measured in MQ2 are high enough (3.1–6.0 wt% NaCl eq.) to either suggest some dissolution of evaporites or connate seawater and evaporated seawater was present in the basin. When considering the elevated T_h measurements, a basin-sourced fluid is the best candidate to account for high-temperature, moderate-salinity fluids (Hanor 1979; Young 2010).

BD

Petrography. BD is typically clear in thin section with a cloudy core and concentric growth bands defined by high abundances of primary fluid

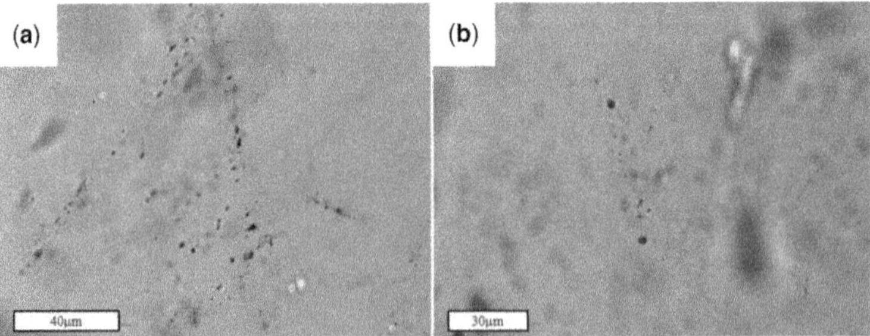

Fig. 10. (**a**) Feather-like distribution of primary FIAs parallel to the direction of growth in MQ2, with variable liquid:gas ratios (sample 5083.7 from Wellington 1-32). (**b**) Fluid inclusions displaying gas-rich and liquid-rich end members in primary FIAs within MQ2 (sample 3-5A (4005.5 ft) from Vulcan).

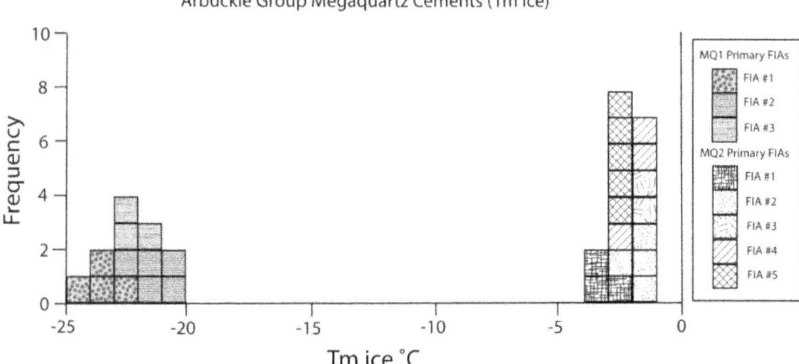

Fig. 11. Frequency distribution of Tm_{ice} of MQ1 and MQ2. The most noticeable trend is the drastic decrease in salinity from early diagenetic cement (MQ1) to late (MQ2).

inclusions (Fig. 13a, b). Fluid inclusions within primary FIAs are typically two-phase inclusions with a small vapour bubble in aqueous liquid. They vary in size and shape; liquid:gas ratios are consistent among fluid inclusions in each FIA (Fig. 13b). All-liquid (aqueous) inclusions can be observed within primary FIAs but are only the relatively small-sized inclusions of the FIA (<5 μm). Pseudosecondary or secondary FIAs in BD in the Vulcan core samples occur as curviplanar, 3D features that cross-cut dolomite crystals. Fluid inclusions within pseudosecondary or secondary FIAs occur as two-phase inclusions (aqueous liquid with vapour bubbles) and all-liquid (aqueous) inclusions. Some pseudosecondary or secondary FIAs contain two-phase inclusions with consistent liquid:gas ratios, whereas others have variable ratios mixed with all-liquid inclusions. Petrographic pairing is not present in primary FIAs but can be clearly observed in pseudosecondary or secondary FIAs where large all-liquid inclusions are present, indicating that necking-down after a phase change may be responsible for the all-liquid inclusions (Goldstein & Reynolds 1994). Additionally, BD contains UV-fluorescent, two-phase hydrocarbon fluid inclusions in pseudosecondary or secondary FIAs entrapped during later recrystallization of BD rhombs; there are also some hydrocarbon inclusions that may be primary in origin (Fig. 13c, d). The hydrocarbon inclusions are clear to light-yellow under transmitted light and fluoresce bright blue under UV illumination (Fig. 13c, d).

No patchiness was observed in BD when using CL, but BSEM analysis revealed extensive evidence of recrystallization along microfractures in earlier growth zones during the precipitation of the latest growth zone (Figs 8a–d & 9). Fluid inclusions are most concentrated in the LDZ, preferentially forming during recrystallization of the earlier growth zones, either along microfractures or vague growth

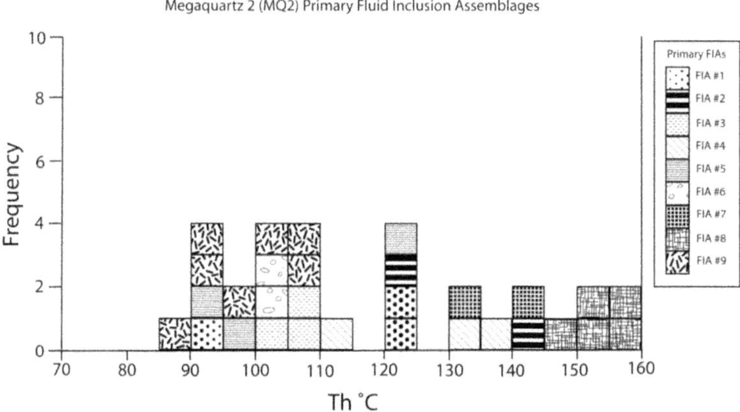

Fig. 12. T_h measured in primary FIAs of MQ2. The range of the data is significant, probably due to the measurement of various end members of inclusions entrapped during heterogeneous conditions.

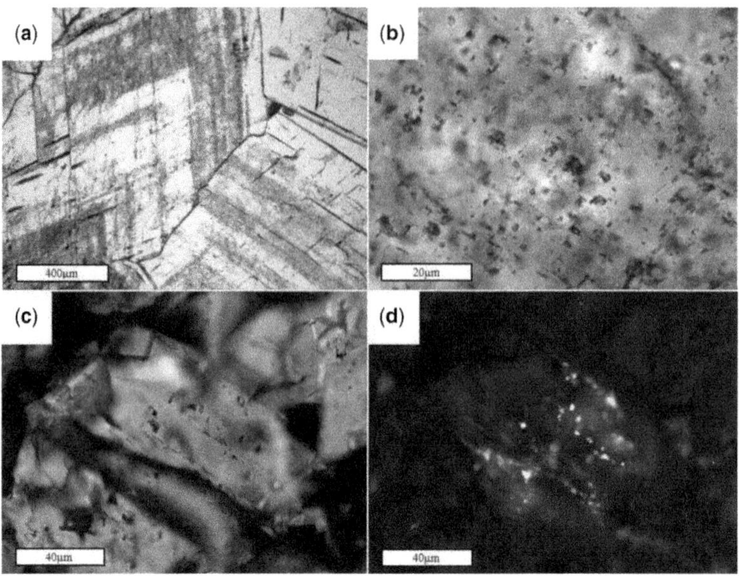

Fig. 13. (**a**) Transmitted light photomicrograph of primary FIAs running perpendicular to growth direction in BD (sample 5063.6A from Wellington 1-32); (**b**) Transmitted light photomicrograph of higher magnification image of 22A, displaying two-phase fluid inclusions within primary FIAs of BD (sample 5063.6A from Wellington 1-32); (**c**) Transmitted light photomicrograph of pseudosecondary, secondary and possibly primary FIAs containing hydrocarbon fluid inclusions in BD crystals (sample 4382.9 from Wellington 1-32); (**d**) same field of view **c**, but illuminated using UV epifluorescence. Photomicrograph of pseudosecondary, secondary and possibly primary FIAs containing hydrocarbon fluid inclusions in BD crystals (sample 4382.9 from Wellington 1-32).

zones within the MLZ (Fig. 9). Although some fluid inclusions can still be observed within the earlier growth zones, most are slightly larger and isolated; this has resulted in the majority of the fluid inclusion data being derived from the latest stage of crystal growth, rather than spanning the entire history of crystal growth.

Microthermometry. T_h data were collected from seventeen consistent primary FIAs and three consistent pseudosecondary or secondary FIAs. Four of the primary FIAs were focused on larger fluid inclusions within earlier growth zones that predate the recrystallization seen in BSEM, and are interpreted to represent early BD precipitation. These gave T_h values ranging from 97.0–113.0°C. The other T_h data were collected from the later stages of BD – and in these growth zones T_h of primary FIAs ranged from 93.0 to 130.6°C and T_h of pseudosecondary or secondary FIAs ranged from 107.0 to 131.0°C (Figs 14 & 15). T_h values collected from the Vulcan core ranged from 97.0 to 124.0°C in primary FIAs and 107.0–128.0°C in pseudosecondary or secondary FIAs, whereas T_h values collected from the Wellington core ranged from 93.0 to 130.6°C in primary FIAs. Several crystal transects were completed, measuring T_h values in primary FIAs from the inner to the outer crystal margin of BD rhombs. T_h values measured along early–late transects did not display any readily observable trends, with temperatures increasing or decreasing from the inner to the outer crystal margin (Fig. 15).

Tm_{ice} data were collected from seven consistent primary FIAs and three consistent psuedosecondary or secondary FIAs. Four of the primary FIAs were interpreted to represent early BD precipitation. Upon warming of the inclusions, the first melting is estimated to have occurred between −42 and −56°C, recognized by the occurrence of an 'orange peel' texture (Goldstein & Reynolds 1994). Tm_{ice} measurements from larger inclusions interpreted as representing earlier BD growth zones ranged from −12.4 to −14.5°C, reflecting salinities of 16.3–18.2 wt% NaCl eq. Tm_{ice} for inclusions in later BD are between −13.2 and −17.3°C, indicating subtly higher salinities of 17.1–20.4 wt% NaCl eq. (Fig. 16).

Interpretation. T_h measurements of two-phase fluid inclusions within consistent FIAs range from 93.0–131.0°C and crystal transects show rises and falls in T_h from the inner to outer margins of individual crystals (Fig. 17). The fact that all-liquid inclusions are small in size suggests that nucleation metastability may play a role in the absence of a gas bubble (Goldstein & Reynolds 1994). If small all-liquid

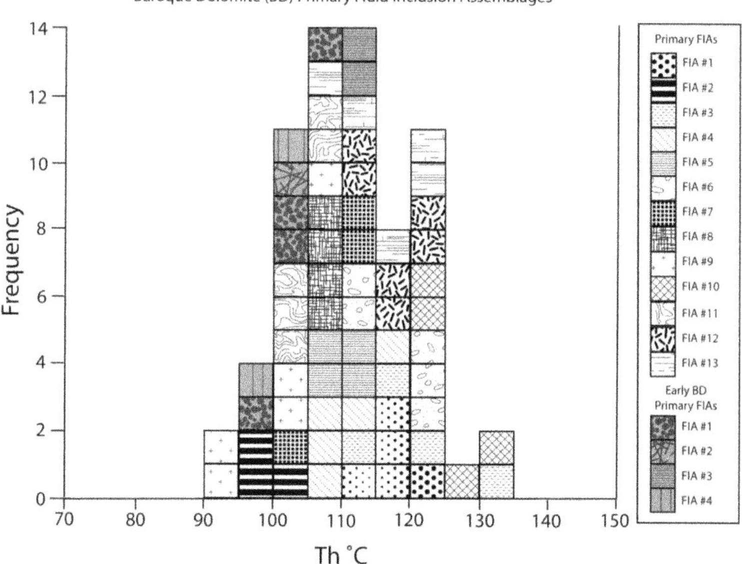

Fig. 14. T_h measured in primary FIAs of BD. The blue data points are values for large inclusions that appear to be related to the earliest stages of BD precipitation. Most measurements fall between 100°C and 125°C.

inclusions are attributed to metastability, then fluid inclusion entrapment most likely occurred during homogeneous conditions at high temperatures. The fact that all FIAs are consistent eliminates the possibility of thermal reequilibration. Variability in T_h could be interpreted to represent either recrystallization at variable high temperatures or homogeneous entrapment during rising and falling high temperatures. BSEM analysis has revealed that recrystallization during the latest stage of crystal growth trapped the majority of the fluid inclusions measured (Fig. 9). As most fluid inclusion data are confined to those recrystallized areas, and FIAs are not trapped in a mixture of recrystallized and unrecrystallized dolomite, homogeneous entrapment of fluid inclusions during multiple fluid migration events at variably high temperatures seems to be the more likely explanation. This is strongly supported by the consistent nature of FIAs. Recrystallization was probably so prevalent that most fluid inclusions

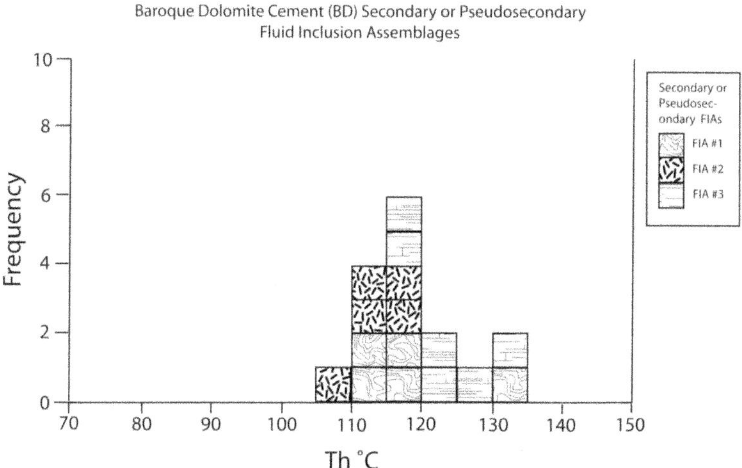

Fig. 15. T_h measured in secondary FIAs of BD. The highest frequency of data points occurs within the same range as primary FIAs, indicating similar conditions during entrapment.

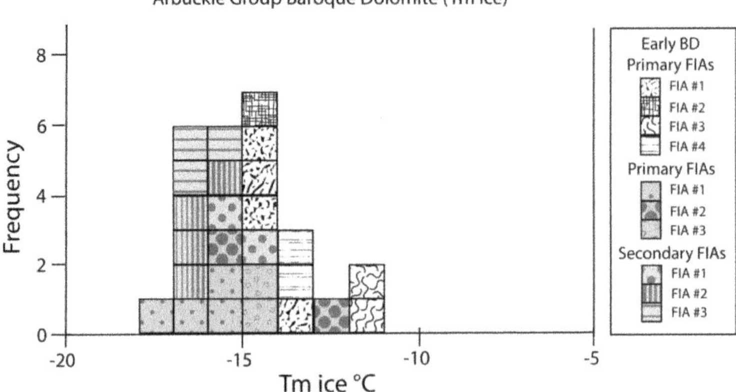

Fig. 16. Frequency distributions showing Tm_{ice} of BD. The most noticeable trend is lower salinity during the earliest stage of BD precipitation.

reflect the conditions during the latest stage of crystal growth. Only the larger inclusions contained in the earlier growth zones represent the earlier conditions, and they are only a minor part of the dataset.

Early BD precipitated from fluids with 16.3–18.2 wt% NaCl eq. and later BD precipitated from fluids with 17.1–20.4 wt% NaCl eq. Such high-salinity fluids suggest the dissolution of evaporites at some point during the fluid migration history or reflux of evaporative fluids. First melting temperatures between −42 and −56°C fall within the observed range of both $NaCl–CaCl_2–H_2O$ and $NaCl–MgCl_2–H_2O$ fluids (Goldstein & Reynolds 1994). When considering the elevated T_h values and the first melting temperatures of the fluids responsible for BD precipitation, a basinal-sourced fluid is the best candidate to account for hydrothermal fluids in the study area (Young 2010).

CC

Petrography. CC is clear in thin section and contains no observable primary FIAs. Secondary FIAs

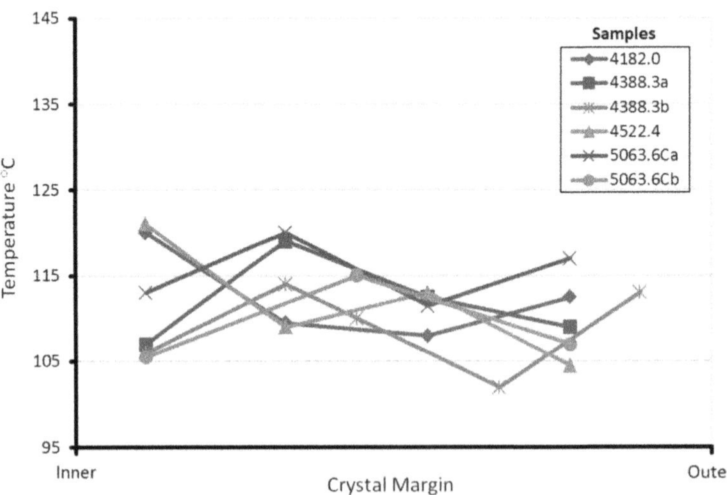

Fig. 17. Transects in BD measuring T_h values progressing from the inner to outer margins of individual crystals. All transects display rises and falls in the temperature of entrapment as crystal growth continued. Transects were conducted in samples spanning the entire thickness of the Arbuckle Group, suggesting that the entire unit was affected by these temperature fluctuations.

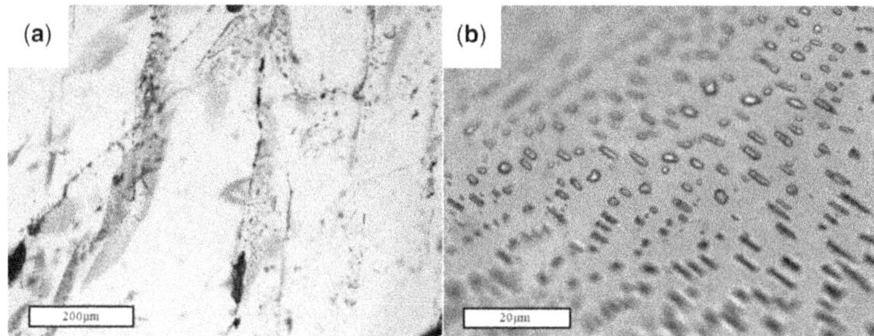

Fig. 18. Transmitted light photomicrographs of (**a**) secondary FIAs cross-cutting CC crystals (sample 3-14 (3979.1 ft) from the Vulcan core) and (**b**) higher magnification image of 22C, displaying two-phase inclusions with consistent liquid:gas ratios within secondary FIAs of CC (sample 3-14 (3979.1 ft) from Vulcan).

occur as curviplanar, 3D features that cut across entire calcite crystals (Fig. 18a). Fluid inclusions within secondary FIAs are generally the same size, rounded negative euhedral in morphology and two-phase (aqueous liquid with small vapour bubbles); liquid:gas ratios in FIAs appear to be consistent (Fig. 18b). All-liquid inclusions can be observed but these are invariably relatively small (<5 μm), suggesting that this is a result of metastability rather than cooler trapping conditions. Petrographic pairing of all-liquid and two-phase inclusions is not observed, indicating that the fluid inclusions did not neck-down after a phase change (Goldstein & Reynolds 1994). No patchiness was observed in CC when using CL or BSEM, indicating that recrystallization is unlikely (Fig. 8).

Microthermometry. T_h values from five consistent secondary FIAs range from 70.5°C to 89.8°C (Fig. 19); these were collected from only the Vulcan core because of the absence of measureable fluid inclusions in the samples from the Wellington core.

Tm_{ice} data were collected from three consistent secondary FIAs with final ice melting occurring between -14.6 and -16.2°C (Fig. 20), indicating salinities ranging from 18.2 to 19.6 wt% NaCl eq.

Interpretation. T_h measurements of two-phase fluid inclusions within consistent secondary FIAs range from 70.5 to 89.8°C. If small all-liquid inclusions are attributed to metastability, then secondary fluid inclusion entrapment most likely occurred during homogeneous conditions at the moderate homogenization temperatures measured. Owing to the lack of primary inclusions, temperature conditions during mineral growth were impossible to derive from fluid inclusion analysis. T_h measurements from secondary FIAs only give the temperature of the fluids during fracture healing after mineral growth,

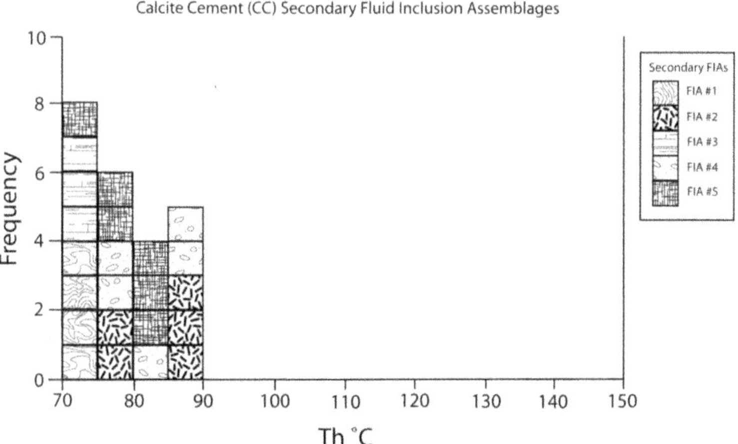

Fig. 19. T_h measured in secondary FIAs of CC. Although inclusions were trapped after mineral growth, they still indicate a decrease in temperature after precipitation of BD.

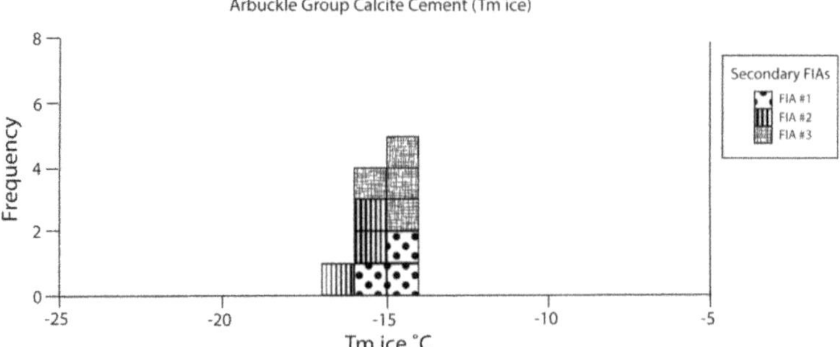

Fig. 20. Frequency distributions showing Tm_{ice} of CC. The most noticeable trend is the slight decrease in salinity from precipitation of BD to the entrapment of secondary fluid inclusions in calcite.

indicating that conditions had cooled down at some point following the precipitation of BD, but remained moderately high following the precipitation of calcite cement.

High-salinity fluids (18.2–19.6 wt% NaCl eq.) could represent the continued migration of basin-sourced fluids.

Further discussion regarding fluid inclusion data is provided later, and a summary of the fluid flow evolution of all analysed cements is provided in Figure 21.

Oxygen and carbon isotopic data

The following section discusses oxygen ($\delta^{18}O$) and carbon ($\delta^{13}C$) isotopic values provided from 44 isotopic analyses of samples from 11 stratigraphic horizons for BD and 4 stratigraphic horizons for calcite. Isotopic values are given in ‰ relative to the Vienna Pee Dee Belemnite (VPDB). BD (event 19) is discussed first, followed by calcite (events 21–23). The range and average of the data, as well as any observable trends, are listed first. Interpretations of $\delta^{18}O$ and $\delta^{13}C$ isotopic values, the spatial variations of the values and the potential fluid flow models are discussed for each precipitate.

BD

A total of 11 stratigraphic horizons containing BD were microsampled and processed for $\delta^{18}O$ and $\delta^{13}C$ isotopic values; all of the microsamples were taken from the Wellington 1-32 core. Ten of the 11 horizons were sampled in transects extending from the inner to the outer margins of individual crystals, providing a total of 27 analyses (Fig. 22). $\delta^{18}O$ ranges from −10.2 to −6.6‰ (average −8.4‰; Fig. 22) and $\delta^{13}C$ ranged from −3.6 to −2.4‰ (average −2.8‰; Fig. 22). Nine of the 27 microsamples were collected from BD present in relatively late fractures (events 14–15) and are noted with sample names in red in the legends for Figures 22–24. In the fractures $\delta^{18}O$ ranges between −9.5 and −7.3‰ (average −8.2‰; Fig. 22) and $\delta^{13}C$ ranges between −3.5 and −2.5‰ (average −2.8‰; Fig. 22).

There are several notable trends that can be observed in the BD data. $\delta^{18}O$ values become more depleted from the base to the top of the Arbuckle Group, with an overall decrease of −3.5‰ (−6.6 to −10.1‰) (Fig. 23). $\delta^{13}C$ values remain fairly constant except for a decrease of −0.9‰, around 300 ft from the top of the Arbuckle Group, to values that remain fairly constant through the upper Arbuckle Group (Fig. 24). Isotopic data from crystal transects with more than two isotopic analyses appear to fluctuate between more negative and more positive values from the inner to the outer crystal margins. The apparent fluctuations in data from transects may support the hypothesis mentioned in the fluid inclusion discussion, which suggests multiple pulses of hydrothermal fluids at variable temperatures. Alternatively, the resolution of isotope sampling may not have been fine enough to distinguish between the different growth zones of BD, leaving potential for the variability in isotopic values to be the result of sampling different proportions of various growth zones.

Interpretation. The $\delta^{18}O$ values range for the BD (−6.6 to −10.2‰), when integrated with petrographic, fluid inclusion and additional geochemical data, are best explained by a high-temperature model (e.g. Gregg & Sibley 1984; Land 1985; Land *et al.* 1988; Goldstein & Reynolds 1994; Davies & Smith 2006). As noted previously, the textures and morphologies of the BD are indicators of elevated temperatures during BD precipitation (Radke & Mathis 1980; Gregg & Sibley 1984; Davies & Smith 2006). Fluid inclusion T_h values of 93.0–131.0°C necessitate a hydrothermal environment to acquire such elevated values in the study area (see discussion of hydrothermal fluids below)

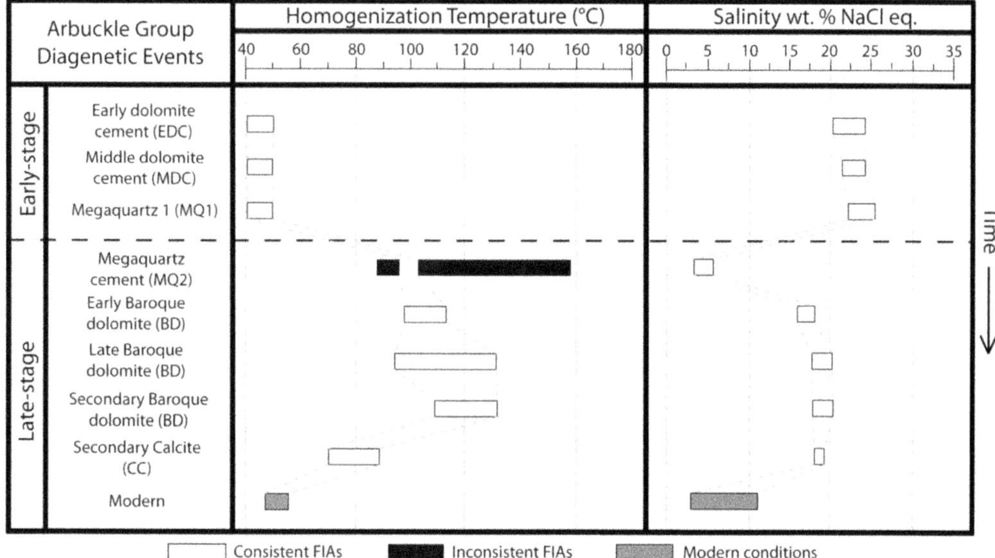

Fig. 21. Fluid evolution illustration that uses fluid inclusion data (T_h and Tm_{ice}) to display changes in reservoir fluid temperature and salinity from early-stage cements through late-stage cements to modern times. Values are fairly consistent during early-stage diagenesis, with temperatures likely representing reservoir temperature due to burial, the geothermal gradient and Ordovician surface temperature. The precipitation of MQ2 signifies a drastic decrease in salinity and increase in temperature, possibly linked to migration of high-temperature connate fluids from the Anadarko Basin. During BD precipitation, salinity increases while temperatures continue to climb; this is thought to represent continued sourcing of hydrothermal fluids from the Anadarko Basin after Permian reflux. Secondary fluid inclusions in CC record a decline in temperature and salinity at some point after the precipitation of CC. Modern values yield temperatures that are significantly lower than what was seen by the reservoir during late-stage diagenesis, but still slightly higher than early-stage temperatures. This is likely to be due to increased burial since the time of early-stage diagenesis. Modern salinity values fall between the range observed in MQ2 and BD, suggesting a continued influence of evaporites along the current fluid migration pathway.

(Goldstein & Reynolds 1994; Goldstein 2012). T_h and $\delta^{18}O_{dolomite}$ data allow for an approximation of $\delta^{18}O_{water}$ (Land 1985):

$$10^3 \ln \alpha = 2.78 \times \frac{10^6}{T^2} + 0.91 \quad (2)$$

where ($10^3 \ln \alpha$) is approximately equal to ($\delta^{18}O_{dolomite} - \delta^{18}O_{water}$) and T is temperature in Kelvin (Land 1985). The $\delta^{18}O_{water}$ PDB was then converted to $\delta^{18}O_{water}$ Vienne Standard Mean Ocean Water (VSMOW) (Arthur *et al.* 1983). The $\delta^{18}O$ values of the fluids responsible for precipitating BD may have ranged from −1.9 to +5.6‰. Enriched $\delta^{18}O_{water}$ values are common in residual evaporite brines (Epstein & Mayeda 1953; Heimstra 2003). Also, extensive rock–water interaction with material in the basin or along the fluid migration pathway could have enriched $\delta^{18}O$ values. High salinities, depleted $\delta^{18}O_{dolomite}$ values and enriched $\delta^{18}O_{water}$ values support the hypothesis of a basin-derived hydrothermal brine migrating through the Arbuckle Group (e.g. Hanor 1979; Land *et al.* 1988; Banner 1995; Davies & Smith 2006). The fluid inclusion and isotopic geochemical data of the fluids imply a basinal source for a hydrothermal brine that either migrated laterally from deeper portions of the basin or along deep-seated faults and fractures, very similar to many models for MVT deposits (e.g. Sverjensky 1986; Garven 1993, 1995; Leach & Sangster 1995; Davies & Smith 2006). The Anadarko basin has been suggested as a primary source for such fluids (e.g. Gao 1990*a*; Gao & Land 1991; Garven 1993; Musgrove & Banner 1993) and encompasses the necessary depth and strata to produce the geochemical data values obtained from the BD (Johnson *et al.* 1988; Gallardo & Blackwell 1999).

If a consistent isotopic composition of the fluid during BD precipitation (2.0‰) is assumed, the range of isotopic compositions yields a temperature range of about 37°C, which is consistent with that of the fluid inclusion homogenization data. Given such an assumption, the values near the top of the Arbuckle Group, at times, may have been 37°C warmer than the base.

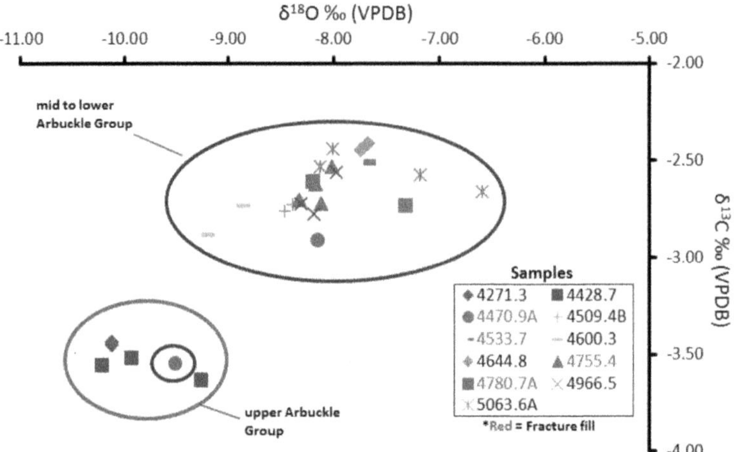

Fig. 22. Cross-plot of the $\delta^{18}O$ values v. the $\delta^{13}C$ values for Arbuckle Group BD. $\delta^{18}O$ and $\delta^{13}C$ values are the most depleted in the same samples, possibly illustrating hotter temperatures that produced depleted $\delta^{18}O$ values, as well as allowing for TSR reactions to slightly deplete $\delta^{13}C$ values. An enhanced porosity zone at the top of the unit or rising lower-density fluids may have produced the observed trend.

The $\delta^{13}C$ values of the BD (range -3.6 to -2.4‰) indicate potential organic influences and the minor variability could be due to differences in fluid–rock interaction along the fluid migration pathway (Arthur *et al.* 1983; Land 1983; Machel 1987, 2001; Davies & Smith 2006); variable fluid

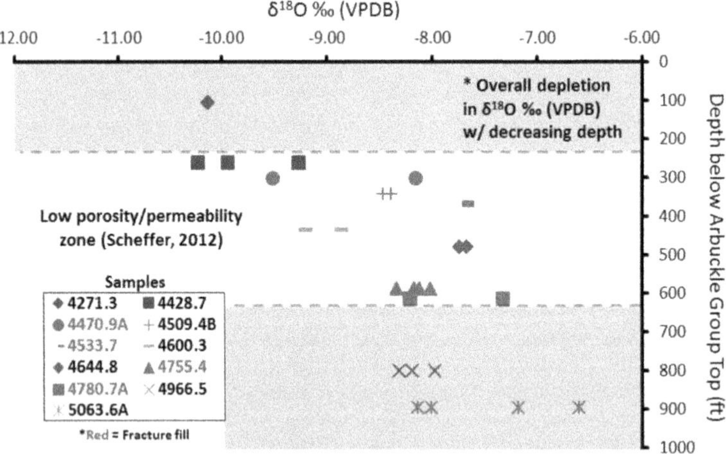

Fig. 23. Cross-plot of the $\delta^{18}O$ values v. the depth of the samples. $\delta^{18}O$ values clearly become more depleted with decreasing depth, probably illustrating higher temperatures that are potentially caused by preferential fluid flow or temperature-controlled density stratification in the Arbuckle Group. The gradual depletion of $\delta^{18}O$ with decreasing depth, especially within Scheffer's (2012) low porosity/permeability zone, supports a temperature-controlled density gradient that affected the entire unit at the time of BD precipitation. The blue boxes represent modern zones of higher porosity/permeability due to solution-enhanced fractures and brecciated zones (Scheffer 2012). The data appear to follow a trend of increasingly depleted values towards the top of the unit, regardless of modern reservoir conditions.

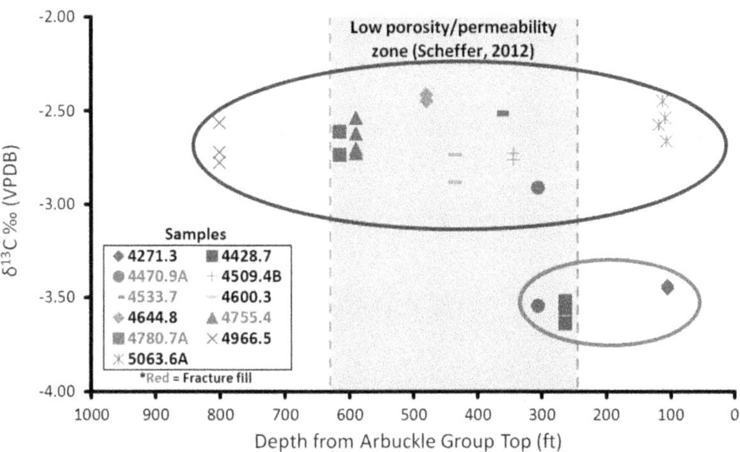

Fig. 24. Cross-plot of the $\delta^{13}C$ values v. the depth of the samples for the Arbuckle Group BD. The upper Arbuckle Group (red oval) displays slightly more depleted $\delta^{13}C$ values than the lower and middle portions of the unit (blue oval). Higher temperatures at the top of the unit may have allowed for TSR to deplete the carbon values. The more depleted carbon values can be found within and outside of Scheffer's (2012) low porosity/permeability zone, again suggesting that fluid flow during BD precipitation may not have been affected by current conditions. The blue boxes represent modern zones of higher porosity/permeability (Scheffer 2012).

sources have already been ruled out above through the evaluation of cement stratigraphy and fluid inclusion data.

Extensive fluid–rock interaction of the brine is plausible, given that even minor dissolution of rock during the migration history of the basinal brine could have a significant impact on the $\delta^{13}C$ values observed (Arthur *et al.* 1983; Lohmann 1988). The spatial variation of $\delta^{13}C$ values can be attributed to fluid–rock interaction because there is a positive correlation between the $\delta^{13}C$ and $\delta^{18}O$ values. The fact that the Arbuckle Group is a reservoir for hydrocarbons, most notably in the upper part of the unit, suggests organic sources in the top portion of the unit could have supplied depleted $\delta^{13}C$ values in the uppermost samples.

Elevated temperatures observed in fluid inclusions indicate the contribution of light carbon may also be influenced by abiotic reactions, presumably thermochemical sulphate reduction (TSR) (Machel 1987, 2001). The majority of T_h values fall within the most common range observed in sites altered by TSR reactions (100–140°C) (Machel 2001). The presence of anhydrite, one of the most prominent sources of sulphate when dissolved in solution, provides the needed catalyst for TSR reactions to exist and the most common products of TSR reactions can be observed throughout the Arbuckle Group (dolomite, calcite, galena, pyrite and sphalerite) (Machel 1987, 2001). The spatial variation could be due to a slight increase in temperature in areas of increased fluid flow in the upper Arbuckle Group, providing conditions conducive to TSR reactions (Machel 2001). The positive correlation between the $\delta^{13}C$ and $\delta^{18}O$ values as one progresses up in the section from the lower to the upper Arbuckle Group supports a greater TSR influence in the upper portion of the unit and is also likely controlled by temperature.

Another notable observation is the fact that both oxygen and carbon isotopic values do not appear to correlate with a low porosity/permeability zone that Scheffer (2012) hypothesized to be responsible for modern separation of fluids between the upper and lower Arbuckle Group and the rest of the unit (Figs 23 & 24); this may mean migration of fluids was not affected by the same low porosity/permeability zone present today.

CC

Four stratigraphic horizons containing calcite were microsampled and processed for $\delta^{18}O$ and $\delta^{13}C$ isotopic values; three samples were taken from the Wellington 1-32 core and one was taken from the Vulcan core. All of the samples were taken in transects extending from the core to outer margins of individual crystals, providing a total of 17 isotopic analyses (Fig. 25). $\delta^{18}O$ ranges from −9.8 to −7.0‰ (average −8.7‰; Fig. 25) and $\delta^{13}C$ ranges from −18.5 to −3.7‰ (average −11.5‰; Fig. 25).

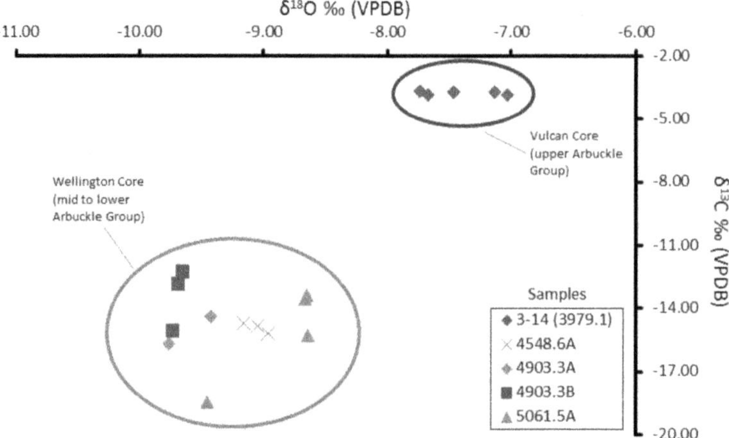

Fig. 25. Cross-plot of the $\delta^{18}O$ values v. the $\delta^{13}C$ values for the Arbuckle Group CC. Samples from the Wellington core display more depleted isotopic values, requiring variability in fluids between the two locations.

The most readily observable trend in the data is the drastic difference in the isotopic compositions of calcite between the Vulcan and Wellington 1-32 cores, with the calcite from Wellington 1-32 having significantly lighter $\delta^{18}O$ and $\delta^{13}C$ values (Figs 25–27). Transects in the calcite crystals resulted in what appears to be overall depletion of $\delta^{18}O$ from the beginning to the culmination of calcite precipitation.

Interpretation. If one assumes that original calcite precipitation occurred at temperatures below the highest temperatures recorded in the fluid inclusions in BD (131.0°C) and above the highest temperatures

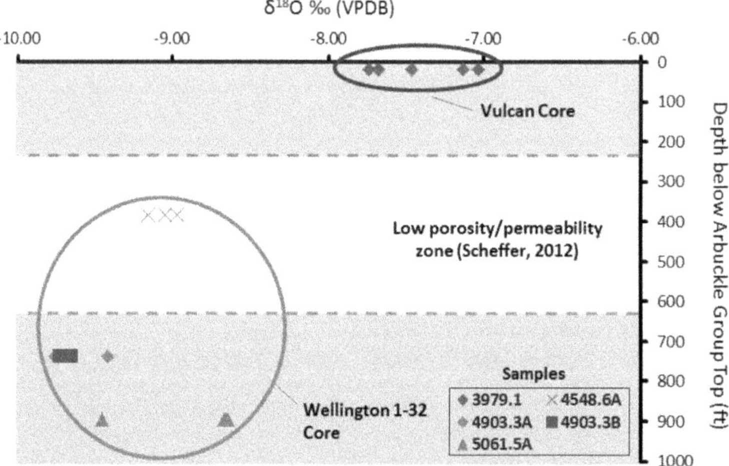

Fig. 26. Cross-plot of the $\delta^{18}O$ values v. the depth of the samples. The only noticeable trend is the more depleted nature of the Wellington core. Also, the $\delta^{18}O$ values do not appear to be influenced by preferential fluid flow or a density gradient, as with BD; this may provide support for the influence of a fracture-controlled fluid flow system affecting the unit at the time of calcite precipitation. Scheffer's (2012) low porosity/permeability zone does not appear to affect the isotopic values of calcite throughout the Arbuckle Group. The blue boxes represent zone of higher porosity/permeability in the upper and lower Arbuckle Group (Scheffer 2012).

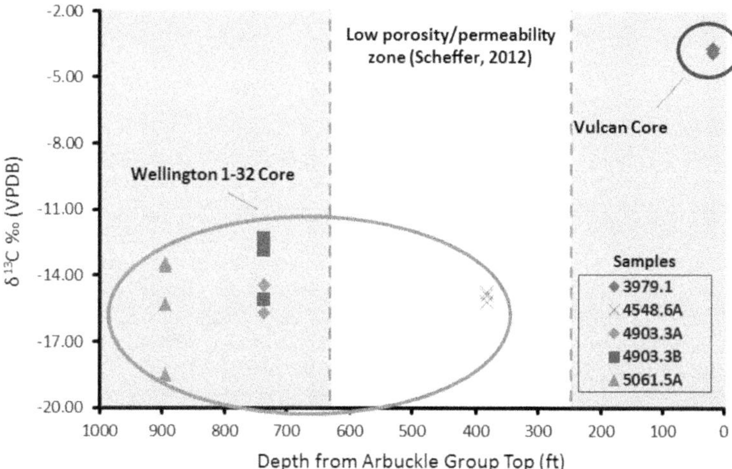

Fig. 27. Cross-plot of the $\delta^{13}C$ values v. the depth of the samples. The only noticeable trend is the more depleted nature of the Wellington core samples compared with the Vulcan core sample. As with the $\delta^{18}O$ data, the $\delta^{13}C$ data do not appear to correlate with the modern zone of low porosity/permeability proposed by Scheffer (2012). The blue boxes represent zones of higher porosity/permeability in the upper and lower Arbuckle Group (Scheffer 2012).

recorded in secondary FIAs in calcite (89.8°C), then temperatures during calcite precipitation would have probably been higher than can be explained by normal burial conditions or geothermal gradients. Using the temperature range estimated from the fluid inclusion data, and the measured calcite $\delta^{18}O$ data, the $\delta^{18}O_{water}$ of fluids responsible for calcite precipitation can be calculated. O'Neil *et al.* (1969) provide the equation:

$$10^3 \ln \alpha = 2.78 \times \frac{10^6}{T^2} - 2.89 \qquad (3)$$

where ($10^3 \ln \alpha$) is approximately equal to ($\delta^{18}O_{calcite} - \delta^{18}O_{water}$) and T is temperature in Kelvin (Land 1985). Using the temperature range 89.9–131.0°C and a $\delta^{18}O_{calcite}$ range −7.0 to −9.8‰, the calculated $\delta^{18}O_{water}$ for original calcite precipitation ranges from about +1.9‰ to +9.1‰. This range of fluid compositions is even more enriched in comparison with the fluids responsible for BD precipitation, suggesting a potential link in the fluid source or migration pathway. Given an assumed constant isotopic composition of the fluid (5.0‰), the temperature would have ranged by approximately 22°C.

As with the BD, the combination of petrographic observations, fluid inclusion data and geochemical data can most adequately be explained by the continued influence of high temperatures. The late-stage precipitation of calcite (potentially the latest event (23) in the paragenesis), correlates well with other authors' interpretations that calcite precipitated in the later stages of cement precipitation in MVT deposits (Hagni & Grawe 1964; Voss *et al.* 1989; Garven 1993; Brannon *et al.* 1996*a*, *b*; Coveney *et al.* 2000). Further discussion regarding potential fluid flow models is addressed in the discussion.

Limited sample depths of calcite inhibit the interpretation of the spatial variations of the $\delta^{18}O$ values and fluid flow, but a couple points can be speculatively made. One potential hypothesis is that calcite precipitation was affected by a fracture-controlled system that occurred after BD precipitation. If fracturing after BD did play a role in fluid flow, then it is possible that the spatial variability could be associated with the location relative to fractures or faults.

The $\delta^{13}C$ values range from the most negative value of −18.5‰ to the most positive value of −3.7‰, with an average of −11.5‰. There is significant depletion of $\delta^{13}C$ values from the Wellington 1-32 core (−12.3 to −18.6‰) when compared with the Vulcan core values (−3.7 to −3.9‰), with variability between $\delta^{13}C$ values from samples in the Wellington core. Similar to $\delta^{18}O$ data, transects through calcite crystals display increasingly depleted $\delta^{13}C$ values as crystal precipitation progressed.

The previously established presence of elevated temperature conditions indicates that the contribution of light carbon may also be influenced by abiotic reactions, presumably TSR (Machel 1987, 2001). Alternatively, the fact that $\delta^{13}C$ values are much more depleted than the values observed in BD may also illustrate the impact that organic sources can have on the depletion of carbon. Petroleum migration occurred during and possibly some time after the precipitation of BD, probably providing an abundance of organics in the reservoir when calcite precipitated. Another hypothesis could be that a fracture-controlled system influenced the precipitation of calcite, in which case the more negative $\delta^{13}C$ values could represent less rock–water interaction and the dominance of flow along fractures.

Alternatively, previous studies have shown evidence for multiple generations of calcite cement near the study area (Coveney 1992; Wojcik *et al.* 1992, 1994, 1997). If the calcite sampled is actually representative of different generations of calcite precipitation, which could not be deduced from this study, then more research is required to shed light on the isotopic values and data trends.

Strontium isotopic data

This section discusses strontium concentrations and isotopic ratios ($^{87}Sr/^{86}Sr$) derived from the analysis of nine BD samples and four calcite samples. BD (event 19) is discussed first, followed by calcite (events 21–23). The range and average of the data, as well as any observable trends, are presented first. Discussions of observed values and trends are offered at the end of each section. For comparison purposes, all values mentioned from previous authors' work have been normalized to National Bureau of Standards (NBS) 987 = 0.710250.

BD

Nine samples of BD were taken from the Wellington 1-32 core; all but one were also paired with carbon and oxygen isotope analyses. Samples ranged from depths of 4190.5 to 5063.6 ft, nearly encompassing the entire extent of the Arbuckle Group. Strontium concentrations varied from 34.68 to 68.33 ppm with an average of 49.11 ppm, whereas $^{87}Sr/^{86}Sr$ ranged from 0.70908 to 0.71019 with an average of 0.70948. $^{87}Sr/^{86}Sr$ values become more radiogenic and variable as depth decreases, with the most radiogenic value (0.71019) seen in the shallowest sample (4271.3 ft) (Fig. 28). For the samples where paired Sr and stable isotope data are available, increasing $^{87}Sr/^{86}Sr$ values correspond with increasingly depleted $\delta^{18}O$ and $\delta^{13}C$ values (Figs 29 & 30).

Interpretation. $^{87}Sr/^{86}Sr$ isotope compositions of Early Ordovician seawater range from approximately 0.70864 to 0.70892 (Denison *et al.* 1998). Every BD sample is more radiogenic than unmodified Ordovician seawater (Fig. 28). In fact, the entire data range (0.70908–0.71019) is more radiogenic

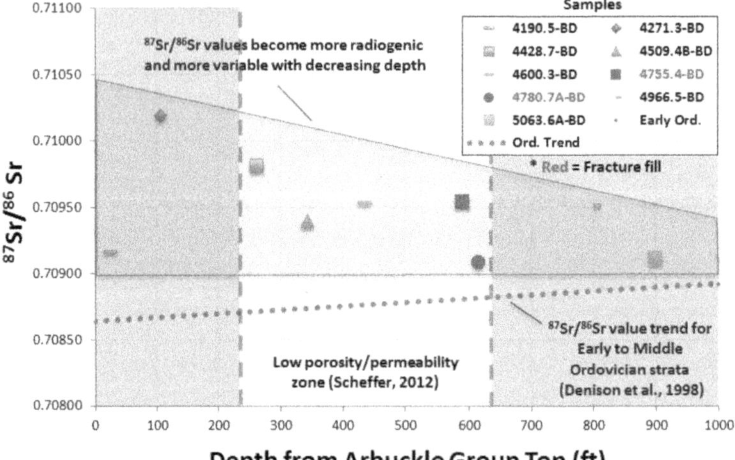

Fig. 28. Cross-plot of the $^{87}Sr/^{86}Sr$ ratio v. the depth of the samples from Arbuckle Group BD. Estimated Early Ordovician samples are also provided from Denison *et al.* (1998). Values become more radiogenic and more variable with decreasing depth, supporting the idea of enhanced fluid and temperature towards the top of the unit. As with the isotopic data, the $^{87}Sr/^{86}Sr$ data do not appear to be affected by Scheffer's (2012) low porosity/permeability zone. The blue boxes represent zones of higher porosity/permeability in the upper and lower Arbuckle Group.

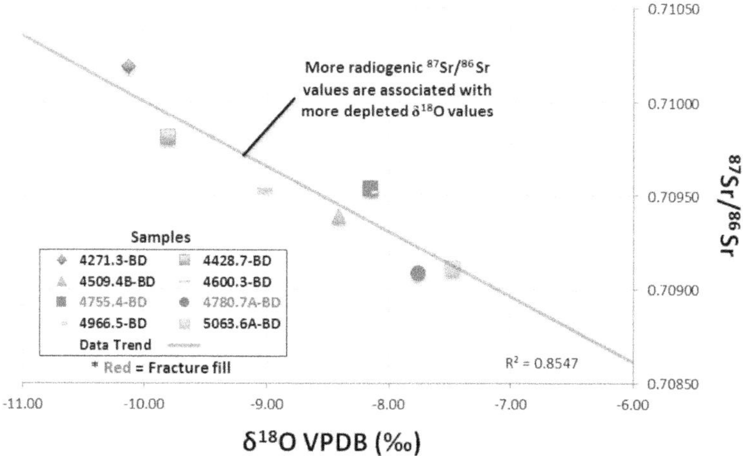

Fig. 29. Cross-plot of the $^{87}Sr/^{86}Sr$ ratio v. $\delta^{18}O$ values for Arbuckle Group BD. More radiogenic $^{87}Sr/^{86}Sr$ values correlate with more depleted $\delta^{18}O$ values.

than any seawater values recorded from the Cambrian (0.70904) to modern (0.70896) (Denison *et al.* 1998). The radiogenic nature of the BD samples suggests rock–water interaction with siliciclastic material or basement at some point during the fluid migration history (Banner 1995; Davies & Smith 2006). The $^{87}Sr/^{86}Sr$ values provide additional support to the hypothesis that the fluids responsible for BD precipitation are basinal.

The spatial variability in the data provides two informative trends that further support previous interpretations. The presence of hydrothermal fluids migrating through the Arbuckle Group has been well established through petrographic observations,

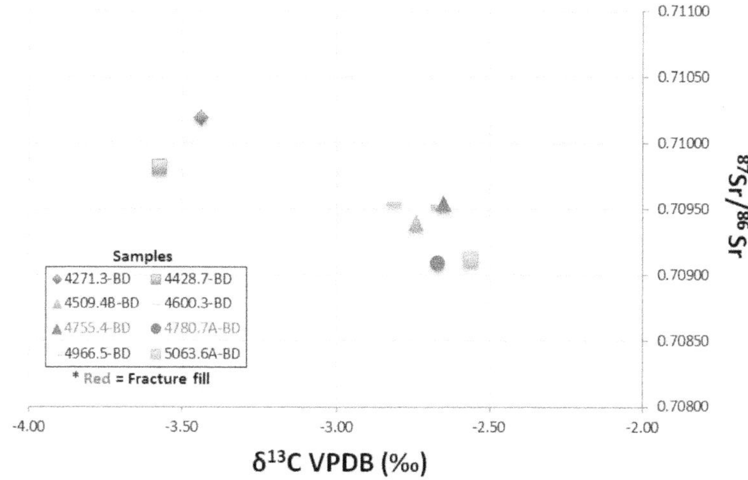

Fig. 30. Cross-plot of the $^{87}Sr/^{86}Sr$ ratio v. $\delta^{13}C$ values for Arbuckle Group BD. More radiogenic $^{87}Sr/^{86}Sr$ values correlate with more depleted $\delta^{13}C$ values. The data support the idea of higher-temperature fluids preserving more radiogenic values and being capable of sustaining TSR, which in turn depletes the $\delta^{13}C$ values.

fluid inclusion microthermometry and carbon and oxygen isotopic analysis. The highly porous and permeable upper portion of the Arbuckle Group has been hypothesized as a preferential fluid flow conduit (this study; Jorgensen *et al.* 1993; Scheffer 2012), but the depletion of $\delta^{18}O$ with decreasing depth supports the idea of a temperature-controlled density gradient affecting the unit as well. $^{87}Sr/^{86}Sr$ values become more radiogenic but also more variable up through the Arbuckle Group; this trend appears to support the idea of a temperature-controlled density gradient with hotter, less-dense fluids towards the top of the unit, whereas the enhanced porosity and permeability at the top of the unit may have allowed for preferential fluid flow at the same time. The enhanced fluid flow could result in less rock–water interaction at the top of the Arbuckle Group. The $^{87}Sr/^{86}Sr$ from this portion of the unit would yield water-dominated values (more radiogenic) whereas ratios obtained from lower in the section would yield rock-dominated values (less radiogenic). Cross-plots of $^{87}Sr/^{86}Sr$ values with $\delta^{13}C$ and $\delta^{18}O$ data provide additional support to the structure of hydrothermal fluid flow. The correlation between radiogenic $^{87}Sr/^{86}Sr$ and depleted $\delta^{18}O$ values (Fig. 29) provides support for preferential fluid flow or a density gradient producing higher temperatures and less rock–water interaction higher in the section. The correlation between radiogenic $^{87}Sr/^{86}Sr$ and depleted $\delta^{13}C$ (Fig. 30) supports the idea of higher temperatures and less rock–water interaction towards the top of the Arbuckle Group, with the depletion in $\delta^{13}C$ from TSR.

Calcite cement

Three locations were sampled from the Wellington 1-32 core and one was sampled from the Vulcan core; all samples were also paired with carbon and oxygen isotope analyses. Sample depths are discussed relative to the top of the Arbuckle Group because they were taken from more than one core. Sample depths ranged from 17.5 to 895.5 ft below the top of the Arbuckle Group, providing data points that span nearly the entire extent of the unit. Strontium concentrations varied from 50.09 to 92.47 ppm with an average of 73.80 ppm, whereas $^{87}Sr/^{86}Sr$ ranged from 0.71257 to 0.71716 with an average of 0.71546. The Wellington 1-32 core samples are more radiogenic than Vulcan core samples and there is no trend relative to depth. As seen in the BD, increasing $^{87}Sr/^{86}Sr$ values correspond with more depleted $\delta^{18}O$ and $\delta^{13}C$ values (Figs 31 & 32), although the correlation between Sr isotope composition and $\delta^{13}C$ is not as strong, generally showing that the Wellington 1-32 core has more radiogenic values associated with the more depleted $\delta^{13}C$ values (Fig. 32).

Interpretation. The radiogenic $^{87}Sr/^{86}Sr$ values (0.71257–0.71716) produced from the calcite are well above any value measured from seawater in any part of the Phanerozoic, clearly suggesting

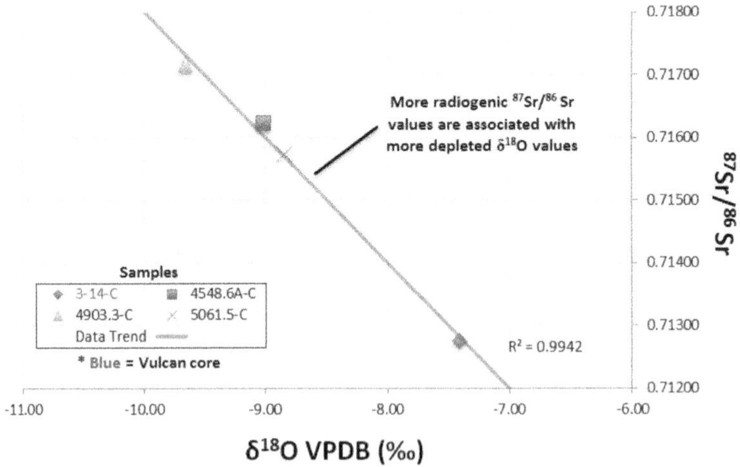

Fig. 31. Cross-plot of the $^{87}Sr/^{86}Sr$ ratio v. $\delta^{18}O$ values of Arbuckle Group calcite. The plot displays more radiogenic values correlating with more depleted $\delta^{18}O$. The data support the idea of higher-temperature fluids preserving more radiogenic values (the Vulcan core sample is highlighted in blue in online version).

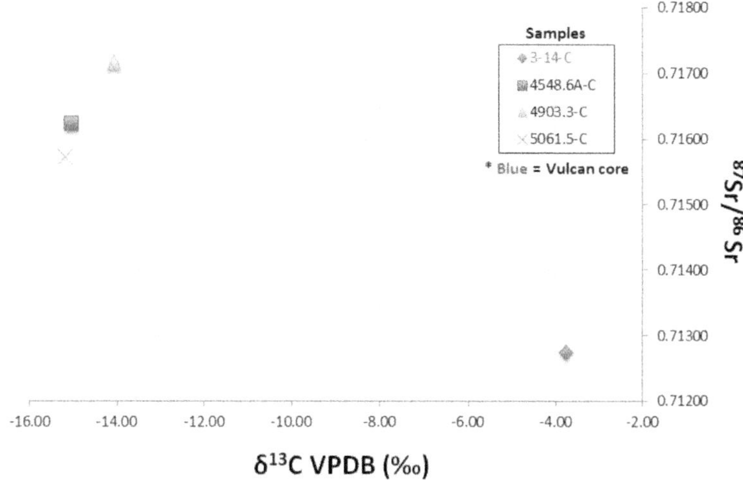

Fig. 32. Cross-plot of the $^{87}Sr/^{86}Sr$ ratio v. $\delta^{13}C$ values of Arbuckle Group calcite. Plot displays more radiogenic values correlating with more depleted $\delta^{13}C$ values. The plot also displays the difference in $\delta^{13}C$ values between the Wellington 1-32 and Vulcan cores (highlighted in blue in online version).

extensive rock–water interaction with siliciclastic or basement material during the fluid migration history. When comparing the calcite values with those observed in BD (0.70908–0.71019), the more radiogenic nature of the calcite suggests that the fluids responsible for calcite precipitation in each of the two localities may have followed different fluid migration pathways than the dolomite-precipitating fluids. If calcites in the two locations are the same age, their different geochemical compositions suggest localized fluid flow rather than regional advective flow. One possible hypothesis could be that tectonism that occurred after BD precipitation may have opened different fluid conduits via additional open fracture networks, possibly allowing for changes in rock–water interaction along the fluid migration pathway.

The linkage between the most depleted $\delta^{18}O$ (−9.6‰ VPDB) and the most radiogenic $^{87}Sr/^{86}Sr$ (0.71716) (Fig. 31) demonstrates that areas with higher temperatures (more depleted $\delta^{18}O$ values) experienced less rock–water interaction with host carbonates. The differences in $\delta^{13}C$ and $^{87}Sr/^{86}Sr$ between the two areas probably reflects preferential fluid flow in the area of the Wellington 1-32 core, allowing for less rock–water interaction and therefore more radiogenic $^{87}Sr/^{86}Sr$ and lighter $\delta^{13}C$, than in the area of the Vulcan core.

Alternatively, the correlation between $^{87}Sr/^{86}Sr$ and $\delta^{18}O$ may represent two-component mixing (Fig. 31) of two end-member calcites – one with depleted $\delta^{18}O$ and radiogenic $^{87}Sr/^{86}Sr$ and the other with enriched $\delta^{18}O$ and less radiogenic $^{87}Sr/^{86}Sr$. The potential clearly exists for sampling of multiple generations of calcite (e.g. Coveney 1992; Wojcik *et al.* 1992, 1994, 1997). If this is the case, interpretations based on this dataset would need to be reconsidered following additional research on calcite cements in the study area. Alternatively, the correlation may represent mixing of two different fluids at the time of calcite precipitation. If fluid mixing produced the linear relationship between $^{87}Sr/^{86}Sr$ and $\delta^{18}O$, then it is possible that one fluid that was slightly hotter and more radiogenic was mixing with a slightly lower-temperature fluid with lower radiogenic values. This type of mixing could mean that fluids were being sourced from multiple sources or along different fluid migration pathways at the time of calcite precipitation.

Discussion

Fluid flow conduits

Primary and secondary porosity, as well as extensive brecciation, in the lower Arbuckle Group form a zone of high porosity and permeability. Another zone of high porosity and permeability exists towards the top of the Group, reflecting dissolution at the culmination of deposition of the Arbuckle Group (see the Paragenesis section). As discussed earlier, trends in $\delta^{18}O$ data suggest that fluid temperatures may have increased from the base to the top of the Arbuckle Group by as much as 37°C, with the highest temperatures existing in the upper section of the unit (see isotope interpretation sections). BD can be observed from the base to the top of the unit, but with more depleted $\delta^{18}O$ values at the top suggesting higher temperatures. This probably means that the modern low porosity/permeability zone in the middle of the section (Scheffer 2012) was not important at the time of fluid migration – and that a temperature-controlled density gradient that may have also been influenced by enhanced fluid flow in the upper section of the unit existed at the time of BD precipitation. High temperature and high salinity, coupled with radiogenic $^{87}Sr/^{86}Sr$ and enriched $\delta^{18}O_{water}$ values, support a regional advective fluid flow system with fluids sourced from the nearby Anadarko basin.

Conversely, $\delta^{18}O$ data from calcite cements do not follow the same kind of depth-dependent trend. Assuming that the $\delta^{18}O$ range observed in calcite (−7.0 to −9.8‰) is the result of elevated temperature values that ranged between those of the inclusion values in the BD (93.0–131.0°C) and secondary inclusions in calcite (70.0–89.9°C) data, temperatures remained too high to be accounted for by burial conditions or geothermal gradients during calcite precipitation. Highly radiogenic $^{87}Sr/^{86}Sr$ values suggest a different fluid source or migration pathway than fluids responsible for BD precipitation. It is unlikely that the same system, with fluids being sourced from the Anadarko basin, would be able to produce the more radiogenic values without at least a change in the migration pathway. Tectonic activity following the precipitation of BD could have altered migration pathways by creating fluid flow through newly created fracture networks. Fracturing in the study area could have altered the connectivity between the Arbuckle Group and underlying stratigraphic units, possibly enabling interaction with the radiogenic fluids present in the underlying Cambrian Reagan sandstone or basement rock. Although the Reagan sandstone is absent in the study area, the basal sand could have acted much like the Arbuckle Group for the regional advective fluid flow of hydrothermal fluids out of the Anadarko basin, with later fracturing allowing for the migration of fluids in underlying units into the overlying Arbuckle Group. It is also possible that the fluids responsible for calcite precipitation may have been sourced from basement rock, using fractures that connected the basement with overlying strata. If the fluids responsible for calcite precipitation did use fracture networks as fluid conduits then calcite precipitation could be expected to be an event controlled by structural deformation.

Hydrothermal fluids

The late-stage mineral assemblage of the Arbuckle Group, including megaquartz, BD, calcite, galena and sphalerite, is similar to MVT deposits believed by most to be the result of basin-derived hydrothermal fluids (Cathles & Smith 1983; Garven & Freeze 1984*a*, *b*; Leach & Rowan 1986; Oliver 1986; Sverjensky 1986; Bethke & Marshak 1990; Wojcik *et al.* 1992; Garven 1993; Leach & Sangster 1995; Young 2010). Moreover, overlying units in the study area have been interpreted as being modified by multiple migration events of hydrothermal fluids sourced from nearby basins (Wojcik *et al.* 1992; Young 2010; Ramaker *et al.* 2014). The existence of basin-derived hydrothermal fluid flow within the Arbuckle Group has been methodically established through petrographic observations, fluid inclusion analysis and geochemical data.

The presence of BD with a xenotopic-C texture is in itself an indicator of elevated temperatures during cement precipitation (Radke & Mathis 1980; Gregg & Sibley 1984; Davies & Smith 2006). Fluid inclusion analysis produced elevated T_h values in late diagenetic cements (≥87°C for megaquartz and as high as 131°C for BD). Depleted $\delta^{18}O$ values recorded in BD and CC also suggest thermal fractionation that resulted from high temperatures commonly associated with the heating of basin-derived brines (Davies & Smith 2006). According to Newell (1997), a burial history model using a reasonable geothermal gradient would produce a maximum formation temperature of around 74°C in counties just outside the study area; this value is 20°C below the T_h values measured in MQ2 and nearly 50°C below those measured in BD. Newell (1997) also explored the possibility of both higher geothermal gradients and Cretaceous thicknesses than were thought to be geologically reasonable. Even with the unreasonably high values, temperature estimates remain approximately 20–30°C lower than the T_h values observed in BD. Moreover, T_h values and $\delta^{18}O$ data recorded in BD suggest multiple pulses of hydrothermal fluids during precipitation; fluctuations in the burial depth or geothermal gradient cannot reasonably account for short-term pulses in reservoir temperature. More depleted $\delta^{18}O$ values near the top of the Arbuckle Group indicate the inverse of what would normally be expected for a normal burial system. During BD precipitation, temperatures at the top of the Arbuckle were 37°C warmer than the base, evidence against a normal burial system and in favour of a hydrothermal system. In calcite cement, depleted $\delta^{18}O$ values also support thermal fractionation resulting from temperatures close to the values recorded in fluid inclusion data from BD (93.0–131.0°C) (Davies & Smith 2006). The only reasonable explanation for temperatures higher than can be modelled from burial, repeated rises and falls in temperature and higher temperatures at the top of the Arbuckle is the pulsed injection of hydrothermal fluids.

Potential sources

The nearby Anadarko basin would seem to be the most likely source for hydrothermal fluids in the study area, as has been suggested by others in the past (Gao 1990; Gao & Land 1991; Musgrove & Banner 1993). High salinities have commonly been attributed to reflux of evaporative brines into deeper portions of basins and radiogenic $^{87}Sr/^{86}Sr$ values support the presence of basinal brines that have undergone extensive rock–water modification during contact with siliciclastic material or basement rock (Hanor 1979; Banner 1995). The Anadarko basin has been proven to contain the necessary siliciclastic strata to produce such fluid compositions (Johnson *et al.* 1988) and Permian-age evaporation is thought to be a potential source for reflux of highly saline fluids into deeper portions of midcontinent basins (Wojcik *et al.* 1994; Young 2010). Alternatively, evaporite dissolution either in the basin or along the migration pathway may have also played an important role in producing the high-salinity fluids; Silurian evaporites or rift-salts in the Anadarko basin associated with Cambrian tectonic activity are potential candidates. Thermal studies of the Anadarko basin have proven that substantial heating of such fluids could easily be attained through burial heating, producing temperatures ranging from 175°C to 220°C at the top of the Arbuckle Group in the deepest portions of the basin (Gallardo & Blackwell 1999); these conditions would provide the necessary environment for the generation of hydrothermal brines. Additionally, the presence of hydrocarbon fluid inclusions within BD further supports the idea that the fluids were sourced from the Anadarko basin because hydrocarbons present in the Arbuckle Group are thought to have been primarily sourced from the nearby basin (Price 1980).

Whereas fluids from the Anadarko basin would seem to account for the majority of the characteristics observed in the late diagenetic fluids, the more radiogenic nature of the calcite samples necessitates a change in the source or migration pathway from the time of BD to calcite precipitation. As suggested in previous sections, an alternative source for such radiogenic fluids may be fluids migrating vertically out of underlying sand units or basement rock through fracture networks. Previous studies have shown evidence for Ouachita and Laramide tectonism reactivating basement faults and extending them into overlying Paleozoic strata (Tweto 1980; Jorgensen *et al.* 1993; Elebiju *et al.* 2011).

Multiple migration events

The geochemical characteristics of calcite and BD have already been interpreted as requiring different fluid sources or migration pathways, clearly demanding more than one fluid migration event. In addition, microthermometric data observed in MQ2 (event 18) and BD (event 19) also supply observations worthy of discussion when considering the potential for multiple migration events. T_h displays fluctuations in temperature during BD precipitation (Fig. 17), suggesting multiple pulses of hydrothermal fluids. Numerous authors have been able to date multiple fluid migration events producing MVT deposits that are thought to be genetically linked to the hydrothermal fluids responsible for late cementation (Table 1) (Brannon *et al.* 1996*b*; Coveney *et al.* 2000; Blackburn *et al.* 2008). Also, Tm_{ice} values produce salinities that increase from the precipitation of MQ2 (3.1–6.0 wt% NaCl eq.) to the precipitation of BD (16.3–20.4 wt% NaCl eq.). It would be unlikely for a single fluid to increase by at least 10.0 wt% NaCl eq. during a single migration event. Similar to the precipitation of calcite, either a different fluid source or migration pathway seems to be necessary for such an increase in salinity to occur. Multiple migration events of hydrothermal fluids – fluids that probably followed different migration pathways or were sourced from different areas – most adequately explain the mineralogy, fluid inclusion data and geochemical data related to late-stage hydrothermal activity.

Hydrologic models

Potential hypotheses regarding fluid flow mechanisms must account for multiple migration events of fluids sourced from areas that would produce high temperatures, variably high salinities and radiogenic $^{87}Sr/^{86}Sr$ values, as well as the known tectonic and stratigraphic history of the study area. Previous authors have explored several hydrologic models that may account for regional fluid migration, including: (1) compaction-driven pore fluid flow and free convection, (2) tectonically controlled, episodic dewatering of basins and (3) gravity-driven fluid flow (Dozy 1970; Sharp 1978; Cathles & Smith 1983; Garven & Freeze 1984*a*, *b*; Bethke 1986; Leach & Rowan 1986; Oliver 1986; Kupecz & Land 1991; Garven 1993, 1995).

Hypothesis (1) is the least likely scenario, given that it involves relatively slow fluid flow and is not thought to be capable of the level of heating necessary to produce the observed fluid inclusion temperatures or the volume of fluid needed for regionally extensive deposits (Bethke 1986; Kupecz & Land 1991; Garven 1993).

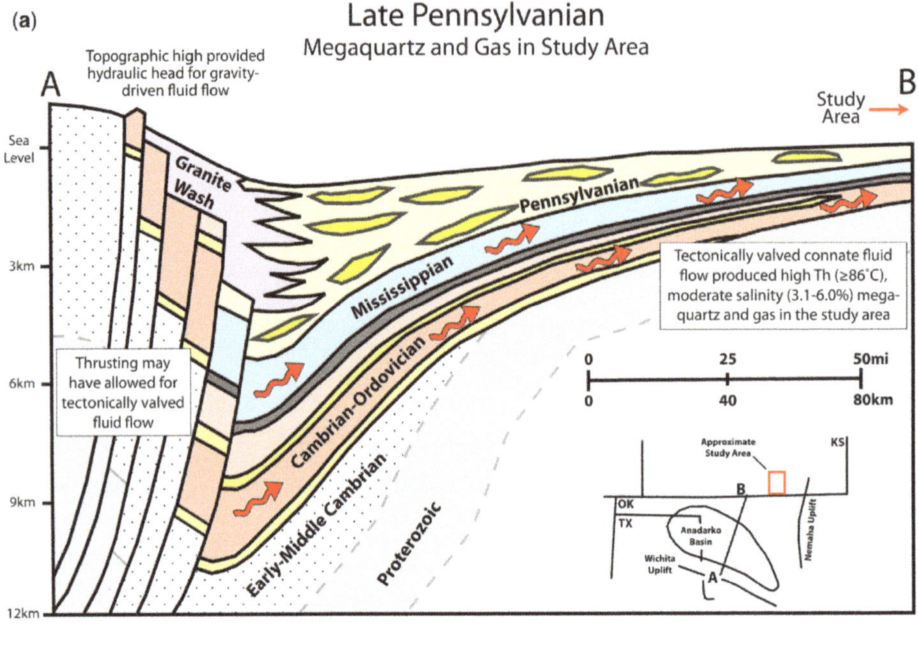

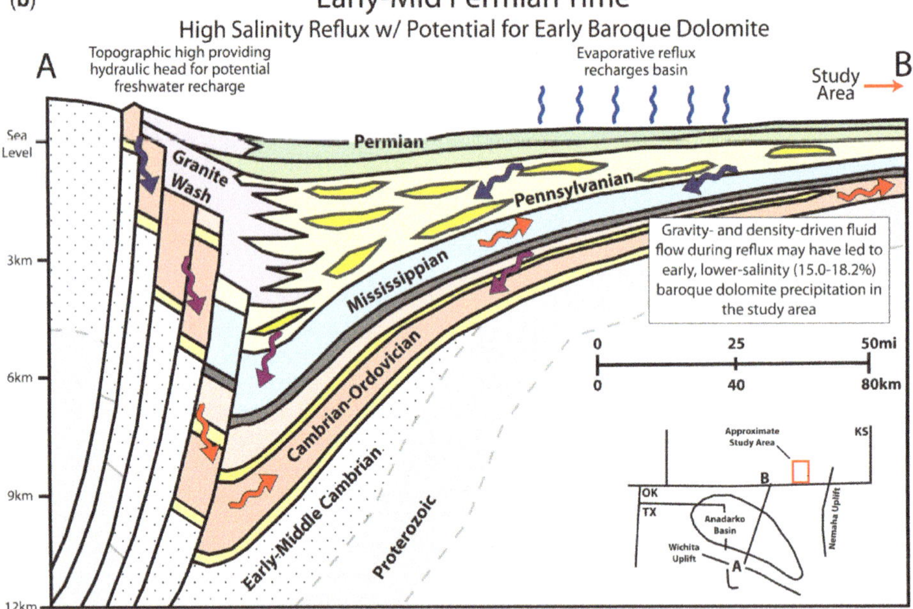

Fig. 33. Cross-sections displaying the fluid sources and migration pathways for fluids responsible for late-stage cementation in the Arbuckle Group. (**a**) Initial expulsion of connate fluids from the Anadarko Basin strata may have occurred with the onset of activity related to the Ouachita orogeny during the Late Pennsylvanian. Tectonic activity could have valved fluids or provided a topographic high and set the scene for gravity-driven fluid flow out of the basin. This initial stage of fluid migration would have consisted of 3.1–6.0 wt% NaCl eq. connate fluids at temperatures $\geq 87°C$, along with gas, migrating into the study area and precipitating MQ2. (**b**) During the Permian evaporitic conditions would have provided highly saline fluids refluxing through permeable strata and fractures associated with the earlier orogenic activity; this fluid could have actively recharged the basin with saline fluids. Besides tectonic valving and gravity-driven fluid flow, density-driven fluid flow may also have played an active role in fluid migration, with high-density fluids migrating downwards and pushing out lower-density fluids.

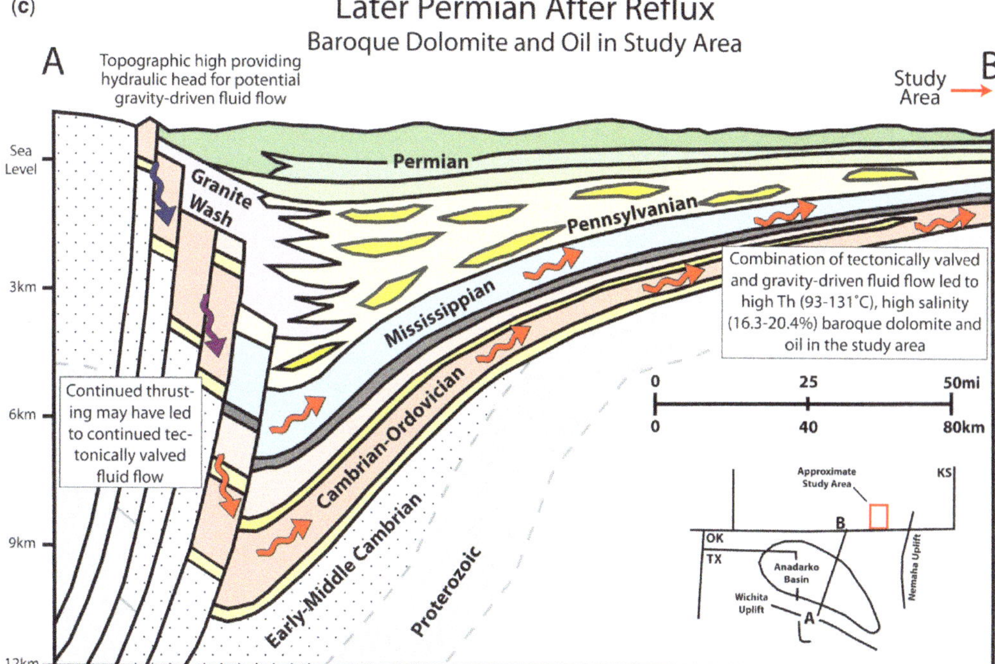

Fig. 33. (*Continued*) (**c**) After Permian reflux, continued tectonic activity, gravity-driven fluid flow or even density-driven fluid flow could have caused continued fluid migration out of the basin and into the study area during the Permian. This stage of fluid migration would have consisted of the 16.3–20.4 wt% NaCl eq. fluid (with temperatures ranging from 93 to 131°C) that was responsible for BD precipitation and petroleum migration in the study area. Modified from Gallardo & Blackwell (1999).

For hypothesis (2), episodic, structurally controlled drivers are capable of expelling hot fluids (Sharp 1978; Cathles & Smith 1983; Oliver 1986; Kupecz & Land 1991), but are also believed to produce fluid volumes that are too low to account for high T_h values in regionally extensive hydrothermal deposits (Leach & Rowan 1986; Garven 1993). Faulting can both episodically valve and pump fluids (Sibson 1981, 1987, 2003; Sibson *et al.* 1988) with short-lived locally high fluid flux (Cox 2010; Lupi *et al.* 2011). For the study area, Young (2010; Ramaker *et al.* 2014) proposed that initial shallow deformation in basins to the south was followed by deeper deformation during the Ouachita event. This explained the succession from low-salinity and moderate-temperature hydrothermal megaquartz to high-salinity and high-temperature hydrothermal dolomite, either through progressive tapping of increased salinity fluids at increased depths or an increasing depth of deformation before and after the charging of the basin with brines in the Permian. As megaquartz and dolomite hydrothermal fluid flow appears to be advective and regional in scale, however, fault pumping and valving cannot be the only driver for fluid flow; the localized and short-lived nature of this mechanism is inconsistent with the observations. For the later calcite on the other hand, this driver may be the perfect explanation. The highly radiogenic $^{87}Sr/^{86}Sr$ and distinctive compositions suggest flow directly out of basement or the underlying Reagan Sandstone that is highly localized rather than regional and advective. Localized faulting or fracturing could have pumped fluids and given rise to the hydrothermal event associated with the calcite. Both Ouachita and Laramide orogenies are thought to have produced strike-slip faulting, probably linked with reactivation of basement faults (Tweto 1980; Jorgensen *et al.* 1993; Gay 2003*a*, *b*; Ohlmacher & Berendsen 2005; Watney *et al.* 2008; Elebiju *et al.* 2011; Wharton 2012; Watney *et al.* 2013). Because of the calcite's late position in the paragenesis, it is hypothesized that the Laramide deformation may be the best explanation for fluid flow associated with the calcite

Hypothesis (3), gravity-driven fluid flow resulting from uplift of adjacent mountains, could also account for basin-sourced hydrothermal fluids facilitating the precipitation of late-stage quartz and dolomite (Garven & Freeze 1984*a*, *b*; Garven

1993, 1995). Tectonic uplift during the Ouachita orogeny was associated with stratigraphic and structural thickening in the Anadarko and Arkoma basins (Figs 1 & 33) (Burgess 1976; Ye *et al.* 1996; Gallardo & Blackwell 1999). Adjacent mountains would have provided the hydraulic head for regional hydrothermal fluid migration out of the basin and into the study area (e.g. Leach & Rowan 1986; Garven 1993, 1995; Leach & Sangster 1995). Ouachita deformation and uplift began before Permian reflux and highly saline basinal brines derived from evaporated seawater (e.g. Shelton *et al.* 2009) might not have been available until late in the Permian. Only lower-salinity connate fluids would have been available before this (Fig. 33a) and after (Fig. 33b, c) the hydraulic head from the mountains could have continued driving hydrothermal fluids northwards. Heated fluids would have migrated laterally out of the basin through multiple carbonate aquifers (Mississippian as well as Cambrian–Ordovician Arbuckle Group) (Fig. 33) (Leach & Rowan 1986; Garven 1993, 1995; Leach & Sangster 1995). The aquifer configuration was such that conditions were warmer at the top of the

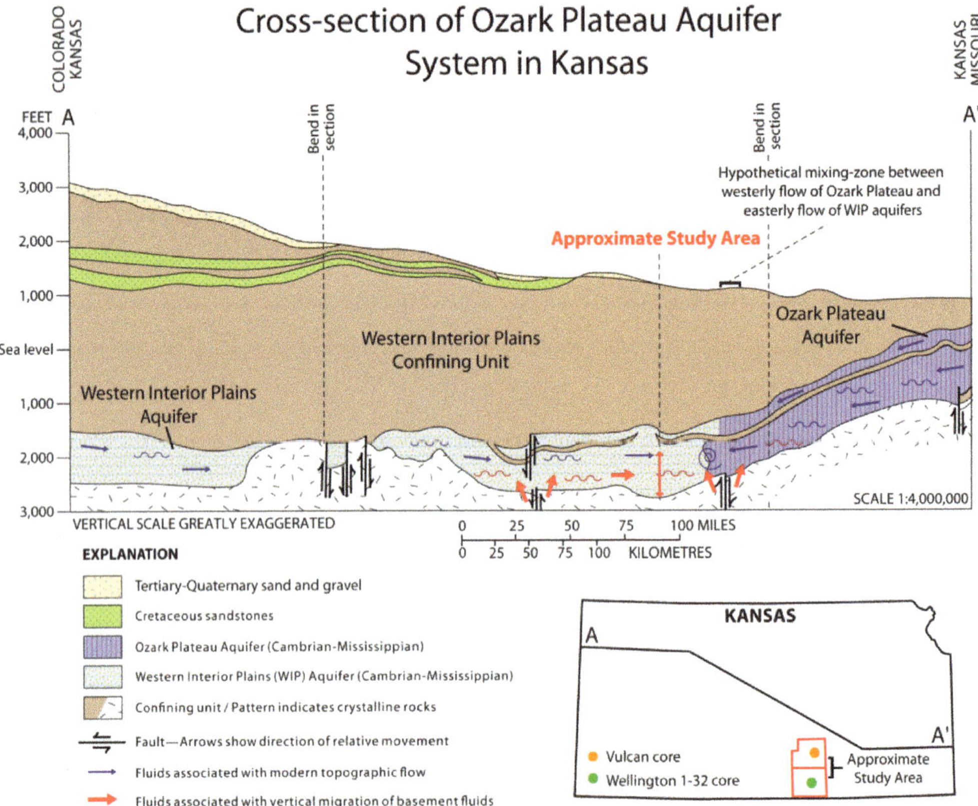

Fig. 34. Cross-section of the OPAS, showing the potential sources and migration pathways of fluids associated with late calcite cementation in the Cambrian–Ordovician Arbuckle Group. The Arbuckle Group is the basal component of the OPAS system (Carr *et al.* 2005). The illustration depicts easterly flow associated with the OPAS throughout the majority of Kansas, as well as the potential mixing zone between the Western Interior Plains aquifer and westerly flow from the Ozark Plateau aquifer in the eastern part of the state (Carr *et al.* 2005). Following strike-slip faulting (hypothesized as occurring during Ouachita or Laramide tectonic activity) and the extension of faults into overlying strata, high-temperature, highly radiogenic Reagan sandstone or basement fluids migrate along fractures into overlying strata (illustrated by red symbols). The proximity of specific areas to fractures may have resulted in enhanced fluid flow in the areas directly affected by fracturing (the Wellington 1-32 core in the case of this study), whereas areas located further from the fractures would display lower temperatures and more rock-dominated isotopic values (the Vulcan core in the case of this study). The injected fluids precipitated high-temperature, highly radiogenic CC in the Arbuckle Group. Fluid flow associated with tectonic activity could mark the transition from an advective system with fluids sourced from the Anadarko Basin to a fracture-controlled fluid flow system with fluids sourced from underlying sandstone or basement rock. Modified from Miller & Appel (1997).

Arbuckle Group rather than at the base. The gravity-driven hydrologic model can account for both the initial low-salinity fluids (connate fluids) associated with MQ2 (Fig. 33a) and later high-salinity fluids (acquired from Permian evaporite reflux; Fig. 33b) associated with BD (Fig. 33c). Repeated deformation could have given rise to pulses of fluid flow by episodically affecting the regional aquifer – essentially adjusting the flow repeatedly, akin to adjusting a valve. Again, the radiogenic compositions and localized nature of calcite samples require a different source and more localized fluid migration pathway than regionally advective flow driven from a hydraulic head in the mountains.

Timing of hydrothermal events

The fact that the fluids responsible for megaquartz precipitation recorded low salinities requires migration prior to the hypothesized reflux of evaporative Permian fluids (Fig. 33a). Either during or after Permian reflux (Fig. 33b), the migration of the highly saline fluids responsible for BD precipitation occurred (Fig. 33c). Lower salinity values in early BD suggest that fluid migration may have occurred during the Early–Middle Permian, with early reflux fluids potentially mixing with lower-salinity connate fluids. Precipitation of the highest-salinity BD may represent migration after the basin was completely flushed by refluxing fluids later in the Permian. Strike-slip faulting occurring after BD precipitation in the study area may have pumped underlying fluids into overlying strata and allowed for vertical migration of the high-temperature and highly radiogenic fluids responsible for late-stage calcite precipitation (Fig. 34). The fact that calcite precipitates after BD – and the fact that fluids responsible for BD precipitation are hypothesized as being derived from Permian-age reflux – implies that the Ouachita orogeny may be an unlikely timing for the strike-slip faulting because, for the most part, Ouachita deformation ended by the Early Permian (Flawn *et al.* 1961; Frezon & Dixon 1975; Kluth & Coney 1981; Marshak *et al.* 2003). Coveney *et al.* (2000) summarized radiometric dates of multiple calcite cements in the nearby Tri-State Area that included 251 $\pm$ 11, 137 $\pm$ 3, 66 $\pm$ 2 and 39 $\pm$ 2 Ma (Blasch & Coveney 1988; Brannon *et al.* 1996*a*, *b*). 251 $\pm$ 11 and 137 $\pm$ 3 Ma both appear to fall between the Ouachita and Laramide orogenies. One of the older dates suggests that calcite could be related to Mesozoic rifting of the Gulf of Mexico. The two youngest dates (66 $\pm$ 2 and 39 $\pm$ 2 Ma) coincide with the estimated timing of the Laramide orogeny (Tweto 1980; Coveney *et al.* 2000). If the calcite observed in the Arbuckle Group is associated with the younger dates then strike-slip faulting or reactivation of basement faults during the Laramide orogeny could be a plausible mechanism for fluid flow. Further research on the calcite cements is required to obtain a more complete understanding of its origin.

Conclusions

Transmitted light, UV epifluorescence and CL microscopy reveal distinctive features within the Arbuckle Group mineral assemblage, enabling the generation of a paragenesis that can be separated into early- and late-stage diagenetic events based on the timing relative to compaction. At some point after the commencement of fracturing hydrothermal brines were injected into the unit, initially resulting in the dissolution of silica phases and carbonates and then precipitation of late-stage cements. T_h values measured in late-stage MQ2 ($\geq$87°C), BD (93.0–131.0°C) and CC (70.5–89.8°C) reveal entrapment temperatures higher than would be expected from elevated geothermal gradients or normal burial conditions, necessitating a hydrothermal influence during late-stage cementation. Low salinities measured in MQ2 (3.1–6.0 wt% NaCl eq.) mark the beginning of late-stage cementation and an increase in the salinity of subsequent BD (16.3–20.4 wt% NaCl eq.) and secondary inclusions in CC (18.2–19.6 wt% NaCl eq.) implies interaction with Permian-age evaporated fluids or evaporite dissolution within the Anadarko basin during the later stages of the fluid migration history. Both high temperatures and high salinities support a basinal or basement source for the hydrothermal brines.

Geochemical data further support the migration of basin-derived hydrothermal fluids within Arbuckle Group strata. Depleted $\delta^{18}O$ values for BD and CC can be attributed to thermal fractionation produced from elevated fluid temperatures, whereas the enrichment of $\delta^{18}O_{water}$ supports evaporative brines as the origin. $\delta^{18}O_{dolomite}$ values become more depleted with decreasing depth, supporting higher temperatures with decreasing depth in the Arbuckle Group, temperature-controlled density stratification and possibly some preferential fluid flow in the upper Arbuckle Group. Slightly more depleted $\delta^{13}C$ values in the upper Arbuckle Group may record TSR that could have persisted along with the higher temperatures produced in the upper portion of the unit. Radiogenic $^{87}Sr/^{86}Sr$ values imply rock–water interaction with siliciclastics or basement rock, probably during migration through various stratigraphic sections of a sedimentary basin before migration into the study area or, if the fluids were basement derived, interaction with basement rock before vertical migration into overlying strata.

The most likely source for the hydrothermal basinal brines responsible for MQ2 and BD precipitation appears to be the Anadarko basin. The believed association of hydrocarbon production in the basin with hydrocarbons present in the Arbuckle Group of Kansas supports the migration of fluids from the basin into the study area. The immense depth of the basin, especially of the Cambrian–Ordovician strata (*c.* 9 km (*c.* 29 500 ft) in the deepest portions of basin), would easily provide the heat necessary for the most elevated T_h values (131°C) and prevalent siliciclastic material within the basin could have caused the radiogenic $^{87}Sr/^{86}Sr$ values. Initial discharge of lower-salinity (3.1–6.0 wt% NaCl eq.) connate fluids would account for MQ2. Reflux of evaporative fluids generated during the Permian could have recharged the basin, accounting for discharge of high-salinity (16.3–20.4 wt% NaCl eq.) fluids that are responsible for the later precipitation of BD and CC.

A combination of tectonically valved and gravity-driven fluid flow most adequately accounts for the petrographic observations, fluid inclusion analysis and geochemical data derived from MQ2 and BD, especially when linked to the stratigraphic and tectonic history of the Anadarko basin and the study area. Fracturing during the Ouachita orogeny would have allowed for tectonic valving of basin fluids and uplifts would have provided a topographic high adjacent to the deep Anadarko basin, setting the stage for gravity-driven fluid flow. Migration probably began with fracturing of basinal strata, discharging connate fluids first followed by the continued reflux and recharge of evaporative fluids during the Permian.

The precipitation of CC appears to mark a transition from an advective fluid-flow system, with fluids being sourced from the Anadarko basin, to a vertical fluid-flow system with fluids being sourced from the underlying Cambrian Reagan sandstone or the basement. $^{87}Sr/^{86}Sr$ values recorded in late CC are much more radiogenic than the values observed in BD, necessitating an alternative source or fluid migration pathway, and $\delta^{18}O$ values in calcite do not appear to be dependent on depth or the enhanced porosity zone at the top of the Arbuckle Group. Fracturing related to the Laramide orogeny may have reactivated fractures present in basement rock and extended those fractures into overlying strata, providing conduits for high-temperature, highly radiogenic fluids from the Reagan sandstone or basement to migrate into the Arbuckle Group.

This project was largely completed as the MSc thesis research of Bradley King, which was supported by the U.S. Department of Energy (DOE) National Energy Technology Laboratory (NETL) under Grant No. DEFE0000002056. The project was managed and administered by the Kansas Geological Survey/KUCR, W. L. Watney, PI, and funded by DOE/NETL and cost-sharing partners. Additional support was provided by the sponsors of the KICC.

References

ARNE, D.C. 1992. Evidence from apatite fission-track analysis for regional Cretaceous cooling in the Ouachita mountain fold belt and Arkoma Basin of Arkansas. *AAPG Bulletin*, **76**, 392–402.

ARNE, D.C., DUDDY, I.R. & SANGSTER, D.F. 1990. Thermochronologic constraints on ore formation at the Gays River Pb-Zn deposit Nova Scotia, Canada, from apatite fission track analysis. *Canadian Journal of Earth Sciences*, **27**, 1013–1022.

ARTHUR, M.A., ANDERSON, T.F., KAPLAN, I.R., VEIZER, J. & LAND, L.S. 1983. *Stable Isotopes in Sedimentary Geology*. SEPM Short Course, 10.

BACHTADSE, V., VAN DER VOO, R., HAYNES, F.M. & KESLER, S.E. 1987. Late Paleozoic magnetism of mineralized and unmineralized Ordovician carbonates from East Tennessee: evidence for post-ore chemical event. *Journal of Geophysical Research*, **92**, 14165–14176.

BAKKER, R.J. 2008. *AqSoVir software package fluids*. Fluid Inclusion Laboratory Leoben, Leoben, http://fluids.unileoben.ac.at

BANNER, J.L. 1995. Application of the trace element and isotope geochemistry of strontium to studies of carbonate diagenesis. *Sedimentology*, **42**, 805–824.

BETHKE, C.M. 1986. Hydrologic constraints on the genesis of the upper Mississippi valley mineral district from Illinois Basin brines. *Economic Geology*, **81**, 233–249.

BETHKE, C.M. & MARSHAK, S. 1990. Brine migrations across North America – the plate tectonics of groundwater. *Annual Review of Earth and Planetary Sciences*, **18**, 287–315.

BLACKBURN, T.J., STOCKLI, D.F., CARLSON, R.W. & BERENDSEN, P. 2008. (U-Th)/He dating of kimberlites- A case study from north-eastern Kansas. *Earth and Planetary Science Letters*, **275**, 111–120.

BLASCH, S.R. & COVENEY, R.M. 1988. Goethite-bearing brine inclusions, petroleum inclusions, and the geochemical conditions of ore deposition at the Jumbo mine, Kansas. *Geochimica et Cosmochimica Acta*, **52**, 1007–1017.

BODNAR, R.J. 1993. Revised equation and table for determining the freezing point depression of H_20-NaCl solutions. *Geochimica et Cosmochimica Acta*, **57**, 683–684.

BRANNON, J.C., COLE, S.C., PODOSEK, F.A., RAGAN, V.M., COVENEY, R.M., JR, WALLACE, M.W. & BRADLEY, A.J. 1996*a*. Th-Pb and U-Pb dating of ore-stage calcite and Paleozoic fluid flow. *Science*, **271**, 491–493.

BRANNON, J.C., PODOSEK, F.A. & COLE, S.C. 1996*b*. Radiometric dating of Mississippi Valley-type ore deposits. *In*: SANGSTER, E.D. (ed.) *Carbonate-Hosted Lead-Zinc Deposit*. Society of Economic Geologists, Special Publications, **4**, 546–554.

BURGESS, W.J. 1976. Geologic evolution of the mid-Continent and Gulf Coast areas – a plate tectonics view. *Transactions-Gulf Coast Association of Geological Societies*, **26**, 132–143.

Carr, T.R., Merriam, D.C. & Bartley, J.D. 2005. Use of relational databases to evaluate regional petroleum accumulation, groundwater flow, and CO_2 sequestration potential in Kansas. *AAPG Bulletin*, **89**, 1607–1627.

Cathles, L.M. & Smith, A.T. 1983. Thermal constraints on the formation of the Mississippi Valley-type lead-zinc deposits and their implications for episodic basin dewatering and deposit genesis. *Economic Geology*, **78**, 983–1002.

Choquette, P.W. & Pray, L.C. 1970. Geologic nomenclature and classification of porosity in sedimentary carbonates. *AAPG Bulletin*, **54**, 207–250.

Coveney, R.M., Jr. 1992. Evidence for expulsion of hydrothermal fluids and hydrocarbons in the Midcontinent during the Pennsylvanian. *In*: Johnson, K.S. & Cardott, B.J. (eds) *Source Rocks in the Southern Midcontinent, 1990 Symposium*. Oklahoma Geological Survey Circular, **93**, 133–143.

Coveney, R.M., Jr, Ragan, V.M. & Brannon, J.C. 2000. Temporal benchmarks for modeling Phanerozoic flow of basinal brines and hydrocarbons in the southern Midcontinent based on radiometrically dated calcite. *Geology*, **28**, 795–798.

Cox, S.F. 2010. The application of failure mode diagrams for exploring the roles of fluid pressure and stress states in controlling styles of fracture-controlled permeability enhancement in faults and shear zones. *Geofluids*, **10**, 217–233.

Davies, G.R. & Smith, T. 2006. Structurally controlled hydrothermal dolomite reservoir facies: an overview. *AAPG Bulletin*, **90**, 1641–1690.

Denison, R.E., Koepnick, R.B., Burke, W.H. & Hetherington, E.A. 1998. Construction of the Cambrian and Ordovician seawater $^{87}Sr/^{86}Sr$ curve. *Chemical Geology*, **152**, 325–340.

Dozy, J.J. 1970. A geological model for the genesis of lead-zinc ores of the Mississippi valley, U.S.A. *Transactions Applied Earth Science*, **79**, B163–B170.

Elebiju, O.O., Matson, S., Keller, G.R. & Marfurt, K.J. 2011. Integrated geophysical studies of the basement structures, the Mississippian chert, and the Arbuckle Group of Osage County region, Oklahoma. *AAPG Bulletin*, **95**, 371–393.

Epstein, S. & Mayeda, T. 1953. Variation of ^{18}O content of waters from natural sources. *Geochimica et Cosmochimica Acta*, **4**, 213–224.

Flawn, P.T., Goldstein, A., Jr, King, P.B. & Weaver, C.E. 1961. The Ouachita System. *In*: *Bureau of Economic Geology*. Report no. 6120. The University of Texas, Austin.

Folk, R.L. 1965. Some aspects of recrystallization in ancient limestones. *In*: Pray, L.C. & Murray, R.C. (eds) *Dolomitization and Limestone Diagenesis: A Symposium*. SEPM Special Publications, **13**, 14–48.

Franseen, E.K. 2000. A review of Arbuckle Group strata in Kansas from a sedimentologic perspective: insights for future research from past and recent studies: the Compass. *Journal of Earth Sciences Sigma Gamma Epsilon*, **75**, 68–89.

Franseen, E.K., Byrnes, A.P., Cansler, J., Steinhauff, D.M. & Carr, T.R. 2004. The Geology of Kansas-Arbuckle Group: current research in earth sciences. *Kansas Geological Survey Bulletin*, **250(part 2)**, 43p, http://www.kgs.ku.edu/Current/2004/franseen/index.html

Frezon, S.E. & Dixon, G.H. 1975. Paleotectonic investigations of the Pennsylvanian System in the United States, Texas Panhandle and Oklahoma. *In*: McKee, E.D. & Crosby, E.J. (eds) *Paleotectonic investigations of the Pennsylvanian System in the United States*. U.S. Geological Survey Professional Paper, **853**, 177–195.

Gallardo, J. & Blackwell, D.D. 1999. Thermal structure of the Anadarko Basin. *AAPG Bulletin*, **83**, 333–361.

Gao, G. 1990. Geochemical and isotopic constraints on the diagenetic history of a massive stratal, late Cambrian (Royer) dolomite, lower Arbuckle Group, Slick Hills, SW Oklahoma, USA. *Geochimica et Cosmochimica Acta*, **54**, 1979–1989.

Gao, G. & Land, L.S. 1991. Early Ordovician Cool Creek Dolomite, middle Arbuckle Group, Slick Hills, SW Oklahoma, U.S.A.: origin and modification. *Journal of Sedimentary Petrology*, **61**, 161–173.

Gao, G., Land, L.S. & Elmore, R.D. 1995. Multiple episodes of dolomitization in the Arbuckle Group, Arbuckle Mountains, south-central Oklahoma. Field, petrographic, and geochemical evidence. *Journal of Sedimentary Research*, **A65**, 321–331.

Garven, G. 1993. Genesis of stratabound ore deposits in the Midcontinent basins of North America; 1, the role of regional groundwater flow. *The American Journal of Science*, **293**, 497–568.

Garven, G. 1995. Continental-scale groundwater flow and geologic processes. *Annual Review of Earth and Planetary Sciences*, **23**, 89–117.

Garven, G. & Freeze, R.A. 1984*a*. Theoretical analysis of the role of groundwater flow in the genesis of stratabound ore deposit: 1 . Mathematical and numerical model. *The American Journal of Science*, **284**, 1085–1124.

Garven, G. & Freeze, R.A. 1984*b*. Theoretical analysis of the role of groundwater flow in the genesis of stratabound ore deposits: 2. Quantitative results. *The American Journal of Science*, **284**, 1125–1174.

Gay, S.P., Jr, 2003*a*. The Nemaha trend – a system of compressional thrust-fold, strike- slip structural features in Kansas and Oklahoma, Part 1. *Shale Shaker*, **53**, 9–17.

Gay, S.P., Jr, 2003*b*. The Nemaha trend – a system of compressional thrust-fold, strike- slip structural features in Kansas and Oklahoma, Part 2. *Shale Shaker*, **54**, 39–49.

Goldstein, R.H. 2003. Ch. 2: petrographic analysis of fluid inclusions. *In*: Samson, I., Anderson, A. & Marshall, D. (eds) *Fluid Inclusions: Analysis and Interpretation*. Mineralogical Association of Canada Short Course, **32**, 9–53.

Goldstein, R.H. 2012. Fluid inclusion geothermometry in sedimentary systems: from paleoclimate to hydrothermal. *In*: Harris, N. (ed.) *Analyzing the Thermal History of Sedimentary Basins: Methods and Case Histories*. SEPM Special Publications, **103**, 45–63.

Goldstein, R.H. & Reynolds, T.J. 1994. *Systematics of Fluid Inclusions in Diagenetic Minerals*. SEPM Short Course, 31.

GREGG, J.M. & SIBLEY, D.F. 1984. Epigenetic dolomitization and the origin of xenotopic dolomite texture. *Journal of Sedimentary Petrology*, **54**, 908–931.

HAGNI, R.D. & GRAWE, O.R. 1964. Mineral paragenesis in the Tri-state district Missouri, Kansas, Oklahoma. *Economic Geology*, **59**, 449–457.

HANOR, J.S. 1979. The sedimentary genesis of hydrothermal fluids. *In*: BARNES, H.L. (ed.) *Geochemistry of Hydrothermal Ore Deposits*. 2nd edn, John Wiley, New York, 137–172.

HEIMSTRA, E.J. 2003. *The diagenesis and fluid migration history of the Indian Basin field, Eddy County, New Mexico*. MSc thesis, University of Kansas.

JOHNSON, K.S., AMSDEN, T.W. *ET AL*. 1988. Southern Midcontinent region. *In*: SLOSS, L.L. (ed.) *Sedimentary Cover – North American craton*. US Geological Society of America, The Geology of North America, **D-2**, 307–359.

JORGENSEN, D.G., HELGESEN, J.O. & IMES, J.L. 1993. *Regional aquifers in Kansas, Nebraska, and parts of Arkansas, Colorado, Missouri, New Mexico, Oklahoma, South Dakota, Texas, and Wyoming – Geohydrologic framework*. U.S. Geological Survey Professional Paper **1414-B**.

KERANS, C. 1988. Karst-controlled reservoir heterogeneity in Ellenburger Group carbonates of West Texas. *AAPG Bulletin*, **72**, 1160–1183.

KLUTH, C.F. & CONEY, P.J. 1981. Plate tectonics of the Ancestral Rocky Mountains. *Geology*, **9**, 10–15.

KUPECZ, J.A. & LAND, L.S. 1991. Late-stage dolomitization of the Lower Ordovician Ellenburger Group, West Texas. *Journal of Sedimentary Petrology*, **61**, 551–574.

LAND, L.S. 1983. The application of stable isotopes to studies of the origin of dolomite and to problems of diagenesis of clastic sediments. *In*: ARTHUR, M.A. (ed.) *Stable Isotopes in Sedimentary Geology*. SEPM Short Course, **10**, 4-1–4-21.

LAND, L.S. 1985. The origin of massive dolomite. *Journal of Geological Education*, **33**, 112–125.

LAND, L.S., MACPHERSON, G.L. & MACK, L.E. 1988. The geochemistry of saline formation waters, Miocene, offshore Louisiana. *Transactions-Gulf Coast Association of Geological Societies*, **38**, 503–511.

LANGE, S., CHAUDHURI, S. & CLAUER, N. 1983. Strontium isotopic evidence for the origin of barites and sulfides from the Mississippi Valley-type ore deposits in southeast Missouri. *Economic Geology*, **78**, 1255–1261.

LEACH, D.L. & ROWAN, E.L. 1986. Genetic link between Ouachita foldbelt tectonism and the Mississippi valley-type lead-zinc deposits of the Ozarks. *Geology*, **14**, 931–935.

LEACH, D.L. & SANGSTER, D.F. 1995. Mississippi Valley-type lead-zinc deposits. *In*: KIRKHAM, R.V., SINCLAIR, W.D., THORP, R.I. & DUKE, J.M. (eds) *Mineral Deposit Modeling*. Geological Association of Canada Special Papers, **40**, 289–314.

LEACH, D.L., BRADLEY, D., LEWCHUK, M.T., SYMONS, D.T.A., MARSILY, G. & BRANNON, J. 2001. Mississippi Valley-type lead-zinc deposits through geological time: implications from recent age-dating research. *Mineralium Deposita*, **36**, 711–740.

LEWCHUCK, M.T. & SYMONS, D.T.A. 1996. Paleomagnetism of Paleozoic carbonates in the Illinois-Kentucky-Tennessee area: the recording of fluid flow events. *In*: SANGASTER, D.F. (ed.) *Carbonate-Hosted Lead-Zinc Deposits*. Society of Economic Geologists Special Publications, **4**, 555–566.

LOHMANN, K.C. 1988. Geochemical patterns of meteoric diagenetic systems and their application to studies of paleokarst. *In*: JAMES, N.P. & CHOQUETTE, P.W. (eds) *Paleokarst*. Ch. 2, Springer-Verlag, New York, 58–80.

LUPI, M., GEIGER, S. & GRAHAM, C.M. 2011. Numerical simulations of seismicity – induced fluid flow in the Tjornes Fracture Zone, Iceland. *Journal of Geophysical Research*, **116**, B07101, https://doi.org/10.1029/2010JB007732

MACHEL, H.G. 1987. Saddle dolomite as a by-product of chemical compaction and thermochemical sulfate reduction. *Geology*, **15**, 936–940.

MACHEL, H.G. 1999. Effects of groundwater flow on mineral diagenesis, with emphasis on carbonate aquifers. *Hydrogeology Journal*, **7**, 94–107.

MACHEL, H.G. 2001. Bacterial and thermochemical sulfate reduction in diagenetic settings - old and new insights. *Sedimentary Geology*, **140**, 143–175.

MARSHAK, S., NELSON, W.J. & MCBRIDE, J.H. 2003. Phanerozoic strike-slip faulting in the continental interior platform of the United States. Examples from the Laramide Orogen, Midcontinent, and Ancestral Rocky Mountains. *In*: STORTI, F., HOLDSWROTH, R.E. & SALVINI, F. (eds) *Intraplate Strike-Slip Deformation Belts*. Geological Society, London, Special Publications, **210**, 159–184.

MILLER, J.A. & APPEL, C.L. 1997. Kansas, Missouri, Nebraska. *In*: *Groundwater Atlas of the United States*. United States Geological Survey, H730-D, http://pubs.usgs.gov/ha/ha730/ch_d/D-text.html

MONTANEZ, I.P. & READ, J.F. 1992. Fluid-rock interaction history during stabilization of early dolomites, upper Knox Group (Lower Ordovician), U.S. Appalachians. *Journal of Sedimentary Petrology*, **62**, 753–778.

MOUNTJOY, E.W., MACHEL, H.G., GREEN, D., DUGGAN, J. & WILLIAMS-JONES, A.E. 1999. Devonian matrix dolomites and deep burial carbonate cements. A comparison between the Rimbey Meadowbrook reef trend and the deep basin of west-central Alberta. *Bulletin of Canadian Petroleum Geology*, **47**, 487–509.

MUSGROVE, M. & BANNER, J.L. 1993. Regional groundwater mixing and the origin of saline fluids: Midcontinent, United States. *Science*, **259**, 1877–1882.

NAKAI, S., HALLIDAY, A.N., KESLER, S.F. & JONES, H.D. 1990. Rb-Sr dating of sphalerites from Tennessee and the genesis of Mississippi Valley-type ore deposits. *Nature*, **346**, 354–357.

NAKAI, S., HALLIDAY, A.N., KESLER, S.F., JONES, H.D., KYLE, J.R. & LANE, T.E. 1993. Rb-Sr dating of sphalerites from Mississippi Valley-type (MVT) ore deposits. *Geochimica et Cosmochimica Acta*, **57**, 417–427.

NEWELL, K.D. 1997. Comparison of maturation data and fluid inclusion homogenization temperatures to simple thermal models: implications for thermal history and fluid flow in the Midcontinent. *Current Research in Earth Sciences, Kansas Geological Survey Bulletin*, **240(part 2)**, 13–27, http://www.kgs.ku.edu/Current/1997/newell/newell1.html

OHLMACHER, G.S. & BERENDSEN, P. 2005. Kinematics, mechanics, and potential earthquake hazards for faults in Pottawatomie County, Kansas, USA. *Tectonophysics*, **396**, 227–244.

OLIVER, J. 1986. Fluids expelled tectonically from orogenic belts: their role in hydrocarbon migration and other geologic phenomena. *Geology*, **14**, 99–102.

O'NEIL, J.R., CLAYTON, R.N. & MAYEDA, T.K. 1969. Oxygen isotope fractionation in divalent metal carbonates. *Journal of Chemical Physics*, **51**, 5547–5558.

PAN, H., SYMONS, D.T.A. & SANGSTER, D.F. 1990. Paleomagnetism of Mississippi Valley-type ore and host rocks in the northern Arkansas and Tri-State districts. *Canadian Journal of Earth Sciences*, **27**, 923–931.

PRICE, L. 1980. Shelf and shallow basin oil as related to hot-deep origin of petroleum. *Journal of Petroleum Geology*, **3**, 91–116.

RADKE, B.M. & MATHIS, R.L. 1980. On the formation and occurrence of saddle dolomite. *Journal of Sedimentary Petrology*, **50**, 1149–1168.

RAMAKER, E.M., GOLDSTEIN, R.H., FRANSEEN, E.K. & WATNEY, W.L. 2014. What controls cherty fine-grained carbonate reservoir rocks? Impact of stratigraphy, unconformities, structural setting and hydrothermal fluid flow: Mississippian, southeast Kansas. *In*: AGAR, S. & GEIGER, S. (eds) *Fundamental Controls on Fluid Flow in Carbonates: Current Workflows to Emerging Technologies*. Geological Society, London, Special Publications, **406**, 179–208, https://doi.org/10.1144/SP406.2

ROSS, C.A. & ROSS, J.R. 1988. Late Paleozoic transgressive-regressive deposition. *In*: WILGUS, C.K., HASTINGS, B.S., KENDALL, C.G., POSAMENTIER, H.W., ROSS, C.A. & VAN WAGONER, J.C. (eds) *Sea Level Changes – An Integrated Approach*. Tulsa, OK, SEPM Special Publications, **42**, 227–247.

SCHEFFER, A. 2012. *Geochemical and microbiological characterization of the Arbuckle saline aquifer, a potential CO_2 storage reservoir; implications for hydraulic separation and caprock integrity*. MSc thesis, University of Kansas.

SHARP, J.M. 1978. Energy and momentum transport model of the Ouachita basin and its possible impact on formation of economic mineral deposits. *Economic Geology*, **73**, 1057–1068.

SHELTON, K.L., GREGG, J.M. & JOHNSON, A.W. 2009. Replacement dolomites and ore sulfides as recorders of multiple fluids and fluid sources in the Southeast Missouri Mississippi Valley-Type District. Halogen-$^{87}Sr/^{86}Sr$-$\delta^{18}O$-$\delta^{34}S$ Systematics in the Bonneterre Dolomite. *Economic Geology*, **104**, 733–748.

SIBSON, R.H. 1981. Fluid flow accompanying faulting. Field evidence and models. *In*: SIMPSON, D.W. & RICHARDS, P.G. (eds) *Earthquake Prediction: An International Review*. American Geophysical Union, Maurice Ewing Series, **4**, 593–603.

SIBSON, R.H. 1987. Earthquake rupturing as a hydrothermal mineralizing agent. *Geology*, **15**, 701–704.

SIBSON, R.H. 2003. Brittle-failure controls on maximum sustainable overpressure in different tectonic regimes. *AAPG Bulletin*, **87**, 901–908.

SIBSON, R.H., ROBERT, F. & POULSEN, K.H. 1988. High-angle reverse faults, fluid-pressure cycling, and mesothermal gold-quartz deposits. *Geology*, **16**, 551–555.

SIMO, G.L. & SMITH, J.A. 1997. Carbonate diagenesis and dolomitization of the lower Ordovician Prairie du Chien Group. *Geoscience Wisconsin*, **16**, 1–16.

SVERJENSKY, D.A. 1986. Genesis of Mississippi valley-type lead-zinc deposits. *Annual Review of Earth and Planetary Sciences*, **14**, 177–199.

SYMONS, D.T.A. & SANGSTER, D.F. 1991. Paleomagnetic age of the Central Missouri barite deposits and its genetic implications. *Economic Geology*, **86**, 1–12.

SYMONS, D.T.A. & STRATAKOS, K.K. 2000. Palaeomagnetic dating of dolomitization and Mississippi Valley-type zinc mineralization in the Mascot-Jefferson City district of eastern Tennessee. A preliminary analysis. *Journal of Geochemical Exploration*, **69–70**, 373–376.

SYMONS, D.T.A., LEWCHUK, M. & LEACH, D.L. 1998*a*. Age and duration of the Mississippi Valley-type mineralizing fluid flow events in the Viburnum Trend, southeast Missouri, USA, from paleomagnetism. *In*: PARNELL, J. (ed.) *Dating and Duration of Fluid Flow and Fluid-Rock Interaction*. Geological Society, London, Special Publications, **144**, 27–39.

TWETO, O. 1980. Summary of Laramide orogeny in Colorado. *In*: KENT, H.C. & PORTER, K.W. (eds) *Colorado Geology*. Rocky Mountain Association of Geologists Symposium on Colorado, Denver, CO, 129–134.

VOSS, R.L., HAGNI, R.D. & GREGG, J.M. 1989. Sequential deposition of zoned dolomite and its relationship to sulfide mineral paragenetic sequence in the Vibernum Trend, southeast Missouri. *Carbonates and Evaporites*, **4**, 195–210.

WATNEY, W.L., FRANSEEN, E.K., BYRNES, A.P. & NISSEN, S.E. 2008. Contrasting styles and common controls on Middle Mississippian and Upper Pennsylvanian carbonate platforms in the Northern midcontinent, U.S.A. *In*: LUKASIK, J. & SIMO, A. (eds) *Controls on Carbonate Platform and Reef Development*. SEPM Special Publications, **89**, 125–145.

WATNEY, W.L., YOULE, J. ET AL. 2013. Sedimentologic and stratigraphic effects of episodic structural activity during the Phanerozoic in the Hugoton Embayment, Kansas USA. *AAPG Annual Meeting*, 21 May 2013, Pittsburgh, PA.

WHARTON, G.C. 2012. *Transitional tectonics: early Laramide strike-slip deformation of the northeastern Front Range, Colorado*. MSc thesis, University of Texas-Austin.

WISNIOWIECKI, M.J., VAN DER VOO, R., MCCABE, C. & KELLY, W.C. 1983. A Pennsylvanian paleomagnetic pole from the mineralized Late Cambrian Bonneterre Formation, southeast Missouri. *Journal of Geophysical Research*, **88**, 6540–6548.

WOJCIK, K.M., MCKIBBEN, M.E., GOLDSTEIN, R.H. & WALTON, A.W. 1992. Diagenesis, thermal history, and fluid migration, Middle and Upper Pennsylvanian rocks, southeastern Kansas. *Oklahoma Geological Survey Circular*, **93**, 144–159.

WOJCIK, K.M., GOLDSTEIN, R.H. & WALTON, A.W. 1994. History of diagenetic fluids in a distant foreland area, Middle and Upper Pennsylvanian, Cherokee Basin, Kansas, USA. Fluid inclusion evidence. *Geochimica et Cosmochimica Acta*, **58**, 1175–1191.

WOJCIK, K.M., GOLDSTEIN, R.H. & WALTON, A.W. 1997. Regional and local controls of diagenesis driven by basin-wide flow system: Pennsylvanian sandstones and limestones, Cherokee basin, southeastern Kansas. *In*: MONTANEEZ, I.P., GREGG, J.M. & SHELTON, K.L. (eds) *Basin-Wide Diagenetic Patterns; Integrated Petrologic, Geochemical, and Hydrologic Considerations*. SEPM Special Publications, **57**, 235–252.

YE, H., ROYDEN, L., BURCHFIEL, C. & SCHUEPBACH, M. 1996. Late Paleozoic deformation of interior North America: the Greater Ancestral Rocky Mountains. *AAPG Bulletin*, **80**, 1397–1492.

YOUNG, E.M. 2010. *Controls on reservoir character in carbonate-chert strata, Mississippian (Osagean-Meramecian), southeast Kansas*. MSc thesis, University of Kansas.

Enhanced porosity preservation by pore fluid overpressure and chlorite grain coatings in the Triassic Skagerrak, Central Graben, North Sea, UK

STEPHAN STRICKER* & STUART J. JONES

Department of Earth Sciences, Durham University, South Road, Durham, DH1 3LE, UK

**Correspondence: stephan.stricker@durham.ac.uk*

Abstract: Current understanding of porosity preservation in deeply buried sandstone reservoirs tends to be focused on how diagenetic grain coatings of clay minerals and microquartz can inhibit macroquartz cementation. However, the importance of overpressure developed during initial (shallow) burial in maintaining high primary porosity during subsequent burial has generally not been appreciated. Where pore fluid pressures are high, and the vertical effective stress is low, the shallow arrest of compaction can allow preservation of high porosity and permeability at depths normally considered uneconomic. The deeply buried fluvial sandstone reservoirs of the Triassic Skagerrak Formation in the Central Graben, North Sea, show anomalously high porosities at depths greater than 3500 metres below sea floor (mbsf). Pore pressures can exceed 80 MPa in the upper part of the Skagerrak Formation at depths of 4000–5000 mbsf, where temperatures are above 140°C. The Skagerrak reservoirs commonly have high primary porosities of up to 35%, little macroquartz cement and variable amounts of diagenetic chlorite grain coats. This research sheds light on the complex controls on reservoir quality in the fluvial sandstones of the Skagerrak Formation by identifying the role of shallow overpressure in arresting mechanical compaction and the importance of chlorite detrital grain coatings in inhibiting macroquartz cement overgrowth as temperature increases during progressive burial.

Deeply buried sandstone reservoirs are the cumulative product of depositional processes and diagenesis during burial both at shallow depths, and at greater depths where temperatures are higher. Simple porosity–depth trends can offer some useful guidance but are not always successful in predicting reservoir porosity. Reservoirs with anomalously high porosities are common in several hydrocarbon basins, e.g. Central Graben, North Sea, UK (Osborne & Swarbrick 1999; Nguyen *et al.* 2013; Grant *et al.* 2014); the Gulf of Mexico, USA (Taylor *et al.* 2004; Ehrenberg *et al.* 2008; Ajdukiewicz *et al.* 2010); the Santos Basin, Brazil (Anjos *et al.* 2003) and the Indus Basin, Pakistan (Berger *et al.* 2009). Current understanding of porosity preservation in deeply buried sandstone reservoirs (>4000 mbsf, metres below sea floor) tends to be focused on how coatings of clay and microquartz on detrital grains can inhibit macroquartz cementation. There are many studies where deep reservoir porosity is linked to early diagenetic clay or microquartz grain coats (Pittman *et al.* 1992; Ehrenberg 1993; Aase *et al.* 1996; Bloch *et al.* 2002; Berger *et al.* 2009; Ajdukiewicz & Lander 2010; Ajdukiewicz & Larese 2012; French *et al.* 2012; Worden *et al.* 2012; Bahlis & De Ros 2013). These studies have proven that quartz-rich sandstones with robust and continuous diagenetic clay or microquartz grain coats contain a much lower volume of macroquartz cement than expected (Berger *et al.* 2009; Ajdukiewicz & Lander 2010).

The role played by fluid overpressure in porosity preservation during the burial of sandstones has often been overlooked or considered less significant (e.g. Audet & McConnell 1992; Gaarenstroom *et al.* 1993; Giles 1997; Bloch *et al.* 2002; Taylor *et al.* 2010). The preservation of enhanced secondary porosity has also been attributed to high overpressures and increased porosity at depth (e.g. Wilkinson *et al.* 1997; Haszeldine *et al.* 1999). Increasing vertical effective stress (VES) caused by sediment loading is the major driver of mechanical compaction and porosity reduction during shallow burial. Pore fluid overpressure reduces the stress on intergranular and cement–grain contacts and inhibits both mechanical compaction and pressure dissolution (Swarbrick & Osborne 1998). The shallow onset of pore fluid overpressure enhances porosity preservation, as noted for the Triassic Skagerrak Formation of the Central North Sea (Nguyen *et al.* 2013; Grant *et al.* 2014).

The Triassic Skagerrak Formation is one of the main reservoirs in the high-pressure high-temperature (HPHT) province in the Central

From: Armitage, P. J., Butcher, A. R., Churchill, J. M., Csoma, A. E., Hollis, C., Lander, R. H., Omma, J. E. & Worden, R. H. (eds) 2018. *Reservoir Quality of Clastic and Carbonate Rocks: Analysis, Modelling and Prediction.* Geological Society, London, Special Publications, **435**, 321–341.
First published online January 5, 2016, https://doi.org/10.1144/SP435.4

Graben, North Sea. Pore pressures can exceed 80 MPa in the upper part of the Skagerrak Formation at depths of 4000–5000 mbsf where temperatures are in the range 166–200°C (Swarbrick *et al.* 2000; di Primio & Neumann 2008; Nguyen *et al.* 2013). Overpressures are widespread in the Mesozoic reservoirs in the Central North Sea, a region that since the early Cretaceous has experienced almost continuous sedimentation of dominantly fine-grained lithologies (Osborne & Swarbrick 1999; Swarbrick *et al.* 2000; Yardley & Swarbrick 2000).

Typical porosity–depth trends for siliciclastic reservoirs, show porosities of 10–15% at depths of 4000–5000 mbsf (Ehrenberg *et al.* 2009). However, the Skagerrak reservoir sandstones retain remarkably good porosity, up to 35%, and a low degree of compaction with respect to their present-day depths of burial. It is these higher-than-expected porosities that form the focus of this study. Here we present new results for the diagenesis and pore pressure evolution of the Triassic Skagerrak Formation in the Josephine High (J-Block), located within the Central Graben in UK Quad 30, and in the Heron field located farther north in UK Quad 22.

Basin analysis and burial history modelling have been undertaken to investigate the role of pore fluid overpressure evolution in the maintenance of high primary porosity during shallow burial (<2500 m burial), where mechanical compaction dominates. The basin model was then compared to petrographic data to investigate the role of palaeopressures in the maintenance of high primary porosity at depth. Detailed SEM and petrographic analysis was then applied to investigate the role of clay mineral grain coatings in maintaining high primary porosity in the high-temperature and diagenetic-controlled deep burial phase (>2500 m burial). This simple approach yields results that allow important inferences to be made about the controls on porosity preservation in HPHT reservoirs. These results complement previous studies in the Central Graben, North Sea (e.g. Nguyen *et al.* 2013; Grant *et al.* 2014) and provide important insights into the controls on reservoir quality of deeply buried sandstones.

Geological setting

The Central Graben of the North Sea is part of the NW–SE-trending southern extension of a trilete rift system (i.e. an incipient ridge–ridge triple junction), with the Viking Graben as the northern arm and the Inner and Outer Moray Firth as the western arm. The North Sea Central Graben is 70–130 km wide with an approximate length of 550 km. The rift system separates the Norwegian basement in the east from the UK continental shelf in the west. The North Sea Central Graben consists of the West and the East Central Graben, divided by the Forties–Montrose and Josephine Ridge horst blocks and flanked by marginal platform areas (Fig. 1). The rift system developed in at least two major rifting phases, one during the Permian–Triassic (290–210 Ma) and the second during the Late Jurassic (155–140 Ma) (Gowers & Sæbøe 1985; Glennie 1998). The geological history has commonly been divided into pre-rift, syn-rift and post-rift phases (Clark *et al.* 1999). Syn-rift sediments are mainly siliciclastic Triassic and Jurassic sediments of up to 2000 m in thickness. The post-rift sediments from the Cretaceous to Holocene are up to 4500 m thick (Fig. 2). Post-rift sediments are mainly siliciclastic rocks dominated by shale, sandstone, silty sandstone and a thick Upper Cretaceous Chalk section (Goldsmith *et al.* 2003). Importantly, Upper Cretaceous Chalk units including the Ekofisk, Tor and Hod Formations (Fig. 2) are the main reservoir seals for the sub-Chalk reservoirs in the Central Graben, North Sea (Mallon & Swarbrick 2002, 2008; Swarbrick *et al.* 2010). These highly cemented and mechanically compacted Chalk units have the potential to seal high overpressure in the underlying highly pressured reservoirs (Mallon *et al.* 2005; Mallon & Swarbrick 2008; Swarbrick *et al.* 2010). After the North Atlantic Ocean opened, the Eocene and younger Hordaland and Nordland groups, comprising up to 2500 m of predominantly siltstone and shale, were deposited.

This study focuses on two key HPHT areas in the Central Graben, North Sea: the Heron Cluster in UK quadrant 22, part of the Eastern Trap Area Project (ETAP) area at the southern end of the Forties–Montrose High; and the J-Block area in UK quadrant 30, located on the Josephine Ridge. Both areas are part of a wider HPHT province that includes the Triassic strata of the Central Graben and the southern part of the Viking Graben (Goldsmith *et al.* 2003; Fig. 1).

Triassic Skagerrak stratigraphy

The Triassic Skagerrak Formation in the Central Graben, North Sea comprises 500–1000 m of predominantly continental braided and meandering fluvial systems and terminal fluvial fans with lacustrine facies (McKie & Audretsch 2005; De Jong *et al.* 2006). The Skagerrak sediments accumulated in a series of fault- and salt-controlled mini-basins, or pods, within the overall rift basin (Smith *et al.* 1993; Matthews *et al.* 2007). The thick Zechstein salt, of late Permian age, strongly influenced sedimentation by forming withdrawal basins due to a combination of localized loading and structural extension (Smith *et al.* 1993; Bishop 1996; Matthews *et al.* 2007). Pod development was active in the study area throughout the Triassic and provided

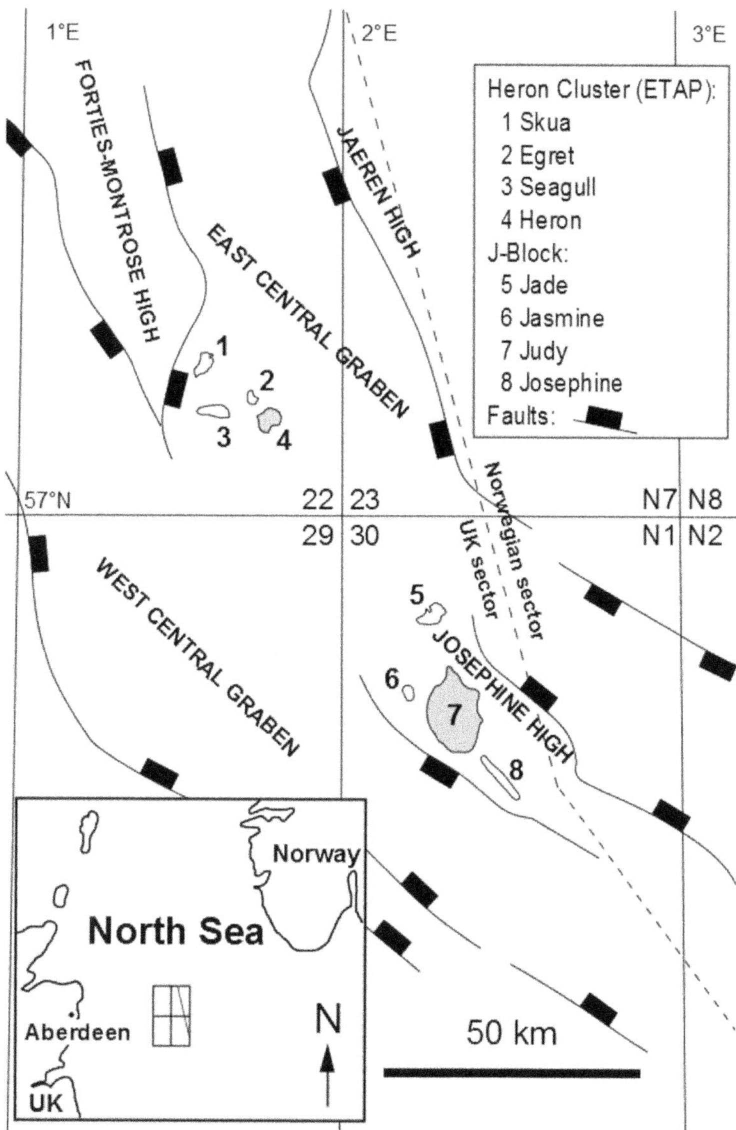

Fig. 1. Location map showing major structural elements of the Central Graben, North Sea (East Central Graben, Forties-Montrose High, Josephine High and West Central Graben) and some Triassic targets within the UK quadrant 22 (Heron Cluster) and UK quadrant 30 (J-Block). The wells chosen are from fields shaded grey.

localized depocentres for the south-easterly flowing Skagerrak fluvial system. The pods were largely responsible for the preservation of the Skagerrak Formation in the study area. However, the salt pods do create facies variability between intra- and inter-pod deposits and influenced reservoir thickness and diagenetic cementation (Nguyen *et al.* 2013).

The stratigraphic nomenclature of the Triassic for the Central Graben was defined by Goldsmith *et al.* (1995), based on detailed biostratigraphic and lithostratigraphic correlation. The Central Graben Triassic succession consists of sediments belonging to the Early Triassic Smith Bank Formation and the Middle to Late Triassic Skagerrak Formation (Fig. 2). The Triassic Skagerrak Formation is subdivided into three sand-dominated units (Judy, Joanne and Josephine) and three mud-dominated units (Julius, Jonathan and Joshua). The sand-dominated units include sheetflood deposits and multistorey

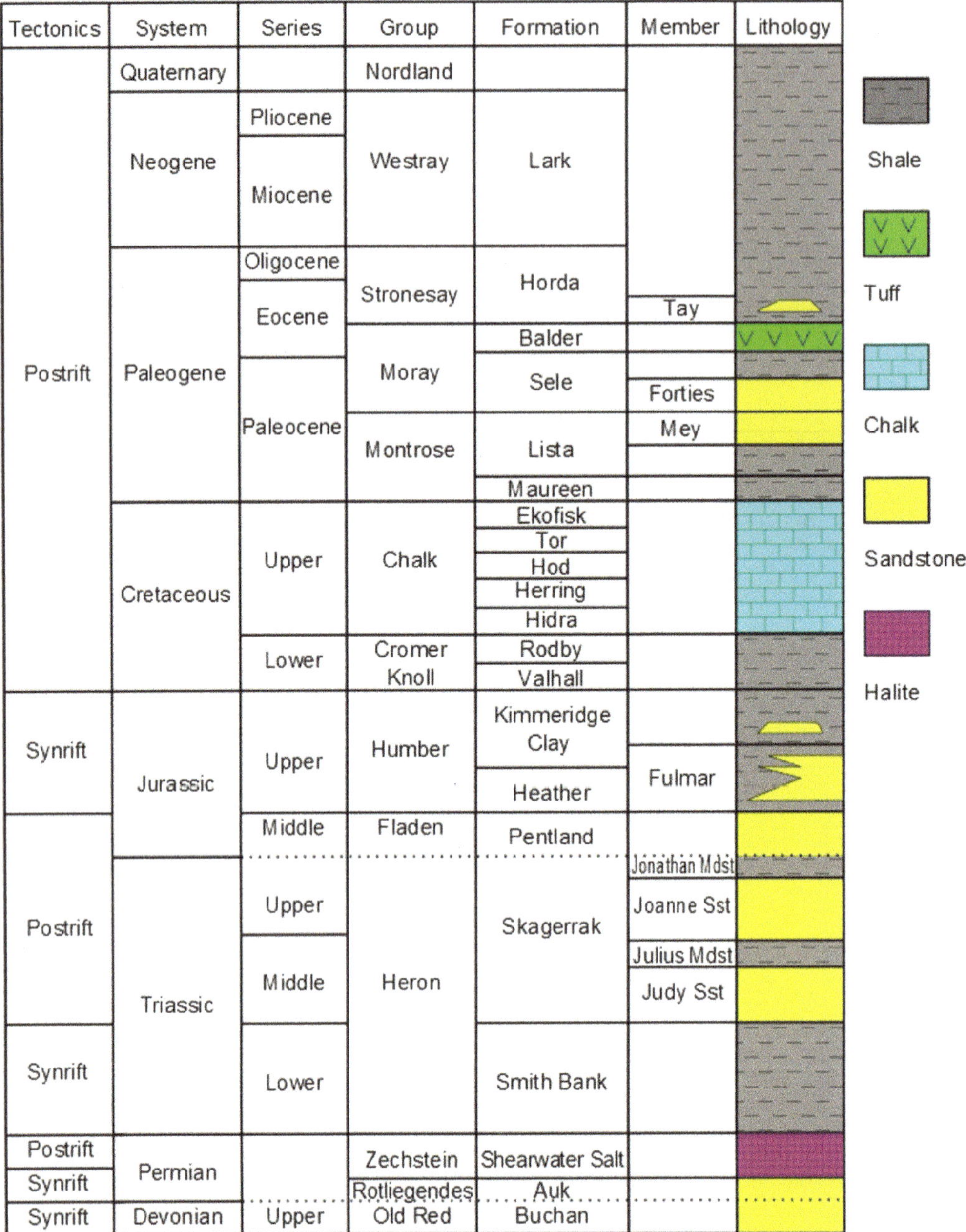

Fig. 2. Regional stratigraphy of the Central Graben, North Sea. North Sea stratigraphic nomenclature based on Knox & Cordey (1992).

stacked channel sandbodies (Goldsmith *et al.* 1995; McKie & Audretsch 2005), whereas the mud-dominated units include a variation of non-marine, basin-wide floodplain and playa deposits. The thick and laterally extensive mud-dominated units provide the main correlative units for the Triassic Skagerrak in the Central Graben (McKie & Audretsch 2005).

The resultant Triassic stratigraphy in UK quadrants 22 and 30 is incompletely preserved due to deep erosion during the Middle and Late Jurassic (Erratt *et al.* 1999). In the J-Block area, the Skagerrak Formation includes both Judy and Joanne Sandstone Members, whereas in the Heron Cluster area only the Judy Sandstone Member is preserved (Figs 1 & 2).

Methodology

Sampling

Core samples and thin sections from six wells in the Judy field (30/7a-7; 30/7a-8; 30/7a-9; 30/7a-11Z; 30/7a-P3; 30/13-5) and one well in the Heron field (22/29-5) have been examined in this study. The 74 Judy core samples were chosen from 3400 to 4000 mbsf to cover the main reservoir fluvial facies for the Joanne and Judy Sandstone Members (Fig. 2). The 136 Heron core samples were chosen from 4300 to 4500 mbsf to cover the main reservoir facies of the Judy Sandstone Member (Fig. 2).

Petrography

Optical porosity, grain size and the fraction of chlorite-coated grains were measured for this study. Optical porosity was measured by using the jPOR digital image analysis technique (Grove & Jerram 2011) on blue epoxy-impregnated thin sections. Grain-size distribution was analysed by using the Leica QWin (V. 3.5.0) software on thin section micrographs and the fraction of chlorite-coated grains was measured by point counting with 300 counts per thin section. The resulting data were used to select samples for additional petrographic analysis (i.e. intergranular volume (IGV) (Paxton *et al.* 2002), total cement volume, porosity loss by mechanical compaction (COPL) and porosity loss by cementation (CEPL) (Lundegard 1992). The samples for further petrographic analysis were also selected by grain size (0.1–0.15 mm) to exclude the effects of grain size on the porosity and intergranular volumes. The intergranular volume and the total cement volume were measured by point counting with 300 counts per thin section.

All thin sections were highly polished to 30 μm and coated with carbon prior to analysis by a Hitachi SU-70 field emission scanning electron microscope (SEM) equipped with an energy-dispersive detector (EDS). Scanning electron microscope analyses of thin sections and bulk rock samples were conducted at acceleration voltages of 5–20 kV with beam currents of 1 and 0.6 nA, respectively. Point analyses had an average duration of 2 min, whereas line analyses were dependent on length. The SEM–EDS assembly was used for rapid identification of chemical species and orientation on the sample.

One-dimensional basin modelling

One-dimensional burial modelling was used to model pore pressure, as it provides a good insight into the pore pressure built up by disequilibrium compaction and pore fluid expansion due to increasing temperatures. Nevertheless one-dimensional models are limited in terms of integrating pore pressure mechanisms such as fluid flow, diagenetic processes, and hydrocarbon charging or gas generation. The one-dimensional burial history simulations were undertaken using Schlumberger's PetroMod (v. 2012.2) software. The software is based on a forward-modelling approach to calculate the geological evolution of a basin and burial history. The one-dimensional burial models are set up from the present-day well stratigraphy, well log lithology and lithological description of the modelled units (Table 1). To create optimum one-dimensional models, it is essential have accurate palaeoheat flow models. Several heat flow models have been published for the North Sea and especially for the Central Graben and can be subdivided into two main types: constant heat flow models (e.g. Schneider & Wolf 2000) and thermal upwelling models (e.g. Swarbrick *et al.* 2000; Carr 2003; di Primio & Neumann 2008). In this study we used the palaeo-basement heat flow and palaeo-surface temperature history published by Swarbrick *et al.* (2000). These data provide the best fit to the rifting events of the Central Graben, North Sea. The lithological unit types used in this model are mainly PetroMod default lithology types or mixed default lithology types based on well log descriptions and core analysis reports for the Central Graben lithologies. The only exceptions are the modelled Chalk units, which were modified to match the specific characteristics of the North Sea non-reservoir chalk (Mallon & Swarbrick 2002; Mallon *et al.* 2005; Swarbrick *et al.* 2010).

Petrography and diagenesis of the Skagerrak Formation

The present-day reservoir quality of the Triassic Skagerrak samples is a cumulative product of depositional attributes, mechanical compaction and diagenesis during early and later stages of burial.

Grain size and porosity distribution

The 136 investigated samples of the Heron field show a wide range of porosity from below 1% up to a maximum of 31% (Fig. 3). The porosity distribution shows a very high fraction (>90% of the samples) at porosities below 15%, which leads to an average porosity of 3.9% for the Heron field

Table 1. *Model parameters, with estimated depositional periods in millions of years, layer thicknesses in metres, modelled erosion (E) in metres and lithology (Sh: Shale and Sst: Sandstone) for the Heron and Judy 1D burial history models*

Time				Heron			Judy		
Start [Ma]	End [Ma]	Formation	Subformation	Thickness [m]	E [m]	Lithology	Thickness [m]	E [m]	Lithology
53	0	Hordaland							
10	0		Hordaland 2	1407		Shale	1424		Shale
53	10		Hordaland 1	1396		Shale	1357		Shale
54.1	53	Tay Sand		15		Silty sh			
54.3	54.1	Balder		18		Shale	17		Silty sh
54.8	54.3	Sele				Shale	54		Silty sh
54.45	54.3		Upper Sele	7					
54.65	54.45		Rogerland	16					
54.8	54.65		Lower Sele	8					
56.1	54.8	Forties		187		Sst			
58.5	56.1	Lista		49		Silty sh	16		Shale
60	58.5	Andrew		51		Silt			
59	58.5		Andrew Clay				18		Shale
59.1	59		Andrew Sand				8		Sst
59.7	59.1		Andrew Silt				50		Silty sh
60	59.7		Andrew Clay				13		Shale
61	60	Maureen		82		Marl			
60.5	60		Maureen Melange				37		Sst
61	60.5		Maureen Marl				55		Marl
65	61	Ekofisk		94		Chalk	28		Chalk
74	65	Tor		459		Chalk	226	0	Chalk
93.5	74	Hod		335		Chalk	154		Chalk
98.9	93.5	Herring		9		Chalk			
136.5	127	Valhall					22	0	Sandy shale
129.5	127		Upper Valhall	43	15	Marl			
136.5	129.5		Lower Valhall	20		Marl			
160	144	Upper Jurassic		0	50	Sandy shale	0	50	Sandy shale
180	160	Mid Jurassic		0	20	Shale	0	20	Shale
188	184	Lias					3		Shale
205.7	195	Lower Jurassic		0	150	Sst	0	150	Sst
241.7	205.7	Skagerrak							
211	205.7		Joshua	0	50	Silty shale	0	50	Silty shale
214	211		Josephine	0	100	70% sst 30% shale	0	100	70% sst 30% shale
220.7	214		Jonathan	0	40	Silty shale	38		Silty shale
234.3	220.7		Joanne	23	375	70% sst 30% shale	469		70% sst 30% shale
237	234.3		Julius	41		Silty shale	140		Silty shale
241.7	237		Judy	339		70% sst 30% shale	385		70% sst 30% shale
251.2	241.7	Smith Bank		200		Silty shale	200		Silty shale
259	251.2	Zechstein		208		Salt	208		Salt

Sst, sandstone.

samples. The grain sizes vary from coarse silt to medium-grained sand, with the majority of the samples between very fine and fine grained (Fig. 4), and an average grain size of 0.147 mm. The Heron field samples have a narrow range of compositions with most in the range of arkosic to lithic–arkosic arenites (McKie *et al.* 2010; Nguyen *et al.* 2013).

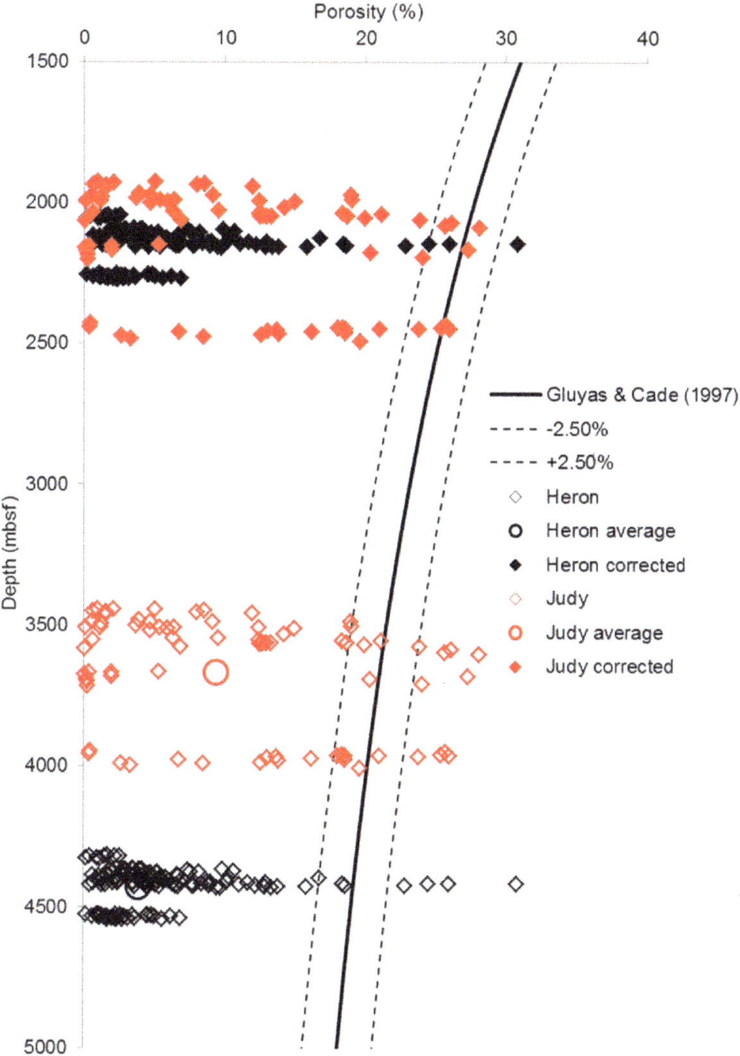

Fig. 3. Porosity–depth measurements with original sample depths and the corrected sample depths and porosity–depth relationship. After Gluyas & Cade (1997).

The 74 samples from the six Judy field wells show porosities ranging from below 1% up to a maximum of 28% (Fig. 3). The porosity distribution of the Judy field samples shows two-thirds of the porosities below 15% porosity, which leads to an average porosity of 9.4%. The grain size distribution of the Judy field samples range from coarse silt to medium-grained sand, with the majority of the investigated samples being very fine to fine grained (Fig. 4), and an average grain size of 0.131 mm. Compositionally, the Judy field samples show a similar narrow range from arkosic to lithic–arkosic arenites (McKie *et al.* 2010; Nguyen *et al.* 2013).

Mechanical compaction

Mechanical compaction in sandstones can be identified by bending of weak grains (mica), deformed soft lithic grains, local fracturing and dissolution at grain contacts, which produces concavo–convex or sutured/stylolitic grain contacts.

The Heron field samples show features of a low degree of mechanical compaction such as deformation of soft lithic grains, slight bending of micas and concavo–convex grain contacts of some detrital quartz (4391 m; Fig. 5a). In contrast, the Judy field samples show no mechanical compaction features,

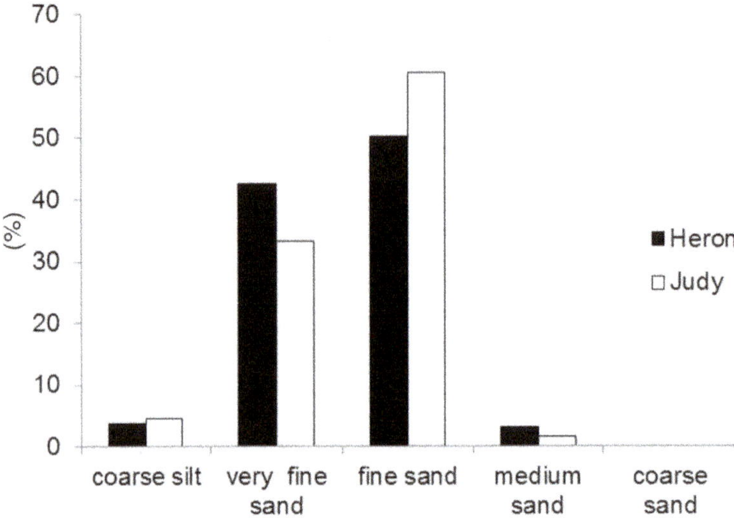

Fig. 4. Grain size distribution for samples from the Heron (136) and the Judy (74) fields.

and the samples appear to be undercompacted at their current depth of burial (3558 m; Fig. 5b).

Diagenetic cement and grain coatings

The diagenesis of the Skagerrak Formation has been the subject of a number of research papers (e.g. Smith *et al.* 1993; Weibel 1998; Kape *et al.* 2010; Nguyen *et al.* 2013). These have highlighted the complex diagenetic history of the sandstone reservoirs and how grain size and facies play an important role. The main diagenetic cements recognized include quartz, localized carbonate (ferroan dolomite) and feldspar, and early precipitates of halite

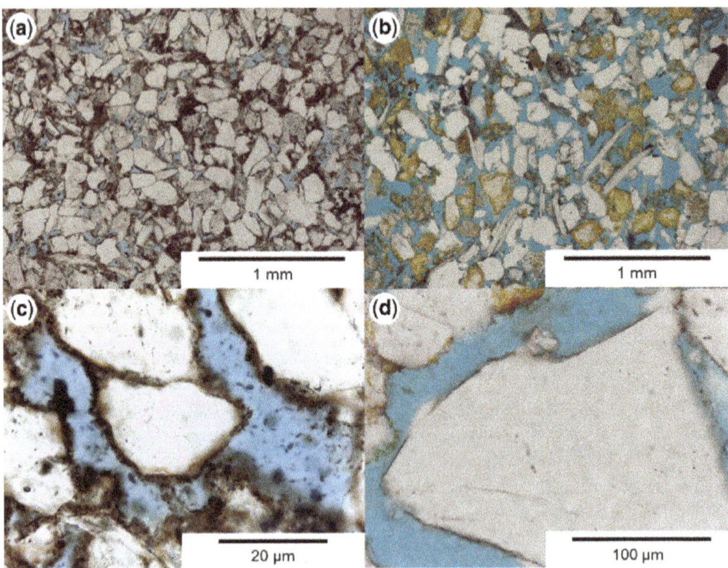

Fig. 5. Micrographs from (**a**) the Heron field sample at 4391 m with mechanical compaction features (e.g. bended mica grain, concavo-convex grain); (**b**) the Judy field at 3558 m with no mechanical compaction features and high porosity; (**c**) the Heron field at 4393 m with thick and complete covering chlorite coatings, and (**d**) the Judy field at 3548 m with thin, partly covering chlorite coatings.

cement as identified by Nguyen *et al.* (2013). Detrital grain coatings such as chlorite are common and their presence has been correlated with low volumes of quartz cement. The major cement types and detrital grain clay coatings important for reservoir quality in this study are discussed in more detail below.

Quartz cements

The quartz cements recognized in this study are common as two diagenetic stages: either an earlier microquartz overgrowth, below and in between the chlorite grain coatings, or a later macroquartz overgrowth, present at non-chlorite coated grain surfaces or at breaks in the grain coatings. The term microquartz overgrowth is used here for polycrystalline growth patterns of individual micro-sized (1–5 μm in length) quartz crystals, which are in optical continuity or discontinuity with the detrital quartz grain. In comparison the macroquartz overgrowth is defined as syntaxial quartz overgrowth larger than 20 μm in optical continuity with the detrital quartz grain.

Minor amounts of macroquartz ($\leq$1% bulk volume) are recognized in the Heron Cluster dataset and typically are only present at rare gaps or breaks in the chlorite grain coatings. Microquartz is more common and can be observed between the detrital quartz grains and the well-developed chlorite grain coatings. The microquartz overgrowths tend to fill small cavities in the detrital quartz grain surface and infill the void space between the detrital surface and the chlorite crystals. Hydrothermal experiments on fluvial–deltaic sediments at similar temperatures to those found in the Skagerrak Formation (~160°C) have identified that quartz cement can fill significant microporosity within diagenetic chlorite coats, potentially affecting mechanical and petrophysical rock properties (Ajdukiewicz & Larese 2012).

Quartz cements play a more significant role in the Judy field datasets, where macroquartz overgrowths commonly occur in all wells but are particularly common in the deepest wells (Fig. 5b). In the deeper-buried sandstones macroquartz overgrowths commonly occur where there are breaks in the thinner, less-well-developed chlorite coatings. Furthermore, small syntaxial quartz overgrowths were also observed on grain surfaces beneath chlorite grain coatings.

Chlorite grain coatings and cement. Detailed SEM analysis identified that chlorite is the most common clay grain coating for all datasets in this study. Chlorite coatings represent over 95% of the clay mineral coatings recognized but frequently occur in close association with illite and mixed-layer chlorite–smectite or illite–smectite. A complex structural pattern for the chlorite grain coatings is observed in all cases, regardless of coating thickness. A root zone (Pittman *et al.* 1992; Ehrenberg 1993) with platey crystals parallel to the surface of the detrital grain is superseded by well-defined platey to curly crystals that grew perpendicular to the detrital surface (Fig. 6a).

The Heron field samples show in general a high fraction (80–100%) of chlorite-coated grains (Fig. 7). Grain coatings are very well developed and range from 1 μm to >20 μm thick, with an average of 11 μm. The coatings are generally absent at grain–grain contacts, but show a very high detrital grain surface coverage of around 95% (Fig. 5c).

The fraction of the chlorite-coated grains in the Judy field samples is more variable, ranging between 10% and 70% (Fig. 7). Chlorite grain coatings vary in thickness from 1–15 μm, with average thicknesses of ~7 μm. The coatings are absent at grain–grain contacts and commonly incomplete in their surface coverage of detrital quartz grains (Fig. 5d). Furthermore, there appears to be a direct relationship between the measured porosity of the sandstone datasets and the fractions of chlorite-coated grains (Fig. 8). The Heron dataset

Fig. 6. SEM images from (**a**) the Heron field sample at 4393 m with a clay coating root zone parallel to the detrital grain surface and a second layer of well-defined chlorite crystals oriented perpendicular to the detrital surface, and (**b**) the Heron field sample at 4417 m with a pore-filling clay mixture of chlorite and illite.

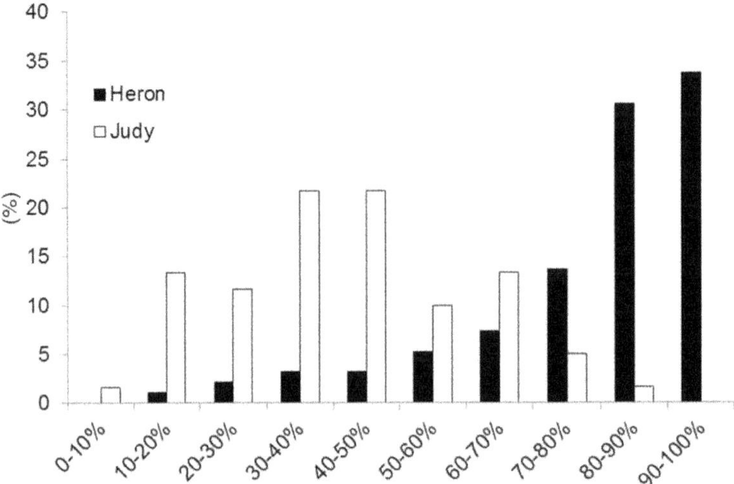

Fig. 7. Fraction of chlorite-coated detrital grains per 300 counts for the 136 Heron samples and the 74 Judy samples.

demonstrates a consistently lower porosity (<15%) for lower fractions of coated grains (<15%), whereas high porosities of up to 30% for higher fractions (>50%) of chlorite-coated grains are common. The Judy dataset does not illustrate such a close relationship between porosity and percentage of coated grains (Fig. 8).

Authigenic chlorite cements are the most important and effective grain coating in terms of limiting early extensive quartz cementation in the Skagerrak sandstones. However, pore-filling chlorite, illite and chlorite–illite mixes commonly occur in the sandstones from the Heron samples. The pore-filling cements comprise small chlorite plates that are orientated parallel and arranged with a denser packing than seen for grain coatings (Fig. 6b). Chlorite-coated grains are locally covered by pore-filling chlorite cements. They represent between 10% and 15% of the bulk rock sample and can fill up to 50% of the remaining intergranular volume (Fig. 5a). In the Skagerrak sandstones from the Judy field authigenic pore-filling chlorite is rarely

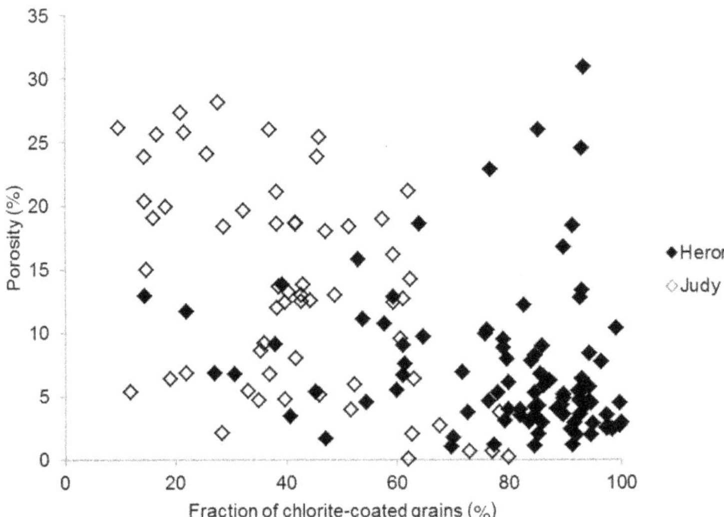

Fig. 8. Measured porosity v. fraction of chlorite-coated detrital grains for the Heron and the Judy samples.

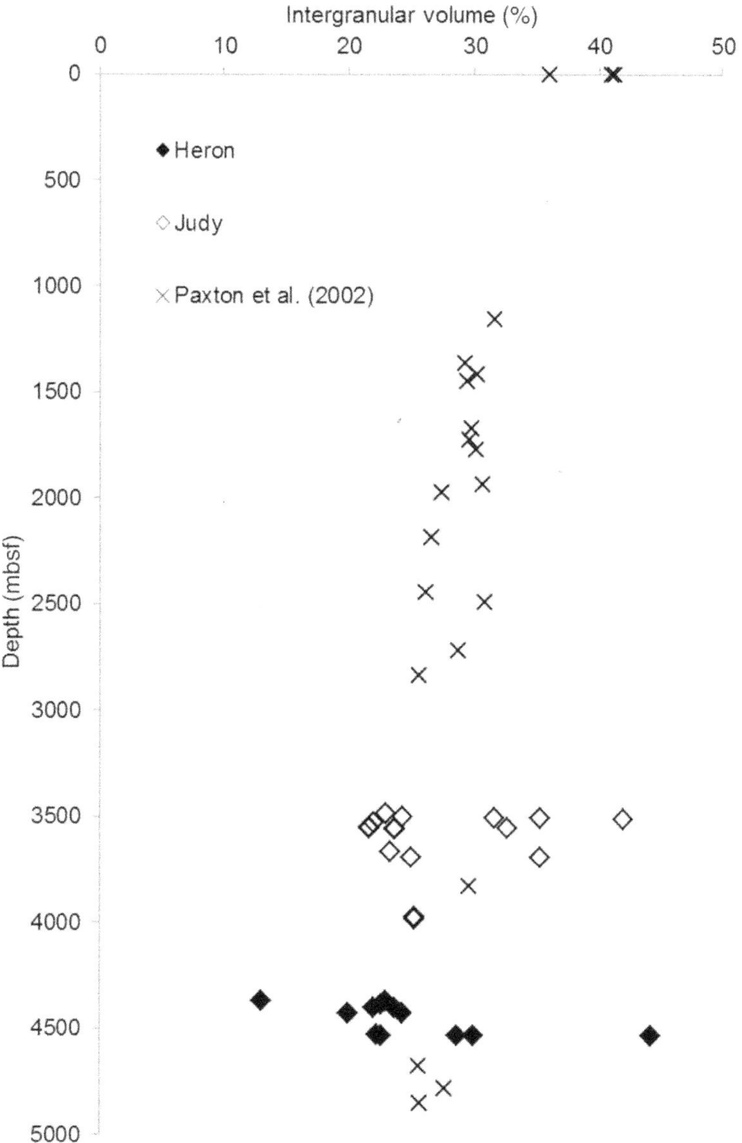

Fig. 9. Intergranular volume (IGV) v. depth for the Heron samples and the Judy samples, with observed reference IGV from Paxton *et al.* (2002).

identified (<5%), (Fig. 5b). This may in part reflect the less-well-developed chlorite grain coatings and occurrence in the Judy compared with the Heron fields (McKie *et al.* 2010; Nguyen *et al.* 2013).

K-feldspar dissolution

K-feldspar dissolution and alteration can be observed in both sample sets. K-feldspar dissolution occurs generally at a late burial stage, after the precipitation of the chlorite grain coating. This is indicated by chlorite grain coatings that preserve the original shape of the partly or fully dissolved K-feldspar grain.

K-feldspar dissolution can be observed in the Heron field samples, but is less common than in the Judy field samples (Fig. 5). Partial dissolution of detrital K-feldspar grains is observed from depths greater than 3200 m (Fig. 5).

Table 2. *Total cement volume (C), measured porosity (P_o), minus-cement porosity (P_{mc}) or intergranular volume (IGV), porosity-loss by mechanical compaction (COPL) and porosity-loss by cementation (CEPL) all in % for the Heron and Judy sample sets*

Field	C [%]	P_o [%]	P_{mc}/IGV [%]	COPL [%]	CEPL [%]
Heron	17.00	6.00	23.00	28.57	12.14
	11.00	2.00	13.00	36.78	6.95
	15.00	7.67	22.67	28.88	10.67
	11.67	10.33	22.00	29.49	8.23
	18.00	5.67	23.67	27.95	12.97
	8.33	11.67	20.00	31.25	5.73
	14.33	10.00	24.33	27.31	10.42
	21.00	1.33	22.33	29.18	14.87
	19.67	3.00	22.67	28.88	13.99
	28.67	1.33	30.00	21.43	22.52
	37.00	7.33	44.33	1.20	36.56
	28.67	0	28.67	22.90	22.10
Average	19.19	5.53	24.72	26.15	14.76
Judy	11.00	12.00	23.00	28.57	7.86
	5.00	19.33	24.33	27.31	3.63
	20.33	11.33	31.67	19.51	16.37
	35.67	6.33	42.00	5.17	33.82
	5.67	16.00	21.67	29.79	3.98
	15.67	17.00	32.67	18.32	12.80
	13.00	10.33	23.33	28.26	9.33
	7.33	17.67	25.00	26.67	5.38
	13.00	22.33	35.33	14.95	11.06
	20.00	15.33	35.33	14.95	17.01
	8.67	13.33	22.00	29.49	6.11
	15.67	6.00	21.67	29.79	11.00
	4.33	19.33	23.67	27.95	3.12
	5.67	18.00	23.67	27.95	4.08
	10.00	18.67	28.67	22.90	7.71
	4.00	15.67	19.67	31.54	2.74
	10.00	15.33	25.33	26.34	7.37
	14.00	11.33	25.33	26.34	10.31
Average	12.17	14.74	26.91	24.21	9.65

Intergranular volume and total cement volume

Total cement and IGV measurements were undertaken on samples with grain sizes between 0.1 and 0.15 mm to reduce grain size variations. The IGV, or minus-cement porosity, is the sum of intergranular pore space, intergranular cement and depositional matrix (Paxton *et al.* 2002). The results of the IGV measurements show wide variations for both sample sets. The Heron field IGVs vary from 13% to 44%, with the majority between 20% and 30% and a mean of 22.8%. The Judy field IGVs range between 20% and 42%. The Judy field IGVs plot in two clusters, between 20–25% and 31–42%, with an overall mean of 24.6% (Fig. 9). The total cement volume represents the volume of all pore-filling diagenetic cements, including quartz, chlorite, illite and other clay mineral cements. The total cement volume is used to calculate the porosity loss by mechanical compaction and by cementation (Lundegard 1992). The total cement volume varies for the Heron samples from 8% up to 37%, with an average of 19%. The Judy field samples show total cement volumes from 4% to 35%, with a lower average of 12% (Table 2). The measured total cement volume (C) can be used to calculate the porosity-losses caused by compaction (COPL) and cementation (CEPL) using the following equations after Lundegard (1992):

$$\mathrm{COPL} = P_i - \left(\frac{(100 - P_i)P_{mc}}{100 - P_{mc}}\right) \tag{1}$$

$$\mathrm{CEPL} = (P_i - \mathrm{COPL})\left(\frac{C}{P_{mc}}\right) \tag{2}$$

where P_i is the initial or depositional porosity and P_{mc} is the IGV or minus-cement porosity calculated from by subtracting the total cement volume, C from the total optical porosity, P_o (Table 2). The calculated COPL and CEPL are only accurate if three conditions are met. First, the assumed initial porosity P_i is correct. Second, the amount of cement derived by local grain dissolution is negligible or known. And third, the amount of framework mass exported by grain dissolution is negligible or known (Lundegard 1992). The initial or depositional porosity assumed for the Skagerrak sandstones samples is 45% (Lundegard 1992). The COPL–CEPL results show porosity loss for both sample sets was predominantly by mechanical compaction (Fig. 10), with averages of 26% and 24% for the Heron and the Judy sample sets, respectively. Furthermore, the results show a CEPL of 15% for the Heron sample set, 5% higher CEPL than for the Judy sample set (Table 2).

Burial history modelling

The one-dimensional burial history models show that the Triassic Skagerrak units are at maximum burial depth and maximum temperature at present day (Fig. 11). The Skagerrak Formation experienced a long shallow burial phase (<1800 mbsf) between deposition and 80 Ma, followed by a phase of rapid burial to their present maximum burial depths. During this rapid burial phase, overpressure started to build up at around 80 Ma ago in the Heron field (Fig. 12a) and at around 70 Ma ago in the Judy field (Fig. 12b). The onset of overpressure in both fields occurred during shallow burial at around 650 m burial depth for the Heron field (Fig. 12a) and around 500 m burial depth for the Judy field (Fig. 12b). The pore fluid pressure

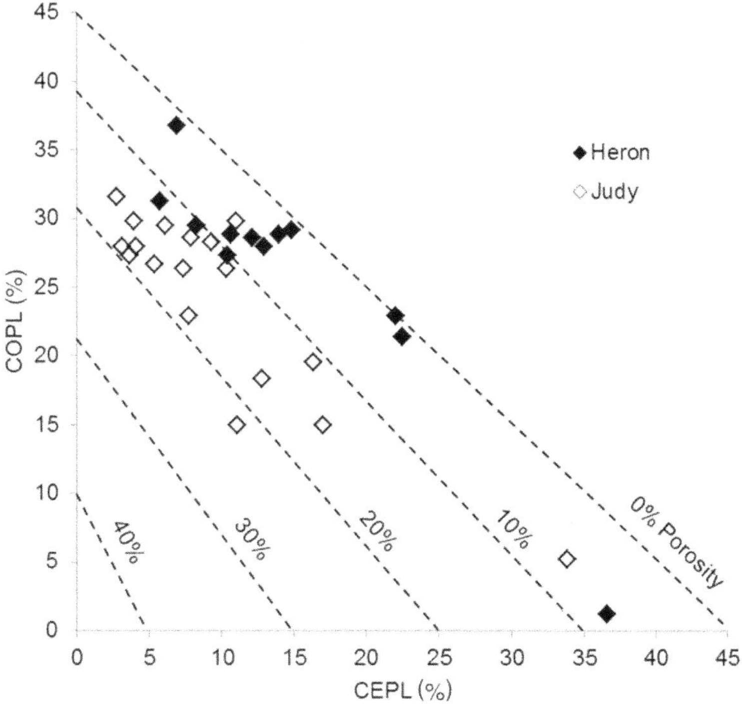

Fig. 10. Mechanical compaction (COPL) and cementational (CEPL) porosity loss for the Heron and the Judy sample set. After Lundegard (1992).

increased during ongoing burial to a present-day maximum for the whole study area. The increase in pore fluid pressure causes a reduction in vertical effective stress with ongoing burial (Fig. 12). The present-day pore fluid pressures for both wells increase downwards through the Chalk units (Ekofisk, Tor and Hod formations) to the Triassic Skagerrak units (Fig. 13). In this model, the only mechanisms for generating overpressure that have been taken into account are disequilibrium compaction and fluid expansion, which explains the lower-than-measured pore fluid pressures for the Heron field and the Judy field. Nevertheless, the models for the Judy field, based on the well 30/7a-9 (Fig. 13a), and the Heron field, based on the well 22/29-5 (Fig. 13b), show pore fluid pressure trends in agreement with measured trends from repeat formation tester (RFT) pressure data.

Overpressure-depth correction

The modelled pore fluid overpressure, generated by disequilibrium compaction, can be used to estimate the equivalent depth where the measured porosity would be found on the normal compaction curve, i.e. if the pore pressure were hydrostatic. For the normal compaction curve, we used the porosity–depth relationship of Gluyas & Cade (1997), which is based on experimental data from laboratory compaction experiments for uncemented sandstones under hydrostatic burial:

$$\varphi = 50\ \exp\left(\frac{-10^{-3}z}{2.4 + 5 \cdot 10^{-4}z}\right) \quad (3)$$

where φ is porosity expressed as a percentage and z is depth in metres. The depth and the rate of mechanical compaction are corrected by scaling the porosity to vertical effective stress, rather than to depth (Gluyas & Cade 1997). The compaction-only porosity–depth relationship is:

$$z' = z - \left(\frac{u}{(\rho_r - \rho_w)g\left(1 - \varphi_{\varepsilon\sum}\right)}\right) \quad (4)$$

where z' is the equivalent depth in metres, z is the depth in metres, u the overpressure in pascals, ρ_r is the rock density in kilograms per cubic metre (typically 2650 kg m^{-3}), ρ_w is the water density in kilograms per cubic metre (typically 1050 kg m^{-3}), g is the gravity in metres per second squared (typically 9.8 m s^{-2}), and $\varphi_{\varepsilon\Sigma}$ is the average

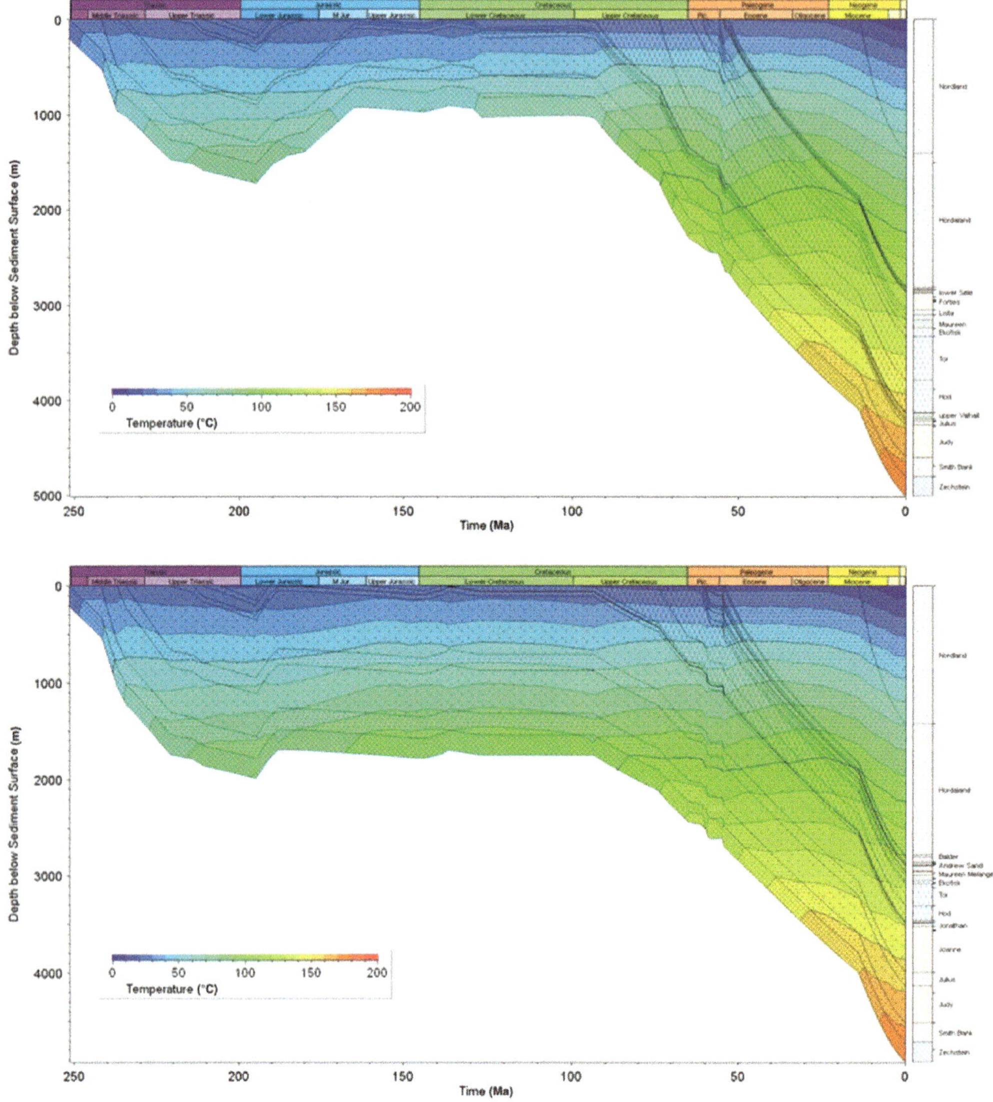

Fig. 11. 1D burial temperature plot for the Heron well (22/29-5) (top) and the Judy well (30/7a-9) (bottom) with the associated stratigraphic columns.

porosity of the overburden expressed as a fraction (typically = 0.2). The corrected depths for the Heron field samples are around 2200 m shallower, corrected from ~4400 m depth to ~2200 m depth (Fig. 3). The corrected depths for the Judy field samples are around 1500 m shallower, corrected from ~3500 to ~2000 m depth (Fig. 3). The high porosities of 25–30% in the Skagerrak sandstones, previously considered anomalous, are within the range of the porosity–depth relationship introduced by Gluyas & Cade (1997) at their corrected depths.

Discussion

Interpretation and diagenetic development

All of the observed petrographic features can be linked into a relative sequence for both reservoirs. The eodiagenetic, dolomitic cements were precipitated early during the initial phase of mechanical compaction and decrease of IGV. This is indicated by high IGVs (close to depositional IGV), in the dolomitic cemented samples in both samples sets. The dolomite cement is likely to be sourced from

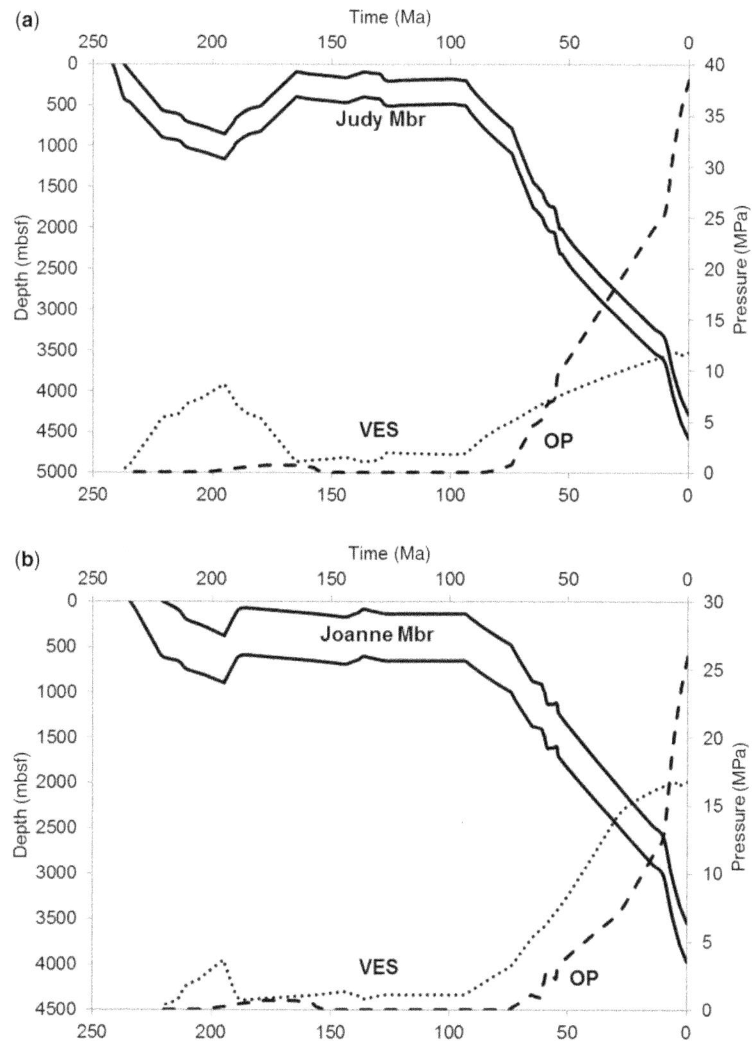

Fig. 12. Burial history (solid lines), pore fluid overpressure (OP; dashed lines) and vertical effective stress (VES; dotted lines) development for (**a**) the Judy Sandstone Member, the uppermost succession within the Skagerrak Formation in the Heron well (22/29-5) and (**b**) the Joanne Sandstone Member, the uppermost succession within the Skagerrak Formation in the Judy well (30/7a-9).

reworked calcrete or dolocrete fragments (McKie & Audretsch 2005); this is indicated by the association of the dolomitic cements with slightly deformed clay clasts. Mechanical compaction had the strongest impact during shallow burial (<2500 m), despite the shallow onset of the pore fluid overpressure which reduces the VES and hence reduces the effects of mechanical compaction. The dolomitic cements were followed by the early microquartz overgrowth, which is proven by the lack of microquartz on detrital grains surrounded by dolomitic cement. The following phase of authigenic grain coating with chlorite is probably the most important diagenetic phase for the Heron field and the Judy field. The authigenic chlorite covers the earlier microquartz on the detrital quartz grains. The authigenic chlorite coatings possibly developed during early mesodiagenesis, at temperatures of 60–100°C. The chlorite probably originates from precursor clay minerals such as smectite, berthierine and kaolinite (Grigsby 2001; Worden & Morad 2003; Berger *et al.* 2009) or the dissolution and reprecipitation of syndepositional chlorite (Anjos *et al.* 2003). These authigenic chlorite grain coatings inhibited a further stage of quartz cementation above temperatures of 70°C (Worden & Morad

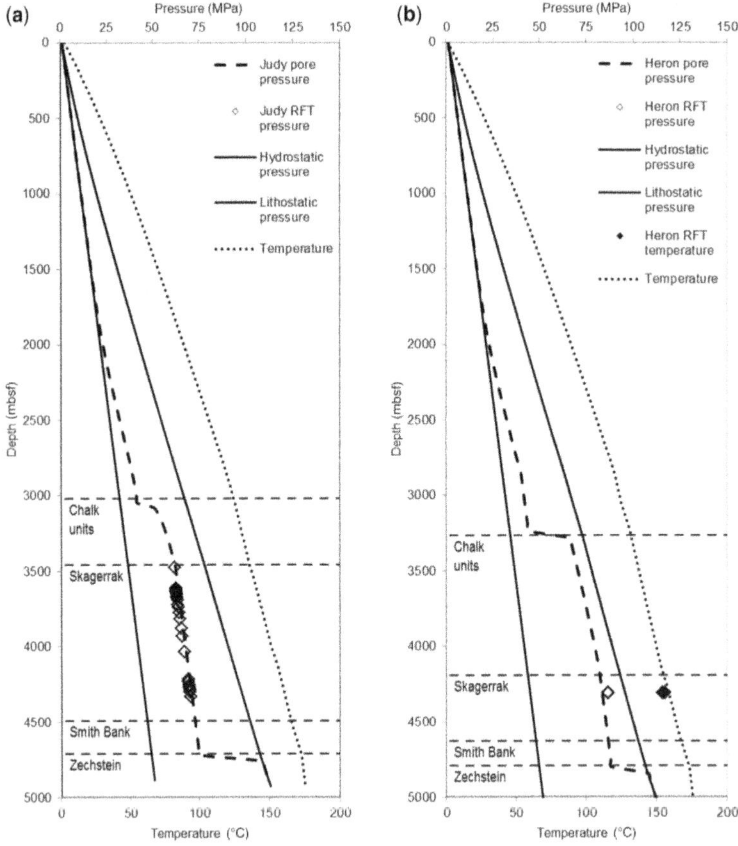

Fig. 13. Modelled formation pressures caused by disequilibrium compaction with hydrostatic pressure, lithostatic pressure and formation temperature, measured RFT pressure and temperature data for (**a**) the Judy field (30/7a-9) and (**b**) the Heron field (22/29-5). Pore fluid pressure can exceed 80 MPa at depths of between 4000 and 5000 mbsf for both wells.

2000) in the Heron field and partly inhibit it in the Judy field. The later stage macroquartz cement is present on the non-covered detrital quartz grain surfaces. Due to the incompleteness of the authigenic coatings the amount of macroquartz cement is higher in the Judy field sample. Deposition of the chlorite grain coats was also followed by a stage of extensive feldspar dissolution in the Judy field, less prominent in the Heron field. The feldspar dissolution phase occurred with, or was closely followed by, a phase of secondary precipitation of pore-filling chlorite and illite, which is likely to have been synchronous with the feldspar alteration and dissolution (Worden & Morad 2003).

Role played by chlorite grain coatings

The occurrence of quartz overgrowths is closely related to chlorite grain coatings (Pittman & Lumsden 1968; Dixon *et al.* 1989; Pittman *et al.* 1992; Ehrenberg 1993; Anjos *et al.* 2003; Berger *et al.* 2009; Ajdukiewicz & Lander 2010; Bahlis & De Ros 2013); therefore the role played by the chlorite grain coatings is important in terms of maintaining primary porosity. This role could not be more different in the two investigated reservoirs.

The Heron field chlorite coatings are thickly developed and, in most cases, cover the detrital quartz grain surfaces completely (Fig. 5a, c). The full coverage of the robust chlorite grain coatings, with their well-developed root zone and their second layer of well-defined perpendicular oriented crystals, seems to be the key in inhibiting extensive macroquartz cement overgrowth at temperatures above 70°C (Worden & Morad 2000). Similar effects on quartz overgrowth have been described by Dixon *et al.* (1989), Pittman *et al.* (1992), Ehrenberg (1993), Anjos *et al.* (2003), Berger *et al.* (2009), Bahlis & De Ros (2013) and Ajdukiewicz & Lander (2010) for other formations with thick and fully covering chlorite grain coatings. The inhibition of quartz cement is potentially the cause for the correlation between porosity and the fraction of coated grains in the Heron field samples (Fig. 8).

The Judy field samples show thin chlorite grain coatings with major gaps and macroquartz overgrowths (Fig. 5d). The impact of grain coatings on the maintenance of porosity is highly dependent on their completeness (Ajdukiewicz & Lander 2010; Ajdukiewicz & Larese 2012). Despite their incomplete coverage, the Judy field coatings are still inhibiting extensive quartz cement precipitation on around 80% of the detrital surfaces. The main nucleation points for the quartz overgrowths are the non-covered gap areas of the detrital quartz grain surfaces where macroquartz overgrowth can be observed (Fig. 5d). Nevertheless, macroquartz overgrowth is less extensive than expected due to the incomplete coverage in the Judy field samples. This could be due to a low concentration of dissolved silica in the pore fluid. The Judy reservoir experienced pore fluid overpressure at relatively shallow depths, which leads to a semi-closed system, low VES, less mechanical compaction and less pressure dissolution at grain contacts. Therefore less silica is dissolved in the reservoir and, due to the pore fluid overpressure, lateral transfer of pore fluid is less likely. This could be a possible explanation for a lower ratio of dissolved silica in the Judy field reservoir.

Nevertheless, it has been reported that thick and extensive chlorite coatings could reduce the reservoir quality by decreasing the porosity and, probably more significantly, the permeability, by bridging and closing of pore throats (Anjos *et al.* 2003; Bahlis & De Ros 2013). Therefore it seems that the thickly developed chlorite coatings in the Heron reservoir have also had contrary impacts on the porosity by bridging the gap between one grain coating to another across open pore space, and filling that pore space with chlorite cement. Some of the thick chlorite coatings actually have a net effect of decreasing the overall porosity, and also reduce the permeability of the reservoir sandstones instead of increasing it. This observation is confirmed by the 5% higher CEPL in the Heron samples compared with the Judy samples (Fig. 10). This porosity reduction is not caused by quartz cementation but by bridging and pore-filling chlorite.

Effect of pore fluid overpressure on porosity preservation

The positive effect of pore fluid pressure on the maintenance of primary porosity has been well known since Terzaghi's introduction of the effective stress concept: overpressured reservoirs are often associated with higher porosities. However, a combination of the magnitude and timing of onset of overpressure needs to be considered in reservoirs with enhanced primary porosity preservation. Pore fluid overpressure can arrest and slow down mechanical compaction, but cannot increase porosity. Therefore the timing of the onset of overpressure is the crucial factor in maintaining primary porosity by overpressure to depth. Furthermore, overpressured reservoirs are often closed or at least semi-closed systems with minimal or no external fluid exchange, which allows them to evolve in their own fluids (Jeans 1994). The exclusion of external sources reduces the number of possible cementational processes in the reservoir and allows better predictions of these processes.

The onset of overpressure in the Heron field was around 80 Ma years ago, which provides an onset burial depth of around 650 m for the investigated samples (Fig. 12a). The pore fluid overpressure then started to ramp up significantly due to the rapid burial and the massive overlaying packages. The VES on the Heron field grain framework further increased during the rapid burial of the Triassic Skagerrak units, but the rapidly increasing pore fluid overpressure reduced the rate of VES increase dramatically (Fig. 12). This slower-than-expected VES increase from the shallow burial depth of 650 m reduced the effects of mechanical compaction on the grain framework significantly. This reduced mechanical compaction can be seen in the investigated samples, where characteristics of a medium compaction state are present (slight bending of mica grains, grain contact dissolution with concavo–convex contacts), but there are no features of a high compaction state (sutured or stylolitic grain contacts and grain fracturing) (Fig. 5a) that would be expected in this deep burial stage.

The overpressure onset in the Judy field was around 70 Ma years ago, at around 500 m burial depth for the investigated samples (Fig. 12b). The pore fluid pressure started to ramp up, which led to a reduced rate of increase for the VES. The lower-than-expected VES led to less mechanical compaction in the Judy field; this can be observed in the investigated samples. There are no common mechanical compaction features such as bent mica grains, fractured grains, pressure dissolution or concavo–convex or sutured grain contacts for the Judy field samples (Fig. 5b). Consequently, greater primary porosity was maintained in the Judy field.

Nevertheless, in both case studies the overpressure only had an impact on the porosity preservation because of its shallow onset, in the case of a deeper onset most of the preserved primary porosity would have been lost due to ongoing mechanical compaction. There would have been almost no effect if overpressure onset occurred at greater depths (>2500 mbsf), when the sandstones had become sufficiently lithified that mechanical compaction was no longer taking place (Paxton *et al.* 2002).

The porosity-preserving impact of the overpressure is shown by the overpressure–depth correction after Gluyas & Cade (1997). The anomalously high porosities for deeply buried sandstone reservoirs, above 20%, are well within the expected range of the hydrostatic porosity–depth relationship after the overpressure–depth correction (Fig. 3).

Furthermore, the IGV and the total cement measurements show that even if most of the porosity loss in both reservoir sandstones has been due to mechanical compaction (Fig. 10), the impact of pore fluid overpressure on porosity maintenance seems to have been considerable in sandstones reservoirs. However, it is not just the overpressure or the magnitude of overpressure that are important and control the VES, the depth of onset is also crucial: the shallower the onset, the greater the potential for preserving high primary porosity.

Implications for deeply buried sandstone reservoirs

Anomalously high porosities can be found in deeply buried sandstone reservoirs where a combination of porosity-preserving mechanisms have been operating. The onset of shallow (<2000 m) pore fluid overpressure reduces the VES on the grain framework and hence reduces the amount of mechanical compaction. Nevertheless, the impact of pore fluid overpressure on the precipitation of cements seems to be negligible, given that temperature and availability of dissolved material is the main driver of cementation (Walderhaug 1994; Oelkers *et al.* 1996; Osborne & Swarbrick 1999).

Therefore, clay mineral or microquartz coatings are important for inhibiting extensive quartz cementation at temperatures above 70°C (Worden & Morad 2000) during deeper burial, where porosity loss is dominated by cementation processes. The effect of grain coatings on inhibiting quartz cement growth is highest in quartz-rich sandstones with silica-rich pore fluids.

It has clearly been shown by the two Triassic Skagerrak case studies that a shallow onset of pore fluid overpressure has a high potential for preserving primary porosities to depth. However, mechanisms that inhibit extensive quartz cementation are also important for porosity preservation during deeper burial (Pittman *et al.* 1992; Jeans 1994; Ajdukiewicz & Lander 2010). Primary porosity maintained by overpressure during shallow burial can be reduced significantly by extensive quartz cementation during later burial. Therefore, a combination of shallow overpressure onset and a high fraction of grains with chlorite coatings, without the development of pore-filling chlorite, is the best possible scenario for maintaining high primary porosities to depth.

Conclusions

(1) High porosities (up to 35%) at depths of >3500 mbsf and temperatures around 150°C are found in the Triassic Skagerrak fluvial sandstones of the J-Block and the Heron Cluster fields in the Central Graben, North Sea.
(2) The rate of porosity decline with increasing burial depth has been significantly arrested by a combination of shallow overpressure development and chlorite coatings of detrital grains.
(3) Timing of overpressure development was crucial as later overpressure development would probably have resulted in poorer reservoir quality for the Skagerrak Formation sandstones. It seems that the isolated nature of the intrapod reservoir units in both areas contributed to the onset of overpressure at shallow burial depth.
(4) Authigenic chlorite coatings maintain porosity by inhibiting quartz cement overgrowth at temperatures >70°C (Worden & Morad 2000). However, the chlorite coatings also reduce porosity due to their very presence, especially when their growth bridges between detrital quartz grains and fills the intervening pore space.
(5) This research has shed light on the controls on reservoir quality in the complex fluvial sandstones of the Skagerrak Formation. It clearly identifies the need to understand the timing of overpressure generation for arresting mechanical compaction and the importance of chlorite detrital grain coatings in inhibiting quartz cement overgrowth as temperature increases during progressive burial.
(6) Compaction remains active until the present day, and its deleterious effects are only too evident where chlorite grain coatings are absent, even where the VES is very low.

The industry funded research consortium GeoPOP sponsored by BG, BP, Chevron, ConocoPhillips, DONG Energy, E.ON, ENI, Petrobras, Petronas, Statoil, Total and Tullow Oil, at Durham University is thanked for funding this research. We acknowledge support from the BGS (British Geological Survey) for access to core material from the Heron Cluster wells. We thank the Special Publications editors P. Armitage and R. Worden for their suggestions and reviewers S. Gier and P. Kukla for their constructive reviews to help improve the manuscript.

References

AASE, N. E., BJORKUM, P. A. & NADEAU, P. H. 1996. The effect of grain-coating microquartz on preservation of reservoir porosity. *AAPG Bulletin*, **80**, 1654–1673.

AJDUKIEWICZ, J. M. & LANDER, R. H. 2010. Sandstone reservoir quality prediction: the state of the art. *AAPG Bulletin*, **94**, 1083–1091.

AJDUKIEWICZ, J. M. & LARESE, R. E. 2012. How clay grain coats inhibit quartz cement and preserve porosity in deeply buried sandstones: observations and experiments. *AAPG Bulletin*, **96**, 2091–2119.

AJDUKIEWICZ, J., NICHOLSON, P. & ESCH, W. 2010. Prediction of deep reservoir quality using early diagenetic process models in the Jurassic Norphlet Formation, Gulf of Mexico. *AAPG Bulletin*, **94**, 1189–1227.

ANJOS, S., DE ROS, L. & SILVA, C. 2003. Chlorite authigenesis and porosity preservation in the Upper Cretaceous marine sandstones of the Santos Basin, Offshore Eastern Brazil. *In*: WORDEN, R. H. & MORAD, S. (eds) *Clay Mineral Cements in Sandstones*. International Association of Sedimentologists Special Publications, **34**, 291–316.

AUDET, D. M. & MCCONNELL, J. D. C. 1992. Forward modelling of porosity and pore pressure evolution in sedimentary basins. *Basin Research*, **4**, 147–162.

BAHLIS, A. B. & DE ROS, L. F. 2013. Origin and impact of authigenic chlorite in the Upper Cretaceous sandstone reservoirs of the Santos Basin, eastern Brazil. *Petroleum Geoscience*, **19**, 185–199.

BERGER, A., GIER, S. & KROIS, P. 2009. Porosity-preserving chlorite cements in shallow-marine volcaniclastic sandstones: evidence from Cretaceous sandstones of the Sawan gas field, Pakistan. *AAPG Bulletin*, **93**, 595–615.

BISHOP, D. J. 1996. Regional distribution and geometry of salt diapirs and supra-Zechstein Group faults in the western and central North Sea. *Marine and Petroleum Geology*, **13**, 355–364.

BLOCH, S., LANDER, R. H. & BONNELL, L. 2002. Anomalously high porosity and permeability in deeply buried sandstone reservoirs: origin and predictability. *AAPG Bulletin*, **86**, 301–328.

CARR, A. 2003. Thermal history model for the South Central Graben, North Sea, derived using both tectonics and maturation. *International Journal of Coal Geology*, **54**, 3–19.

CLARK, J. A., CARTWRIGHT, J. A. & STEWART, S. A. 1999. Mesozoic dissolution tectonics on the West Central Shelf, UK Central North Sea. *Marine and Petroleum Geology*, **16**, 283–300, https://doi.org/10.1016/S0264-8172(98)00040-3

DE JONG, M., SMITH, D., NIO, S. D. & HARDY, N. 2006. Subsurface correlation of the Triassic of the UK southern Central Graben: new look at an old problem. *First Break*, **24**, 103–109.

DI PRIMIO, R. & NEUMANN, V. 2008. HPHT reservoir evolution: a case study from Jade and Judy fields, Central Graben, UK North Sea. *International Journal of Earth Sciences*, **97**, 1101–1114.

DIXON, S., SUMMERS, D. & SURDAM, R. 1989. Diagenesis and preservation of porosity in Norphlet Formation (Upper Jurassic), southern Alabama. *AAPG Bulletin*, **73**, 707–728.

EHRENBERG, S. N. 1993. Preservation of anomalously high porosity in deeply buried sandstones by grain-coating chlorite: examples from the Norwegian continental shelf. *AAPG*, **77**, 1260–1286.

EHRENBERG, S., NADEAU, P. & STEEN, Ø. 2008. A megascale view of reservoir quality in producing sandstones from the offshore Gulf of Mexico. *AAPG Bulletin*, **92**, 145–164.

EHRENBERG, S. N., NADEAU, P. H. & STEEN, O. 2009. Petroleum reservoir porosity v. depth: influence of geological age. *AAPG Bulletin*, **93**, 1281–1296.

ERRATT, D., THOMAS, G. & WALL, G. 1999. The evolution of the central North Sea Rift. *In*: FLEET, A. J. & BOLDY, S. A. R. (eds) *Petroleum Geology of Northwest Europe: Proceedings of the 5th Conference*. Geological Society, London, 63–82.

FRENCH, M. W., WORDEN, R. H., MARIANI, E., LARESE, R. E., MUELLER, R. R. & KLIEWER, C. E. 2012. Microcrystalline quartz generation and the preservation of porosity in sandstones; evidence from the Upper Cretaceous of the Subhercynian Basin, Germany. *Journal of Sedimentary Research*, **82**, 422–434.

GAARENSTROOM, L., TROMP, R. A. J., DE JONG, M. C. & BRANDENBURG, A. M. 1993. Overpressures in the Central North Sea: implications for trap integrity and drilling safety. *In*: PARKER, J. R. (ed.), *Petroleum Geology of Northwest Europe: Proceedings of the 4th Conference*. Geological Society, London, 1305–1313.

GILES, M. R. 1997. *Diagenesis: A Quantitative Perspective*. Kluwer Academic, Dordrecht.

GLENNIE, K. W. 1998. *Petroleum Geology of the North Sea: Basic Concepts and Recent Advances*. Blackwell, Oxford.

GLUYAS, J. & CADE, C. A. 1997. Prediction of porosity in compacted sands. *In*: KUPECZ, J. A., GLUYAS, J. G. & BLOCH, S. (eds) *Reservoir Quality Prediction in Sandstones and Carbonates*. American Association of Petroleum Geologists Memoir, American Association of Petroleum Geologists, Tulsa, USA, **69**, 19–27.

GOLDSMITH, P. J., HUDSON, G. & VAN VEEN, P. 2003. Triassic. *In*: EVANS, D., GRAHAM, C., ARMOUR, A. & BATHURST, P. (eds) *The Millennium Atlas: Petroleum Geology of the Central and Northern North Sea*. Geological Society, London, 105–127.

GOLDSMITH, P. J., RICH, B. & STANDRING, J. 1995. Triassic correlation and stratigraphy in the south Central Graben, UK North Sea. *In*: BOLDY, S. A. R. (ed.) *Permian and Triassic Rifting in Northwest Europe*. Geological Society, London, Special Publications, **91**, 123–143.

GOWERS, M. B. & SÆBØE, A. 1985. On the structural evolution of the Central Trough in the Norwegian and Danish sectors of the North Sea. *Marine and Petroleum Geology*, **2**, 298–318.

GRANT, N. T., MIDDLETON, A. J. & ARCHER, S. 2014. Porosity trends in the Skagerrak Formation, Central Graben, UKCS: the role of compaction and pore pressure history. *AAPG Bulletin*, **98**, 1111–1143.

GRIGSBY, J. D. 2001. Origin and growth mechanism of authigenic chlorite in sandstones of the lower Vicksburg Formation, south Texas. *Journal of Sedimentary Research*, **71**, 27–36.

Grove, C. & Jerram, D. A. 2011. jPOR: an ImageJ macro to quantify total optical porosity from blue-stained thin sections. *Computers & Geosciences*, **37**, 1850–1859.

Haszeldine, R. S., Wilkinson, M. *et al.* 1999. Diagenetic porosity creation in an overpressured graben. *In*: Fleet, A. J. & Boldy, S. A. R. (eds) *Petroleum Geology of Northwest Europe: Proceedings of the 5th Conference*. Geological Society, London.

Jeans, C. 1994. Clay diagenesis, overpressure and reservoir quality: an introduction. *Clay Minerals*, **29**, 415–424.

Kape, S., De Souza, O. D., Bushnaq, I., Hayes, M. & Turner, I. 2010. *Predicting Production Behaviour from Deep HPHT Triassic Reservoirs and the Impact of Sedimentary Architecture on Recovery*. Geological Society, London, Petroleum Geology Conference series, 405–417.

Knox, R. W. O'B. & Cordey, W. G. (eds) 1992. *Lithostratigraphic Nomenclature of the UK North Sea*. British Geological Survey, Nottingham.

Lundegard, P. D. 1992. Sandstone porosity loss; a 'big picture' view of the importance of compaction. *Journal of Sedimentary Petrology*, **62**, 250–260.

Mallon, A. J. & Swarbrick, R. E. 2002. A compaction trend for non-reservoir North Sea Chalk. *Marine and Petroleum Geology*, **19**, 527–539.

Mallon, A. J. & Swarbrick, R. E. 2008. Diagenetic characteristics of low permeability, non-reservoir chalks from the Central North Sea. *Marine and Petroleum Geology*, **25**, 1097–1108.

Mallon, A. J., Swarbrick, R. E. & Katsube, T. J. 2005. Permeability of fine-grained rocks: new evidence from chalks. *Geology*, **33**, 21–24.

Matthews, W. J., Hampson, G. J., Trudgill, B. D. & Underhill, J. R. 2007. Controls on fluviolacustrine reservoir distribution and architecture in passive salt-diapir provinces: insights from outcrop analogs. *AAPG Bulletin*, **91**, 1367–1403.

McKie, T. & Audretsch, P. 2005. Depositional and structural controls on Triassic reservoir performance in the Heron Cluster, ETAP, Central North Sea. *In*: Doré, A. G. & Vining, B. A. (eds) *Petroleum Geology: North-West Europe and Global Perspectives – Proceedings of the 6th Petroleum Geology Conference*. Geological Society, London, 285–297.

McKie, T., Jolley, S. & Kristensen, M. 2010. Stratigraphic and structural compartmentalization of dryland fluvial reservoirs: Triassic Heron Cluster, Central North Sea. *In*: Jolley, S. J., Fisher, Q. J. *et al.* (eds) *Reservoir Compartmentalization*. Geological Society, London, Special Publications, **347**, 165–198.

Nguyen, B. T. T., Jones, S. J., Goulty, N. R., Middleton, A. J., Grant, N., Ferguson, A. & Bowen, L. 2013. The role of fluid pressure and diagenetic cements for porosity preservation in Triassic fluvial reservoirs of the Central Graben, North Sea. *AAPG Bulletin*, **97**, 1273–1302.

Oelkers, E. H., Bjorkum, P. A. & Murphy, W. M. 1996. A petrographic and computational investigation of quartz cementation and porosity reduction in North Sea sandstones. *American Journal of Science*, **296**, 420–452.

Osborne, M. J. & Swarbrick, R. E. 1999. Diagenesis in North Sea HPHT clastic reservoir – Consequences for porosity and overpressure prediction. *Marine and Petroleum Geology*, **16**, 337–353.

Paxton, S., Szabo, J., Ajdukiewicz, J. & Klimentidis, R. 2002. Construction of an intergranular volume compaction curve for evaluating and predicting compaction and porosity loss in rigid-grain sandstone reservoirs. *AAPG Bulletin*, **86**, 2047–2067.

Pittman, E. D. & Lumsden, D. N. 1968. Relationship between chlorite coatings on quartz grains and porosity, Spiro Sand, Oklahoma. *Journal of Sedimentary Petrology*, **38**, 668–670.

Pittman, E. D., Larese, R. E. & Heald, M. T. 1992. Clay coats: occurrence and relevance to preservation of porosity in sandstones. *In*: Pittman, E. D. (ed.) *Origin, Diagenesis, and Petrophysics of Clay Minerals in Sandstones*. SEPM Special Publications, Tulsa, USA, **47**, 241–255.

Schneider, F. & Wolf, S. 2000. Quantitative HC potential evaluation using 3D basin modelling: application to Franklin structure, Central Graben, North Sea, UK. *Marine and Petroleum Geology*, **17**, 841–856.

Smith, R. I., Hodgson, N. & Fulton, M. 1993. Salt control on Triassic reservoir distribution, UKCS Central North Sea. *In*: Parker, J. R. (ed.) *Petroleum Geology of Northwest Europe: Proceedings of the 4th Conference*. Geological Society, London, 547–557.

Swarbrick, R. E. & Osborne, M. J. 1998. Mechanisms that generate abnormal pressures: an overview. *In*: Law, B. E., Ulmishek, G. F. & Slavin, V. I. (eds) *Abnormal Pressures in Hydrocarbon Environments*. American Association of Petroleum Geologists Memoir, Tulsa, USA, **70**, 13–34.

Swarbrick, R. E., Osborne, M. *et al.* 2000. Integrated study of the Judy field (Block 30/7a) – an overpressured central North Sea oil/gas field. *Marine and Petroleum Geology*, **17**, 993–1010.

Swarbrick, R. E., Lahann, R. W., O'Connor, S. A. & Mallon, A. J. 2010. Role of the Chalk in development of deep overpressure in the Central North Sea. *In*: Vining, B. A. & Pickering, S. C. (eds) *Petroleum Geology: From Mature Basins to New Frontiers – Proceedings of 7th Petroleum Geology Conference*. Geological Society, London, **1**, 493–507.

Taylor, T., Stancliffe, R., Macaulay, C. & Hathon, L. 2004. High temperature quartz cementation and the timing of hydrocarbon accumulation in the Jurassic Norphlet Sandstone, offshore Gulf of Mexico, USA. *In*: Cubitt, J. M., England, W. A. & Larter, S. R. (eds) *Understanding Petroleum Reservoirs: Towards an Integrated Reservoir Engineering and Geochemical Approach*. Geological Society, London, Special Publications, **237**, 257–278.

Taylor, T. R., Giles, M. R. *et al.* 2010. Sandstone diagenesis and reservoir quality prediction: models, myths, and reality. *AAPG Bulletin*, **94**, 1093–1132.

Walderhaug, O. 1994. Temperatures of quartz cementation in Jurassic sandstones from the Norwegian continental shelf – evidence from fluid inclusions. *Journal of Sedimentary Research*, **64**, 311–323.

Weibel, R. 1998. Diagenesis in oxidising and locally reducing condition – an example from the Triassic Skagerrak Formation, Denmark. *Sedimentary Geology*, **121**, 259–276.

WILKINSON, M., DERBY, D., HASZELDINE, R. S. & COUPLES, G. D. 1997. Secondary-porosity generation during deep burial associated with overpressure leak-off: fulmar Formation, UKCS. *AAPG Bulletin*, **81**, 803–813.

WORDEN, R. H. & MORAD, S. 2000. Quartz cementation in oilfield sandstones: a review of the key controversies. *In*: WORDEN, R. H. & MORAD, S. (eds) *Quartz Cement in Sandstones*. International Association of Sedimentologists Special Publications, **29**, 1–20.

WORDEN, R. H. & MORAD, S. 2003. *Clay Minerals in Sandstones: Controls on Formation, Distribution and Evolution*. Blackwell, Oxford.

WORDEN, R. H., FRENCH, M. W. & MARIANI, E. 2012. Amorphous silica nanofilms result in growth of misoriented microcrystalline quartz cement maintaining porosity in deeply buried sandstones. *Geology*, **40**, 179–182.

YARDLEY, G. & SWARBRICK, R. 2000. Lateral transfer: a source of additional overpressure Marine and Petroleum. *Geology*, **17**, 523–537.

Deciphering multiple controls on reservoir quality and inhibition of quartz cement in a complex reservoir: Ordovician glacial sandstones, Illizi Basin, Algeria

MARTIN WELLS[1]*, PHILIP HIRST[1], JON BOUCH[2], EMMA WHEAR[2] & NIGEL CLARK[1]

[1]*BP Exploration, Chertsey Road, Sunbury-on-Thames, Middlesex TW16 7LN, UK*

[2]*Pore Scale Solutions, Riverside Cottage, Carwynnen, Camborne, Cornwall TR14 9LR, UK*

**Correspondence: Martin.Wells2@uk.bp.com*

Abstract: Late Ordovician (*c.* 445 Ma) glacial sandstones form important gas reservoirs in the Illizi Basin, SW Algeria. These reservoirs have a high degree of depositional and diagenetic complexity, such that understanding and predicting reservoir quality (RQ) presents a major challenge to their economic development. Porosity is typically 1–10%, but reaches up to 15% and permeability is typically $<10^{-15}$ m^2 (<1 mD), but locally reaches $>10^{-13}$ m^2 (>100 mD). The key questions addressed herein concern the development and distribution of this RQ variability, specifically why has good RQ been locally preserved?

Primary depositional fabric exerts a strong control on RQ. Muddy sandstones are either highly compacted or pervasively cemented by quartz and microporous illite, and have very poor RQ. Only fine- to medium-grained, moderately well sorted, clean sandstones can contain good RQ, but texturally and mineralogically similar sandstones span a wide range of porosity and permeability. This range is primarily driven by the degree of quartz cementation, with incomplete cementation resulting in the best RQ. Quartz overgrowths in incompletely cemented clean sandstones are patchy and non-luminescent in scanning electron microscopy with cathodoluminescence (SEM-CL), possibly indicating slow growth rates. There is tentative evidence to link incomplete quartz cementation with oil charging of the reservoir. An alternative or additional explanation of RQ preservation may be related to limited silica supply in the centres of the thickest, stacked, clean sandstones, where the better RQ tends to reside.

The results of this study imply that sustained high-energy depositional processes, coupled with an early oil charge, are prerequisites for retaining the best RQ. This has important implications for the exploration and development of Late Ordovician glacial sandstones in the Illizi Basin, and potentially similar plays elsewhere.

Lower Palaeozoic sandstones host important hydrocarbon resources across North Africa and the Middle East (e.g. Le Heron *et al.* 2009). Sandstones deposited during the Late Ordovician (Hirnantian) glaciation form one of the most important reservoirs. These reservoirs display considerable depositional complexity (e.g. Hirst *et al.* 2002) and have been subject to more than 400 myr of burial diagenesis. Consequently, there are many reservoir quality (RQ) controls that interact in complex ways and are often difficult to predict; the ability to unravel the various depositional and diagenetic RQ controls is of considerable scientific and economic significance. Previous RQ studies of Ordovician glacial sandstones from North Africa have highlighted the importance of depositional fabrics and clay content (Hirst *et al.* 2002), syn-sedimentary emplacement of clay coats by pressurized glacial meltwater (Tournier *et al.* 2010), and eo- and meso-diagenetic cements (Hirst *et al.* 2002; El-ghali *et al.* 2006*a*, *b*; Tournier *et al.* 2010).

This paper documents a case study from the In Amenas and Bourarhat Sud area of the Illizi Basin, Algeria (Fig. 1), where Late Ordovician gas-bearing sandstones typically have porosities of <10% and permeabilities of $<10^{-15}$ m^2 (<1 mD). However, sandstones recording porosities up to 15% and permeabilities of $>10^{-13}$ m^2 (>100 mD) exist and have dominated production to date. This study integrates detailed sedimentology, petrography and routine core analysis data with results from other analyses including X-ray diffraction (XRD), scanning electron microscopy (SEM), SEM with cathodoluminescence (SEM-CL) and fluid-inclusion analysis from a large subsurface dataset to address two key questions:

- What are the depositional and diagenetic features that resulted in the typically poor RQ?
- What depositional and diagenetic features have caused the local preservation of moderate–good RQ?

From: Armitage, P. J., Butcher, A. R., Churchill, J. M., Csoma, A. E., Hollis, C., Lander, R. H., Omma, J. E. & Worden, R. H. (eds) 2018. *Reservoir Quality of Clastic and Carbonate Rocks: Analysis, Modelling and Prediction.* Geological Society, London, Special Publications, **435**, 343–372.
First published online December 11, 2015, https://doi.org/10.1144/SP435.6

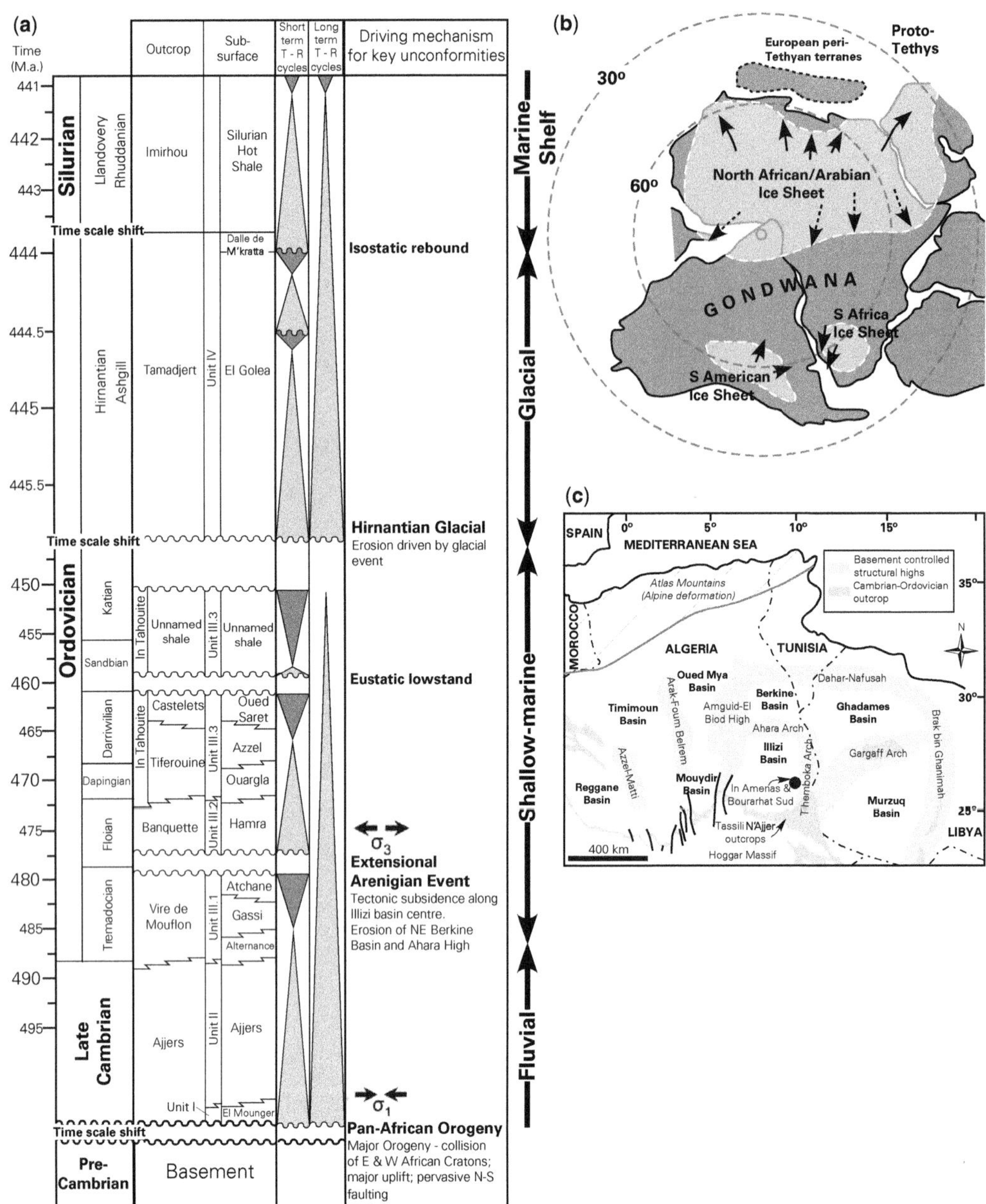

Fig. 1. (**a**) Lower Palaeozoic stratigraphic column for the Illizi Basin, Algeria (compiled from Galeazzi *et al.* 2010; Hirst 2012). Strata belonging to the Hirnantian glacial episode are the focus of this study. (**b**) Palaeogeographical reconstruction of Gondwana at approximately 440 Ma showing the distribution of Hirnantian ice sheets (after Le Heron *et al.* 2009). (**c**) Palaeozoic basins and arches of Algeria and Libya, showing the location of the Illizi Basin and In Amenas–Bourarhat Sud study area (after Hirst 2012). T–R cycles, transgressive–regressive cycles

The studies cited above illustrate that subtly different RQ controls appear to have operated in different areas of North Africa. However, it is hoped that the outcome of this study may shed further light on RQ controls in other Lower Palaeozoic reservoirs, both across North Africa and worldwide.

The geological setting of the Late Ordovician glacial deposits of the Illizi Basin, Algeria is first summarized before the various methods employed are reviewed. A large dataset is then interrogated to better understand the likely controls on RQ. Finally, two hypotheses regarding the preservation of RQ in these sandstones are outlined, including hydrocarbon inhibition and silica budgets.

Geological setting

Regional context

The Palaeozoic of North Africa records mainly fluvial and shallow-marine sedimentation on a low-relief passive-margin ramp more than 1000 km wide on the northern edge of the Gondwana supercontinent (Fig. 1) (Scotese & McKerrow 1990; Scotese *et al.* 1999; Le Heron *et al.* 2004; Ghienne *et al.* 2007; Craig *et al.* 2010; Eschard *et al.* 2010; Galeazzi *et al.* 2010). Reactivation of basement lineaments in the Cambrian–Lower Ordovician led to the segregation of the passive-margin ramp into areas of lower and higher subsidence (arches and basins, respectively: Fig. 1). The Illizi Basin was flanked by the Amguid-el Biod High to the west, the Ahara Arch to the north and the Tihemboka Arch to the east (Fig. 1). By the Late Ordovician, when the sediments that are the subject of this study were deposited, subsidence rates equalized across the basins and arches marking a period of tectonic quiescence that extended into Silurian times (Boote *et al.* 1998; Eschard *et al.* 2010).

During the Lower Palaeozoic, Gondwana was located over the South Pole, and a short-lived (*c.* 1–2 myr) glacial episode is recorded at the end of the Ordovician (Hirnantian, *c.* 445 Ma) (Fig. 1) (Beuf *et al.* 1971; Hambrey & Harland 1981; Vaslet 1990; Moreau *et al.* 2005; Saltzman & Young 2005; Ghienne *et al.* 2007; Le Heron & Dowdeswell 2009). The Late Ordovician ice sheet covered a large part of North Africa and Arabia (McClure 1978; Vaslet 1990; Powell *et al.* 1994; Ghienne & Deynoux 1998; Senalp & Al Laboun 2000; Ghienne *et al.* 2007; Le Heron & Dowdeswell 2009), with glacial influence extending as far north as Spain at the glacial maximum (Guitierrez-Marco *et al.* 2010). Algeria and Libya were situated towards the margin of the North African ice sheet, with glacially transported sediment generally directed north to northwestwards towards the ocean (Beuf *et al.* 1971, pp. 77, 165–170).

Algeria and Libya are interpreted to have been extensively ice-covered during the ice maximum (Le Heron & Craig 2008). During this time and the subsequent ice retreat, a series of glacial valleys were carved into the pre-glacial semi-consolidated substrate (Hirst *et al.* 2002; Moreau *et al.* 2005; Le Heron & Craig 2008; Hirst 2012). The valleys were largely filled during phases of overall ice-sheet retreat by a complex assemblage of glaciogenic and glacially influenced sediments (e.g. Hirst *et al.* 2002; Hirst 2012; Lang *et al.* 2012).

Late Ordovician glacial deposits: facies scheme and depositional model

Eleven descriptive lithotypes have been identified from outcrop and core studies (Table 1) (Hirst 2012). They include:

- fine- to coarse-grained, bedded and cross-bedded sandstones (Sb and CSx);
- unstructured to faintly laminated sandstones with water-escape features (Su and SMu);
- heterolithics displaying various portions of Bouma sequences (HS and HM);
- laminated mudstones with rare outsized grains (MM and MMg);
- very poorly sorted massive to highly deformed sandstones and sandy mudstones with floating outsized grains (SSg, SMg and MSg).

These lithotypes have been consistently applied on a cm-scale to the more than 3 km of core in the In Amenas–Bourarhat Sud study area. The distinction of 'clean' v. 'muddy' sandstones (e.g. Su v. SMu) is made based on the colour of the core (muddy sandstones tend to be a darker grey) and integration with wireline log data (especially gamma ray, where Su is typically <30 g API and SMu >30 g API).

Five facies associations have been identified (Table 2) (Hirst *et al.* 2002; Hirst 2012). FA1 (fine- to coarse-grained, bedded and cross-bedded sandstones) typically occur towards the base of glacial valley fills (i.e. towards the base of Unit IV: Fig. 1a) and have limited lateral extent at outcrop and in the subsurface (<200 m perpendicular to palaeoflow). They are interpreted as subglacial channel-fills to ice-proximal proglacial outwash fans (Fig. 2) (Hirst *et al.* 2002; Dixon *et al.* 2008; Hirst 2012). FA2 (massive to low-angle cross-laminated, dewatered sandstones) are more common towards the top of glacial valley fills and in strata that overtops the valleys to form a regionally extensive blanket (i.e. towards the top of Unit IV: Fig. 1a). The sandstones occur in both laterally extensive (10 s km) amalgamated sheets and channels, up to tens of metres thick, and, more rarely, in sinuous ribbon channels, which are typically 50–400 m wide and 4–12 m thick, capping the glacial stratigraphy (Beuf *et al.* 1971). The presence of transcritical bedforms and dewatering fabrics distinguish these sandstones from FA1. FA2 is interpreted as the deposits of relatively ice-distal proglacial high-density turbidites (Fig. 2) deposited by glacier meltwater events, possibly related to

Table 1. *Summary of the lithotypes identified in the subsurface cores used in this study (after Hirst 2012)*

Lithotype code	Lithotype description	Lithology qualifier based on visual estimation from core	Depositional process
CSx	Coarse sandstone, some cross-beds; locally conglomeratic		Sediments with tractional bedforms
Sb	Fine to coarse bedded sandstone		
Su	Fine- to very-fine-grained sandstone; massive to subtle low-angle bedding, water-escape structures	Very light grey, low clay content (<10%)	High-density turbidites
SMu		Light grey, moderate clay content (10–20%)	
HS	Heterolithics, sandstone-dominated	>50% sandstone layers	Low-density turbidites
HM	Heterolithics, mudstone-dominated	<50% sandstone layers	
MM	Mudstone and siltstone, laminated with some granule layers		Pelagic fines
MMg	Mudstone with granule or pebble layers	Mudstone-dominated	Ice-rafted debris
MSg	Argillaceous, very poorly sorted, granular sandstones ('diamictites')	Dark grey, high clay content (>50%)	Debris-flow deposits
SMg		Medium grey, moderate to high clay content (10–50%)	
SSg		Light-grey, moderately low clay content (<10%)	

periodic catastrophic draining of subglacial lakes (Hirst 2012). FA3 (heterolithic deposits with localized deformation) and FA4 (laminated mudstones and siltstones with rare granules) are interpreted as ice-distal deposits comprising the distal expression of meltwater turbidites and hemipelagic sedimentation with iceberg rain-out material (Fig. 2). Finally, FA5 (poorly sorted argillaceous sandstones with granules and deformed fabrics) are interpreted as the deposits of debris flows shed from the unstable ice-margin and subaqueous valley walls (Fig. 2).

The proglacial turbidite sandstones are the dominant reservoir facies in the subsurface: hence, understanding their distribution is highly valuable. Their extensive development in the uppermost glacial stratigraphy is thought to be related to stillstands during the final, relatively rapid retreat of the marine-terminating ice sheet. The glacial valley network and the end-glacial ribbon channels both occur as segments, tens of kilometres long, with abrupt beginnings and ends (e.g. Le Heron & Craig 2008; Girard *et al.* 2012). Mega-regional mapping of the glacial sediment isopach reveals local thickening orientated perpendicular to the ends of these valley trends (Le Heron & Craig 2008). These large-scale patterns are consistent with a phased ice retreat, with the development of grounding lines during periods of ice stillstand (Le Heron & Craig 2008). Grounding-line fans would be expected to form in front of these grounding lines (e.g. Lønne *et al.* 2001; Powell & Cooper 2002; Dowdeswell *et al.* 2015), providing a model to predict the presence of reservoirs in the subsurface.

No grounding-zone wedges have been explicitly identified in the outcrops of southern Algeria and Libya; however, candidate grounding-zone wedges have been interpreted from 3D seismic surveys to the north (Decalf *et al.* 2013). The mapped grounding lines in the subsurface are spaced at approximately 30 km, consistent with the spacing of valley and ribbon-channel segments observed elsewhere.

Methods

With more than 3 km of core available, covering a geographical area of around 5500 km^2, there is a large dataset on which to draw. In order to unravel the controls on reservoir quality, the detailed sedimentological descriptions outlined above have been integrated with:

- routine core analysis (RCA) from 3519 core plugs from 36 wells;
- detailed qualitative and quantitative petrography on 235 thin sections from 25 wells;
- X-ray diffraction (XRD);
- scanning-electron microscopy (SEM);
- scanning electron microscopy with cathodoluminescence (SEM-CL);
- fluid-inclusion data.

Routine core analysis (RCA)

The RCA database is a combination of data acquired using a CMS-300 (Core Measurement System) machine since 2008 and historical data (from the 1970s to 2000s) collected using older machinery.

The majority of the data used here (*c*. 70%) were collected using the CMS-300: an automated computer-directed, unsteady state pressure decay permeameter and porosimeter with a sleeve pressure of 5.5 MPa (800 psi). Owing to the different vintages of equipment used to collect the data presented here, and the lack of measurements at different effective stresses for all of these machines, no overburden correction has been applied and measurements should be treated as unstressed. Porosities and permeabilities, especially at the lower end of the scale, will be lower at reservoir conditions. However, unstressed data are sufficient for the purpose of this study: that is, identifying causes in the variability of RQ rather than highly quantitative petrophysical rock typing, for example.

The RCA database was first quality controlled to remove fractured plugs and low-permeability samples analysed with older machinery that was unable to resolve permeabilities below about 10^{-16} m^2 (0.1 mD). This left 3519 plugs with useable data. The sedimentary lithotypes were interpolated onto the RCA data, together with other pertinent information such as true stratigraphic thickness below top reservoir seal.

Petrography

The petrographical dataset comprises textural analysis and point-count data (200 mineral point counts + porosity) from 25 wells, with 235 out of a total of 342 thin sections sampling the Late Ordovician glacial stratigraphy. The remainder of the thin sections sample the pre-glacial Cambrian–Ordovician stratigraphy and the Late Ordovician–Early Silurian post-glacial stratigraphy. Grain size was categorized on the Udden–Wentworth grain-size scale with Ø upper and lower divisions. Sorting was calculated as the standard deviation from the mean length (expressed as Ø), and is categorized according to the scheme of Folk & Ward (1957).

Porosity measurements derived from RCA are unable to distinguish macroporosity from microporosity: however, these different pore architectures have an important impact on permeability. An attempt has been made to quantify macroporosity v. microporosity in all of the petrographical samples. Macroporosity is derived from modal analysis and microporosity is calculated as: [measured routine core analysis porosity]–[point counted total macroporosity].

Petrographical data have been collected by several different petrographers. Therefore, caution must be exercised when integrating all of the data, as there is likely to be a degree of operator bias. Of the 235 thin sections, 212 have accompanying porosity–permeability data. A review of the petrography samples shows that they cover the range of porosity and permeability in the larger RCA database (Fig. 3f). This adds confidence to conclusions drawn from the petrography database regarding RQ controls. There are detailed proprietary petrographical reports for four of the wells and two proprietary overview reports covering 15 of the remaining wells. The detailed qualitative observations included in these reports have been instrumental in understanding the RQ controls.

X-ray diffraction (XRD)

Whole-rock and <2 μm clay fraction XRD data has been collected using standard practices (Appendix 1) from 89 of the petrographical samples, covering a range of lithotypes (NB SSg, SMg and MSg lithotypes are not represented), and has assisted in the identification of the major mineral phases present.

Scanning electron microscopy (SEM)

Backscattered and secondary electron SEM imaging of rough surfaces was used primarily to investigate the morphologies, compositions, mineralogy and paragenetic relationships of the clay minerals. SEM and SEM-CL imaging of polished thin sections was primarily used for detailed analysis of quartz cements and compaction, but was also employed for observation of the clays and their paragenetic relationships.

Fluid inclusions

Fluid inclusions were first identified and characterized using routine petrographical techniques in plane polarized and ultraviolet (UV) light to determine a primary or secondary origin, and to define an aqueous or hydrocarbon content. The fluid inclusions were almost exclusively aqueous, with rare hydrocarbon-filled examples, and of primary origin: trapped in quartz either at grain–cement boundaries (dust rims) or within the main bodies of the quartz overgrowths (typically along crystal-growth-zonation boundaries). Rare secondary aqueous and hydrocarbon fluid inclusions were observed: trapped along microfractures within grains of detrital quartz. Ultraviolet epifluorescence was used to identify and differentiate different types of hydrocarbon based on the colour of any fluorescence. Microthermometric analysis was carried out to determine temperatures and salinities of the pore-filling fluid phase at the time of entrapment (Appendix 1).

Results

Plots of porosity v. permeability shaded by lithotype show a strong relationship between lithotype and RQ (Fig. 3), highlighting the importance of

Table 2. *Summary of the facies associations identified at outcrop and in the subsurface cores used in this study (compiled following the facies scheme of Hirst 2012)*

Facies association	Dominant lithotypes	Subordinate lithotypes	Description	Example core photograph
FA1	CSx, Sb	Su	Light-grey, clean, moderately to poorly sorted, fine- to coarse-grained sandstones. Well-defined, sharp-based beds range from 0.1 to >10 m in thickness and comprise trendless or gross upward-fining grain-size profiles. Internally, the sandstones may be decimetre-scale cross-stratified to structureless. Grain size is typically variable from bed to bed. Beds can contain a high proportion of granules and pebbles, which sometimes occur as basal lags	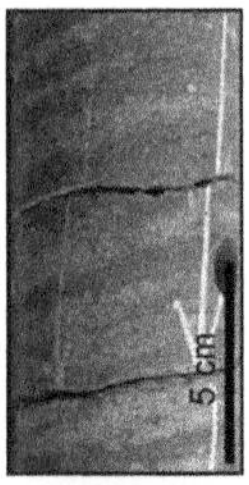 5 cm
FA2	Su, SMu	HS, HM, SSg, SMg, MM, MMg	Light- to medium-grey, very clean to moderately clean, well-sorted to moderately sorted, very-fine- to medium-grained sandstones. Poorly defined beds range from 0.5 to 6 m thick with trendless to upward-fining grain-size profiles. Erosive bed boundaries can be observed. Internally, the sandstones are typically structureless, with rare faint mm- to cm-scale planar lamination to low-angle cross-lamination. Diffuse variations in argillaceous content on a cm-scale are locally common. Dewatering fabrics including subhorizontal consolidation laminae, dish structures and subvertical sheets or pipes are often observed. Beds may rarely contain outsized 'floating' granules with a variable concentration	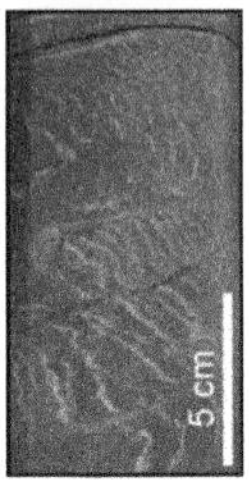 5 cm
FA3	HS, HM	MSg, MM, MMg	Well-defined cm-scale interbedded very-fine- to fine-grained sandstones, siltstones and mudstones. Often stacked, forming sequences of sharp-based fining-upward cycles. Internally, the sandstones often represent various portions of Bouma-type cycles, including structureless fabrics together with mm-scale planar argillaceous lamination and ripple and climbing cross-lamination. The siltstones and mudstones are structureless to mm-scale planar laminated. Planolites burrows and graptolites are sparsely present. Rare, outsized 'floating' granules can be observed	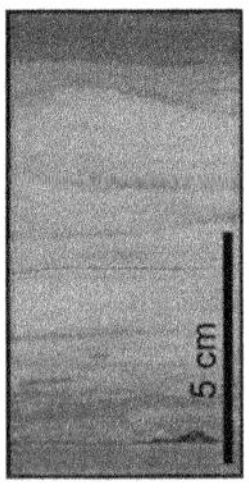 5 cm
FA4	MM	MMg	Millimetre-scale planar laminated to structureless, dark-grey silty mudstones or muddy siltstones with rare graptolites, and single instances of trilobite and brachiopod remains. Coarser-grained laminae, often granules and rarely isolated pebbles are sparsely present and can, in some instances, be observed to indent the underlying laminae	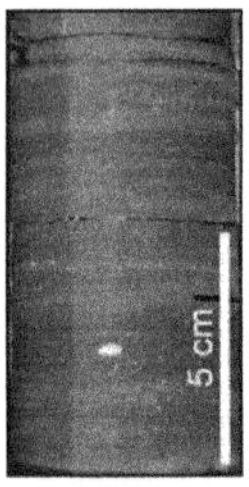 5 cm
FA5	SSg, SMg, MSg	Su, Sb, HS, HM, MM, MMg	Light- to medium-dark-grey, variably argillaceous, poorly to very poorly sorted, fine- to medium-grained sandstones and sandy mudstones. Characterized by common outsized 'floating' granules to pebbles and deformed intraformational clasts. Poorly defined beds range from <1 to >15 m in thickness with sharp bases and tops. Thinner-bedded deposits comprise intensely deformed cm-scale heterolithics. Thicker-bedded sediments are typically structureless, although brittle and ductile deformation fabrics are locally common. Dewatering features are not observed.	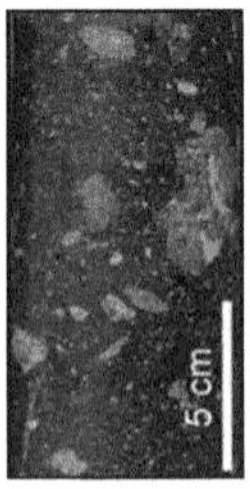 5 cm

Characteristic wireline log signature	Example wireline log signature	Probable depositional setting and processes
• Low (*c.* 20–40 gAPI), blocky to slightly serrate gamma ray. Packages often bedded at 1–5 m scale, with sharp bases and an upward increasing gamma-ray profile • Moderate sand-style separation between the neutron-density logs • Cross-bedding observed in image logs	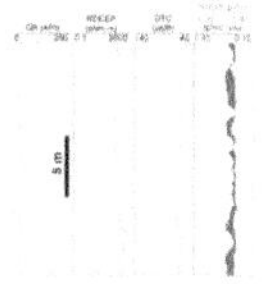	Deposition by tractional flows, probably unidirectional, of waxing and waning flow events, periodically high energy. Interbedding with turbidite, debrite and heterolithic packages suggests a marine setting. Deposition probably occurred in a confined setting such as a channel or valley base, possibly at or near a glacial tunnel valley mouth.
• Low to relatively low (*c.* 10–50 gAPI), blocky to upward-increasing gamma-ray log profile. Packages often bedded at 1–5 m scale, with sharp bases and subtle increases in gamma ray towards the tops • Sand-style separation between the neutron-density logs • Massive to subtle flat- to low-angle lamination with dewatering fabrics observed in image logs		Rapid deposition as metre-scale 'event' beds. Low-angle cross-stratified sandstones resemble antidune (transcritical flow) bedforms at outcrop and indicate deposition from sustained, high-velocity flows where flow thickness $\ll$ water depth. Diffuse cm-scale variations in argillaceous content are interpreted to represent flow banding, further supporting deposition from sustained (quasi-steady) flows. Combined, these attributes suggest deposition from sustained, hyperpycnal high-density turbidites (concentrated density flows) in a pro-glacial marine shelf setting
• Highly serrate gamma ray (*c.* 50–150 gAPI), rapid changes on a metre-scale • Alternating sand-style and mud-style cross-overs between the neutron and density logs at a metre-scale • 'Stripy' appearance with some localized soft-sediment deformation observed in image logs	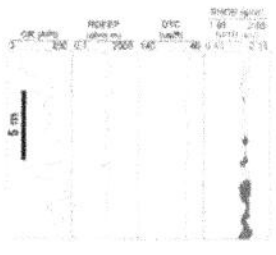	Relatively low-energy, ice-distal marine shelf setting characterized by hemipelagic settling of mud and deposition of sand-prone event beds (commonly turbiditic in appearance). The coarser-grained elements (outsized grains) are interpreted as ice-rafted debris and reworked ice-rafted debris
• Slightly serrate gamma-ray (>150 gAPI), subtle increasing and decreasing trends on a 10 m-scale • Mud-style separation between the neutron-density logs.	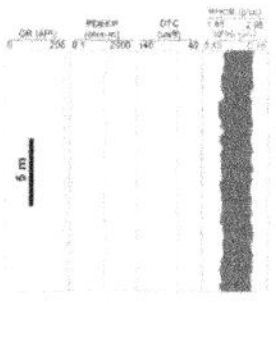	Low-energy marine shelf setting (presence of Graptolites) dominated by hemipelagic settling of mud. The coarser-grained elements (outsized grains) are interpreted as ice-rafted debris. These sediments were deposited in an ice-distal setting
• Blocky to serrate gamma ray (*c.* 50–100 gAPI) often with an increasing or decreasing trend over a 10 m scale • Coincident neutron-density logs or a mud-style separation • 'Speckled' appearance with soft-sediment deformation fabrics readily observed in image logs		Deposition from cohesive debris flows to hyperconcentrated density flows. The presence of abundant outsized grains suggests that these debris flows had a different provenance to the high-density turbidites and did not evolve into them. They were likely sourced from a combination of muddy rainout/meltout diamictites and interbedded sand prone turbidites and were probably triggered by slope instability immediately ahead of the ice front. Interbedded tractional deposits and/or thin turbidites are interpreted to represent small perennial outflows or turbidite events.

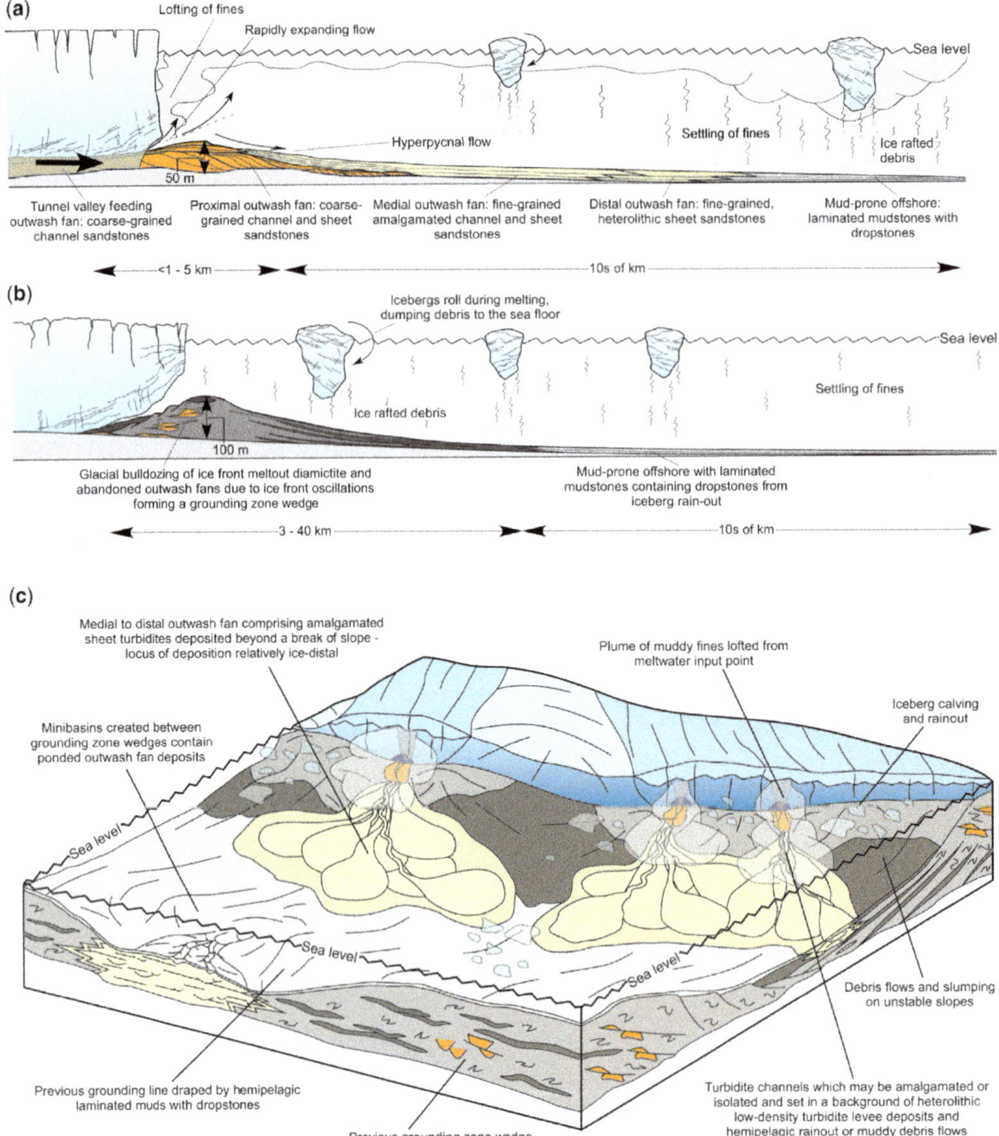

Fig. 2. Depositional model of pro-glacial outwash-fan-related deposits, Late Ordovician of the Illizi Basin, Algeria. Schematic depositional dip profile axial (**a**) and off-axis (**b**) to a pro-glacial outwash fan. (**c**) Block diagram illustrating a late deglaciation depositional model. Colour scheme represents facies associations, which are summarized in Table 2. Based on the models of Lønne *et al.* (2001), Powell & Cooper (2002), Russell & Arnott (2003) and Hornung *et al.* (2007).

depositional textures and sedimentary structures on RQ. Two broad trends are observed, both of which are well represented in the petrographical dataset:

- A ‘high-gradient’ porosity-permeability trend defined by lithotypes CSx, Sb and Su (defined in Table 1). These are all clean to very clean sandstones (light grey in core, low gamma-ray log values).
- A ‘low-gradient’ porosity-permeability trend defined by lithotypes SMu, SSg, SMg, HS and HM (defined in Table 1). These are muddier sandstones (darker grey in core, higher gamma-ray log values).

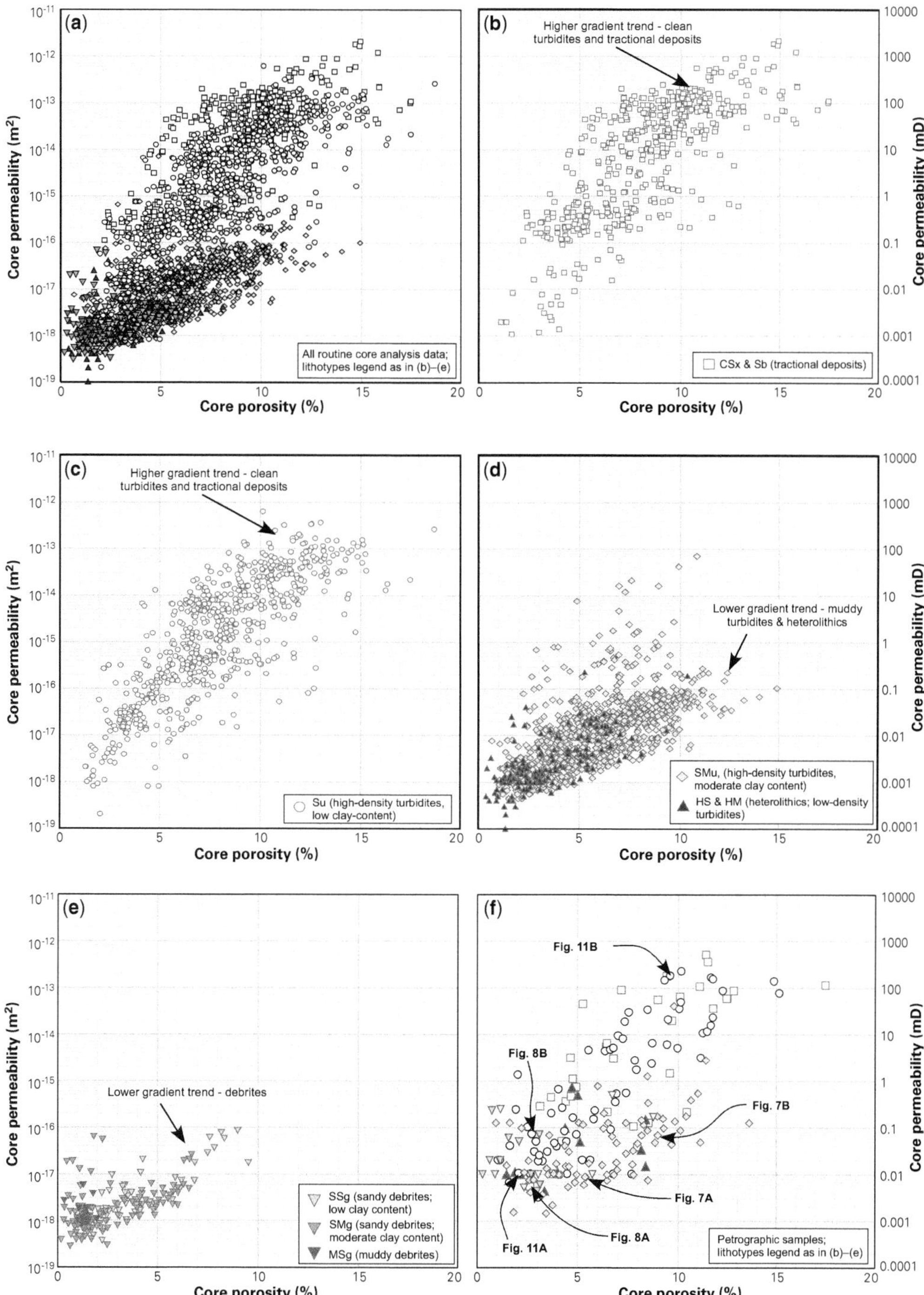

Fig. 3. (**a**) Porosity–permeability cross-plot for the cores used in this study, shaded by lithotype (see Table 1). The data fall into two broad trends: a higher gradient trend comprising lithotypes CSx, Sb and Su (**b** & **c**); and a lower gradient comprising lithotypes SMu, HS, SSg, SMg and MSg (**d** & **e**). (**f**) Petrographical samples with RCA data cover a representative range of lithotypes and porosity–permeability compared to the larger RCA dataset. The locations of four samples chosen to illustrate various controls on RQ in Figures 7 and 11 are shown.

There is a broad range in porosity and permeability for each lithotype. Therefore, although lithotype controls which porosity–permeability trend a given rock will occur on, additional factors control the position on that trend.

The following subsections provide systematic descriptions of the detrital and diagenetic characteristics of the sediments, with the objective of better understanding the porosity–permeability cross-plot (i.e. RQ controls).

Mineralogy and sandstone classification

Petrographical point-count data and XRD analysis illustrate that, in general, these are highly chemically mature sediments (Fig. 4). The detrital mineralogy is typically dominated by monocrystalline quartz, with only minor lithic material, feldspar and detrital clays. The lithic assemblage is dominated by polycrystalline quartz with lesser amounts of typically highly altered grains of unidentifiable original composition. There has been some minor (<1% by volume) secondary porosity generation through feldspar dissolution: however, depositional abundances of plagioclase and K-feldspar are considered to have typically been very low. Given the low volumes of porosity generated, the causes of this dissolution have not been explored in more detail. Detrital clay is relatively scarce, and has been partially altered to authigenic illite and, more locally, chlorite. Illite and chlorite, and rarely kaolinite, also occur as discrete authigenic cements

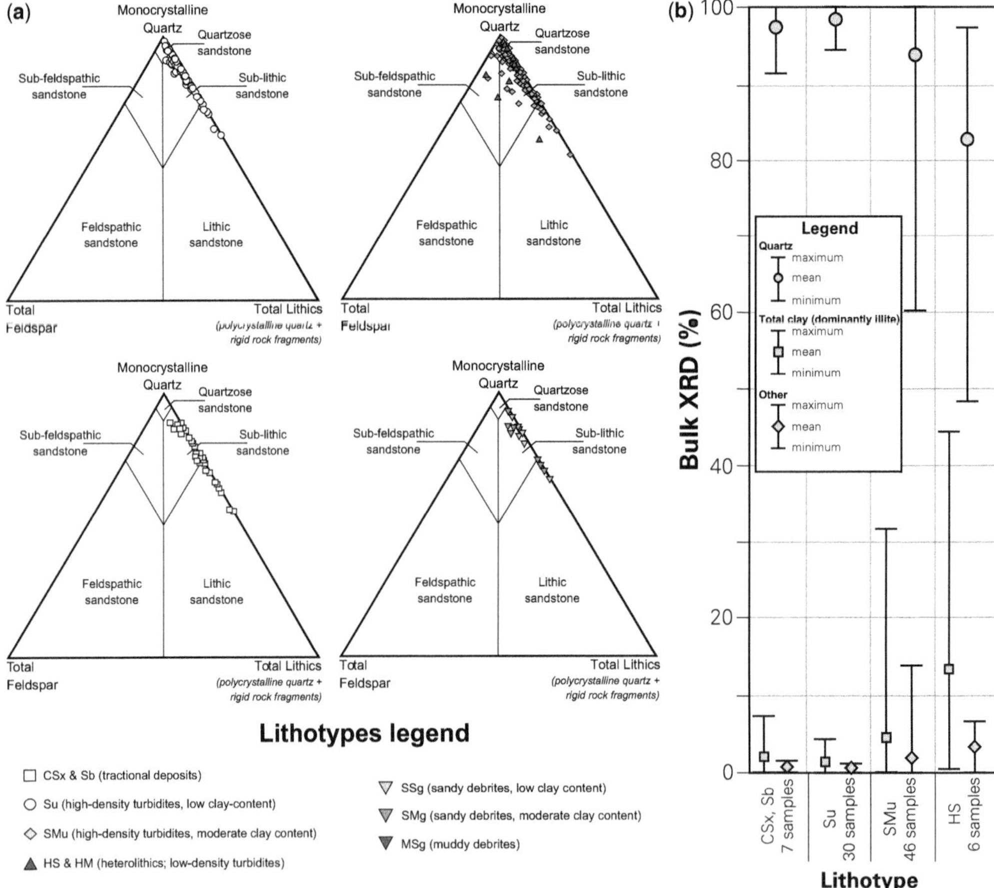

Fig. 4. (**a**) Sandstone classification for Late Ordovician sandstones derived from point-count data. All of the lithotypes plot in the same area of the quartz–feldspar–lithics diagram and are strongly quartz-dominated with negligible feldspar. (**b**) Summary of whole-rock XRD data categorized by lithotype. 'Clay' includes illite + illite–smectite, kaolinite and chlorite. 'Other' typically includes minor calcite, siderite, dolomite and K-feldspar. These data reinforce the quartz-dominated nature of the deposits, but illustrate that clay contents are subtly higher in lithotypes SMu and HS compared to lithotypes CSx, Sb and Su.

and grain-replacive components, which makes it difficult to assess original detrital clay content.

Sandstones classify as either quartzose, sublithic or lithic sandstones (Fig. 4a). All of the lithotype groupings occupy broadly the same space on the sandstone classification diagram (Fig. 4a), suggesting similar derivation. This is consistent with erosion and mixing of the relatively thick (>200 m) underlying pre-glacial Cambro-Ordovician stratigraphy. In places, Late Ordovician syn-glacial sandstones directly overlay crystalline basement, proving that glacial erosion did reach basement locally (e.g. Hirst 2012). This may be a source for the more chemically stable lithics (e.g. polycrystalline quartz) observed in thin sections (local basement typically comprises meta-granite in the study area).

XRD data confirms that all of the sandstone lithotypes are dominated by quartz (from 49 to 100% bulk XRD, with a mean of 95%: Fig. 4b). Lithotypes CSx, Sb and especially Su have the lowest total clay content (mean <2% bulk XRD), followed by SMu (mean 4.4% bulk XRD), then HS (mean 13.5% bulk XRD: Fig. 4). All other identified minerals (most commonly K-feldspar, plagioclase, calcite, siderite, dolomite, ferroan dolomite, pyrite, barite and muscovite and minor anatase, analcime, sphalerite, halite, hornblende, and marcasite) are very rare (mean <1% bulk XRD) in CSx, Sb and Su lithotypes, becoming slightly more common, although still rare, in the muddier lithotypes (Fig. 4). These data corroborate the more qualitative metrics used to define the lithotypes (core colour and gamma-ray log values: Table 1).

The <2 μm clay fraction is dominated by illite (45–100% clay fraction XRD with a mean of 80%) and mixed layer illite + illite–smectite (0–23% clay fraction XRD, with a mean of 11%). Chlorite is typically minor by comparison (0–33% clay fraction XRD, with a mean of 6%). The dominance of illite is interpreted to reflect both the cold-climate weathering origin of the detrital clays and extensive recrystallization during prolonged diagenesis (e.g. Worden & Morad 2003).

Grain size and sorting

There is a broad relationship between grain size, sorting and lithotype (Fig. 5). Lithotypes SMu and HS tend to have the finest grain sizes and are the most well sorted. Lithotypes Su, Sb and CSx overlap the SMu and HS trend, but extend to coarser grain sizes and poorer sorting. Lithotypes SSg and SMg tend to have poorer sorting for a given grain size compared with the other lithotypes. This is consistent with the core-based observation of scattered outsized granules in SSg and SMg lithotypes that are absent in lithotypes Su, SMu and HS. This difference in grain size and sorting suggest that the debris flows (SSg, SMg lithotypes) were unlikely to have evolved into the turbidity currents (Su, SMu and HS lithotypes) inasmuch as high-energy turbidity currents should be able to entrain the granule-grade material if it were available (Hirst 2012).

Permeability and, especially, porosity relationships with grain size and sorting are more difficult to discern, as might have been expected from the porosity–permeability cross-plot. However, it appears that the finest grained, most-sorted sandstones tend to have the worst permeabilities. The coarsest-grained, least-sorted sandstones have moderate permeabilities, and the highest permeabilities are associated with sandstones of intermediate grain size (medium-grained) and sorting (moderately well sorted to moderately sorted: Fig. 5). However, medium-grained, moderately well-sorted to moderately sorted sandstones can also have low porosities and permeabilities. This indicates that, in spite of diagenetic overprinting exerting a very strong control on RQ, an underlying primary depositional texture control is still locally evident.

Paragenetic sequence and burial history

The main diagenetic degraders of RQ in these rocks are strong compaction together with quartz and illite cementation (Fig. 6a). Other diagenetic minerals such as kaolinite, siderite and chlorite are present in lesser abundance. Poor RQ is to be expected considering the burial history of these reservoirs (Fig. 6b). There is considerable uncertainty surrounding the burial history of the Ordovician reservoirs since Carboniferous rocks currently crop out at the surface, leaving an approximately 300 Ma unconformity. However, the preferred burial history model takes into account the major tectonic events and is calibrated using fluid-inclusion data. Although burial history modelling suggests that the reservoirs have not been especially deeply buried (maximum burial depth of *c.* 2.5–3 km), they have been at moderate burial depths from the Carboniferous (*c.* 330–320 Ma) to present day. Modelled sustained effective stress of approximately 5500–6500 psi is consistent with strong compaction, and modelled sustained temperatures of >90°C is consistent with strong quartz cementation. Indeed, a more pertinent question might be why do these reservoirs have any porosity and permeability remaining at all? Before returning to this question, the factors impacting compaction and cementation are examined using the datasets available.

Ductile content (detrital and authigenic clay): impact on compaction diagenesis

A variety of datasets suggests a strong relationship between ductile content and RQ. Muddier lithotypes

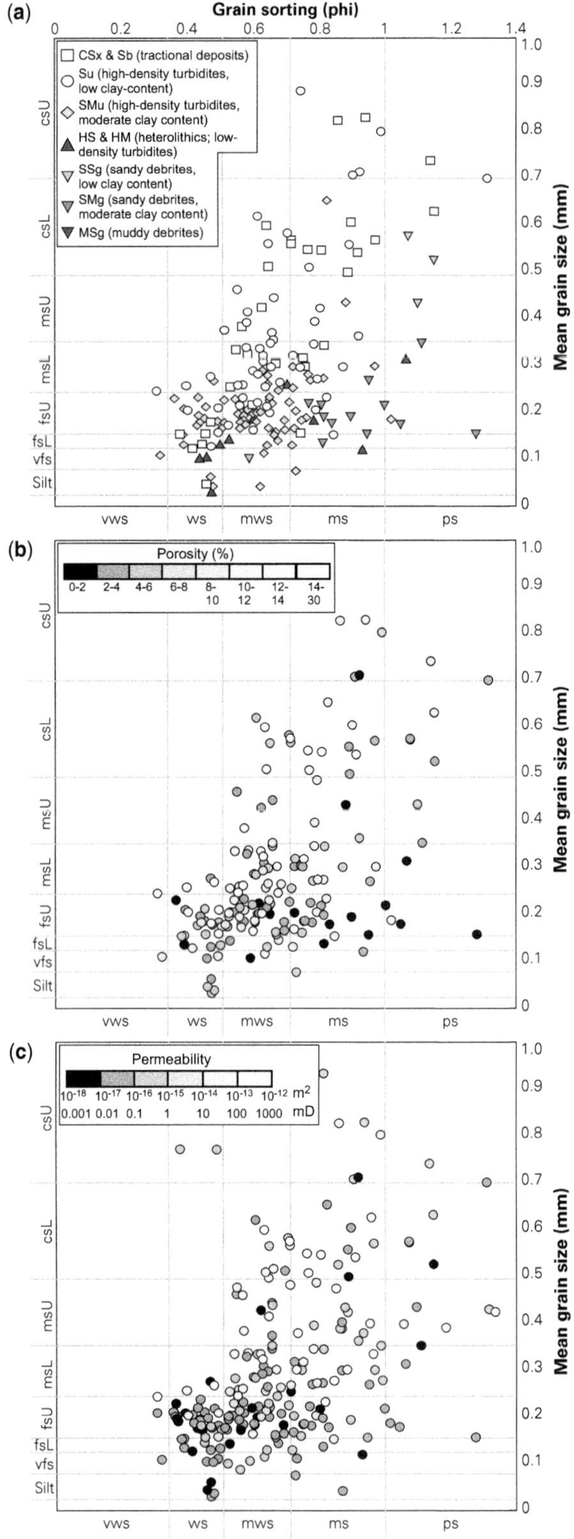
(a)
Grain sorting (phi)
0 0.2 0.4 0.6 0.8 1 1.2 1.4
CSx & Sb (tractional deposits)
Su (high-density turbidites, low clay-content)
SMu (high-density turbidites, moderate clay content)
HS & HM (heterolithics; low-density turbidites)
SSg (sandy debrites, low clay content)
SMg (sandy debrites, moderate clay content)
MSg (muddy debrites)
csU
csL
msU
msL
fsU
fsL
vfs
Silt
Mean grain size (mm)
1.0 0.9 0.8 0.7 0.6 0.5 0.4 0.3 0.2 0.1 0
vws ws mws ms ps
(b)
Porosity (%)
0-2 2-4 4-6 6-8 8-10 10-12 12-14 14-30
(c)
Permeability
10-18 10-17 10-16 10-15 10-14 10-13 10-12 m2
0.001 0.01 0.1 1 10 100 1000 mD

(SMu and HS: as defined by core observations and gamma-ray logs, and corroborated by XRD and modal analysis data) have poorer RQ (Fig. 3). Cleaner lithotypes (Su, Sb and CSx) can have better RQ (Fig. 3).

Two principal habits of clay are observed in thin sections: pore-filling detrital and authigenic clay (Fig. 7); and grain-rimming detrital/recrystallized clay (Fig. 8). XRD data confirm that the dominant clay mineral is illite. Other clay minerals are rare, but can include kaolinite and chlorite. Pore-filling, finely crystalline, loosely packed illite cements with significant intercrystalline microporosity are relatively common in the muddier sandstones (Fig. 7b).

The most ductile-rich samples (>15% total ductiles: lithotypes SMu, HS, SSg and SMg) have no macropores, moderate–negligible microporosity that decreases with increasing ductile content (Fig. 9), and low–negligible permeability ($<2 \times 10^{-16}$ m^2 (<0.2 mD)). Thin-section observations illustrate that these samples are strongly compacted, facilitated by the high ductile content. Ductile clays have deformed between rigid quartz framework grains destroying all macroporosity and preventing the precipitation of microporous authigenic illite in significant quantities (Fig. 7a).

Sandstones with 5–15% total ductiles (lithotype SMu) have limited–no macroporosity and are typically dominated by microporosity (Fig. 9), with microporosity increasing as total ductile content increases. Rare macropore-dominated samples can have good permeability (up to 10^{-13} m^2 (100 mD)), but micropore-dominated samples have poor–negligible permeability ($<10^{-15}$ m^2 (<1 mD): Fig. 9). This suggests that the microporosity does not provide an effective flow path, even for gas. Stylolites are often observed at the core to thin-section scale and a strongly compacted fabric is observed in thin sections. Thin-section observations suggest that slightly lower initial detrital clay volumes, including less-well-developed grain-rimming clays (see below), relative to the most ductile-rich sandstones resulted in a slightly more open fabric post-compaction, allowing the precipitation of microporous illite cements (Fig. 7b).

The cleanest sandstones (<5% total ductiles: lithotypes Su, Sb and CSx) tend to be dominated by macropores, with lower microporosity (Fig. 9), and these are the only sandstones where good permeabilities are encountered ($>10^{-13}$ m^2 (>100 mD)). However, even in these sandstones, permeability can range from $>10^{-13}$ to $<10^{-17}$ m^2 (from >100 to <0.01 mD: Fig. 9).

Quartz cements

Quartz cements are typically the volumetrically dominant authigenic phase in these sandstones and are responsible for occluding much of the post-compaction porosity.

Cross-plotting total quartz cement against other variables, such as permeability, can give a distorted view on the importance of quartz cements on RQ as the absolute abundance of quartz cement is limited by the amount of space available for precipitation. For example, the muddier sandstones may have lower volumes of quartz cement than the cleaner sandstones because there was limited space for quartz cements to precipitate following strong compaction and/or authigenic illite precipitation. Making the simplification that the majority of the quartz cements were the last authigenic phase to precipitate, and thus post-date the main phase of compaction, the degree of quartz cementation can be calculated from the point-count petrographical data: [(present-day macroporosity + total quartz cement)/total quartz cement]. There is a reasonably strong negative correlation between degree of quartz cementation and permeability; higher permeabilities are associated with less complete quartz cementation (Fig. 10). The factors controlling the degree of quartz cementation are clearly complex and challenging to consistently relate back to depositional fabrics (lithotypes, grain size, sorting and ductile content: Fig. 10). A deeper understanding of the variability in quartz cementation was achieved through the use of SEM-CL to reveal multiple quartz-cement generations and their habit, not consistently differentiable using optical microscopy.

An earlier generation of quartz is overlain by kaolinite, illite, siderite, residual oil and, locally, barite. It forms volumetrically scarce–minor (<3%: based on point-counting of SEM-CL images from selected samples), thin, variably continuous, non-luminescent (dark) overgrowths, which post-date minor–moderate grain-contact elongation. These overgrowths tend to be less well developed in highly compacted samples, but are thicker and more continuous in less-compacted samples. Furthermore,

Fig. 5. Cross-plots of grain size v. sorting shaded by (**a**) lithotype (see Table 1); (**b**) porosity; and (**c**) permeability. The data illustrate that lithotypes SMu and HS (muddy turbidites) tend to be the finest-grained, most-sorted sandstones. Lithotypes SSg, SMg and MSg (debrites) are the least-sorted deposits. Lithotypes CSx, Sb and Su (tractional deposits and clean turbidites) have intermediate grain size and sorting. There is a high degree of scatter in the grain size and sorting v. porosity and permeability data, but, generally, the finest-grained, most-well-sorted sandstones have the lowest porosity and permeability. Very-fine- to medium-grained, moderately well-sorted sandstones retain the best porosity and permeability, but can also have very low porosity and permeability.

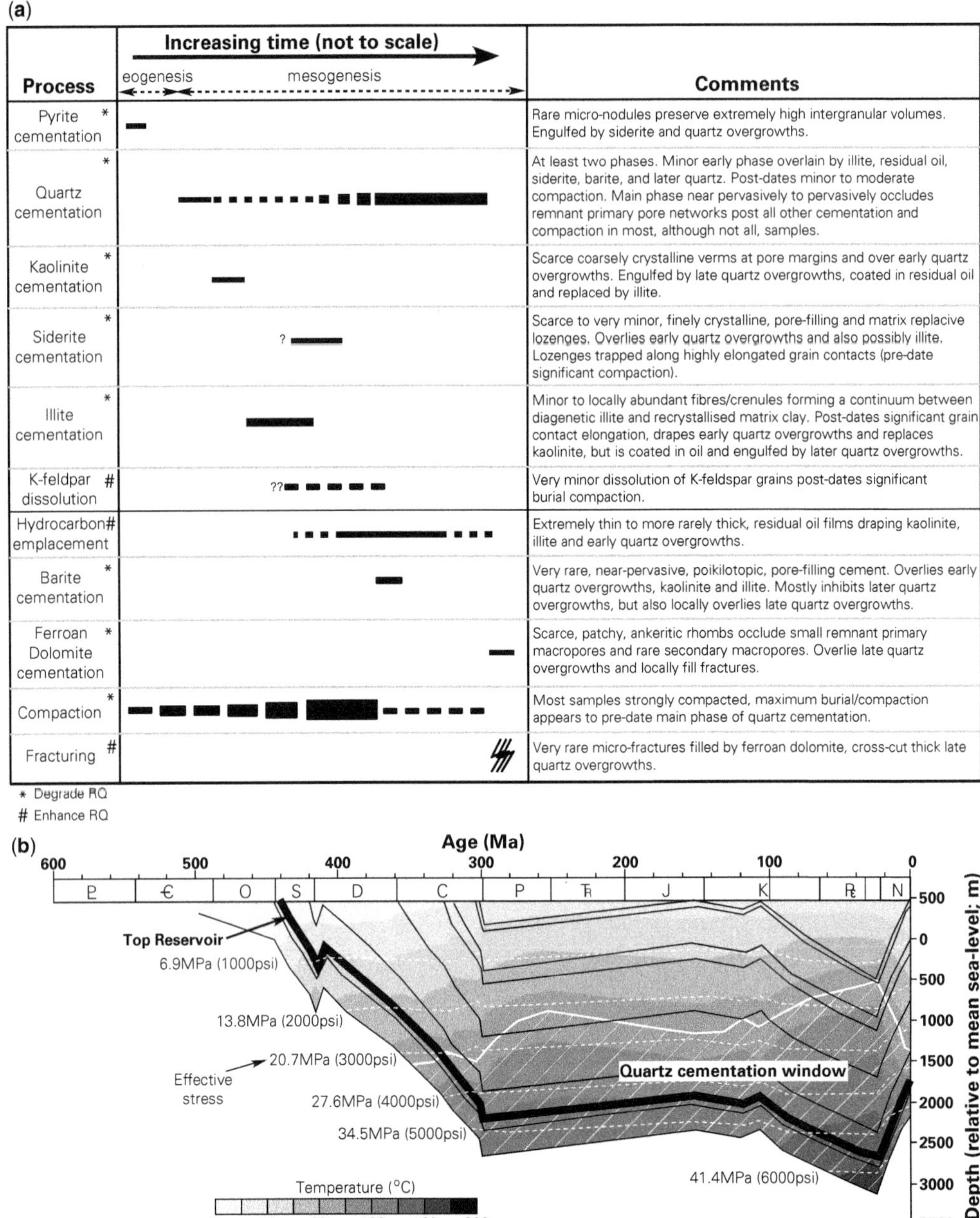

Fig. 6. (**a**) Generalized paragenetic sequence based on thin-section observations. (**b**) Typical burial history plot for the study area. Modelled temperatures are denoted by the shaded fill, with the 'quartz cementation window' (i.e. above 80°C) marked by diagonal white hatching. Isolines of modelled effective stress are shown in dashed white lines. Note that the reservoir (the top is marked by the bold black line) has been at elevated temperatures (inside the 'quartz cementation window') and at moderately high effective stress for a long period of time (>300 myr).

the early overgrowths appear to form more readily in exceptionally clean sandstones with thin, discontinuous clay rims around detrital quartz grains (Fig. 8b). Conversely, thicker, more continuous clay rims appear to have inhibited the early phase of quartz cement (Fig. 8a).

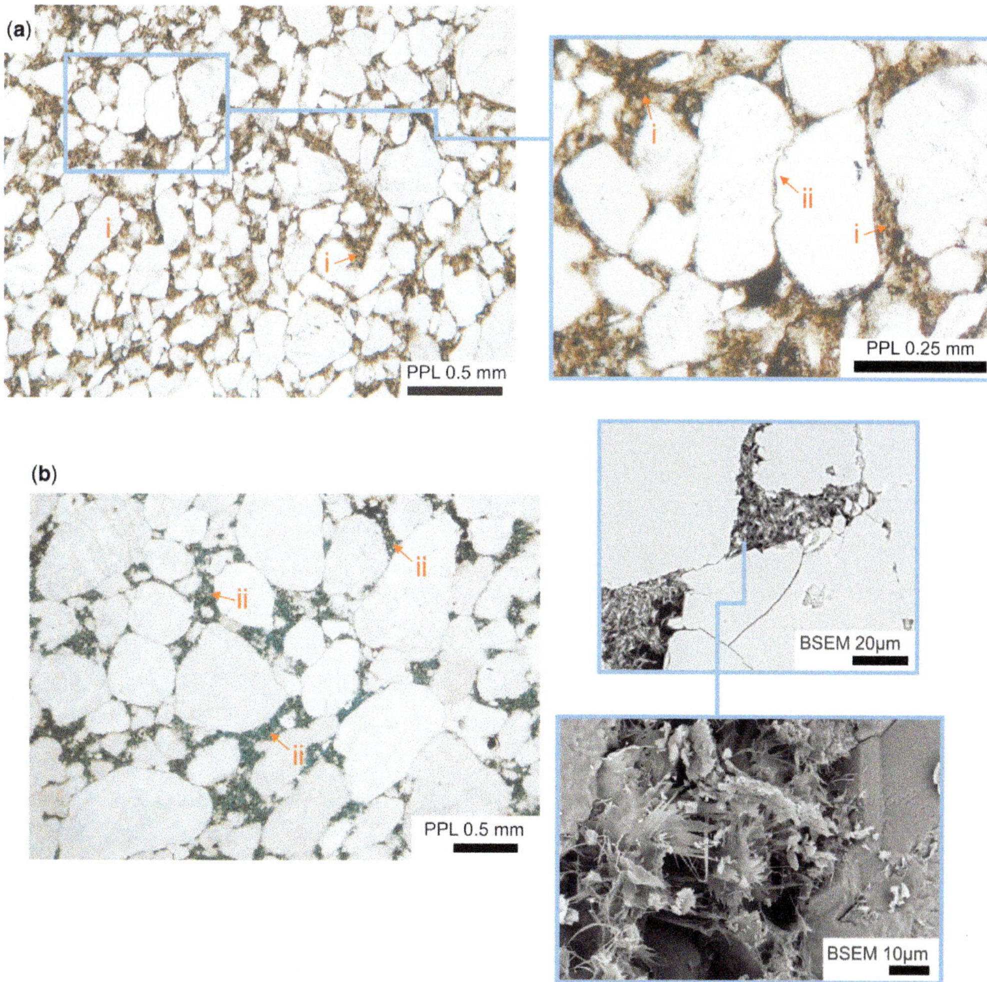

Fig. 7. Thin-section photomicrographs of pore-filling clay habits. (**a**) The relatively high clay content (24%) in this sandstone has resulted in strong compaction, as shown by the tightly compressed matrix clay (i), together with numerous elongate, sutured and stylotized (ii) grain contacts. No macroporosity has been preserved and only minor (3.1%) microporosity is present. Consequently, porosity (5.4%) and permeability (9×10^{-18} m^2 (0.009 mD)) is negligible (Fig. 3f). (**b**) The slightly lower clay content (14%) in this sandstone comprises finely crystalline, loosely packed, webby to fibrous authigenic illite (ii). The presence of significant intercrystalline microporosity (7.8%) leads to a green tinge in plane polarized light (PLL) due to partial resin impregnation. The illite cements are best viewed in polished and rough surface backscattered electron microscopy (BSEM) images shown to the top and bottom right of the PPL image, respectively. Although the porosity is higher (9%), the permeability (7×10^{-14} m^2 (0.07 mD)) remains negligible (Fig. 3f).

A later generation of quartz is inhibited by, and also engulfs, kaolinite, illite, siderite, residual oil and, locally, barite (Fig. 6). In pervasively quartz-cemented sandstones, this later, dominant phase typically forms thick, well-developed, anhedral, pore-filling overgrowths that occlude the remnant primary pore networks. The overgrowths are typically moderately luminescent and can show growth zonation (Fig. 11a). These moderately luminescent, zoned overgrowths are absent in sandstones with the highest permeabilities, which instead display patchy, non-luminescent overgrowths that, at least in part, post-date significant compaction (Fig. 11b). This may suggest slower quartz-cement precipitation rates in these sandstones, leading to there being fewer of the defects in the crystal lattice that give rise to luminescence in quartz (Götze *et al.* 2001). Given the degree of compaction, the non-luminescent

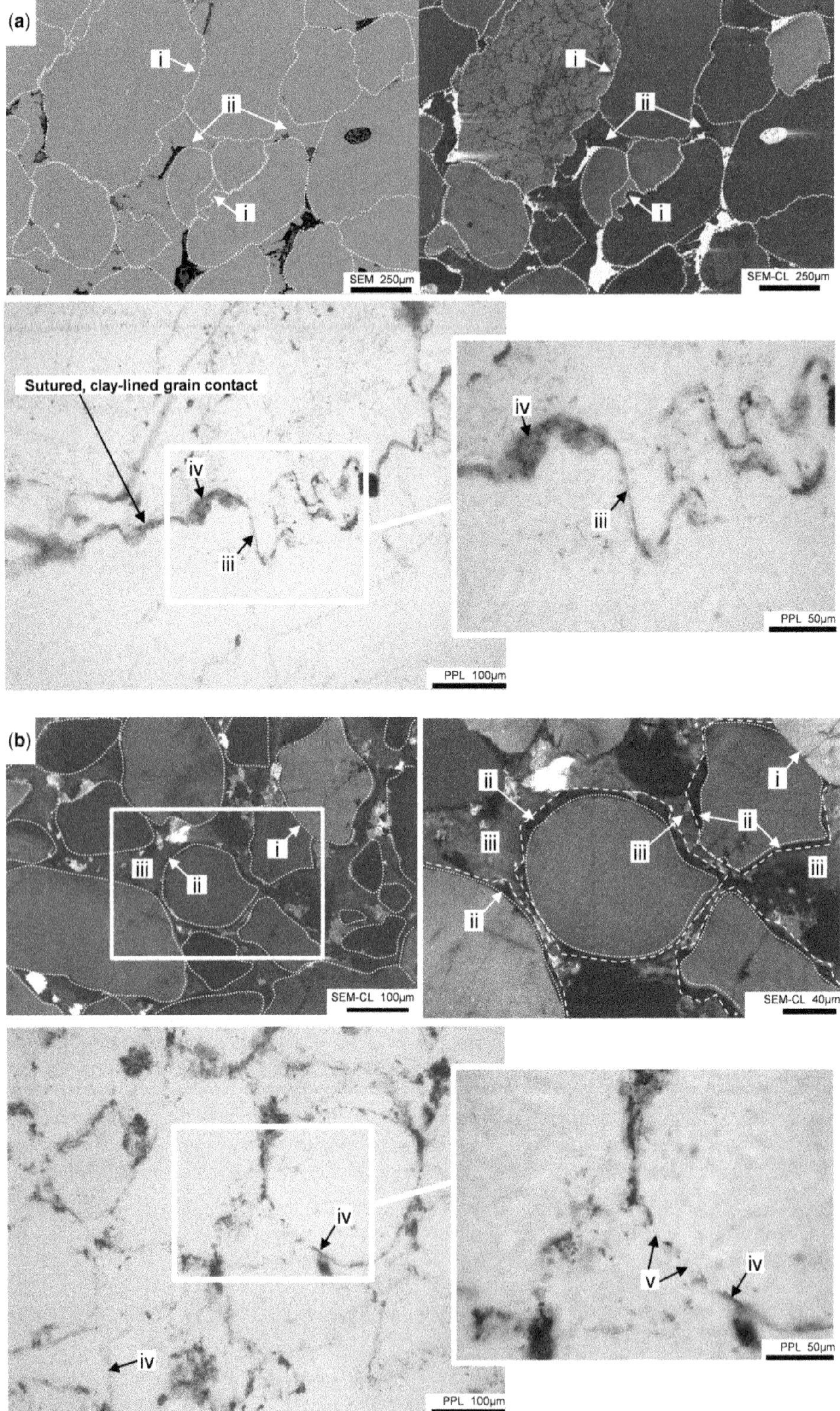
(a)
i
ii
i
SEM 250μm
SEM-CL 250μm
Sutured, clay-lined grain contact
iv
iii
PPL 100μm
PPL 50μm
(b)
iii
ii
i
SEM-CL 100μm
SEM-CL 40μm
v
PPL 50μm

overgrowths are interpreted to represent an amalgamation of the earlier and later generations of quartz cements. In other words, cementation continued in the incompletely cemented sandstones, possibly synchronously with the moderately luminescent, zoned cements in the pervasively cemented sandstones, but at a much slower rate.

Summary: pore evolution pathways

From all of the observations and interpretations given above, four main pore evolution pathways can be identified based on precursor sediment and major diagenetic processes (Fig. 12).

Very argillaceous, fine-grained, well-sorted sandstones (SMg & SMu). Fine-grained sandstones both with (SMg) and without (SMu) outsized grains are prone to very strong compaction, especially where detrital ductile content increases above 15%. Ductile grains and clays are deformed and totally occlude the pore space as the sediment is buried. Macropores are not present and micropores are typically minor; RQ is negligible.

Argillaceous, fine-grained, well-sorted sandstone (SSg & SMu). Typically, fine-grained, well-sorted sandstones (SMu) with 5–15% detrital ductile content are also strongly compacted. Well-developed clay rims around detrital quartz grains prevented an early phase of quartz cement rendering the sandstone more prone to later compaction. However, some post-compaction pore space was available for the precipitation of quartz cements and microporous illite. Macropores are rare and microporosity dominates; RQ is poor, being composed of poorly connected microporosity.

Clean, medium-grained, moderately well-sorted sandstone (Su, Sb & CSx) with pervasive quartz cementation. Very clean (<5% detrital ductiles), medium- to coarse-grained, moderately well-sorted to moderately sorted sandstones (Su, Sb & CSx) are compacted, but often not as strongly as muddier sandstones. Thin, discontinuous clay rims around detrital quartz grains allowed an early phase of quartz cement to precipitate, which appears to have held open the pore network during initial compaction. However, protracted, moderately deep burial (Fig. 6) resulted in pervasive quartz cementation. Both macropores and micropores are rare, and RQ is very poor in these highly cemented sandstones.

Clean, medium-grained, moderately well-sorted sandstones (Su, Sb & CSx) with incomplete quartz cementation. These sandstones appear to have the same fabric and early diagenetic history as the pervasively quartz-cemented very clean sandstones above. However, in this case, some macroporosity has been preserved due to incomplete quartz cementation. These sandstones have the best RQ (up to 15% porosity and 2×10^{-13} m^2 (200 mD) permeability). SEM-CL observations suggest that the quartz cements that do exist were precipitated slowly.

The primary control on RQ is, therefore, depositional fabric: most importantly, the detrital ductile content. The muddy sandstones always exhibit poor RQ, due either to very strong compaction or to strong compaction followed by illite and quartz cementation. The clean sandstones have typically preserved more porosity through the compaction phase of diagenesis, but pervasive quartz cementation has often occluded this porosity. Yet, against expectations from burial history modelling (Fig. 6b), some clean sandstones have preserved good RQ due to incomplete quartz cementation. The degree of quartz cementation in these cleanest sandstones does not appear to be related to grain size or to total ductile content. SEM-CL observations suggest that the highest permeability sandstones have

Fig. 8. Thin-section photomicrographs illustrating the impact of grain-coating clays on compaction and quartz cementation. (**a**) A clean, moderately sorted, coarse-grained sandstone that has experienced pronounced compaction, with highly elongated, often interpenetrating grain contacts (i). Quartz cement is relatively minor (8%), and is dominated by a single generation forming highly discontinuous, yet thick, pore-filling overgrowths and outgrowths (ii), post-dating pronounced grain-contact elongation. Earlier, thin, non-luminescent quartz overgrowths are only rarely observed. Although the sample only has 2.5% grain-coating clay, these coats are generally continuous and typically from 1 μm thick (iii) to 20 μm thick (iv). They appear to have inhibited earlier quartz overgrowths, which in turn allowed for strong compaction. Strong compaction means that porosity (2.8%) and permeability (7×10^{-14} m^2 (0.07 mD)) is negligible. (**b**) A clean, well-sorted, fine-grained sandstone that has preserved a very high intergranular volume (32.5%). Compaction is weak, with only minor, patchy grain-contact elongation observed (i). Quartz cement is abundant (22%), with two generations identifiable. The earliest generation is volumetrically minor, forming non-luminescent, fairly continuous, thin (<15 μm thick) syntaxial overgrowths (ii), which coat the surface of most quartz grains. The later generation is volumetrically dominant, forming thick, pore-filling overgrowths, which pervasively occlude the primary pore network (iii). The sample has minor grain-rimming detrital clay (3%) that forms very thin (<1 μm thick), discontinuous coats around most detrital quartz grains (iv). These clay rims do not appear to have inhibited an early phase of quartz cement, which stabilized the grain framework, and inhibited compaction. Nonetheless, pervasive quartz cementation has resulted in negligible porosity (2.8%) and permeability (6×10^{-18} m^2 (0.006 mD)).

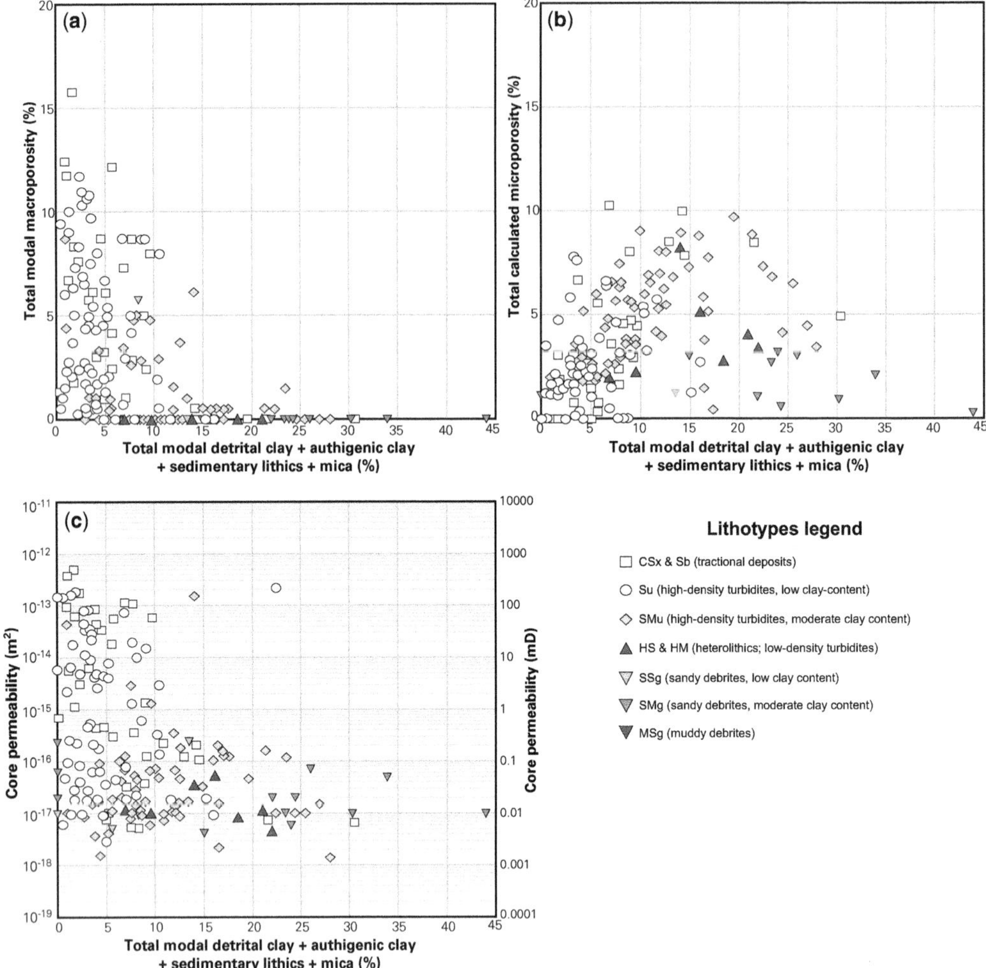

Fig. 9. Cross-plots from petrographical samples illustrating the impact of detrital ductiles and authigenic clays on RQ: (**a**) detrital ductiles and authigenic clay v. macroporosity from modal analysis; (**b**) detrital ductiles and authigenic clay v. calculated microporosity; and (**c**) detrital ductiles and authigenic clay v. permeability from RCA. Only the cleanest sandstones have significant macroporosity and permeability remaining. However, equally clean sandstones can have negligible macroporosity and permeability. Intermediate ductile and clay content are associated with microporosity-dominated systems with negligible permeability. At higher ductile and clay content, not even microporosity is preserved.

patchy, non-luminescent overgrowths. Understanding the preservation of good RQ, therefore, becomes a question of what is preventing pervasive quartz cementation by slowing down or stopping the precipitation of quartz cements.

Discussion: preservation of RQ in the cleanest sandstones

The rate of quartz cementation is a function of the rate of silica supply from various sources, the rate of diffusion from the source to the cementation site and the rate of precipitation at the cementation site (e.g. Walderhaug 1996). Three theories to explain the preservation of RQ in the cleanest sandstones are investigated here: surface area available for quartz cementation; hydrocarbon inhibition of quartz cementation; and reservoir-architecture control on quartz cementation.

Surface area available for quartz cementation

Surface area available for cementation is suggested to be an important control on quartz cementation

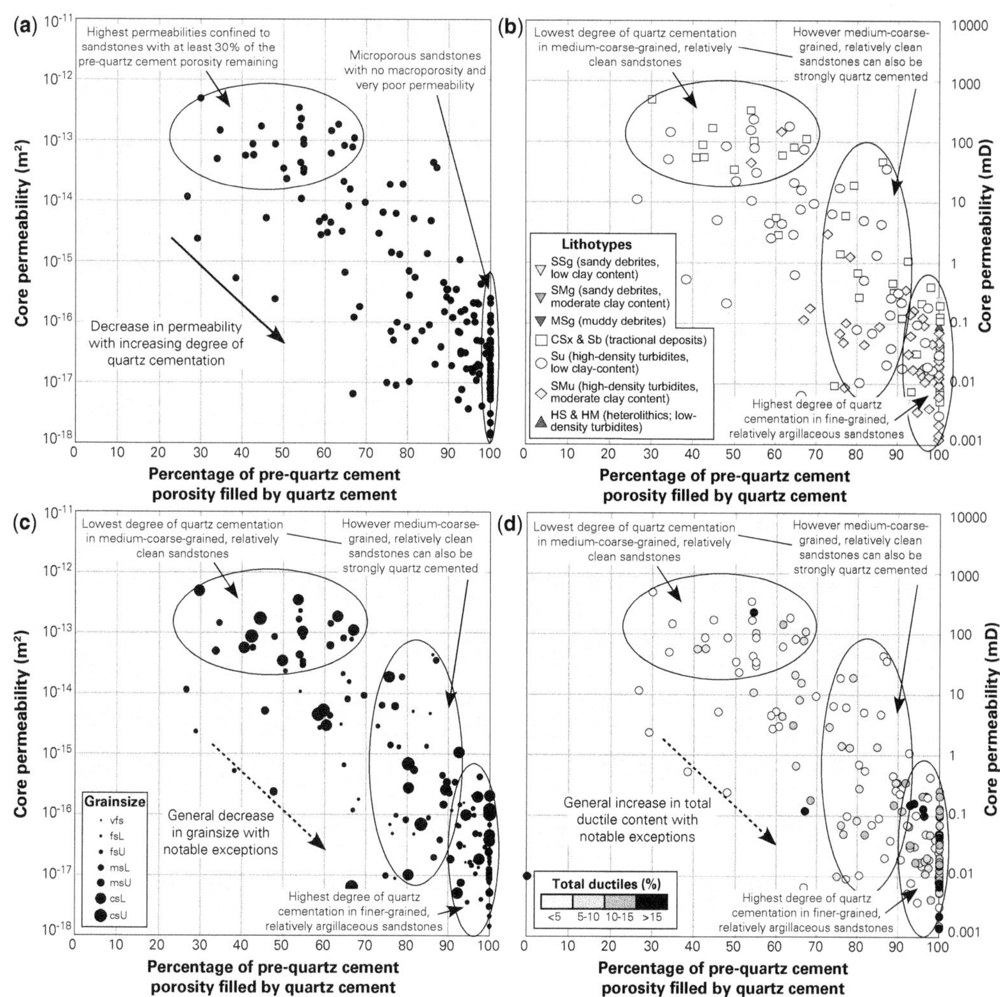

Fig. 10. (**a**) Degree of quartz cementation v. core permeability; shaded by (**b**) depositional lithotype; (**c**) grain size; and (**d**) total ductile content. Degree of quartz cementation is calculated as: [total quartz cement/(total macroporosity + total quartz cement)]. Permeability decreases as the degree of quartz cementation increases. The most quartz-cemented sandstones tend to be fine-grained and argillaceous, reflecting both a large surface area for quartz-cement nucleation, and limited pore space available post-compaction and illite cementation. The least quartz-cemented sandstones tend to be medium- to coarse-grained and less argillaceous, reflecting both a smaller surface area for quartz-cement nucleation, and more pore space available due to less compaction and less illite cementation. However, notable exceptions to these generalizations exist; medium- to coarse-grained, less argillaceous sandstones can also be strongly quartz-cemented with very poor permeability.

(e.g. Walderhaug 1996). Finer-grained sandstones have a larger surface area for quartz cements to nucleate and so are expected to be more cemented than coarser-grained sandstones, all else being equal and if precipitation rate is the rate-controlling step (Walderhaug 1996). This is broadly corroborated by the data in this study: the finer-grained, more detrital-clay-rich sandstones are typically pervasively quartz-cemented and have poor RQ (Fig. 10b–d). Whilst the absolute volume of quartz cement in these sandstones may be less than in some of the cleaner sandstones, this reflects lower pre-quartz-cement porosity in the muddier sandstones (i.e. there was less space available for quartz cementation). As a function of space available, they are more strongly quartz-cemented. Conversely, in terms of space availability, the least quartz-cemented sandstones tend to be medium- to coarse-grained and have a low–very low ductile content (Fig. 10b–d). This can be explained by

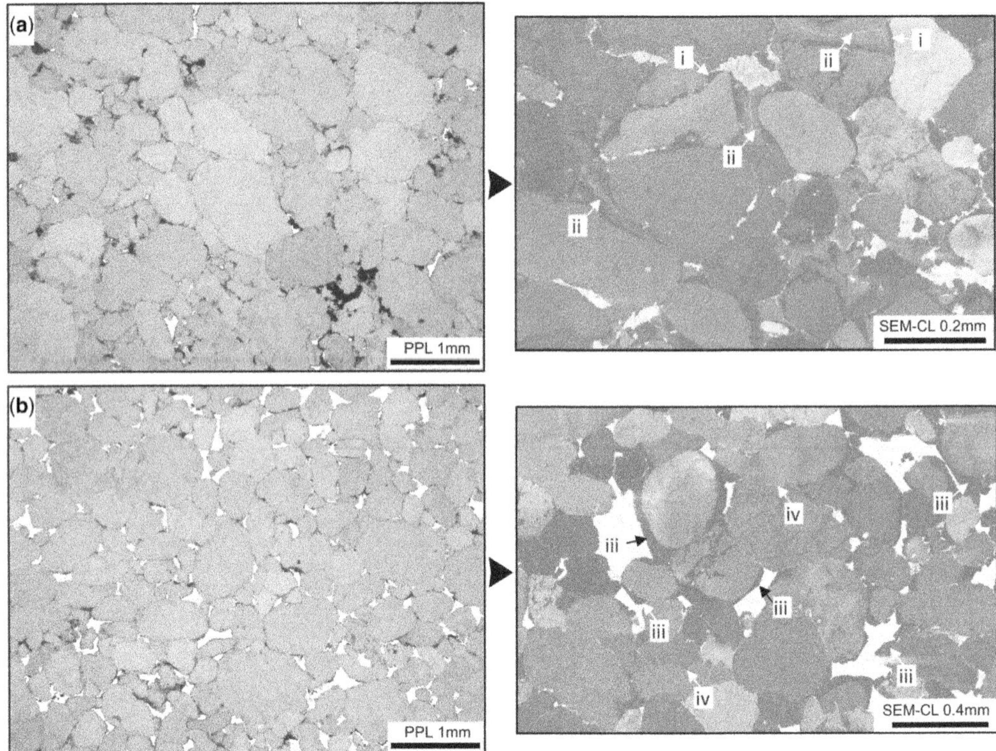

Fig. 11. Plane polarized light (PPL) and SEM-CL photomicrographs illustrating typical quartz-cement habits in strongly cemented v. incompletely cemented sandstones. Pores are shaded white in the PPL and SEM-CL images. (**a**) A clean, well-sorted, fine-grained sandstone that has been pervasively quartz-cemented (24.5%), with two distinct generations. The earlier is thinly developed and non-luminescent (i). The later phase is more abundant, well-developed (thick and often continuous/interlocking), moderately luminescent and with zoned overgrowths (ii). The pervasive quartz cementation has resulted in negligible porosity (2.1%) and permeability (10^{-17} m^2 (0.01 mD)) (Fig. 3f). (**b**) A clean, moderately well-sorted, medium-grained sandstone that has been moderately quartz-cemented (9%). The quartz cement comprises patchy, euhedral, non-luminescent (dark) overgrowths (iii) that, at least in part, post-date significant compaction/grain interpenetration (note the stylolitized grain contacts (iv). This sample has good porosity (9.7%) and permeability (1.8×10^{-13} m^2 (183.3 mD): Fig. 3f) due to the absence of the moderately luminescent quartz seen in (a).

less compaction, which itself might be a function of discontinuous clay coats and the precipitation of a volumetrically minor early quartz cement that held the pore space open during compaction (Fig. 8b), and less illite cementation. However, there are almost as many exceptions to this rule as there are applications. Although some of the strongly quartz-cemented clean sandstones may be explained by their smaller grain size, there are also medium- to coarse-grained, clean sandstones that have been strongly quartz-cemented (Fig. 10b–d).

Surface area available for quartz cementation is also controlled by the presence and distribution of grain coats, such as clays or microcrystalline quartz (e.g. Heald & Larese 1974; Pittman *et al.* 1992; Aase *et al.* 1996; Walderhaug 1996; Bloch *et al.* 2002; Taylor *et al.* 2010). No microcrystalline quartz has been observed in these sandstones, consistent with a lack of volcanic glass, tuffs and sponge spicules or interbedded radiolarian-bearing mudstones, which tend to be precursors to microcrystalline quartz (e.g. Vagle *et al.* 1994; Taylor *et al.* 2010; Weibel *et al.* 2010). In this study, the samples with the thickest, most continuous clay coats are typically highly compacted with poor reservoir quality (Fig. 8). Thin, discontinuous clay rims in the cleanest sandstones appear to have allowed the precipitation of an early phase of quartz cement that resisted strong compaction. Presumably, these early overgrowths would provide ideal nucleation sites for the main phase of quartz cementation. In other words, discontinuous clay coats appear to be required to preserve some porosity pre-main-phase quartz cementation. However, these sandstones have a high risk of subsequent pervasive quartz cementation.

Surface area available for quartz cementation therefore appears to explain the preservation of good RQ in some, but not all, medium-grained, moderately sorted, clean sandstones.

Hydrocarbon inhibition of quartz cementation

Hydrocarbons may be able to inhibit (prevent or slow down) the transport of aqueous silica and the precipitation of quartz cement by lowering the water saturation in a reservoir and preventing aqueous silica reaching quartz grain surfaces (e.g. Worden *et al.* 1998). This is a contentious concept and has been the subject of debate since the 1990s (e.g. the summary in Worden & Morad 2000). In oil-wet sandstones, quartz cementation could theoretically be halted by preventing aqueous silica reaching the quartz-grain surfaces. However, sandstone reservoirs are often water-wet and a thin film of water remains around quartz grains, meaning that quartz cementation could continue, possibly at a reduced rate.

In this study of gas reservoirs, residual oil is observed in thin sections typically as oil-impregnated illite fibres and as very thin (<1 μm thick) discontinuous drapes on the surfaces of framework grains and, rarely, incompletely developed quartz cements (Fig. 13). Rare hydrocarbon-bearing primary and secondary fluid inclusions are also observed (Fig. 13c). All samples have been cleaned in toluene before RCA and thin-section preparation: therefore, further residual oil may well have been removed, leading to an underestimation of the residual oil. There appears to be a relationship between permeability preservation and the presence of residual oil in very clean sandstones (<5% total ductiles: Fig. 13). Seventy per cent of very clean sandstones with permeabilities $>10^{-15}$ m^2 (>1 mD) contain at least traces of residual oil, which has been identified in thin section. Conversely, residual oil has only been identified in 30% of very clean sandstones with $<10^{-15}$ m^2 (<1 mD) permeability (Fig. 13).

Fluid-inclusion data can be used to tentatively place the timing of hydrocarbon charge relative to quartz cementation (Fig. 14). Fluid inclusions are very rare and small in the early quartz cements and dominantly occur in the later phase of quartz cement. Hydrocarbon fluid inclusions have only been observed in about the top 20 m of the reservoir beneath the Silurian Hot Shale top seal. Aqueous primary fluid inclusions suggest that quartz-cement growth commenced at around 80°C, which is consistent with models of quartz kinetics (e.g. Walderhaug 1996), and, according to the preferred burial history, was attained by the formation during the Carboniferous (Fig. 14). The majority of the aqueous fluid inclusions span the temperature range 120–140°C, attained by the formation post-Carboniferous (*c.* 300 Ma). Homogenization temperatures of fluid and gas phases from primary and secondary aqueous inclusions cogenetic with hydrocarbon inclusions indicate oil emplacement above about 120–125°C throughout the sample set, up to approximately 155°C (Fig. 14). Only eight of the measured 146 hydrocarbon fluid inclusions imply emplacement below 120°C. The hydrocarbon fluid inclusions exhibit yellow-green–blue-green fluorescence and have spectrometrically determined API gravities of 26–43°. Minor populations of wet-gas inclusions were found and possible dry-gas inclusions were observed in the outer part of some quartz overgrowths. Raman spectrometry indicates that the wet-gas contains only hydrocarbons, and has a composition of approximately 67 mol% CH_4, 19 mol% C_2H_6 and 14 mol% $C_3H_8^+$.

These data have several implications. First, it seems probable that these reservoirs were oil charged before a later gas charge displaced the oil and/or the oil was cracked to gas in the reservoir, leaving the observed gas charge. Second, the inferred modal hydrocarbon (light oil) emplacement temperatures are consistent with higher temperatures that the reservoir would have attained at the time the immediately overlying Silurian Hot Shale was oil mature. This supports a component of direct downward migration as the main charging mechanism, as opposed to long-distance lateral migration. Third, it suggests that quartz cement had already started to precipitate before significant volumes of light oil reached the reservoir (Fig. 14). Given the sustained moderately high thermal stress that the reservoir has been subjected to and the current gas charge, the likely source of the residual bitumen rarely observed impregnating illite fibres and coating quartz overgrowths (Fig. 13a, b) is through thermal cracking of the earlier oil charge or possibly gas-induced asphaltene flocculation, as opposed to lower temperature biodegradation (e.g. see Rogers *et al.* 1974; Blanc & Connan 1994). Clearly, this late-stage development of grain- and cement-coating residual oil relative to the onset of quartz cementation impacts the extent to which oil can have inhibited quartz cementation by preventing aqueous silica reaching grain and cement surfaces. Furthermore, the mere presence of, albeit very rare, hydrocarbon fluid inclusions implies that the initial oil charge failed to completely stop quartz cementation by lowering the water saturation. However, both factors could possibly have slowed the rate of precipitation of quartz cement in the later stages of burial history. This is consistent with the apparent relationship between the presence of

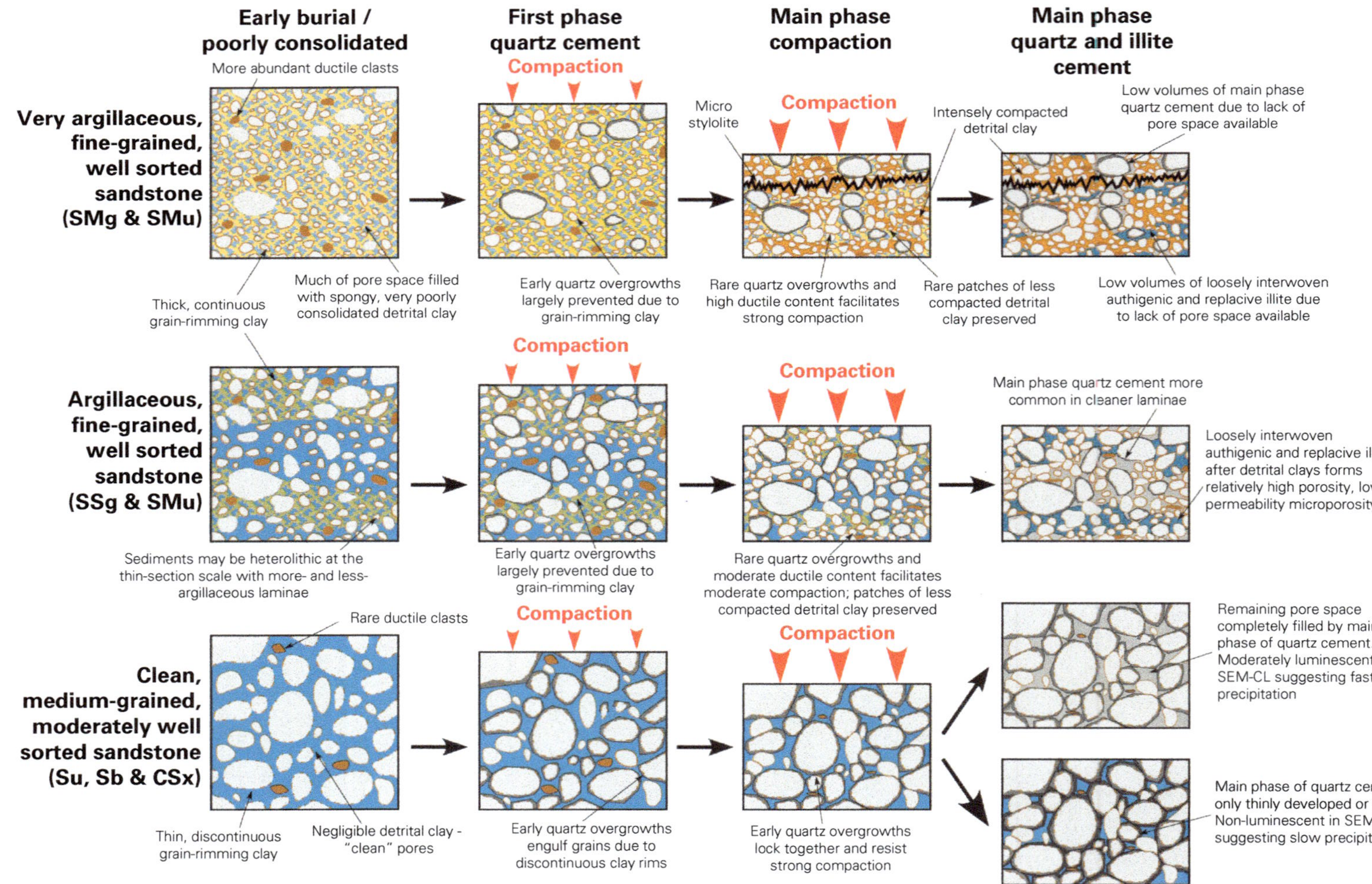

Early burial / poorly consolidated
First phase quartz cement
Main phase compaction
Main phase quartz and illite cement
More abundant ductile clasts
Very argillaceous, fine-grained, well sorted sandstone (SMg & SMu)
Compaction
Micro stylolite
Intensely compacted detrital clay
Low volumes of main phase quartz cement due to lack of pore space available
Thick, continuous grain-rimming clay
Much of pore space filled with spongy, very poorly consolidated detrital clay
Early quartz overgrowths largely prevented due to grain-rimming clay
Rare quartz overgrowths and high ductile content facilitates strong compaction
Rare patches of less compacted detrital clay preserved
Low volumes of loosely interwoven authigenic and replacive illite due to lack of pore space available
Argillaceous, fine-grained, well sorted sandstone (SSg & SMu)
Main phase quartz cement more common in cleaner laminae
Loosely interwoven authigenic and replacive illite after detrital clays forms relatively high porosity, low permeability microporosity
Sediments may be heterolithic at the thin-section scale with more- and less-argillaceous laminae
Early quartz overgrowths largely prevented due to grain-rimming clay
Rare quartz overgrowths and moderate ductile content facilitates moderate compaction; patches of less compacted detrital clay preserved
Rare ductile clasts
Clean, medium-grained, moderately well sorted sandstone (Su, Sb & CSx)
Remaining pore space completely filled by main phase of quartz cement. Moderately luminescent in SEM-CL suggesting faster precipitation
Main phase of quartz cement only thinly developed or absent. Non-luminescent in SEM-CL suggesting slow precipitation
Thin, discontinuous grain-rimming clay
Negligible detrital clay - "clean" pores
Early quartz overgrowths engulf grains due to discontinuous clay rims
Early quartz overgrowths lock together and resist strong compaction

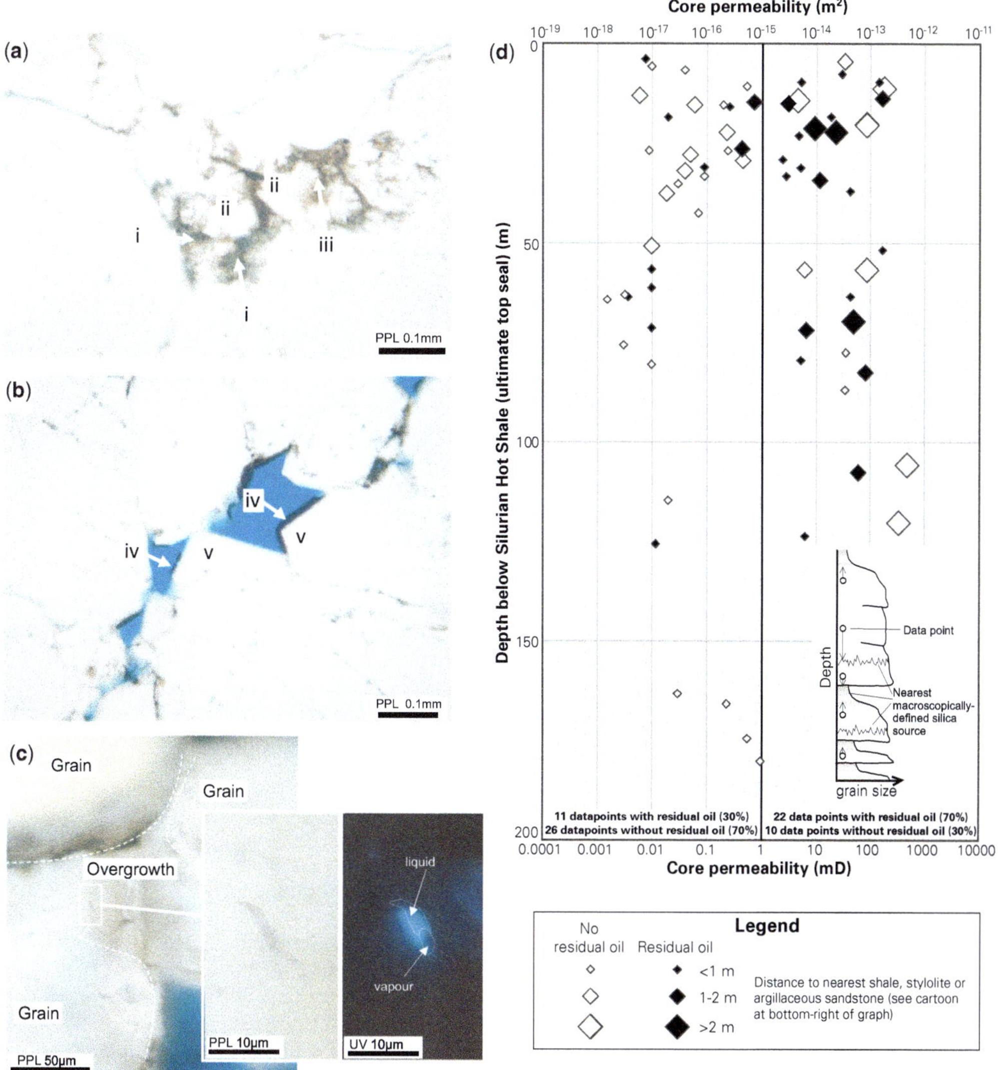

Fig. 13. Observations of residual oil and possible implications on RQ. (**a**) Loosely packed, finely crystalline, wispy, pore-lining to pore-bridging illite fibres (i), engulfed by thick, pore-filling quartz overgrowths (ii). The illite is locally coated in residual oil and appears brownish to black in colour (iii). (**b**) Thin, discontinuous films of residual oil (iv) locally draping well-developed quartz overgrowths (v). (**c**) Hydrocarbon-filled fluid inclusion seen within authigenic quartz; note the strong blue fluorescence in ultraviolet light. (**d**) Cross-plot of permeability v. depth below the Silurian Hot Shale top seal (Fig. 1). Only the cleanest sandstones (<5% total ductiles) are plotted and they are shaded by the presence of residual oil as derived from thin-section observations. Data points are sized according to the distance to the nearest macroscopically defined likely silica source: shales, stylolites and argillaceous sandstones.

Fig. 12. Pore evolution diagram summarizing the main controls on RQ. Four end members can be recognized: very argillaceous, fine-grained, well-sorted sandstones are strongly compacted; argillaceous, fine-grained, well-sorted sandstones are strongly compacted with the remaining pore space filled by authigenic illite and quartz; clean, medium-grained, moderately well-sorted sandstones with pervasive quartz cements; and clean, medium-grained, moderately well-sorted sandstones with patchy, incomplete quartz cementation.

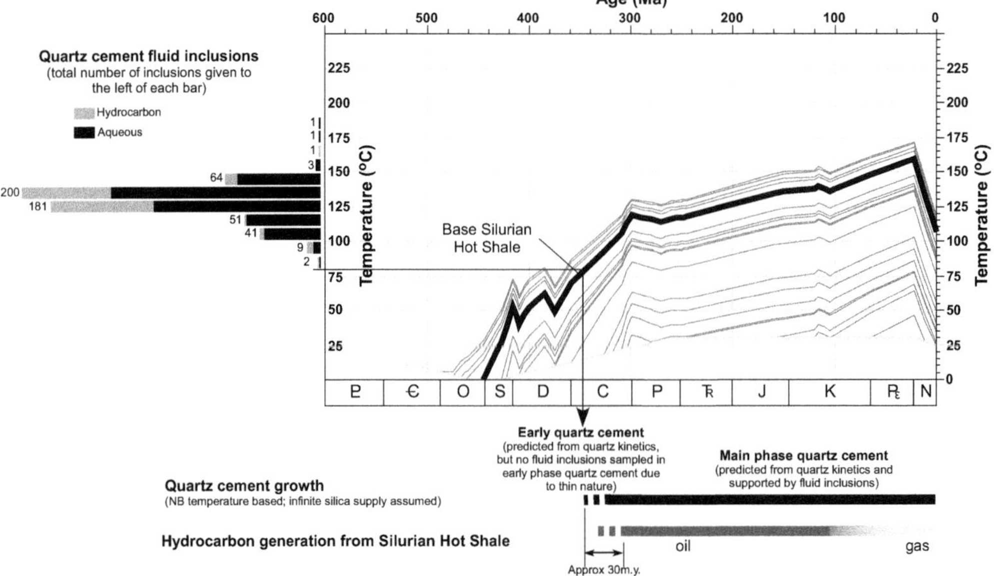

Fig. 14. Thermal history (derived from burial history: Fig. 6b) plotted alongside quartz-cement homogenization temperatures from fluid inclusions. Quartz cementation commenced at about 80°C, consistent with kinetic models (e.g. Walderhaug 1996), and, according to the preferred burial and thermal history model, occurred during the early Carboniferous. Significant hydrocarbon fluid inclusions are not noted until significantly higher homogenization temperatures (*c.* 120–140°C). This is consistent with direct downward migration of hydrocarbons from the Silurian Hot Shale source rock immediately above the reservoir. It also implies that, although hydrocarbons were present in the pore system at the time of quartz cementation, they did not arrive in significant quantities until after quartz cementation had commenced.

residual oil and preservation of permeability (Fig. 13d), and the petrographical observation that the least-cemented clean quartz sandstones have non-luminescent, presumably slowly precipitated, quartz cements (Fig. 11). The present-day gas charge may be continuing to inhibit quartz cementation in the remaining macroporous sandstones that otherwise remain in the quartz-cementation temperature window (at *c.* 100°C).

Reservoir architecture and silica budgets

Aqueous silica in quartz-cemented sandstones may be sourced internally (relatively locally) or externally (basin-scale convection currents). The latter would appear to be less likely given that 'the abundant evidence that quartz cementation typically takes place at advanced burial depths and high temperatures where flow of meteorically derived groundwaters does not take place' (Walderhaug 1996, p. 732). Internal silica sources include:

- dissolution of quartz grains at grain contacts and stylolites due to pressure solution (chemical compaction), which can be clay- or mica-catalysed (e.g. Walderhaug 1996);
- clay reactions (e.g. smectite to chlorite or illite; kaolinite to chlorite or illite: Worden & Morad 2003);
- feldspar reactions (e.g. K-feldspar to illite: Worden & Morad 2003);
- biogenic silica (e.g. sponge spicules or radiolarians; Vagle *et al.* 1994; Weibel *et al.* 2010);
- volcanic glass and tuffs (e.g. Taylor *et al.* 2010).

No volcanic glass, tuff, sponge spicules or radiolarians have been observed in any of the studied thin sections: hence, a volcanogenic or biogenic source of silica is deemed unlikely in these Late Ordovician glacial sandstones. Similarly, negligible feldspars (Fig. 4) or secondary porosity from dissolved feldspars have been observed, suggesting an insignificant contribution of silica from feldspar reactions. Conversely, the interpreted thermal history (Fig. 6b) and observation of illite as the dominant clay mineral (Fig. 7) suggests that diagenetic clay reactions have taken place, providing a potential silica source. Quartz-grain dissolution features are relatively common at the core- and thin-section-scale, including stylolites, microstylolites, grain-contact elongation and suturing, and grain interpenetration. These features are products of the sustained mechanical and thermal stresses that these sandstones have encountered, and are deemed a significant source of the quartz cements observed. Interpenetrative grain contacts and stylolites are observed to be more common in the muddier sandstones (SMu) than in the cleanest sandstones (Su). This illustrates the impact of detrital clays in facilitating strong mechanical compaction and catalysing chemical compaction.

It is difficult at best, and sometimes impossible, to definitively quantify the amount of silica produced at stylolites, but an attempt has been made to quantify the amount of silica generated by grain-contact dissolution using SEM-CL in a few selected samples following the methodology of Tournier *et al.* (2010). Comparison of quartz cements (measured from SEM-CL images) with this estimation of internal silica supply illustrates that most sandstones, even moderately argillaceous ones, are silica importers at the thin-section-scale (Fig. 15). In other words, an additional source of silica is typically required to explain the amount of quartz cement in these samples. This is a common observation in quartz-cemented sandstones, with nearby (within tens of cm) stylolites often supposed to supply the remaining silica in kinetic models of quartz cementation (Walderhaug 1996). Stylolites are present in the Late Ordovician rocks of this study, and clay transformations in interbedded mudstones may provide some additional silica. However, neither of these potential silica sources are uniformly distributed. The distance from each clean sandstone sample (<5% total ductiles) to the nearest mudstone, stylolite or argillaceous sandstone (i.e. a macroscopically identified potential silica source) was measured and the results are plotted in Figure 13d. These data have their limitations inasmuch as stylolites may have been missed in the core description, microstylolites and grain interpenetration are omitted, and shale, stylolite or argillaceous sandstone frequencies are not accounted for. For example, a clean sandstone sample may plot <1 m from a stylolite, but this could be the only stylolite for many metres. The amount of silica being exported in this case would clearly be less than in an interval of more frequent stylolites. Nonetheless, the data provide a useful, readily measureable proxy for the amount of silica available for cementation.

In thin-bedded, heterolithic packages undergoing strong compaction, stylolites and grain dissolution in interbedded muddy sandstones may produce more silica than can be accommodated, thus leading to silica export (Fig. 16a). Any given clean sandstone bed within such a heterolithic package will be relatively close (within tens of cm or less) to a potential supply of aqueous silica and, hence, prone to a higher degree of cementation (Fig. 16a). This is supported by the generally lower permeability encountered in clean sandstones <2 m away from a macroscopically identified potential silica source (Fig. 13d) and by the porosity–permeability profile of the example well in Figure 16a.

By comparison, in thick-bedded, stacked, clean–very clean sandstone successions, any given bed will be further away from a potential silica source (Fig. 16). Stylolites are rare (several metres apart) to absent in these clean sandstones and an early phase of quartz cement appears to have held the pore space open during compaction, leading to few interpenetrative grain contacts. Viable silica sources may be confined to underlying and overlying muddy sandstones and mudstones. A scenario could theoretically develop whereby sandstones closest to a silica source at the top and base of the stacked sandstone became preferentially cemented, leading to silica starvation in the centre of the clean sandstone package (Fig. 16). It is somewhat difficult to conclusively prove this hypothesis, as within stacked sandstones subtle variations in grain size, sorting and clay content complicate the picture. However, clean sandstones >2 m away from a macroscopically identified potential silica source tend to record good permeabilities (Fig. 13d), and the best RQ is typically encountered in thick (>50 m), stacked, clean sandstone successions (Fig. 16b).

Conclusions

A detailed study of >3 km of core from the In Amenas–Bourarhat Sud area, Illizi Basin, Algeria has helped to elucidate the main controls on reservoir quality (RQ). The primary reservoir target comprises Late Ordovician glaciogenic sandstones that have been subject to moderate mechanical and thermal stresses for a prolonged period of time (>400 myr).

Integration of sedimentological and petrographical techniques shows that a primary depositional fabric exerts a strong control on RQ:

- Sandstones with a high detrital clay content (>15%) have been intensely compacted, and have negligible porosity and permeability.
- Very-fine-grained, well-sorted sandstones with a moderate detrital clay content (5–15%) have been strongly compacted, but some post-compaction pore space remained. Macroporosity is typically pervasively occluded by quartz cements derived from readily available local silica sources, including clay-catalysed dissolution of quartz grains at grain contacts and along stylolites. The fine grain size provided a large surface area over which the quartz cements can precipitate. Variable amounts of microporosity are present in fibrous illite cements;

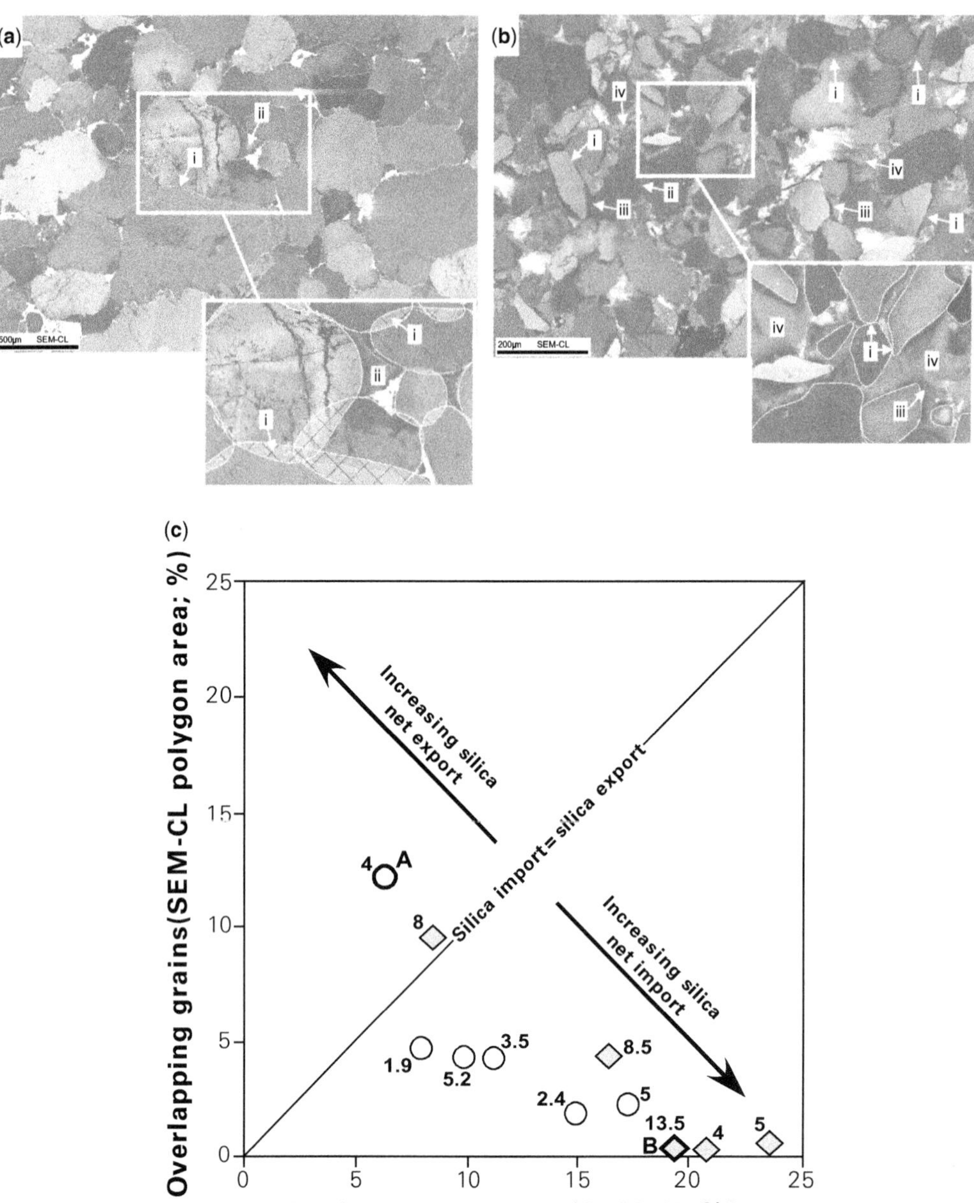

Fig. 15. Quantification of silica import v. export using detailed SEM-CL analysis. (**a**) A very heavily compacted (pre-quartz-cement porosity = 10%) sandstone with sutured, strongly interpenetrating grain contacts (i). Quartz cement (ii) is relatively scarce and the earlier generation evident in (b) is not present. (**b**) A relatively lightly compacted (pre-quartz-cement porosity = 31%) sandstone, with predominantly tangential to elongate grain contacts (i) and relatively minor grain interpenetration (ii: overlapping areas highlighted on the higher magnification inset). Quartz cement is very abundant (17.5%) and two generations are easily differentiable due to differences in their luminescence (iii, older, non-luminescent; and iv, younger, moderately luminescent). (**c**) Cross-plot of total quartz cement (silica import: point-counted from SEM-CL images) v. overlapping grains (proxy for silica export: areas measured from SEM-CL images). Data points corresponding to the samples shown (a) and (b) are highlighted in bold. Total ductile content (point-counted volume %) is shown next to the data points.

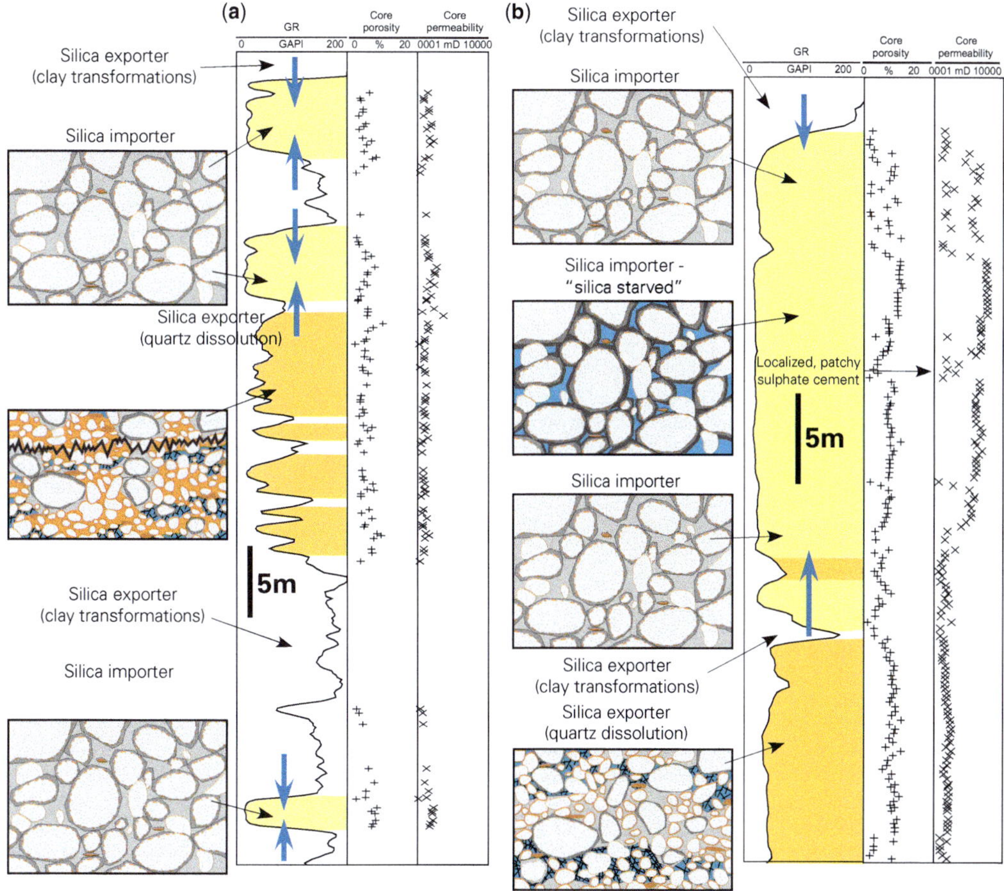

Fig. 16. Cartoon illustrating the possible control of bed thickness and reservoir heterogeneity on silica budgets and the amount of quartz cementation. (**a**) In thin-bedded clean sandstones, local silica supply is at least the same, if not greater than, the pre-quartz-cement porosity, and strong quartz cementation is likely. (**b**) In thick-bedded clean sandstones, the local silica supply may be less than the pre-quartz-cement porosity, especially in the centre of the sandstone package and this may result in less pervasive quartz cementation. Core plug porosity and permeability data corresponding to the selected intervals are shown to the right of the gamma-ray log profile. Core porosity is plotted on a linear scale, while core permeability is plotted on a logarithmic scale.

however, low permeabilities show that the microporosity does not provide an effective flow pathway, even for gas.

- Very-fine- to medium-grained, moderately well-sorted, very clean sandstones (<5% detrital clay) preserve the best RQ. They have been strongly compacted and variably quartz-cemented. The best RQ occurs in the least-cemented sandstones. Such low quartz-cement volumes are at odds with the burial (temperature) history. Close inspection of SEM-CL photomicrographs shows that the quartz overgrowths in the most permeable sandstones are non-luminescent – interpreted to be due to slow quartz-precipitation rates.

The best RQ sandstones are typically, but not exclusively, observed to occur in thick (>50 m) stacked sandstones at the top of the reservoir interval immediately below the regionally extensive lowermost Silurian Hot Shale seal and source rock. Although downward charge from the overlying Silurian Hot Shale occurred after quartz cementation had begun, there are observations to suggest that this hydrocarbon charge may have slowed the rate of quartz cementation and preserved RQ. Consideration of silica budgets suggests that the centres of thick, stacked sandstones may become silica-starved as they are more distant from silica sources, in contrast to thinner-bedded, more heterolithic intervals.

In reality, both of these processes are likely to be, at least in part, responsible for the observed RQ variations in this sedimentologically and diagenetically complex system. However, this study provides some simple concepts that can be tested in the search for better-quality reservoirs in the late Ordovician play in North Africa: that is, that the thicker, clean glaciogenic outwash sandstones close to the overlying Silurian Hot Shale should be targeted.

We would like to take this opportunity to thank those colleagues who have contributed to the data presented and to discussions on the contents and ideas herein, in particular Dorothy Payne, Norman Oxtoby, Richard Dixon, Barnaby Roome, Chereé Stover and Ed Jones. We also thank BP and Sonatrach for permission to publish this material. Finally, thanks to Richard Worden and two anonymous reviewers for their insightful comments on an earlier version of this manuscript, and to Pete Armitage and James Churchill for editorial comments.

Appendix 1

X-ray diffraction (XRD) analysis standard practices

A Hiltonbrooks generator with a Philips PW1050 goniometer was used to carry out all XRD measurements. Samples were first cleaned, where necessary, using distilled water and then toluene, and then gently crushed until fragments <5 mm in diameter were obtained. Whole-rock samples were disaggregated using a pestle and mortar: 2 g of material was then placed into a McCrone micronizing mill for 5 min. The slurry was dried, and the resulting powder tightly packed and reverse-mounted into an aluminium sample holder. The whole-rock powders were run from 5° to 60°2θ, with a step size of 0.05°, and were exposed to X-ray radiation from a copper anode at 40 kV and 30 mA.

Clay fraction (<2 μm) samples were totally disaggregated using a pestle and mortar: 5 g of the sample was mixed with 50 ml of distilled water and 5 ml of Calgon (sodium hexametaphosphate and sodium carbonate solution). The samples were then placed in an ultrasonic bath, followed by mixing table, ultrasonic bath and centrifuge (×2) before decanting 200 ml into a beaker. The total weight of clay extracted was obtained by removing 20 ml of the clay suspension, and evaporating this to dryness. A clay mount was made using the 'Millipore filter' (filtering the clay suspension until an even clay layer is achieved) before mounting into an aluminium XRD slide. Analysis of the clay-fraction mount was carried out in the following sequence: (1) after air-drying; (2) supersaturated in ethylene glycol overnight, using a temperature of 65°C; (3) heated for 2 h at 380°C; and (4) further heated at 550°C for 1 h. The clay-fraction samples are run from 3° to 32°2θ, with a step size of 0.05°, and were exposed to X-ray radiation from a copper anode at 40 kV and 30 mA.

Identification of minerals in both whole-rock and clay-fraction traces was made by comparison of their diffraction profiles with standard traces.

Fluid-inclusion analysis standard practices

For the determination of all phase changes during microthermometry, a 'cycling' protocol, following the general methodology for the study of fluid inclusions in diagenetic cements outlined by Goldstein & Reynolds (1994), was applied. The errors in the determination of phase changes were therefore controlled by the temperature increment used during the approach to the end point of the phase change being measured. Typical errors for final ice melting (T_{ice}) are between ±0.2 and ±0.1°C, and, for homogenization temperatures (T_h), between ±2 and ±5°C. No pressure corrections were applied to the homogenization temperatures.

Salinities were modelled in the system $NaCl-H_2O$ using 'Calcicbrine' (Naden 1996). Very limited first ice-melting temperature data indicate that this is a simplification of the reality and that other divalent cations are also present. However, owing to the generally very small size of the inclusions, it was not possible to derive data on the temperatures of other phase changes (i.e. hydrate melting, etc.) that could be used to model salinities in a more sophisticated manner.

References

Aase, N. E., Bjorkum, P. A. & Nadeau, P. H. 1996. The effect of grain-coating microquartz on preservation of reservoir porosity. *American Association of Petroleum Geologists Bulletin*, **80**, 1654–1673.

Beuf, S., Biju-Duval, B., de Charpal, O., Rognon, P., Gariel, O. & Bennacef, A. 1971. *Les Grès du Paleáozoïque Inférieur au Sahara*. Editions Technip, Paris.

Blanc, P. & Connan, J. 1994. Preservation, degradation, and destruction of trapped oil. *In*: Magoon, L. B. & Dow, W. G. (eds) *The Petroleum System – from Source to Trap*. American Association of Petroleum Geologists, Memoirs, **60**, 237–247.

Bloch, S., Lander, R. H. & Bonnell, L. M. 2002. Anomalously high porosity and permeability in deeply buried sandstone reservoirs: origin and predictability. *American Association of Petroleum Geologists Bulletin*, **86**, 301–328.

Boote, D. R. D., Clark-Lowes, D. D. & Traut, M. W. 1998. Palaeozoic petroleum systems of North Africa. *In*: MacGregor, D. S., Moody, R. T. J. & Clark-Lowes, D. D. (eds) *Petroleum Geology of North Africa*. Geological Society, London, Special Publications, **132**, 7–68, https://doi.org/10.1144/GSL.SP.1998.132.01.02

Craig, J., Grigo, D., Rebora, A., Serafini, G. & Tebaldi, E. 2010. From Neoproterozoic to Early Cenozoic: exploring the potential of older and deeper hydrocarbon plays across North Africa and the

Middle East. *In*: Vining, B. A. & Pickering, S. C. (eds) *Petroleum Geology: from Mature Basins to New Frontiers – Proceedings of the 7th Petroleum Geology Conference*. Geological Society, London, 673–705, https://doi.org/10.1144/0070673

Decalf, C., Fugelli, E. & Dowdeswell, J. A. 2013. The seismic architecture and geometry of glacial grounding-zone wedges from Greenland and North Africa. (Abstract.) *American Association of Petroleum Geologists 3P Arctic Polar Petroleum Potential Conference & Exhibition*, 15–18 October 2013, Stavanger, Norway, AAPG Search and Discovery Article #30310.

Dixon, R. J., Patton, T. L., Hirst, J. P. P. & Diggens, J. 2008. Transition from subglacial to proglacial depositional systems: Implications for reservoir architecture, Illizi Basin, Algeria. (Abstract.) *American Association of Petroleum Geologists Annual Convention*, 20–23 April 2008, San Antonio, Texas, USA, AAPG Search and Discovery Article #50095.

Dowdeswell, J. A., Hogan, K. A., Arnold, N. S., Mugford, R. I., Wells, M., Hirst, J. P. P. & Decalf, C. 2015. Sediment-rich meltwater plumes and ice-proximal fans at the margins of modern and ancient tidewater glaciers: observations and modelling. *Sedimentology*, **62**, 1665–1692, https://doi.org/10.1111/sed.12198

El-ghali, M. A. K., Mansurbeg, H., Morad, S., Al-Aasm, I. & Ramseyer, K. 2006*a*. Distribution of diagenetic alterations in glaciogenic sandstones within a depositional facies and sequence stratigraphic framework: evidence from the Upper Ordovician of the Murzuq Basin, SW Libya. *Sedimentary Geology*, **190**, 323–351.

El-ghali, M. A. K., Tajori, K. G., Mansurbeg, H., Ogle, N. & Kalin, R. M. 2006*b*. Origin and timing of siderite cementation in Upper Ordovician glaciogenic sandstones from the Murzuq basin, SW Libya. *Marine and Petroleum Geology*, **23**, 459–471.

Eschard, R., Braik, F., Bekkouche, D., Ben Rahuma, M., Desaubliaux, G., Deshcamps, R. & Proust, J. N. 2010. Palaeohighs: their influence on the North African Paleozoic petroleum systems. *In*: Vining, B. A. & Pickering, S. C. (eds) *Petroleum Geology: from Mature Basins to New Frontiers – Proceedings of the 7th Petroleum Geology Conference*. Geological Society, London, 707–724, https://doi.org/10.1144/0070707

Folk, R. L. & Ward, W. 1957. Brazos River bar: a study in the significance of grain size parameters. *Journal of Sedimentary Petrology*, **27**, 3–26.

Galeazzi, S., Point, O. N., Haddadi, N., Mather, J. & Druesne, D. 2010. Regional geology and petroleum systems of the Illizi–Berkine area of the Algerian Saharan Platform: an overview. *Marine and Petroleum Geology*, **27**, 143–178.

Ghienne, J.-F. & Deynoux, M. 1998. Large-scale channel fill structures in late Ordovician glacial deposits in Mauritania, western Sahara. *Sedimentary Geology*, **119**, 141–159.

Ghienne, J.-F., Le Heron, D. P., Moreau, J. & Deynoux, M. 2007. The Late Ordovician glacial sedimentary system of the West Gondwana platform. *In*: Hambrey, M. J., Christofferson, P., Glasser, N. F. & Hubbard, B. (eds) *Glacial Sedimentary Processes and Products*. International Association of Sedimentologists, Special Publications, **39**, 295–319.

Girard, F., Ghienne, J. F. & Rubino, J.-L. 2012. Channelized sandstone bodies ('cordons') in the Tassili N'Ajjer (Algeria & Libya): snapshots of a Late Ordovician proglacial outwash plain. *In*: Huuse, M., Redfern, J., Le Heron, D. P., Dixon, R. J., Moscariello, A. & Craig, J. (eds) *Glaciogenic Reservoirs and Hydrocarbon Systems*. Geological Society, London, Special Publications, **368**, 355–379, https://doi.org/10.1144/SP368.3

Goldstein, R. H. & Reynolds, T. J. 1994. *Systematics of Fluid Inclusions in Diagenetic Minerals*. SEPM Short Course, **31**. Society for Sedimentary Geology (SEPM), Tulsa, OK.

Götze, J., Pötze, M. & Habermann, D. 2001. Origin, spectral characteristics and practical applications of the cathodoluminescence (CL) of quartz – a review. *Mineralogy and Petrology*, **71**, 225–250.

Guitierrez-Marco, J. C., Ghienne, J.-F., Bernardez, E. & Hacar, M. P. 2010. Did the Late Ordovician ice sheet reach Europe? *Geology*, **38**, 279–282.

Hambrey, M. J. & Harland, W. B. 1981. *Earth's Pre-Pleistocene Glacial Record*. Cambridge University Press, Cambridge.

Heald, M. T. & Larese, R. E. 1974. Influence of coatings on quartz cementation. *Journal of Sedimentary Petrology*, **44**, 1269–1274.

Hirst, J. P. P. 2012. Ordovician proglacial sediments in Algeria: insights into the controls on hydrocarbon reservoirs in the Amenas field, Illizi Basin. *In*: Husse, M., Redfern, J., Le Heron, D. P., Dixon, R. J., Moscariello, A. & Craig, J. (eds) *Glaciogenic Reservoirs and Hydrocarbon Systems*. Geological Society, London, Special Publications, **368**, 319–353, https://doi.org/10.1144/SP368.17

Hirst, J. P. P., Benbakir, A., Payne, D. F. & Westlake, I. R. 2002. Tunnel valleys and density flow processes in the Upper Ordovician glacial succession, Illizi Basin, Algeria: influence on reservoir quality. *Journal of Petroleum Geology*, **25**, 297–324.

Hornung, J. J., Asprion, U. & Winsemann, J. 2007. Jet-efflux deposits of a subaqueous ice-contact fan, glacial Lake Rinteln, northwestern Germany. *Sedimentary Geology*, **193**, 167–192.

Lang, J., Dixon, R. J., Le Heron, D. P. & Winsemann, J. 2012. Depositional architecture and sequence stratigraphic correlation of Upper Ordovician glaciogenic deposits, Illizi Basin, Algeria. *In*: Huuse, M., Redfern, J., Le Heron, D. P., Dixon, R. J., Moscariello, A. & Craig, J. (eds) *Glaciogenic Reservoirs and Hydrocarbon Systems*. Geological Society, London, Special Publications, **368**, 293–317, https://doi.org/10.1144/SP368.1

Le Heron, D. P. & Craig, J. 2008. First-order reconstructions of a Late Ordovician Saharan ice sheet. *Journal of the Geological Society, London*, **165**, 19–29, https://doi.org/10.1144/0016-76492007-002

Le Heron, D. P. & Dowdeswell, J. A. 2009. Calculating ice volumes and ice flux to constrain the dimensions of a 440 Ma North African ice sheet. *Journal of the Geological Society, London*, **166**, 277–281, https://doi.org/10.1144/0016-76492008-087

Le Heron, D., Sutcliffe, O., Bourgig, K., Craig, J., Visentin, C. & Whittington, R. 2004. Sedimentary architecture of Upper Ordovician tunnel valleys, Gargaf Arch, Libya: implications for the genesis of a hydrocarbon reservoir. *GeoArabia*, **9**, 137–160.

Le Heron, D. P., Craig, J. & Etienne, J. L. 2009. Ancient glaciations and hydrocarbon accumulations in North Africa and the Middle East. *Earth-Science Reviews*, **93**, 47–76.

Lønne, I., Nemec, W., Blikra, L. H. & Lauritsen, T. 2001. Sedimentary architecture and dynamic stratigraphy of a marine ice-contact system. *Journal of Sedimentary Research*, **71**, 922–943.

McClure, H. A. 1978. Early Palaeozoic glaciation in Arabia. *Palaeogeography, Palaeoclimatology, Palaeoecology*, **25**, 315–326.

Moreau, J., Ghienne, J.-F., Le Heron, D., Rubino, J.-L. & Deynoux, M. 2005. A 440 Ma old ice stream in North Africa. *Geology*, **33**, 753–756.

Naden, J. 1996. CalcicBrine 1.5: a Microsoft Excel 5.0 Add-in for calculating salinities from microthermometric data in the system $NaCl–CaCl_2–H_2O$. *In*: Brown, P. E. & Hagemann, St. G. (eds) *PACROFI VI: Sixth Biennial Pan-American Conference on Research on Fluid Inclusions: Program and Abstracts*. University of Wisconsin Press, Madison, WI, 97–98.

Pittman, E. D., Larese, R. E. & Heald, M. T., 1992. Clay coats: occurrence and relevance to preservation of porosity in sandstones. *In*: Houseknecht, D. W. & Pittman, E. D. (eds) *Origin, Diagenesis, and Petrophysics of Clay Minerals in Sandstones*. Society for Sedimentary Geology (SEPM), Special Publications, **47**, 241–264.

Powell, J. H., Khalil Moh'd, B. & Masri, A. 1994. Late Ordovician–Early Silurian glaciofluvial deposits preserved in palaeovalleys in South Jordan. *Sedimentary Geology*, **89**, 303–314.

Powell, R. D. & Cooper, J. M. 2002. A glacial sequence stratigraphic model for temperate, glaciated continental shelves. *In*: Dowdeswell, J. A. & Ó Cofaigh, C. (eds) *Glacier-Influenced Sedimentation on High-Latitude Continental Margins*. Geological Society, London, Special Publications, **203**, 215–244, https://doi.org/10.1144/GSL.SP.2002.203.01.12

Rogers, M. A., McAlary, J. D. & Bailey, N. J. L. 1974. Significance of reservoir bitumens to thermal-maturation studies, Western Canada Basin. *American Association of Petroleum Geologists Bulletin*, **5**, 1806–1824.

Russell, H. A. J. & Arnott, R. W. C. 2003. Hydraulic-Jump and Hyperconcentrated-flow deposits of a glacigenic subaqueous fan: Oak Ridges Moraine, southern Ontario, Canada. *Journal of Sedimentary Research*, **73**, 887–905.

Saltzman, M. R. & Young, S. A. 2005. Long-lived glaciation in the Late Ordovician? Isotopic and sequence stratigraphic evidence from western Laurentia. *Geology*, **33**, 109–112.

Scotese, C. R. & McKerrow, W. S. 1990. Revised world maps and introduction. *In*: McKerrow, W. S. & Scotese, C. R. (eds) *Palaeozoic Palaeogeography and Biogeography*. Geological Society, London, Memoirs, **12**, 1–21, https://doi.org/10.1144/GSL.MEM.1990.012.01.01

Scotese, C. R., Boucot, A. J. & McKerrow, W. S. 1999. Gondwana palaeogeography and palaeoclimatology. *Journal of African Earth Sciences*, **28**, 99–114.

Senalp, M. & Al Laboun, A. 2000. New evidence on the Late Ordovician glaciation in Central Saudi Arabia. *Saudi Aramco Journal of Technology*, (Spring 2000), 11–40.

Taylor, T. R., Giles, M. R. *et al.* 2010. Sandstone diagenesis and reservoir quality prediction: models, myths, and reality. *American Association of Petroleum Geologists Bulletin*, **94**, 1093–1132.

Tournier, F., Pagel, M., Portier, E., Wazir, I. & Fiet, N. 2010. Relationship between deep diagenetic quartz cementation and sedimentary facies in a Late Ordovician glacial environment (Sbaa Basin, Algeria). *Journal of Sedimentary Research*, **80**, 1068–1084.

Vagle, G. B., Hurst, A. & Dypvik, H. 1994. Origin of quartz cement in some sandstone from the Jurassic of the Inner Moray Firth (UK). *Sedimentology*, **41**, 363–377.

Vaslet, D. 1990. Upper Ordovician glacial deposits in Saudi Arabia. *Episodes*, **13**, 147–161.

Walderhaug, O. 1996. Kinetic modeling of quartz cementation and porosity loss in deeply buried sandstone reservoirs. *American Association of Petroleum Geologists Bulletin*, **80**, 731–745.

Weibel, R., Friis, H., Kazerouni, A. M., Svendsen, J. B., Stokkendal, J. & Poulsen, M. L. K. 2010. Development of early diagenetic silica and quartz morphologies – examples from the Siri Canyon, Danish North Sea. *Sedimentary Geology*, **228**, 151–170.

Worden, R. H. & Morad, S. 2000. Quartz cementation in oil field sandstones: a review of the key controversies. *In*: Worden, R. H. & Morad, S. (eds) *Quartz Cementation in Sandstones*. International Association of Sedimentologists, Special Publications, **29**, 1–20.

Worden, R. H. & Morad, S. 2003. Clay minerals in sandstones: controls on formation, distribution and evolution. *In*: Worden, R. H. & Morad, S. (eds) *Clay Mineral Cements in Sandstones*. International Association of Sedimentologists, Special Publications, **34**, 3–41.

Worden, R. H., Oxtoby, N. H. & Smalley, P. C. 1998. Can oil emplacement prevent quartz cementation in sandstones? *Petroleum Geoscience*, **4**, 129–137, https://doi.org/10.1144/petgeo.4.2.129

Trace element composition of authigenic quartz in sandstones and its correlation with fluid–rock interaction during diagenesis

THOMAS GÖTTE

Institute for Geosciences, Goethe-University Frankfurt, Altenhöferallee 1, D-60438 Frankfurt am Main, Germany

goette@em.uni-frankfurt.de

Abstract: The properties of sandstones are strongly affected by formation of minerals in the pore space during diagenesis. In many sandstones, quartz overgrowths are the most important pore-filling cements and show a characteristic trace element composition. The most important impurities are Al, Li, H and Ge.

The Al concentration of quartz may reach up to several 1000 μmol mol^{-1} and is assumed to reflect the composition of the porewater. Geochemical modelling of the activity ratio of Al and Si revealed a minimum at nearly neutral to slightly acid conditions. This minimum shifts to lower pH with increasing temperatures. Due to complexing, organic acids may strongly affect the Al solubility especially in acidic water at low temperatures. There is a linear correlation between Al and Li, indicating their incorporation in a combined [AlO_4|Li] centre. Since Li only accounts for 10–30% of the Al concentration, the remaining Al needs to be balanced by H. The ratio of Li/H-compensated Al centres seems to depend on the Li activity and the pH in the aqueous solution. Germanium concentrations in quartz cements are slightly higher than the crustal average and they show a weak correlation with Al. The excess of Ge in authigenic quartz requires pre-enrichment, probably by formation of kaolinite. Possible applications of trace element analyses of authigenic quartz include discrimination of different sources that contribute to the supply of silica, enhanced understanding of inhomogeneities that are related to cementation and possible tracking of fluid migration.

The diagenesis of sediments and sedimentary rocks is known to directly affect the petrophysical properties of source and reservoir rocks for oil and gas. For that reason, the understanding and prediction of diagenetic processes is of great interest for the exploration and production of hydrocarbons and has been an important aspect of sedimentary research for more than 40 years (McDonald & Surdam 1984; Crossey *et al.* 1996). Recently the focus of economic interest has increasingly shifted to unconventional and tight reservoirs, for which the prediction of diagenetic processes has become even more important than for sandstone reservoirs. It is relevant to utilize geochemical signals in quartz for the reconstruction of diagenesis in reservoir sandstones for two reasons:

First, quartz is one of the most common authigenic minerals and is the most important pore-filling cement in sandstones, especially those buried to depths where the temperature is greater than about 80–100°C. Its great influence on the quality of hydrocarbon reservoirs has led to a large number of studies on general and specific aspects of quartz cementation (cf. compilation of Worden & Morad 2000) and contributed to the development of models that help to predict cementation of sandstones (Walderhaug 1996; Lander *et al.* 2008). Although this approach is more or less successful in various cases, many facets of the formation of quartz cements are still unclear concerning the sources, mobility and transfer rates of silica and other relatively immobile elements. The key questions that need to be answered in any study are: (1) what processes trigger the formation of quartz cement; and (2) is silica derived from internal sources or does it require mass transfer from adjacent rocks (Worden & Morad 2000; Milliken 2005)? Of course, general rules can be derived from numerous studies, but there are so many exceptions that they should not be accepted as universally valid *a priori*.

The second reason is that quartz, once it is formed, is chemically so stable that it does not change even on geological timescales (in contrast to many other minerals, such as carbonates and sulphates). This means that the conditions under which quartz crystallizes are permanently archived, provided that they leave traces in the isotopic or trace element composition or the defect structure. While isotope studies have largely been restricted to $\delta^{18}O$ to date, allowing statements about the origin of the porewater and/or the formation temperatures of the cements (Rezaee & Tingate 1997; Williams *et al.* 1997; Marchand *et al.* 2002; Hiatt *et al.* 2007; Kelly *et al.* 2007), the trace element composition reflects the concentrations or activities of certain elements in the aqueous solution which

From: Armitage, P. J., Butcher, A. R., Churchill, J. M., Csoma, A. E., Hollis, C., Lander, R. H., Omma, J. E. & Worden, R. H. (eds) 2018. *Reservoir Quality of Clastic and Carbonate Rocks: Analysis, Modelling and Prediction.* Geological Society, London, Special Publications, **435**, 373–387.
First published online January 19, 2016, https://doi.org/10.1144/SP435.2

primarily depends on the pH-Eh conditions, temperature, and diagenetic reactions that take place (Lehmann *et al.* 2011; Götte *et al.* 2013). Thus, decoding the processes that control the trace element incorporation in quartz should provide information on the porewater evolution, help to determine the timing and regional pattern of quartz cementation, and elucidate the diagenetic reactions that trigger the formation of quartz. Overall, this will result in enhanced models for reactive transport and diagenesis in sandstones and for the prediction of reservoir quality.

The very limited volume of material available in an individual quartz overgrowth and the low concentrations of impurities, however, require highly evolved techniques for the analysis that have very good spatial resolution and extremely low detection limits. With the development of mass spectrometric methods, especially in the last 10–15 years, these chemical analyses of such small sample volumes are possible. This article reviews the characterisation of the trace element inventory of authigenic quartz on the basis of previous studies, most of which are already published. It also presents some ideas on the processes, the physicochemical conditions and diagenetic reactions that mobilise elements that can be incorporated in the quartz lattice.

Geological background of the case studies of trace elements in quartz cement

The data sets upon which this discussion is based, are the following (Table 1):

(1) The Witten-Formation (Upper Carboniferous) north of the Rhenish Massif (Western Germany); this is an alternating sequence of fluvial sandstones, mudstones, coal measures and marine marls. The feldspathic litharenites and subarkoses consist mainly of quartz, various lithic fragments, K-feldspar, plagioclase and detrital mica (Füchtbauer 1974; Büker 1996; Götte *et al.* 2013). After strong mechanical compaction, quartz overgrowths and illite cements were formed due to feldspar-breakdown reactions and as a consequence of pressure dissolution. Cathodoluminescence revealed three distinct generations of overgrowths of which the second and the third are volumetrically important (Götte *et al.* 2013). The second generation forms euhedral overgrowths and is assumed to have been accompanied by kaolinite which should have been formed during early stages of diagenesis, while the third one is accompanied by pore-filling illite. Some late baryte cements post-date the quartz cements. Additional to alteration of feldspars, pressure dissolution of quartz is an important source of silica. Modern porewater has been analysed by Wedewardt (1995) who found low salinity ground-water in surficial aquifers and high-salinity porewater at greater depths.

(2) The Solling-Formation (Lower Triassic, Central Germany) consists of red to beige fluvial sandstones which can be mainly classified as subarkoses or lithic subarkoses (Weber & Ricken 2005). The predominantly sandy sequence is only rarely interrupted by mudstones which only are some centimetres to tens of centimetres thick. Early diagenetic incrustations of hematite are common and predate quartz overgrowths. Three phases of quartz cementation can be distinguished by their CL-properties. The first one may be relatively early or correspond to the 'anhedral growth' of Lander *et al.* (2008) while the second one forms euhedral overgrowths. The paragenesis of these early cements remains unclear, but they might be accompanied by kaolinite and carbonate cements (Weber & Ricken 2005). The feldspars are altered and partly dissolved. Plagioclase grains are albitized and discrete albite overgrowths on relics of feldspar grains have been recognised. The latest generation of quartz overgrowths and albite cements are accompanied by illite cements (Götte *et al.* 2013). Pressure dissolution of quartz grains has been recognised but seems not to be the dominant process for the supply of silica. Analyses of recent porewaters from various depth revealed several aquifers that contain low-salinity water near the surface and high-salinity brines from greater depths (Kloppmann *et al.* 2001; Möller *et al.* 2008).

(3) In contrast to all other examples, the Rockenberg-Formation (Neogene) of the Wetterau-Basin (Central Germany) has never been deeply buried. It consists of predominantly uncemented sands which are underlain by fine-grained sediments and overlain by basaltic rocks of the Vogelsberg-volcano. The sands of the Rockenberg-Formation are nearly pure quartz sands with >98% quartz and quartzose lithic fragments. Particularly noteworthy is that Al-bearing minerals are lacking. Lenses of tightly cemented sandstones occur in the Rockenberg-Formation which are several meters thick and extend over some tens of square-meters. Quartz overgrowths and Fe-sesquioxides are the predominant cements, additionally more unstable SiO_2-modifications like micro-quartz and chalcedony have been found which post-date the overgrowths and the Fe-phases (Götte *et al.* submitted).

Table 1. *Summary of the diagenetic properties of the sandstones and chemical composition of quartz cements*

	Paragenesis	*T* (°C)	Li	B	Na	Al	Ti	Fe	Ge	AlOH	LiOH	H_2O
Witten Fm.												
Quartz overgrowth 2	Quartz (+kaolinite?)		92	6,3	16	750	57	20	1,4			
Standard deviation			*76*	*0,8*	*10*	*430*	*50*	*13*	*0,5*			
Quartz overgrowth 3	Quartz + illite	150–180	430	0	19	2100	1,5	65	1,9	600	49	2400
Standard deviation			*170*	*0*	*7*	*370*	*0,5*	*27*	*0,6*	*330*	*81*	*1100*
Solling Fm.												
Quartz overgrowth 2	Quartz (+hematite + kaolinite?)		83	18,5	69	304	26	359	0,99			
Standard deviation			*31*	*2,1*	*84*	*125*	*20*	*452*	*0,45*			
Quartz overgrowth 3	Quartz + illite + albite	*c.* 100	19,5	0,58	31	140	16	159	0,77	86	11	1100
Standard deviation			*10*	*0,49*	*38*	*143*	*11*	*159*	*0,17*	*53*	*2*	*790*
Rockenberg Fm.												
Quartz overgrowths	Quartz (+limonite)	<50	280	17	n.d.	810	0,5	60	1,9			
Standard deviation			*270*	*13*	*n.d.*	*790*	*0,5*	*57*	*0,65*			
Chalcedony	Quartz + limonite	<50	3,2	15	580	370	0,24	59	0,44			
Standard deviation			*1,4*	*4,5*	*500*	*130*	*0,15*	*50*	*0,14*			
Late quartz	Quartz (+limonite)	<50	3,4	11	1000	250	0,41	95	0,4			
Standard deviation			*0,56*	*3,8*	*1100*	*180*	*0,32*	*82*	*0,21*			
Micro-quartz	Quartz + kaolinite + limonite	<50	150	270	410	5900	10 200	1060	1,7			
Standard deviation			*20*	*197*	*160*	*2700*	*8000*	*910*	*0,62*			
Haushi Group												
Quartz overgrowths	Quartz + illite (kaolinite) + dolomite	85–135	180		23	1570	77	164	1,2			
Standard deviation			*160*		*69*	*1340*	*570*	*530*	*0,77*			
Keuper Sdst.												
Quartz overgrowths	Quartz + illite		62			1080						
Standard deviation			*74*			*1100*						

All element/compound concentrations are in μmol mol^{-1}.
Data are from Lehmann *et al.* 2011 (Haushi Group); Büker 1996, Götte *et al.* 2013 (Witten Formation); Weber & Ricken 2005; Götte *et al.* 2013 (Solling Formation); and Demars *et al.* 1996 (Keuper Sanstones).

Pressure dissolution or other internal sources have not been recognised. Therefore, advective transport in an open system is required for the formation of these sandstones. The composition of the recent low-salinity porewater is dominated by the weathering of the basaltic rocks with Ca and Mg being the most important cations. However, there have been local influxes of deep thermal water with increased NaCl-concentrations (Leßmann *et al.* 2001).

(4) The sandstones of the Haushi-Group (Carboniferous to Permian; Oman) are of glacial and fluvial origin in the deeper part of the section and of fluvial to coastal origin in the shallower interval. The Haushi-Group sandstones consist of quartz, plagioclase, K-feldspar and rock-fragments. An early diagenetic sequence of cements comprises hematite and pedogenic carbonates. The burial diagenetic paragenesis consist of quartz overgrowths, kaolinite, dolomite, anhydrite and illite. The quartz cements content increases with increasing depths due to stronger pressure dissolution which seems to be a predominant source of silica (Lehmann *et al.* 2011). Only two porewater analyses have been published from the Oman (Ramseyer *et al.* 2004), both of which have relatively high salinities but low pH.

(5) The Keuper-sandstones (Upper Triassic) of the Paris Basin are fluvial arkoses and subarkoses that contain varying amounts of lithic fragments. Evaporites occur beneath these sediments in the eastern part of the basin. Dolomite is the predominant cement and is accompanied by anhydrite. Additionally, quartz overgrowths that post-dates the carbonates, albite, K-feldspar, illite and traces of pyrite and anatase, have been found (Demars *et al.* 1996). The recent porewater in the Paris Basin has been analysed by Azaroual *et al.* (1997) who found low to high-salinity aqueous solutions with a pH near to neutral in aquifers of Dogger and Keuper-sandstones.

Methods

The electron microprobe (EMP) is the most established technique for chemical analysis with high spatial resolution. It has been applied to magmatic and hydrothermal quartz in several studies (Ramseyer *et al.* 1988; Perny *et al.* 1992; Müller *et al.* 2003) but much less frequently to quartz cements in sandstones (e.g. Demars *et al.* 1996; Kraishan *et al.* 2000; Fischer *et al.* 2013). Quartz is very sensitive to ionizing irradiation and damage of the crystal lattice increases with decreasing beam diameter and increasing beam current (Kronz *et al.* 2012). Therefore, the beam should be defocused to a spot size of at least 5 μm. The electron energy is commonly about 15–20 keV. The method has two great disadvantages with regards to the geochemistry of quartz: first, the detection limits are not low enough (50–100 μmol mol^{-1} at best) for most impurities since quartz is known to be a very pure mineral and incorporates less than 10 ppm of most foreign cations. The second disadvantage is that other elements that are present in relatively high concentrations, such as H, Li or B, cannot be analysed due to the Be-windows in front of the X-ray detectors. The list of elements that can be analysed by EMP in quartz is restricted to Al, K, Ti, and Fe (Kronz *et al.* 2012), of which only Al is present above the detection limit in authigenic quartz samples. For these reasons, EMP has only achieved partial success in chemical analyses of quartz cements.

To reduce detection limits, Bruhn *et al.* (1996) used a proton microprobe (proton-induced X-ray Emission, PIXE) for the analysis of trace elements in authigenic quartz. The proton beam, with an energy of 3 MeV, was produced in a Dynamitron Tandem Accelerator and had a diameter of approximately 10 μm. Using this technique, it is possible to quantify concentrations as low as 5–10 ppm of some cations, e.g. Ti and Fe. Unfortunately, the detection limits of Al are not lower than those obtained with the EMP (Götte 2004). The limitations concerning H, Li or B are the same as described above for the EMP and so prevented this method being applied to a large number of studies of sandstones.

Mass-spectrometric techniques can have very low detection limits and are therefore the optimal choice for the analysis of trace elements in quartz. Generally, two different techniques are available: Secondary ion mass spectrometry (SIMS) or inductively coupled plasma mass spectrometry (ICP-MS) combined with laser ablation. The first one requires great technical effort and has been found to be not optimal for trace element analysis (Müller *et al.* 2003; Lehmann *et al.* 2009). The greatest advantage is the ability to detect H and the analyses of the isotopic composition, which has become the preferred application of this technique (Rezaee & Tingate 1997; Williams *et al.* 1997; Marchand *et al.* 2002; Hiatt *et al.* 2007; Kelly *et al.* 2007).

Laser ablation techniques (LA-ICP-MS) were established in the last decade and are ideal for the analysis of quartz with high spatial resolution and low detection limits (Müller *et al.* 2003). The development of the ArF-excimer laser provided optimal conditions for the ablation of quartz. The analyses are commonly calibrated with the NIST 612 standard and ^{29}Si was used as the internal standard for the quantification. Isotopes that are suitable for the analyses comprise ^{6}Li (0.2 μmol mol^{-1}), ^{7}Li (0.1 μmol mol^{-1}), ^{11}B (0.3 μmol mol^{-1}), ^{22}Na (1 μmol mol^{-1}), ^{23}Na (1 μmol mol^{-1}), ^{27}Al (1 μmol mol^{-1}), ^{39}K (4 μmol mol^{-1}), ^{48}Ti (0.01 μmol mol^{-1}), ^{52}Ti (0.1 μmol mol^{-1}), ^{55}Mn (0.1 μmol mol^{-1}), ^{56}Fe (2 μmol mol^{-1}), ^{57}Fe (1 μmol mol^{-1}), ^{72}Ge (0.01 μmol mol^{-1}) and ^{74}Ge (0.02 μmol mol^{-1}; detection limits for a pit with a diameter of 32 μm are in brackets; Götte *et al.* 2011, 2013; Lehmann *et al.* 2011). The isotopes that can be used depend on the setting of the instrument and the gases that are used during the analysis (for technical details see Pettke 2008). The absolute values of the detection limits vary with the pit size and become higher at smaller beam diameters. Changing mass fractions can be recognized in the chromatogram during the ablation process and can be put down to zoning within the quartz cement. Each growth zone should be evaluated separately.

The most important disadvantage of LA-ICP-MS is that it is not possible to quantify hydrogen. This element is one of the most important impurities in quartz. It is most efficiently analysed by Fourier transform infrared spectroscopy (FTIR) and not by SIMS, which is a very expensive and time-consuming technique. The analyses are carried out on 50–100 μm thick sections in transmitted infrared light (that can be used in polarized or non-polarized mode). In the case of sandstones, non-polarized light should be used because quartz grains are randomly oriented and the results obtained by polarized light cannot be interpreted. Generally, the detection limits decrease with increasing spot size, but the latter is limited by the small size of quartz overgrowths and by the fact that the entire pathway through the sections must be in the quartz cement because only the integrated signal is detected. The absorption bands of OH-defects in quartz are in the spectral range of 2500–4000 cm^{-1}. After deconvolution of the spectra, the concentration of H can be calculated as H_2O by the Lambert–Beer law $\alpha = \varepsilon\, c\, d$ with the absorbance α, sample thickness d (in cm), integral absorption coefficient ε and H_2O-concentration c (in mol l^{-1}, cf. compilation of Rossman 2006). For the absorption coefficient, either mineral-specific (Thomas *et al.* 2009) or wavelength-specific parameters can be used (Libowitzky & Rossman 1997).

Trace element composition of authigenic quartz

The published data on the trace element geochemistry of quartz cements are based on a relatively small number of studies. All quartz cements have a characteristic composition in common. Under diagenetic conditions, only a few elements are incorporated in concentrations that are large enough to be quantified and interpreted. These include Al, B, and Ge, which all substitute for Si on a tetrahedral site, as well as the alkali metal elements (Li, Na, K) and H. Titanium is incorporated in quartz only at higher temperatures (>400°C, Wark & Watson 2006), and Fe is generally present only at very low concentrations.

Substituting high-field cations

Aluminium is by far the most common substitution that is found in diagenetic quartz, and it reaches concentrations of up to several thousand μmol mol^{-1} (Fig. 1a–d). In cathodoluminescence (CL) images and element maps it is apparent that zoning in quartz cement is the rule rather than the exception; CL-zoning is at least partially related to trace element chemistry (Lehmann *et al.* 2009). In most cases, there are distinct cement generations, or individual growth zones, that have increased Al concentrations (Lehmann *et al.* 2011; Götte *et al.* 2013). The sharp zoning that is visible both in the CL images as well as in the trace element distributions, indicates episodic formation of quartz cementation where the distinct growth zones are separated by episodes of reduced or transiently stopped growth of quartz. Thus, each generation needs a trigger that initiates growth and each generation may vary in the trace element composition due to changes in the porewater chemistry and diagenetic reactions.

The highest Al concentrations have been measured in the quartz cements of the Keuper Sandstones (Fig. 1a, Demars *et al.* 1996, the highest two might be outliers) and of the Haushi Group Sandstones (Fig. 1a, Lehmann *et al.* 2011). However, different generations of quartz cement have not been considered in these studies and much of the large scattering may be put down to growth zoning. In the case of fluvial sandstones from the Witten and Solling formations, three generations of quartz cement have been found (Büker 1996; Götte *et al.* 2013). The respective second cement generation in both the Witten and Solling formations have moderate Al contents of 200–1000 μmol mol^{-1}, while the third generation differs significantly in both cases from the previous one and also from each other (Fig. 1b, c). While in the sandstones of the Witten Formation, the Al- contents increased significantly to approximately 2000 μmol mol^{-1} or more (Fig. 1b), only very low concentrations of approximately 100 μmol/mol^{-1} are measurable in the latest generation quartz of the Solling Formation (Fig. 1c).

Similarly, low concentrations of Al have been found in the metastable chalcedony cements from the meteoric sandstones in the Rockenberg

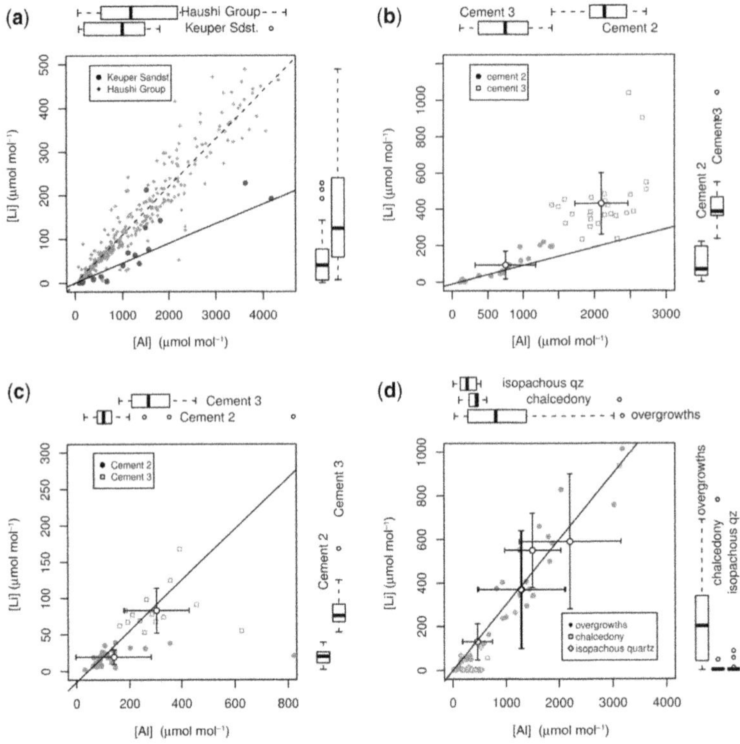

Fig. 1. Crossplots of Al and Li concentrations of silica cements of different case studies. The regression lines are fitted to the lower concentrations: (**a**) Keuper Sandstones (data from Demars *et al.* 1996) and sandstones from the Haushi Group (data from Lehmann *et al.* 2011). (**b**) Witten Formation (data from Götte *et al.* 2013); three generations of quartz-cements have been found of which the second and third were analysed, circles with error bars represent averages and standard deviations (1 s) of each generation. (**c**) Solling Formation (data from Götte *et al.* 2013; cf. b). (**d**) Rockenberg Formation from four locations (data from Götte *et al.* submitted); open circles represent averages and standard-deviations (1 s) for the quartz overgrowths in each sample.

Formation, which contain less than 500 μmol mol^{-1} Al compared with 500–3000 μmol/mol in the overgrowths (Fig. 1d).

Germanium is the second most abundant element substituted for Si in authigenic quartz. The concentrations are generally low (0.5–5 μmol mol^{-1}), but it has been found in all studies that have been published so far. The concentrations are slightly above the crustal average of 1 μmol mol^{-1} in quartz. However, the variability is too large to unambiguously reveal trends or differences between distinct cement generations. Commonly, Ge is weakly correlated to Al (Fig. 2a–d).

Among the substituting cations, B is that one that shows the most curious behaviour. It is commonly below the detection limit but has been repeatedly found in different samples with concentrations up to 20 μmol mol^{-1}. In particular, it has been found in the meteoric-derived quartz cements of the Rockenberg Formation, in which there are apparently no internal sources. Locally, the sediments might be under the influence of saline thermal water that rises in proximity to the Vogelsberg volcano (Leßmann *et al.* 2001). However, B is not correlated to any other element, especially not to Al, except in the case of chalcedony cements, which show a quite good correlation between Al and B.

Alkali elements on substitutional sites

Lithium, Na and K are potential charge-compensating elements that can be incorporated in quartz on interstitial sites. However, the concentration of K is commonly below the detection limit (approx. 5–10 μmol mol^{-1}) even using very sensitive techniques, although there are indications that K may be incorporated into quartz at elevated temperatures (Landtwing & Pettke 2005). The same may be valid for Na, which can be detected in most quartz overgrowths, but there are very high variations and no correlation to Al (Lehmann *et al.* 2011). Commonly,

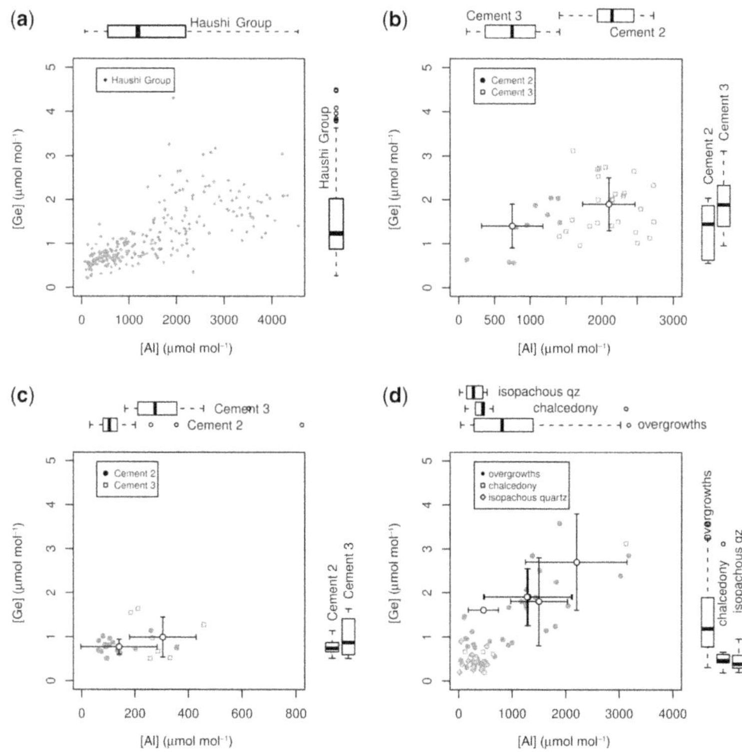

Fig. 2. Crossplots of Al and Ge concentrations in silica cements from different case studies. For details cf. Figure 1. Demars *et al.* (1996) did not report values for Ge.

Na is absent in most measurements, but some have Na contents of up to 70 μmol mol^{-1}. It has therefore been suggested to be present in sub-microscopic secondary mineral and fluid inclusions and is not incorporated in the quartz lattice (Lehmann *et al.* 2011; Götte *et al.* 2013).

The remaining alkali element, Li, is incorporated on interstitial sites and compensates the charge imbalance that results from the substitution of Al^{3+} for Si^{4+}. This interpretation is supported by the observation that, in most cases, there is a very good correlation between these elements (Fig. 1a–d). Li concentrations are between 5 and 1000 μmol mol^{-1}, which is well below those of Al. Thus, only 10–30% of Al is compensated by Li and the excess of Al needs to be balanced by H^+. In the late quartz overgrowths of the sandstones from the Witten Formation, there is no correlation between Li and Al; Li is enriched relative to the earlier quartz cement phases (Fig. 1a, b, Götte *et al.* 2013). Similar results have been reported from the Paris Basin (Demars *et al.* 1996), however, the results were not originally interpreted in this way by the authors. A second deviation from the linear correlation between Li and Al exists in the samples from the Rockenberg Formation, where chalcedony and late, isopachous quartz cements are strongly depleted in Li (Fig. 1d).

Discussion

Trace element incorporation in quartz

The discussion of the trace element composition focuses on Al, Li, H and Ge, which are the most important impurities in quartz. The relatively high degree of covalent bonding in the Si—O bond inhibits the formation of a dense ion package. As a consequence, large channels run parallel to the *c*-axis through the crystal lattice (cf. compilation of Preusser *et al.* 2009). Generally, there are three possible ways to incorporate trace elements into this lattice: as substitution for Si on a tetrahedral site, on interstitial sites in the large channels (Fig. 3) or by a combination of both.

Substitutions. Cations that substitute for Si on a tetrahedral site are restricted to high-field, tri- to pentavalent ions (e.g. Al^{3+}, B^{3+}, Ge^{4+}, Ti^{4+}, P^{5+}; Larsen *et al.* 2004). Isovalent substitution is only possible by Ge^{4+} or Ti^{4+}. The relatively simple

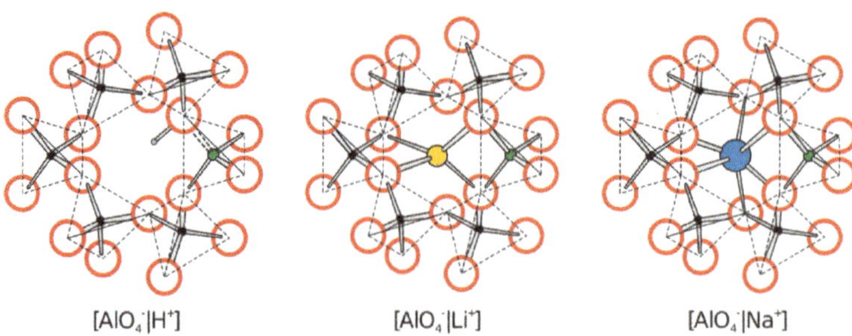

Fig. 3. Part of the quartz lattice with large channels parallel to the *c*-axis (projection of (0001)). Red circles represent the position of O, black dots the position of Si. Tetrahedral sites may be occupied by Al, Ge or B (green), interstitial sites can be occupied by H (grey), Li (yellow) or Na/K (blue), respectively, in slightly different positions (redrawn and modified from Rossman 1994; Preusser *et al.* 2009).

exchange of one cation for another should be controlled only by temperature, pressure and the composition of the parental fluid. This has been already observed for Ti (Wark & Watson 2006). Heterovalent substitution needs to be compensated either by monovalent cations (in case of a trivalent cation) or by a coupled substitution of a tri- and a pentavalent cation. The latter is the only possible way to incorporate, for example, P^{5+}, which is closest to Si^{4+} in terms of ion radius, but seems to be relatively rare due to problems with charge balance. As well as the physicochemical conditions, the concentration of certain defects is influenced by the activities of all elements involved in the crystal and in the aqueous solution as well as by the crystallographic properties of the defect (i.e. the defect's stability).

Interstitial impurities and stuffed derivatives. The large channels parallel to the *c*-axis provide interstitial sites that are large enough to be occupied by alkali ions (Li, Na, or K; cf. compilation of Preusser *et al.* 2009). Lithium, the smallest alkali ion, commonly occupies a special site on the *a*-axis that is coordinated by four O atoms in a distorted tetrahedron (Rossman 1994). Commonly Li is adjacent to an Al substitution and forms an $[AlO_4|Li]$ centre, which is balanced in respect to the electric charge (Fig. 3). The large cations, Na^+ and K^+, have a higher coordination that is slightly below the *a*-axis and forms a distorted octahedron in $[AlO_4|Na]$ centres (Fig. 3). Porewater always contains greater concentrations of Na and K than Li, whereas Li is the most abundant element in quartz and seems to be preferentially incorporated under diagenetic conditions. However, Na and K have been detected in hydrothermal quartz that was formed at higher temperatures (Bambauer 1961; Landtwing & Pettke 2005).

Hydrogen in quartz is incorporated in OH groups. Due to the very small size of single protons, they can nearly always be bound to O without occupying a lattice or interstitial site. If the O is on a lattice site, then the OH bond commonly has a preferred orientation (Rossman 1994). Hydrogen is typically one of the most important impurities in quartz and may either be incorporated together with Al in $[AlO_4|H]$ centres (Fig. 3), or replacing Si^{4+} with four hydrogen ions, as H_2O in the channels along the *c*-axis or as hydrated Si—O bond (Stenina 2004; Rosa *et al.* 2005). In authigenic quartz from the Witten and Solling formations molecular water has been found to be the most important species, but $[AlO_4|H]$ centres are also common (Table 1, Götte *et al.* 2013).

Mobilization of aluminium and silica

The Al concentration of authigenic quartz has been found to reflect the composition of the parental solution (i.e. the activity ratio of Al/Si in the aqueous solution) but is primarily not a function of temperature (Rusk *et al.* 2008; Götte *et al.* 2011). Similarly, the incorporation of Al in authigenic quartz has been proposed to depend on the activities of Al and Si in the porewater (Lehmann *et al.* 2011; Götte *et al.* 2013). However, it should be mentioned that the substitution of Si^{4+} by Al^{3+} generates an excess of negative charge that needs to be compensated either by incorporation of a monovalent cation on an interstitial site or by substitution of P^{5+}. The distribution of Al and Si between aqueous solution and quartz may be written as follows (Merino *et al.* 1989):

$$\frac{x_{[\mathrm{Al|M}]}}{x_{[\mathrm{Si}]}} = K' \frac{a_{\mathrm{M}} a_{\mathrm{Al(OH)_4}}}{a_{\mathrm{Si(OH)_4}}} \quad (1)$$

with $x_{[Al|M]}$ and $x_{[Si]}$ being the activities of the distinct species in quartz and a_M, $a_{Al(OH)4}$ and $a_{Si(OH)4}$ are the activities in aqueous solution. The constant of proportionality, K', results from the mass action law (Merino *et al.* 1989).

Aluminium is an immobile element that occurs in high concentrations in mudrocks and in matrix- or feldspar-rich sandstones because in sedimentary environments it is predominantly found in clay-minerals and feldspars. Since feldspars tend to be reactive under typical diagenetic conditions, they are commonly a source of Al while clay minerals and quartz are formed (McBride 1989; Milliken 2005) and Al-bearing quartz cements have been linked to feldspar dissolution (Kraishan *et al.* 2000). If equilibrium is assumed, the Al activity in aqueous solution is controlled by the stable clay mineral and the activity of $Si(OH)_4$ is controlled by equilibrium with quartz. At low ionic strength, pH and temperature, kaolinite will be the stable Al phase, while in most burial diagenetic settings it is replaced by illite. In the samples that this article refers to, illite is the stable clay mineral with exception of the Rockenberg Formation, which is in the stability field of kaolinite.

To illustrate the mobility of Al and Si and the evolution of the Al/Si-ratio at different temperatures, salinities and pH values, the total activities have been calculated for equilibrium with illite and quartz in the system $NaOH—KCl—HCl—AlCl_3—SiOH_4—H_2O$ (Figs 4 & 5; thermodynamic data are from Castet *et al.* 1993; Wesolowski & Palmer 1994; Stumm & Morgan 1996). The concentration of K has been set to 0.01 of the Na concentration and the chemical activities of the electrolytes have been calculated with the Specific Ion Interaction Theory (SIT) model (Grenthe *et al.* 1997).

Regardless of the Al mineral that is formed, the total activity of all Al species (i.e. Al^{3+}, $AlOH^{2+}$, $Al(OH)_2^{+}$, $Al(OH)_3^{0}$, $Al(OH)_4^{-}$) shows a strong minimum at nearly neutral to slightly acid pH at low temperatures. This is a consequence of the high dependency of Al species on pH (e.g. de Jong *et al.* 1983; Merino *et al.* 1989). At low pH, Al^{3+} is surrounded by six H_2O ligands in an octahedral

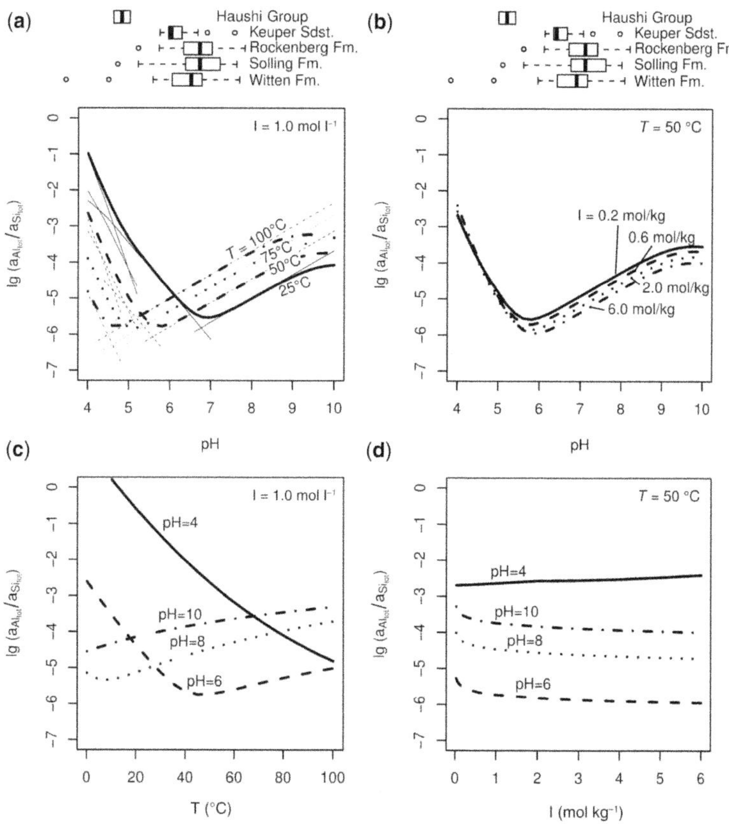

Fig. 4. Calculation of the logarithm (lg) of the ratio of total Al and Si activities at different pH, temperatures and ionic strengths (I). Thin lines in (a) represent activity ratios of different Al species with $Si(OH)_4^0$. The boxplots above the upper diagrams indicate the pH range of the recent porewaters (Haushi Group: Ramseyer *et al.* 2004; Keuper Sandstones: Azaroual *et al.* 1997; Rockenberg Formation: Leßmann *et al.* 2001, Solling Formation: Kloppmann *et al.* 2001; Witten Formation: Wedewardt 1995). I, definition of logarithm of the lg ionic strength.

coordination. With increasing pH the water molecules are replaced one by one by OH ligands, which form much stronger, partly covalent bonds with the central Al cation. In alkaline solutions, tetrahedrally coordinated $Al(OH)_4^-$ is the dominant species (de Jong *et al.* 1983). In contrast, $Si(OH)_4^0$ is the dominant species at acid and neutral pH and only in more alkaline solutions (pH > 10) species like $SiO(OH)_3^-$ and $SiO_2(OH)_2^{2-}$ become important and influence the solubility of quartz. The Al/Si-ratio in porewater therefore reflects the aqueous geochemistry of Al with the minimum of solubility of Al-minerals at nearly neutral pH (Fig. 4). With increasing temperatures, the total activities of Al and Si rise so that both elements are present at higher concentrations. However the minimum value of the Al/Si ratio shifts to lower pH values with increasing temperatures (Fig. 4a, b). While this minimum value is at a pH of about six at standard conditions, it lies at a pH of about 4.5 at 100°C. In contrast, the ionic strength has only a weak effect on the Al/Si ratio. Neither the general form nor the position of the minimum of the Al/Si ratio varies greatly over a large range of ionic strengths (Fig. 4c, d). The most important effect of salinity is a change in mineralogy, with kaolinite being stable at very low salinities while illite becomes stable with increasing K activities that commonly go hand in hand with increasing salinities (Hanor 2001).

Organic acids have repeatedly been shown to form soluble complexes with Al. The best studied complex-forming components are acetic acid (Wesolowski & Palmer 1994) and oxalic acid (Fein & Hestrin 1994). Both of them mainly form Al complexes at low pH. The most important are $Al(CH_3COO)^{2+}$, $Al(CH3COO)_2^+$ for the acetate complexes (Wesolowski & Palmer 1994) and $Al((COO)_2)^+$ and $Al((COO)_2)_3^{3-}$ for oxalate complexes (Fein & Hestrin 1994). The effect of oxalic acid on the Al/Si ratio is several orders of magnitude greater than that of acetic acid (Fig. 5). The solubility of Al is especially increased at low to neutral pH and low temperatures. At higher pH, the effect of organic acids is low. At higher temperatures, the effectiveness of oxalic acid decreases because the stability field of the complex shifts towards lower pH and becomes less important near the minimum of the Al solubility. However, oxalic complexing of Al may play an important role for the mobilization and transport of Al in acid, organic-rich water at lower temperatures. If the complexes decay either due to oxidation of the organic molecules or due to thermal degradation, then Al is released to the porewater solution. In this case, the Al/Si ratio can temporarily rise until the excess of Al is removed from the water. Oxidation of complex-forming organic components might be effective in surficial environments of the Rockenberg Formation where high Al concentrations have been observed in sandstone lenses that might be classified as groundwater silcretes, while thermal degradation of organic acids may be important in the burial diagenetic environments of the Witten Formation and the Haushi Group sandstones. It is questionable whether the low pH that has been found in the recent porewater of the latter formation is favourable for mobilization of Al, because at higher temperatures the minimum solubility shifts to lower pH even at high activities of organic acids (Fig. 5c, d). Highly saline porewater that is poor in organic components will provide only low Al concentrations at low temperatures and nearly neutral pH, as has been found in the quartz overgrowths of the Solling Sandstone.

Mobilization of Li and the Li/Al-ratio

Lithium is generally not an abundant element in porewater and is commonly present only at a sub-ppm level. Its concentration is positively correlated with salinity but it is enriched with respect to Cl and Na in many porewaters compared with seawater (Fontes & Matray 1993; Kloppmann *et al.* 2001; Chan *et al.* 2002). This behaviour has been put down to two processes: it behaves conservatively during the formation of evaporitic minerals and remains in the solution (Fontes & Matray 1993) and it may be released to the porewater by diagenetic reactions (Kloppmann *et al.* 2001; Chan *et al.* 2002). The latter authors attributed the Li excess in porewaters to exchange reactions in smectite. Since Li commonly substitutes for Mg, it is also proposed that smectite–illite reactions will release Li (Williams & Hervig 2005). The high Li activity in porewater has been linked to a high Li/Al ratio in authigenic quartz according to the distribution equation formulated by Merino *et al.* (1989, see above), because these elements are incorporated together in $[AlO_4|Li]$ centres. Therefore, the Li concentration in quartz may be used as proxy for either migration of pre-evaporated brines or for clay mineral reactions. However, the Li concentrations reach only 30% of those of Al even in very Li-rich authigenic quartz (Fig. 1a–d) and the excess Al needs to be compensated by H, which commonly accounts for 70–100% of Al.

In Al defects, H competes with Li for the charge-compensating site adjacent to the substituting Al ion (Müller & Koch-Müller 2009; Götte *et al.* 2011). Hydrothermal quartz typically displays a good balance between Al and the charge-compensating ions Li and H. Only extremely fast-grown quartz might have an imbalance and electron–holes linked to the substituting Al (smoky quartz centres) due to a lack of monovalent cations (Götte *et al.* 2011).

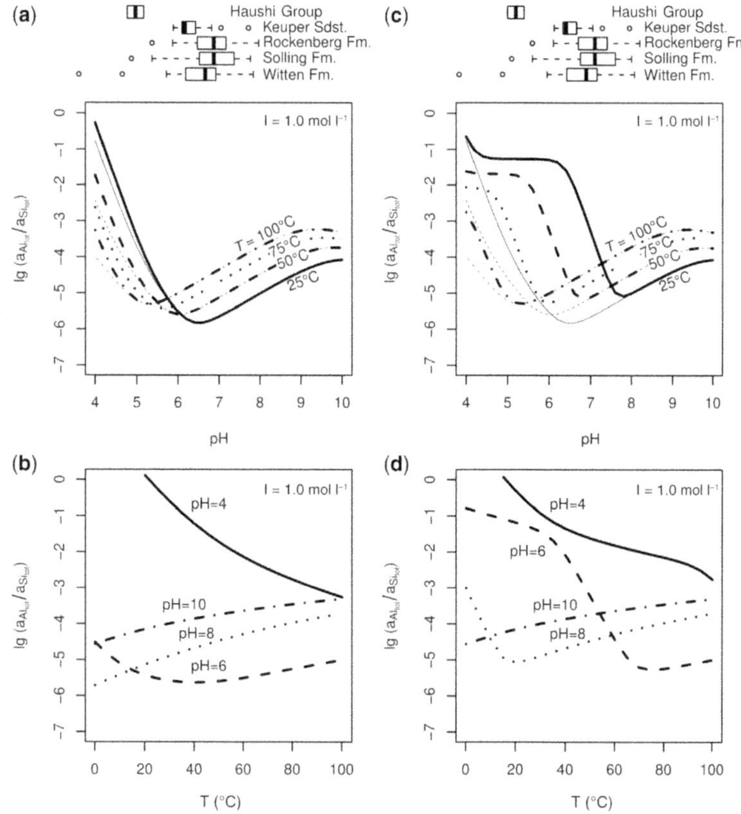

Fig. 5. Calculation of the ratio of total Al and Si activities at different pH and temperatures with a concentration of 10^{-3} mol kg^{-1} acetic acid (**a**, **b**) or 10^{-4} mol kg^{-1} oxalic acid (**c**, **d**), respectively. Thin lines in a and c represent activity ratios of inorganic Al species with Si. The boxplots above the upper diagrams indicate the pH range of the recent porewaters (Haushi Group: Ramseyer *et al.* 2004; Keuper Sandstones: Azaroual *et al.* 1997; Rockenberg Formation: Leßmann *et al.* 2001, Solling Formation: Kloppmann *et al.* 2001; Witten Formation: Wedewardt 1995).

Analyses of H defects in quartz overgrowths by FTIR revealed high concentrations of H defects in Al rich cements (Götte *et al.* 2013) and SIMS analyses confirmed enrichment of H in distinct growth zones (Lehmann *et al.* 2009). The FTIR measurements have a high error, because the quartz grains in sandstones are randomly oriented. This feature of sandstones negatively affects the quality of the analysis (which might be up to three times too low) but the analysed concentrations seem to be high enough to compensate the Al (Table 1). The ratio of H- and Li-compensated Al impurities at the surface in equilibrium with the aqueous solution can be described by:

$$[AlO_4|Li]_{qz} + H^{+}{}_{aq} \longrightarrow [AlO_4|H]_{qz} + Li^{+}{}_{aq}. \quad (2)$$

The activity ratios of $[AlO_4|Li]$ and $[AlO_4|H]$ in quartz ($x_{[AlO4|Li]}$ and $x_{[AlO4|H]}$) are related to the ratio of the activities of the aqueous species, a_{Li} and a_H, by the proportionality constant K'':

$$\frac{x_{[AlO_4|Li]}}{x_{[AlO_4|H]}} = K'' \frac{a_{Li}}{a_H}. \quad (3)$$

Increasing Li activities, either due to influx of pre-evaporated brines or due to clay mineral transformations, will shift the equilibrium to the left side of the equation while low pH should shift it to the right side. Thus, the Li/H ratio will not only be controlled by the Li activity but also by the pH of the porewater at the time the quartz cement formed. For that reason, the Li/H ratio of the quartz overgrowths in the Solling Formation might result from a high pH that is also found in the recent porewater of the North German Basin (Kloppmann *et al.* 2001; Möller *et al.* 2008), while the low Li/H ratio in quartz cements of the Haushi Group and the Witten Formation formed in more acid, organic-rich water. The absolute Al concentration is additionally affected by other parameters (see above).

Germanium systematics

Concentrations of Ge in silicate minerals are commonly at the ppm level. In particular, Ge concentrations in minerals that form detrital components in sandstones are commonly low and near the crustal average of 1.0–1.2 μmol mol^{-1} (Bernstein 1985). In contrast, they are enriched by up to a factor of two in quartz cements in all case studies reported here, except for the Solling Formation where the Ge concentrations are slightly below the crustal average. An additional source for Ge is therefore needed; the correlation of Ge with Al indicates that both elements are mobilized together either due to the same diagenetic reaction or at least at the same diagenetic stage. Lehmann *et al.* (2011) found slightly higher Ge concentrations in feldspars than in quartz, and concluded that this may be the source of Al and Ge in quartz cement. However, two other mechanisms might be much more effective for a pre-enrichment of Ge. First, kaolinite is commonly relatively rich in Ge compared with other minerals indicating a high distribution coefficient (Bernstein 1985). If feldspar is dissolved and kaolinite is formed Ge remains preferentially in the solid phase. Subsequent degradation of kaolinite and replacement by illite can release this excess Ge to the porewater and result in an enrichment of Ge in the quartz- cement. This process has been described from weathering of igneous rocks, where kaolinite is dissolved but not replaced by another clay mineral (Kurtz *et al.* 2002). This process may also work during burial diagenesis. Second, formation of Ge-poor quartz cements would result in enrichment of Ge in the porewater. Subsequent formation of quartz would produce cements that display increasing Ge concentrations. This process has been proposed to be responsible for high Ge/Si ratios in thermal water (Evans & Derry 2002). However, if no advection occurs during burial diagenesis, the same process may take place *in situ*. In the referenced studies on the Haushi Group and the Witten and Solling formations, the replacement of kaolinite by illite seems to be more probable, especially because kaolinite has been assumed to be part of the earliest diagenetic paragenesis in all cases (Table 1; Weber & Ricken 2005; Lehmann *et al.* 2011; Götte *et al.* 2013). Furthermore, the Ge concentration decreases in the quartz cements of the Witten Formation from the second to the third generation of quartz cements.

Applications and future perspectives

It is of general interest for the prediction of the reservoir quality to improve the understanding of the processes that lead to the formation of pore-filling cements (Ajdukiewicz & Lander 2010). In addition, trace element analyses can particularly contribute to finding solutions to some special issues:

(1) Although quartz formation is commonly considered to be reaction-controlled, zoning in cements indicate that cement formation is not strictly uniform (Lehmann *et al.* 2009; Götte *et al.* 2013). Analysis of trace elements may help to clarify this variability and help to differentiate distinct sources for the silica.
(2) Inhomogeneities in reservoirs, which are related to the cementation, can probably be

Table 2. *Most abundant impurities in quartz and the processes that affect their occurrence in quartz cements*

Cation	Defect	Sources
Substituting cations		
Al^{3+}	$[AlO_4\|M]$	Feldspar dissolution, dissolution of Al-bearing unstable minerals equilibrium with clay minerals (kaolinite, illite) oxidation or thermal degradation of complexing organic ligands
Ge^{4+}	$[GeO_4]$	Feldspar dissolution kaolinite dissolution (pre-enrichment of Ge) enrichment in the porewater after formation of Ge-poor quartz
B^{3+}	$[BO_4\|H]$	smectite to illite transformation migrating brines hydrothermal water
(Ti^{4+})	$[TiO_4]$	not in authigenic quartz, only incorporated at high temperature
Interstitial cations		
H^+	$[AlO_4\|H]$	porewater, influenced by diagenetic reactions, Controlled by the prevailing buffering system
	H_2O	water molecules in sub-microscopic inclusions and on interstitial sites
Li^+	$[AlO_4\|Li]$	clay mineral reactions (smectite to illite transformation) migration of pre-evaporated brines
Na^+	$[AlO_4\|Na]$,	migration of brines, porewater
	Fluid inclusions	probably not incorporated in the quartz lattice, but in inclusions
(K^+)	$[AlO_4\|K]$	same as Na^+

better understood by using trace element analyses. Moreover, internal or external sources of silica for formation of quartz can be described.
(3) Where the cementation is influenced by fluid migration, this is likely to be reflected by the trace elements of the quartz overgrowths.

All these issues cannot be addressed by analysing only one proxy element, and thus not only one single method should be used. In most cases, a combination of imaging techniques (CL, element mapping) and trace element quantification by using mass spectrometry promises the most success. Further progress may be obtained when the trace element analyses are combined with the determination of isotope ratios. In addition to O isotopes, Li or Si isotopes are also relevant in analysing authigenic quartz. They could provide information on the origin of silicic acid, migration of fluids and formation temperatures.

A point that is poorly understood is the influence of Al on the reaction rates of quartz. High Al concentrations in aqueous solutions have been found to reduce the dissolution (and potentially precipitation) rate significantly (Bickmore *et al.* 2006) and will also result in quartz cements with high Al contents. Considering this effect will further improve the models that are used for reservoir quality prediction.

Conclusions

The trace element composition of authigenic quartz in sandstones is mainly controlled by the composition of the porewater and the crystal chemistry of quartz. Thus, diagenetic reactions that take place during quartz cementation leave markers in its chemical composition (Table 2). The key points of the trace element chemistry of quartz formed in sandstones are summarized as follows:

(1) Aluminium is incorporated in relatively high concentrations in quartz and forms combined defects with Li and H. Due to its wide occurrence, it is commonly the most important impurity that substitutes for Si.
(2) According to the mass action law, Al concentrations in quartz are linked to the activities in the porewater. Near-neutral to slightly acid pH values, which commonly prevail in natural porewater, are not favourable for high Al-concentrations because the minimum Al solubility lies in this pH range.
(3) Organic ligands can form soluble complexes with Al. Oxalic acid, for example, may increase the Al/Si ratio by several orders of magnitude. Due to oxidation or thermal degradation of the ligands, larger quantities of Al can be released to the porewater and produce a temporarily high Al/Si-ratio.
(4) Lithium and H compete for the charge-balancing interstitial site adjacent to an Al impurity. The activity ratio in the aqueous solution and the diagenetic conditions control their ratio in quartz cements. Transformation of smectite to illite and migration of pre-evaporated brines are the most prominent sources for Li, while the H activity (i.e. the pH) depends on the buffering system that prevails during diagenesis.
(5) Germanium in overgrowths is commonly slightly enriched compared with the crustal average in quartz. A weak correlation between Al and Ge can be frequently observed and is considered to indicate their common origin or their mobilization in the same diagenetic stage. The presence of excess of Ge in quartz requires pre-enrichment either by the formation of Ge-rich kaolonite during early stages of diagenesis and subsequent dissolution or the formation of Ge-poor quartz that results in an enrichment of Ge in the aqueous solution.

I thank K. Ramseyer (Bern) for helpful comments on the geochemistry of quartz and T. Pettke (Bern) and M. Seitz (Frankfurt) for discussions on technical details. Furthermore, I am grateful to R. Worden for all his efforts with the editorial handling and reviewing of this article and two referees for helpful and constructive comments that significantly improved the manuscript.

References

AJDUKIEWICZ, J. M. & LANDER, R. H. 2010. Sandstone reservoir quality predictions: the state of the art. *AAPG Bulletin*, **94**, 1083–1091.

AZAROUAL, M., FOUILLAC, C. & MATRAY, J. M. 1997. Solubility of silica polymorphs in electrolyte solutions, II. Activity of aqueous silica and solid silica polymorphs in deep solutions from the sedimentary Paris-Basin. *Chemical Geology*, **140**, 167–179.

BAMBAUER, H. U. 1961. Spurenelementgehalte und γ-Farbzentren in Quarzen aus Zerrklüften der Schweizer Alpen. *Schweizerische Mineralogisch und Petrographische Mitteilungen*, **41**, 335–369.

BERNSTEIN, L. R. 1985. Germanium geochemistry and mineralogy. *Geochimica et Cosmochimica Acta*, **49**, 2409–2422.

BICKMORE, B. R., NAGY, K. L., GRAY, A. K. & BRINKERHOFF, A. R. 2006. The effect of $Al(OH)_4^-$ on the dissolution rate of quartz. *Geochimica et Cosmochimica Acta*, **70**, 290–305.

BRUHN, F., BRUCKSCHEN, P., MEIJER, J., STEPHAN, A., RICHTER, D. K. & VEIZER, J. 1996. Cathodoluminescence investigations and trace-element analysis of quartz by micro-PIXE: implications for diagenetic and provenance studies in sandstone. *Canadian Mineralogist*, **34**, 1223–1232.

BÜKER, C. 1996. *Absenkungs-, Erosions- und Wärmeflußgeschichte des Ruhr-Beckens und des nordöstlichen Rechtsrheinischen Schiefergebirges*. PhD thesis, Forschungszentrum Jülich.

CASTET, S., DANDURAND, J.-L., SCHOTT, J. & GOUT, R. 1993. Boehmit solubility and aqueous aluminium speciation in hydrothermal solutions (90–350°C): experimental study and modelling. *Geochimica et Cosmochimica Acta*, **57**, 4869–4884.

CHAN, L.-H., STARINSKI, A. & KATZ, A. 2002. The behaviour of lithium and its isotopes in oil field brines: evidence from the Heletz-Kokhov field, Israel. *Geochimica et Cosmochimica Acta*, **66**, 615–623.

CROSSEY, L. J., LOUCKS, R. & TOTTEN, M. W. 1996. Siliciclastic diagenesis and fluid flow: concepts and applications. *SEPM Special Publication*, **55**, III.

DE JONG, B. H. W. S., SCHRAMM, C. M. & PARZIALE, V. E. 1983. Polymerisation of silicate and aluminate tetrahedra in glasses, melts, and aqueous solutions. IV. Aluminium coordination in glasses and aqueous solutions and comments on the aluminium avoidance principle. *Geochimica et Cosmochimica Acta*, **47**, 1223–1236.

DEMARS, C., PAGEL, M., DELOULE, E. & BLANC, P. 1996. Cathodoluminescence of quartz from sandstones: interpretation of the UV-range by determination of trace element distributions and fluid-inclusion P-T-X properties in authigenic quartz. *American Mineralogist*, **81**, 891–901.

EVANS, M. J. & DERRY, L. A. 2002. Quartz control of high germanium/silicon ratios in geothermal waters. *Geology*, **30**, 1019–1022.

FEIN, J. B. & HESTRIN, J. E. 1994. Experimental studies of oxalate complexation at 80°C: gibbsite, amorphous silica, and quartz solubilities in oxalate-bearing fluids. *Geochimica et Cosmochimica Acta*, **58**, 4817–4829.

FISCHER, C., WALDMANN, S. & VON EYNATTEN, H. 2013. Spatial variation in quartz cement type and concentration: An example from the Heidelberg formation (Teufelsmauer outcrops), Upper Cretaceous Subhercynian Basin, Germany. *Sedimentary Geology*, **29**, 48–61.

FONTES, J. CH. & MATRAY, J. M. 1993. Geochemistry and origin of formation brines from the Paris Basin, France. 1. Brines associated with Triassic salts. *Chemical Geology*, **109**, 149–175.

FÜCHTBAUER, H. 1974. Zur Diagenese fluviatile Sandsteine. *Geologische Rundschau*, **63**, 904–925.

GÖTTE, TH. 2004. *Petrographische und geochemische Untersuchungen zu den postvariszischen Mineralisationen des nordwestlichen Rechtsrheinischen Schiefergebirges unter besonderer Berücksichtigung der Kathodolumineszenz*. PhD thesis, Ruhr-University.

GÖTTE, TH., PETTKE, TH., RAMSEYER, K., KOCH-MÜLLER, M. & MULLIS, J. 2011. Cathodoluminescence properties and trace element signature of hydrothermal quartz: a fingerprint of growth dynamics. *American Mineralogist*, **96**, 802–813.

GÖTTE, TH., RAMSEYER, K., PETTKE, TH. & KOCH-MÜLLER, M. 2013. Implications of trace element composition of syntaxial quartz cements for the geochemical conditions during quartz precipitation in sandstones. *Sedimentology*, **60**, 111–117.

GRENTHE, I., PLYASUNOV, A. V. & SPAHIU, K. 1997. Estimation of medium effects on thermodynamic data. *In*: GRENTHE, I. & PUIGDOMENECH, I. (eds) *Modeling in Aquatic Chemistry*. OECD Nuclear Energy Agency, Issy-les-Moulineaux, France, 325–426.

HANOR, J. S. 2001. Reactive transport involving rock-buffered fluids of varying salinity. *Geochimica et Cosmochimica Acta*, **65**, 3721–3732.

HIATT, E. E., KYSER, T. K., FAYEK, M., POLITO, P., HOLK, G. J. & RICIPUTI, L. R. 2007. Early quartz cements and evolution of paleohydraulic properties of basal sandstones in three Paleoproterozoic continental basins: evidence from in situ ^{18}O analysis of quartz cements. *Chemical Geology*, **238**, 19–37.

KELLY, J. L., FU, B., KITA, N. T. & VALLEY, J. W. 2007. Optically continuous silcrete quartz cements of the St. Peter Sandstone: high precision oxygen isotope analysis by ion microprobe. *Geochimica et Cosmochimica Acta*, **71**, 3812–3832.

KLOPPMANN, W., NÉGREL, PH., CASANOVA, J., KLINGE, H., SCHELKES, K. & GUERROT, C. 2001. Halite dissolution derived brines in the vicinity of a Permian salt dome (N German Basin). Evidence from boron, strontium, oxygen and hydrogen isotopes. *Geochimica et Cosmochimica Acta*, **65**, 4087–4101.

KRAISHAN, G. M., RAZAEE, M. R. & WORDEN, R. H. 2000. Significance of trace element composition in quartz cement as a key to reveal the origin of silica in sandstones: an example from the Cretaceous of the Barrow Sub-basin, Western Australia. *In*: WORDEN, R. H. & MORAD, S. (eds) *Quartz Cements in Sandstones*. IAS, Special Publications, John Wiley & Sons, New York, **29**, 317–331.

KRONZ, A., VAN DEN KERKHOF, A. M. & MÜLLER, A. 2012. Analysis of Low Element Concentrations in Quartz by Electron Microprobe. *In*: GÖTZE, J. & MÖCKER, R. (eds) *Quartz: Deposits, Mineralogy and Analytics*. Springer, Heidelberg, 191–217.

KURTZ, A. C., DERRY, L. A. & CHADWICK, O. A. 2002. Germanium-silicon fractionation in the weathering environment. *Geochimica et Cosmochimica Acta*, **66**, 1525–1537.

LANDER, R. H., LARESE, R. E. & BONNELL, L. M. 2008. Toward more accurate quartz cement models: the importance of euhedral v. noneuhedral growth rates. *AAPG Bulletin*, **92**, 1537–1563.

LANDTWING, M. R. & PETTKE, T. 2005. Relationships between SEM-cathodoluminescence response and trace-element composition of hydrothermal vein quartz. *American Mineralogist*, **90**, 122–131.

LARSEN, R. B., HENDERSON, I., IHLEN, P. M. & JACAMON, F. 2004. Distribution and petrogenetic behaviour of trace elements in granitic pegmatite quartz from South Norway. *Contributions to Mineralogy and Petrology*, **147**, 615–628.

LEHMANN, K., BERGER, A., GÖTTE, TH., RAMSEYER, K. & WIEDENBECK, M. 2009. Growth related zonations in authigenic and hydrothermal quartz characterized by SIMS-, EPMA-, SEM-CL- and SEM-CC-imaging. *Mineralogical Magazine*, **73**, 633–643.

LEHMANN, K., PETTKE, TH. & RAMSEYER, K. 2011. Significance of trace elements in syntaxial quartz cement, Haushi Group sandstones, Sultanate of Oman. *Chemical Geology*, **280**, 47–57.

LESSMANN, B., WIEGAND, K. & SCHARPFF, H.-J. 2001. Die Hydrogeologie des vulkanischen Vogelsberges. *Geologische Abhandlungen Hessen*, **108**, 3–143.

LIBOWITZKY, E. & ROSSMAN, G. R. 1997. An IR absorption calibration for water in minerals. *American Mineralogist*, **82**, 1111–1115.

MCBRIDE, E. F. 1989. Quartz cementation in sandstones: a review. *Earth Science Reviews*, **26**, 69–112.

MCDONALD, D. A. & SURDAM, R. C. 1984. Clastic diagenesis. *AAPG Memoir*, **37**, IX.

MARCHAND, A. M. E., MACAULAY, C. K., HASZELDINE, R. S. & FALLICK, A. E. 2002. Pore water evolution in oilfield sandstones: constraints form oxygen isotope microanalyses of quartz cement. *Chemical Geology*, **191**, 285–304.

MERINO, E., HARVEY, C. & MURRAY, H. H. 1989. Aqueous-chemical control of the tetrahedral-aluminium content of quartz, halloysite, and other low-temperature silicates. *Clays and Clay Minerals*, **37**, 135–142.

MILLIKEN, K. L. 2005. Late diagenesis and mass transfer in sandstone-shale sequences. *In*: MACKENZIE, F. (ed.) *Sediments, Diagenesis, and Sedimentary Rocks*. Treatise on Geochemistry, **7**, 159–190.

MÖLLER, P., WEISE, S. M., TESMER, M., DULISKI, P., PEKDEGER, A., BAYER, U. & MAGRI, F. 2008. Salinisation of groundwater in the North German Basin: results from conjoint investigation of major, trace element and multi-isotope distribution. *International Journal of Earth Sciences*, **97**, 1057–1073.

MÜLLER, A. & KOCH-MÜLLER, M. 2009. Hydrogen speciation and trace element contents of igneous, hydrothermal and metamorphic quartz from Norway. *Mineralogical Magazine*, **73**, 569–583.

MÜLLER, A., WIEDENBECK, M., VAN DEN KERKHOFF, A. M., KRONZ, A. & SIMON, K. 2003. Trace elements in quartz a combined electron microprobe, secondary ion mass spectrometry, laser-ablation ICP-MS, and cathodoluminescence study. *European Journal of Mineralogy*, **15**, 747–763.

PERNY, B., EBERHARDT, P., RAMSEYER, K., MULLIS, J. & PANKRATH, R. 1992. Microdistribution of Al, Li and Na in a quartz: possible causes and correlation with short-lived cathodoluminescence. *American Mineralogist*, **77**, 534–544.

PETTKE, T. 2008. Analytical protocols for element concentration and isotope ratio measurements in fluid inclusions by LA-(MC)-ICP-MS. *In*: SYLVESTER, P. (ed.) *Laser Ablation ICP-MS in the Earth Sciences: Current Practices and Outstanding Issues*. Mineralogical Association of Canada, Vancouver, **40**, 189–218.

PREUSSER, F., CHITHAMBO, M. L. *ET AL*. 2009. Quartz as a natural luminescence dosimeter. *Earth-Science Reviews*, **97**, 196–226.

RAMSEYER, K., BAUMANN, J., MATTER, A. & MULLIS, J. 1988. Cathodoluminescence colours of α-quartz. *Mineralogical Magazine*, **52**, 669–677.

RAMSEYER, K., AMTHOR, J. E., SPÖTL, CH., TERKEN, J. M. J., MATTER, A., VROON-TEN HOVE, M. & BORGOMANO, J. R. F. 2004. Impact of basin evolution, depositional environment, pore water evolution and diagenesis on reservoir-quality of Lower Paleozoic Haima Supergroup sandstones, Sultanate of Oman. *GeoArabia*, **9**, 107–138.

REZAEE, M. & TINGATE, P. R. 1997. Origin of quartz cement in Tirrawarra Sandstone, Southern Cooper Basin, South Australia. *Journal of Sedimentary Research*, **67**, 168–177.

ROSA, A. L., EL-BARBARY, A. A., HEGGIE, M. I. & BRIDDON, P. R. 2005. Structural and thermodynamic properties of water related defects in a-quartz. *Physics and Chemistry of Minerals*, **32**, 323–331.

ROSSMAN, G. R. 1994. Colored Varieties of Silica Minerals. *In*: HEANEY, P. J., PREWITT, C. T. & GIBBS, G. V. (eds) *Silica – Physical behaviour, Geochemistry and materials Application*. Reviews in Mineralogy, Mineralogical Society of America, Washington, DC, **29**, 433–367.

ROSSMAN, G. R. 2006. Analytical Methods for measuring water in nominally anhydrous minerals. *Reviews in Mineralogy and Geochemistry*, **62**, 1–28.

RUSK, B. G., LOWERS, H. A. & REED, M. H. 2008. Trace elements in hydrothermal quartz: relationships to cathodoluminescence textures and insights into vein formation. *Geology*, **36**, 547–550.

STENINA, N. G. 2004. Water-related defects in quartz. *Bulletin of Geosciences*, **79**, 251–268.

STUMM, W. & MORGAN, J. J. 1996. *Aquatic Chemistry: Chemical Equilibria and Rates in Natural Waters*. Wiley, Cambridge.

THOMAS, S.-M., KOCH-MÜLLER, M., REICHART, P., RHEDE, D., THOMAS, R., WIRTZ, R. & MATSYUK, S. 2009. IR-calibrations for water determination in olivin, r-GeO_2 and SiO_2-polymorphs. *Physics and Chemistry of Minerals*, **35**, 489–509.

WALDERHAUG, O. 1996. Kinetic modeling of quartz cementation and porosity loss in deeply buried sandstone reservoirs. *AAPG Bulletin*, **80**, 731–745.

WARK, D. A. & WATSON, E. B. 2006. TitaniQ: a titanium in quartz geothermometer. *Contributions to Mineralogy and Petrology*, **152**, 743–754.

WEBER, J. & RICKEN, W. 2005. Quartz cementation and related sedimentary architecture of the Triassic Solling Formation, Reinhardswald Basin, Germany. *Sedimentary Geology*, **175**, 459–477.

WEDEWARDT, M. 1995. Hydrochemie und Genese der Tiefenwässer im Ruhr-Revier. *DMT-Berichte aus Forschung und Entwicklung*, **39**, 1–170.

WESOLOWSKI, D. J. & PALMER, D. A. 1994. Aluminium speciation and equilibria in aqueous solution: V. Gibbsite solubility at 50°C and pH 3–9 in 0.1 molal NaCl solutions (a general model for aluminium speciation; analytical methods). *Geochimica et Cosmochimica Acta*, **58**, 2947–2969.

WILLIAMS, L. B. & HERVIG, R. L. 2005. Lithium and boron isotopes in illite-smectite: the importance of crystal size. *Geochimica et Cosmochimica Acta*, **69**, 5705–5716.

WILLIAMS, L. B., HERVIG, R. L. & BJORLYKKE, K. 1997. New evidence for the origin of quartz cements in hydrocarbon reservoirs revealed by oxygen isotope microanalyses. *Geochimica et Cosmochimica Acta*, **61**, 2529–2538.

WORDEN, R. H. & MORAD, S. 2000. Quartz cementation in oil field sandstones: a review of the key controversies. *In*: WORDEN, R. H. & MORAD, S. (eds) *Quartz Cementation in Sandstones*. IAS, Special Publications, John Wiley & Sons, New York, **29**, 1–20.

Comparing clay mineral diagenesis in interbedded sandstones and mudstones, Vienna Basin, Austria

SUSANNE GIER[1]*, RICHARD H. WORDEN[2] & PETER KROIS[3]

[1]*Department of Geodynamics and Sedimentology, University of Vienna, 1090 Vienna, Austria*

[2]*Department of Earth, Ocean and Ecological Sciences, University of Liverpool, Liverpool L69 3GP, UK*

[3]*OMV Exploration and Production GmbH, 1020 Vienna, Austria*

**Correspondence: susanne.gier@univie.ac.at*

Abstract: There is no consensus about the rate and style of clay mineral diagenesis in progressively buried sandstones v. interbedded mudstones. The diagenetic evolution of interbedded Miocene sandstones and mudstones from the Vienna Basin (Austria) has therefore been compared using core-based studies, petrography, X-ray diffraction and X-ray fluorescence. There was a common provenance for the coarse- and fine-grained sediments, and the primary depositional environment of the host sediment had no direct effect on illitization. The sandstones are mostly lithic arkoses dominated by framework grains of quartz, altered feldspars and carbonate rock fragments. Sandstone porosity has been reduced by quartz overgrowths and calcite cement; their pore-filling authigenic clay minerals consist of mixed-layer illite–smectite, illite, kaolinite and chlorite. In sandstones, smectite illitization progresses with depth; at 2150 m there is a transition from randomly interstratified to regular interstratified illite–smectite. The overall mineralogy of mudstones is surprisingly similar to the sandstones. However, for a given depth, feldspars are more altered to kaolinite, and smectite illitization is more advanced in sandstones than in mudstones. The higher permeability of sandstones allowed faster movement of material and pore fluid necessary for illitization and feldspar alteration than in mudstones. The significance of this work is that it has shown that open-system diagenesis is important for some clay mineral diagenetic reactions in sandstones, while closed-system diagenesis seems to operate for clay mineral diagenesis in mudstones.

Although there have been numerous separate studies of both sandstone and mudstone burial diagenesis, relatively little research has been carried out or documented on the comparative clay mineralogical relationships of interbedded sandstones and mudstones. For example, existing publications have not revealed a systematic pattern for the occurrence and rates of smectite illitization in interbedded sandstones and mudstones (McKinley *et al.* 2003). Three different smectite illitization scenarios have been proposed: (i) illitization occurs at a similar depth in sandstones and mudstones (Hugget 1996); (ii) illitization occurs at a slower rate in sandstones than in mudstones (Boles & Franks 1979; Howard 1981; Niu & Ishida 2000); and (iii) illitization occurs at a faster rate in sandstones than in mudstones (Hillier *et al.* 1996; Ko & Hesse 1998). The last scenario has been mainly attributed to higher permeabilities in sandstones compared to mudstones and/or the effects of circulating higher temperature fluids, suggesting the need to understand the effects of an open system v. a closed system on changes in clay mineralogy.

The illitization of smectite is controlled by a number of factors, including temperature, pressure, porosity and permeability (McKinley *et al.* 2003). Illitization also requires the availability of potassium in the pore fluid (Weibel 1999), which typically is assumed to have been derived from the dissolution or alteration of K-feldspar or muscovite (Hower *et al.* 1976). Differences in reaction rates of specific smectite compositions are also considered to play an important role (McKinley *et al.* 2003).

One of the products of the smectite to illite transformation in mudstones is quartz, which can be precipitated as microcrystalline quartz within the mudstone matrix (Peltonen *et al.* 2009; Thyberg *et al.* 2010; Thyberg & Jahren 2011) in closed systems or as cements in adjacent sandstones in open systems (Lynch *et al.* 1997; Thyne 2001).

The primary composition of both the sandstones and the mudstones plays an important role not only for illitization of smectites but also for the authigenesis of kaolinite. Arkosic sandstones have a higher potential to produce kaolinite cement than quartz arenites (Worden & Morad 2003).

From: Armitage, P. J., Butcher, A. R., Churchill, J. M., Csoma, A. E., Hollis, C., Lander, R. H., Omma, J. E. & Worden, R. H. (eds) 2018. *Reservoir Quality of Clastic and Carbonate Rocks: Analysis, Modelling and Prediction.* Geological Society, London, Special Publications, **435**, 389–403.
First published online November 20, 2015, https://doi.org/10.1144/SP435.9

Understanding the reaction from smectite to illite is of economic importance, especially for the petroleum industry, because sandstones that contain smectite normally have a poor reservoir quality. In some cases, sandstone reservoir quality might be improved by the illitization of smectite because illite has a lower specific surface area than smectite (Bjørlykke & Jahren 2010). In fine-grained facies, the change from smectite to illite also causes an increase in rock density and seismic velocity, and this constitutes the first seismically important mineral reaction in mudstones. The smectite to illite reaction has several implications for mudstones. Weak smectite is replaced by stiffer illite. The reaction creates free water and the volume of solids is decreased because illite has a higher density than smectite. Quartz is produced as a by-product and porosity may be reduced by chemical compaction (Avseth 2010). All of these have an effect on seismic signatures, and must be considered as input controls and parameters for rock physics modelling.

Extensive geological research has been carried out in the Vienna Basin (Fig. 1) for more than 150 years (Sauer *et al.* 1992). Oil and gas exploration activities started in the early twentieth century and the first hydrocarbon discoveries were made in the early 1930s. Since then, more than 6000 wells have been drilled (Arzmüller *et al.* 2006) and significant parts of the basin have been covered with two-dimensional (2D) and, more recently, 3D seismic data. This extensive dataset over the basin has permitted detailed studies of the structural evolution and stratigraphy of the basin and its depositional environments, and also the diagenesis of the sedimentary basin fill.

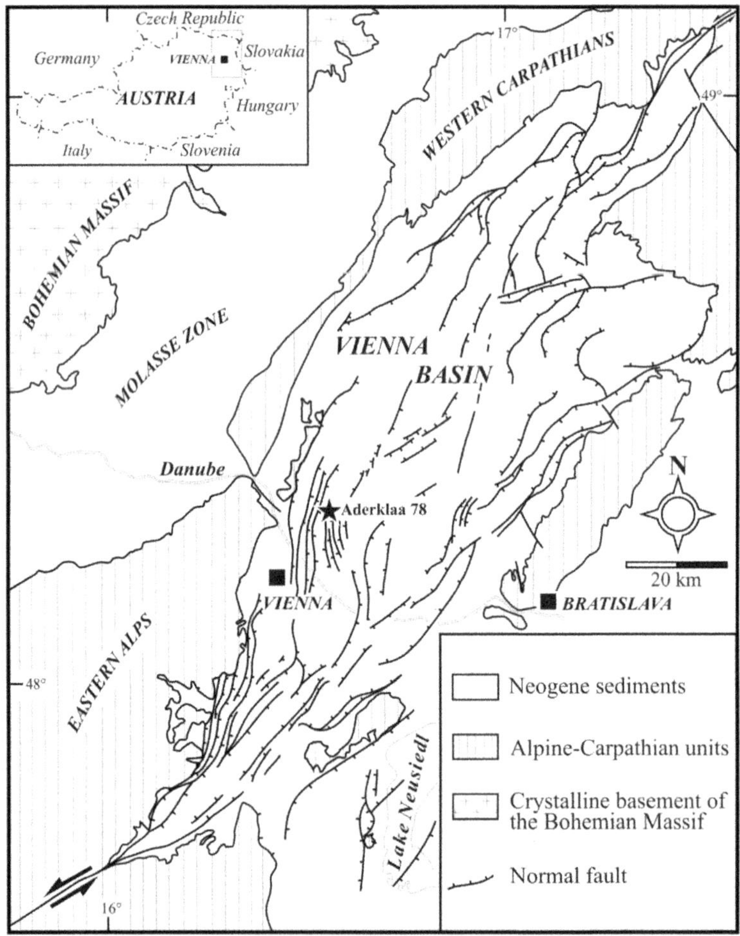

Fig. 1. Location map of the Vienna Basin and surrounding areas (adapted from Wagreich & Schmid 2002). The studied Aderklaa 78 well is highlighted.

The Aderklaa 78 well (Figs 1 & 2), utilized in this study, was drilled by OMV Aktiengesellschaft in 1958. This well had a comprehensive data acquisition programme, which, in addition to wireline logging, included the coring of long intervals. The Aderklaa 78 well reached its final depth of 2851 m in Upper Triassic dolomites of the Alpine–Carpathian fold and thrust belt. Extensive coring was carried out in the Neogene basin fill, which here has allowed a detailed study of the diagenetic evolution of the interbedded sandstones and mudstones, and an assessment of the controlling parameters, such as temperature, depth, depositional environment and primary lithologies.

The specific objectives of this study are to: (1) characterize the diagenesis of the sandstones and mudstones; (2) compare clay mineral evolution in progressively buried interbedded sandstones and mudstones; and (3) document lithological influences and the control of the depositional environment on progressively modified clay minerals, especially on mixed-layer illite–smectite.

Geology

The Vienna Basin (Fig. 1) is located in the NE part of Austria, extending into Slovakia and the Czech Republic. The NE–SW-trending Vienna Basin has a rhomboidal shape, and is approximately 200 km long and 40 km wide. It is a classic pull-apart basin and formed along a sinistral fault system during the lateral extrusion of the Eastern Alps (Royden 1985). Maximum sediment thickness in the depocentres is up to 5500 m. The present-day geothermal gradient in the Vienna Basin is 30°C km^{-1} (Sachsenhofer 2001).

During the first phase of its evolution in the Early Miocene, the Vienna Basin formed as a piggy-back basin on top of the thrusted and imbricated nappes of the Alpine–Carpathian fold and thrust belt. The second phase of basin formation, the classic pull-apart phase, started in the Middle Miocene and continued until the Late Miocene, where east–west compression led to basin inversion (Decker & Peresson 1996). This was followed

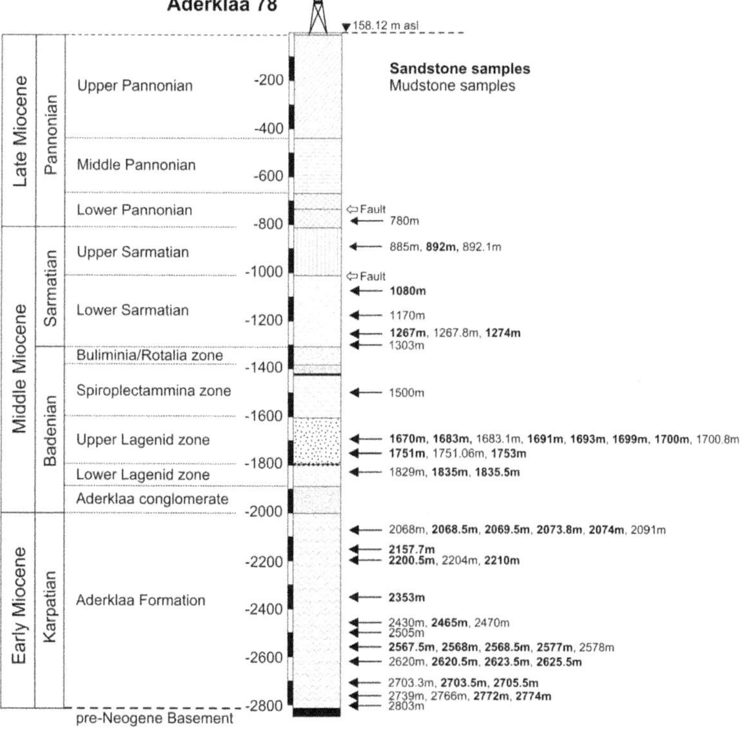

Fig. 2. Stratigraphic sequence of Miocene sediments from the Aderklaa 78 well. Depositional environments range from limnic-fluviatile in the Karpatian, marine in the Badenian, brackish in the Sarmatian to limnic in the Pannonian. These are stages of the Central Paratethys and correspond to the Burdigalian–Tortonian stages of the international timescale. The third column from the left gives the local lithostratigraphic units. The positions of studied sandstone samples are indicated in bold; mudstone samples in normal font. Depths are given as relative depth in m (below kelly bushing).

by east–west extension from the Pleistocene to present times.

The Neogene basin, and the thrusted and imbricated units of the Northern Calcareous Alps and the Flysch, are underlain by an autochthonous, passive-margin sequence. There, crystalline basement is overlain by a predominantly Jurassic sedimentary section. Thin Cretaceous and Palaeogene sediments unconformably overlie this section. The main hydrocarbon source rock of the Vienna Basin, marls of the Malmian Mikulov Formation, belongs to this sequence. Triggered by thrusting, the source rock entered the oil window in the Early Miocene. Its further maturation was controlled by local subsidence of the Neogene basin. Depending on its location, the present-day maturity of the source rock ranges from being in the oil window to overmature. The present-day depth of oil generation is between 4000 and 6000 m (Ladwein 1988).

Neogene sedimentation in the Aderklaa 78 well started with the Lower Miocene, Karpatian sediments of the Aderklaa Formation. These sediments are predominantly mudstones and marls, and are interpreted to be of limnic-fluviatile to lacustrine origin (Seifert 1996). An increase in sand content towards the top of the Aderklaa Formation reflects the progradation of a delta system. An inversion phase, coinciding with a change from piggy-back to rhombic pull-apart basin development at the end of the Early Miocene, stopped sedimentation and caused uplift and erosion (Hölzel *et al.* 2010). Subsequently, braided river sediments of the 'Aderklaa Conglomerate' were deposited on the peneplained surface.

Afterwards, fully marine conditions were established as the Vienna Basin connected with the Paratethys via a seaway through the Pannonian Basin. Throughout most of the Middle Miocene (Badenian), fully marine conditions predominated. Depositional environments alternated from shaly and marly basin floors to sandy delta-front and slope environments. Prograding and retrograding delta systems caused the alternation of sandstones and mudstones. A regional sea-level highstand was reached during 'Spiroplectammina' times, when carbonate deposition took place on regional highs. In Aderklaa 78, the lower two units of the Badenian, the Upper and Lower Lagenid zones, reveal a significant influx of sand, whereas the upper two units, the Buliminia Rotalia and the Spiroplectammina zones, are dominated by shaly and marly sediments. Regionally, delta progradation culminated at the Badenian–Sarmatian boundary. Depositional environments for the Sarmatian remained basinal to delta front and slope. However, conditions changed from open marine to restricted marine and brachyhaline. The basin reached its maximum extent and subsidence slowed down (Seifert 1996). During the Pannonian, the basin got completely cut off from the Paratethys and an extensive freshwater lake developed. Pannonian sediments were deposited in delta-front to delta-plain environments.

Net sand in Aderklaa 78, derived from a spontaneous potential (SP) log, reaches maximums in the Upper and Lower Lagenid zones with 22 and 65%, respectively. In contrast, sediments of the Aderklaa Formation, the Sarmatian and the lower Pannion have sandstone–mudstone ratios of 8–13%. No sandstones can be identified on the SP log in the Buliminia Rotalia and the Spiroplectammina zones.

Materials and methods

Samples were taken from cores that were cut from Miocene sediments at depths of between 780 and 2803 m in the Aderklaa 78 well (Fig. 2). Resistivity and SP logs were available for the whole well. Temperature measurements were taken from the surface to 1999 m. In addition, macroscopic core descriptions and compiled cutting descriptions could be accessed.

Thirty-four sandstone samples and 23 samples from intercalated mudstone intervals were available. Special emphasis was given to 23 samples of sandstone–mudstone pairs where both lithologies were in close vertical proximity, with distances varying between 0.1 and 1 m. The samples were compared using thin-section microscopy, X-ray diffraction (XRD) analysis, X-ray fluorescence analysis and scanning electron microscopy (SEM).

Thin sections of the sandstones and mudstones were impregnated with blue resin to highlight porosity. In addition, they were stained with Alizarin Red S and K-ferricyanide to aid carbonate mineral detection.

To separate the clay fraction (<2 μm) from the sandstones and mudstones, the inner part of the core was first disaggregated manually and the organic material removed with H_2O_2, and then treated with a 400 W ultrasonic probe for 3 min. The outer part of the core was discarded to avoid contamination from the drilling mud. Atterberg cylinders were used to separate the <2 μm fraction; the <0.2 μm fraction was obtained by centrifugation (Tanner & Jackson 1947).

Bulk and clay fraction mineralogies of the powdered samples were ascertained by XRD using a Panalytical X'Pert PRO diffractometer (Cu Kα radiation, 40 kV, 40 mA, step size 0.0167, 5 s per step). Semi-quantitative mineral estimates of the bulk sample were made using the method of Schultz (1964), which has error limits of $\pm 10\%$. Orientated samples of the <2 μm fraction for XRD were made by dispersing 8 mg of clay in 1 ml of water, pipetting 1 ml of suspension onto a glass slide and drying at room temperature. The orientated mounts were analysed in an air-dried state and after vapour solvation with ethylene glycol (EG) at 60°C for 12 h. The proportions of smectite and illite in the illite–smectite phase were determined using the 2-theta method described by Moore & Reynolds (1997).

The major-element composition of the sandstone and mudstone samples was analysed using a Philips PW 2400 X-ray fluorescence spectrometer.

Morphology of authigenic minerals was studied with a Philips XL 30 ESEM scanning electron microscope on fractured, gold-coated surfaces of sandstones.

Results

For the present paper, results of studies of sandstone (Gier *et al.* 2008) and mudstone diagenesis (Kurzweil & Johns 1981; Horton *et al.* 1985) of the Aderklaa 78 well were integrated with new results from 23 samples of sandstone–mudstone pairs from similar depths.

Sandstones

Mineralogy and petrology. The fine- to coarse-grained sandstones (Fig. 3b, c) can be classified from point-count data as predominately lithic arkoses to feldspathic litharenites, with subordinate subarkoses (Gier *et al.* 2008) and an average sandstone composition of Q_{60} F_{22} L_{18}. Detrital grains comprise monocrystalline quartz, feldspar, limestone, dolomite and some ductile lithic grains. The feldspars are mostly K-feldspars and show variable degrees of alteration. The replacement of feldspars by kaolinite is common. Intergranular pores are partly filled by calcite cement, quartz overgrowths, clay minerals (illite, illite–smectite, kaolinite and chlorite) and some minor pyrite (Gier *et al.* 2008).

Fig. 3. Thin-section photomicrographs, all plane-polarized light: porosity is indicated in light blue. (**a**) Mudstone sample 892 m containing a silt layer: flakes of mica, grains of quartz, calcite (red) and Fe-dolomite (blue) within the brownish clayey matrix and the silt layer are discernible. (**b**) Sandstone sample 2620.5 m predominantly composed of quartz, feldspar and calcite (red) grains. The calcite grains dissolve and reprecipitate *in situ* forming a patchy cement. (**c**) Mudstone intercalation within sandstone (sample from 1274.1 m). (**d**) SEM image of a pore-filling kaolinite (sample from 2353 m). Qtz, quartz; Glt, glauconite; Cal, calcite; Kln, kaolinite.

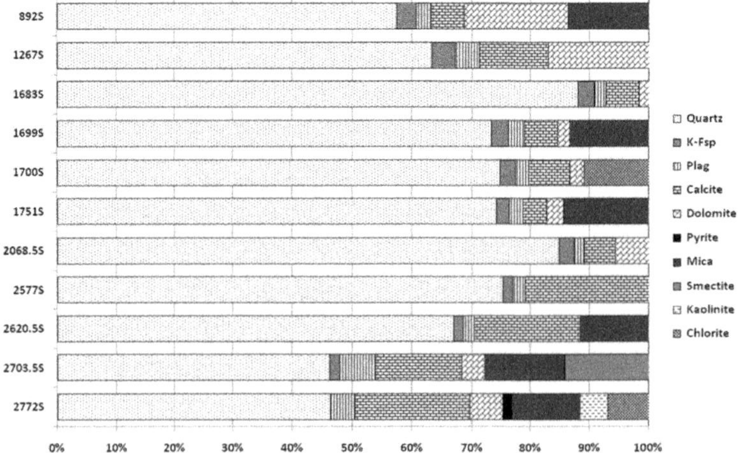

Fig. 4. Bulk mineralogy of sandstone samples from XRD: S at the end of the depth value stands for sandstone. K-Fsp, K-feldspar; Plag, plagioclase.

Bulk mineralogy from XRD (Fig. 4) gives a range of 46–88% quartz, 4–8% feldspar, 4–21% calcite, 0–18% dolomite and 0–28% clay minerals. The majority of the bulk samples contain mica (0–14%), and only some bulk samples contain smectite, kaolinite and chlorite.

In the following, we use the term mica for well-crystallized detrital 10 Å minerals (mostly muscovite) occurring in the bulk samples and illite for the authigenic 10 Å mineral found in the clay fraction (less than 2 μm).

Authigenic clay cements in the pores of the sandstones are illite, mixed-layer illite–smectite, kaolinite and chlorite (Fig. 5). Expandable smectite, as part of the mixed-layer illite–smectite, decreases with depth, from 76% in the shallowest (892 m) to 20% in the deepest sample (2774 m). The quantification of the clay minerals (Fig. 6a–d) gives 8–54% illite–smectite, 33–71% illite, 3–37% kaolinite and 5–21% chlorite. Kaolinite shows increasing abundance at depths greater than 2000 m (Figs 3d, 5 & 6c), in accordance with kaolinite data collected with Fourier transform infrared spectroscopy (FTIR) (Gier *et al.* 2008, fig. 8d). The illitization of smectite is also expressed by the transition of a randomly interstratified (R0) to regular interstratified (R1 ordering) I/S mixed layer at a depth of 2150 m. The percentage of illite layers in the illite–smectite mixed layer increases, as in the mudstones, from around 24% at 892 m to 80% at 2774 m (Fig. 5).

Chemistry. The bulk chemical composition of the sandstones (Fig. 7) reflects their mineralogy (compare Figs 4–7). They contain 55–84% SiO_2, 2–8% Al_2O_3, 1.3–3.6% Fe_2O_3, 0.5–4.9% MgO, 4–13% CaO, 0.4–1.34% Na_2O and 0.7–1.4% K_2O. The sandstones have been geochemically classified based on Herron (1988). Most sandstone samples plot into the fields of sublitharenites, with lesser subarkoses and litharenites (Fig. 8): this is broadly in agreement with results from point-counting. It has to be mentioned that geochemical data were only available for sandstones from the sandstone–mudstone pairs and that they only constitute a subset of the overall sandstone samples.

Mudstones

Mineralogy and petrology. By grain size, the mudstones can be characterized as clayey siltstones (Kurzweil & Johns 1981). Bedding-parallel muscovite flakes and silt-sized quartz and carbonate grains can be identified in thin sections of mudstones (Fig. 3a, c). Bulk mineralogy of mudstones was determined by XRD (Fig. 9). The constituent minerals are broadly similar to the sandstones but the relative abundance of individual minerals is different. As in the sandstones, the main components are quartz (16–37%), feldspar (3–10%), calcite (7–20%), dolomite (2–13%) and clay minerals (35–69%). Amongst these, mica is the most abundant (17–32%), followed by smectite (0–30%), chlorite (0–15%) and kaolinite (0–9.4%).

The clay fraction (<2 μm) of the mudstones is broadly similar to the sandstones at any given depth and the clay fraction shows similar changes with depth (Fig. 6). The only notable exception is kaolinite (Fig. 6c). The mudstones contain 5–62% illite–smectite, 26–68% illite, 3–13% kaolinite and 3–16% chlorite. The fine clay fraction (<0.2 μm) of the mudstones is dominated by

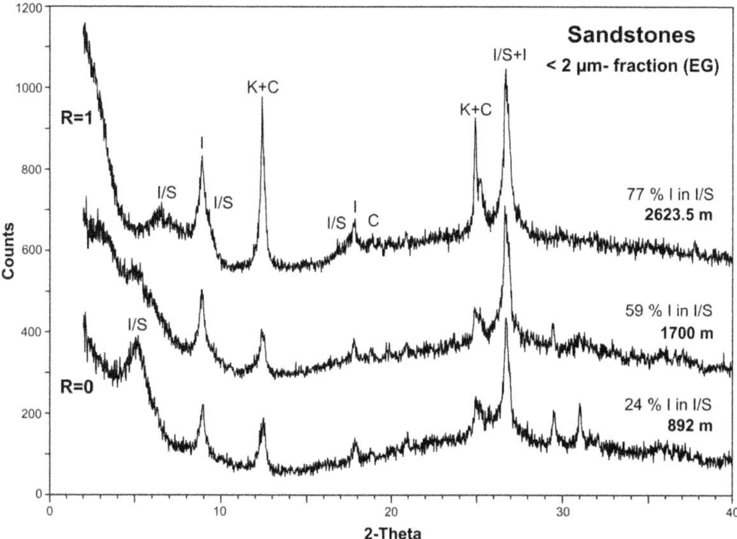

Fig. 5. X-ray diffraction patterns of orientated, ethylene glycol (EG)-saturated clay fractions of sandstones from different depths. Illitization of smectite is increasing with depth, kaolinite is dominant in the deeper samples. I/S, illite–smectite; I, illite; K, kaolinite; C, chlorite; R0, randomly interstratified; R1, ordered.

mixed-layer illite–smectite. The main clay mineral transformation with depth, the change from smectite to illite involving mixed-layer illite–smectite intermediates, is also observed. The percentage of illite layers in the illite–smectite mixed layer increases, as in the sandstones, from 25% at 780 m to 84% at 2803 m (Fig. 10).

Chemistry. In contrast to their mineralogical composition (Fig. 9), the bulk chemical composition of the mudstones is broadly homogeneous (Fig. 11). They contain 45–54% SiO_2, 10–17% Al_2O_3, 4–6% Fe_2O_3, 2.3–4.4% MgO, 6–12% CaO, 0.7–1.2% Na_2O and 2.2–3.5% K_2O. The loss on ignition varies between 10 and 18%. When geochemically classified, the mudstones plot into the shale and wacke fields (Herron 1988) (Fig. 8), and show only a minor scatter.

Based on literature data, the total organic carbon (TOC) values of the mudstones vary from 0.22 to 2.04%, with a mean value of 0.81% (Johns & Hoefs 1985).

Discussion

Depth-related variations in bulk mineralogy and geochemistry

A number of indices for sandstones and mudstones have been calculated and plotted to better highlight changes in mineralogy and chemistry. The majority of them show no significant variations with depth in either the sandstones or the mudstones (e.g. $TiO_2/(TiO_2 + Al_2O_3)$, $TiO_2/(TiO_2 + Fe_2O_3)$, quartz/(quartz + K-feldspar + plagioclase), $Fe_2O_3/(Fe_2O_3 + Al_2O_3)$ and $MnO/(MnO + Fe_2O_3)$).

Al_2O_3 and TiO_2 are refractory oxides and are highly resistant to weathering. They are also particularly immobile during diagenesis because they are relatively insoluble in common pore fluids (Andersson & Worden 2004; Andersson *et al.* 2004). Changes in the ratios of these elements might, thus, be indicative of changes in sediment provenance. Absolute amounts of TiO_2 and Al_2O_3 may vary because of dilution by quartz, calcite or dolomite. For this reason, a ratio of the two has been used. The $TiO_2/(TiO_2 + Al_2O_3)$ ratio is nearly constant for the studied mudstones (Fig. 12a), indicating a broadly uniform sediment provenance. Variations of this ratio in the sandstones are caused by local differences in the feldspar content that have an effect on Al_2O_3 (Fig. 12a).

Similarly, the ratio of quartz/(quartz + K-feldspar + plagioclase) (Fig. 12b) remains constant with depth and supports the concept of a uniform sediment provenance.

The $MgO/(MgO + Fe_2O_3)$ index shows a significant increase upsection (with decreasing depth) in both mudstones and sandstones (Fig. 12c). This coincides with a decrease in the calcite/(calcite + dolomite) ratio in both sandstones and mudstones (Fig. 12d). This relative decrease of calcite v. dolomite upsection indicates both a change in sediment input and a diagenetic overprint. This is

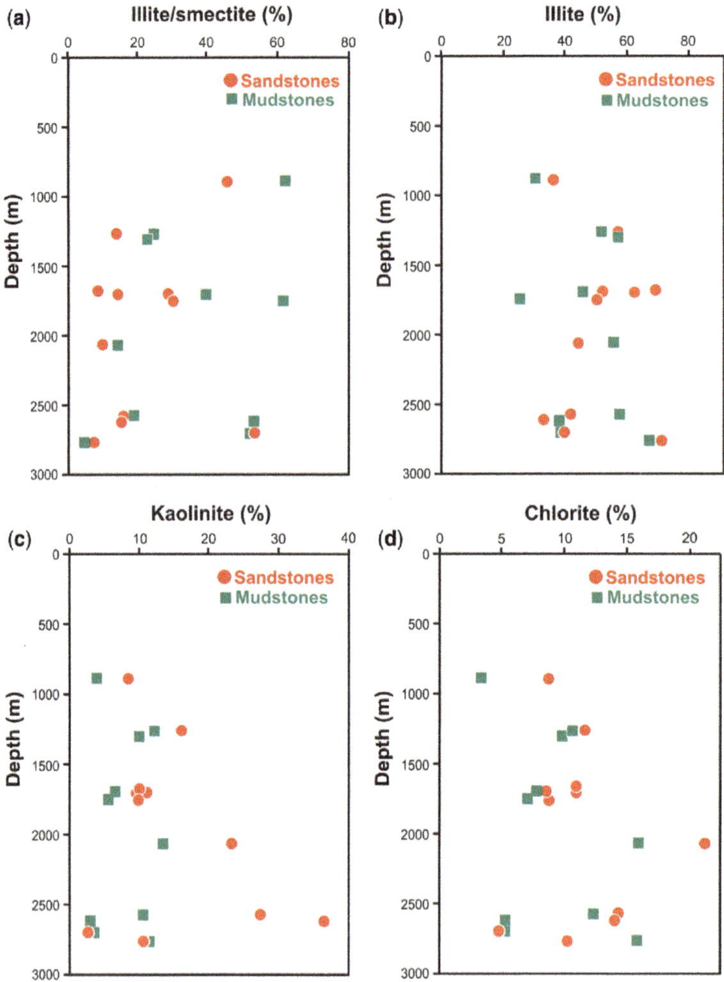

Fig. 6. Comparison of clay mineral distribution in sandstones (red dots) and mudstones (green squares) with depth: (**a**) illite–smectite; (**b**) illite; (**c**) kaolinite; and (**d**) chlorite.

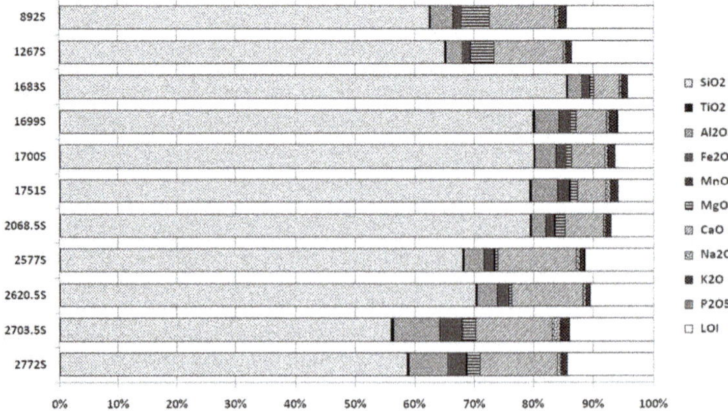

Fig. 7. Bulk chemistry of sandstone samples. LOI, loss on ignition; S after the depth value stands for sandstone.

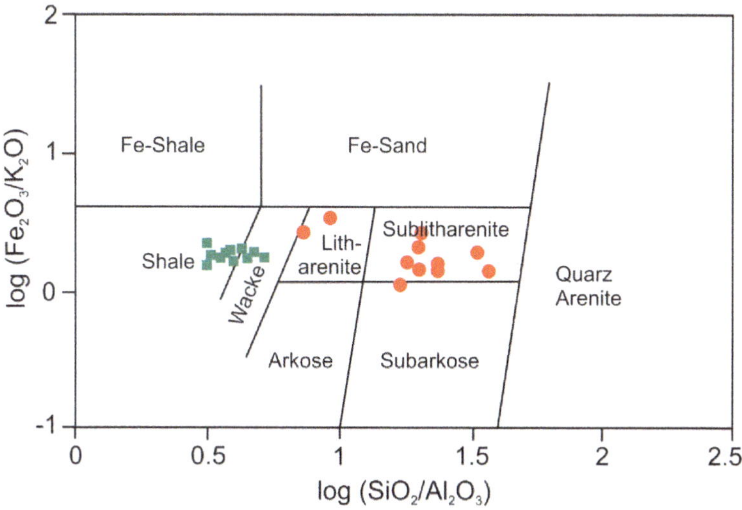

Fig. 8. Geochemical classification of the Aderklaa samples after Herron (1988). The mudstones (squares) plot in the shale to wacke fields; the sandstones (circles) in the sublitharenite to litharenite fields.

in accordance with microscopic observations in the sandstones, where a larger amount of dolomite clasts relative to calcite clasts can be seen in the shallower samples. The deeper samples contain more calcite than dolomite clasts and, in addition, are calcite cemented.

From the base of the section, the $Na_2O/(Na_2O + K_2O)$ ratio for sandstone decreases upwards, while the ratio increases upwards in mudstones, up to about 1750 m (Fig. 12e). In the shallower section, the $Na_2O/(Na_2O + K_2O)$ ratio stays constant, and is approximately the same for the sandstones and mudstones (Fig. 12e). The sandstones seem to have lost K_2O in the deeper section relative to the shallow section (Fig. 12f), presumably because of the dissolution of K-feldspar in the older and high-temperature sediments. This change coincides with the depth interval where the authigenic kaolinite is seen to increase (Fig. 6c). In order to establish whether this K_2O had an effect on the illitization of smectites in adjacent mudstones, the ratios of $Na_2O/(Na_2O + Al_2O_3)$ and $K_2O/(K_2O + Al_2O_3)$ were used (Fig. 12f). While Na_2O increases only slightly upsection in the mudstones and stays

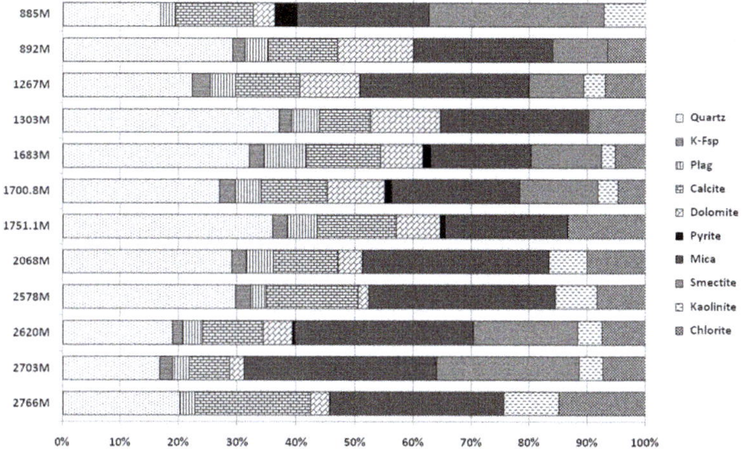

Fig. 9. Bulk mineralogy of mudstones from XRD; M after the depth value stands for mudstone. K-Fsp; K-feldspar; Plag, plagioclase.

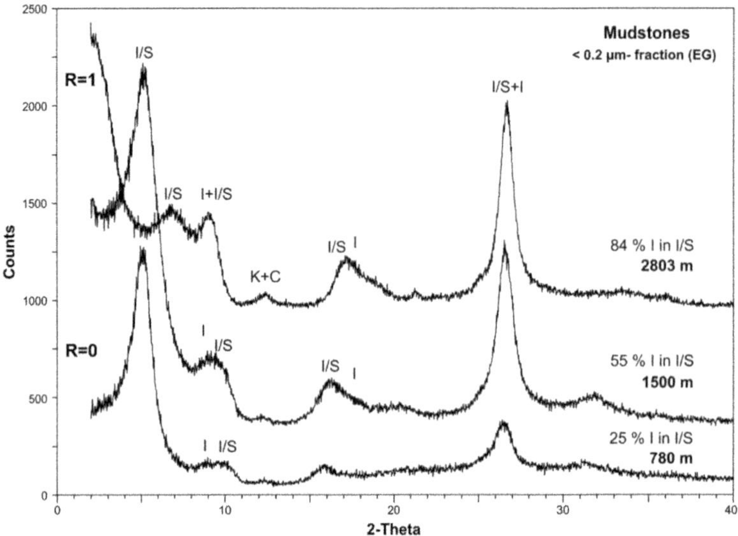

Fig. 10. X-ray diffraction patterns of orientated, ethylene glycol (EG)-saturated fine clay fractions of mudstones from different depths. Illitization of smectite is increasing with depth. I/S, illite–smectite; I, illite; K, kaolinite; C, chlorite; R0, randomly interstratified; R1, ordered.

constant with depth for the sandstones, the increase in K_2O upsection in the sandstones is significant. However, K_2O is roughly constant with depth for the mudstones (Fig. 12f). This suggests that the sandstones did not contribute K_2O to the mudstone for illitization.

The chemical index of alteration (CIA $= (Al_2O_3/(Al_2O_3 + CaO^* + Na_2O + K_2O))$: Nesbitt & Young 1982) can be an indicator of major-element changes resulting from weathering and the conversion of feldspars and other labile components to clay minerals. For mudstones, it has previously been used to characterize the degree of weathering of the sediment-source region and varying hinterland terrains (Andersson & Worden 2004). The CIA can also be interpreted to represent the degree of diagenetic alteration and material loss from the sandstones during diagenesis. To remove the effects of the high carbonate content of the Aderklaa sediments, the CIA has been calculated without CaO. The Aderklaa sandstone samples show decreased alteration upsection (Fig. 12g). In unburied sediments that are not diagenetically altered, this would be interpreted as a loss of alkali elements by increased weathering: in our case, this 'CIA' index is interpreted to reflect the greater degree of

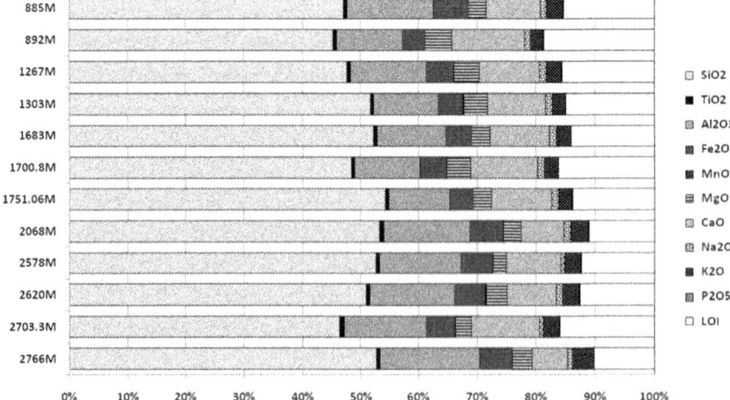

Fig. 11. Bulk chemistry of mudstone samples. LOI, loss on ignition; M after the depth value stands for mudstone.

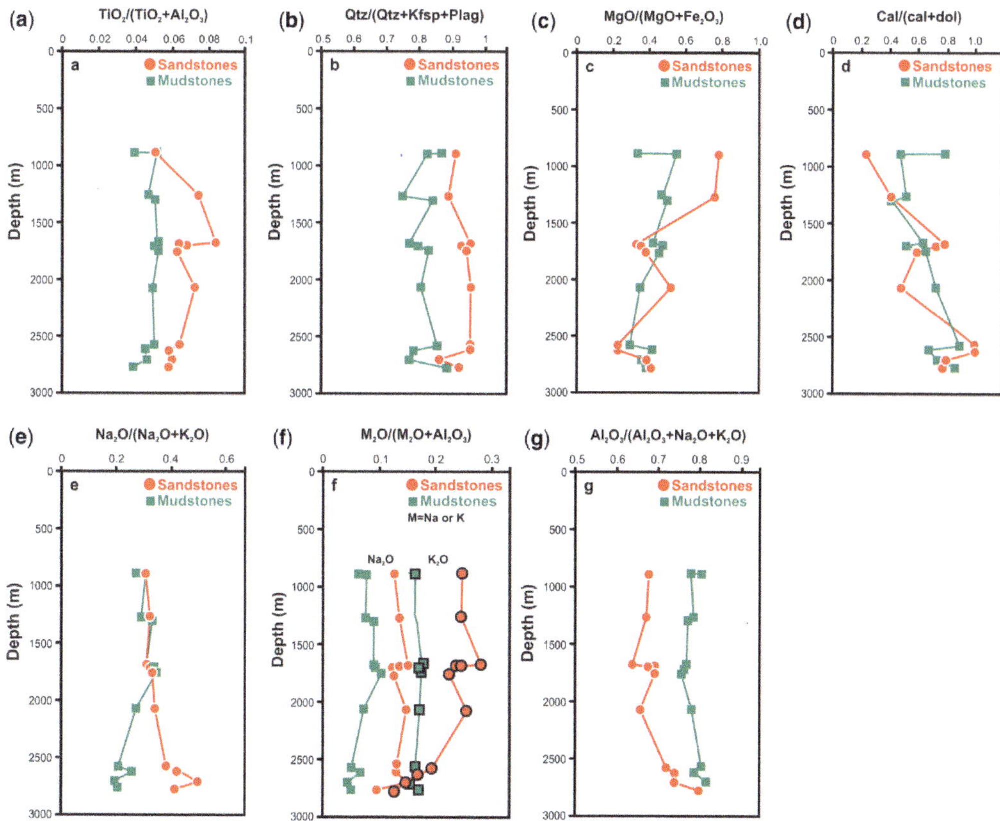

Fig. 12. (**a**) $TiO_2/(TiO_2 + Al_2O_3)$, (**b**) quartz/(quartz + K-feldspar + plagioclase), (**c**) $MgO/(MgO + Fe_2O_3)$, (**d**) calcite/(calcite + dolomite), (**e**) $Na_2O/(Na_2O + K_2O)$, (**f**) $Na_2O/(Na_2O + Al_2O_3)$ and $K_2O/(K_2O + Al_2O_3)$ and (**g**) $Al_2O_3/(Al_2O_3 + Na_2O + K_2O)$ (also known as the chemical index of alteration (CIA)) ratios from sandstones and mudstones.

diagenetic alteration of feldspar in the deeper sandstones than in the shallower sandstones.

Comparison of clay mineralogy and illitization in sandstones and mudstones

Smectite illitization in sandstones can both be similar or dissimilar to the equivalent process in adjacent mudstones (McKinley *et al.* 2003). Key controls are the initial rock mineralogy and composition, the local water geochemistry and the higher permeability of sandstones compared to mudstones (McKinley *et al.* 2003).

Hugget (1996) showed the detrital and authigenic mineralogies of Paleocene sandstones and mudstones of the Central North Sea to be quite similar, the only difference being the relative proportions of the clays.

Niu & Ishida (2000) and Niu *et al.* (2000) reported a slower rate of illitization in sandstones compared to mudstones. This is in line with observations by, for example, Boles & Franks (1979) and Howard (1981) from Palaeogene sandstones in the Gulf of Mexico. They showed a higher smectite content in the illite–smectite mixed-layer clay in the sandstones. The slower illitization rate in the sandstones could possibly be explained by different initial types of smectite in the sandstones and mudstones: a low-charged and a high-charged type, respectively. The smectites in the sandstones were formed by authigenesis during burial, and thus grew in equilibrium with the pore fluids, whereas the smectites in the mudstones could be largely derived from weathered illite (McKinley *et al.* 2003). The smectites derived from weathered illite in the mudstones are higher charged than the authigenic smectites in the sandstones. For higher charged smectite, the uptake of potassium is easier (Gier *et al.* 1998) and, thus, the illitization process might be faster in the mudstones.

In a hydrothermal setting, sandstones displayed a higher rate of illitization compared to adjacent mudstones (Hillier *et al.* 1996). There, higher permeability in the sandstones allowed the circulation of hydrothermal fluids and resulted in higher temperatures in the sandstones. A higher rate of illitization in sandstones caused by higher permeability and porosity was also reported by Ko & Hesse (1998).

In the Aderklaa sandstone–mudstone sequence, the illitization of the illite–smectite mixed-layer mineral has progressed faster (more extensively) in the sandstones than in the mudstones (Fig. 13). We explain this as being a consequence of the higher porosity and permeability of the sandstones that allowed faster advection of the pore fluid (or faster diffusion within the better-connected pore fluid). In comparison, there is less smectite but more illite, chlorite and kaolinite in the sandstone. The faster fluid flow (or diffusion) in the sandstones relative to the mudstones is also documented in the comparative clay mineral plots of illite–smectite, illite, chlorite and kaolinite (Fig. 6a–d). Kaolinite authigenesis has progressed more in the sandstones compared to the mudstones. In Aderklaa 78, this increase in kaolinite content coincides exactly with the depth interval with the largest amount of K-feldspar (Gier *et al.* 2008). In this sandstone interval with an elevated K-feldspar content, alteration because of fluid flow or diffusion is demonstrably greater than in the mudstones. The dissolution of K-feldspar also provides K- and Al-ions, which are necessary for the illitization of smectite (Hower *et al.* 1976). This is another possible factor accelerating the illitization in the sandstones. The lower fluid flow or diffusion rate in the mudstone samples of Aderklaa 78 is also supported by their bulk

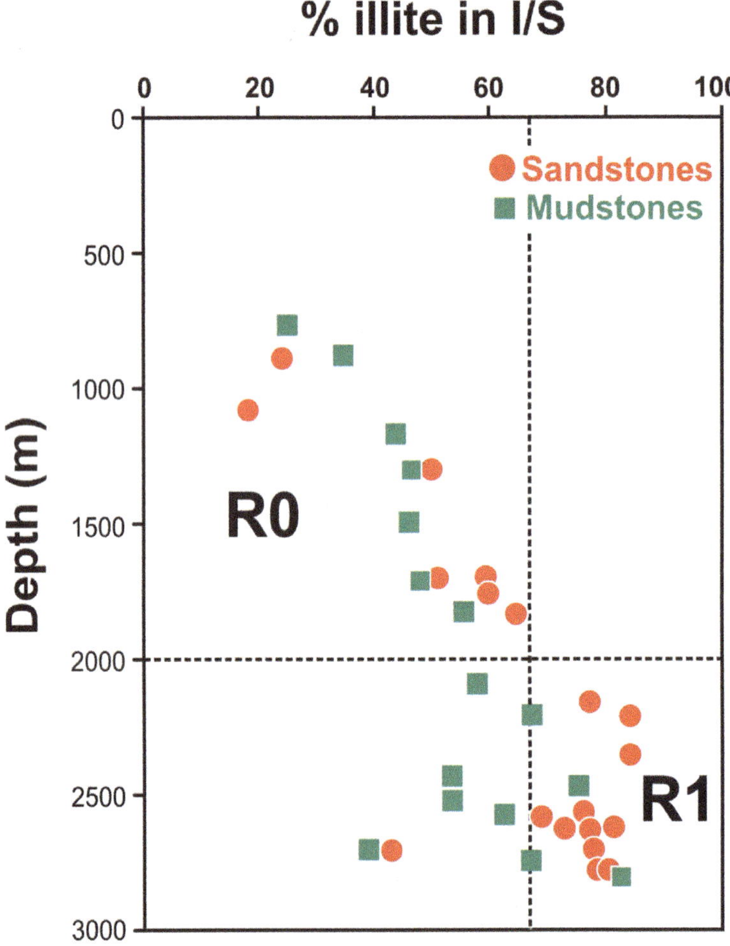

Fig. 13. Comparison of illitization trends of interbedded sandstones and mudstones. R0, randomly interstratified; R1, ordered.

composition, which shows no significant variation with depth (Figs 11 & 12) even though their clay mineralogy is quite different (Fig. 9). This indicates a closed system for the diagenetic changes within the mudstones. The process of conversion of smectite into illite layers is, thus, dependent on lithology because it has progressed by a transformation process in the mudstones within a restricted chemical system and by a dissolution–precipitation process in the sandstones within a more open chemical system (Clauer *et al.* 1999).

There is a reversal in the illitization trend at a depth of around 2700 m (Fig. 13). Kurzweil & Johns (1981) interpreted this to be the result of the variation in composition of the subsurface pore fluids. In areas adjacent to Aderklaa 78, pore fluids within underlying Triassic dolomites have an unusually high Mg content that increases upwards and also within the basal sediments of the Neogene basin fill (Kurzweil & Johns 1981). Generally, illitization with depth is a function of temperature and the availability of K-ions supplied by pore waters (Hower *et al.* 1976). A high Mg content within pore-fluids would promote the stability of smectite relative to illite. Another explanation for the reversal in illitization could be early carbonate cementation. Samples from the lowermost part of the Aderklaa Formation (>2500 m) show a high degree of carbonate cementation. This coincides with a deviation in the illitization trend (Fig. 13). Carbonate cementation causes a reduction in porosity and permeability, and results in a decrease in supply of the K and Al ions necessary for illitization. The substantial influence of permeability on illitization was also shown by Ramseyer & Boles (1986), who described early carbonate cementation in sandstones having retarded the illitization process, probably by impeding the transport of Al and K ions to and from the reaction sites.

Effects of different depositional environments on clay mineral assemblage and illitization

Three factors are particularly important in determining the clay mineral assemblages exhibited by alternating mudstones and sandstones (Jeans 1989): (1) the mineralogical and chemical nature of the detrital and colloidal sediment load; (2) the extent to which individual components within this load are segregated by variations in the depositional energy of the environment at the point of sedimentation and by the effects of differential flocculation; and (3) the variations in the chemistry of the depositional fluids and the diagenetic formation water.

Sediments in Aderklaa 78 were deposited in a variety of environments ranging, from the bottom to the top of the sedimentary sequence, from limnic-fluviatile to lacustrine, fully marine, to restricted marine and brachyhaline, and back to limnic. Based on regional geology, mineralogy (Fig. 12b) and geochemistry (Fig. 12a), a common and rather uniform source area can be interpreted for the alternating mudstones and sandstones in Aderklaa 78. Regionally, the source area did not fundamentally change during Miocene sedimentation in the Vienna Basin (Kurzweil & Johns 1981). From heavy mineral analyses, Wieseneder (1953) and Wieseneder & Maurer (1958) interpreted only minor changes of the sediment provenance (source area) for the Vienna Basin. In general, the heavy mineral assemblages can be related to erosional processes within the Alps and, to a lesser extent, the Bohemian Massif. The mineralogical decrease in calcite (Fig. 12d) and the increase in the $MgO/(MgO + Fe_2O_3)$ ratio (Fig. 12c) upsection, in both sandstones and mudstones, can be explained by a primary depositional control combined with a diagenetic overprint, where the younger sediments have a higher input of dolomite clasts, and the older sediments are richer in limestone clasts and, in addition, are calcite cemented.

The higher K-feldspar content in the limnic-fluviatile Karpatian Aderklaa Formation (Gier *et al.* 2008) can presumably be attributed to a higher sediment input from the Bohemian Massif, where weathering and erosion of granitic and gneissic rocks took place. Alteration of this detrital K-feldspar in the sandstones leads to the authigenesis of kaolinite, possibly by interaction with acidic water from thermal maturation of organic matter (Marfil *et al.* 2003). This explains the higher kaolinite content in the clay fraction of the sandstones at depths greater than 2000 m. However, the different depositional environments identified in the core do not have a major influence on the overall illitization trend of smectite.

Conclusions

- In relation to depth, the illitization of the illite–smectite mixed-layer mineral has progressed faster in the sandstones than in the mudstones.
- Illitization of smectite took place in a relatively open system in the sandstones and in a closed system in the mudstones. The sandstones lost bulk K_2O with increasing depth, in contrast to the mudstones where bulk K_2O remains roughly constant with depth. The sandstones apparently did not donate K_2O to the mudstones for illitization.
- The higher permeability of the sandstones compared to the mudstones is also expressed by the

higher kaolinite content, which was caused by accelerated alteration of feldspar in the sandstones compared to the mudstones.
- Based on regional geology, sedimentary petrology, heavy mineral analyses and geochemistry, a common source area is interpreted for the studied sediments. The younger sediments had a higher input of dolomite clasts, while the older sediments are richer in limestone clasts and K-feldspar.
- The variety of depositional environments, from limnic-fluviatile to marine, did not have a major influence on the illitization process.

The authors thank OMV Exploration & Production for providing the core samples of the Aderklaa 78 well and also for the permission to publish this work. S. Gier acknowledges the financial support of the 'Hochschuljubiläumsstiftung der Stadt Wien (Project H-942/2004)'. We thank P. Nagl and J. Weber for their assistance on the X-ray fluorescence and SEM. The comments of two anonymous reviewers are highly acknowledged. Special thanks go to S. Gier's clay mineral teachers, the late W.D. Johns and H. Kurzweil, who initiated the diagenesis studies on the Aderklaa 78 well.

References

ANDERSSON, P. O. D. & WORDEN, R. H. 2004. Mudstones of the Tanqua Basin, South Africa: an analysis of lateral and stratigraphic variations within mudstones and a comparison of mudstones within and between turbidite fans. *Sedimentology*, **51**, 479–502.

ANDERSSON, P. O. D., WORDEN, R. H., HODGSON, D. M. & FLINT, S. 2004. Provenance evolution and chemostratigraphy of a Palaeozoic submarine fan complex: Tanqua Karoo Basin, South Africa. *Marine and Petroleum Geology*, **21**, 555–577.

ARZMÜLLER, G., BUCHTA, Š., RALBOVSKÝ, E. & WESSELY, G. 2006. The Vienna Basin. *In*: GOLONKA, J. & PICHA, F. J. (eds) *The Carpathians and Their Foreland: Geology and Hydrocarbon Resources*. American Association of Petroleum Geologists, Memoirs, **84**, 191–204.

AVSETH, P. 2010. Explorational rock physics – the link between geological processes and geophysical observables. *In*: BJØRLYKKE, K. (ed.) *Petroleum Geoscience: From Sedimentary Environment to Rock Physics*. Springer, Berlin, 403–426.

BJØRLYKKE, K. & JAHREN, J. 2010. Sandstone and sandstone reservoirs. *In*: BJØRLYKKE, K. (ed.) *Petroleum Geoscience: From Sedimentary Environment to Rock Physics*. Springer, Berlin, 113–140.

BOLES, J. R. & FRANKS, S. G. 1979. Clay diagenesis in Wilcox sandstones of southwest Texas: implications of smectite diagenesis on sandstone cementation. *Journal of Sedimentary Petrology*, **49**, 55–70.

CLAUER, N., RINCKENBACH, T., WEBER, F., SOMMER, F., CHAUDHURI, S. & O'NEIL, J. R. 1999. Diagenetic evolution of clay minerals in oil-bearing Neogene sandstones and associated shales, Mahakam delta basin, Kalimantan, Indonesia. *American Association of Petroleum Geologists Bulletin*, **83**, 62–87.

DECKER, K. & PERESSON, H. 1996. Tertiary kinematics in the Alpine–Carpathian–Pannonian system: links between thrusting, transform faulting and crustal extension. *In*: WESSELY, G. & LIEBL, W. (eds) *Oil and Gas in Alpidic Thrustbelts and Basins of Central and Eastern Europe*. European Association of Geoscientists & Engineers (EAGE), Special Publications, **5**, 69–77.

GIER, S., OTTNER, F. & JOHNS, W. D. 1998. Layer-charge heterogeneity in smectites of *I/S* phases in pelitic sediments from the Molasse Basin, Austria. *Clays and Clay Minerals*, **46**, 670–678.

GIER, S., WORDEN, R. H., JOHNS, W. D. & KURZWEIL, H. 2008. Diagenesis and reservoir quality of Miocene sandstones in the Vienna Basin, Austria. *Marine and Petroleum Geology*, **25**, 681–695.

HERRON, M. M. 1988. Geochemical classification of terrigenous sands and shales from core or log data. *Journal of Sedimentary Petrology*, **58**, 820–829.

HILLIER, S., SON, B. K. & VELDE, B. 1996. Effects of hydrothermal activity on clay mineral diagenesis in Miocene shales and sandstones from the Ulleung (Tsushima) backarc basin, East Sea (Sea of Japan), Korea. *Clay Minerals*, **31**, 113–126.

HORTON, R. B., JOHNS, W. D. & KURZWEIL, H. 1985. Illite diagenesis in the Vienna Basin, Austria. *TMPM Tschermaks Mineralogische und Petrographische Mitteilungen*, **34**, 239–260.

HÖLZEL, M., DECKER, K., ZAMOLYI, A., STRAUSS, P. & WAGREICH, M. 2010. Lower Miocene structural evolution of the central Vienna Basin (Austria). *Marine and Petroleum Geology*, **27**, 666–681.

HOWARD, J. J. 1981. Lithium and potassium saturation of illite–smectite clays from interlaminated shales and sandstones. *Clays and Clay Minerals*, **29**, 136–142.

HOWER, J., ESLINGER, E. V., HOWER, M. E. & PERRY, E. A. 1976. Mechanism of burial metamorphism of argillaceous sediment: 1. Mineralogical and chemical evidence. *Geological Society of America Bulletin*, **87**, 725–737.

HUGGET, J. M. 1996. Aluminosilicate diagenesis in a tertiary sandstone-mudrock sequence from the central North Sea, UK. *Clay Minerals*, **31**, 523–536.

JEANS, C. V. 1989. Clay diagenesis in sandstones and shales: an introduction. *Clay Minerals*, **24**, 127–136.

JOHNS, W. D. & HOEFS, J. 1985. Maturation of organic matter in Neogene sediments from the Aderklaa oilfield, Vienna Basin, Austria. *TMPM Tschermaks Mineralogische und Petrographische Mitteilungen*, **34**, 143–158.

KO, J. & HESSE, R. 1998. Illite/smectite diagenesis in the Beaufort–Mackenzie Basin, Arctic Canada: relation to hydrocarbon occurrence? *Bulletin of Canadian Petroleum Geology*, **46**, 74–88.

KURZWEIL, H. & JOHNS, W. D. 1981. Diagenesis of Tertiary marlstones in the Vienna Basin. *TMPM Tschermaks Mineralogische und Petrographische Mitteilungen*, **29**, 103–125.

LADWEIN, H. W. 1988. Organic geochemistry of Vienna Basin: model for hydrocarbon generation in overthrust belts. *American Association of Petroleum Geologists Bulletin*, **72**, 586–599.

LYNCH, F. L., MACK, L. E. & LAND, L. S. 1997. Burial diagenesis of illite/smectite in shales and the origins of authigenic quartz and secondary porosity in sandstones. *Geochimica et Cosmochimica Acta*, **61**, 1995–2006.

MARFIL, R., DELGADO, A., ROSSI, C., LA IGLESIA, A., RAMSEYER, K. 2003. Origin and diagenetic evolution of kaolin in reservoir sandstones and associated shales of the Jurassic and Cretaceous, Salam Field, Western Desert (Egypt). *In*: WORDEN, R. H. & MORAD, S. (eds) *Clay Mineral Cements in Sandstones*. International Association of Sedimentologists, Special Publications, **34**, 319–342.

MCKINLEY, J. M., WORDEN, R. H. & RUFFELL, A. H. 2003. Smectite in sandstones: a review of the controls on occurrence and behaviour during diagenesis. *In*: WORDEN, R. H. & MORAD, S. (eds) *Clay Mineral Cements in Sandstones*. International Association of Sedimentologists, Special Publications, **34**, 109–128.

MOORE, D. M. & REYNOLDS, R. C. 1997. *X-Ray Diffraction and the Identification and Analysis of Clay Minerals*. Oxford University Press, Oxford.

NESBITT, H. W. & YOUNG, G. M. 1982. Early Proterozoic climates and plate motions inferred from major element chemistry of lutites. *Nature*, **299**, 715–717.

NIU, B. & ISHIDA, H. 2000. Different rates of smectite illitization in mudstones and sandstones from the Niigata Basin, Japan. *Clay Minerals*, **35**, 163–173.

NIU, B., YOSHIMURA, T. & HIRAI, A. 2000. Smectite diagenesis in Neogene marine sandstone and mudstone of the Niigata Basin, Japan. *Clays and Clay Minerals*, **48**, 26–42.

PELTONEN, C., MARCUSSEN, Ø., BJØRLYKKE, K. & JAHREN, J. 2009. Clay mineral diagenesis and quartz cementation in mudstones: the effects of smectite to illite reaction on rock properties. *Marine and Petroleum Geology*, **26**, 887–898.

RAMSEYER, K. & BOLES, J. R. 1986. Mixed-layer illite/smectite minerals in Tertiary sandstones and shales, San Joaquin Basin, California. *Clays and Clay Minerals*, **34**, 115–124.

ROYDEN, L. H. 1985. The Vienna Basin: a thin-skinned pull-apart basin. *In*: BIDDLE, K. T. & CHRISTIE-BLICK, N. (eds) *Strike-Slip Deformation, Basin Formation and Sedimentation*. Society of Economic Paleontologists and Mineralogists (SEPM), Special Publications, **37**, 319–339.

SACHSENHOFER, R. F. 2001. Syn- and post-collisional heat flow in the Cenozoic Eastern Alps. *International Journal of Earth Sciences*, **90**, 579–592.

SAUER, R., SEIFERT, P. & WESSELY, G. 1992. Guidebook to excursions in the Vienna Basin and the adjacent Alpine-Carpathian thrustbelt in Austria. *Mitteilungen der Österreichischen Geologischen Gesellschaft*, **85**, 1–264.

SCHULTZ, L. G. 1964. *Quantitative Interpretation of Mineralogical Composition from X-ray and Chemical Data for the Pierre Shale*. United States Geological Survey, Professional Paper, **0391-C**.

SEIFERT, P. 1996. Sedimentary–tectonic development and Austrian hydrocarbon potential of the Vienna Basin. *In*: WESSELY, G. & LIEBL, W. (eds) *Oil and Gas in Alpidic Thrustbelts and Basins of Central and Eastern Europe*. European Association of Geoscientists & Engineers (EAGE), Special Publications, **5**, 331–341.

TANNER, C. B. & JACKSON, M. L. 1947. Nomographs of sedimentation times for soil particles under gravity or centrifugal acceleration. *Soil Science Society of America Proceedings*, **11**, 60–65.

THYBERG, B. & JAHREN, J. 2011. Quartz cementation in mudstones: sheet-like quartz cement from clay mineral reactions during burial. *Petroleum Geoscience*, **17**, 53–63, https://doi.org/10.1144/1354-079310-028

THYBERG, B., JAHREN, J., WINJE, T., BJØRLYKKE, K., FALEIDE, J. I. & MARCUSSEN, Ø. 2010. Quartz cementation in Late Cretaceous mudstones, northern North Sea: changes in rock properties due to dissolution of smectite and precipitation of micro-quartz crystals. *Marine and Petroleum Geology*, **27**, 1752–1764.

THYNE, G. 2001. A model for diagenetic mass transfer between adjacent sandstone and shale. *Marine and Petroleum Geology*, **18**, 743–755.

WAGREICH, M. & SCHMID, H. P. 2002. Backstripping dip-slip fault histories: apparent slip rates for the Miocene of the Vienna Basin. *Terra Nova*, **14**, 163–168.

WEIBEL, R. 1999. Effects of burial on the clay assemblage in the Triassic Skagerrak Formation, Denmark. *Clay Minerals*, **34**, 619–635.

WORDEN, R. H. & MORAD, S. 2003. Clay minerals in sandstones: controls on formation, distribution and evolution. *In*: WORDEN, R. H. & MORAD, S. (eds) *Clay Mineral Cements in Sandstones*. International Association of Sedimentologists, Special Publications, **34**, 3–41.

WIESENEDER, H. 1953. Zur Diagenese klastischer Sedimente im Wiener Becken. *TMPM Tschermaks Mineralogische und Petrographische Mitteilungen*, **3**, 142–153.

WIESENEDER, H. & MAURER, I. 1958. Ursachen der räumlichen und zeitlichen Änderung des Mineralbestandes der Sedimente des Wiener Beckens. *Eclogae Geologicae Helvetiae*, **51**, 1155–1172.

The relevance of dawsonite precipitation in CO_2 sequestration in the Mihályi-Répcelak area, NW Hungary

CSILLA KIRÁLY[1], ESZTER SENDULA[1], ÁGNES SZAMOSFALVI[2], RÉKA KÁLDOS[1], PÉTER KÓNYA[2], ISTVÁN J. KOVÁCS[2], JUDIT FÜRI[2], ZSOLT BENDŐ[1] & GYÖRGY FALUS[2]*

[1]*Lithosphere Fluid Research Lab, Department of Petrology and Geochemistry, Eötvös University, Budapest, Hungary. H-1117, Budapest, Pázmány Péter sétány 1/c, Hungary*

[2]*Geological and Geophysical Institute of Hungary, H-1143 Budapest, XIV., Stefánia út 14, Hungary*

**Correspondence: falus.gyorgy@mfgi.hu*

Abstract: A natural CO_2 reservoir system with a sandstone lithology in NW Hungary has been studied due to its similarities to a large saline reservoir formation that is widespread in the the Pannonian Basin (Central Europe) and is suggested to be one of the best candidates for industrial CO_2 storage. A range of analytical techniques has been used on core samples from CO_2-containing sandstone layers that represent a wide range of pressures (90–155 bar), temperatures (79–95°C) and pore fluid compositions (total dissolved solids between 18 000 and 50 700 mg l^{-1}) to identify the mineralogy and textural characteristics of the natural reservoir.

The only clear CO_2-related feature in the studied lithology was the occurrence of dawsonite ($NaAlCO_3(OH)_2$) in a close textural relationship with albite. This is in clear agreement with our geochemical modelling results, which also underline the presence of albite as a precondition for the crystallization of dawsonite at the given $P–T–X$ conditions. Our results suggest that, at least in the Pannonian Basin, dawsonite may be an important mineral for the safe sequestration of industrial CO_2 in the subsurface.

Supplementary material: Raman spectra of dawsonite are available at https://doi.org/10.6084/m9.figshare.c.3268322

A fundamental prerequisite for safe subsurface CO_2 storage is the availability of a suitable reservoir. The required features of a reservoir are described in detail by Bachu *et al.* (2007). The main reservoir parameters are volume, porosity, reservoir depth and the presence of an impermeable cap rock. Reservoir productivity (permeability) is also important.

In addition, the effect of CO_2 on the reservoir porewater–rock system must be well understood before an industrial CO_2 storage project commences (Bachu *et al.* 2007; Arts *et al.* 2008). Reactions between CO_2 and the porewater–rock system will necessarily occur, however, the rate of chemical reactions can vary and reactivity may compromise the long-term safety of storage. For these reasons, geochemical reactivity must be studied early during the site selection procedure.

In order to understand and estimate the effect of CO_2 on the reservoir four main techniques can be applied: (1) laboratory experiments (Kaszuba *et al.* 2003; Carroll & Knauss 2005; Berta *et al.* 2012; Armitage *et al.* 2013), (2) numerical models (Xu *et al.* 2004; Allen *et al.* 2005; Kampman *et al.* 2014), (3) the study of natural CO_2 reservoirs (Moore *et al.* 2005; Pearce *et al.* 2005; Worden 2006; Gao *et al.* 2009), and (4) field trials and EOR, such as study of Krechba field (Worden & Smith 2004; Armitage *et al.* 2010). The study of natural CO_2 reservoirs has two benefits compared with the other three approaches. Firstly, CO_2 has been stored in these systems for geological timescales. Thus studying samples from natural systems could help in understanding the long-term reactions that might not be observed in laboratory experiments due to their limited duration. Secondly, samples from natural systems have experienced a range of geological processes that cannot be reproduced in laboratory experiments or by geochemical modelling.

The formation of dawsonite is one of the most disputed geochemical reactions related to the processes that take place during/after industrial CO_2 storage. Dawsonite is only rarely precipitated during laboratory experiments using natural reservoir samples because the kinetic rate of the growth of

From: Armitage, P. J., Butcher, A. R., Churchill, J. M., Csoma, A. E., Hollis, C., Lander, R. H., Omma, J. E. & Worden, R. H. (eds) 2018. *Reservoir Quality of Clastic and Carbonate Rocks: Analysis, Modelling and Prediction.* Geological Society, London, Special Publications, **435**, 405–418.
First published online June 20, 2016, https://doi.org/10.1144/SP435.15

dawsonite is slow (e.g. Hellevang *et al.* 2005). However, most of the geochemical models show that dawsonite should precipitate under reservoir conditions (e.g. Xu *et al.* 2004; Knauss *et al.* 2005). It is stable at high CO_2 partial pressures and becomes unstable when the CO_2 partial pressure drops as a consequence of CO_2 dissolution into the porewater (Hellevang *et al.* 2005). The stability conditions of dawsonite, such as pressure, temperature and water salinity, are still not completely defined, therefore detailed analyses of natural CO_2 reservoirs that contain dawsonite are expected to better define the conditions under which this mineral is stable.

The Mihályi-Répcelak area (Little Hungarian Plain, NW Hungary) is the oldest known well-sealed natural CO_2 reservoir (Fig. 1). The exploration of this area started in 1933, with production since the 1940s to the present day and 28 CO_2 reservoirs are now known (Mészáros *et al.* 1979). The original gas in place is approximately 25 Mt of CO_2, from which approximately 100 000 t are produced annually. As a consequence there is a great deal of publicly available data about these reservoirs, regarding their geophysical characteristics, structure and reservoir geological parameters (Mészáros *et al.* 1979). Moreover rock samples are accessible from CO_2-bearing strata. The geological facies of the CO_2-bearing strata are similar to a widespread saline aquifer formation (the Szolnok Fm; central-eastern Hungary; Fig. 1.), which is the best candidate for industrial CO_2 storage in the region.

The main aim of this paper is to identify CO_2-related features from samples of the CO_2-bearing reservoir sandstones. These features are assumed to represent a first-order approximation of the reactions that can be expected to take place as a result of any future industrial CO_2 injection in the Szolnok Fm.

Geological background

The Neogene Basin of the Little Hungarian Plain subsided and was filled by a prograding delta system from the NW during the Late Miocene – (from 10 Ma) (Magyar *et al.* 2013) (Fig. 1). The composition of the porewater that is assumed to have developed during the filling of the Pannonian Basin is 18 000–50 700 mg l^{-1} (Mészáros *et al.* 1979). The reservoir rock is composed of fine-grained sandy turbidites interlayered with argillaceous beds in the deepest part of the basin. A *c.* 300 m thick argillaceous succession was deposited on a slope setting, providing regional coverage for the sandy turbidite. The mineral composition, thickness and low permeability of the argillaceous overlying strata make it a suitable, leakage-safe cap rock for the storage system. The reservoir sandstone formation is also underlain by argillaceous rocks that were formed in the basin far from sediment input, whereas the whole sedimentary sequence is overlain by an interfingering siltstone, sandstone and claystone succession that formed in delta and shoreline environments and in the alluvial plain. The thickness of these sediments is approximately 800 m in this area (Mészáros *et al.* 1979; Juhász 1991).

As recent studies suggest, oil was generated in the Middle Miocene source rock during subsidence, and migrated into the pore spaces of the Pannonian sediments from 8 Ma ago (Csizmeg *et al.* 2012). Subsequently (between 5 and 4 Ma ago) the arriving

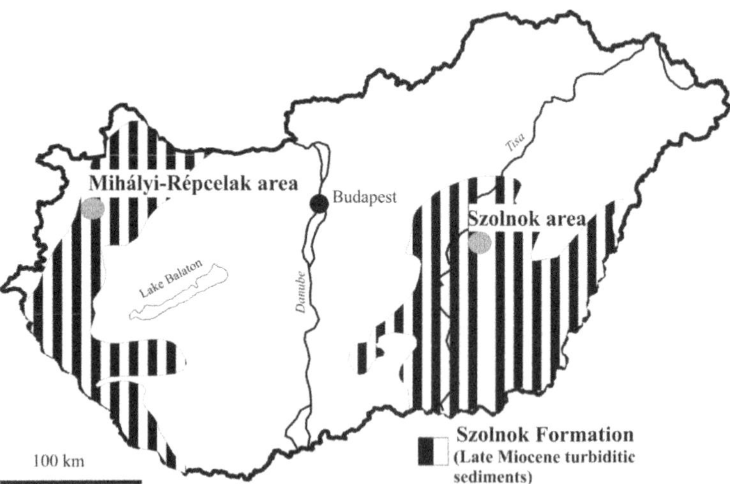

Fig. 1. Map of the Pannonian Basin, showing the turbiditic sandstones of Szolnok Fm (Little Hungarian Plain and Great Hungarian Plain). The grey dot shows the Mihályi-Répcelak area (modified from Kubus 2009).

CO_2 replaced the oil (Vető *et al.* 2014). The origin of the CO_2 is contested, a previous study suggested that the CO_2 has a metamorphic origin (Clayton *et al.* 1990). However, the newest results indicate a magmatic origin based on the isotopic ratio of $\delta^{13}C$ and $\delta^{18}O$ (Palcsu *et al.* 2014; Vető *et al.* 2014). During the migration of the CO_2 the deeper reservoirs were filled first, followed by the overlying ones. The accumulation of CO_2 is suggested to be a short process in this area (Vető *et al.* 2014). The natural CO_2-bearing succession (69–98 vol% (volumetric percentage); other gases present include CH-gases and N_2, Vető *et al.* 2014) in the Mihályi-Répcelak area was deposited in the same geological facies as saline reservoirs in the Great Hungarian Plain (central-eastern Hungary; Fig. 1) that bear with the largest estimated CO_2 storage potential in Hungary (Fancsik *et al.* 2007). The calculated storage capacity of the saline reservoirs in the Great Hungarian Plain, based on volumetric approach, is approximately 10^9 tons of CO_2 (Szamosfalvi *et al.* 2011).

Analytical methods

Petrographic method

The studied rock samples were analysed by a NIKON Eclipse LV100 POL polarization microscope equipped with a NIKON DS-Fi1 digital camera at the Geological and Geophysical Institute of Hungary (MFGI). A scanning electron microscope (SEM) was also used for the analysis at the Institute of Geography and Earth Sciences, Eötvös University (ELTE). The instrument is an AMRAY 1830 I/T6 equipped with EDAX PV 9800 energy dispersive X-ray spectrometer. During the analyses the accelerating voltage was 20 kV and the primary sample current was 1–2 nA. Thin sections and artificially fractured surfaces were examined. Backscattered electron (BSE) images and secondary electron (SE) images were taken from the thin sections and artificially fractured surfaces, respectively. Using this method the texture, the mineralogical environment of dawsonite, the type and structure of the carbonate minerals, plagioclase and clay minerals were observed. Grain size distribution measurements were carried out with a Horiba LA-950V2 laser particle size analyser at the ELTE Faculty of Science, Research and Instrument Core Facility, Budapest – ELTE FS-RICF. The samples were disaggregated by the freezing–melting method in a sodium pyrophosphate solution. Subsequently, an ultrasonic treatment (for 5 min to 1 h) was used to complete the dispersion. Three repeated measurements were taken from each sample to ensure homogeneity of the grain size distribution.

X-ray diffraction (XRD)

All samples were analysed for whole rock XRD at MFGI (Table 1) using a Phillips PW 1730 diffractometer (Cu cathode, 40 kV and 30 mA tube current, graphite monochromator, goniometer speed 2 min^{-1}). A semi-quantitative assessment of the relative concentrations of the phases was performed using the relative intensity ratios and half widths of specific reflections of minerals with the XDB Powder Diffraction Phase Analytical software (version 2.7).

Thermal Analysis

Samples RM17-4R, RM32-5RI, RM19-6R, RM6-7R1 and RM46-7/2R were studied by thermal analysis at MFGI in order to verify the results of the XRD measurements. The instrument is a computer controlled and evaluated Derivatograph-PC simultaneous thermogravimetry, derivative thermogravimetry, differential thermal analysis unit. The measurements are carried out in a ceramic crucible. The heating rate was 10°C min^{-1} from room temperature to 1000°C. The thermal analysis method is

Table 1. *Estimated pressure and temperature values for the studied samples as calculated from local gradients (100 bar km^{-1}, 50°C km^{-1}; Dövényi* et al. *1983)*

Sample	Depth (m)	Pressure (bar)	Temperature (°C)
RM6-2R	1280–1287	128.7	75.4
RM17-4R	1271–1288	128.8	75.4
RM17-5RI	1288–1305	130.5	76.3
RM32-5R	1375–1392	139.2	80.6
RM6-4R	1375–1392.6	139.3	80.7
RM19-6R	1394–1415	141.5	81.8
RM6-7R1	1417–1425.4	142.5	82.3
RM6-7R2	1417–1425.4	142.5	82.3
RM6-7R3	1417–1425.4	142.5	82.3
RM46–4R	1461–1478	147.8	84.9
RM46-7/2R	1626.5–1643.5	164.4	93.2

primarily sensitive to organic matter, carbonate minerals and the crystallization rate of clay minerals.

Fourier transform infrared (FTIR) and Raman spectrometry

Dawsonite was identified by a Jobin Yvon LabRAM HR confocal Raman microspectrometer (at ELTE FS-RICF). Furthermore, all samples were analysed by an FTIR spectrometer (Bruker Vertex 70) with a single pass attenuated total reflectance (ATR) cell (Bruker Platinum diamond ATR cell) at MFGI. An MCT (mercury–cadmium–telluride) detector was used to obtain ATR spectra in the mid-infrared spectral range (400–4000 cm^{-1}). The samples were powdered and dried for 30 min at 80°C in an oven immediately prior to the ATR FTIR measurements. This process ensured that the majority of the adsorbed water was removed (Udvardi *et al.* 2014). The relatively low temperature of the heat treatment guarantees that it does not cause any first-order alterations in the structure of the clay minerals but significantly simplifies the interpretation of the ATR spectra (Tóth *et al.* 2012). This analysis is sensitive to the presence of clay minerals and dawsonite as these minerals contain OH^-. The dawsonite peak is at 3280 cm^{-1}, whereas the clay mineral peaks are between 3696 cm^{-1} and 3621 cm^{-1} in the OH^- region.

Kinetic modelling

In order to better understand the mineral reaction processes in the studied natural CO_2 reservoir we constructed a kinetic model to describe the type and relative timing of the processes taking place. We assumed that petrographic and mineralogical observations in the studied sandstone samples from the Mihályi-Répcelak area are the end-products from a reaction between porewater, injected CO_2 and a potential CO_2-barren reservoir lithology due to the same geological facies and physical conditions.

Input data used in the model

For the input mineral composition to the model, a CO_2-barren sandstone composition was used as CO_2-free end member based on XRD data (same geological formation as the Mihályi-Répcelak area, similar recent pressure and temperature conditions). This assumption is based on the fact that mineral compositions and diagenetic features of the studied sandstone unit are ubiquitous in the Pannonian Basin (Mátyás & Matter 1997). Ten per cent porosity and water saturated pores were assumed, giving a water–rock mass ratio of approximately 1:25.

The representative porewater chemical composition that was applied in the model is assumed to be characteristic for this geological formation and this depth interval. As a first step in the geochemical modelling, the porewater data were equilibrated with the reservoir rock composition. The results obtained from this equilibration (Tables 2 & 3) were then applied to modelling the effect of pure CO_2 on the properties of the reservoir rock. Thus, the reactions taking place as a result of chemical disequilibrium between the initial rock and porewater data were excluded and the model clearly shows the geochemical effect of CO_2.

The pressure ($p = 158$ bar) and temperature ($T = 88$°C) values used in the simulation were calculated from a depth of 1558 m assuming a hydrostatic pressure gradient, in agreement with earlier studies (i.e. Mészáros *et al.* 1979) and a 50°C km^{-1} geothermal gradient at the studied area (Dövényi *et al.* 1983). The partial pressure of pure CO_2 used in the model was assumed to be equal to the calculated hydrostatic pressure. The rate constants at 298.15 K and the activation energies used in equation (2) are taken from the literature (Table 2).

Specific surface area (SSA), one of the controlling factors of rock reactivity, was estimated by assuming spherical grains (according to previous studies, e.g. Amin *et al.* 2014) with two typical grain diameters (based on grain size analysis and petrographic observations): 30 μm for carbonates and 180 μm for other silicates (Table 2). In the case of kaolinite, an SSA value of 13.1 $m^2\ g^{-1}$ was used, as reported by Dogan *et al.* (2006). For kinetic calculations the surface area is required in units of m^2 per kg of H_2O for each mineral phase.

Modelling approach and main equations

A closed-system batch kinetic model was run, applying the PHREEQC-3 code (Parkhurst & Appelo 2012).

A general form of rate law given by Lasaga (1984) was used to express the reaction rate for mineral dissolution and precipitation:

$$\text{rate}_m = \text{SSA}_m k(T)_m (a_{\text{H}+})^n \left| \left[\left(\frac{Q_m}{K_m} \right)^{\mu} - 1 \right] \right|^{\nu} \quad (1)$$

where rate_m is the dissolution/precipitation rate (positive values indicate dissolution, negative values indicate precipitation) of mineral *m*, SSA is the specific surface area per kg H_2O, $k(T)$ is the temperature-dependent rate constant, $a_{\text{H}+}$ is the activity of H^+ and *n* is the empirical reaction order ($0 \leq n \leq 1$). *K* is the equilibrium constant written for the dissolution of 1 mol of *m* in the

Table 2. *Kinetic parameters used for the primary and secondary phases, as well as the amount present for each mineral*

Minerals	Amount present after equilibration with pore water (mol per kg H_2O)	SSA* (m^2 g^{-1})	$k^†$ (mol m^{-2} s^{-1})	$E_a^‡$ (kJ mol^{-1})	Source of parameters k, E_a and SSA
Albite	5.66	1.91E−01	1.10E−12	6.50E + 01	Brantley 2008
Ankerite	1.85E + 01	3.28E−02	7.98E−08	5.01E + 01	k and E_a values assumed to be intermediate between dolomite/siderite
Calcite	2.86E + 01	3.69E−02	1.55E−06	2.35E + 01	Palandri & Kharaka (2004)
Dawsonite	0.00	4.13E−02	1.80E−09	6.38E + 01	Hellevang *et al.* (2010)
Dolomite	7.19	3.88E−02	2.95E−08	5.22E + 01	Palandri & Kharaka (2004)
Illite	4.43E − 07	1.36E−02	3.10E−14	5.82E + 01	set to muscovite rate in Oelkers *et al.* (2008)
Kaolinite	1.45E + 01	1.31E + 01	1.20E−13	2.93E + 01	k, E_a: Yang & Steefel (2008) SSA: Dogan *et al.* (2006)
K-feldspar	2.21	1.95E−01	2.30E−13	5.17E + 01	k: Brantley (2008) E_a: Palandri & Kharaka (2004)
Pyrite	1.43	9.98E−02	9.60E−11	5.69E + 01	Palandri & Kharaka (2004)
Quartz	1.94E + 02	1.90E−01	4.50E−14	7.20E + 01	Rimstidt & Barnes (1980)
Siderite			1.30E−07	4.80E + 01	Golubev *et al.* (2009)

*Specific surface area.
†Reaction rate coefficient at 25°C.
‡Activation energy.

mineral–water reaction, Q is the reaction quotient and μ and v are two experimentally determined positive constants that are usually taken to be equal to unity (e.g. Xu *et al.* 2005; Amin *et al.* 2014).

Table 3. *Fluid composition of the reservoir formation after equilibration with the initial reservoir mineralogy*

Parameter	Value	Elements	Concentration (mol l^{-1})
Temperature (°C)	88	Na	2.40E−01
Ionic strength	0.2267	K	1.54E−03
pH	6.899	Ca	3.35E−04
		Mg	1.21E−05
		Cl	2.03E−01
		S	1.05E−03
		C	8.50E−02
		Al	4.46E−07
		Fe	2.03E−06
		Si	6.86E−04

Because data from literature are insufficient and often inconsistent regarding n in the rate law, we used a value of 0.5 for all minerals according to previous studies (Gaus *et al.* 2005; Amin *et al.* 2014).

The temperature dependence of k in equation (1) was considered using the Arrhenius relation (Gaus *et al.* 2005; Amin *et al.* 2014):

$$k = k_{25}\left[\frac{-E_a}{R}\left(\frac{1}{T} - \frac{1}{298.15}\right)\right] \quad (2)$$

where k_{25} is the rate constant at 298.15 K (in mol m^{-2} s^{-1}) (Table 2), E_a is the activation energy (in J mol^{-1}), R is the gas constant (8.314 J mol^{-1} K^{-1}) and T is the temperature (in K).

Reservoir heterogeneity (e.g. the spatial distribution of different minerals) was not taken into account. In addition, our model considered closed-system behaviour, hence it does not assume the fluid flow through the reservoir. However, despite these simplifications, it can predict the direction (dissolution and/or precipitation) of the mineral reactions caused by CO_2.

Description of the CO_2-bearing reservoir rocks

Table 1 shows the depth of origin as well as the physical conditions ($P-T$) of the studied samples. The samples are light grey in colour and very fine to fine grained sandstones. The rocks show a bimodal grain size distribution with an average grain size of 12.44 ± 0.23 μm for the matrix and 153.02 ± 52.19 μm for the grains. All samples are grain supported.

Detrital composition

The most important detrital components in the sandstones are monocrystalline and polycrystalline quartz, plagioclase feldspar, K-feldspar, mica, dolomite and both sedimentary and metamorphic rock fragments. Rutile, apatite, zircon and matrix clays occur in minor or trace amounts. Quartz, the most abundant mineral phase in the studied lithology, is characterized by an average grain size of 150 μm. Mica generally occurs in oriented patches subparallel to the overall lamination observed in the sandstone samples. As a consequence of compaction some of the mica crystals are bent on quartz or feldspar grains.

Diagenetic minerals

The authigenic mineral assemblages are dominated by carbonates (calcite, ankerite, siderite and dawsonite), quartz and clay minerals (mostly kaolinite). Minor amount of K-feldspar overgrowth and albitization of plagioclase feldspar grains are observed. Authigenic quartz is common and occurs in two forms, either as a very fine grained cement or as syntaxial overgrowth on detrital quartz grains (Fig. 2a, b).

Kaolinite, the dominant clay mineral phase in the studied sandstones occurs either as a fine grain coating on detrital minerals or as large 'book shaped' crystals in textural equilibrium with syntaxial quartz and dawsonite in the pore space (Fig. 2c). Kaolinite flakes are also frequently found in close vicinity to the mica patches, oriented subparallel to the micas (Fig. 2d). Other clay minerals (e.g. illite) occur only rarely.

Carbonates, the volumetrically most significant authigenic minerals occur in a wide range of textural varieties in the studied rock samples. Polarized light microscopy identified micritic, fibrous and idiomorphic carbonate occurrences. Detailed analysis by SEM showed that dolomites, most likely original detrital grains, are overgrown by multiple ankerite generations with variable Fe^{2+} content (Fig. 2e). Iron content increases toward the rims and ankerite zones with lower Fe-content exhibit dissolution features. In some of the samples the Fe-zoning can be reversed. Ankerite also occurs as cement or individual idiomorphic grains in the pore space. Calcite generally occurs as residual cement in sealed pores or as late stage idiomorphic grains in the pore space. Siderite is found in all studied samples and occurs only as a fine grained cement (Fig. 2e) intermixed with ankerite and kaolinite.

Dawsonite is a fibrous pore-filling mineral in the studied sandstones (Fig. 2b). It generally occurs in a close textural relationship with corroded albite grains (Fig. 2f). Dawsonite is found in all but one of the samples selected for the current study. Dawsonite forms thin fibrous bundles in samples with low dawsonite concentrations. In samples with high abundance, dawsonite fibers are thick. In some cases dawsonites are found in textural relationships with K-feldspars. Nevertheless, K-feldspars only rarely exhibit any dissolution textures. In some samples the dissolution of detrital K-feldspar grains is observed, as well as the formation of diagenetic K-feldspar.

The textural relation of the cements in the sandstones is also strongly influenced by the abundance of dawsonite. In sample RM6-9R, which has the highest abundance of dawsonite (16%), the images taken at artificially fractured surfaces show that ankerite dissolution/resorption predated dawsonite formation (Fig. 2b) in some parts of the rock. Whereas in other areas ankerite resorption textures are missing and dawsonite fibers are in contact with euhederal ankerite crystals (Fig. 2e). Moreover, dawsonite crystals display textural equilibrium with fine diagenetic quartz overgrowths and book-shaped kaolinite crystals that formed in the pore space (Fig. 2a, 2c). Dawsonite crystals are never coated by fine grain kaolinite. However, kaolinite and mica patches can be completely surrounded by dawsonite.

Other studied samples exhibit an even more complex crystallization history of carbonates. Nevertheless, dawsonite in these rocks always occurs in the pores of resorbed albite crystals (Fig. 2f).

Results of analytical measurements

XRD

The results show that the average mineral composition of the samples is quartz (*c.* 46.6%), dolomite + ankerite (*c.* 16.1%), muscovite (*c.* 12.1%), kaolinite (*c.* 10.7%), calcite (*c.* 4.8%), siderite (*c.* 2.9%), dawsonite (*c.* 2.3%), illite (*c.* 1.8%), K-feldspar (*c.* 1.2%), plagioclase (*c.* 0.9%), gypsum (*c.* 0.4%) and goethite (*c.* 0.4%).

Thermal analysis

The carbonate minerals in the analysed samples are calcite, dolomite, ankerite, siderite and dawsonite.

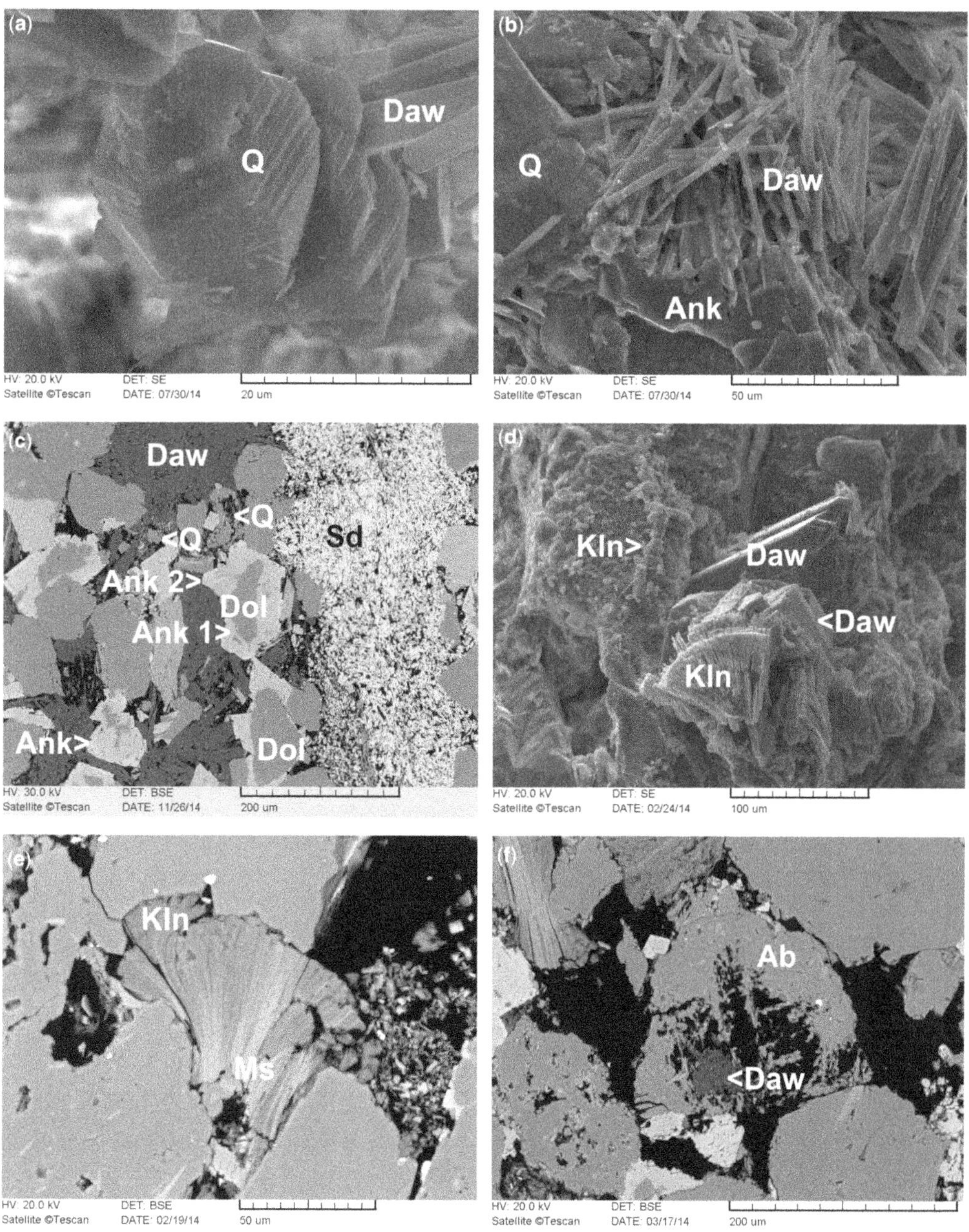

Fig. 2. SE (**a**–**c**) and BSE (**d**–**f**) images from the natural CO_2-bearing reservoir samples. Daw, dawsonite; Ab, albite; Ank, ankerite; Q, quartz; Kln, kaolinite; Ms, muscovite; Dol, dolomite; Sd, siderite; Ank 1, Fe-poor ankerite; Ank 2, Fe-rich ankerite. (**a**) Diagenetic quartz overgrowth with dawsonite (sample: RM6-9R). (**b**) Fibrous dawsonite precipitation as a pore-filling mineral phase with ankerite cement and quartz overgrowth (sample: RM6-9R). (**c**) Diagenetic book-shape kaolinite in dawsonite and kaolinite coating (sample: RM19-6R). (**d**) Kaolinite formation in the close vicinity of muscovite (sample: RM19-6R). (**e**) Occurrence of carbonates (sample RM6-9R). (**f**) Dawsonite precipitation in the secondary pore of an albite relict (sample: RM46-7/2R).

The weight loss 'peaks' of dawsonite are separated from the other carbonates and OH^--bearing phases – the peak of CO_2 loss appears at 346°C and the peak of OH^- loss is at 318°C in dawsonite (Fig. 3) (Földvári 2011).

FTIR and Raman measurements

The results of Raman measurements confirm that the fibrous carbonate minerals are dawsonites, because their spectra are very similar to those of the synthetic dawsonites reported by Serna *et al.* (1985) and Frost & Bouzaid (2007). The ATR FTIR is sensitive to the presence of molecular H_2O or OH^-, therefore it is particularly effective for the identification of sheet silicates and dawsonite in these mineralogical assemblage. The main band of dawsonite appears at 3280 cm^{-1} (Serna *et al.* 1985). The results show that dawsonite could, in fact, be present in more samples than is indicated by XRD. Composition of the samples, based on the ATR FTIR, include kaolinite, dawsonite, carbonates and silicates (Fig. 4).

Results of the kinetic model

As a consequence of the addition of CO_2, the model suggests dissolution of albite, calcite (early stage), dolomite, illite and small amounts of pyrite. The minerals precipitating are quartz, dawsonite, kaolinite, ankerite and calcite (late stage). In case of K-feldspar, dissolution and precipitation also occur. The result of the model shows that the pH immediately decreases from 6.89 to 4.72 after the addition of CO_2 into the model system. However, at the end of the simulation it reaches a value of 6.84 as a consequence of the dissolution and precipitation of different mineral phases (Fig. 5f). The mineral dissolution/precipitation reactions may lead to changes in the porosity and permeability of the reservoir rock. The most sensitive minerals, calcite and dolomite, begin to dissolve rapidly as the pH drops (Fig. 5c, D). The dissolution of these carbonates increases the pH and the porosity. Subsequently, the precipitation of ankerite was observed (Fig. 5b), which consumes the Fe^{2+} coming from the simultaneous slight dissolution of pyrite. Following this, albite and small amount of K-feldspar dissolve and kaolinite, quartz and dawsonite precipitate according the model (Fig. 5a, e, g, h), which is in agreement with the petrographic observations of the CO_2-bearing reservoir rocks in the Mihályi-Répcelak area (Fig. 2a, c). The precipitation of K-feldspar following its dissolution is also supported by some of the studied core samples. There is no petrographic evidence for the dissolution of kaolinite that is indicated by the model in the late stage. This may be explained by its small grain size. Furthermore, the model suggests the dissolution of illite, which was initially present in the model rock composition in small amounts. The model also indicates the precipitation of late-stage calcite (Fig. 5d), which is also observed in some of the samples from the Mihályi-Rápcelak area. However, we should note that the change in the amount of carbonate minerals (calcite, dolomite and ankerite) in the model is negligible compared

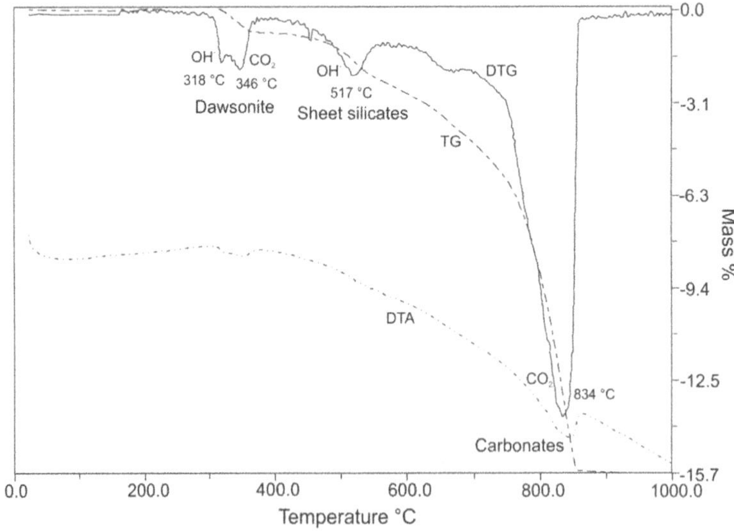

Fig. 3. The result of the thermal analysis shows peak OH^- loss at 318°C and peak CO_2 loss at 346°C for dawsonite. DTA, differential thermal analysis; DTG, derivative thermogravimetry; TG, thermogravimetry.

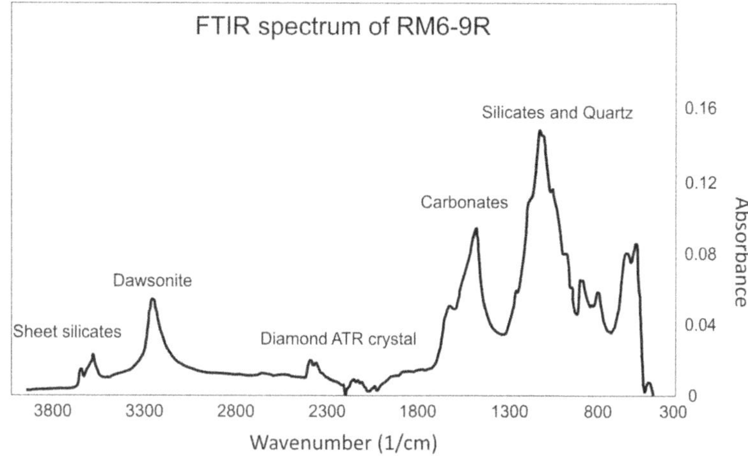

Fig. 4. FTIR spectra of RM6-9R, where the volume ratio of dawsonite is approximately 16%. The main peak of dawsonite is at 3280 cm^{-1}.

with the amount of dawsonite formed, which can also trap CO_2 in a solid phase.

The time required for the reactions highly depends on the kinetic rates of mineral dissolution and precipitation, as well as the SSA values of the minerals (Xu *et al.* 2005; Hellevang & Aagaard 2013). Therefore the timing of the reactions reported here is uncertain, but the relative timing of the dissolution/precipitation processes can be considered as a good approximation of the reaction order because the by petrographic observations confirm these results.

Discussion

Effect of CO_2 on the reservoir

According to our petrographic observations and other studies dealing with the target turbiditic sandstone formation (e.g. Mátyás & Matter 1997), the first type of cement in the system was a carbonate (calcite, ankerite) cement, similar in composition to the inner zones of carbonates found on exogenic dolomite grains. When the CO_2 plume reached the reservoir and dissolved in the pore water, the pH decreased. The result of this process was that the carbonate cement and the overgrowth on dolomite grains were completely or partially dissolved. Furthermore albite grains started to dissolve in this lower pH environment. When the porewater became oversaturated in Na^+, Al^{3+} and CO_3^{2-} the dawsonite could precipitate. We assume that this precipitation was a local process, because Al^{3+} is an immobile ion in water (Hayes & O'Driscoll 1990). During these mineral reaction processes the pH increased; as a result the permeability of the reservoir may have significantly changed.

Kinetic model

The results of the kinetic model are in good agreement with the petrographic observations in the CO_2-bearing sandstone samples from the Mihályi-Répcelak area. The dissolution of albite, early calcite, dolomite and K-feldspar are observed, as are the precipitation of quartz, dawsonite, kaolinite, ankerite and late calcite.

The model indicates that the most important impact of CO_2 is the significant dissolution of albite and the subsequent formation of dawsonite, suggesting that albite plays a key role in dawsonite formation, acting as a source for the necessary Al^{3+} (and Na^+) (Fig. 5a). A similar explanation for the albite–dawsonite relationship was given by Amin *et al.* (2014). This result is well supported by our petrographic observations, namely that dawsonite is texturally intimately related to dissolving albite grains (Fig. 2f). According to the model, the contribution of other carbonates to CO_2 mineral trapping is insignificant compared with dawsonite, at least in the studied area.

CO_2 trapping capacity of dawsonite

Based on the maximum concentration of dawsonite (16%; sample RM6-9R), we have performed a rough estimation of its CO_2 trapping capacity – assuming that dawsonite is the major mineral phase that entraps CO_2 during the reaction between reservoir and injected carbon dioxide in the studied system.

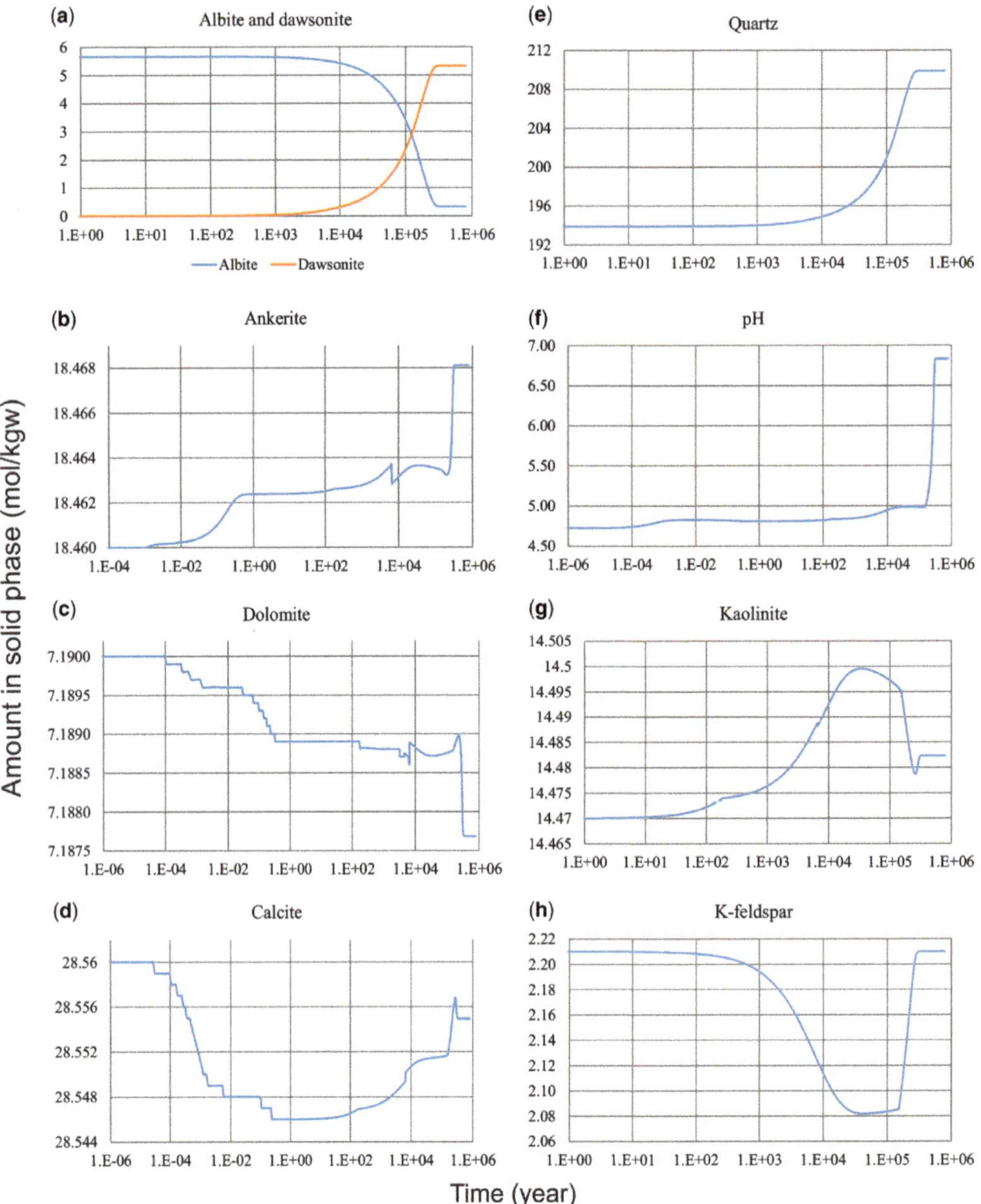

Fig. 5. Results of kinetic batch modelling showing the dissolution and precipitation of reservoir mineral secondary phases as the consequence of the reaction with CO_2. (**a**) dissolution of albite and the subsequent formation of dawsonite; (**b**) precipitation of ankerite; (**c**): dissolution of dolomite; (**d**) early dissolution and late-stage precipitation of calcite; (**e**) precipitation of quartz; (**f**) pH decreases immediately after the addition of CO_2 into the model from 6.89 to 4.72 and subsequent increases up to 6.84 as a consequence of the mineral reactions; (**g**) kaolinite precipitation following by its late-stage dissolution; (**h**) dissolution and subsequent formation of K-feldspar. kgw, kg of water.

The reaction that describes the precipitation of dawsonite is shown below:

$$\underset{\text{albite}}{NaAlSi_3O_8} + CO_2 + H_2O \longleftrightarrow \underset{\text{dawsonite}}{NaAlCO_3(OH)_2} + 3SiO_2 \quad (3)$$

Taking a volume of 1 m^3 with 10% porosity, the mass of rock is approximately 2.4 t according to the density and the volume of rock (2.7 g cm^{-3}, 0.9 m^3). The mass of dawsonite in this system is thus 0.4 tonnes (2.4 tonnes × 16%). The molar mass of dawsonite is 143.98 g mol^{-1}, so the amount of dawsonite (0.4/143.98 g mol^{-1}) is 2.7×10^3 mol. For this reason the CO_2 trapping capacity is also 2.7×10^3 mol (v. equation 3), which corresponds to 0.12 t of CO_2 per m^3 (according to the molar mass of CO_2). Where the dawsonite concentration is lower the CO_2 trapping capacity of dawsonite is also lower. In some samples the dawsonite concentration is lower than 1% (below the detection limit of XRD), so the applied complex analytical method was important in detecting this mineral. The FTIR analysis is sensitive to the OH^- content of dawsonite, and in some samples the dawsonite was detected only with SEM.

Stability of dawsonite

According to the kinetic modelling dawsonite plays a significant role in trapping CO_2 in the mineral phase (Hellevang *et al.* 2011 and references therein). Many of the natural CO_2 reservoirs contain dawsonite (Baker *et al.* 1995; Moore *et al.* 2005; Worden 2006; Gao *et al.* 2009), however, dawsonite-barren natural CO_2 reservoirs are also known (Pearce *et al.* 1996). The stability of dawsonite depends on the partial pressure of CO_2, porewater composition and temperature but the precise definition of the stability zone of dawsonite is incomplete and is estimated only by geochemical models (Hellevang *et al.* 2005; Hellevang *et al.* 2011).

We summarized the published physical and geochemical conditions of dawsonite-bearing natural CO_2 reservoirs worldwide (Table 4). From these data the stability interval of dawsonite can be estimated. Dissolution of the dawsonite present in these reservoirs was not observed, therefore we may assume stability of the mineral at these conditions. The minimum partial pressure of CO_2 in these systems is approximately 5 bar, calculated assuming hydrostatic pressure gradients (0.1 bar m^{-1}), minimum depth (160 m) and minimum CO_2 concentration (30%) (160 m × 0.1 bar m^{-1} × 0.3 = 4.8 bar) in the Bowen-Gunnedah area, Sydney Basin. The minimum temperature is 20°C and the minimum TDS in adjacent porewater is 4210 mg l^{-1} (Baker *et al.* 1995). The maximum partial pressure of CO_2 is approximately 188.7 bar calculated with the maximum depth, hydrostatic pressure and CO_2 concentration (Hailer Basin, Gao *et al.* 2009). The reported maximum temperature is approximately 110°C in the Upper Cretaceous reservoir of the Songliao Basin, China (Me *et al.* 2003). The maximum TDS is found in the Triassic Lam Fm in the Shabwa Basin, Yemen (Worden 2006), which is 240 000 mg l^{-1}. The partial pressure range of CO_2 in the whole Mihályi-Répcelak area is between 128 bar and 155 bar and the temperature is 75–95°C (Mészáros *et al.* 1979). The TDS in the porewater falls between 18 000 mg l^{-1} and 50 700 mg l^{-1}. Therefore, the study area fits well within these dawsonite stability conditions.

Table 4. *Physical and chemical properties of natural dawsonite-bearing CO_2 reservoirs*

Area	Depth (m)	CO_2 concentration (vol%)	Temperature (°C)	TDS (mg l^{-1})	References
Late Miocene Mihályi-Répcelak area, Hungary	1225–1643.5	69–98	72.8–93.2	18 000–50 700	This paper
Late Mesozoic–Cenozoic Hailaer Basin, China	1331.4–1949.5	80.2–96.8	Not reported	Not reported	Gao *et al.* (2009)
Cretaceous Izumi Group, Japan Okuyama *et al.* (2013)	Not	reported	Not reported	80–85	Not reported
Triassic Lam Fm, Shabwa Basin, Yemen	2000–3000	*c.* 20	85–100	>240 000	Worden (2006)
Upper Cretaceous reservoir, Songliao Basin, China Liu *et al.* (2011); Me *et al.* (2003)	1256–	1599.15	Not reported	*c.* 110	7600–10 000
Permian Springerville-St. Johns Field, USA	463–495	*c.* 90	20–30	4210	Moore *et al.* (2005)
Triassic Bowen-Gunnedah-Sydney Basin. Australia	160–1401	>30	25–75	Not reported	Baker *et al.* (1995)
Early Permian Fizzy accumulation, Southern North Sea, UK	2300–2400	50	80–85	9000	Wilkinson *et al.* (2009)

Dawsonite seems to be stable over a wide pressure, temperature and TDS range and it may very well be that other factors, such as the availability of Al^{3+} (and Na^{+}) or the presence/lack of source minerals, are also important limiting factors for the growth of dawsonite. Furthermore, the analytical tools used with an emphasis on ATR FTIR proved to be very sensitive to the occurrence of dawsonite. Its growth and presence even as a trace in natural CO_2 reservoirs may be more widespread than previously thought.

Conclusion

(1) Using multiple analytical techniques we have demonstrated that dawsonite is present in every studied reservoir rock sample in Mihályi-Répcelak area. Dawsonite is detected as the most visible effect of CO_2 emplacement in the samples.
(2) Dawsonite precipitation is strongly related to albite dissolution, which is probably the most important source of Al^{3+} (and Na^{+}) in the studied area for the formation of dawsonite. In most cases this is a local (sub-mm scale) reaction as aluminium is a highly immobile ion at the given conditions.
(3) The Mihályi-Répcelak natural CO_2 reservoir is an excellent study area to better understand the long-term geochemical processes taking place in a sandstone reservoir. The leakage-free storage of CO_2 for geological timescales implies that, at least in the Pannonian Basin, areas with similar geological characteristics are potential candidates for long-term safe CO_2 storage.
(4) Basic kinetic modelling was capable of closely describing most of the mineral reactions in the studied reservoirs and proved to be an effective tool for predicting the changes taking place in a reservoir after CO_2 injection, even on geological timescales.

The authors are grateful to Csaba Szabó for the consultations and his help in the organization of measurements. The authors give thanks to the Hungarian Geological and Geophysical Institute and to Faculty of Science Research and Instrument Core Facility at Eötvös University (ELTE FS-RICF) for the use of analytical equipment. The authors are very grateful to the two anonymous reviewers and Peter Armitage for thoughtful critique and detailed and constructive comments and to Jenny Omma for her patience in improving the style and English of the paper. This research was supported by KMOP Project No. 4.2.1/B-10-2010-002 of the European Union, carried out in agreement between ELTE and MFGI (TTK 2461/1/2013 and MFGI 206-114/2013) and K 115927 115927 (PI: Gy. Falus). This is the 74th publication of the Lithosphere Fluid Research Lab at Eötvös University.

References

Allen, D.E., Strazisar, B.R., Soong, Y. & Hedges, S.W. 2005. Modelling carbon dioxide sequestration in saline aquifers: significance of elevated pressures and salinities. *Fuel Processing Technology*, **86**, 1569–1580.

Amin, S.M., Weiss, D.J. & Blunt, M.J. 2014. Reactive transport modelling of geologic CO_2 sequestration in saline aquifers: the influence of pure CO_2 and of mixtures of CO_2 with CH_4 on the sealing capacity of cap rock at 37°C and 100 bar. *Chemical Geology*, **367**, 39–50.

Armitage, P.J., Worden, R.H., Faulkner, D.R., Aplin, A.C., Butcher, A.R. & Iliffe, J. 2010. Diagenetic and sedimentary controls on porosity in Lower Carboniferous fine-grained lithologies, Krechba field, Algeria: a petrological study of a caprock to a carbon capture site. *Marine and Petroleum Geology*, **27**, 1395–1410.

Armitage, P.J., Faulkner, D.R. & Worden, R.H. 2013. Caprock corrosion. *Nature Geoscience*, **6**, 79–80.

Arts, R., Beaubien, S. *et al.* 2008. *What does CO_2 Storage Really Mean?* CO2GeoNet European Network of excellence, Orléans, France, 1–20.

Bachu, S., Bonijoly, D., Bradshaw, J., Burruss, R., Holloway, S., Christensen, N.P. & Mathiassen, O.M. 2007. CO_2 storage capacity estimation: methodology and gaps. *International Journal of Greenhouse Gas Control*, **1**, 430–443.

Baker, J.C., Bai, G.P., Hamilton, P.J., Golding, S.D. & Keene, J.B. 1995. Continental-scale magmatic carbon dioxide seepage recorded by dawsonite in the Bowen-Gunnedah Sydney Basin system, Eastern Australia. *Journal of Sedimentary Research*, **A65**, 522–530.

Berta, M., Király, Cs., Lévai, Gy., Falus, Gy. & Szabó, Cs. 2012. Carbon Capture and Sequestration: a preliminary study on its determinant geochemical processes on Pannonian sedimentary formations. *Hungarian Geophysics*, **4**, 258–264 [in Hungarian].

Brantley, S.L. 2008. Kinetics of mineral dissolution. *In*: Brantley, S.L., Kubicki, J.D. & White, A.F. (eds) *Kinetics of Water-Rock Interactions*. Springer, New York, 151–196.

Carroll, S.A. & Knauss, K.G. 2005. Dependence of labradorite dissolution kinetics on CO_2(aq), Al(aq), and temperature. *Chemical Geology*, **217**, 213–225.

Clayton, J.L., Spencer, C.W., Koncz, I. & Szalay, A. 1990. Origin and migration of hydrocarbon gases and carbon dioxide, Békés Basin, southeastern Hungary. *Organic Geochemistry*, **15**, 233–247.

Csizmeg, J., Márton, B., Szalay, Á., Vető, I., Peffer, M., Varga, G. & Pogácsás, Gy. 2012. Neogene hydrocarbon potential and inert gas risk in the Hungarian part of Danube Basin. Paper presented at the 29th IAS Meeting of Sedimentology, Sedimentology in the Heart of Alps, 10–13 September, 2012, Schladming, Austria, 343.

Dogan, A.U., Dogan, M., Onal, M., Sarikaya, Y., Aburub, A. & Wurster, D.E. 2006. Baseline studies of the Clay Minerals Society source clays: specific surface area by the Brunauer Emmett Teller (BET) method. *Clays and Clay Minerals*, **54/1**, 62–66.

Dövényi, P., Horváth, F., Liebe, P., Gálfi, J. & Erki, I. 1983. Geothermal conditions of Hungary. *Geophysical Transactions*, **29**, 3–114.

Fancsik, T., Török, K., Törökné Sinka, M., Szabó, Cs. & Lenkey, L. 2007. *Long Term Storage Options of Industrial Carbon Dioxide in Hungary*. Research Reports by the Hungarian Academy of Sciences and the Prime Minister's Office 89–119 [in Hungarian].

Földvári, M. 2011. *Handbook of Thermogravimetric System of Minerals and its Use in Geological Practice*. Occasional Papers of the Geological Institute of Hungary, Budapest, **213**, 1–179.

Frost, R.L. & Bouzaid, J.M. 2007. Raman spectroscopy of dawsonite $NaAl(CO_3)(OH)_2$. *Journal of Raman Spectroscopy*, **38/7**, 873–879.

Gao, Y., Liu, L. & Hu, W. 2009. Petrology and isotopic geochemistry of dawsonite-bearing sandstones in Hailaer basin, northeastern China. *Applied Geochemistry*, **24**, 1724–1738.

Gaus, I., Azaroual, M. & Czernichowski-Lauriol, I. 2005. Reactive transport modelling of the impact of CO_2 injection on the clayey cap rock at Sleipner (North Sea). *Chemical Geology*, **217**, 319–337.

Golubev, S.V., Bénézeth, P., Schott, J., Dandurand, J.L. & Castillo, A. 2009. Siderite dissolution kinetics in acidic aqueous solutions from 25 to 100°C and 0 to 50 atm pCO_2. *Chemical Geology*, **265**, 13–19.

Hayes, J.P. & O'Driscoll, C.F. 1990. *Regional Geological Setting and Alteration within the Eastern Avalon High-Alumina Belt, Avalon Peninsula, Newfoundland*. Newfoundland Department of Mines and Energy, Geological Survey Branch Report **90/1**.

Hellevang, H. & Aagaard, P. 2013. Can the long-term potential for carbonatization and safe long-term CO_2 storage in sedimentary formations be predicted? *Applied Geochemistry*, **39**, 108–118.

Hellevang, H., Aagaard, P., Oelkers, E.H. & Kvamme, B. 2005. Can dawsonite permanently trap CO_2? *Environmental Science & Technology*., **39**, 8281–8287.

Hellevang, H., Declercq, J., Kvamme, B. & Aagaard, P. 2010. The dissolution rates of dawsonite at pH 0.9 to 5 and temperature of 22, 60 and 77°C. *Applied Geochemistry*, **25**, 1575–1586.

Hellevang, H., Declercq, J. & Aagaard, P. 2011. Why is dawsonite absent in CO_2 charged reservoirs? *Oil & Gas Science and Technology*, **66**, 119–135.

Juhász, Gy. 1991. Lithostratigraphic and sedimentological framework of the Pannonian (s.l.) sedimentary sequence in the Hungarian Plain (Alföld), Eastern Hungary. *Acta Geologica Hungarica*, **34**, 53–72.

Kampman, N., Bickle, M., Wigley, M. & Dubacq, B. 2014. Fluid flow and CO_2-fluid-mineral interactions during CO_2-storage in sedimentary basins. *Chemical Geology*, **369**, 22–50.

Kaszuba, J.P., Janecky, D.R. & Snow, M.G. 2003. Carbon dioxide reaction processes in a model brine aquifer at 200°C and 200 bars: implication for geologic sequestration of carbon. *Applied Geochemistry*, **18**, 1065–1080.

Knauss, K.G., Johnson, J.W. & Steefel, C.I. 2005. Evaluation of the impact of CO_2, co-contaminant gas, aqueous fluid and reservoir rock interactions on the geologic sequestration of CO_2. *Chemical Geology*, **217**, 339–350.

Kubus, P. 2009. CO_2 storage possibilities in Hungary. *MOL Group Scientific Magazine (SD & HSE Special Issue)*, 120–132.

Lasaga, A.C. 1984. Chemical kinetics of water–rock interactions. *Journal of Geophysical Research*, **89**, 4009–4025.

Liu, N., Qu, X., Yang, H., Wang, L. & Zhao, S. 2011. Genesis of authigene carbonate minerals in the Upper Cretaceous reservoir, Honggang Anticline, Songliao Basin: a natural analogue for mineral trapping of natural CO_2 storage. *Sedimentary Geology*, **237**, 166–178.

Magyar, I., Radivojević, D., Sztanó, O., Synak, R., Ujszászi, K. & Pócsik, M. 2013. Progradation of the paleo-Danube shelf margin across the Pannonian Basin during the Late Miocene and Early Pliocene. *Global and Planetary Change*, **103**, 168–173.

Mátyás, J. & Matter, A. 1997. Diagenetic indicators of meteoric flow in the Pannonian Basin, Southeast Hungary. In: *Basin-Wide Diagenetic Patterns: Integrated Petrologic, Geochemical and Hydrologic Considerations*, SEPM Special Publications, Tulsa, USA, **57**, 281–296.

Me, X.N., Jiao, J.J. & Cheng, J.M. 2003. Regional variation of formation water chemistry and diagenesis reaction in underpressured system: example from Shiwu depression of Songliao Basin, NE China. *Journal of Geochemical Exploration*, **78-80**, 585–590.

Mészáros, L., Dallos, E. et al. 1979. *Final Report on the Exploration Phase and OGIP-Calculation of Non-Combustible Mixed Gas Reservoirs from the Mihályi Exploration Area.*, Országos Kőolajipari Tröszt, 1–116 [in Hungarian].

Moore, J., Adams, M., Allis, R., Lutz, S. & Rauzi, S. 2005. Mineralogical and geochemical consequences of the long-term presence of CO_2 in natural reservoirs: an example from the Springerville-St. Johns Field, Arizona, and New Mexico, U.S.A. *Chemical Geology*, **217**, 365–385.

Oelkers, E.H., Schott, J., Gauthier, J.-M. & Herrero-Roncal, T. 2008. An experimental study of the dissolution mechanism and rates of muscovite. *Geochimica et Cosmochimica Acta*, **72**, 4948–4961.

Okuyama, Y., Funatsu, T., Fujii, T. & Take, S. 2013. Dawsonite and other carbonate veins in the Cretaceous Izumi Group, SW Japan: a natural support for fracture self-sealing in mudstone caprock in CGS. *Procedia Earth and Planetary Science*, **7**, 6363–6639.

Palandri, J.L. & Kharaka, Y.K. 2004. *A Compilation of Rate Parameters of Water–Mineral Interaction Kinetics for Application to Geochemical Modelling*. U.S. Geological Survey Water-Recourses Investigations Report **04/1068**.

Palcsu, L., Vető, I., Futó, I., Vodila, G., Papp, L. & Major, Z. 2014. In-reservoir mixing of mantle-derived CO_2 and metasedimentary CH_4-N_2 fluids – noble gas and stable isotope study of two multistacked fields (Pannonian Basin System, W-Hungary). *Marine and Petroleum Geology*, **54**, 216–227.

Parkhurst, D.L. & Appelo, C.A.J. 2012. *Description of Input and Examples for PHREEQC Version 3–A Computer Program for Speciation, Batch-Reaction, One-Dimensional Transport, and Inverse Geochemical Calculations*. U.S. Geological Survey Open-File Report **490**.

PEARCE, J.M., HOLLOWAY, S., WACKER, H., NELIS, M.K., ROCHELLE, C. & BATEMAN, K. 1996. Natural occurrences as analogues for the geological disposal of carbon dioxide. *Energy Conversion and Management*, **37**, 1123–1128.

PEARCE, J.M., SHEPHERD, T.J. ET AL. 2005. *Natural analogues for the geological storage of* CO_2. NASCENT project, Nottingham, 1–115.

RIMSTIDT, J.D. & BARNES, H.L. 1980. The kinetics of silica-water reactions. *Geochimica et Cosmochimica Acta*, **44**, 1683–1699.

SERNA, C.J., GARCIA-RAMOS, J.V. & PENA, M.J. 1985. Vibrational study of dawsonite type compounds $MAl(OH_2)CO_3$ ($M = Na$, K, NH_4). *Spectrochimica Acta*, **41A**, 697–702.

SZAMOSFALVI, Á., FALUS, Gy. & JUHÁSZ, Gy. 2011. The potential options of storing CO_2 in saline reservoir in Hungary. *Hungarian Geophysics*, **52**, 95–106.

TÓTH, J., UDVARDI, B., KOVÁCS, I., FALUS, Gy., SZABÓ, Cs., TROSKOT-ČORBIĆ, T. & SLAVKOVIĆ, R. 2012. Analytical development in FTIR analysis of clay minerals. *MOL Group Scientific Magazine*, **1**, 52–61.

UDVARDI, B., KOVÁCS, I.J. ET AL. 2014. Application of attenuated total reflectance Fourier transform infrared spectroscopy in the mineralogical study of a landslide area, Hungary. *Sedimentary Geology*, **313**, 1–14.

VETŐ, I., CSIZMEG, J. & SAJGÓ, Cs. 2014. Accumulation and mixing of magmatic CO_2 and hydrocarbon – nitrogen gas in the southern Danube Basin. *Central European Geology*, **57/1**, 53–69.

WILKINSON, M., HASZELDINE, R.S., FALLICK, A.E., ODLING, N., STOKER, S.J. & GATLIFF, R.W. 2009. CO_2-mineral reaction in a natural analogue for CO_2 storage – implications for modeling. *Journal of Sedimentary Research*, **79/7**, 486–494.

WORDEN, R.H. 2006. Dawsonite cement in the Triassic Lam Formation, Shabwa Basin, Yemen: a natural analogue for a potential mineral product of subsurface CO_2 storage for greenhouse gas reduction. *Marine and Petroleum Geology*, **23**, 61–77.

WORDEN, R.H. & SMITH, L.K. 2004. CO_2 injection into oil fields for enhanced oil recovery (EOR): lessons for the geological sequestration of CO_2 in the subsurface. *In*: BAINES, S. & WORDEN, R.H., (eds) *Geological Storage of Carbon Dioxide*. Geological Society, London, Special Publications, **233**, 211–224, https://doi.org/10.1144/GSL.SP.2004.233.01.14

XU, T., APPS, J.A. & PRUESS, K. 2004. Numerical simulation of CO_2 disposal by mineral trapping in deep aquifers. *Applied Geochemistry*, **19**, 917–936.

XU, T., APPS, J.A. & PRUESS, K. 2005. Mineral sequestration of carbon dioxide in a sandstone-shale system. *Chemical Geology*, **217**, 295–318.

YANG, L. & STEEFEL, C.I. 2008. Kaolinite dissolution and precipitation kinetics at 22°C and pH 4. *Geochimica et Cosmochimica Acta*, **72**, 99–116.

Reactive transport modelling of compacting siliciclastic sediment diagenesis

C. GELONI*, A. ORTENZI & A. CONSONNI

Department of GEOLAB, eni SpA, via Maritano 26, 20097, San Donato Milanese, Italy

**Correspondence: claudio.geloni@eni.com*

Abstract: A reactive transport model (RTM) is used to simulate the diagenetic evolution of a siliciclastic reservoir with a known burial and thermal history. The diagenetic phenomena occurring in two temperature regimes are simulated: kaolinite/chlorite formation at low/medium temperatures by means of isothermal zero-dimensional (0D) flush models, and smectite illitization and quartz precipitation at high temperatures by means of 0D and 1D non-isothermal models.

Zero-dimensional models show that at 30°C kaolinite forms only in freshwater from K-feldspar and quartz. In river or seawater, muscovite is stable instead of kaolinite. Calcite formation depends on pH and total inorganic carbon. At 50°C, seawater promotes Mg-chlorite formation from mica alterations. At 70–110°C, evaporated seawater favours smectite–illite transformation and quartz precipitation.

The non-isothermal 1D models are used to simulate the diagenesis of a compacting clay expelling fluid into an underlying sandstone.

An sensitivity analysis of the clays' thermodynamic and kinetic parameters is carried out to assess the possibility of modelling the transformation of smectite into illite in the clay, with the concomitant formation of a quartz cement in the adjacent sandstone. The results of the numerical simulations point out that the extent of the smectite–illite conversion and quartz precipitation is dependent primarily on the availability of potassium: temperature, however, does not seem to play a major role.

The diagenesis of sand is characterized by the complexity of the dissolution/precipitation reactions (Primmer *et al.* 1997), and by the interaction between chemical and mechanical compaction. The primary depositional characteristics of sand, including grain size, sorting and mineral composition, directly affect the diagenetic processes that turn sand into sandstone.

Any attempt to predict the diagenetic history of a sandstone has to take into consideration the aforementioned complexity and the evolution of the waters that fill the pore space between grains, in addition to the thermal and stress history of the rock. An educated guess at the initial composition of the porewaters may be made on the basis of the environment of deposition and constrained by the present-day chemical composition, generally known from analyses, especially in oil fields. The evolution of porewater composition between deposition and the present day is poorly constrained, even though it is possible to make some assumption about the original chemical composition on the basis of the petrographical observations on early diagenetic processes. Major alterations of the sandstone and its porewaters begin when the temperature rises above 80°C. At the burial depths corresponding to such temperatures, water flux is usually reduced to very low rates. This is why there is a debate on the relative effects of fluid flow v. diffusion in deeply buried sandstones.

When modelling the diagenetic history, two approaches can be considered. The first approach considers a limited range of phenomena (such as mechanical compaction and quartz cementation) using the Arrhenius formulation for the kinetics of mineral precipitation. This approach can be considered as a zero-dimensional (0D) model, with no inherent spatial scale. These models can be predictive, as they are calibrated to measured petrographical data, and to the burial and thermal history of the reservoir (Walderhaug 2000; Lander *et al.* 2008; Lander & Bonnell 2010).

The other approach, known as reactive transport modelling (RTM), is based on a mechanistic formulation of water–rock interactions. The reactive behaviour of the system is described by thermodynamic and kinetics laws (for the precipitation/dissolution reactions), and is constrained by mass balances of the reactive species. The transport can be simulated in a multidimensional grid, where fluid flow, pressure and temperature fields are defined or computed on the basis of the thermophysical properties of the system. To perform such simulations, a considerable amount of input data are needed. The most difficult input data to specify are the thermodynamic parameters of the clay minerals

From: Armitage, P. J., Butcher, A. R., Churchill, J. M., Csoma, A. E., Hollis, C., Lander, R. H., Omma, J. E. & Worden, R. H. (eds) 2018. *Reservoir Quality of Clastic and Carbonate Rocks: Analysis, Modelling and Prediction*. Geological Society, London, Special Publications, **435**, 419–439.
First published online December 10, 2015, https://doi.org/10.1144/SP435.7

and the chemical composition of the interstitial waters.

The high variability of these initial constraints can severely affect the predictive character of this kind of modelling, which is, instead, used as a means of describing the phenomenology of the processes occurring in the investigated geochemical system (possibly giving a formal description in terms of a series of strictly balanced chemical reactions) and to assess the effects induced by the migration of the silica released by the processes itself. It is also an attempt to evaluate the relative weights of the various factors driving the fate of the system, mainly focusing on temperature and the availability of components needed to trigger the mineralogical transformations (i.e. K^+ and $SiO_{2(aq)}$).

The aim of this work is to test the capability of the RTM approach in simulating multiple diagenetic scenarios during burial with a particular focus on the diagenesis of a clay layer where smectite–illite transformation takes place, acting as a source of silica for the adjacent sandstone cementation.

TOUGHREACT (Xu *et al.* 2006) is the simulator used to solve the numerical models. This is a widely used integral finite-difference code that couples a non-isothermal multiphase flow model (TOUGH2: Pruess *et al.* 1999) to a reactive transport module. The code has been modified in order to take into account the effect of a uniaxial compaction of the simulation grid.

Experimental data for model initialization

The case study is a North African Cretaceous sandstone reservoir characterized by relatively simple mineralogy and diagenesis (Fig. 1). The detrital composition is quartzose, and the main diagenetic event was the precipitation of quartz overgrowths during the intermediate–late stages of burial. The other diagenetic products observed during the petrographical analysis were early diagenetic pore-filling kaolinite derived from feldspar alterations and early- to intermediate-stage chlorite sourced from almost completely dissolved unstable components, the remnants of which show a biotitic composition at the microprobe. Intermediate- to late-stage fibrous illite is also present in small amounts. The sediments were deposited in a continental fluvial–lacustrine setting. Shortly after deposition, a marine transgression occurred, followed by a period of restricted circulation that resulted in the formation of thick salt deposits (Ahlbrandt 2001).

The burial and thermal history of the reservoir is relatively simple (Fig. 2) and was reconstructed by means of an eni proprietary petroleum system model. The quantitative calibration was carried out based on classical geochemical data, such as vitrinite reflectance.

The present-day water composition (Table 1) is characterized by a high salinity, and a significant content of calcium and magnesium. The use of

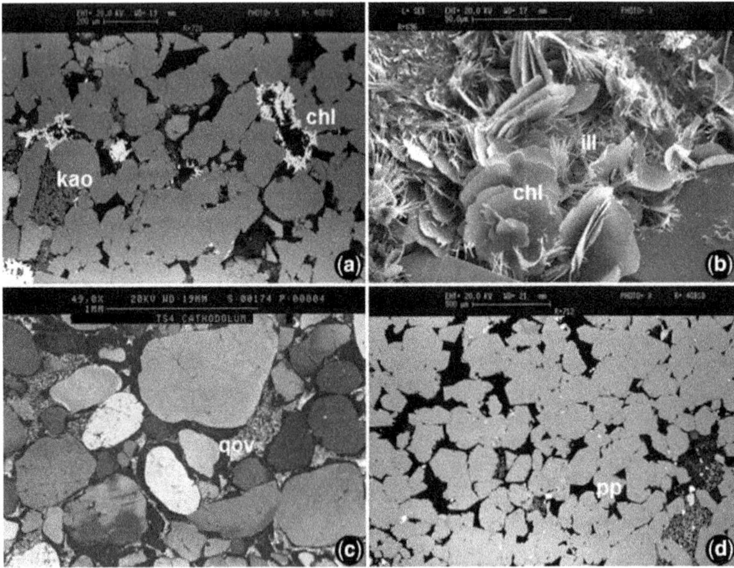

Fig. 1. Backscattered electron images of the sandstone reservoir. The diagenetic events are represented by kaolinite formation ('kao' in **a**), chlorite ('chl' in a & b) and fibrous illite precipitation ('ill' in **b**), and by quartz overgrowths ('qov' in **c**, CL image). Primary porosity is still preserved ('pp' in **d**).

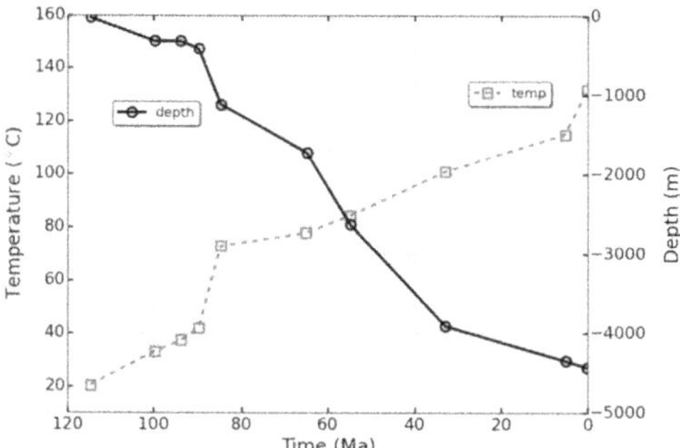

Fig. 2. Burial and thermal history of the sandstone reservoir.

activity models developed for non-ideal electrolytic solutions is mandatory when modelling brines with such a high salinity (Pitzer & Mayorga 1973; Helgeson & Kirkham 1974*a*, *b*). In the simulations, the activity coefficients of charged aqueous species are computed using the extended Debye–Huckel equation and parameters derived by Helgeson *et al.* (1981).

Model set-up

In this paper, three diagenetic events are simulated. The temperature range and the time span of the simulations are constrained by integrating the diagenetic events into the burial history (Fig. 2).

The three diagenetic events are associated, in the conceptual models, to three burial stages, each characterized by different thermodynamic conditions, mineral and water compositions, and hydrological regimes. This is a type of 'spot simulation' approach followed to investigate the most representative condition defining the system during its burial history, where the end of a given stage is not necessarily used as the starting point of the following one.

Table 1. *Average composition of the interstitial water**

pH	5.14
Al	3.53×10^{-6}
B	1.23×10^{-2}
Ba	5.31×10^{-4}
Br	5.23×10^{-3}
C	2.70×10^{-3}
Ca	3.14×10^{-1}
Cl	3.15
F	2.23×10^{-4}
Fe	3.48×10^{-3}
K	1.01×10^{-2}
Li	1.68×10^{-3}
Mg	4.16×10^{-2}
Mn	1.33×10^{-3}
N	5.86×10^{-3}
Na	2.42
S	2.07×10^{-3}
Si	1.16×10^{-3}
Sr	7.95×10^{-3}

*Concentrations are given in terms of molality, pH in pH units.

The three burial stages that can be schematically identified by the major mineral transformation occurring within the sediment are:

- Stage 1 – kaolinitization of K-feldspar: low-temperature (20–30°C) alteration of a quartzose sandstone with 10% K-feldspar. In order to consider the environment of deposition and the subsequent evolution of pore-fluid composition, the composition of freshwater, river water and seawater were all tested.
- Stage 2 – chloritization of biotite: medium–low temperatures (50°C). During this stage, the sandstone alteration is driven by moderately strong flushing of dense hypersaline fluids.
- Stage 3 – illitization of smectite: high temperature (70–110°C). Late burial diagenesis under a very-low-flow hydrological regime of a hypersaline water is assumed to be approximately in equilibrium with the sediments.

Water composition reconstruction

As pointed out in the previous discussion, the water composition plays a key role in driving the overall reactivity of the rock–fluid system. The present-day water composition was used as a guide to reconstruct the main chemical characteristics of the fluid during most of the burial history. One straightforward approach to reconstructing the origin and evolution of the porewater is to use the Cl/Br ratio of water in log–log plots (Fig. 3) with respect to the seawater evaporative trajectory (SET: Carpenter 1978). If the Cl/Br ratio is greater than (above) the SET, this indicates that the brine was in direct contact or mixed with water of evaporitic origin. Ratios less than (below) the SET are formed by dilution with less saline water after halite deposition. Analytical data for porewaters from the case-study reservoir lie on the SET (Fig. 3) just at the onset of sulphate precipitation and before halite formation. This is consistent with the hypothesis of a brine formed by evaporation of seawater in a subaerial environment. A preliminary geochemical model, performed with the PHREEQC2 code (Parkhurst & Appelo 1999), was used to model the chemical composition for the flushing fluids used in stages 2 and 3.

Kinetics and thermodynamic uncertainties

In RTM, especially when dealing with clay minerals, the formal stoichiometric description of each

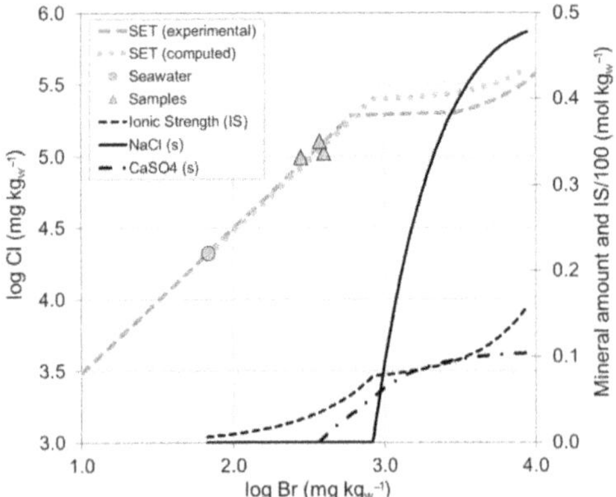

Fig. 3. Seawater evaporation trajectory (SET) and mineral (gypsum and halite) precipitation curves in the Cl–Br domain. Evaporation is simulated with PHREEQC2 by removing water from the system up to very high ionic strength. Analysed water samples fall on the SET.

phase and the accuracy of the thermodynamic parameters describing the stability of the mineral phases as a function of temperature are critical.

The stability of a mineral phase can be quantified in terms of its dissociation constant, K, and of the activities of the species forming the mineral according to the law of mass action (equation 1):

$$\log K = \sum_{i}^{N} v_i \log a_i \qquad (1)$$

where N is the number of components forming the mineral, and v_i and a_i are the stoichiometric coefficient and the activity of the i-th component, respectively.

The temperature dependence of K can be defined in terms of thermochemical laws and can be formally described by equation (2):

$$\log K(T) = a + b\,T + \frac{c}{T} + d\ln(T) + \frac{e}{T^2} \qquad (2)$$

where T is the absolute temperature, and a, b, c, d and e are regression coefficients tabulated in the thermodynamic databases that greatly affect the reactivity of the system.

A first source of uncertainty in the value of log K is related to the computation of the Gibbs free energy, enthalpy and entropy of the mineral species that determine the log K and its dependence on temperature (Garrels 1984; Vieillard 2000).

A second source of uncertainty is the natural variability in the chemical composition of many clay minerals, resulting in a continuous variation of the stoichiometric coefficients. This variability is not captured by RTM, which uses mineral phases with fixed chemical compositions.

As a consequence of these uncertainties, several different sets of parameters exist in the literature to describe the thermodynamic stability of a given mineral phase. This is particularly true for clay minerals such as smectite and illite (Fig. 4). Thermodynamic data for minerals and aqueous species are mostly taken from the EQ3/6 V7.2b database of Wolery (1992). The original database has been integrated to account for recent published revisions in thermodynamic properties of rock-forming minerals (especially clays: Blanc *et al.* 2012).

However, the formal description of the kinetics of mineral dissolution/precipitation reactions is even more challenging. This is why many semi-empirical approaches have been proposed to represent the kinetics for carbonate, silicate and clay reactions (Sverdrup & Warfvinge 1988; Morse & Arvidson 2002). In TOUGHREACT, the kinetics are implemented following the Lasaga general formulation (Lasaga *et al.* 1994), as reported in equation (3):

$$r = \pm kA|1 - \Omega^{\theta}|^{\eta} \qquad (3)$$

where r is the reaction rate (mol s^{-1}: a positive value indicates dissolution, a negative value indicates precipitation), k is the kinetic rate constant (mol m^{-2} s^{-1}) that is dependent on temperature, A (m^2) is the reactive mineral surface area computed from the specific reactive areas and the quantity of each mineral, assuming that the reactive surface is proportional to the mineral volume fraction (Xu *et al.* 2006), the term $1-\Omega$ accounts for the reaction rate decrease as equilibrium is approached (Ω is the mineral saturation ratio), and θ and η are empirical exponents, usually but not always, taken to be equal to 1. The temperature dependence of the reaction rate constant is expressed via the Arrhenius equation (equation 4):

$$k = k_{25}\exp\left(-\frac{E_a}{R}\left(\frac{1}{T} - \frac{1}{298.15}\right)\right) \qquad (4)$$

where k_{25} is the kinetic constant at 25°C (mol m^{-2} s^{-1}), E_a the activation energy (kJ mol^1), R is the universal gas constant and T is the absolute temperature. In the simulations, k_{25}, E_a and the values of the specific reactive areas are based on widely accepted published data (Palandri & Kharaka 2004; Xu *et al.* 2005). Some of the most relevant kinetic parameters are listed in Table 2. During simulations, a sensitivity analysis was performed for k_{25} and the activation energies of all the mineral species, and for the specific reactive surface area of

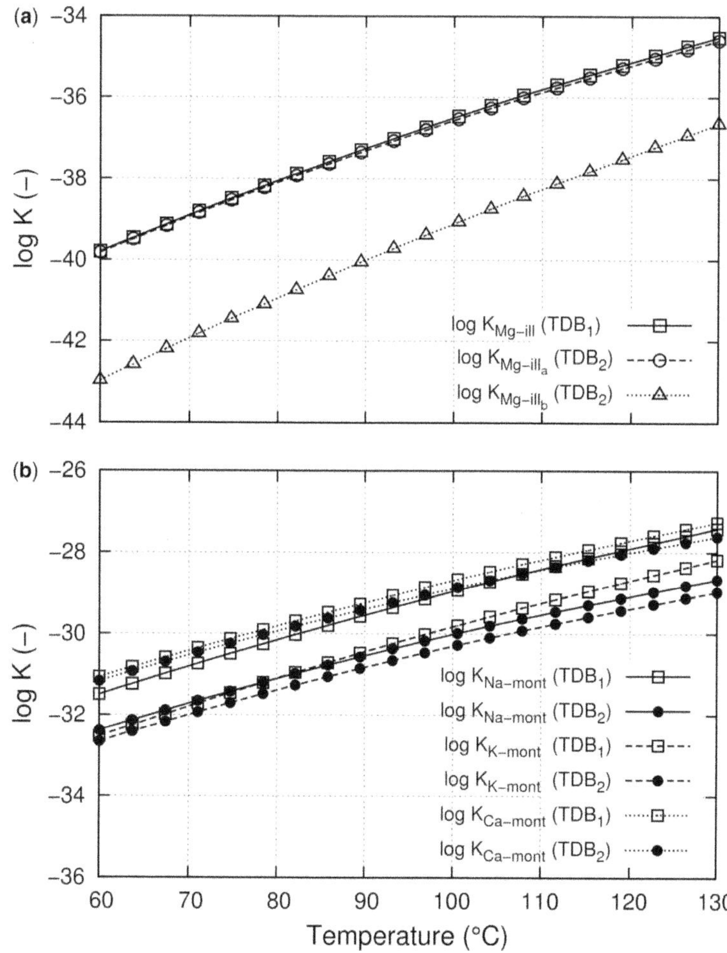

Fig. 4. Temperature dependence of the stability constants of (**a**) illites and (**b**) montmorillonites from different thermodynamic databases. TDB1 is compiled by the BRGM (Blanc *et al.* 2012); TDB2 is the standard database released with the TOUGHREACT simulator. A wide spread of stability can be noted for both minerals.

Table 2. *Kinetic rate constants (k_{25}), activation energies (E_a) and specific reactive areas (Sra) for the minerals taken into consideration in our simulations*

Mineral	k_{25} (mol m^{-2} s^{-1})	E_a (kJ mol^{-1})	Sra (cm^2 g^{-1})
Calcite		Set to equilibrium	
Anhydrite		Set to equilibrium	
Quartz	1.00×10^{-12}/ 1.00×10^{-14}	87.70	10.0/3.0
Kaolinite	1.50×10^{-12}	22.20	20.0
K-feldspar	3.90×10^{-13}	38.00	10.0
Illite	1.70×10^{-13}	35.00	150.0
Smectite	1.70×10^{-13}	35.00	150.0
Chlorite	3.00×10^{-13}	50.00	100.0
Muscovite	1.00×10^{-12}	38.00	150.0
Phlogopite	1.00×10^{-12}	22.20	150.0

Since the reactivity of calcite and anhydrite are orders of magnitude faster than the other minerals, they are considered at thermodynamic equilibrium with the porewater.

quartz in order to evaluate the influence of these parameters on the overall reactivity of the system.

The discussion on the validity of kinetic parameters for a given mineral, especially at temperatures >25°C, is beyond the scope of this study as the thermodynamic parameters are not rigorously determined for many of the mineral phases involved in the reaction. The sequence of the reaction rate for illite/smectite, feldspars, kaolinite, carbonates and sulphates is, however, respected in terms of orders of magnitude.

Simulation strategy

Model development has been carried out in stages of increasing complexity, starting from geometrically and chemically simple models (0D 'flush' models: Fig. 5a) to accommodate the uncertainties discussed above. More complex non-isothermal 1D models (Fig. 5b) are then used to reproduce the dynamics of the diagenetic events on timescales of the order of tens of millions of years under non-isothermal conditions.

Simulation of diagenetic Stage 1

The reactions occurring at low temperature (20–30°C) are verified in different conditions by means of 0D flush models, considering three water types and two different CO_2 partial pressures, according to the scheme summarized in Table 3. During this stage, a constant mass injection rate of 1×10^{-7} kg s^{-1} was used. This corresponds approximately to a residence time of fluids in the system of 100 years, consistent with the lower end of the range of fluid fluxes expected during very early diagenesis.

Representative compositions for meteoric, river and marine waters are taken from, respectively, Stumm & Morgan (1996), Gabelich *et al.* (2002) and Ball & Nordstrom (1991), with the integration of Turekian (1968) for bromides (details in Table 4).

Meteoric and river waters are considered to have been present during sediment deposition. Meteoric water was initially equilibrated with kaolinite to fix the hexogenic silica and aluminum species. Meteoric water is undersaturated with respect to K-feldspar, muscovite, quartz, montmorillonite and illite. Colorado River water, with an alkaline pH, is oversaturated with respect to the same mineral phases. Seawater probably infiltrated the sediment slightly later during burial. For this reason, the fluid was equilibrated with quartz and kaolinite in order to fix the activities of Si and Al. The resulting composition is oversaturated with respect to clinochlore, phlogopite and calcite, while it is undersaturated with respect to K-feldspar, muscovite, montmorillonite and illite.

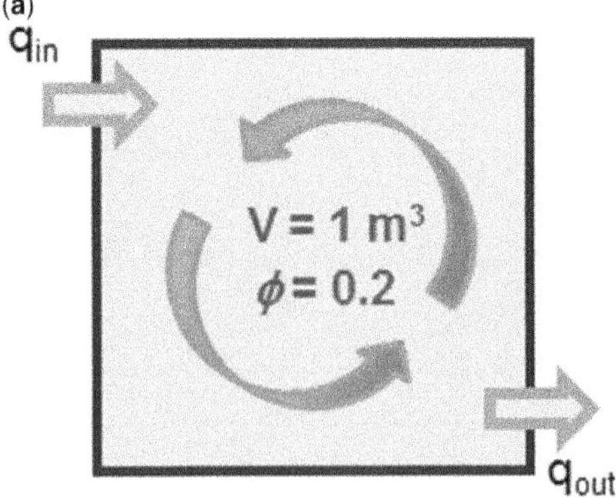

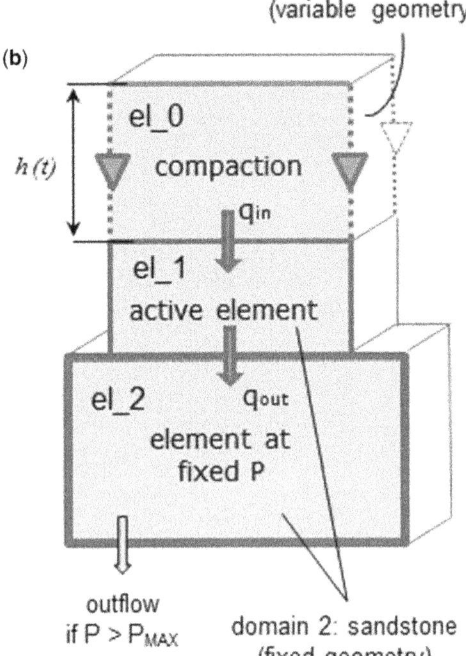

Fig. 5. Conceptual models used in the simulations. The 0D models (**a**) are set up for isothermal and non-isothermal fluid flow. The 1D model (**b**) is used to simulate non-isothermal compaction. It consists of a compacting element with a volume of 2000 m^3, an active element of 100 m^3 and a fixed-pressure-element of 10 000 m^3. The contact surface area is set to 10 m^2. The nodal distance el_0–el_1 is initially set to 1.1 m, the contact surface being placed 1 m from the el_0 node. The nodal distance el_1–el_2 is set to 14 m, the contact surface being placed 9 m from the el_1 node.

In the following, the mineral trends are reported as variations in moles per cubic metre (Δmol m^{-3}) of porous medium, since it makes easier to derive

Table 3. *Stage 1 case study with different mineral assemblages, flushing fluids types and CO_2 partial pressure*

Initial mineralogy of solid volume (%)		Porosity (%)	Flushing fluids (mass injection rate $=1 \times 10^{-7}$ kg s^{-1})	CO_2 partial pressure (atm)
Quartz	82	35	Meteoric, river, seawater	0.01, 0.001
K-feldspar	6			
Phlogopite	12			
Kaolinite	tr			
Quartz	86			
K-feldspar	12			
Muscovite	1			
Kaolinite	1			

balanced chemical reactions. The main model results for Stage 1 showed that, with meteoric water, K-feldspar is dissolved, and kaolinite and quartz are precipitated (Fig. 6a). When river water or seawater are used instead, K-feldspar (and kaolinite if present) dissolves, quartz forms, but muscovite is stable rather than kaolinite (Fig. 6b). The substitution of phlogopite (biotite Mg-end member) for muscovite in the initial mineral assemblage makes the system much more reactive, as phlogopite begins to dissolve very quickly and delays the K-feldspar dissolution. The system evolution is, therefore, more complex as phlogopite dissolution induces an transient calcite formation and a significant precipitation of muscovite (Fig. 6c). Calcite formation depends on the water's pH and total dissolved inorganic carbon. Low P_{CO_2} results in a continuous precipitation of calcite (Fig. 6d) in the time frame of the simulation. All reactions are supposed to produce crystalline quartz.

In meteoric fluids, the reactivity of the system can be described by the following independent reactions related, respectively, to the kaolinitization of feldspar and of muscovite:

$$2\text{K-feldspar} \rightarrow \text{kaolinite} + 4\,\text{quartz} \quad \text{(r1)}$$

$$2\,\text{muscovite} \rightarrow 3\,\text{kaolinite}. \quad \text{(r2)}$$

More details and fully balanced reactions are given in the Appendix. The overall reactivity is a

Table 4. *Water compositions used in 0D flush models of Phase 1*

Meteoric water		Colorado River		Seawater	
pH	4.4	pH	8.2	pH	8.22
Al*	5.65×10^{-6}	Al	1.97×10^{-6}	Al‡	4.23×10^{-7}
Ca	1.70×10^{-5}	B	1.48×10^{-5}	B	4.31×10^{-4}
Cl	2.66×10^{-5}	Ba	4.37×10^{-7}	Br	8.73×10^{-4}
Fe†	1.86×10^{-8}	Br	1.25×10^{-6}	Ca	1.07×10^{-2}
K	2.13×10^{-6}	Ca	1.25×10^{-3}	Cl	5.66×10^{-1}
Mg	4.79×10^{-6}	Cl	1.78×10^{-3}	F	7.09×10^{-5}
N(−3)	9.05×10^{-5}	F	1.11×10^{-5}	Fe	1.86×10^{-8}
N(5)	6.39×10^{-5}	Fe	1.79×10^{-7}	K	1.06×10^{-2}
Na	5.33×10^{-6}	K	8.90×10^{-5}	Mg	5.51×10^{-2}
S	4.69×10^{-5}	Mg	8.64×10^{-4}	Mn	1.89×10^{-9}
Si*	5.65×10^{-6}	N	2.74×10^{-5}	N(−3)	1.72×10^{-6}
Alkalinity	1.89×10^{-5}	Na	2.87×10^{-3}	N(5)	4.85×10^{-6}
		S	1.61×10^{-3}	Na	4.86×10^{-1}
		Si	1.60×10^{-4}	O(0)	4.64×10^{-4}
		Sr	7.42×10^{-6}	S	2.93×10^{-2}
		Alkalinity	2.06×10^{-3}	Si‡	1.58×10^{-4}
				Sr	9.35×10^{-5}
				Alkalinity	2.41×10^{-3}

Concentrations are given in terms of molality, pH in pH units.
*Equilibrium with kaolinite.
†Set to seawater value.
‡Equilibrium with kaolinite and quartz.

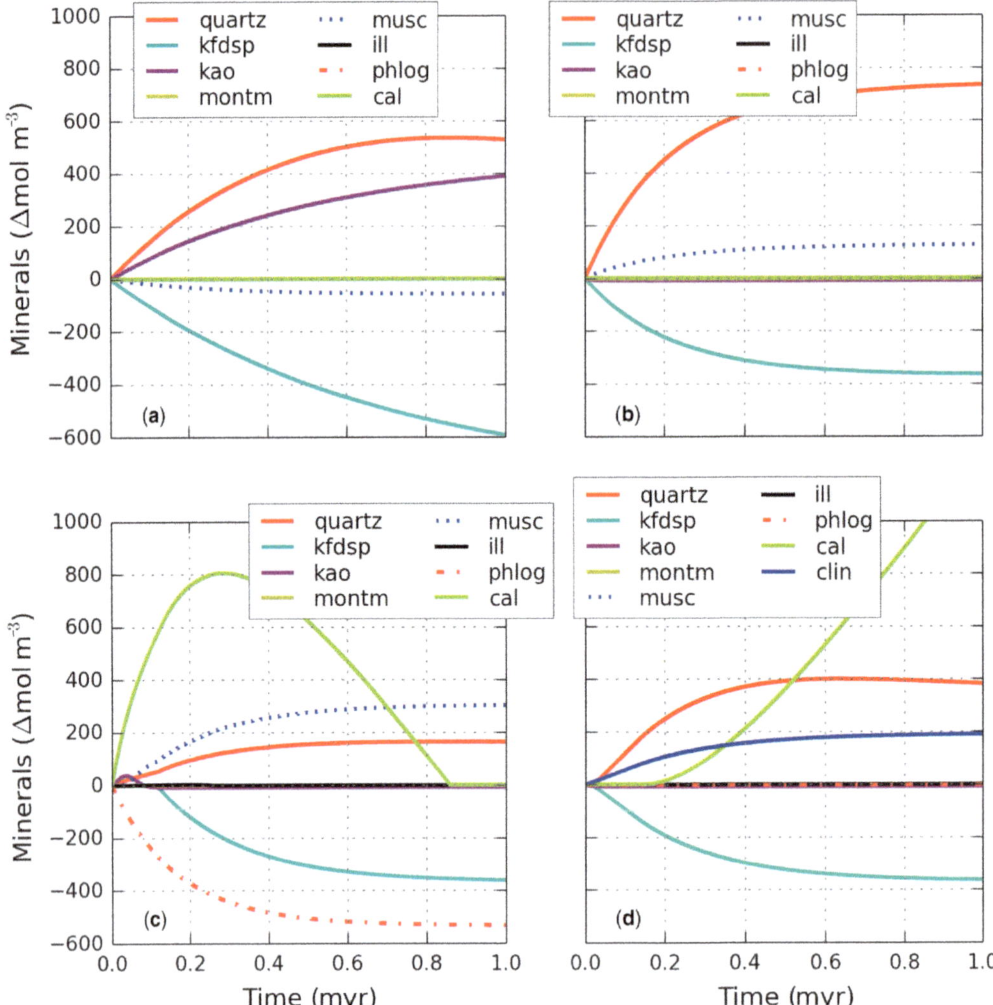

Fig. 6. Mineral evolution of 0D flush models for the isothermal diagenetic Stage 1 at 30°C: (**a**) meteoric water, $P_{CO_2} = 0.01$ atm; (**b**) river/seawater, $P_{CO_2} = 0.01$ atm; (**c**) seawater with $P_{CO_2} = 0.01$ atm with phlogopite in the initial mineral assembly; (**d**) similar to (c) but with $P_{CO_2} = 0.001$ atm. The magnitude of the precipitation and dissolution reactions is given in terms of variations in moles per cubic metre (mol m^{-3}) of rock. kfdsp, K-feldspar; kao, kaolinite; montm, montmorillonite; musc, muscovite; ill, illite; phlog, phlogopite; cal, calcite.

combination of the two reactions, the second one accounting for more or less one-tenth of the total.

When river water or seawater are used, in the early stages of the simulation, two different reactive paths are observed depending on the initial mineralogy. If muscovite is present, a single reaction takes place (r3):

$$3\,\text{K-feldspar} \rightarrow \text{muscovite} + 6\,\text{quartz}. \quad \text{(r3)}$$

If, however, phlogopite is the main mica phase, after a short cycle of kaolinite formation and destruction, which consumes many protons (r5: see also the Appendix), phlogopite transforms to muscovite. This reaction acts as a solid buffer favouring calcite precipitation, using calcium from flushing water:

$$2\,\text{phlogopite} \rightarrow \text{kaolinite} + 4\,\text{quartz} \quad \text{(r4)}$$

$$\text{phlogopite} + \text{kaolinite} + \text{H}^+ \rightarrow \text{muscovite} + 2\,\text{quartz}. \quad \text{(r5)}$$

The reduction in CO_2 partial pressure inhibits phlogopite dissolution, as protons are not available

Table 5. *Stage 2 case study with different illite reactive behaviour and mass injection rate*

Initial mineralogy of solid volume (%)		Porosity (%)	Illite precipitation	Flushing fluid		CO_2 partial pressure (atm)
				Mass injection rate (kg s^{-1})	Type	
Quartz	83	25	Yes	1×10^{-7} residence time (100 years)	Equilibrated seawater	0.75
K-feldspar	1					
Phlogopite	3					
Annite	3					
Kaolinite	4		No	1×10^{-9} residence time (10 kyr)		
Muscovite	3					
Montmorillonite	3					

for the reaction (r5). The main source for muscovite formation remains K-feldspar, as noted in (r3).

In all cases, the reactivity of the system is driven by the dissolution of K-feldspar and biotite, if present (the K-bearing mineral phases). The results are consistent with the well-known stability diagrams for K-feldspar and its weathering products (Garrels & Mackenzie 1967; Helgeson *et al.* 1969; Tardy 1971), in which the formation of kaolinite from K-feldspar is possible only when using meteoric water or, broadly speaking, waters with low K^+/H^+ ratios. However, in waters characterized by high K^+/H^+ ratios or in the presence of phlogopite in the initial mineralogy, muscovite forms as a product of feldspar or kaolinite alteration (Rex & Martin 1966). Carbonates should also be taken into account, as calcite precipitation can be effective in the reduction of porosity if sufficient carbon can enter the system either as dissolved solute in porewater or as gas migrating into the reservoir.

Simulation of diagenetic Stage 2

The medium-low temperature (50°C) diagenetic Stage 2 is defined to simulate the formation of chlorite from biotitic micas. The fluid fluxes in this relatively shallow burial condition are poorly constrained, so two mass flow rates were used (Table 5): one corresponding to a relatively short residence time (representative of a system still connected to a surface circulation); and the other representing a more closed system.

Geological evidence suggests that, at this burial stage, the reservoir was flushed by more saline fluids that substituted the modified waters resulting from the end of Stage 1. The flushing water used in this stage is, therefore, reconstructed based on the assumption that evaporation is the primary mechanism of concentration up to a chlorinity of 100 g l^{-1} (Fig. 3). This stage is modelled as a two-step procedure. A first stage of isothermal (25°C) evaporation at atmospheric conditions was performed, reaching the onset of gypsum precipitation. The resulting water was then heated to 50°C in the presence of a mineral assemblage similar to that used in Stage 1 under reducing conditions and at a P_{CO_2} of 0.0316 atm to constrain the pH (Table 6).

The initial mineralogy (Table 5) considers the increase in kaolinite due to feldspar alteration. Annite and phlogopite were used to represent the end members of the biotites group added to the initial mineralogy in order to promote the formation of chlorite. Chlorite is modelled as an ideal solution, with clinochlore and ripidolite representing, respectively, the Mg-end member and an intermediate term with a Fe:Mg ratio of 1:1. Illite is not present in the initial mineralogy, and two scenarios are considered according to the different reactive behaviour of illite (Table 5): (a) precipitation is allowed; and (b) precipitation is suppressed in order to represent

Table 6. *Water compositions used in 0D flush models of Stage 2 under reducing conditions (pE = −1.86)*

Flushing seawater $T = 50°C$	
pH	5.72
Al	0.1245×10^{-6}
B	0.1823×10^{-2}
Ba	0.8685×10^{-6}
Br	0.3688×10^{-2}
C	0.5498×10^{-2}
Ca	0.4034×10^{-1}
Cl	0.2391×10^{1}
F	0.2997×10^{-3}
Fe	0.1569×10^{-6}
K	0.4787×10^{-1}
Mg	0.2327
Na	0.2051×10^{1}
Si	0.3530×10^{-3}
Sr	0.3948×10^{-3}

Concentrations are given in terms of molality, pH in pH units.

a type of kinetic meta-stability condition. A quantitative and balanced relationship between the mineral phases involved in the reaction series can be made by considering the ratio between the moles of the dissolved and precipitated minerals given by the simulation (see the Appendix for full details of the reactions).

Considering the case with a high fluid flux and no illite precipitation, four reactive regimes can be recognized over time (Fig. 7a):

(i) 0–15 kyr: during this period K-montmorillonite completely dissolves. At the beginning of the period (0–3 kyr), annite and phlogopite

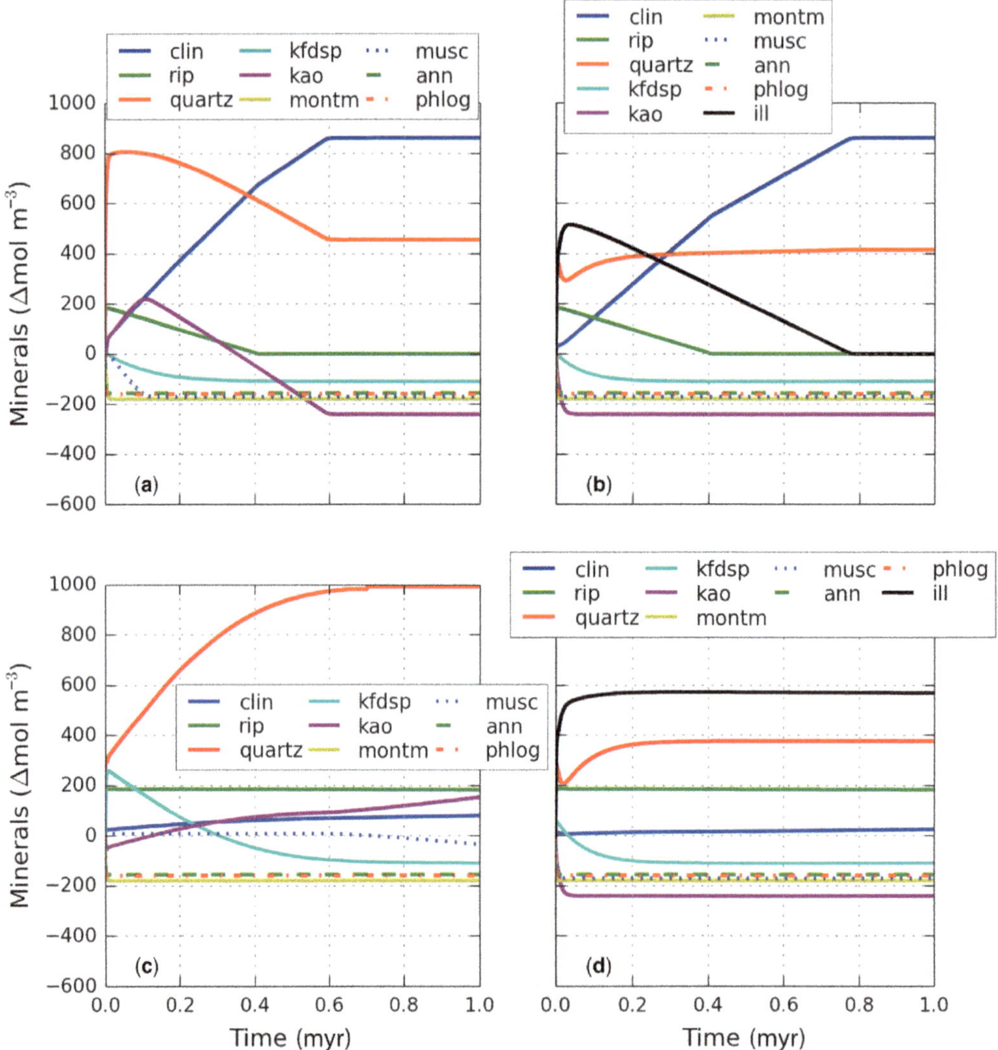

Fig. 7. Mineral evolution of 0D flush models for the isothermal diagenetic Stage 2 at 50°C. (**a**) Flow rate of 1×10^{-7} kg s^{-1} with inhibited illite precipitation. (**b**) Flow rate as in (a) but with illite precipitation allowed. (**c**) Flow rate of 1×10^{-9} kg s^{-1} with inhibited illite precipitation. (**d**) Flow rate as in (c) but with illite precipitation allowed. The magnitude of the precipitation and dissolution reactions is given in terms of variations in moles per cubic metre (mol m^{-3}) of rock. Variations in terms of mineral volumes are reported in the following: quartz 86 → 88.25; muscovite 1.2 → 0; K-feldspar 12 → 6.3; kaolinite 1.2 → 8.8; porosity 35 → 36.8. clin, clinochlore; rip, ripidolite; kfdsp, K-feldspar; kao, kaolinite; montm, montmorillonite; musc, muscovite; ann, annite; phlog, phlogopite.

(Fe and Mg biotites) also disappear, promoting the formation of quartz, ripidolite and minor clinochlore:

$$\text{annite} + 0.7\ \text{phlogopite} \rightarrow 0.9\ \text{ripidolite} + 2.6\ \text{quartz}. \quad (r6)$$

From 3 to 15 kyr: the smectite dissolves to combine with a very small amount of ripidolite to form clinochlore, kaolinite and quartz:

$$\text{K-montmorillonite} \rightarrow 0.4\ \text{clinochlore} + 0.4\ \text{kaolinite} + 1.9\ \text{quartz}. \quad (r7)$$

(ii) 15–100 kyr: when the supply of smectite is exhausted, muscovite, ripidolite and K-feldspar dissolve to form kaolinite and clinochlore:

$$\text{muscovite} + 0.4\ \text{K-feldspar} + 0.25\ \text{ripidolite} \rightarrow \text{kaolinite} + \text{clinochlore}. \quad (r8)$$

(iii) 100–600 kyr: from 100 to 400 kyr, kaolinite, quartz, K-feldspar and ripidolite react to form clinochlore:

$$\text{ripidolite} + 1.4\ \text{quartz} + 2\ \text{kaolinite} + 0.4\ \text{K-feldspar} \rightarrow 3.2\ \text{clinochlore}. \quad (r9)$$

When ripidolite and K-feldspar are exhausted, kaolinite and quartz continue to dissolve to form clinochlore by obtaining Mg from flushing water:

$$\text{kaolinite} + \text{quartz} + 5\ \text{Mg} \rightarrow \text{clinochlore}. \quad (r10)$$

(iv) 600–1000 kyr: after all kaolinite has been consumed, the system is stable.

The precipitation of illite influences the reactivity of the system (Fig. 7b). As in the previous case, four reactive periods may be recognized:

(i) 0–3 kyr: this period is characterized by concurrent dissolution of K-montmorillonite, annite, phlogopite and muscovite, resulting in the precipitation of illite, quartz, ripidolite and minor clinochlore. During this period kaolinite and K-feldspar begin to dissolve, contributing to the overall reactivity:

$$\text{annite} + \text{phlogopite} + 0.8\ \text{muscovite} + 0.75\ \text{K-montmillonite} + 1.3\ \text{kaolinite} + 0.2\ \text{K-feldspar} \rightarrow 1.9\ \text{quartz} + 2.7\ \text{illite} + \text{ripidolite} \quad (r11)$$

(ii) 3–25 kyr: kaolinite and quartz react to give illite, while minor ripidolite dissolution produces a small amount of clinochlore:

$$\text{kaolinite} + 1.9\ \text{quartz} + K^{+} \rightarrow \text{illite}. \quad (r12)$$

(iii) 25–785 kyr: from 25 to 400 kyr, ripidolite and illite react to form clinochlore and quartz:

$$\text{ripidolite} + 1.4\ \text{illite} + 0.4\ \text{K-feldspar} \rightarrow 2.8\ \text{clinochlore} + 0.4\ \text{quartz} \quad (r13)$$

When ripidolite has been entirely consumed, illite, just like kaolinite in (r10), is converted to clinochlore by taking Mg from porewater:

$$\text{illite} + 0.1\ \text{quartz} + 5\ Mg^{2+} \rightarrow 1.1\ \text{clinochlore}. \quad (r14)$$

During this period, at around 400 kyr, K-feldspar dissolution ends.

(iv) 785–1000 kyr: after illite consumption the system is stable.

The reduction in fluid fluxes in the absence of illite precipitation (Fig. 7c) has an impact on the reactivity, the main differences being early K-feldspar formation and ripidolite stability. The low fluid flow rate limits the Mg^{2+} supply and, consequently, the formation of clinochlore. Initially (4 kyr), quartz precipitation is limited by K-feldspar formation, but later on all K-feldspar dissolves to produce quartz and kaolinite. The total amount of authigenic quartz produced is by far the largest obtained during the simulation. When illite is allowed to precipitate, it dominates the overall reactivity, inhibiting the formation of quartz and clinochlore (Fig. 7d). Kaolinite is completely dissolved at 20 kyr, consuming the quartz and feldspar formed during the early brief period of muscovite, smectite, annite and phlogopite dissolution to form illite:

$$\text{kaolinite} + 0.3\ \text{K-feldspar} + 0.3\ \text{quartz} \rightarrow \text{illite}. \quad (r15)$$

The stability of the system is substantially reached at 400 kyr.

The results of Stage 2 confirmed that in these conditions of temperature, fluid composition, fluid flux regimes and initial mineralogy, biotites completely alter with respect to chlorites, whether ferroan or magnesian. Since the aim of this stage is not to fully match the observed mineralogy, no sensitivity analysis on input parameters was carried out for a quantitative tuning of the results.

Simulation of diagenetic Stage 3

Diagenetic Stage 3 is characterized by the precipitation in sandstone of quartz cement and illite during a period of burial lasting for about 60–70 myr. During this period, the temperature increases from around 70 to 110°C. Because of this wide temperature range, the simulations cannot be isothermal, as in the geologically 'instantaneous' previous stages. Also, the fluid flow cannot be assumed constant during this long period, and model flow rates are considered to be decreasing over time to simulate the transition from advective to diffusive regimes of mass transport.

In this stage, 0D models were used to test the interactions of clays with evaporated seawater at constant composition, allowing a sensitivity analysis to be carried out of the results of uncertainties in the thermodynamic parameters of the clays. One-dimensional models were set up to describe the diagenetic evolution of the sandstone, taking into account the compaction and diagenetic alteration of the adjacent clays, silica expulsion, solute diffusion and sandstone quartz cementation.

0D flush models

In this model, different temperatures and flow regimes are tested (Table 7). A constant heat flow is imposed on the system to achieve a linear increase of temperature from 70 to 110°C. Mass injection rates are decreased from an initial value of 1×10^{-9} kg s^{-1} to 0 at 60 Ma (Fig. 8) with a sharp decrease in the first 5 myr.

The flushing seawater was reconstructed by following two steps in a procedure similar to the one used in Stage 2: (i) isothermal seawater evaporation at 25°C at atmospheric pressure until reaching the chlorinity at which precipitation of gypsum begins; and (ii) heating of the resulting water to 70°C, attaining equilibrium with a quartz-rich initial mineralogy under reducing conditions and P_{CO_2} at 1 bar. The incoming water (Table 8) is undersaturated with respect to Ca-montmorillonite and

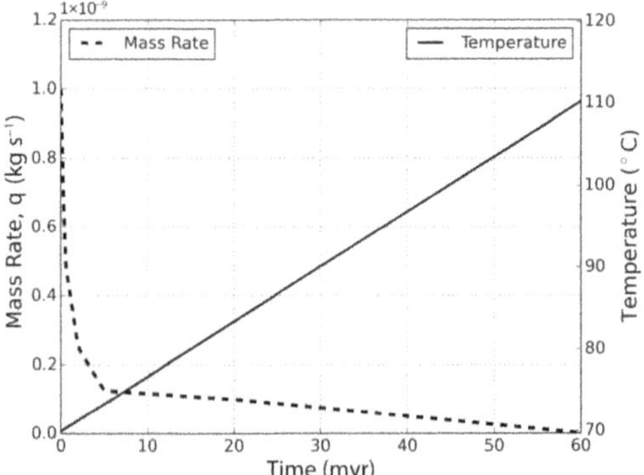

Fig. 8. Heating schedule and mass injection rate history used in the 0D models for the non-isothermal Stage 3.

calcite, oversaturated relative to illite and clinochlore, and near equilibrium for quartz, K-feldspar, smectite and anhydrite.

Taking into account that the no-flux models show only a transient and negligible mineral–water re-equilibration under both isothermal and non-isothermal conditions, the magnitude of the fluid flux seems to play a major role in mineral alteration: the major reactive process being the conversion of montmorillonite to illite and quartz. The reactivity of the system is clearly driven by a common ion effect, since K^+ is present in K-feldspar, illite, montmorillonite and the flushing water.

Without considering other factors not captured by these kind of models, such as crystalline structure changes, the heating seems to have a secondary role on the overall reactivity (Fig. 9) and is in contrast with the formation of illite. Simulations performed at lower temperature confirmed that the illitization of smectite are driven by potassium availability for all of the smectite–illite pairs tabulated in the thermodynamic database.

The conversion of montmorillonite to illite and quartz has a very small impact on the porosity, with a decrease of 2.5 percentage points in the

Table 7. *Stage 3 0D models case study with different mass injection rate and temperature conditions*

Initial mineralogy of solid volume (%)		Porosity (%)	Flushing fluid		Temperature
			Mass injection rate (kg s^{-1})	Type	
Montmorillonite	82	20	No flux or decaying flux (Fig. 8) (initial residence time = 10 myr)	Equilibrated seawater	Isothermal (70°C)
Quartz	15				Linearly increasing from 70 to 110°C (Fig. 8)
K-feldspar	1				
Illite	1				
Kaolinite	1				

Table 8. *Water compositions used in 0D flush and 1D compacting models of Stage 3 under reducing conditions (pE = −0.5)*

Flushing seawater $T = 70°C$	
pH	5.77
Al	4.232×10^{-7}
B	2.151×10^{-3}
Br	4.353×10^{-3}
C	1.661×10^{-2}
Ca	2.522×10^{-2}
Cl	2.821
F	3.536×10^{-4}
Fe	1.851×10^{-7}
K	6.062×10^{-2}
Mg	2.747×10^{-1}
Na	2.421
S	1.236×10^{-1}
Si	5.601×10^{-4}

Concentrations are given in terms of molality, pH in pH units

isothermal case and of 1 percentage point in the non-isothermal case. In terms of volume percentages, the dissolution of 45% of the initial montmorillonite (isothermal case) furnishes 32% of the illite and 10% of the quartz formed in this reaction series.

K-montmorillonite is used in the simulation. Owing to the high Na^+/K^+ ratio of the flushing water (Table 8), a Na-montmorillonite was also tested. As expected with this clay mineral suite, the system was stable with very limited precipitation or dissolution processes.

1D compaction models

Zero-dimensional (0D) flush models demonstrate that it is possible to generate authigenic quartz and illite in shale sediments. Such models are not capable of quantifying the amount of diagenetic products observed in the samples. In a more geologically plausible model, which considers the compaction of clays, the water expulsion in the sandstone and the diffusion of solutes in the porewater is necessary in order to improve the quantitative estimate of such products. In this model, geometric parameters (nodal distances, and the contact areas between nodes and volumes: see Fig. 5b for details) influence the specific amount of the minerals precipitating or dissolving.

The large uncertainties related to many input data (initial mineralogy, water chemistry, thermodynamics and kinetic parameters) and the lack of calibration against petrographical data do not allow this model to be used as a predictive tool.

Since these models are able to handle multiple diagenetic processes they can be used, at the present stage, as diagnostic tools that allow a sensitivity analysis of the input parameters and of different geological and diagenetic scenarios.

A version of TOUGHREACT developed in-house was used in which a geometric compaction module is coupled to the original reactive transport module.

The model consists of three elements (Fig. 5b): (i) a compacting clay element (el_0), which initially (70°C) is considered to be a Na-smectite with minor

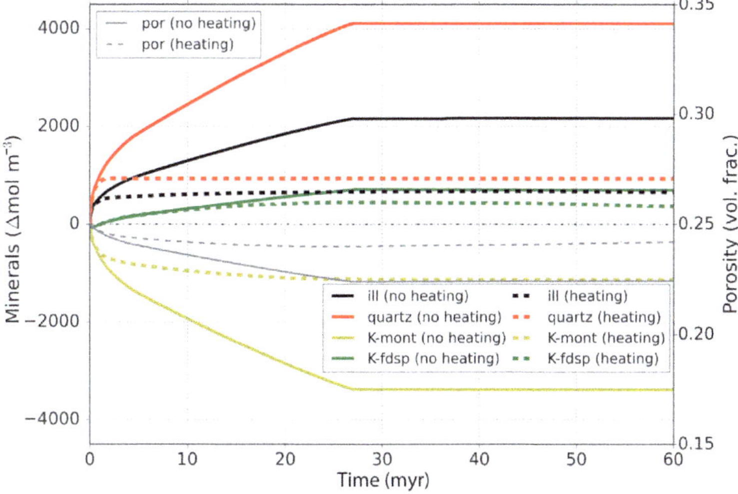

Fig. 9. Mineral evolution of 0D models for Stage 3. The dashed line indicates the result of the non-isothermal simulation; the solid line the results of the isothermal stage at 70°C. The magnitude of the precipitation and dissolution reactions is given in terms of variations in moles per cubic metre (mol m^{-3}) of rock. ill, illite; K-mont, K-montmorillonite; K-fdsp, K-feldspar.

illite and minor K-feldspar with an initial porosity of 27%; (ii) an active sandstone element (el_1) with a fixed geometry made of an almost pure quartz arenite with K-feldspar and an initial porosity of 30%; and (iii) a sandstone element (el_2) similar to el_1 that acts as a sink for fluxes in order to avoid a build-up of pressure. The compaction is simulated by reducing the thickness of the el_0 according to a history of porosity evolution (from 27 to 6%) extracted from the burial curve for the time period of diagenetic Stage 3. Initially, all elements are saturated with the evaporated seawater already used in the 0D models of this stage (Table. 8). A constant heat flow is imposed at the base of the system (el_2) to achieve a linear increase in temperature over the range of 70–115°C (Fig. 10). This set-up also causes a very small difference in temperature between the top and the bottom according the petrophysical parameters assigned to the rocks. Fluxes are computed by the simulator on the basis of the compaction history (Fig. 10) and compete with diffusive transport throughout the entire simulation (the diffusion coefficient is $1.5 \times 10^{-9}\ m^2\ s^{-1}$ for all species, a value close to the average of values reported by Oelkers & Helgeson 1988).

Since this 1D model consists of only three single elements with no further internal discretization, a sensitivity study concerning the impact of the nodal distances between el_0 and el_1 has indicated that this distance, strongly affecting the travel time of silica released from the clay, influences the amount of quartz precipitation 'visible' at the node representing el_1. A finer discretization of the domains (clays and sandstone) should minimize this issue.

The type of illite/smectite, the K-feldspar content, the reactive specific surface areas of quartz and the residence time of silica in the compacting clay element (el_0) are the major factors affecting the reactivity of the system from a quantitative point of view. The residence time depends on the interplay of compacting and diffusive fluxes, and is also affected by the internodal distance of the compacting and active elements.

Mg-illite and Na-smectite are the phases where the properties are compiled in the standard TOUGHREACT database. The K-montmorillonite of the 0D models is not used as it is more stable than the Mg-illite at high temperature, with an unrealistic transformation of illite back to montmorillonite above 90–100°C.

While all the 0D models are focused on the thermodynamic aspects of some selected diagenetic reactions, the more geologically realistic 1D models are also used to inspect the effects of kinetic parameters, K-feldspar content and the reactive specific surface area of quartz.

Two sets of kinetic parameters are employed, differing mainly for the reaction rate constant of quartz and K-feldspar, one set being globally less reactive than the other (Table 2).

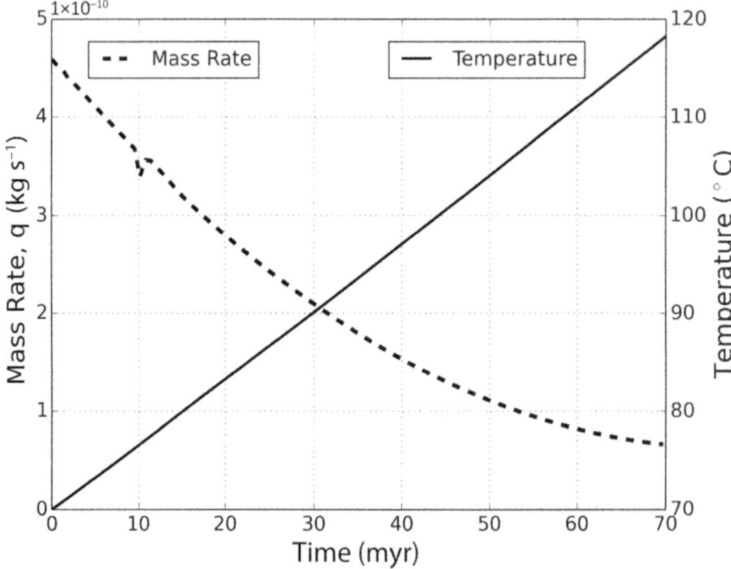

Fig. 10. Heating schedule and mass injection rate history used in the 1D models for the non-isothermal Stage 3. The fluxes are calculated according to the burial history.

For the 'fast' kinetic set, in order to avoid the precipitation of silica in the compacting element (el_0), the reactive specific surface area of quartz is reduced by a factor of 30 from the commonly used value of $10\ cm^2\ g^{-1}$, based on the results of a sensitivity study on this parameter. This tuning in the simulation is a numerical device to reproduce a condition of kinetic inhibition for quartz formation justified by the geological evidence that illite-rich, deeply buried shales are not usually heavily silicified. For the 'slow' kinetic set, this inhibition is not required.

Two different sandstone compositions are considered to assess the impact of potassium availability both on the amount of illite and quartz precipitation, and on the reaction dynamics. In the first scenario, the K-feldspar content is 2% (by volume) in all elements. In the second scenario, the K-feldspar amount is increased to 4% in el_1 and to 7% in el_2.

The main results given by the 'slow' kinetic model are illustrated in Figure 11, where the mineral volumes in the compacting element are related to the rock fraction of the porous medium. The reactivity of the system is primarily driven by the availability of K released by feldspar dissolution in all the elements, while kinetic parameters affect the amounts of quartz formed at the beginning of the simulation (in these simulations, the reactive specific surface area of quartz is set to the more commonly used value of $10\ cm^2\ g^{-1}$).

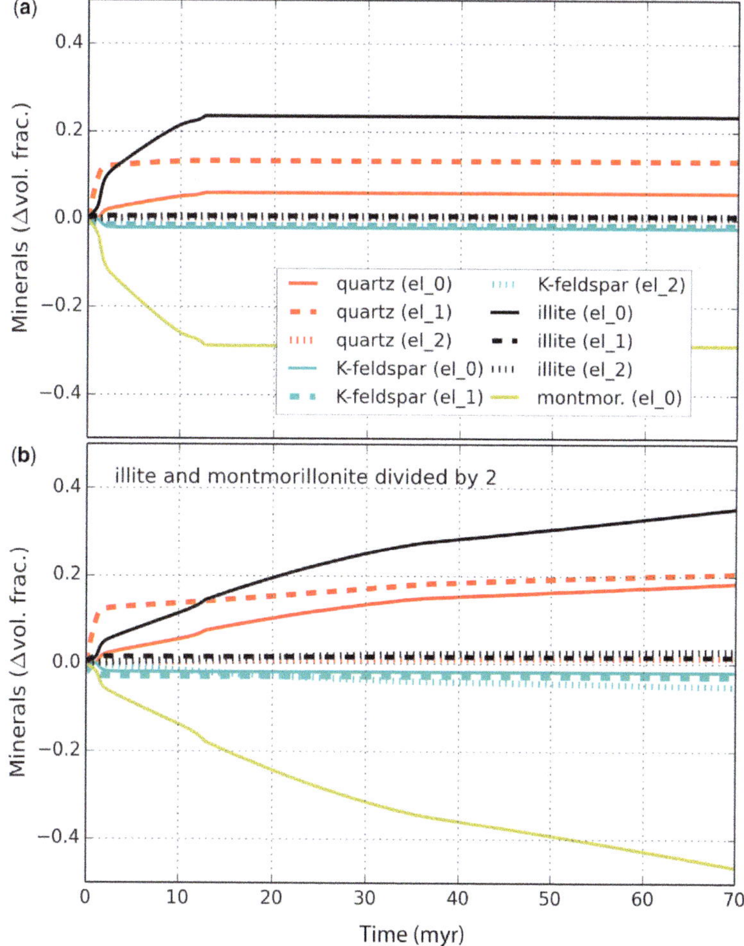

Fig. 11. Mineral evolution of 1D compacting non-isothermal models for Stage 3. Plot (**a**) reproduces the low K-feldspar case, where the reactivity of the system ends at around 12–13 Ma. Plot (**b**) represents the high K-feldspar case; the reactivity is almost stopped after 70 Ma due to the complete conversion of smectite to illite. The effects of porosity changes due to compaction are not considered in the computation of the mineral amounts. montmor., montmorillonite.

Figure 11a refers to the K-feldspar-poor case. The system reaches its geochemical stability after 12–13 Ma, when all the K-feldspar of el_2 is consumed.

In the K-feldspar-rich case (Fig. 11b), the system is still reactive after 70 Ma as K-feldspar of el_2 is not completely dissolved. The first reactions occurring in this system are (Fig. 12a) the dissolution of K-feldspar and the precipitation of quartz and illite in el_1 (see the Appendix for the fully balanced equations):

$$\text{K-feldspar} \rightarrow 0.4\,\text{illite} + 1.5\,\text{quartz}. \qquad \text{(r16)}$$

The K released by the process diffuses into the clay and triggers the dissolution of smectite, with the formation of incipient illite. The same phenomenon takes place in the el_2, providing additional K.

The acceleration of illite formation in the clay element is due to K-feldspar dissolution within the element itself:

$$\text{K-feldspar} + 4.6\,\text{Na-smectite} + 1.4\,K^{+} \rightarrow 4\,\text{illite} + 7.3\,\text{quartz}. \qquad \text{(r17)}$$

This reaction accounts for most of the illite and quartz precipitation in the element over the time interval of 1–2.5 Ma (Fig. 12a).

From about 2.5 Ma to the end of the simulation, the ongoing reactivity of the system is ensured by the dissolution of K-feldspar in the el_2:

$$\text{Na-smectite} + 0.5\,K^{+} \rightarrow 0.8\,\text{illite} + 1.3\,\text{quartz}. \qquad \text{(r18)}$$

In the K-feldspar-poor case at the end of K-feldspar dissolution in the el_2 (at *c.* 10 Ma:

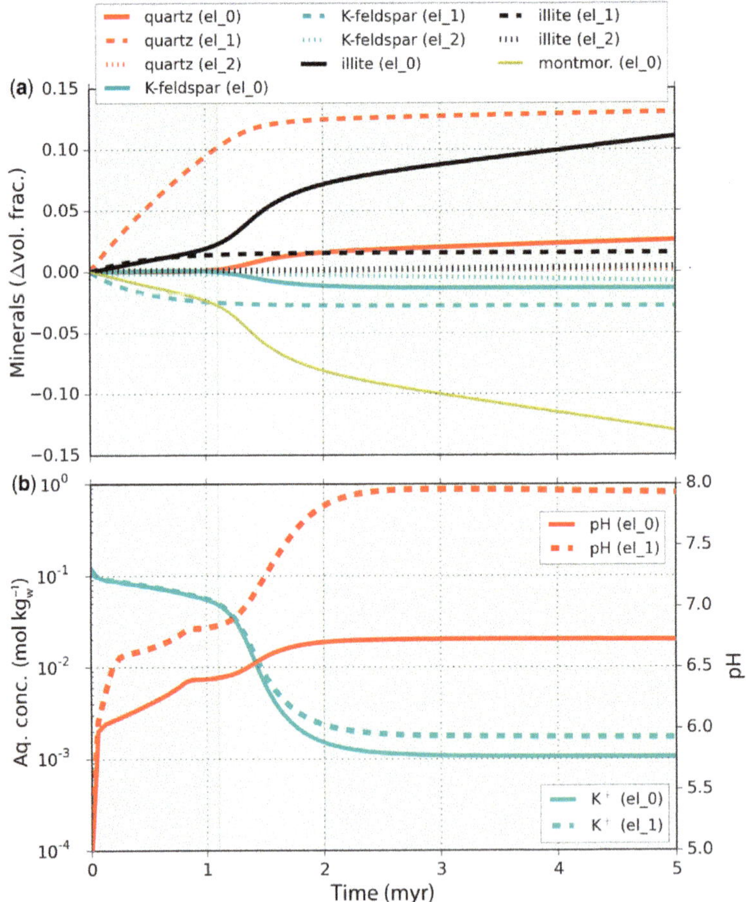

Fig. 12. Close up of the mineral variations in the first 5 myr (**a**) and the consequent chemical evolution of the water composition (**b**). Three successive reactive regimes, differentiate by colour, can be identified over time according to the main mineral reactions. The sharp drop (from 1.1 to 2.0 Ma) in the K^{+} ion in water is related to the process described in (r18) (see text). montmor., montmorillonite.

Fig. 11a), a further smectite–illite conversion occurs from 10 to 13 Ma. This process is promoted by the diffusion of K from el_2 to el_0. Considering that the distance between the sink and the compacting element nodes is of 15 m, the net velocity of K migration is around 1.5×10^{-6} m s^{-1}, corresponding to the difference between the compaction (see Fig. 10) and diffusive fluxes.

The reactions alter the water chemical composition by increasing pH from almost neutral to mildly basic condition in all elements (Fig. 12b).

The volumes of minerals precipitated in the elements change in the two K-feldspar scenarios. In the low-feldspar case, around 13.5% quartz and 1% illite form in the active element el_1 (Fig. 11a). Owing to the limited amount of K in the system, only 29% of the smectite is transformed to illite (23%) plus quartz (6%) in el_0. In element el_1, the bulk of quartz-cement precipitation takes place during the first 1.5 myr, and around 10% (volume) of the total quartz cement is of internal origin (promoted by the K-feldspar dissolution), while the remaining 90% is of external origin, mostly from the illitization of smectite in el_0.

In the high-feldspar case, around 21% quartz and 2% illite are present at the end of the simulation in el_1. In el_0, almost all smectite (92%) is transformed to illite (70%) and quartz (19%). In this scenario, one-third of the quartz of el_1 is formed over a long time span (from 1.5 to 70 Ma).

The results of this sensitivity study suggest that the original composition of the sandstone of the case study should be feldspar-poor, as only about 11.5% of quartz cement is present.

In the simulation, the quartz precipitated in el_1 during the first 1 myr accounts for 92 and 60%, respectively (Fig. 11). During this relatively short time span, temperature increases from 70 to 71°C. This quick quartz cementation depends on the reactivity of the rock–water system in all the elements and on the flux history. Four types of fluxes are active contemporaneously in the system: (i) at the very beginning (from 0 to 0.1 Ma), the hydrodynamical regime is dominated by pressure-driven flows induced by porosity changes related to the mineralogical reactions occurring in the elements; (ii) diffusive flows, active throughout the entire simulation, maintain a continuous supply of solutes through the elements; (iii) compaction-driven flows, induced by the mechanical reduction of porosity in el_0 (Fig. 10); and (iv) weak thermal flows (about two order of magnitude less than compaction flows) generated by the small gradient of density due to the temperature difference between el_0 and el_2.

The quartz cementation is initially promoted by the chemical reactions related to the equilibration of the rock–water system that was only partially equilibrated at the beginning of the simulation. Pressure and diffusive flows maintain this transient reactivity until the fluxes are stabilized following the compaction curve (Fig. 10), and the diagenetic transformation of smectite to illite becomes the major process to drive the quartz cementation.

Tests carried out with initial more stable systems showed that the contribution of the pressure-driven flow of the aforementioned point (i) to this transient period can be actually reduced. In these more stable conditions, smoother transients are still present depending on the compaction history, the diffusivity of the solutes and the kinetic parameters of the mineral phases. Temperature seems to play a secondary role.

Conclusions

A RTM based on the TOUGHREACT simulator has been developed to investigate the diagenetic evolution of a Cretaceous sandstone reservoir with a known present-day mineral and water composition, and a well-defined burial and thermal history.

Three distinct diagenetic stages – two of them isothermal at low temperatures (30°C and 50°C) and one non-isothermal at high temperatures (70–115°C) – are simulated using 0D flushing models for all three stages and a 1D compacting model for the third late diagenetic episode.

The first isothermal stage involved the formation of kaolinite from K-feldspar at low temperature. K-feldspar alters to kaolinite only in freshwater, while in river water and seawater muscovite is produced. Calcite precipitation can be promoted or inhibited by varying the CO_2 partial pressure within the sediment. An important role for calcite formation is played by the presence of mineral phases with a high buffering capacity, such as phlogopite.

The second isothermal episode consisted of the alteration of micas to chlorites in an evaporated seawater in different flow regimes. High fluid fluxes led to the formation of a significant amount of Mg-chlorite (clinochlore); at lower flushing rates, Mg–Fe-chlorite (ripidolite) is favoured.

During both stages, small amounts (<2% by volume) of quartz also form.

The 0D flush models set up to simulate the non-isothermal stage deal with the transformation of smectite to quartz and illite under decreasing fluxes of evaporated seawater. The simulations showed that the smectite species used (K-montmorillonite) transforms to quartz and illite, but the transformation, already effective at 70°C, is inhibited by increasing temperature.

The 1D non-isothermal models were set up to mimic the diagenesis and compaction of clays, and the subsequent fluid expulsion into an underlying sandstone, providing a source of silica for quartz cementation.

The models are not set up in order to be predictive and do not take into account the quartz cement sourced by internal mechanisms such as chemical or pressure solution. The model indicates that quartz formation is promoted both by the K-feldspar dissolution in all the elements of the system, by the smectite to illite transformation occurring in the compacting element, and by the capability of the solutes to reach the reactive sites. These models indicate that it is also possible to completely convert the smectite to illite in 70 myr.

The limiting factor constraining the illitization of smectite and, by consequence, the quartz cementation in the sandstone does not seem to be temperature but, rather, the availability of potassium and silica in the reaction sites. Potassium and silica may be supplied by *in situ* dissolution of K-feldspar or by migration from other elements: in this last case, the mobility of the species seems to be the most important limiting factor (regime of diffusion-controlled kinetics).

As temperature is commonly thought to play a significant role in smectite illitization (Freed & Peacor 1989; Pytte & Reynolds 1989), further investigations are required to fully assess the relationships between the structural changes in the clays and the conceptual thermochemical framework on which the RTM is based.

All model runs show that the system resolves the transient reactivity in a relatively short time with respect to the simulation time (up to 10 myr for the most stable systems). Episodes of calcite precipitation and dissolution are predicted by some of the models. Such episodes may play a role on the porosity evolution of the reservoir.

The RTM approach followed in this work allows the evaluation of the impact of factors such as the compaction history, diffusivity of the solutes, kinetic parameters of the mineral phases, temperature and the chemical composition of formation water on the dynamics of the smectite–illite transformation v. quartz precipitation to be evaluated.

The 1D model discussed here is the first step towards the numerical simulation of the observed diagenetic events. In order to improve the accuracy of the model in describing the whole diagenetic evolution, it will be necessary to have a better geometrical discretization of the system, a reliable description of the thermodynamic and kinetic behaviour of the specific clay minerals actually present in the sediment, and a better constraint on the transport mechanism of the solutes dissolved in the fluids.

Eni is kindly acknowledged for according permission to publish these preliminary results of the R&D project on the modelling of siliciclastic sediment diagenesis. All the reviewers are also sincerely acknowledged for their valuable contribution to the improvement of the manuscript.

Appendix

Even if not representative of the exact processes occurring within the sediment (stoichiometric coefficients have been included in the following reactions only to honour mass balance constraints), these equations do represent the main mineralogical transformations completely.

The reactions proposed here synthesize in a compact form the complex multicomponent reactive processes occurring in the different diagenetic stages. Their validity is supported by inspection of the results of the numerical simulation.

Stage 1 reactions:

$$2\,\underbrace{\mathrm{KAlSi_3O_8}}_{\text{K-feldspar}} + 2\,\mathrm{H^+} + \mathrm{H_2O} = \underbrace{\mathrm{Al_2Si_2O_5(OH)_4}}_{\text{kaolinite}} + 2\,\mathrm{K^+} + 4\,\underbrace{\mathrm{SiO_2}}_{\text{quartz}} \qquad \text{(r1)}$$

$$2\,\underbrace{\mathrm{KAl_2(AlSi_3O_{10})(OH)_2}}_{\text{muscovite}} + 2\,\mathrm{H^+} + 3\,\mathrm{H_2O} = 3\,\underbrace{\mathrm{Al_2Si_2O_5(OH)_4}}_{\text{kaolinite}} + 2\,\mathrm{K^+} \qquad \text{(r2)}$$

$$3\,\underbrace{\mathrm{KAlSi_3O_8}}_{\text{K-feldspar}} + 2\,\mathrm{H^+} = \underbrace{\mathrm{KAl_2(AlSi_3O_{10})(OH)_2}}_{\text{muscovite}} + 2\,\mathrm{K^+} + 6\,\underbrace{\mathrm{SiO_2}}_{\text{quartz}} \qquad \text{(r3)}$$

$$2\,\underbrace{\mathrm{KMg_3AlSi_3O_{10}(OH)_2}}_{\text{phlogopite}} + 14\,\mathrm{H^+} = \underbrace{\mathrm{Al_2Si_2O_5(OH)_4}}_{\text{kaolinite}} + 7\,\mathrm{H_2O} + 2\,\mathrm{K^+} + 6\,\mathrm{Mg^{2+}} + 4\,\underbrace{\mathrm{SiO_2}}_{\text{quartz}} \qquad \text{(r4)}$$

$$\underbrace{\mathrm{KMg_3AlSi_3O_{10}(OH)_2}}_{\text{phlogopite}} + \underbrace{\mathrm{Al_2Si_2O_5(OH)_4}}_{\text{kaolinite}} + 6\,\mathrm{H^+} = \underbrace{\mathrm{KAl_2(AlSi_3O_{10})(OH)_2}}_{\text{muscovite}} + 5\,\mathrm{H_2O} + 3\,\mathrm{Mg^{2+}} + 2\,\underbrace{\mathrm{SiO_2}}_{\text{quartz}}. \qquad \text{(r5)}$$

Stage 2 reactions:

$$\underbrace{\mathrm{KFe_3AlSi_3O_{10}(OH)_2}}_{\text{annite}} + 0.71\,\underbrace{\mathrm{KMg_3AlSi_3O_{10}(OH)_2}}_{\text{phlogopite}} + \mathrm{H^+} = 0.86\,\underbrace{\mathrm{Fe_{2.5}Mg_{2.5}Al_2Si_3O_{10}(OH)_8}}_{\text{ripidolite}} + 2.57\,\underbrace{\mathrm{SiO_2}}_{\text{quartz}} + 1.71\,\mathrm{K^+} + 0.86\,\mathrm{Fe^{2+}} \qquad \text{(r6)}$$

$$\underbrace{K_{0.33}Mg_{0.33}Al_{1.67}Si_4O_{10}(OH)_2}_{\text{K-montmorillonite}} + 3.22H_2O + 1.84\,Mg^{2+} = 0.40\,\underbrace{Al_2Si_2O_5(OH)_4}_{\text{kaolinite}} + 0.43\,\underbrace{Mg_5Al_2Si_3O_{10}(OH)_8}_{\text{clinochlore}} + 1.90\,\underbrace{SiO_2}_{\text{quartz}} + 0.33\,K^+ + 3.36\,H^+ \quad (r7)$$

$$\underbrace{KAl_2(AlSi_3O_{10})(OH)_2}_{\text{muscovite}} + 0.38\,\underbrace{KAlSi_3O_8}_{\text{K-feldspar}} + 0.25\,\underbrace{Fe_{2.5}Mg_{2.5}Al_2Si_3O_{10}(OH)_8}_{\text{ripidolite}} + 4.38\,Mg^{2+} + 6.94\,H_2O = 0.94\,\underbrace{Al_2Si_2O_5(OH)_4}_{\text{kaolinite}} + \underbrace{Mg_5Al_2Si_3O_{10}(OH)_8}_{\text{clinochlore}} + 6.12\,H^+ + 1.38\,K^+ + 0.62\,Fe^{2+} \quad (r8)$$

$$\underbrace{Fe_{2.5}Mg_{2.5}Al_2Si_3O_{10}(OH)_8}_{\text{ripidolite}} + 1.4\,\underbrace{SiO_2}_{\text{quartz}} + 2.0\,\underbrace{Al_2Si_2O_5(OH)_4}_{\text{kaolinite}} + 0.4\,\underbrace{KAlSi_3O_8}_{\text{K-feldspar}} + 21.6\,H^+ = 3.2\,\underbrace{Mg_5Al_2Si_3O_{10}(OH)_8}_{\text{clinochlore}} + 15.6\,H_2O + 0.4\,K^+ + 3.0\,Mg^{2+} + 8.0\,Fe^{2+} \quad (r9)$$

$$\underbrace{Al_2Si_2O_5(OH)_4}_{\text{kaolinite}} + \underbrace{SiO_2}_{\text{quartz}} + 5\,Mg^{2+} + 7\,H_2O = \underbrace{Mg_5Al_2Si_3O_{10}(OH)_8}_{\text{clinochlore}} + 0.4\,K^+ + 10\,H^+ \quad (r10)$$

$$\underbrace{KFe_3AlSi_3O_{10}(OH)_2}_{\text{annite}} + \underbrace{KMg_3AlSi_3O_{10}(OH)_2}_{\text{phlogopite}} + 0.75\,\underbrace{K_{0.33}Mg_{0.33}Al_{1.67}Si_4O_{10}(OH)_2}_{\text{K-montmorillonite}} + 0.8\,\underbrace{KAl_2(AlSi_3O_{10})(OH)_2}_{\text{muscovite}} + 1.28\,\underbrace{Al_2Si_2O_5(OH)_4}_{\text{kaolinite}} + 0.19\,\underbrace{KAlSi_3O_8}_{\text{K-feldspar}} + 1.45\,H^+ = 2.7\,\underbrace{K_{0.85}Mg_{0.25}Al_{2.35}Si_{3.4}O_{10}(OH)_2}_{\text{illite}} + 1.03\,\underbrace{Fe_{2.5}Mg_{2.5}Al_2Si_3O_{10}(OH)_8}_{\text{ripidolite}} + 1.88\,\underbrace{SiO_2}_{\text{quartz}} + 15.6\,H_2O + 0.4\,K^+ + 0.94\,K^+ + 0.43\,Fe^{2+} \quad (r11)$$

$$\underbrace{Al_2Si_2O_5(OH)_4}_{\text{kaolinite}} + 1.88\,\underbrace{SiO_2}_{\text{quartz}} + 0.05\,\underbrace{Fe_{2.5}Mg_{2.5}Al_2Si_3O_{10}(OH)_8}_{\text{ripidolite}} + 0.8\,K^+ = 0.94\,\underbrace{K_{0.85}Mg_{0.25}Al_{2.35}Si_{3.4}O_{10}(OH)_2}_{\text{illite}} + 0.05\,\underbrace{Mg_5Al_2Si_3O_{10}(OH)_8}_{\text{clinochlore}} + 1.37\,H_2O + 0.24\,H^+ + 0.13\,Fe^{2+} + 0.16\,Mg^{2+} \quad (r12)$$

$$\underbrace{Fe_{2.5}Mg_{2.5}Al_2Si_3O_{10}(OH)_8}_{\text{ripidolite}} + 1.40\,\underbrace{K_{0.85}Mg_{0.25}Al_{2.35}Si_{3.4}O_{10}(OH)_2}_{\text{illite}} + 0.40\,\underbrace{KAlSi_3O_8}_{\text{K-feldspar}} + 11.38\,Mg^{+2} + 14.06\,H_2O = 2.84\,\underbrace{Mg_5Al_2Si_3O_{10}(OH)_8}_{\text{clinochlore}} + 0.42\,\underbrace{SiO_2}_{\text{quartz}} + 16.16\,H^+ + 1.59\,K^+ \quad (r13)$$

$$\underbrace{K_{0.85}Mg_{0.25}Al_{2.35}Si_{3.4}O_{10}(OH)_2}_{\text{illite}} + 0.125\,\underbrace{SiO_2}_{\text{quartz}} + 5.625\,Mg^{2+} + 8.90\,H_2O = 1.175\,\underbrace{Mg_5Al_2Si_3O_{10}(OH)_8}_{\text{clinochlore}} + 10.40\,H^+ + 0.85\,K^+ \quad (r14)$$

$$\underbrace{Al_2Si_2O_5(OH)_4}_{\text{kaolinite}} + 0.35\,\underbrace{KAlSi_3O_8}_{\text{K-feldspar}} + 0.35\,\underbrace{SiO_2}_{\text{quartz}} + 0.25\,Mg^{2+} + 0.50\,K^+ = \underbrace{K_{0.85}Mg_{0.25}Al_{2.35}Si_{3.4}O_{10}(OH)_2}_{\text{illite}} + H^+ + 0.50\,H_2O. \quad (r15)$$

Stage 3 reactions:

$$\underbrace{KAlSi_3O_8}_{\text{K-feldspar}} + 0.109\,Mg^{2+} + 0.522\,H^+ + 0.174\,H_2O = 0.435\,\underbrace{K_{0.6}Mg_{0.25}Al_{2.30}Si_{3.50}O_{10}(OH)_2}_{\text{illite}} + 1.478\,\underbrace{SiO_2}_{\text{quartz}} + 0.739\,K^+ \quad (r16)$$

$$\underbrace{KAlSi_3O_8}_{\text{K-feldspar}} + 4.6\ \underbrace{Na_{0.29}Mg_{0.26}Al_{1.77}Si_{3.97}O_{10}(OH)_2}_{\text{Na-smectite}} + 1.385\,K^+ + 0.354\,H^+ = 3.975\ \underbrace{K_{0.6}Mg_{0.25}Al_{2.30}Si_{3.50}O_{10}(OH)_2}_{\text{illite}} + 7.35\ \underbrace{SiO_2}_{\text{quartz}} + 1.334\,Na^+ + 0.202\,Mg^{2+} + 0.802\,H_2O \quad \text{(r17)}$$

$$\underbrace{Na_{0.29}Mg_{0.26}Al_{1.77}Si_{3.97}O_{10}(OH)_2}_{\text{Na-smectite}} + 0.462\,K^+ = 0.770\ \underbrace{K_{0.6}Mg_{0.25}Al_{2.30}Si_{3.50}O_{10}(OH)_2}_{\text{illite}} + 1.277\ \underbrace{SiO_2}_{\text{quartz}} + 0.290\,Na^+ + 0.068\,Mg^{2+} + 0.035\,H^+ + 0.212\,H_2O. \quad \text{(r18)}$$

References

Ahlbrandt, S. T. 2001. *The Sirte Basin Province of Libya – Sirte-Zelten Total Petroleum System*. United States Geological Survey, Bulletin, **2002-F**.

Ball, J. W. & Nordstrom, D. K. 1991. *WATEQ4F – User's Manual with Revised Thermodynamic Data Base and Test Cases for Calculating Speciation of Major, Trace and Redox elements in Natural Waters*. United States Geological Survey, Open-File Report, **90-129**.

Blanc, Ph., Lassin, A., Piantone, P., Azaroual, M., Jacquemet, N., Fabbri, A. & Gaucher, E. C. 2012. Thermoddem: a geochemical database focused on low temperature water/rock interactions and waste materials. *Applied Geochemistry*, **27**, 2107–2116.

Carpenter, A. B. 1978. Origin and chemical evolution of brines in sedimentary basins. *Oklahoma Geological Survey Circular*, **79**, 78–88.

Freed, R. L. & Peacor, D. R. 1989. Variability in temperature of the smecite/illite reaction in Gulf Coast sediments. *Clay Minerals*, **24**, 171–180.

Gabelich, C., Yun, T. I., Green, J. F., Suffet, I. H. & Chen, W. R. 2002. *Evalution of Precipitative Fouling for Colorado River Water Desalination using Reverse Osmosis. Agreement No. 99-FC-81-0185*. Desalination and Water Purification Research and Development Program Report, DWPR No. 85. United States Department of the Interior, Bureau of Reclamation, Denver Office, Denver, CO.

Garrels, R. M. 1984. Montmorillonite/Illite stability diagrams. *Clay and Clay Minerals*, **32**, 161–166.

Garrels, R. M. & Mackenzie, F. T. 1967. Origin of the chemical composition of the some springs and lakes. *In*: Stumm, W. (ed.) *Equilibrium Concepts in Natural Water Systems*. Advances in Chemical Series, **67**. American Chemical Society, Washington, DC, 222–242.

Helgeson, H. C. & Kirkham, D. H. 1974*a*. Theoretical prediction of the thermodynamic behavior of aqueous electrolytes at high pressures and temperatures: I. Summary of the thermodynamic/electrostatic properties of the solvent. *American Journal of Science*, **274**, 1089–1198.

Helgeson, H. C. & Kirkham, D. H. 1974*b*. Theoretical prediction of the thermodynamic behavior of aqueous electrolytes at high pressures and temperatures: II. Debye–Huckel parameters for activity coefficients and relative partial molal properties. *American Journal of Science*, **274**, 1199–1261.

Helgeson, H. C., Garrels, R. M. & Mackenzie, F. T. 1969. Evaluation of irreversible reactions in geochemical processes involving minerals and aqueous solutions. *Geochimica et Cosmochimica Acta*, **34**, 569–592.

Helgeson, H. C., Kirkham, D. H. & Flowers, D. C. 1981. Theoretical prediction of the thermodynamic behavior of aqueous electrolytes at high pressures and temperatures: IV. Calculation of activity coefficients, osmotic coefficients, and apparent molal and standard and relative partial molal properties to 600°C and 5 kb. *American Journal of Science*, **281**, 1249–1516.

Lander, R. H. & Bonnell, L. M. 2010. A model for fibrous illite nucleation and growth in sandstones. *American Association of Petroleum Geologists Bulletin*, **94**, 1161–1187.

Lander, R. H., Larese, E. L. & Bonnel, L. M. 2008. Toward more accurate quartz cement models: the importance of euhedral v. noneuhedral growth rates. *American Association of Petroleum Geologists Bulletin*, **92**, 1537–1563.

Lasaga, A. C., Soler, J. M., Ganor, J., Burch, T. E. & Nagy, K. L. 1994. Chemical weathering rate laws and global geochemical cycles. *Geochimica et Cosmochimica Acta*, **58**, 2361–2386.

Morse, J. W. & Arvidson, R. S. 2002. The dissolution kinetics of major sedimentary carbon mineral. *Earth-Science Reviews*, **58**, 51–84.

Oelkers, E. H. & Helgeson, H. C. 1988. Calculations of the thermodynamic and transport properties of aqueous species at high pressures and temperatures: aqueous tracer diffusion coefficients of ions to 1000°C and 5 kb. *Geochimica et Cosmochimica Acta*, **52**, 63–85.

Palandri, J. L. & Kharaka, Y. K. 2004. *A Compilation of Rate Parameters of Water–Mineral Interaction Kinetics for Application to Geochemical Modeling*. United States Geological Survey, Open-File Report, **1068**.

Parkhurst, D. & Appelo, C. A. J. 1999. *Users's Guide to PHREEQC (version 2) – A Computer Program for Speciation, Batch-Reaction, One Dimensional Transport, and Inverse Geochemical Calculations*. United States Geological Survey, Water-Resources Investigations Report, **99-4259**.

Pitzer, K. S. & Mayorga, G. 1973. Thermodynamics of electrolytes. II. Activity and osmotic coefficients for strong electrolytes with one or both ions univalent. *Journal of Physical Chemistry*, **77**, 2300–2307.

Primmer, T. J., Cade, C. A. *et al.* 1997. Global patterns in sandstone diagenesis: their application to reservoir quality prediction for petroleum exploration.

In: Kupecz, J. A., Gluyas, J. & Bloch, S. (eds) *Reservoir Quality Prediction in Sandstones and Carbonates*. American Association of Petroleum Geologists, Memoirs, **69**, 61–77.

Pruess, K., Oldenburg, C. & Moridis, G. 1999. *TOUGH2 User's Guide, Version 2.0 Report*, LBNL-43134. Lawrence Berkeley National Laboratory, Berkeley, CA.

Pytte, A. M. & Reynolds, R. C. 1989. The thermal transformation of smectite to illite. *In*: Naeser, N. D. & McCulloh, T. H. (eds) *Thermal History of Sedimentary Basins. Methods and Case Histories*. Springer, Berlin, 134–140.

Rex, R. W. & Martin, B. D. 1966. Clay mineral formation in seawater by submarine weathering of K-feldspar. *In*: Bailey, S. W. (ed.) *Clays and Clay Minerals. Proceedings of the Fourteenth National Conference, Berkeley, California*. Pergamon Press, Oxford, 235–240.

Stumm, W. & Morgan, J. J. 1996. *Aquatic Chemistry: Chemical Equilibria in Natural Waters*, 3rd edn. Wiley, New York.

Sverdrup, H. U. & Warfvinge, P. 1988. Weathering of primary silicate minerals in the natural soil environment in relation to a chemical weathering model. *Water, Air, & Soil Pollution*, **38**, 387–408.

Tardy, Y. 1971. Characterization of the principal weathering types by the geochemistry of waters from some Europen and African crystalline massifs. *Chemical Geology*, **7**, 253–271.

Turekian, K. K. 1968. *Oceans*. Prentice-Hall, Dunwoody, GA.

Vieillard, P. 2000. A new method for the prediction of Gibbs free energies of formation of hydrated clay minerals based on the electronegativity scale. *Clays and Clay Minerals*, **48**, 459–473.

Walderhaug, O. 2000. Modeling quartz cementation and porosity loss in Middle Jurassic Brent Group sandstones of Kvitebjorn field, norther North Sea. *American Association of Petroleum Geologists Bulletin*, **84**, 1325–1339.

Wolery, T. J. 1992. *Software Package for Geochemical Modeling of Aqueous Systems: Package Overview and Installation Guide (Version 7.2)*. Lawrence Livermore National Laboratory Report UCRL-MA-10662 PT I, Livermore, California.

Xu, T., Apps, J. A. & Pruess, K. 2005. Mineral sequestration of carbon dioxide in a sandstone-shale system. *Chemical Geology*, **217**, 295–318.

Xu, T., Sonnenthal, E., Spycher, N. & Pruess, K. 2006. TOUGHREACT-A simulation program for non-isothermal multiphase reactive geochemical transport in variably saturated geologic media: applications to geothermal injectivity and CO_2 geological sequestration. *Computers & Geosciences*, **32**, 145–165.

Index

Page numbers in *italics* refer to Figures. Page numbers in **bold** refer to Tables.